Mechanics of Swelling

From Clays to Living Cells and Tissues

NATO ASI Series

Advanced Science Institutes Series

A series presenting the results of activities sponsored by the NATO Science Committee, which aims at the dissemination of advanced scientific and technological knowledge, with a view to strengthening links between scientific communities.

The Series is published by an international board of publishers in conjunction with the NATO Scientific Affairs Division

A Life Sciences B Physics	Plenum Publishing Corporation London and New York
C Mathematical and Physical Sciences D Behavioural and Social Sciences E Applied Sciences	Kluwer Academic Publishers Dordrecht, Boston and London
F Computer and Systems Sciences G Ecological Sciences H Cell Biology I Global Environmental Change	Springer-Verlag Berlin Heidelberg New York London Paris Tokyo Hong Kong Barcelona Budapest

NATO-PCO DATABASE

The electronic index to the NATO ASI Series provides full bibliographical references (with keywords and/or abstracts) to more than 30000 contributions from international scientists published in all sections of the NATO ASI Series. Access to the NATO-PCO DATABASE compiled by the NATO Publication Coordination Office is possible in two ways:

- via online FILE 128 (NATO-PCO DATABASE) hosted by ESRIN, Via Galileo Galilei, I-00044 Frascati, Italy.
- via CD-ROM "NATO-PCO DATABASE" with user-friendly retrieval software in English, French and German (© WTV GmbH and DATAWARE Technologies Inc. 1989).

The CD-ROM can be ordered through any member of the Board of Publishers or through NATO-PCO, Overijse, Belgium.

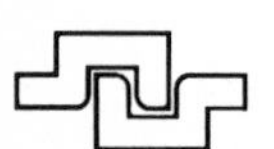

Series H: Cell Biology, Vol. 64

Mechanics of Swelling

From Clays to Living Cells and Tissues

Edited by

Theodoros K. Karalis

Democritos University of Thrace
Department of Civil Engineering
67100 Xanthi
Greece

Springer-Verlag
Berlin Heidelberg New York London Paris Tokyo
Hong Kong Barcelona Budapest
Published in cooperation with NATO Scientific Affairs Division

Proceedings of the NATO Advanced Research Workshop on Mechanics of Swelling:
From Clays to Living Cells and Tissues held at Corfu (Greece) from July 1–6, 1991

ISBN 3-540-54607-3 Springer-Verlag Berlin Heidelberg New York
ISBN 0-387-54607-3 Springer-Verlag New York Berlin Heidelberg

Library of Congress Cataloging-in-Publication Data
Mechanics of Swelling : from clays to living cells and tissues / edited by Theodoros K. Karalis.
(NATO ASI series. Series H, Cell biology ; vol. 64)
"Proceedings of the NATO Advanced Research Workshop on Mechanics of Swelling: from Clays to Living Cells and Tissues held at Corfu (Greece) from July 1-6, 1991." Includes bibliographical references and index.
ISBN 0-387-54607-3
1. Edema--Congesses. 2. Tissues--Mechanical properties--Congresses. 3. Swelling soils--Congresses. I. Karalis, Theodoros K., 1940- . II. NATO Advanced Research Workshop on Swelling Mechanics: From Clays to Living Cells and Tissues (1991 : Kerkyra, Greece) III. Series. RB 144.M43 1992 574.19'1--dc20

Printed in Germany

Typesetting: Camera ready by author
31/3145 - 5 4 3 2 1 0 - Printed on acid-free paper

It is the work of an educated man to look for precision in each class of things insofar as the nature of the subject admits

Aristotle

Souvenir photograph of the NATO ARW participants held in Corfu 1–6 July, 1990. By numbers: *(1) Wendy Silk, (2) Pierre–Gilles de Gennes, (3) John R. Philip, (4) Giovanni Pallotti, (5) Alex Silberberg, (6) Evan A. Evans, (7) Theodoros K. Karalis, (8) Alice Maroudas, (9) Francoise Brochard–Wyart, (10) de Barios, (11) Pedro Verdugo, (12) Peter J. Basser, (13) Dennis Pufahl, (14) Jacob Israelachvili, (15) Stefan Marcelja, (16) Sidney A. Simon, (17) J.P.G Urban, (18) Peter R. Rand, (19) S.D. Tyerman, (20) Paul Janmey, (21) Joe Wolfe, (22) Paolo Bernardi, (23) James S. Clegg, (24) Claude Lechene, (25) John Passioura, (26) Larry L. Boersma, (27) Thanassis Sambanis, (28) Kenneth R. Spring, (29) Philippe Baveye, (30) Avinoam Nir, (31) Pierre Cruiziat, (32) Kostas Gavrias, (33) Panayiotis Kotzias, (34) Frank A. Meyer, (35) Wayne Comper, (36) J. M. A. Snijders, (37) Yoram Lanir, (38) Rolf K. Reed, (39) Adrian Parsegian, (40) Aris Stamatopoulos, (41) George F. Oster, (42) Leonid B. Margolis, (43) Makoto Suzuki, (44) A. P. Halestrap, (45) Charles A. Pasternak, (46) Jersy Nakielski, (47) Louis Hue*

Preface

Mechanics of Swelling is crucial in any decision process in a diversity of problems in engineering and biological practice. It is part of the control steps following identification of certain materials preceding the organization that should be made before an industrial action is undertaken. It includes research on all aspects of osmotic phenomena in clays, plants, cells and tissues of living systems, gels and colloidal systems, vesicle polymeric systems, forces between surfactants, etc. It embraces also research on oedema, obesity, tumours, cancer and other related diseases which are connected with oncotic variations in the parts of a living system.

The ancient Greeks investigating our physical universe, by wrestling with logical reflections, continuously asking and asking again themselves, they found a certain exodos and in this particular problem too. *Tumor, dolor, calor, rubor*, was a maxim already known by Hippocrates and later–on by the Romans. Presently, major scientific achievement has been attained in the last decades by biologists, medics doctors and physicists, but it still remains difficult to explain the cause, the evolution and the various details concerning oncotic variations in our animate world. Presumably, different generic errors and environmental factors operate at different stages in the development of these events through multiplication, differentiation, aggregation, localized proliferation and cell death; each of which is itself a complex system of biomedical activities. Oedema, obesity, swelling of cells and tissues, teratogenic mechanisms, etc. remain of intense clinical importance and the contributions to explain these diseases have been the subjects of widespread research and observations.

From the inanimate world, the fact that certain materials swell was already known in antiquity. Ceramic manufacture could not be achieved without knowledge of the used clays' swelling potential. The Egyptians isolated stones from the rocks by carefully filling holes in the rock with dried wood and then pouring water over it. The force developed by the swelling wood made the stone burst. Among other examples the swelling of certain resins in aqueous solutions and of soils at different moisture content was known earlier.

Silicates and certain crystalline substances swell and/or lose water without ceasing to behave like crystals. Haemoglobin crystals are capable of swelling taking up and losing water without any change in their apparent (microscopic) homogeneity. Hair and textile fibres swell when placed in water. The way they swell depends on their chemical constitution which affects their behaviour during dyeing and finishing. Knowledge of fibre swelling behaviour would be valuable, but it has proved hard to ascertain because of their small dimensions. Few textile fibres are shorter than about a centimetre in length whilst many are much longer and few fibres are more than one thousandth of a centimetre in breadth and many are smaller. These dimensions make it comparatively easy to measure longitudinal swelling and difficult to measure transverse swelling. Fibres are allotropic and transverse swelling cannot be directly inferred from the longitudinal value. If swelling values could be determined in both directions their ratio would be a valuable index in measuring their allotropy.

Keratin or high molecular weight materials when placed in a suitable solvent swell to an equilibrium value determined by the solvent, the temperature and the nature of the polymer. Furthermore, for a long time now certain clay minerals (montmorillonite) have attracted interest and found large uses because of their ability to adsorb large amounts of water. The role of Polysaccharides in Corneal swelling is also substantial. The problem of hydration of the cornea has received considerable attention in the past because of its correlation with corneal transparency. Cornea stroma isolated and immersed in an aqueous solution swells excessively and consequently loses its transparency.

The systematic and scientific study of the dimension changes of solids was first discussed by Cauchy in 1828. His outstanding masterpiece was entitled "Sur l' equilibre et le mouvement d' un system des points materiels sollicités par les forces d'attraction ou de répulsion mutuelle". However, Cauchy was concerned only with explaining dilation and/or condensation of a solid when it was solicited by external forces. His objective was to study elastic tension and/or condensation of a solid subjected to bending and torsion which were pointed out earlier by Leonard Euler (1755) and much earlier by Galilei (1564–1642), Mariote (1620–1684), Leibnitz (1646–1716), Robert Hooke (1635 – 1703), Jacob Bernoulli (1654–1705), Thomas Young (1773–1829) and many others.

All these studies concern changes in the dimensions of a solid and not with the interaction of the solid with fluids, specifically when an amount of liquid is taken up by certain materials. The systematic work from this point of view was done by the earlier botanists during the period 1860–1880. Deluc (1791) made some valuable contributions on this subject but it is only through the work of Vide Nageli (1862), Reinke, Pfeffer, Hugo de Vries and others that swelling really started to be modelled. According to them, a solid is said to swell when it takes up a liquid whilst at the same time: (*i*) it does not lose its apparent homogeneity, (*ii*) its dimensions are enlarged and (*iii*) its cohesion is diminished; the latter statement outlining the fact that a material instead of being hard and brittle becomes soft and flexible. Therefore, the flow through a material imbibing water and producing swelling is clearly distinct from capillary imbibition, such as is shown by a solid having many fine capillary canals *e.g.,* a piece of birch etc. Such a solid taking up liquids, remains clearly microscopically inhomogeneous but its dimensions do not change and their cohesion is not drastically diminished through imbibition of the liquid.

However, despite the remarkable research by the earlier botanists and other contemporary workers, it still remains difficult to have a complete picture of swelling via phase interaction, phase transition, mechanical and thermal considerations. It is difficult to understand why certain materials swell taking up liquids and still not lose its cohesion entirely but only partially. This and other facts are the reason why swelling was discussed in all the fields of our natural world in this workshop.

Throughout the meeting our aim was to elucidate the physical meaning and implications of the concepts which already apply to swelling behaviour in most of the materials and systems of our natural world. We tried to promote the mechanics of swelling entering upon a new stage, which is less empirical and where the experimental study of better defined objects was guided rather by more quantitative theories than by qualitative "rules" or working hypotheses. The mechanics of swelling as was discussed in this workshop, may serve as an example of this development reviewing most of the crucial aspects of swelling in nature and to develop as far as possible a quantitative theory giving as a result a clear, concise and relatively complete treatment. The material in this volume is arranged in six parts which cover swelling in soils, plants, cells, tissues and gels. Developments in various techniques are drawn in part six. The booklet ends with a subject index.

It was the general opinion of all who attended the Conference that this Advanced Research Workshop was important and valuable to their on–going research projects and everyone saw the need of having further exchanges on the same subject. The lectures and the short papers presented here are of exceptional quality and I would like to thank all the contributors for their efforts. As an editor I had the pleasurable opportunity of becoming familiar with all the contributions and to interact with their authors. It is with great pleasure that I extend my sincerest congratulations along with those of the participants of this ARW to Pierre–Gilles de Gennes who has become a Nobel prize laureate. Furthermore, let me acknowledge the gratitude of all the participants to the NATO Scientific Committee for its generous support and worthwhile goal of bringing together scientists from many countries. Also I which to extend a word of appreciation to the Greek Institutions for their financial support and to my wife Photini for her efficient secretarial work.

I also feel very proud to have had the confidence of all the participants who supported the organization of this Advanced Research Workshop held in Corfu from 1–6 July 1991 and I hope that the papers presented in this Springer Verlag's edition will lead to further advances in the Mechanics of Swelling.

Theodoros K. Karalis
ARW Director

Contents

Contributors

John Agortsas, MD., Venizelou 100, 67100 Xanthi, Greece.

Peter J. Basser, National Institute of Health, NIH Building 13, Rm 3W13, Bethesda, MD 20892, USA

Philippe Baveye, Cornell University, College of Agriculture and Life Sciences, Bradfield and Emerson Halls, Ithaca, New York 14853, USA

Paolo Bernardi, MD, Universita degli Studi di Padova, Instituto Di Patologia Generale, Via Trieste, 75 35121 Padova, Italy

Larry L. Boersma, Soil Science Department, Oregon State University, Corvallis, OR 97331–2213, USA

Francoise Brochard-Wyart, Université Pierre et Marie Curie (Paris VI), Structure et Reactivité aux interfaces, Batiment Chimie–Physique 11, rue Pierre et Marie Curie 75231, Paris Cedex 05, France

James S. Clegg, University of California, Bodega Marine Laboratory, PO Box 247, Bodega Bay, California 94923, USA

Wayne Comper, Department of Biochemistry, Monash University, Clayton, Melbourne, Victoria, 3168, Australia

Pierre Cruiziat, Centre de Recherches Agronomiques du Massif Central, Lab. de Bioclimatologie, Domaine de Crouelle, 63039 Clermont–Ferrand Cedex, France

Evan A. Evans, University of British Columbia, Department of Pathology, Faculty of Medicine, 2211 Wesbrook Mall, Vancouver, B.G. V6T 1W5, Canada

Kostas Gavrias, Civil Engineer, Amphitrionos 4, 41336 Larisa, Greece

Pierre-Gilles de Gennes, College de France, Physique de la Matiere Condensee, 11, Place Marcelin-Berthelot, 75231 Paris CEDEX 05, France

A. P. Halestrap, University of Bristol, Department of Biochemistry, School of Medical Sciences, University Walk, Bristol BS8 1TD, United Kingdom

Dieter Haussinger, Medizin Klinik, Universitat Freiburg, Hugstetterstrasse 55, D-7800 Freiburg im BR, Germany

Louis Hue, Unite Horemones et Metabolism Research, UCL 7529, Avenue Hippocrate 75, B-1200 Bruxelles, Belgium

Jacob Israelachvili, University of California, Santa Barbara, Department of Chemical and Nuclear Engineering, Santa Barbara, California 93106, USA

Theodoros K. Karalis, School of Civil Engineering, Democritos Univercity of Thrace, 67100, Greece

Panayiotis Kotzias, President of the Hellenic Society of Soil Mechanics, and Engineering Foundation, Isavron 5, 11471 Athens, Greece

Paul Janmey, Massachusetts General Hospital Division of Hematology & Oncology, Bulding 149, 13tb Street, MGH-West, 8tb Floor, Cbarlestown, MA 02129, USA

Yoram Lanir, Technion-Israel Institute of Technology, Department of Bio-Medical Engineering, The Julius Silver Institute of Bio-Medical Engineering Sciences, Technion City, Haifa 32000, Israel

Claude Lechene, Harvard Medical School, Brigham and Women's Hospital, Laboratory of Cellular Physiology, 221 Longwood Avenue Boston, MA 02115, USA

Stefan Marcelja, The Australian National University, Research School of Physical Sciences, Department of Applied Mathematics, GPO Box 4, Canberra ACT 2601, Australia

Leonid B.Margolis, Moscow State University, A.N. Belozersky Laboratory of Molecular Biology and Bioorganic Chemistry, 119899, Moscow, Bldg. A", USSR

Alice Maroudas, Technion–Israel Institute of Technology, Department of Biomedical Engineering, Technion City, Haifa 32000, Israel

Frank A. Meyer, Tel Aviv–Elias Sourasky Medical Center, Department of Rheumatology, Ichilov Hospital, 6 Weizmann St., Tel–Aviv 64239, Israel.

Jersy Nakielski, Department of Biophysics and Cell Biology, Silesian University, ul. Jagiellonska 28, 40–032 Katowice Poland

Avinoam Nir, Technion–Israel Institute of Technology, Technion City–Haifa, Israel

George F. Oster, University of California, 201 Wellman Hall, Berkley, CA 94720, USA

Giovanni Pallotti, Professor of Medical Physics, Faculty of Medicine and Surgery, Department of Physics, University of Bologna, Via Irnerio 46, 40126 Bologna, Italy.

Adrian Parsegian, Physical Sciences Laboratory, Division of Computer Research and Technology, National Institutes of Health, Building 12A, Room 2007, Bethesda, Maryland 20205, USA

John Passioura, CSIRO, Division of Plant Industry, GPO Box 1600, Canberra, ACT 2601, Australia

Charles A. Pasternak, St. George's Hospital Medical School, University of London, Department of Cellular & Molecular Sciences, Division of Biochemistry, Granmer Terrace, London SW17 ORE, United Kingdom

John R. Philip, Division of Enviromental Mechanics, Centre for Environmental Mechanics, Black Maountain, Canberra, GPO Box 821, Canberra, ACT 2601, Australia

Sergey Popov, Department of Biological Sciences, Columbia University, Fairchild Building, Room 915, New York, NY 10027, USA

Dennis Pufahl, University of Saskatchewan, Department of Civil Engineering, Saskatoon, Canada S7N OWO

Peter R. Rand, Brock University, Department of Biological Sciences, St. Catharines, Ontario, Canada L2S 3A1

Rolf K. Reed, School of Medicine, University of Bergen, Department of Physiology, Arstadveien 19 N–5009 Bergen, Norway.

Thanassis Sambanis, Chemical Engineering Department, Georgia Institute of Technology, Atlanta Georgia 30332–0100, USA

Sidney A. Simon, Duke University Medical Center, Department of Neurobiology, Durham, Box 3209, North Carolina 27710, USA

Alex Silberberg, The Weizmann Institute of Science, Department of Polymer Research, Rehovot 76100, Israel

Wendy Silk, Department of Land, Air and Water Resources, Hoagland Hall, University of California, Davis CA 95616, USA

J. M. A. Snijders, Rijksuniversiteit Limburg, Department of Movement Sciences, P.O.BOX 616, Holland

Kenneth R. Spring, Section on Transport Physiology, Laboratory of Kidney and Electrolyte Metabolism, National Institutes of Health, National Heart, Lung, and Blood Institute, Bethesda, Maryland 20892, USA

David E. Smiles, CSIRO Division of Soils, GPO Box 639, Canberra ACT 2601, Australia

Aris Stamatopoulos, Isavron 5, 11471 Athens, Greece

Makoto Suzuki, Mechanical Engineering Lab., Agency of Industrial Science and Thechnology, Ministry of International Trade and Industry, Namiki 1–2, Tsukuba, Ibazak 305, Japan

Toyoichi Tanaka, Massachusetts Institute of Technology, Department of Physics, 77 Massachusetts Avenue, Cambridge, MA 02139, Room 13–2153, USA

Perikles S. Theocaris, Member of the Academy of Athens, Professor of N.T.U., PO. Box 77230, 17510, Athens, Greece

S.D. Tyerman, The Flinders University of South Australia, School of Biological Sciences, Bedford Park, South Australia 5042.

J.P.G Urban, Department of Physiology, University of Oxford, Park Road, Oxford OX1 3 PT, United Kingdom

Pedro Verdugo, Bioengineering, University of Washington, Seattle WA, USA

Joe Wolfe, School of Physics, University of New South Wales, PO Box 1, Kensington, 2033 Australia

Part 1

Swelling in Soils

FLOW AND VOLUME CHANGE IN SOILS AND OTHER POROUS MEDIA, AND IN TISSUES

J.R. Philip
CSIRO Centre for Environmental Mechanics
GPO Box 821
Canberra ACT 2601
Australia

Some 140 years ago Edward Lear, the English artist and humorist, visited Corfu and wrote the following:

> There was an old man in Corfu
> Who never knew what he should do,
> So he rushed up and down
> Till the sun made him brown,
> That bewildered old man of Corfu.

I do not doubt that many of you, too, will let the sun make you brown; but, unlike that old man, you will all know perfectly well what to do: namely, to make the most of this unique research workshop. Gathered together here are scientists from 16 countries and from a great diversity of disciplines from agronomy to molecular physics, from civil engineering to physiology and biochemistry, all concerned in one way or another with the common theme of the mechanics of swelling.

We owe this unusual and valuable event to the imagination, tenacity, and energy of one man, Professor Theodoros Karalis. He has created this Workshop virtually single-handed (though he tells me his charming wife Photini helped with some practical details).

In November 1989 Professor Karalis wrote to tell me of his hopes of bringing his Workshop into being. I saw immediately the compelling logic behind his vision of a meeting bringing together research scientists from diverse disciplines, but united in their concern with problems of volume change. Much illumination may follow from examining both what swelling phenomena in these various fields have in common, and the ways in which they differ.

Over my research life I have been somewhat involved in volume-change problems, both in porous medium physics and in physiology. Most of those 40 years have been spent on simpler systems innocent of the complications

NATO ASI Series, Vol. H 64
Mechanics of Swelling
Edited by T. K. Karalis

of volume change, but I have always found my excursions into swelling and associated problems of equilibrium and flow fascinating and challenging. My own experience and, I suppose disposition, leads me to concentrate here on what swelling processes across the board might have in common, and on their phenomenological characterization and analysis at an appropriate macroscopic scale. This review of swelling mechanics thus makes no claim to be encyclopaedic.

1. The Scales of Discourse

In treating the mechanics of porous media (and of cell aggregations), we must recognise clearly three distinct and separate scales of discourse (Raats 1965, Philip 1972a, Philip and Smiles 1982): 1) the molecular scale; 2) the microscopic (for fluids, the Navier-Stokes) scale; 3) the macroscopic (for porous media, the Darcy) scale. Our *molecular* scale is precisely that called 'mass-point' or 'microscopic' in continuum mechanics (e.g. Truesdell and Noll, 1965; Sedov, 1971); and in that sub-discipline our *microscopic* scale is confusingly designated 'macroscopic' or 'phenomenological'. Discourse on our *macroscopic* or *phenomenological* scale deals with physical quantities related to averages of analogous quantities on our microscopic scale, the averages being taken over a volume, or cross-section, large compared with that of the individual pores of porous media, or of individual cells of tissues, or large enough to contain many particles in colloid pastes and suspensions. A formal description of a suitable averaging process has been given, for example, by Zaslavsky (1968).

The primary goal of porous medium (and many physiological) studies is theory on this macroscopic scale: physical observations are most readily made on this scale, and the practical concern is with processes on this scale. Usually a full microscopic theory would be needlessly elaborate, even if it happened to be feasible. Of course microscopic studies are important in their own right: well-founded theory on this scale may guide us as to the form macroscopic theory may take, and as to the limits of its applicability. Macroscopic phenomenological theory must be at least consistent with what we know on the microscopic scale. Most of this review concerns the macroscopic scale, but Section 4 on colloid pastes illustrates the interplay between inquiry on the microscopic scale and theory formulation on the macroscopic scale.

In my view, failure to distinguish clearly between the scales of discourse has led to confusion and misplaced effort in many studies. It would be invidious to elaborate. Some examples are enumerated in Philip (1972a).

2. Flow in Unsaturated Nonswelling Porous Media

A prime example of phenomenological theory on the macroscopic scale is the analysis of water movement in unsaturated nonswelling soils and porous media. The underlying physical concepts were understood by Buckingham (1907) and the formulation sharpened by Richards (1931). Much basic work was published in the 1950's (e.g. Childs and Collis-George, 1950; Klute, 1951; Philip, 1954, 1957a, 1957b), and the theory has been in general use in soil physics and hydrology since about 1960. We describe it in some detail here, because we shall go on to show how it is generalized to deal with (at least one-dimensional) problems of equilibrium, flow, and volume change in swelling media.

We begin, then, with an unsaturated nonswelling porous medium made up of three components: the rigid solid matrix, a liquid, and a gas. We outline the theory in the simplification (often justified in the applications) that gas pressure differences may be neglected. It then suffices to examine the flow of the liquid component: details of the gas flow follow from continuity. We deal specifically with the case where the liquid is water and the gas air (including water vapour): the extension to other incompressible Newtonian liquids and other gases will be obvious.

2.1 Darcy's law for unsaturated nonswelling media

We may write Darcy's law for *saturated* media, specialized to water flow, as

$$\underline{\underline{v}} = -K \nabla\Phi \ . \qquad (1)$$

$\underline{\underline{v}}$ is the vector flow velocity, Φ is the total potential, and K is the hydraulic conductivity. Expressing potentials per *unit weight* simplifies our equations and units: Φ then has the dimension [length] and K the dimensions [length] [time]$^{-1}$. We note that

$$\Phi = p/(\rho g) + \Xi ,$$

where p is the pressure, ρ the water density, g the gravitational acceleration, and Ξ the potential of the external forces.

Buckingham (1907) suggested that Darcy's law should hold for *unsaturated* media in a modified form with K a function of θ, the volumetric moisture content. Richards (1931), Moore (1939), Childs and Collis-George (1950), and others confirmed this experimentally and established the general character of $K(\theta)$. For obvious physical reasons (Philip 1954, 1957a) K decreases through as many as 6 decades (Gardner 1960) as θ decreases from its saturation value through the range of interest.

2.2 Total potential and moisture potential of water in nonswelling media

In (water-wet) unsaturated media the water is not free in the thermodynamic sense because of capillarity, adsorption, and electrical double layers (Edlefsen and Anderson 1943, Schofield 1935a). Capillarity is dominant in wet, coarse-textured media, and adsorption assumes its greatest importance in dry media. Double-layer effects may be significant in fine-textured media exhibiting colloidal properties. Buckingham (1907), a keen disciple of Willard Gibbs, was the first to appreciate that the conservative forces governing the equilibrium and flow of soil-water are amenable to treatment through their associated scalar potentials.

We define such potentials relative to the reference state of water (of composition identical to the soil solution) at atmospheric pressure and datum elevation $z = 0$. Here z is the vertical ordinate, conveniently taken positive downward. We then have $\Xi = -z$ and

$$\Phi = \Psi - z . \tag{2}$$

Ψ, the *moisture potential*, is the potential of the forces arising from local interactions between solid and water (Philip 1970b). It is not essential either to know or to specify these forces in detail: it suffices that Ψ can be measured by well-established techniques (Croney, Coleman and Bridge 1952; Richards 1965; Holmes, Taylor and Richards 1967). In water-wet nonswelling media $\Psi = 0$ at saturation and decreases with θ to very large negative values (typically -10^4m) at the dry end of the moisture range of interest.

The partial volumetric Gibbs free energy associated with the local solid-water interaction is $\rho g\Psi$ and it follows that (in the absence of solutes) the liquid and vapour systems are connected at equilibrium by the relation

$$H = \exp g\Psi/RT , \tag{3}$$

where H is relative humidity, R the gas constant for water vapour, and T absolute temperature. We see that the $\Psi(\theta)$ relation presents in different guise exactly the information conveyed by the adsorption isotherm for water in the medium.

2.3 General partial differential equation of flow in unsaturated nonswelling media.

Combining (1) with $K = K(\theta)$ and (2) with the continuity requirement yields

$$\partial\theta/\partial t = \nabla.(K\nabla\Psi) - \partial K/\partial z , \tag{4}$$

where t denotes time. When the relations between K, Ψ, and θ are single-valued, (4) may be rewritten in terms of a single dependent variable. In terms of θ, the equation is

$$\frac{\partial\theta}{\partial t} = \nabla.(D\nabla\theta) - \frac{dK}{d\theta}\frac{\partial\theta}{\partial z} . \tag{5}$$

Both the *moisture diffusivity* D, defined by

$$D = K\, d\Psi/d\theta ,$$

and the coefficient $dK/d\theta$ are, in general, strongly-varying functions of θ.

Richards (1931) developed (4). Childs and George (1948) recognized the diffusion character of (5) for a horizontal one-dimensional system. Klute (1951) explicitly derived (5). Philip (1954, 1955, 1957b) extended the approach to include water transfer in vapour and adsorbed phases in the same formalism. The strong nonlinearity of Fokker-Planck equation (5) cannot be ignored, and progress in unsaturated flow studies has depended centrally on solution of it and related equations (e.g. Philip 1969a, 1988).

We emphasize the macroscopic character and the great generality of this approach. The macroscopic functions $K(\theta)$ and $\Psi(\theta)$ represent a sufficient characterization of the medium for the purposes of analysis and prediction of unsaturated flow phenomena. These two functions may be established directly from routine macroscopic measurements and may be of quite arbitrary functional form. We are not limited to simplifying assumptions about the internal geometry of the medium nor to any molecular or microscopic model of the solid-water interaction. We simply feed the appropriate $K(\theta)$ and $\Psi(\theta)$ functions, whatever their form, into (4) and use its solutions for analysis and prediction as required.

3. Equilibrium, Flow, and Volume Change in Swelling Media

The foregoing developments have provided a fruitful theoretical framework for study of the hydrology of nonswelling soils, but the need for generalization to swelling soils has long been recognized (Philip 1958a). We now consider some modest steps toward the required extension.

I must warn at once that these extensions are limited in character: they are, for the most part, restricted to *one-dimensional* systems; and they do not purport to treat irreversible structural changes which may occur in the presence of large enough stresses. Nevertheless, they have important consequences: they reveal, for example, that many classical concepts of soil- and ground-water hydrology, based on the behaviour of nonswelling media, fail completely for swelling soils.

3.1 The extension to swelling media.

At the outset we note that, since K, Ψ, and the *moisture ratio* ϑ are all free to vary in both saturated and unsaturated swelling media, both are amenable to the same general formulation. ϑ is the ratio of the volume occupied by water to that occupied by particles, and is equal to $\theta(1 + e)$, where e is the *void ratio*, defined as the ratio of void volume to particle volume. For swelling media it is usually more convenient to work with ϑ rather than θ.

The analysis is simpler and more straightforward for saturated or two-component (solid, water) systems, since they necessarily exhibit "normal" volume change (Keen 1931, Marshall 1959) and $e \equiv \vartheta$. Unsaturated or three-component (solid, water, air) systems exhibit "residual" volume change, with e dependent both on ϑ and on the normal stress P. Note that both the particles and the water are taken to be incompressible.

Three new basic elements enter the extension of the flow theory to swelling media:

A. In unsteady swelling systems, the soil particles are, in general, in motion, so that it must be recognized (Gersevanov 1937) that Darcy's law applies to flow relative to the soil particles. We therefore replace (1) with $K = K(\theta)$ by

$$\underline{\underline{v}}_r = -K(\vartheta)\nabla\Phi \ , \tag{6}$$

where $\underline{\underline{v}}_r$ is the vector flow velocity in the local rest frame of the particles.

B. For one-dimensional systems involving self-weight and/or surface loading, (2) must be generalized to include the *overburden potential* Ω (Philip 1969b), so that it becomes

$$\Phi = \Psi + \Omega - z \ . \tag{7}$$

It is convenient to take Ψ as the "unloaded" moisture potential, and Ω is then the contribution to Φ due to the normal stress, P. The measured water pressure in such systems is $\Psi + \Omega$. We discuss overburden potential in greater detail in section 3.2 below.

C. Whereas $K(\theta)$ and $\Psi(\theta)$ provide a sufficient hydrodynamic characterization of non-swelling media (for nonhysteretic processes), we now require for unsaturated systems $K(\vartheta)$, $\Psi(\vartheta)$, and $e(\vartheta,P)$, as well as the particle specific gravity γ_s. Figures 1 and 2 show typical $\Psi(\vartheta)$ and $e(\vartheta,P)$ relations for a swelling soil. For mineral soils $\gamma_s \approx 2.7$. Saturated systems need much less elaborate characterization. They require merely $K(\vartheta)$, $\Psi(\vartheta)$, and γ_s.

Unsteady swelling systems may be subjected either to Eulerian analysis (Prager 1953, Philip 1968) in the physical space coordinate, or to Lagrangian analysis (Hartley and Crank 1949, McNabb 1960, Smiles and Rosenthal 1968) in material coordinate m such that

$$\frac{dm}{dw} = (1 + e)^{-1} \ , \tag{8}$$

when the datum of m is taken in a plane where particles are stationary. Here w is the one-dimensional space coordinate. Analogous material coordinates may be used for two- and three-dimensional axisymmetric systems.

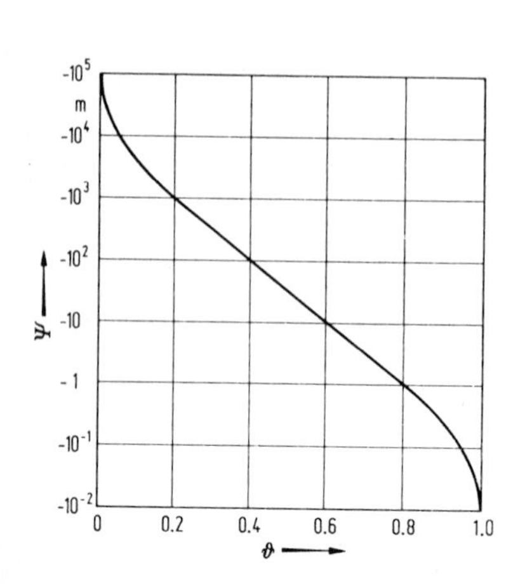

Fig. 1. Typical relation between moisture potential, Ψ, and moisture ratio, ϑ, for a swelling soil (Philip 1969b), adopted for illustrative purposes.

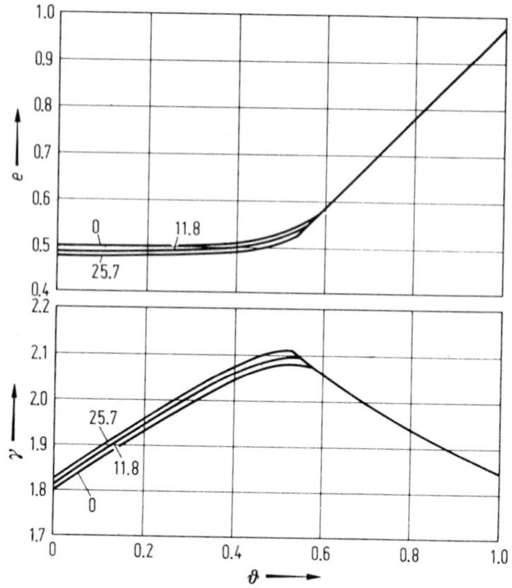

Fig. 2. The functions $e(\vartheta, P)$ and $\gamma(\vartheta, P)$ for the illustrative swelling soil (Philip 1971), e is the void ratio, γ the apparent wet specific gravity, and P the normal stress. Numerals on the curves denote values of P(m).

The Lagrangian analysis in material coordinates is much better adapted to swelling systems and turns out, quite generally, to be simpler and more manageable than the Eulerian analysis. Note that recognition that the particles move in unsteady systems is no mere exercise in pedantry: it is demonstrable that the mass flow with the particles can be of the same order of magnitude as the Darcy flow relative to the particles (Philip 1968).

3.2 Hydrostatics in swelling media.

The overburden potential Ω enters the analysis of vertical and loaded systems. It should be understood that the separation of the non-gravitational contributions to the total potential (i.e. $\Phi + z$) into Ψ and Ω is, in a sense, arbitrary; but it has the practical advantage that it enables us to identify, study, and calculate the separate contributions to water pressure arising from local interaction with the particles (Ψ) and from the normal stress produced by the weight of overlying strata and/or surface loading (Ω). It has been shown by Bolt (Philip 1970c, Groenevelt and Bolt 1972) that

$$\Omega = \alpha P, \quad \text{with } \alpha(\vartheta,P) = P^{-1}\int_0^P \frac{\partial}{\partial\vartheta}\left[e(\vartheta,P)\right]dP \; . \qquad (9)$$

For a vertical column at equilibrium, then, the use of (9) in (7) yields

$$\Phi = \Psi - z + \alpha\left[P(0) + \int_0^z \gamma \, dz\right] = \text{constant} = -Z \; . \qquad (10)$$

Here P(0) is the vertical stress due to loading at the upper surface $z = 0$ and γ is the *apparent wet specific gravity*, defined by

$$\gamma = (\vartheta + \gamma_s)/(1 + e) \; .$$

Figure 2 shows γ for our illustrative swelling soil. In hydrological contexts Z is the "water table depth". With γ_s and characterizing functions $\Psi(\vartheta)$ and $e(\vartheta,P)$ known, (10) may be solved to yield the equilibrium moisture profile corresponding to a given Z.

Experimental data (Talsma 1977a, 1977b) and exploratory calculations indicate that the P-dependence of α is weak and also that γ is a slowly-varying function of P; and it appears that the variation of these functions with P may be neglected without serious error in a P-range of perhaps 5m (corresponding to the top 2.5 m of the soil) (Philip 1971).

In this approximation Eq. (10) reduces to a linear first-order differential equation, for which the solution $z(\vartheta)$ is available in closed form. There are three classes of solution giving three distinct types of equilibrium profile: wet profiles with ϑ greatest at the surface and decreasing as depth increases, separated by a singular profile with ϑ constant and wet apparent specific gravity γ at its maximum value, from dry profiles with ϑ least at the surface and increasing with depth (Philip 1969b). These are the results for the practical case with $\gamma_s > 1$.

Overburden effects in hydrology had previously been recognized by some authors. Schofield (1935b) saw that Φ should include an overburden contribution. Coleman and Croney (1952) introduced a "compressibility factor" α, though their definition, evaluation, and use of α differ from those reviewed here. See also Collis-George (1961) and Rose, Stern and Drummond (1965).

We note that in saturated, two-component, systems $e = \vartheta$ and $\alpha = 1$, so that (10) reduces to the simpler

$$\Phi = \Psi - z + P(0) + \int_0^z \gamma dz = \text{constant} = -Z \; . \tag{11}$$

In this case, with $\gamma > 1$ all solutions are of the "wet" class.

3.3 Steady vertical flows

Combining (6) and (10) gives the equation for vertical flow. With the P-dependence of α and γ neglected, this is

$$v_r = -K\left[\left\{\frac{d\Psi}{d\vartheta} + \frac{d\alpha}{d\vartheta}\left[P(0) + \int_0^z \gamma dz\right]\right\}\frac{\partial\vartheta}{\partial z} + \alpha\gamma - 1\right] \; . \tag{12}$$

For steady flows this reduces to a linear differential equation of the first order in $d\vartheta/dz$ and ϑ, which may be integrated to yield $z(\vartheta)$ in closed form. The solution is singular under certain conditions: steady vertical flows are possible only for certain combinations of values of ϑ_0 and ϑ_∞ (values of ϑ at $z = 0$ and at large z). The details are complicated (Philip 1969d). For two-component systems (12) assumes the simpler form

$$v_r = -K\left[\frac{d\Psi}{d\vartheta} \cdot \frac{\partial\vartheta}{\partial z} + \gamma - 1\right].$$

For steady flows a quadrature gives $z(\vartheta)$ for given v_r and ϑ_0.

3.4 Unsteady flows

Applying continuity to (12) and using (8), we obtain the equation for unsteady flow and volume-change in vertical swelling systems,

$$\frac{\partial}{\partial t} = \frac{\partial\vartheta}{\partial m}\left[\frac{K}{1 + e}\left\{\frac{d\Psi}{d\vartheta} + \frac{d\alpha}{d\vartheta}\left[P(0) + \int_{m(0)}^{m} (\vartheta + \gamma_s)dm\right]\right\}\frac{\partial\vartheta}{\partial m}\right] + \frac{d}{d\vartheta}\left[K(\gamma\alpha - 1)\right]\frac{\partial\vartheta}{\partial m}. \tag{13}$$

$m(0)$ is the m–value corresponding to the soil surface $z = 0$.

The theory of various unsteady processes in vertical systems may be developed through solution of (13) subject to appropriate conditions: we have in mind infiltration, capillary rise, drainage, and evaporation; and also the soil-mechanical processes of consolidation and swelling of layers so thick that self-weight cannot be neglected. Equation (13) is a generalization of the nonlinear Fokker-Planck or convection-diffusion equation: in certain important special cases it reduces to a Fokker-Planck equation. For example, in two-component systems with $e = \vartheta$, (13) becomes

$$\frac{\partial\vartheta}{\partial t} = \frac{\partial}{\partial m}\left[\frac{K}{1+e}\frac{d\Psi}{d\vartheta}\cdot\frac{\partial\vartheta}{\partial m}\right] + (\gamma - 1)\,\frac{dK}{d\vartheta}\cdot\frac{\partial\vartheta}{\partial m}\,, \tag{14}$$

of the same general form as the nonlinear Fokker-Planck equation for infiltration in nonswelling media. Methods developed for studying the nonlinear Fokker-Planck equation promise to be useful in connexion with (13) (Philip 1969d, 1970b). For example, certain terms on the right of (13) are negligibly small at small t for many unsteady phenemona of interest, so that solutions of the nonlinear diffusion equation

$$\frac{\partial\vartheta}{\partial t} = \frac{\partial}{\partial m}\left[D^{**}\,\frac{\partial\vartheta}{\partial m}\right] \tag{15}$$

with

$$D^{**}(\vartheta) = (1+e)^{-1}\,K\{d\Psi/d\vartheta + P(0)\,d\alpha/d\vartheta\}$$

form the basis of perturbation methods of solving the full equation. Equation (15) is the counterpart of (13) for horizontal systems. For two-component and unloaded three-component horizontal systems D^{**} in (15) is replaced by D^*, with

$$D^*(\vartheta) = (1+e)^{-1}\,K\,d\Psi/d\vartheta$$

(Smiles and Rosenthal 1968, Philip 1968, Philip and Smiles 1969). The classical theory of consolidation (Terzaghi 1923) takes K and $d\Psi/d\vartheta$ constant and involves a linear diffusion equation. It is expressed in Eulerian coordinates, so that mass flow is ignored.

3.5 Applications

Although much further work is required, it is clear that hydrologic theory based on Eqs. (10), (12), (13) differs profoundly from the classical theory which takes no account of swelling and neglects the contribution to Φ of the overburden potential Ω. Perhaps the simplest general statement one can offer on the influence of swelling is this: the net effect of gravity on the equilibrium and flow of water in swelling media is approximately $(1 - \gamma\alpha)$ times that in nonswelling ones. For a mineral soil this factor is about -1 in the normal range, increasing to 0 when γ is maximum, and approaching +1 as $\alpha \to 0$ at small values of ϑ. Various "intuitions" of the hydrologist are therefore invalidated.

Equilibrium moisture distributions for a swelling soil are thus totally different in character from those for a nonswelling soil (with $d\vartheta/dz$ necessarily positive and the moisture distribution invariant with respect to the water table). Attempts to interpret equilibria in a swelling soil through classical concepts are therefore doomed to failure (Philip 1969c).

This *bouleversement* carries over to vertical flow processes. For example, the course of infiltration in a swelling soil evidently has analogies with that of capillary rise in a nonswelling one, and vice versa: and evaporation from an initially wet swelling soil does not necessarily exhibit the sharp transition between constant-rate and falling-rate phases characteristic of nonswelling media (Philip 1954, 1957c). These developments have evident practical consequences for groundwater hydrology (Philip 1971) and irrigation technology (Philip 1972b).

There are, of course, other important applications of this approach to swelling media. My long-time colleague Dr David Smiles (see for example, Philip and Smiles 1982, Smiles and Kirby 1991) has been active in applying the foregoing phenomenological theory to chemical engineering processes such as filtration and sedimentation.

3.6 Limitations

Exploration of the full implications of the foregoing analysis to three-component systems has been relatively slow. One impediment has been the experimental difficulty of making detailed measurements of $e(\vartheta, P)$ (e.g., Stroosnijder, 1976), though the work of Talsma both in the field (Talsma 1977a) and in the laboratory (Talsma 1977b) is notable.

A second limitation is that generalization to two- and three-dimensional systems requires that we integrate into the analysis an appropriate macroscopic representation of stress-strain relations in such systems. In the following section 4 we summarize an attempt to develop such a representation.

4. Mechanics of Colloid Pastes

In the late 1960's it was perceived that the preceding concepts and analysis of two-component swelling media carry over to the unsteady behavior of suspensions in such processes as sedimentation, filtration, and centrifigation (Philip, 1970a). Pastes consisting of dense suspensions of colloidal particles are typical systems to which the analysis applies.

The dominant particle-water (strictly particle-electrolyte) interaction in such pastes is that arising from electrical double layers. We may therefore apply the Poisson-Boltzmann equation to such systems, map all relevant microscopic physical quantities, and integrate up to phenomenological results such as the tensorial relations between macroscopic strain and macroscopic stress.

A programe of research along these lines was initiated by Philip, Knight, and Mahony (1985). Much remains to be done, but we indicate briefly here the character of the work and some of its results.

The connexions between anisotropic strain and anisotropic stress arise *inter alia* in the context of swelling soils in the field. An uncracked swelling soil in the field is constrained to change volume in one dimension only, the vertical. The horizontal dimensions of a field of swelling soil do not change with moisture content, but the elevation of its surface does. As the soil dries and (more or less) vertical cracks open, however, the individual monoliths between the cracks are free to change volume three-dimensionally. Change in the constraints on swelling and shrinking evidently produces a change in the energetics, and this is related to the energetics of cracking.

4.1 Questions from soil mechanics and soil physics

A related motivation was the hope that the work might shed light on some questions arising in soil mechanics (and indeed soil physics).

Although the more thoughtful writers on soil mechanics draw distinctions between soils of high colloid content and soils such as sands, there appears to remain a need to test various concepts against what happens on the microscopic scale in a colloidal soil. We list some questions here.

4.1.1 *State of stress of soil-water.* Classical soil mechanics considers the water in soil-water systems to be in a state of isotropic stress; and soil physicists would generally agree. At one time there were arguments in the literature about whether the soil water was in compression or in tension. But double-layer analysis reveals that the water is, in general, in an anisotropic (tensorial) state of stress: in some circumstances and at some points it will be in tension in some directions, and in compression in others.

4.1.2 *Transmission of load. Particle to particle contact.* With honorable exceptions, soil mechanics literature envisages the transmission of load between soil particles as through grain-to-grain contact. As we shall see, this is by no means essential. Load can be transmitted from charged particle to charged particle by means of tensorial electrostatic Maxwell stresses in the water.

4.1.3 *Concepts of "effective stress" and "pore pressure".* The soil-mechanics concepts of "effective stress" and "pore pressure" tend to be obscured and confused by the mental picture of both these stresses as isotropic, and by lack of recognition that particle load transmission doesn't need particle to particle contact.

4.2 The Poisson-Boltzmann equation in homogeneous particle arrays

Our point of departure is the Poisson-Boltzmann equation applied to the diffuse double layer:

$$\nabla^2\psi = -(4\pi\epsilon/D) \sum z_i n_i(0) \exp(-z_i\epsilon\psi/kT) . \qquad (16)$$

∇^2 is the Laplacian in physical space coordinates; ψ is the electrostatic potential; D is the dielectric constant of the solution; $n_i(0)$ is the number per unit volume of ions of species i in regions of the solution at potential 0; z_i is the signed valency; ϵ is the protonic charge; k is the Boltzmann constant; and T is the absolute temperature. (16) combines Poisson's equation for the distribution of potential with Boltzmann's equation for the ionic concentrations. Boltzmann's equation implies that the total number of ions per unit volume at potential ψ,

$$N(\psi) = \sum n_i(0) \exp(-z_i \epsilon \psi / kT).$$

We apply equation (16) to arrays of identical charged particles. Minimum energy considerations require that the equilibrium configuration be a regular particle array. Each particle occupies its own basic cell, bounded by a surface on which the normal component of the electrostatic field strength, $\partial\psi/\partial\nu$, vanishes. All cells are identical, so the problem reduces to solving (16) subject to the conditions

$$\psi = \psi_0 \text{ on } A_0 \; ; \quad \partial\psi/\partial\nu = 0 \text{ on } A_1 \; . \tag{17}$$

Here A_0 is the particle surface and A_1 the surface of its basic cell.

These regular arrays of identical colloidal particles are not just a convenient fantasy. They actually happen, though of course not in any exact sense in soils. Colloid crystals occur in nature: for example ordered arrays of identical spheres of amorphous silica are precursors to opals (Sanders, 1964). In recent years they have become commonplace in the laboratory where colloid scientists produce ordered arrays of identical particles of latex and other polymers (Efremov 1976 , Bartlett, Ottewill and Pusey 1990).

4.3 Gibbs free energy, variational principle, and microscopic and macroscopic stress tensors

With ψ known from solution of (16) subject to (17), it is a relatively straightforward matter to analyze the energetics, and the stress and load distributions in a particle array.

4.3.1 *Gibbs free energy, variational principle.* The Gibbs free energy of interaction per particle is

$$G = -\int_V \frac{D}{8\pi} [\nabla\psi]^2 + kT\ [N(\psi) - N(0)]\ dv\ , \tag{18}$$

where the integral is taken over the volume of the cell external to the particle.

The colloid literature gives elaborate, lengthy, and opaque derivations of (18). In fact (18) follows immediately from recognition that (16) yields a variational principle governing ψ, namely that the integral of (18) must be minimized. Formally, (16) is the Euler-Lagrange equation for $-G$ treated as a functional of ψ (i.e. the differential equation for ψ giving the distribution of ψ minimizing $-G$ defined by (18)).

The first term on the right of (18) is associated with an anisotropic microscopic (Maxwell) stress tensor due to the electric field; and the second is connected with the osmotic pressure due to the ionic excess in the double layer.

4.3.2 *Microscopic stress tensor.* The local microscopic stress tensor $\underline{\underline{\sigma}}$ has as its principal components

$$\sigma_1 = \sigma_2 = \frac{D}{8\pi} [\nabla\psi]^2 + kT\ [N(\psi) - N(0)]\ ;$$

$$\sigma_3 = -\frac{D}{8\pi} [\nabla\psi]^2 + kT\ [N(\psi) - N(0)]\ . \tag{19}$$

σ_3 is the component of $\underline{\underline{\sigma}}$ in the direction of the electric field with σ_1, σ_2 in directions normal to it. Tensile stress is positive. The Maxwell stress (the first term on the right) is positive for σ_1 and σ_2 and negative for σ_3. The osmotic term (second on the right) is positive for all three components.

4.3.3 *Macroscopic stress tensor.* For exploratory calculations we adopt a cuboidal array of particles, with a cuboidal basic cell with sides 2a, 2b, 2c in the principal directions of the particle array. The principal components of the macroscopic stress tensor $\underline{\underline{S}}$ (defined in terms of averages over the relevant cross-sections) are then:

$$S_1 = -V^{-1}\,\partial G/\partial \ell n\ a\ , \quad S_2 = -V^{-1}\,\partial G/\partial \ell n\ b\ ,$$
$$S_3 = -V^{-1}\,\partial G/\partial \ell n\ c\ , \tag{20}$$

with cell volume $V = 8abc$.

4.3.4 *Debye-Hückel approximation.* In the Debye-Hückel (linear) approximation, and with suitable normalization, (18) reduces to the simple

$$G = -\int_V [(\nabla\psi)^2 + \psi^2]\,dv\ . \tag{21}$$

Applying the divergence theorem and using (17), we get the beautiful result

$$G = \int_{A_0} \frac{\partial\psi}{\partial\nu}\,dA_0\ , \tag{22}$$

with the normal derivative taken outward. A nice economy is that if we use orthogonal expansions to solve Debye-Hückel for cylindrical and spherical particles, G turns out to be simply proportional to the coefficient of the leading term.

Results are most economically expressed in dimensionless form. When stresses are normalized so that (20) holds also in the new symbols, (19) is replaced by

$$\sigma_1 = \sigma_2 = (\nabla\psi)^2 + \psi^2\ ;$$
$$\sigma_3 = -(\nabla\psi)^2 + \psi^2\ . \tag{23}$$

4.4 Some illustrative results

Various relevant solutions have been found, and an ongoing program is planned. To date the major body of results is for the very simplest system: a two-dimensional array of cylindrical particles (dimensionless radius 1) satisfying the Debye-Hückel equation. Each particle sits in its own rectangular cell of dimensions $2a \times 2c$. Among numerous ways of presenting the results, the following are perhaps the most significant.

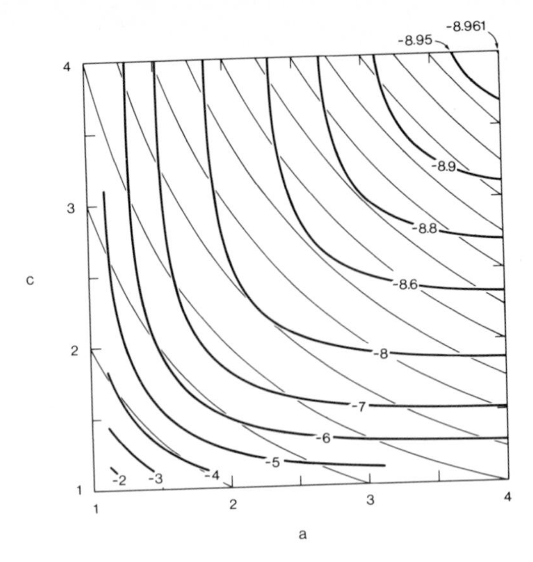

Fig. 3. Dependence of normalized Gibbs free energy per particle, G, on normalized cell dimensions a and c. Values of G are shown on the bold curves. Lighter curves are hyperbolae ac = constant and are thus curves of constant cell volume. Note that G(a,c) = G(c,a).

4.4.1 *Total Gibbs free energy per particle.* Figure 3 shows the variation of total Gibbs free energy per particle as a function of cell dimensions a and c. The plot is necessarily symmetrical about the diagonal from lower left to upper right, since G(a,c) = G(c,a).

Values of G are shown on the bold curves. The lighter curves are various rectangular hyperbolae ac = constant, representing constant cell volume. Note that the hyperbolae are always less curved than the constant G curves. This means that for fixed cell volume G is always more negative on the diagonal (which represents isotropic volume change) than anywhere else on a given hyperbola. For fixed cell-volume the excess of G over its value on the diagonal can be interpreted as strain energy, potentially available to initiate cracking.

4.4.2 *Macroscopic stress tensor.* In this two-dimensional system, S_1 is the principal macroscopic stress component in the direction of variation of a, and S_3 that in the direction of variation of c. Because of the symmetry of G(a,c), $S_3(a,c) = S_1(c,a)$. Figure 4 thus provides the required information on the nonlinear relations between the strain tensor defined by a and c and the macroscopic stress tensor $\underline{S}$.

4.4.3 *Details of the macroscopic stress distribution.* The maps of G and S_1 show in global form macroscopic implications of our solutions. There is

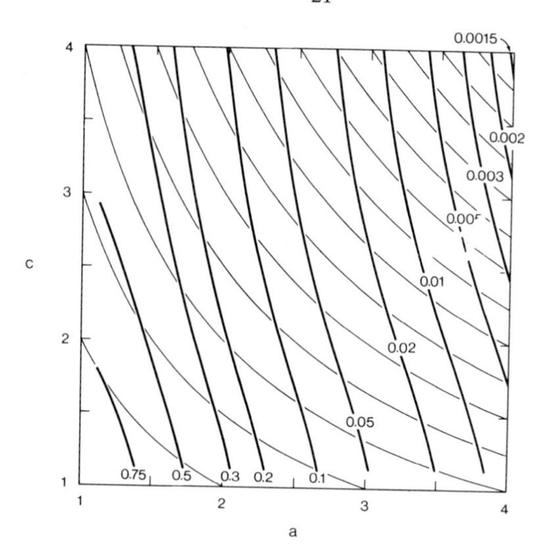

Fig. 4. Dependence of normalized macroscopic principal stress component S_1 (in direction of variation of a) on normalized cell dimensions a and c. Numerals on the curves are values of S_1. Note that $S_1(c,a) = S_3(a,c)$, where S_3 is principal component in direction of variation of c.

also an enormous amount of detail on the microscopic stress distribution around the particle and in the water. We limit ourselves here to just two aspects, the principal component of the macroscopic stress in the direction of the electric field, σ_3, and load sharing between particle and water.

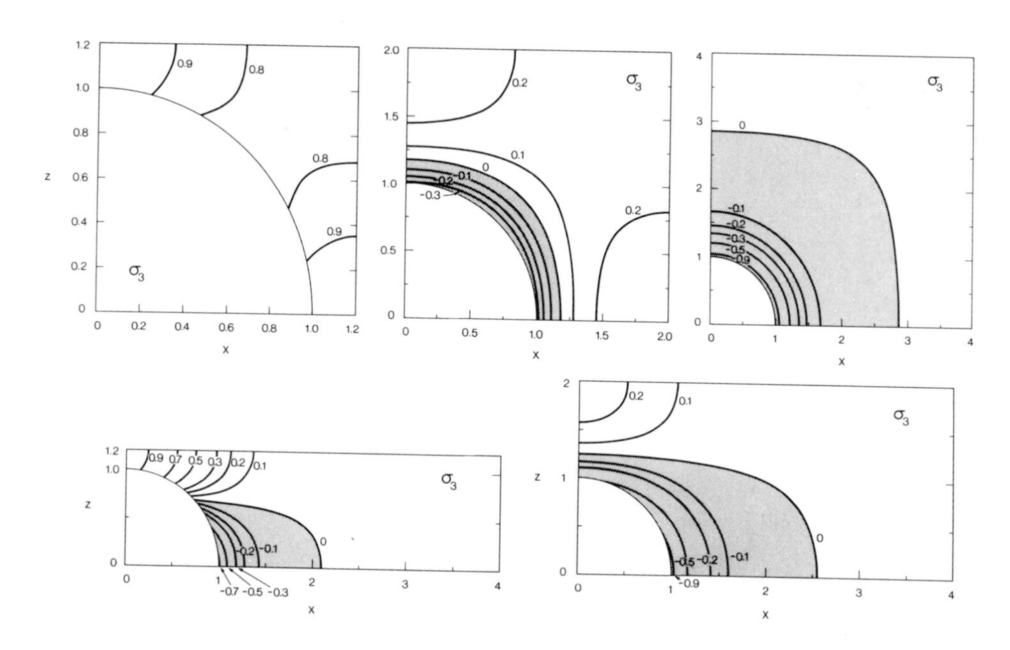

Fig. 5. Normalized microscopic principal stress component σ_3 (in direction of the electric field) for various cell configurations. Numerals on the curves are values of σ_3. Toned regions are regions of negative σ_3 (compression).

The principal component of microscropic stress normal to the electric field (i.e. roughly parallel to the particle surface), σ_1 is everywhere positive (tensile). The component in the direction of the field (i.e. roughly normal to the particle surface) may be either positive or negative (compression). Figure 5 maps σ_3 for some configurations of the basic cell. The upper row are isotropic cells: for the quarter-cell 1.2 × 1.2 there is no compression, for 2 × 2 a small compressive zone, and for 4 × 4 a large one. The two lower maps, plus that for 4 × 4, form a series with the cell swelling one-dimensionally. Compression occurs in regions of large gradient of electrical potential and charge density (regions close to one particle but far from the next particle in that direction).

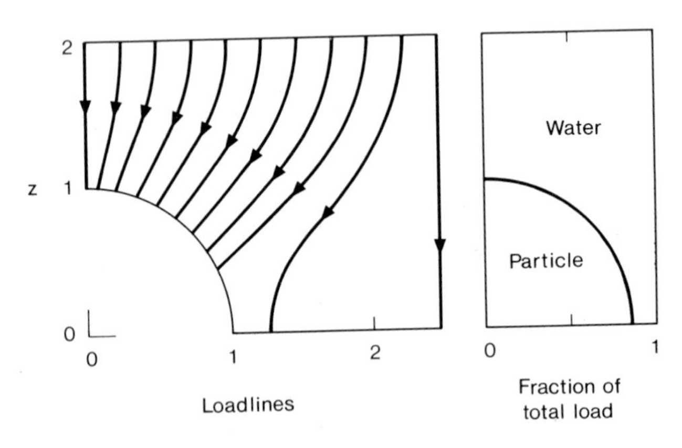

Fig. 6. Left: load lines illustrating how a vertical load is carried by the water and the particle. Right: partition of the load between water and particle.

Figure 6 is a plot of load lines (analogous to streamlines in fluid mechanics) representing how a given vertical load is shared on various horizontal cross-sections between particle and water. Above the particle the load is, of course, taken 100% by the water; but by the equatorial plane of the particle, some 85% of the load is carried by the particle.

4.4.4 *Two remarks*. First we recall that the results presented are based on the linear Debye-Hückel form of the Poisson-Boltzmann equation. We have here a striking example of a linear process on the microscopic scale leading to a strongly nonlinear process on the macroscopic scale.

Second, we observe that, in the unloaded state the water is generally in tension and the particle in compression. Our particle-water system thus behaves essentially as a prestressed composite material. A fanciful analogy is with prestressed concrete, the water behaving like the steel, the particles like the concrete.

5. Equilibrium, Flow, and Volume Change in Cells and Tissues

My first excursion into the mechanics of swelling concerned cells and tissues (Philip 1958b, 1958c, 1958d). Here I relate that and some connected work to concepts treated earlier in this presentation.

5.1 Energetics of osmotic cells and tissues.

5.1.1 *The classical osmotic cell.* Our point of departure is the energetics of the classical osmotic cell (Starling 1896, Höfler 1920). We express this in symbols connected in part to those used here earlier:

$$\Psi(V) = \tau(V) - \omega(V) \quad . \tag{24}$$

The cell water potential Ψ can be regarded as a unique function of cell volume V when the turgor pressure τ and the osmotic pressure of the cell contents are unique functions of V. These conditions will be met when both cell osmotic behaviour and cell deformation are reversible.

5.1.2 *Total potential and gravitational potential in cells and tissue.* In tissues made up of aggregations of cells, equilibrium, flow, and volume change are determined by the distribution of total potential Φ throughout the system.

For systems of this type with the vertical dimension significant, various physiologists (see, for example, Passioura 1982, Kirkham 1990) have recognized that Φ includes a gravitational component, so that

$$\Phi = \Psi(V) + z \quad , \tag{25}$$

where z is the vertical ordinate. It is natural to take z positive upward, unlike in the soils context earlier where it was positive downward. We use potentials per unit weight of water here also, so the gravitational component of Φ is just z. (25) is strictly comparable to (2) for nonswelling porous media.

5.1.3 *Overburden potential in cells and tissues.* So far as I am aware, physiologists have not taken the further step of recognizing that, for tissues free to change volume in the vertical, Φ should include also the overburden component Ω. For such tissues (25) should be replaced by

$$\Phi = \Psi(V) + z + \Omega \quad , \tag{26}$$

just as, for swelling porous media, (7) replaced (2).

Unfortunately a convincing evaluation of Ω in tissues seems very difficult: both estimation of the directional constraints on volume change, and of the vertical stress against which changes occur, present problems. A first guess is that Ω might effectively reduce by 1/3 the variation of Φ due to gravity.

The gravitational effect on the water relations of tall trees may be rather less than supposed; and the same may hold for the expected gravitational effect on blood pressure changes due to changes of (animal and human) posture.

5.2 Propagation of turgor and other properties through tissies.

In 1958 I proposed a macroscopic phenomenological analysis of the propagation of changes of turgor and related quantities such as water potential, osmotic pressure, and water content, through tissue made up of aggregations of osmotic cells (Philip 1958c, 1958d).

The initial formulation was rather simplified, but its basic elements persist in more sophisticated analyses. With the individual cell taken to be small compared with tissue dimensions, I proceeded to the limit with cell volume treated as infinitessimal and derived an unsteady diffusion equation describing the propagation of disturbances through the tissue. The formulation thus proceeded from the microscopic (cell) scale to macroscopic phenomenological theory dealing with quantities averaged over volumes large compared with the individual cell.

5.2.1 *Cell volume V, water content θ, and water ratio ϑ.* Volume V was classically used as the independent variable specifying the state of the osmotic cell. We adopt the convention that each cell is bounded by the mid-surface of each wall or structure it has in common with adjoining cells, and that V is the volume contained within that bounding surface. Within that volume is included, as well as water, the non-aqueous constituents of the cell, of volume V_n, taken to be constant. Then the volumetric moisture content θ is $1 - V_n/V$, and the moisture ratio ϑ is $V/V_n - 1$. These interpretations of θ and ϑ are consistent with their definitions for porous media.

5.2.2 *Dependence of Ψ, τ, and ω on θ or ϑ.* A source of economy in the analysis is the functional interdependence of various properties. As we have seen, Ψ, τ, and ω are all functions of V and, equally, they are functions of either θ or ϑ. The chain rule for differentiation thus transforms the equation for propagation of any one property into that for any other.

5.2.3 *Essential ingredients of the analysis.* One essential ingredient of the analysis is that, at least on relevant time scales, thermodynamic equilibrium holds on the scale of the individual cell. With this requirement met, we may characterize the tissue for our purposes *solely* by the following (or equivalent) functionals:

(1) The function $\Psi(\theta)$ specifying the energetics of the tissue.

(2) The function $K(\theta)$ specifying the water-conducting properties of the tissue. [$K(\theta)$ enters a macroscopic equation for water transfer, analogous to (1).]

These two macroscopic functions (or their equivalents) are all we need to develop the analysis. We may remain ignorant (agnostic) about the geometry and disposition of cell structures determining K, and also about the components of Ψ.

5.2.4 *The Eulerian equations.* For processes with tissue volume change relatively small, we obtain a general nonlinear diffusion equation

$$\frac{\partial \theta}{\partial t} = \nabla \cdot (D_1 \nabla \theta) \tag{27}$$

with

$$D_1(\theta) = K d\Psi/d\theta \quad ,$$

wholly analogous to (5) for nonswelling unsaturated porous media. [The gravitational first-order term is missing: it could be included, but we omit if for simplicity.] Equations for propagation of the other properties follow simply. They are of the heat-conduction, not the diffusion, form.

5.2.5 *The Lagrangian equations.* When tissue volume change cannot be ignored, we proceed to a Lagrangian formulation (Philip, 1969e), just as for swelling porous media. We use ϑ rather than θ, and for one-dimensional systems we employ material coordinate m, defined by

$$\frac{dm}{dw} = (1 + \vartheta)^{-1} \tag{28}$$

and comparable with (8). Axisymmetric two- and three-dimensional systems may be treated similarly.

In one dimension we obtain the equation (Philip, 1969e)

$$\frac{\partial \vartheta}{\partial t} = \frac{\partial}{\partial m}\left[D_2 \frac{\partial \vartheta}{\partial m}\right] \tag{29}$$

with

$$D_2(\vartheta) = (1 + \vartheta)^{-1} K d\Psi/d\vartheta \quad ,$$

in close analogy to (15).

5.3 Discussion

The major burden of this section 5 has been that, with appropriate reinterpretation of various quantities, the macroscopic analysis of flow and volume change in porous media carries over to the same processes in tissues. One facet is the comparable use of Lagrangian coordinates in the two systems, an aspect recognized by Raats (1987).

Strictly, equations such as (27) and (29) are nonlinear, but for tissues the nonlinearities tend to be weaker, and relative volume changes less, than in some swelling media. The consequence is that the simplifications of linear equations and Eulerian coordinates may be justified in some tissue contexts.

Molz (Molz and Klepper 1972, Molz, Klepper, and Browning 1973) successfully applied a linear and Eulerian form of the analysis to the radial system of cotton stem phloem. Molz and Klepper (1972) took account of tissue swelling by allowing the stem radius to be time-dependent. Molz, Klepper, and Browning (1972) used a constant 'average' radius, but recognized that a Lagrangian analysis was desirable.

Later Molz (Molz and Ikenberry 1974, Molz 1976) noted that the original formulation (Philip 1958c) took no explicit account of water transport within the cell-wall structures, and that its success might therefore occasion some surprise. With a cell model including both vacuole and wall structures, he found local thermodynamic equilibrium on the scale of the individual cell and arrived back at the diffusion description. This is of course consistent with the general analysis of propagation of disturbances in tissues described in section 5.2 (and especially 5.2.3) above and first outlined in Philip (1969e).

To this point our treatment has been for tissues in which the solutes are non-diffusible through the cell membranes. Philip (1958b) and Dainty (1963) showed that cell dynamics is much complicated by the presence of diffusible solutes; and the complications carry over to tissue dynamics. Molz (Molz and Hornberger 1973, Molz 1976) has taken the first steps towards generalizing the foregoing analysis to embrace solute diffusion.

References

Bartlett P, Ottewill RH, Pusey PN (1990) Freezing of binary mixtures of colloidal hard spheres. J Chem Phys 93:1299-1312

Buckingham E (1907) Studies on the movement of soil moisture. U.S. Dept Agr Bull 38.

Childs EC, George NC (1948) Soil geometry and soil-water equilibrium. Disc Faraday Soc 3:78-85

Childs EC, Collis-George N (1950) The control of soil water. Advan Agron 2:233-272

Collis-George N (1961) Free energy considerations in the moisture profile at equilibrium and effect of external pressure. Soil Sci 91:306-311

Coleman JD, Croney D (1952) The estimation of the vertical moisture distribution with depth in unsaturated cohesive soils. Road Res Lab Note RN/1709/JDC DC

Croney D, Coleman JD, Bridge PM (1952) The suction of moisture held in soil and other porous materials. Road Res Tech Paper 24

Dainty J (1963) Water relations of plant cells. Adv Bot Res 1:279-326

Edlefsen NE, Anderson ABC (1943) Thermodynamics of soil moisture. Hilgardia 15:31-298

Efremov IF (1976) Periodic colloid structures. Surface and Colloid Science 8:85-192

Gardner WR (1960) Dynamic aspects of water availability in plants. Soil Sci 89:63-73

Gersevanov NM (1937) The foundations of dynamics of soils. 3rd ed Stroiizdat, Moscow Leningrad

Groenevelt PH, Bolt GH (1972) Water retention in soil. Soil Sci 113:238-245

Hartley GS, Crank J (1949) Some fundamental definitions and concepts in diffusion processes. Trans Faraday Soc 45:801-818

Höfler K (1920) Ein Schema für die osmotische Leistung der Pflanzenzelle. Ber deut bot Ges 38:288-298

Holmes JW, Taylor SA, Richards SJ (1967) Measurement of soil water. In: Hagen RM, Haise HR, Edminster TW (eds) Irrigation of agricultural lands, Am Soc Agronomy, Madison pp 547-579

Keen NA (1931) The physical properties of the soil. Longmans London

Kirkham MB (1990) Plant responses to water deficits. In: Stewart BA, Nielsen DR (eds) Irrigation of agricultural crops, Am Soc Agronomy, Madison pp 323-342

Marshall TJ (1950) Relations between water and soil. Comm. Agr. Bureau Farnham Royal

McNabb A (1960) A mathematical treatment of one-dimensional soil consolidation. Quart Appl Math 17:337-347

Molz FJ (1976) Water transport through plant tissue: the apoplasm and symplasm pathways. J Theor Biol 59:277-292

Molz FJ, Hornberger GM (1973) Water transport through plant tissues in the presence of a diffusable solute. Soil Sci Soc Am Proc 37:833-837

Molz FJ, Ikenberry E (1974) Water transport through plant cells and cell walls. Soil Sci Soc Am Proc 38:699-704

Molz FJ, Klepper B (1974) Radial propagation of water potential in stems. Agronomy J 64:469-475

Molz FJ, Klepper B, Browning VD (1973) Radial diffusion of free energy in stem phloem: an experimental study. Agronomy J 65:219-225

Moore RE (1939) Water conduction from shallow water tables. Hilgardia 12:383-426

Passioura JB (1982) Water in the soil-plant-atmosphere continuum. In: Lange OL, Nobel PS, Osmond CB, Ziegler H (eds) Physiological plant ecology II, Springer, Berlin Heidelberg New York, pp 5-33

Philip JR (1954) Some recent advances in hydrologic physics. J Inst Engrs Australia 26:255-259

Philip JR (1955) The concept of diffusion applied to soil water. Proc Nat Acad Sci India (Allahabad) 24A:93-104)

Philip JR (1957a) The physical principles of soil water movement during the irrigation cycle. Trans ICID 3rd Congr Irrig Drain 8.125-8.154

Philip JR (1957b) The theory of infiltration: 1. The infiltration equation and its solution. Soil Sci 83:345-357

Philip JR (1957c) Evaporation, and moisture and heat fields in the soil. J Meteorology 14:356-366

Philip JR (1958a) Physics of water movement in porous solids. Highway Res Board Spec Report 40:147-163

Philip JR (1958b) The osmotic cell, solute diffusibility, and the plant water economy. Plant Physiol 33:264-271

Philip JR (1958c) Propagation of turgor and other properties through cell aggregations. Plant Physiol 33:271-274

Philip JR (1958d) Osmosis and diffusion in tissue: half-times and internal gradients. Plant Physiol 33:275-278

Philip JR (1968) Kinetics of sorption and volume change in clay-colloid pastes. Aust J Soil Res 6:249-267

Philip JR (1969a) Theory of infiltration. Adv Hydroscience 5:215-296

Philip JR (1969b) Moisture equilibrium in the vertical in swelling soils, 1. Basic theory. Aust J Soil Res 7:99-120

Philip JR (1969c) Moisture equilibrium in the vertical in swelling soils, 2. Applications. Aust J Soil Res 7:121-141

Philip JR (1969d) Hydrostatics and hydrodynamics in swelling soils. Water Resour Res 5:1070-1077

Philip JR (1969e) Theory of flow and transport processes in pores and porous media. In: Wolstenholme GEW, Knight J (eds) Circulatory and respiratory mass transport, Churchill, London pp 25-48

Philip JR (1970a) Hydrostatics in swelling soils and soil suspensions: unification of concepts. Soil Sci 109 294-298

Philip JR (1970b) Flow in porous media. Ann Rev Fluid Mech 2:177-204

Philip JR (1970c) Reply to note by E G Youngs and G D Towner on 'Hydrostatics and hydrodynamics in swelling soils'. Water Resour Res 6:1248-1251

Philip JR (1971) Hydrology of swelling soils. In: Talsma T, Philip JR (eds) Salinity and water use, Macmillan, London, pp 95-107

Philip JR (1972a) Flow in porous media. In: Becker E, Mikhailov GK (eds) Theoretical and applied mechanics, Springer, Berlin pp 279-294

Philip JR (1972b) Recent progress in the theory of irrigation and drainage of swelling soils. Trans ICID 8th Congress Irrig Drain, paper C2, 16pp

Philip JR (1988) Quasianalytic and analytic approaches to unsaturated flow. In: Steffen WL, Denmead OT (eds) Flow and transport in the natural environment: advances and applications, Springer, Heidelberg pp 30-47

Philip JR, Smiles DE (1969) Kinetics of sorption and volume change in three-component systems. Aust J Soil Res 7:1-19

Philip JR, Smiles DE (1982) Macroscopic analysis of the behaviour of colloidal suspensions. Adv Colloid Interface Sci 17:83-103

Philip JR, Knight JH, Mahony JJ (1985) Mechanics of colloidal suspensions with application to stress transmission, volume change, and cracking in clay soils. Proc ISS Symp Water and solute movement in heavy clay soils (eds Bouma J Raats PAC) Int Inst Land Reclamation Improvement Wageningen pp 39-44

Prager S (1953) Diffusion in binary systems. J Chem Phys 21:1344-1347

Raats PAC (1965) Development of equations describing transport of mass and momentum in porous media, with particular reference to soils. PhD thesis, University of Illinois, Urbana

Raats PAC (1987) Applications of material coordinates in the soil and plant sciences. Neth J Agr Sci 35:361-370

Richards LA (1931) Capillary conduction of liquids through porous mediums. Physics 1:318-333

Richards LA (1965) Water conducting and retaining properties of soils in relation to irrigation. Desert research, Res Council Israel Spec Report 2:523-546

Rose CW, Stern WR, Drummond JE (1965) Determination of hydraulic conductivity as a function of depth and water content for soil in situ. Aust J Soil Res 2:1-9

Sanders JV (1964) Colour of precious opal. Nature 204: 1151-1153

Schofield RK (1935a) The interpenetration of the diffuse double layers surrounding soil particles. Trans 3rd Cong Soil Sci 2:30-33

Schofield RK (1935b) The pF of water in soil. Trans 3rd Cong Soil Sci 2:37-48

Sedov LI (1971) A course in continuum mechanics vol 1. Wolters-Nordhof, Groningen

Smiles DE Rosenthal MJ (1968) The movement of water in swelling materials. Aust J Soil Res 6:237-248

Smiles DE, Kirby JM (1991) Water movement and volume change in swelling systems. In: Karalis TK (ed) Swelling mechanics: from clays to living cells and tissues. Springer, Berlin pp 000-000

Starling EH (1896) The glomerular functions of the kidney. J Physiol Lond 24:317-330

Stroosnijer L (1976) Infiltratie en herverdeling van water in grond. Versl Landbouwk Onderz 847

Talsma T (1977a) Measurement of the overburden component of total potential in swelling field soils. Aust J Soil Res 15:95-102

Talsma T (1977b A note on the shrinkage behaviour of a clay paste under various loads. Aust J Soil Res 15:275-277

Terzaghi K (1923) Die Berechnung der Durchlässigkeitsziffer des Tones aus dem Verlauf der hydrodynamischen Spannungserscheinungen. Sitsb Akad Wiss (Wien) Abt 2a, 132:125-138

Truesdell C, Noll W (1965) The non-linear field theories of mechanics. Handbuch der Physik III/3. Springer, Berlin

Zaslavsky D (1968) Average entities in kinematics and thermodynamics of porous materials. Soil Sci 106:358-362

WATER MOVEMENT AND VOLUME CHANGE IN SWELLING SYSTEMS

D.E. Smiles and J.M. Kirby
CSIRO Division of Soils
GPO Box 639
Canberra ACT 2601
Australia

INTRODUCTION

Volume change associated with water movement occurs in many areas of science and technology. In civil engineering for example consolidation of soil beneath man-made structures is a well known phenomenon as is consolidation associated with drainage and exploitation of aquifers. Analogous problems arise also in the chemical engineering processes of sedimentation, centrifugation and filtration where liquid is separated from a consolidating solid in two-phase systems.

These processes are still described using different formalisms and terminology and to different degrees of scientific rigour in the identifiably different disciplines of, for example, soil mechanics, chemical engineering and soil science. In fact these differences also exist within disciplines. Reference to reviews by Leu (1986), Shirato et al. (1986) and Wakeman (1986) for example reveal that the processes of filtration, thickening and expression in chemical engineering are still approached differently despite the fact that the basic physical processes of movement of liquid relative to solid are identical.

Furthermore, acceptance of new approaches to an old problem is often prejudiced by the need to displace physically inaccurate theory which has wide acceptance by virtue of usage and often of regulatory blessing. An example is the linearised, small strain, approximate approach to consolidation of Terzaghi (1956). The utility of this approach obscured its inherent approximations (which Terzaghi (1923) recognised), despite apparent physical failure to match the prediction based in this incomplete theory (Smiles and Poulos 1969; Smiles 1973). Only in the past decade has there been significant reappraisal of the assumptions of strain and linearity by the industry in question (Gibson et al., 1981). It is distressing to note that this incomplete theory has recently provided basis for "development" in chemical engineering (Shirato et al. 1986).

This paper reviews a general and physically correct approach to one-dimensional liquid flow and volume change in a two component (solid/liquid) system which is macroscopic (Philip, 1972, 1991, Raats and Klute, 1968a) in the sense that it is based on well defined and measurable properties which represent averages over a material volume which is great relative to the size of the particle or pore. In particular, it requires that the permeability (or hydraulic conductivity) of the system, and the potential of the liquid, be well-defined and measurable functions of the liquid content. The specific gravity of the solid is also required.

NATO ASI Series, Vol. H 64
Mechanics of Swelling
Edited by T. K. Karalis

The paper also cites experiments which provide support for the approach and identifies practically important situations for which solutions to the flow equation have been explored. Finally, the paper identifies areas where the approach presents challenges for theoretical development and practical application.

PHYSICAL CONSIDERATIONS AND THEORY

The theory depends for much of its insight on seminal papers by Raats and Klute (1968a,b) and Philip (1968), as well as extensive discussion with Drs Raats and Philip who provided impetus to many of the experiments described.

We deal exclusively with one-dimensional, two-phase flow. This provides the simplest system fully revealing the basic phenomena involved. It also encompasses a great many practically important problems and is the one most comprehensively studied. In this system, and without loss of generality we deal with water as the liquid component while most of the experiments deal with Wyoming bentonite as the solid component.

In order to describe unsteady water flow in such a system, where both phases are moving relative to an external observer and to each other, one must formulate continuity statements for each component, and formally describe the physical law that describes their relative motion in response to space gradients of force. Bird et al. (1960) explore these issues in a general sense while Raats and Klute (1968a,b) developed these ideas for water flow in soils.

Material Balance Equations

During one dimensional non-steady flow, equations of continuity may be written for the water (1) and for the solid (2), viz.

$$\left(\frac{\partial \theta_w}{\partial t}\right)_z = -\left(\frac{\partial F_w}{\partial z}\right)_t \quad , \qquad (1)$$

$$\left(\frac{\partial \theta_s}{\partial t}\right)_z = -\left(\frac{\partial F_s}{\partial z}\right)_t \quad , \qquad (2)$$

in which θ_w and θ_s are the volume fractions of the water and solid respectively, F_w and F_s are the respective volume fluxes relative to an external observer, and z and t are distance and time, respectively.

In a two-phase system it is evident that

$$\theta_w + \theta_s = 1 \quad . \qquad (3)$$

During non-steady flow the flux of water relative to an observer has a component, u, relative to the solid particles, and a "convective" component associated with the moving particles. F_w may therefore be written

$$F_w = u + \theta_w F_s/\theta_s = u + \vartheta F_s \quad . \tag{4}$$

In (4), F_s/θ_s is the average velocity of the solid relative to an observer, and $\theta_w/\theta_s=\vartheta$ is the moisture ratio (volume of water per unit volume of solid). In a saturated system, ϑ is identical to the void ratio, e: the distinction is maintained to anticipate situations where air may enter the system and $\vartheta<e$ (Philip and Smiles 1969).

Substitution for F_w from (4) in (1), and also $\vartheta\theta_s$ for θ_w yields

$$\left(\frac{\partial(\theta_s\vartheta)}{\partial t}\right)_z = -\left(\frac{\partial u}{\partial z}\right)_t - \left(\frac{\partial(\vartheta F_s)}{\partial z}\right)_t \quad . \tag{5}$$

Differentiation by parts followed by the elimination of two terms using (2), and division by θ_s then yield the equation

$$\left(\frac{\partial\vartheta}{\partial t}\right)_z = -(1/\theta_s)\left(\frac{\partial u}{\partial z}\right)_t - (F_s/\theta_s)\left(\frac{\partial\vartheta}{\partial z}\right)_t \quad . \tag{6}$$

which satisfies both continuity equations (1) and (2), but which focusses on the water component of the system.

Material Coordinates

Philip (1968) and Wakeman (1986) implicitly develop an Eulerian analysis of unsteady flow problems expressed in terms of the space coordinate z consistent with (6). Alternatively, a Lagrangian analysis may be developed recognising that left-hand side of (6), together with the second term on the right, represent the differential of ϑ following the motion of the solid (e.g., Bird et al. 1960), that is

$$\left(\frac{\partial\vartheta}{\partial t}\right)_z + (F_s/\theta_s)\left(\frac{\partial\vartheta}{\partial z}\right)_t = \left(\frac{\partial\vartheta}{\partial t}\right)_m \quad , \tag{7}$$

with m(z,t) a material coordinate defined by the equations

$$\frac{\partial m}{\partial z} = \theta_s \quad \text{and} \quad \frac{\partial m}{\partial t} = -F_s \quad , \tag{8}$$

so that

$$dm = \theta_s\, dz - F_s\, dt \quad . \tag{9}$$

Substitution of (7) in (6) and the use of the first of (8), yields the continuity equation for the water in material space which satisfies material balance for the solid, viz.

$$\left(\frac{\partial\vartheta}{\partial t}\right)_m = -\left(\frac{\partial u}{\partial m}\right)_t \quad . \tag{10}$$

Further development of the theory concentrates on (10), and in particular on the laws of flow necessary to define u.

Darcy's Law for Colloidal Systems

In the system we describe, Darcy's Law describes the volume flux of the water relative to the particles in response to a space gradient of the total potential, Φ, of the water (Zaslavsky 1964). Thus

$$u = -k(\vartheta)\ \nu^{-1}\ \partial\Phi/\partial z \quad . \tag{11}$$

In (11) u has units (ms^{-1}), ν is the kinematic viscosity of water (m^2s^{-1}), and $k(\vartheta)$ is the water content dependent permeability. If Φ is expressed as energy per unit mass of water, with SI units Jkg^{-1}, then the permeability takes units of (m^2). The form of $k(\vartheta)$ in systems that change their volume with ϑ is well-known.

Total Potential of the Water

In a one dimensional (vertical) swelling material, the total potential of the water is given by

$$\Phi = \psi(\vartheta) + \Omega + gz \quad . \tag{12}$$

In (12), g is the acceleration due to gravity, so gz is the gravitational potential of water at z, defined positive upwards relative to a convenient datum; Ω is the overburden potential, the work associated with vertical displacement of the system following addition of unit mass of water at position z; and $\psi(\vartheta)$ is the water content dependent potential that arises as a result of interaction of the water with the solid surfaces and their geometry. $\psi(\vartheta)$ is readily measured (Black 1965).

The overburden potential in a two-phase system is defined (Philip 1969a, Philip 1991) by

$$\Omega(z) = g\left[\int_z^{z_1} \gamma_w\, dz + P(0)\right] = gP \tag{13}$$

where γ_w is the wet specific gravity of the system, $z=z_1$ is its upper surface, P(0) (units L) is any normal surface load.

Combination of (12) and (13) yields

$$\Phi(z) = \psi(\vartheta) + g\left[\int_z^{z_1} \gamma_w \, dz + P(0)\right] + gz$$

$$= p_w(z) + gz \quad . \tag{14}$$

In (14), $p_w(z)$ is the water pressure measured with a manometer fitted with a membrane that permits passage of water but not solid. According to (14), $\psi(\vartheta)$ is the negative of the "effective stress" of civil engineering theory (Aitchison 1961) or the interparticle (solid compressive) pressure of filtration theory (Shirato et al. 1986).

Equation of Unsteady Vertical Flow in a Two-Phase System

Substitution of (12) in (13) and the inclusion of (13) then yields Darcy's Law in the form

$$u = -k(\vartheta)\nu^{-1}\left(\frac{\partial\psi}{\partial z} + g(1-\gamma_w)\right)$$

$$= -\frac{k(\vartheta)}{\nu(1+\vartheta)}\left(\frac{\partial\psi}{\partial m} + g(1-\gamma)\right) \quad . \tag{15}$$

In (15), γ is the specific gravity of the solid component of the system, and we have used (8) to derive the second equality, noting that in a two-phase system $\theta_s=(1+\vartheta)^{-1}$. The group $k(\vartheta)\nu^{-1}(1+\vartheta)^{-1}(s)$ plays the same role in material space as the hydraulic conductivity $K(s)(=k(\vartheta)\nu^{-1})$ plays in physical space.

A general equation of flow now follows if we substitute for u from (15) in (11)

$$\frac{\partial\vartheta}{\partial t} = \frac{\partial}{\partial m}\left(\frac{k(\vartheta)}{\nu(1+\vartheta)}\frac{\partial\psi}{\partial m}\right) - g(\gamma-1)\frac{\partial}{\partial m}\left(\frac{k(\vartheta)}{\nu(1+\vartheta)}\right) \tag{16}$$

which may be written as the non-linear Fokker-Planck equation (Philip 1969b)

$$\frac{\partial\vartheta}{\partial t} = \frac{\partial}{\partial m}\left(D(\vartheta)\frac{\partial\vartheta}{\partial m}\right) - E(\vartheta)\frac{\partial\vartheta}{\partial m} \tag{17}$$

in which the moisture diffusivity D is given by

$$D(\vartheta) = \frac{k(\vartheta)}{\nu(1+\vartheta)}\frac{d\psi}{d\vartheta} \tag{18}$$

and the coefficient E given by

$$E(\vartheta) = g(\gamma-1)\,\frac{d}{d\vartheta}\left[\frac{k(\vartheta)}{\nu(1+\vartheta)}\right] \quad . \tag{19}$$

The moisture diffusivity $D(\vartheta)(m^2s^{-1})$ is a coefficient of consolidation in civil engineering terms, and an expression coefficient in filtration theory (Shirato et al. 1986). The coefficient $E(\vartheta)(ms^{-1})$, which embodies the effects of gravity both directly and through the overburden component of potential, does not generally appear in filtration theory.

The fact that both D and E vary with ϑ complicates the solution of (17). There is, nevertheless, a substantial literature, in particular in soil physics and hydrology, devoted to analytical, quasi-analytical and numerical methods of solving this equation subject to relevant conditions.

In practice, however, this full solution may be unnecessary for many situations because the gravitational component of potential acts in an opposite sense and tends to cancel Ω (Smiles 1974). The importance of gravity also diminishes as the imposed pressure P_i is increased (cf. (14)). In these circumstances, flow approximates closely that of a "gravity free" system for which (17) becomes the non-linear diffusion equation

$$\frac{\partial\vartheta}{\partial t} = \frac{\partial}{\partial m}\left[D(\vartheta)\,\frac{\partial\vartheta}{\partial m}\right] \quad . \tag{20}$$

The major features of constant pressure filtration, for example, are well predicted by solutions to (20). Again, many solutions to (20) have been documented in the literature.

If the material properties are well-defined and the problem well-posed in terms of initial and boundary conditions we proceed to seek a solution to (17) or (20), $\vartheta(m,t)$, in material space. Integration of the first of (8) using $\vartheta(m,t)$ and the fact that $\theta_w=\vartheta/(1+\vartheta)$ in two-phase systems, then permit us to present the solution in Eulerian space as $\theta_w(z,t)$. The procedure is demonstrated for example by Smiles et al. (1982). For three-phase systems where $e>\vartheta$, there remain problems of analysis yet to be resolved. We return to this issue later in the paper.

MATERIAL PROPERTIES AND TEST OF THEORY

Evidently use of (16), (17) and (20) demands that the conductivity function, $k(\vartheta)$, the water potential function, $\psi(\vartheta)$, and the specific gravity of the solid be well-defined. These macroscopic functions, or combinations such as the diffusivity $D(\vartheta)$, are readily measured, but are not generally predictable.

In this section, we present examples of these properties measured for suspensions of bentonite in water with initial water contents $\vartheta_n \simeq 40$.

Water Potential - Water Content Relation, $\psi(\vartheta)$

The $\psi(\vartheta)$ data here were derived from pressure cell experiments in which a thin sample of a clay-water mixture is subjected to a constant gas pressure P_i in excess of atmospheric pressure. The water but not the clay may escape through a filter membrane to a pool of free water at atmospheric pressure in the plane of the pressure cell membrane. Static equilibrium exists when no more water leaves the cell, and when in a thin sample (according to (14)), $\psi=-gP_i$. The sample is then recovered and the water content determined by drying at 105°C.

Figure 1 shows the results of a series of such experiments on suspensions of bentonite in equilibrium with three different NaCl solutions. For each solution concentration, three different temperature regimes were used. Further detail of the chemical properties of the samples of clay used is provided in Smiles et al.(1985). Linear regressions were fitted to these data but it should be noted that the analysis applies generally, and independently of the form of $\psi(\vartheta)$; all that is required is that $\psi(\vartheta)$ be single valued.

It is clear from the figure that while the equilibrium salt concentration significantly affects the $\psi(\vartheta)$ relation, there is no significant temperature effect. Diffuse double layer concepts are qualitatively consistent with these data, but quantitative prediction of the three $\psi(\vartheta)$ relations is not possible.

Conductivity and Diffusivity Relations

Flow experiments are necessary to measure $k(\vartheta)$ and $D(\vartheta)$.

Non-steady Flow Experiments: Non-steady flow experiments are often easier to perform than are those involving steady flow. The simplest impose constant pressure, P, on an initially uniform vertical column of wet clay permitted to drain at its base to a pool of free water at similar elevation through a membrane permeable to water but not to solid. The cumulative outflow of water from the system is measured as a function of time. The evolution of the water content profile is measured in a series of experiments in which the columns are destructively sampled at a range of times. Such experiments were first described by Smiles and Rosenthal (1968) and reviewed by Smiles and Kirby (1988).

It is assumed during these experiments that we seek solutions to (20) subject to the conditions

$$\vartheta = \vartheta_n \qquad m > 0 \qquad t = 0 \tag{21}$$

$$\vartheta = \vartheta_o \qquad m = 0 \qquad t > 0 \tag{22}$$

in which ϑ_n is the initial uniform moisture ratio, and ϑ_o is the value of ϑ at the outflow membrane in equilibrium with $\psi=-gP_i$ at the membrane. (According to (18) if the membrane is freely water permeable and we define as our datum, atmospheric pressure at the external free surface at elevation z=0, then within the column at z=0, $\Phi=0$ and $\psi=-gP_i$.)

The material coordinate, m, is determined by integrating (9) using the fact that at the filter membrane at the base of the column the flux of solid, $F_s=0$. Hence

$$m(z,t) = \int_0^z \theta_s(\tilde{z},t)\, d\tilde{z} - \int_0^t F_s(o,\tilde{t})\, d\tilde{t}$$

$$= \int_0^z (1+\vartheta)^{-1}\, dz \quad . \tag{23}$$

Thus m is simply the cumulative volume of solid per unit area of cross section measured away from the membrane where z=m=0.

If the flow equation (20) is valid and conditions defined by (21) and (22) are realised, then the introduction of $\lambda = mt^{-\frac{1}{2}}$ eliminates m and t from (20), (21) and (22) and (20) may be reduced to

$$D(\vartheta)\, d\vartheta/d\lambda = - \int_{\vartheta_n}^{\vartheta} \lambda/2\, d\vartheta \tag{24}$$

whence we expect cumulative outflow from the column, i, to be linear with $t^{\frac{1}{2}}$ with slope proportional to P which is an integral outcome of the similarity of $\vartheta(\lambda)$ (Smiles and Rosenthal, 1968). We also expect the profiles of ϑ graphed as functions of $\lambda = mt^{-\frac{1}{2}}$ to be unique. These profiles permit calculation of $D(\vartheta)$ according to (24). Smiles and Harvey (1973) provided an alternative method for obtaining $D(\vartheta)$ by differentiating $i(t^{\frac{1}{2}})$ vs gP. This is a useful method for materials that do not permit easy measurement of water content profiles.

Figure 2a shows $\vartheta(\lambda)$ data for the clay in equilibrium with 3mM/l NaCl as shown in Fig. 1. These data were generated in experiments at the same imposed constant pressure, but were terminated at different times to test similarity in λ behaviour required of the theory. Subsets of the experimental set were performed at different temperatures (277, 293, 306K). For a given temperature $\vartheta(\lambda)$ data are effectively unique as required by the analysis. Furthermore, the graph of ϑ against $(\nu/\nu_*)^{\frac{1}{2}}k$ shown in Fig. 2b is also unique. In this figure, ν_* is the kinematic viscosity of water at 293K. This scaling implies that any temperature effects are wholly explained in terms of the temperature dependence of the viscosity of water.

Figure 3 shows $D(\vartheta)$ at 293K calculated using the data of Fig. 2b, and (24), and Fig. 4 shows corresponding values of $K(\vartheta)$ calculated using Figs. 1 and 3 and (18). The conductivity $K(\vartheta)$ varies with equilibrium salt content indicating that structure associated with double layer effects vitiate interpretation of K simply in terms of the macroscopic property ϑ. It is also easily shown that a Kozeny-Carman analysis (Carman 1956) will not fit these curves.

We note that $K(\psi)$ at 293K, shown as Fig. 5, and determined using Fig. 4 and Fig. 1 is effectively unique for all solution salt concentrations; that is, the mobility of the water relative to the solid is apparently simply related to the chemical potential of the water. Philip (1968a) used such a relation for his early illustrative calculations and, of course, it offers significant opportunities for generalising material

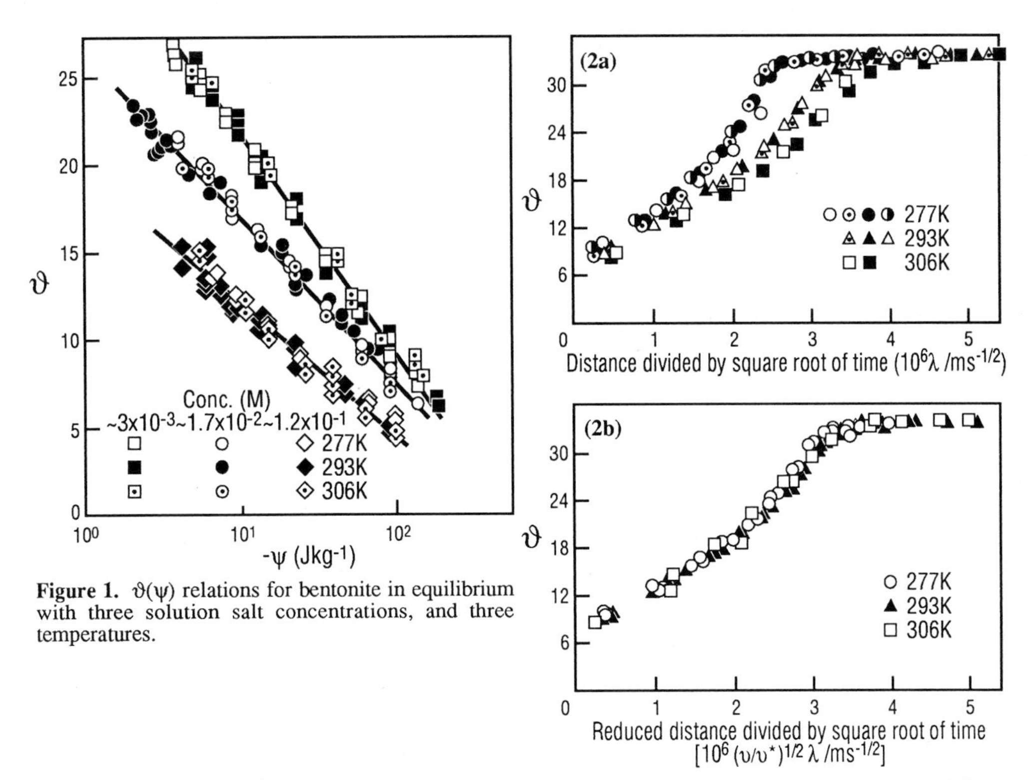

Figure 1. $\vartheta(\psi)$ relations for bentonite in equilibrium with three solution salt concentrations, and three temperatures.

Figure 2a. $\vartheta(\lambda)$ observed during constant pressure desorption of bentonite in equilibrium with 1.2×10^{-1} M NaCl, at three different temperatures. **Figure 2b.** $\vartheta(\{\upsilon/\upsilon^*\}^{1/2}\lambda)$ data, corresponding to figure 2a, demonstrating that temperature differences shown in that figure may be ascribed entirely to the temperature dependence of the kinematic viscosity, υ, of water.

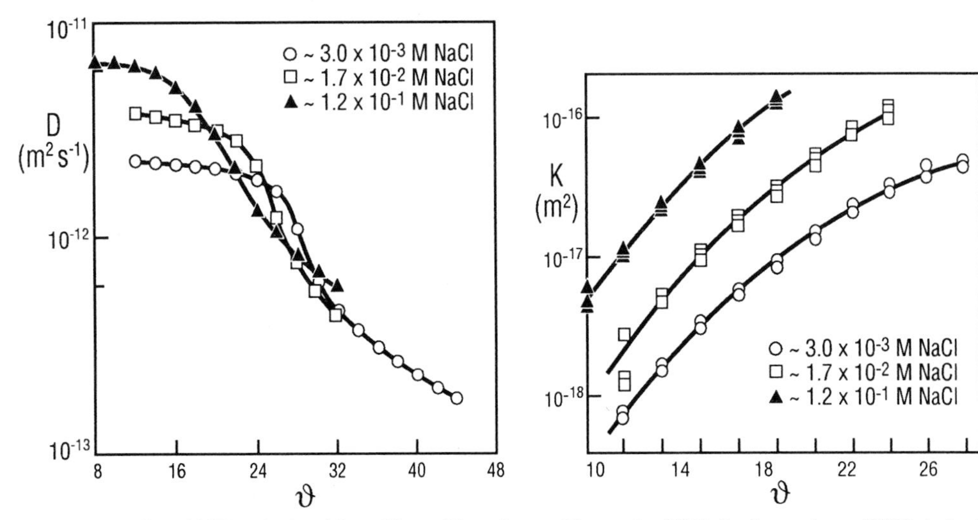

Figure 3. $D(\vartheta)$ at 293K, calculated from Figure 2b, and corresponding data for other salt concentrations shown in Figure 1.

Figure 4. $K(\vartheta)$ for bentonite at 293K derived from data of Figure 1 and Figure 3.

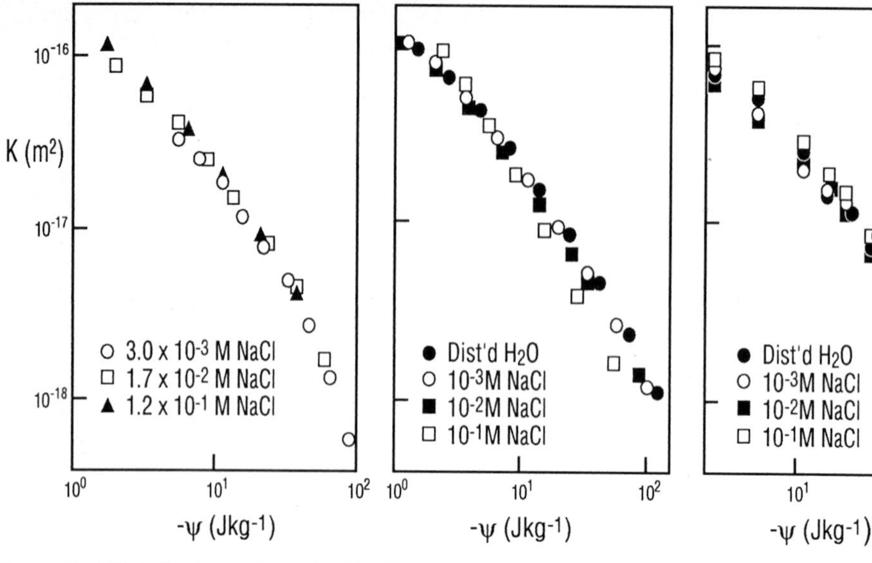

Figure 5. K(ψ) for bentonite at 293K derived from data of Figure 1 and Figure 4.

Figure 6. K(ψ) for bentonite at 293K derived from steady flow experiments.

Figure 7. K(ψ) for bentonite at 293K derived by differentiating outflow data from constant pressure experiments.

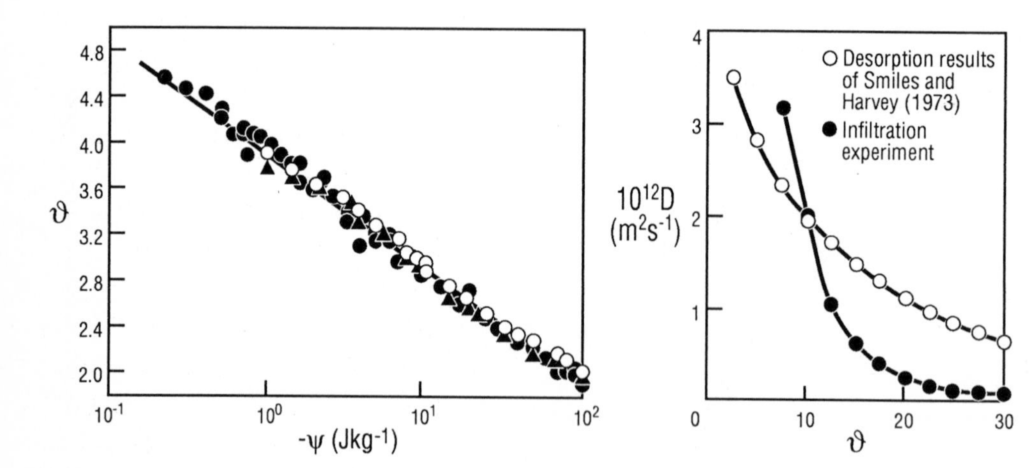

Figure 8. ϑ(ψ) for red mud.

Figure 9. D(ϑ) during sorption and desorption of bentonite.

behaviour where it arises. We are exploring this challenge with other materials.

<u>Steady Flow Experiments</u>: The theory is also tested if $K(\psi)$ determined from transient experiments corresponds to that measured in steady flow experiments. Kirby and Smiles (1988) performed such a series of experiments with clays defined in Fig. 1. Figure 6 shows $K(\psi)$ obtained in these experiments. They compare most satisfactorily with the data of Fig. 5. Kirby and Smiles (1988) also derived $K(\psi)$ independently by differentiating cumulative outflow data using the method of Smiles and Harvey (1973). Figure 7 shows $K(\psi)$ obtained in these experiments and, again, correspondence with data of Fig. 5 is close. These experiments provide strong support for the general approach.

We reiterate that while the $K(\psi)$ relations are not predictable, they are readily and reliably measured using a range of methods that may be matched to the properties of the material of concern. Furthermore, the relation appears to be unique over three orders of magnitude of equilibrium solution concentration, and implies the great practical advantage that only the dependence on electrolyte of $\vartheta(\psi)$ needs to be measured, in addition to $K(\psi)$, at any convenient electrolyte concentration, to fully characterise these particular systems.

EXEMPLARY APPLICATIONS

It is useful to discuss elements of the approach in association with exemplary application of the theory. We follow this procedure dealing first with static equilibrium and then flow problems. We emphasise that many applications of the theory take advantage of solutions to analogous problems in the equilibrium and flow of water in unsaturated, non-swelling soils.

Static Equilibrium

Philip (1969a) explored, in principle, the static equilibrium profile that might be expected in swelling systems. Only the so-called hydric situation has been subject to experiment however, by Smiles (1976) who integrated (14) using the experimentally derived $\vartheta(\psi)$ for red mud shown as Fig. 8 to predict static equilibrium profiles in equilibrium with a free water surface above the surface of the mud, and also for the case where the mud drained to equilibrium with a free water surface at its base. Red mud is an effluent produced by the aluminium industry and its de-watering and stabilisation presents a major management problem in the industry. Theory predicts and experiment confirms that the latter situation results in a drier and more stable material, and leads to the recommendation that provision should generally be made for drainage beneath a tailings dam if the site ultimately has to be restored to a "natural" state.

In passing, also we note in relation to (14) that in its differential form

$$d\psi/dm = - g(\gamma-1) \tag{25}$$

it takes a form exactly analogous to the well-known Buckingham (1907) equation for static equilibrium in a non-swelling porous medium but with the z-coordinate replaced by the material m, and with the gravity term modified by the submerged specific gravity of the solid.

Flow Processes

Many unit operations of chemical engineering which seek to separate liquid from solid, and many consolidation and swelling problems of civil engineering have initial and boundary conditions that permit quasi-analytical or numerical solution of (17) and (20). Furthermore, developments in the solution of the analogues of these equations for water flow in non-swelling soils (Philip 1988) provide a considerable catalogue of methods for dealing with many practical problems in swelling systems.

Thus the processes of constant pressure, and constant rate, filtration for materials of the sort characterised above have been explored using quasi-analytical methods developed in soil science (Smiles 1986, Smiles and Kirby 1988). In addition, aspects of centrifugation and sedimentation have been studied, as well as self-weight filtration. Furthermore, Smiles et al. (1979) examined the case of self-weight filtration during which accretion of slurry occurs at a constant rate at the top of the system. This is important in tailings dam drainage where the dam continues to have effluent added as it drains from beneath.

The effect of a relatively impermeable filter membrane in constant pressure filtration has also been satisfactorily predicted using (20) (Smiles et al. 1982).

RESEARCH NEEDS AND GENERAL DISCUSSION

As Fell and McDonogh (1988) point out, there still remain practical concerns about the approach and the conditions under which it has been tested. In particular they note that in no experiments performed by Smiles and co-workers do imposed pressures exceed 100kPa, and that the approach remains one-dimensional. The former objection is essentially irrelevant insofar as the approach requires only that the macroscopic properties $\vartheta(\psi)$ and $K(\vartheta)$ be well-defined, and Darcy's Law valid, in the range of pressure to be imposed for (17) and (20) to be valid. In those circumstances solutions can be found for any problem appropriately posed in terms of initial and bounday conditions. Also, the criticism is incorrect because Kirby (1985) applied the approach to filtration of industrial pastes to pressures up to 300 bar.

Multi-dimensional Flow

Fell and McDonogh's second criticism is more significant and while we observe that there a great number of practical problems that are essentially one-dimensional, there remain multidimensional problems that require solution.

The most important of these is the rheological one associated with any anisotropic stress field in such a system. In one dimension we assume, and experiment supports the assumption, that the energetics of the system reflect only the movement of the water, and the [one-dimensional] strain in the matrix is wholly reflected in $\vartheta(\psi)$. In more than one dimension the energetics of matrix distortion needs to be taken into account in both the statics and the dynamics (Biot 1941). The problem presents formidable experimental challenges as is evident in the paucity of such experiments.

Three-Phase Systems

A problem that Fell and McDonogh (1988) do not recognise, however, is the restriction of tested theory to two-phase systems. In fact the derivation leading to (10) is applicable to both two- and three-phase systems. Thereafter, we experience problems in the definition of the overburden potential, Ω (equation 13), and in the conversion from m to z and vice versa. The former are discussed in detail by Philip (1972) and arise because in an unsaturated, vertical swelling medium, addition of unit volume of water to the system at any point z is not generally accompanied by unit vertical displacement of the system above z. The overburden potential defined by (13) is reduced by a factor α, and must be written

$$\Omega(z) = g\left[\alpha(\vartheta,P)\int_z^{z_1}\gamma_w\,dz + P(0)\right] = gP \qquad (26)$$

with α calculated according to the equation

$$\alpha(\vartheta,P) = P^{-1}\int_0^P \frac{\partial}{\partial\vartheta}\left(e(\vartheta,P)\right)\,dP \qquad (27)$$

Equation (27) in principle is evaluated using the experimentally determined $e(\vartheta,P)$ relation which describes the way the volume of the system, described by e, depends on ϑ and the surface load, P. No satisfactory methods have been described to measure $e(\vartheta,P)$ however, and no adequate test of (27) performed.

In the three-phase system, $e(\vartheta,P)$ is also required to convert the Lagrangian solution of the three-phase analogue of (20) to Eulerian space using (23), because

$$\frac{\partial m}{\partial z} = \theta_s = 1/(1+e) < 1/(1+\vartheta) \quad . \qquad (28)$$

In principle, Philip and Smiles (1969) treated the issue of three-phase systems for horizontal, one-dimensional flow, developing (20) but with $D(\vartheta)$ replaced by

$$D(\vartheta,e) = \frac{k(\vartheta)}{\nu(1+e)}\frac{d\psi}{d\vartheta} \qquad (29)$$

Effect of Solution Salt Concentrations

It is evident from Figs. 1-5 that the equilibrium concentration of NaCl affects the structure and flow properties of these systems. There is therefore a need to extend these studies to systems of mixed salts, and to examine the effect of pH. Change in pH may substantially change the surface charge density of the clay mineral and associated oxides in colloidal systems, as well as the activity particularly of Al^{3+} which will profoundly affect electrical double layer interaction and hence $\vartheta(\psi)$. This need will be particularly significant in relation to clay membranes used to line dams constructed to retain noxious, and chemically reactive wastes.

Other Materials

Bentonites form stable and very wet suspensions and are ideal for testing the theory developed here. Thus while red mud, phosphate slimes and unconsolidated marine muds display properties consistent with the approach, there remains a need to systematically test application of the approach described here to drier systems and materials with coarser particles.

Issues of Scale

Philip (1991) deals with this issue fully elsewhere in this workshop. It is important however to identify two areas where microscopic considerations unnecessarily complicate a macroscopic approach. They relate to anomalous structure of water at colloidal surfaces, and the possible stress related volume change of the matrix of the system (eg. Tosun 1985). The first effect might be expected to affect the mobility and viscosity of the water in the vicinity of the colloid surface. Our experimental data (Fig 2) reveal no anomalous viscosity effects, and any other structural effects close to the surface are incorporated in the measured and macroscopic $K(\vartheta)$ function. In relation to the second issue, it is readily shown that stress dependent density change in the system matrix is accomodated within $\vartheta(\psi)$.

CONCLUDING REMARKS

The approach we review here is simple and direct, and based on macroscopic, well-defined and measurable material properties. These properties have exact analogues in well-established theory of water movement in non-swelling soil and the non-linear Lagrangian equations (17) and (20) which encapsulate the approach, similarly have exact analogues for those systems. We therefore have access to a considerable catalogue of solutions to (17) and (20) for a wide range of important initial and boundary conditions.

The approach is also well supported by experimental data, although admittedly for a restricted range of materials. It thus provides a suitable base from which to explore other solid-liquid systems. In association with (eg) bentonite it also provides a well-defined and predictable system against which the integrated outcome of theory developed from single particle and molecular considerations, must be tested.

REFERENCES

Aitchison GD (1961) Relationships of moisture stress and effective stress functions in unsaturated soils. In: Conf on pore pressure and suction in soils. Butterworths, London, pp47-52

Bird RB, Stewart WE, Lightfoot EN (1963) Transport phenomena. Wiley, New York

Biot MA (1941) General theory of three-dimensional consolidation. J Appl Phys 12:155-164

Black CA (ed) (1965) Methods of soil analysis. Part 1 Physical and mineralogical properties, including statistics of measurement and sampling. Am Soc Agron Inc, Madison, Wisconsin

Buckingham E (1907) Studies on the movement of soil water. US Dept Agr, Bur Soils, Bull 38

Carman, P.C., 1956. Flow of gases through porous media. London, Butterworths Scientific Publications, 182p.

Fell CJD, McDonogh RM (1988) One dimensional solid-liquid separation: Commentary. In: Steffen WL, Denmead OT (eds) Flow and transport in the natural environment: advances and applications. Springer-Verlag Berlin. pp 343-351

Gibson RE, Schiffman RL, Cargill KW (1981) The theory of on dimensional consolidation of saturated clays II Finite non-linear consolidation of thick homogeneous layers. Can Geotech J 18:280-293

Kirby JM, Smiles DE (1989) Hydraulic conductivity of aqueous bentonite pastes. Aust J Soil Res 26: 561-74

Kirby JM (1985) Compression of slurries - filtration or consolidation? In: Banks PJ (Ed) Filtration and press dewatering: principles andindustrial applications. CSIRO Division of Energy Technology, Melbourne, 64-5

Leu W (1986) Principles of compressible cake filtration. In: Cheremisinoff NP (ed) Encyclopedia of fluid mechanics, vol 5, Slurry flow technology. Gulf Pub Co, Houston, pp 865-904

Philip JR (1968) Kinetics of sorption and volume change in clay-colloid pastes. Aust J Soil Res 6:249-267

Philip JR (1969a) Moisture equilibria in the vertical in swelling soils. I. Basic theory. Aust J Soil Res 7:99-120

Philip JR (1969b) Theory of infiltration. Adv Hydrosci 5:215-296

Philip JR (1972) Hydrostatics and hydrodynamics in swelling media. In: Fundamentals of transport phenomena in porous media. IAHR, Elsevier, Amsterdam pp341-350

Philip JR (1988) Quasianalytic and analytic approaches to unsaturated flow. In: Steffen WL, Denmead OT (Eds) Flow and transport in the natural environment: advances and applications. Springer-Verlag Berlin. pp 30-47

Philip JR (1991) Mechanics of flow and volume change in soils and other porous media and in some biological contexts. In: Karalis TK (ed) Swelling mechanics: from clays to living cells and tissues. Springer-Verlag Berlin. pp ???-???

Philip JR, Smiles DE (1969) Kinetics of sorption and volume change in three component systems. Aust J Soil Res 7:1-19

Raats PAC, Klute A (1968a) Transport in soils: The balance of mass. Soil Sci Soc Am Proc 32:161-166

Raats PAC, Klute A (1968b) Transport in soils: The balance of momentum. Soil Sci Soc Am Proc 32:452-456

Shirato M (1986) Course notes for Workshop on filtration and expression. CSIRO Division of Energy Technology Clayton Australia
Shirato M, Murase T, Iwata M, Kurita T (1986) Principles of expression and design of membrane compression-type filter press operation. In: Cheremisinoff NP (ed) Encyclopedia of fluid mechanics, vol 5, Slurry flow technology. Gulf Pub Co, Houston, pp 905-964
Smiles DE (1973) Examination of settlement data from an embankment of wet light clay. Aust Road Res 5:55-59
Smiles DE (1974) Infiltration into a swelling material. Soil Sci 117:140-147
Smiles DE (1986) Principles of constant pressure filtration. In: Cheremisinoff NP (ed) Encyclopedia of fluid mechanics, vol 5, Slurry flow technology. Gulf Pub Co, Houston, pp 905-964
Smiles DE, Barnes CJ, Gardner WR (1985) Water relations of saturated bentonite: some effects of temperature and solution salt concentration. Soil Sci Soc Amer J 49:66-69
Smiles DE, Colombera PM (1975) The early stages of infiltration into a swelling soil. In: De Vries DA, Afgan NH (eds) Heat and mass transfer in the biosphere: 1. transfer processes in the plant environment. Scripta Book Co Washington. pp77-85
Smiles DE, Harvey AG (1973) Measurement of moisture diffusivity of wet swelling systems. Soil Sci 116:391-399
Smiles DE, Kirby JM (1988) One-dimensional solid-liquid separation. In: Steffen WL, Denmead OT (eds) Flow and transport in the natural environment: advances and applications. Springer-Verlag Berlin. pp 327-342
Smiles DE, Knight JH, Nguyen-Hoan TXT (1979) Gravity filtration with accretion of slurry at constant rate. Spar Sci Technol 14:175-192
Smiles DE, Poulos GH (1969) The one dimensional consolidation of columns of soil of finite length. Aust J Soil Res 7:285-291
Smiles DE, Raats PAC, Knight JH (1982) Constant pressure filtration: the effect of a filter membrane. Chem Eng Sci 37:707-714
Smiles DE, Rosenthal MJ (1968) The movement of water in swelling materials. Aust J Soil Res 237-248
Terzaghi K (1923) Die Berechnung der Durchlassigkeitsziffer des Tones aus dem verlauf der hydrodynamischen Spannungserscheinungen. Sitzer Akad Wiss Wien, Abt IIa, vol 32
Terzaghi K (1956) Theoretical soil mechanics. 8th Ed. Wiley, NY
Tosun I (1985) Mathematical formulation of cake filtration for deformable solid particles. Chem Eng Sci 40:637-674
Wakeman RJ (1986) Theoretical approaches to thickening and filtration. In: Cheremisinoff NP (ed) Encyclopedia of fluid mechanics, vol 5, Slurry flow technology. Gulf Pub Co, Houston, pp 649-684
Zaslavsky D (1964) Saturated and unsaturated flow equation in an unstable porous medium. Soil Sci 98:317-321

THERMODYNAMICS OF SOILS SWELLING NON-HYDROSTATICALLY

Theodoros K. Karalis

Democritos University of Thrace,
67100 Xanthi, Greece

1. INTRODUCTION

The thermodynamics of non–saturated soils swelling non–hydrostatically means knowledge of the free and bounded energy from heat adsorption, fluid imbibition, and fluid–solid interaction at any time and position of the continuum. However, of all the forms of energy, heat is the most disorderly. That is, it is harder to describe soils that differ in heat content alone than soils that differ in the amount of the moisture content or in the varying configurations due to the changes in the force field. The problem becomes difficult when heat flow must be considered in conjunction with fluid imbibition, therefore only isothermal conditions will be considered in this paper.

The purpose of this paper is to develop a solid model in order to simulate both diffusivity and the resulting mechanical properties of a non–saturated soil, swelling non–hydrostatically and in which the deforming of the soil skeleton cannot be calculated according to Terzaghi's (1923), Biot's (1941) or to more recent works based on the theory of consolidation (McNabb, 1960). The development that follows leads to the proposed model which can be viewed in accordance with experimental observations (Karalis, 1090a). Specifically, the analysis deals with the flow equation and with the problem of stress–strain–moisture density time properties of the non–saturated and non–hydrostatically swelling soil.

The flow equation is discussed in the second part of this development. The mechanical model is then presented in the third part in which the implicated parameters are compatible with the thermodynamical properties of the swelling matrix which are developed in the fourth part. Conclusions are drawn in the fifth part.

Specifically, the article proposes both: (*i*) a set of equations governing the

NATO ASI Series, Vol. H 64
Mechanics of Swelling
Edited by T. K. Karalis

simultaneous movement of aqueous and gaseous phases in non–saturated soils swelling non–hydrostatically and (*ii*) a deeper understanding of the behaviour of a swelling non–hydrostatically, non–saturated and non–isotropic soil by considering the geometric representation of swelling in terms of: (*i*) the isotropic p_I part of the stress tensor characterizing the gradient of the forces driving swelling and the deviatoric p_G part of the stress tensor characterizing the gradient of the forces opposing swelling (*ii*) the moisture ratio ϑ defined as the volume of water per volume of the dry soil and (*iii*) the hygrometric state h defined as the ratio of the actual negative water pressure to the maximum possible.

2. WATER FLOW THROUGH A SWELLING CLAY–SOIL

2.1. The conservation of mass

The discussion of the mechanics of swelling is based upon several postulates. One of these is that the rate at which the volume of the soil increases is related to the net rate at which the bounding surface moves in the outward direction. Another is that the derivatives with respect to time (while following the material particles $D/D^m t$) are replaced by the total derivatives with respect to time (while following the fictitious system of particles d/dt) and (*ii*) the velocity vector corresponding to a material system of particles is replaced by the velocity vector corresponding to a fictitious system of particles. The balance of mass reads

$$\frac{d}{dt}\int_{V(m)} \rho_a dV = -\int_{S(m)} (\rho_a \mathbf{v}_a)\cdot\mathbf{n}\, dS + \int_{V(m)} c_a dV \qquad (1)$$

where ρ_a is the microscopic mass density of the phase a, $\boldsymbol{v}_a$ designates the velocity of the phase a, c_a is the rate of production of the phase a per unit volume as a result of the adsorption occurring in the double layer, phase transitions, etc.; $S(m)$ denotes the time–dependent closed bounding surface of the soil (m) on which the unit outwardly normal $\boldsymbol{n}$ is considered, and $V(m)$ is the region of space which contains the soil in its current configuration.

Considering water as one of the fluid phases ($a = w$), the spatial differential form of the mass balance of water reads

$$\frac{d}{dt}\int_{V(m)} \rho_w dV = -\int_{S(m)} (\rho_w \mathbf{v}_w)\cdot\mathbf{n}\, dS + \int_{V(m)} c_w dV \qquad (2)$$

At this point it is important to point out that the intrinsic seepage velocity $\boldsymbol{v}_w$ figured in (2) is different from the macroscopic or mean velocity $\boldsymbol{v}^w$ (defined by $\boldsymbol{v}^w = n^w \boldsymbol{v}_w$ and used for instance in Darcy's law). Considering the macroscopic mean velocity $\boldsymbol{v}^w$ and $\rho_w = \rho^w n^w$ the first term of the *LHS* of (2) results in

$$-\int_{S(s)} (\rho_w \boldsymbol{v}_w).\boldsymbol{n}dS = -\int_{S(s)} (\rho^w n^w)\boldsymbol{v}_w.\boldsymbol{n}dS = -\int_{S(s)} (\rho^w \boldsymbol{v}^w).\boldsymbol{n}dS \tag{3}$$

where $S(s)$ designates the closed surface bounding the swelling system of fictitious particles (s). Equation (2) by virtue of (3) reads

$$-\int_{S(s)} \rho^w \boldsymbol{v}^w.\boldsymbol{n}dS = \frac{d}{dt}\int_{V(s)} \rho^w n^w dV - \int_{V(s)} c_w dV \tag{4}$$

where $V(s)$ designates the region of space currently occupied by the fictitious system of particles, $\rho^w v^w (=f_n^w)$ indicates the mass flux of water per unit area (normal to v^w) and per unit time; the scalar product of $\rho^w v^w$ and $\boldsymbol{n}$ gives the component of $\rho^w v^w$ perpendicular to the surface S^w. Hence:

$$-\int_{S(s)} f_n^w.\boldsymbol{n}dS = \frac{d}{dt}\int_{V(s)} \rho^w n^w dV - \int_{V(s)} c_w dV \tag{5}$$

or in differential form

$$\frac{d}{dt}(\rho^w n^w) + \int_{S(s)} f_n^w \boldsymbol{n}dS - c_w = 0 \quad . \tag{6}$$

<u>*The first term*</u> of (6) can be transformed according to

$$\frac{d}{dt}(\rho^w n^w) = \frac{d}{dp_I}(\rho^w n^w)\frac{dp_I}{du_f}\frac{du_f}{dw}\frac{dw}{dt} \tag{7}$$

where p_I is defined by $p_I=(P_1+P_2+P_3)/3$; P_1, P_2 and P_3 being the normal swelling stresses along the principal directions, u_f is the suction or the negative fluid pressure and w is the moisture density defined by

$$w = \frac{n^w(1+e)\rho^w}{\rho^s} \leftrightarrow n^w = \frac{w\rho^s}{(1+e)\rho^w} \tag{8}$$

ρ^s being the macroscopic mass density of the clay particles and e is the void ratio.

A more convenient expression for (7) may be found if some simple relationship between the variables describing soil structure, such as the void ratio and porosity are involved and if the volumetric water content is replaced by the moisture density. Considering (8) then (7) reads

$$\frac{d}{dt}(\rho^w n^w) = \frac{d}{dp_I}\left(\frac{w\rho^s}{1+e}\right)\frac{dp_I}{du_f}\frac{du_f}{dw}\frac{dw}{dt} =$$
$$= \rho^s\left[\frac{1}{1+e}\frac{dw}{dp_I} - \frac{w}{(1+e)^2}\frac{de}{dp_I}\right]\frac{dp_I}{du_f}\frac{du_f}{dw}\frac{dw}{dt}. \qquad (9)$$

Since

$$\frac{dw}{dp_I} = \frac{dV}{dp_I}\frac{dw}{dV}, \qquad (10)$$

equation (9) reads

$$\frac{d}{dt}(\rho^w n^w) =$$
$$= \rho^s\left\{\frac{1}{1+e}\frac{dw}{dV}\frac{dV}{dp_I} - \frac{w}{(1+e)^2}\frac{de}{dp_I}\right\}\frac{dp_I}{du_f}\frac{du_f}{dw}\frac{dw}{dt}. \qquad (11)$$

Furthermore, the volume of the soil at a point is given by

$$V \equiv \frac{(1+w)\,m^s}{\rho} \qquad (12)$$

where m^s is the mass of the solid phase and ρ is the bulk density. Considering (12), the specific swelling defined as the volume change per unit change of the moisture density at constant swelling pressure (another expression for the specific swelling is discussed elsewhere by Karalis (1990b)) is given by

$$\left(\frac{dV}{dw}\right)_{p_I} = m^s\left\{\frac{d}{dw}\left(\frac{1+w}{\rho}\right)\right\}_{p_I}. \qquad (13)$$

Combining: (13), the compressibility equation given by

$$b = -\frac{1}{V}\left(\frac{dV}{dp_I}\right)_w \qquad (14)$$

and the expression

$$e = -1 + V\frac{\rho^s}{m_s}\;;\;\left(\frac{de}{dp_I}\right)_w = \frac{\rho^s}{m^s}\left(\frac{dV}{dp_I}\right)_w \qquad (15)$$

then (11) yields

simultaneous movement of aqueous and gaseous phases in non–saturated soils swelling non–hydrostatically and (*ii*) a deeper understanding of the behaviour of a swelling non–hydrostatically, non–saturated and non–isotropic soil by considering the geometric representation of swelling in terms of: (*i*) the isotropic p_I part of the stress tensor characterizing the gradient of the forces driving swelling and the deviatoric p_G part of the stress tensor characterizing the gradient of the forces opposing swelling (*ii*) the moisture ratio ϑ defined as the volume of water per volume of the dry soil and (*iii*) the hygrometric state h defined as the ratio of the actual negative water pressure to the maximum possible.

2. WATER FLOW THROUGH A SWELLING CLAY–SOIL

2.1. *The conservation of mass*

The discussion of the mechanics of swelling is based upon several postulates. One of these is that the rate at which the volume of the soil increases is related to the net rate at which the bounding surface moves in the outward direction. Another is that the derivatives with respect to time (while following the material particles $D/D^m t$) are replaced by the total derivatives with respect to time (while following the fictitious system of particles d/dt) and (*ii*) the velocity vector corresponding to a material system of particles is replaced by the velocity vector corresponding to a fictitious system of particles. The balance of mass reads

$$\frac{d}{dt}\int_{V(m)} \rho_a dV = -\int_{S(m)} (\rho_a \mathbf{v}_a) \cdot \mathbf{n} dS + \int_{V(m)} c_a dV \qquad (1)$$

where ρ_a is the microscopic mass density of the phase a, v_a designates the velocity of the phase a, c_a is the rate of production of the phase a per unit volume as a result of the adsorption occurring in the double layer, phase transitions, etc.; $S(m)$ denotes the time–dependent closed bounding surface of the soil (m) on which the unit outwardly normal $\boldsymbol{n}$ is considered, and $V(m)$ is the region of space which contains the soil in its current configuration.

Considering water as one of the fluid phases ($a = w$), the spatial differential form of the mass balance of water reads

$$\frac{d}{dt}\int_{V(m)} \rho_w dV = -\int_{S(m)} (\rho_w \mathbf{v}_w) \cdot \mathbf{n} dS + \int_{V(m)} c_w dV \qquad (2)$$

At this point it is important to point out that the intrinsic seepage velocity v_w figured in (2) is different from the macroscopic or mean velocity v^w (defined by $v^w = n^w v_w$ and used for instance in Darcy's law). Considering the macroscopic mean velocity v^w and $\rho_w = \rho^w n^w$ the first term of the *LHS* of (2) results in

$$\frac{d}{dt}(\rho^w n^w) = \frac{d}{dp_I}\left(\frac{w\rho^s}{1+e}\right)\frac{dp_I}{du_f}\frac{du_f}{dw}\frac{dw}{dt} = \\ =\rho^s\left[\frac{1}{1+e}\frac{dw}{dp_I} - \frac{w}{(1+e)^2}\frac{de}{dp_I}\right]\frac{dp_I}{du_f}\frac{du_f}{dw}\frac{dw}{dt}. \tag{9}$$

Since

$$\frac{dw}{dp_I} = \frac{dV}{dp_I}\frac{dw}{dV}, \tag{10}$$

equation (9) reads

$$\frac{d}{dt}(\rho^w n^w) = \\ =\rho^s\left\{\frac{1}{1+e}\frac{dw}{dV}\frac{dV}{dp_I} - \frac{w}{(1+e)^2}\frac{de}{dp_I}\right\}\frac{dp_I}{du_f}\frac{du_f}{dw}\frac{dw}{dt}. \tag{11}$$

Furthermore, the volume of the soil at a point is given by

$$V \equiv \frac{(1+w)\,m^s}{\rho} \tag{12}$$

where m^s is the mass of the solid phase and ρ is the bulk density. Considering (12), the specific swelling defined as the volume change per unit change of the moisture density at constant swelling pressure (another expression for the specific swelling is discussed elsewhere by Karalis (1990b)) is given by

$$\left(\frac{dV}{dw}\right)_{p_I} = m^s\left\{\frac{d}{dw}\left(\frac{1+w}{\rho}\right)\right\}_{p_I}. \tag{13}$$

Combining: (13), the compressibility equation given by

$$b = -\frac{1}{V}\left(\frac{dV}{dp_I}\right)_w \tag{14}$$

and the expression

$$e = -1 + V\frac{\rho^s}{m_s}\ ;\ \left(\frac{de}{dp_I}\right)_w = \frac{\rho^s}{m^s}\left(\frac{dV}{dp_I}\right)_w \tag{15}$$

then (11) yields

$$\frac{d}{dt}(\rho^w n^w) =$$

$$=\rho^s\left\{\frac{wb}{(1+e)} - \frac{b}{\rho^s}\frac{1}{\frac{d}{dw}\left(\frac{1+w}{\rho}\right)_{p_I}}\right\}\frac{dp_I}{du_f}\frac{du_f}{dw}\frac{dw}{dt}. \tag{16}$$

Considering as known the ratio of the change of the fluid pressure in the interstitial space to the change of the mean swelling pressure (dp_k/du_f) and the change of the fluid pressure in the interstitial space to the change in the moisture density (du_f/dH) and putting

$$M_c=\rho^s\left\{\frac{wb}{(1+e)} - [\frac{b}{\rho^s}\frac{1}{\frac{d}{dw}\left(\frac{1+e}{\rho}\right)}]\right\}\frac{dp_I}{du_f}\frac{du_f}{dw} \tag{17}$$

where M_c is a coefficient qualifying the mobility of the moisture density, the first term of (6) reads

$$\frac{d}{dt}\int_{V(s)}\rho^w n^w dV=M_c\frac{dw}{dt}. \tag{18}$$

<u>The second term</u> of (6) can be transformed considering that water flux figuring in this term can be analyzed primarily in three terms: the flow at constant volume, the influence of gravity and the restriction of the flow due to swelling. The flow at constant volume can be analyzed into two terms: the viscous flow and the flow due to the concentration difference giving rise to a chemical potential difference.

(a) *Flow at constant volume*

i. Viscous flow. The viscous flow at constant volume may be described by the equations

$$f^w_{\mathbf{n},d}=-D\frac{dC}{dw}\frac{\partial w}{\partial \mathbf{n}}\Big|_t \quad ; \quad D=K\frac{d\Psi}{d\eta} \tag{19}$$

where C is the fraction of the moving aqueous phase per unit volume of dry clay, D is the Fickian soil moisture diffusivity, K is the hydraulic conductivity; η is the porosity of the swelling soil, and Ψ is the pressure head.

ii. Flow due to the chemical potential difference. Swelling is strongly dependent on concentration in solutes as shown by Karalis (1990b). At constant volume, the flow of the aqueous phase due to the chemical potential difference, at constant volume, is given by

$$f^{w}_{\mathbf{n},c}=-\overline{DS}\frac{dc}{dw}\frac{\partial w}{\partial \mathbf{n}}\Big|_{t} \tag{20}$$

where t is the time, DS is the product of the self–diffusion constant of water D the effective area S of diffusion; $\delta c/\delta \boldsymbol{n}$ standing for the concentration gradient.

(b) The influence of gravity

The influence of gravity is expressed by

$$f^{w}_{\mathbf{n},g}=-kg\rho^{w}\frac{\partial z}{\partial \mathbf{n}} \tag{21}$$

where k is a proportionality factor.

(c) The influence of swelling

Any compressing load pressure attends to deform the soil structure by relatively displacing the soil particles from their original position; bringing them closer together and resulting in a decrease of the soil volume. Inversely, when the soil swells the pressure driving swelling results in increasing the soil volume and water capacity as well. The influence of swelling in the fluid flow can be expressed by

$$\begin{aligned} f^{w}_{\mathbf{n},p_I}&=-k\frac{dp_I}{dw}\frac{\partial w}{\partial \mathbf{n}}\Big|_{t}=-k\frac{dp_I}{dV}\frac{dV}{dw}\frac{\partial w}{\partial \mathbf{n}}\Big|_{t}= \\ &=\frac{k}{bV}\frac{d}{dw}\left(\frac{1+w}{\rho}\right)\frac{\partial w}{\partial \mathbf{n}}\Big|_{t} \end{aligned} \tag{22}$$

The total flux f_{n}^{w} can be considered by adding (19), (20), (21) and (22) e.g.,

$$f^{w}_{\mathbf{n}}=\left\{-D\frac{dC}{dw}-\overline{DS}\frac{dc}{dw}+\frac{k}{bV}\frac{d}{dw}\left(\frac{1+w}{\rho}\right)\right\}\frac{\partial w}{\partial \mathbf{n}}-kg\rho^{w}\frac{\partial z}{\partial \mathbf{n}} \tag{23}$$

Putting in the equation of flow

$$\frac{dF}{d\mathbf{n}}=\left\{-D\frac{dC}{dw}+\frac{k}{bV}\frac{d}{dw}\left(\frac{1+w}{\rho}\right)-\overline{DS}\frac{dc}{dw}\right\}=T \tag{24}$$

and considering Green's theorem

$$\int_{S(s)} f_n^w \mathbf{n}\, dS = \iiint \frac{\partial f_n^w}{\partial n} dx_1 dx_2 dx_3 = \nabla f_n^w \tag{25}$$

then according to equations (18), (23), (24) and (25), equation (6) yields:

$$\nabla\left(T\frac{\partial w}{\partial \mathbf{n}} - kg\rho^w \frac{\partial z}{\partial \mathbf{n}}\right) - M_c\frac{dw}{dt} - c_w = 0. \tag{26}$$

Furthermore since

$$\frac{dw}{dt} = \frac{\partial w}{\partial t} + \mathbf{v}^s \nabla w, \tag{27}$$

where vs is the velocity vector corresponding to the fictitious system of particles, equation (26) reads

$$\nabla\left(T\frac{\partial w}{\partial \mathbf{n}} - kg\rho^w \frac{\partial z}{\partial \mathbf{n}}\right) - M_c\left(\frac{\partial w}{\partial t} + \mathbf{v}^s \nabla w\right) + c_w = 0. \tag{28}$$

A new frame of reference by the Introduction of the variable Q through the balance of volumes. Stating that the volumes of the dry soil, water and water vapour are additives

$$S\, d\mathbf{n} = \sum_a dm^a / \rho^a \tag{29}$$

and for the system soil–water–water vapour, equation (29) along the vertical reads

$$dx = \{1 + w(\rho^s/\rho^w) + e\} dQ = (1 + \vartheta + e)\, dQ \tag{30}$$

where $dQ = dm^s / S\rho^s$.

Considering (30) and neglecting the influence of gravity the following partial differential yields from (28)

$$\nabla_Q (T_Q \nabla_Q w) = M_c \frac{\partial w}{\partial t} + \mathbf{v}^s M_c \nabla_Q w - c_w \;;\; \nabla_Q \equiv \frac{\partial}{\partial Q} \tag{31}$$

which is the same form as the differential equation already discussed by Philip (1969) if one disregards the second and the third term.

<u>The third term of equation (6).</u> Considering the interaction of the solid particles with water, then

$$C_w = n^w c^w = n^w \frac{\partial \rho^w}{\partial t} = \frac{m^s}{V} \frac{\partial w_a}{\partial t} = \rho^s \frac{Q}{x} \frac{\partial w_a}{\partial t} \quad ; \quad w = w_a + w_b \tag{32}$$

where w_a designates the moisture density fixed around the soil particles (responsible for swelling) and w_b is the moisture density corresponding to the non–viscous (Darcian) flow.

Considering that the flow along the vertical direction and also that the double layer formation are both time and space dependent then

$$\frac{\partial w_a}{\partial t} = a_1 w - a_2 w_a \tag{33}$$

where a_1,a_2 are constants. Hence, the third term in (31) does not imply further complication in the form of the resulting differential equation, which yields

$$\begin{aligned} &\nabla_Q (T_Q \nabla_Q w + k_Q \frac{\partial Q}{\partial n}) = \\ &= M_c \frac{\partial w}{\partial t} + \mathbf{v}^s M_c \nabla_Q w - \rho_s \{1 + w(\rho_s/\rho_w) + e\}^{-1} (a_1 w + a_2 w_a) \end{aligned} \tag{34}$$

where $\nabla_Q = \partial/\partial Q$.

Equation (34) gives the transport of the moisture density through a one–dimensional soil column under steady state conditions in which the change in the figured fluid pressure u_f may be expressed in terms of the change of the mean swelling stress p_I. The solution of the form of the differential equation (34) is discussed by Karalis (1991a).

2.2 *The conservation of momentum*

Momentum from the action of all the exterior phases upon the surface of a soil segment. Let p_I be the mean swelling pressure which at equilibrium equals the load pressure. The swelling force acted upon the surface S_k will be $p_I S_k$ and the force opposing swelling from the actual vapour pressure h_v will be $-h_v S_k$. During the time interval dt the momentum delivered to the surface S_k will be $(p_I - h_v) S_k dt$, or

$$\frac{\partial (p_k - h_v)}{\partial Q} S_k \, dQ \, dt \tag{35}$$

where Q denotes the material coordinate defined with (30). Additionally, the momentum in the swelling segment results from the flowing solid phase corresponding at a particular material coordinate Q, and also from the gradients in the fluid mass flux and velocity at Q.

Momentum from the flowing fluid phase. The volume of the flowing water (entering and leaving through the surface S_k across an elemental length which is considered along the normal to the surface S_k is given by

$$\frac{f_n^W}{\rho^W} S_k dt \tag{36}$$

and the momentum crossing the Eulerian coordinate is

$$f_n^W S_k \mathbf{v}^W dt \tag{37}$$

or in material coordinates

$$\frac{\partial}{\partial Q}(f_n^W \mathbf{v}^W)\, S_k dQ dt. \tag{38}$$

Momentum from the flowing solid phase. Considering that the dry mass in the soil segment is $\rho^s S_k dQ$, the rate of momentum from the mass flux of the solid phase is

$$\rho^s S_k dQ \frac{\partial \mathbf{v}^s}{\partial t}. \tag{39}$$

Conserving momentum, (35), (38) and (39) result in

$$\frac{\partial (p_k - h_v)}{\partial Q} + \frac{\partial}{\partial Q}(f_n^W \mathbf{v}^W) = \rho^s \frac{\partial \mathbf{v}^s}{\partial t} \tag{40}$$

which represent the generalisation of Darcy's law; the first term expressing the external applied forces and the second the drag force acting upon the solid phase.

Since the accumulation of water in the soil segment is

$$S_k \left(\frac{\partial f_n^W}{\partial Q} \right) dQ dt \tag{41}$$

and the mass of the segment may be derived from the definition (30), the rate with which the moisture density varies reads

$$\frac{\partial w}{\partial t} = \frac{1}{\rho^s} \frac{\partial f_n^W}{\partial Q}. \tag{42}$$

Considering (42), then (40) reads

$$\frac{\partial (p_k - h_v)}{\partial Q} + f_n^W \frac{\partial \mathbf{v}^W}{\partial Q} + \rho^s \mathbf{v}^W \frac{\partial w}{\partial t} = \rho^s \frac{\partial \mathbf{v}^s}{\partial t} \tag{43}$$

which is an expression relating both water velocity and solid phase velocity with the rate of change of the moisture density.

Equation (43) expresses the fact that the sum of the momentum transfer from the combined action of swelling pressure and vapour pressure depends not explicitly upon the velocity difference $v^w - v^s$ (e.g., the velocity of water and the velocity of the bounds of the elemental bulk volume of the swollen clay matrix). Moisture density, the flux of water, and compressibility are also important factors contributing to the transfer of momentum.

3. NORMAL STRESSES DRIVING SWELLING

Functional determinants for a non-saturated swelling soil. Considering the volume V of a test piece, the variations of p_I, w and h are not independent. The relation $V(p_I,w,h) = ct$ gives an implicit functional relationship among p_I, w and h which may be solved for one variable (say p_I) in terms of the other two (say w,h). Considering the same for $p_I(V,w,h) = \text{ct}$, these two functions can easily be treated in order to derive certain relationships among their partial derivatives. However, it is convenient at this point to have a name for the curves figuring in the hygrometric state–load pressure diagram, drawn so that the actual negative water pressure h_v is always a constant fraction of the maximum possible h_o. Calling these curves isohygrometric curves, it is also convenient to have a term for the curve giving at different normal swelling stress the same moisture content of the wetted soil holding the weight of the soil constant. These curves are called isoneric curves or shortly isoneres (from *nero*=water); already in use in describing the aneroid state of a material.

However, an important advantage in representing the swelling behaviour in terms of the void ratio e and the moisture ratio ϑ is that $e = \vartheta$ when the soil becomes saturated. In order to transform V and w in terms of e and ϑ the following relations are considered: (*i*)

$$e=-1+\frac{V}{m^s \upsilon^s} \quad \textbf{(44)}$$

where υ^s is the specific volume and (*ii*)

$$w=\frac{\upsilon^s}{\upsilon^w}\vartheta=\frac{\rho^w}{\rho^s}\vartheta \quad \textbf{(45)}$$

where υ^w is the specific volume of the liquid phase. Representing swelling with the following pairs of variables

$$\begin{aligned}
&p_I=\phi_{p_I,V}(w,h)\ ;\ V=\psi_{V,p_I}(w,h)\\
&V=\phi_{V,w}(p_I,h)\ ;\ w=\psi_{w,V}(p_I,h)\\
&p_I=\phi_{p_I,w}(V,h)\ ;\ w=\psi_{w,p_I}(V,h)\\
&h=\phi_{h,V}(p_I,w)\ ;\ V=\psi_{V,h}(p_I,w)\\
&h=\phi_{h,w}(p_I,V)\ ;\ w=\psi_{w,h}(p_I,V)\\
&p_I=\phi_{p_I,h}(w,V)\ ;\ h=\psi_{h,p_I}(w,V)
\end{aligned}\tag{46}$$

where the continuity of the functions and of their first derivatives is proven experimentally, the variables w,h in (46a) be assigned constant values, the equations define two curves and if w,h be assigned a series of such values, then (46a) define a network of curves in some part of the p_I–V plane. However, if there is a functional relation $w=F(h)$ so that the isoneres coincide with the isohygrometrics (i.e., if the Jacobian of (46a) vanishes identically) a constant value of h implies a constant value of w and hence the locus for which w is constant; i.e., the set of h–curves coincides with the set of w–curves and no true network is formed. This case is uninteresting since as it is proved experimentally (Karalis, 1990b) these curves do not coincide and the Jacobian does not vanish for any point (p_I,V) of the region of the p_I–V plane. Equations (46a) may then be solved for p_I,V in terms of w,h at any point of the region; the latter forming a pair of curves passing through each point. Furthermore, it is sometimes proper to consider (w,V) as coordinates. To any of these points correspond not only the rectangular coordinates (p_I,h) but also the curvilinear coordinates (w,V). The w,V may be called the coordinates of extensity and p_I,h the corresponding coordinates of intensity.

The equations connecting the rectangular and curvilinear coordinates may be taken in either of the two forms

$$h=\chi(w,V)\ ;\ p_I=\zeta(w,V)\ ;\ w=\omega(p_I,h)\ ;\ V=\xi(w,h)\tag{47}$$

each of which are the solutions of the other. Assigning a definite value of the void ratio V to the set of the *3* independent variables p_I,w and h (*called the coordinates of a generic point of the swelling system*) the values so obtained determine a point in a *3*–Dimensional Manifold; the *3*–variables being the co–ordinates of the Manifold and

$$V=V(p_I,w,h)\ .\tag{48}$$

However, designating with V, w and h the independent variables, the expression for the pressure driving swelling takes the form

$$p_I=p_I(V,w,h)\tag{49}$$

Disregarding the coupling terms the equations transforming the void ratio V, or the isotropic part p_I of the swelling stress tensor, can be obtained by differentiating (48) and (49) respectively i.e.,

$$dV=\left(\frac{\partial V}{\partial p_I}\right)_{w,h}dp_I+\left(\frac{\partial V}{\partial w}\right)_{p_I,h}dw+\left(\frac{\partial V}{\partial h}\right)_{p_I,w}dh$$
$$dp_I=\left(\frac{\partial p_I}{\partial V}\right)_{w,h}dV+\left(\frac{\partial p_I}{\partial w}\right)_{V,h}dw+\left(\frac{\partial p_I}{\partial h}\right)_{V,w}dh \quad (50)$$

The coefficients $(\partial i/\partial j)_{n,k}$ (where any one of i,j,k,n may be any of w,h,V,p_I) indicate that the implied functional relations between i and j are those determined by the constancy of n and k. Rewriting the coefficients of (50) in a more explicit form, in order to express

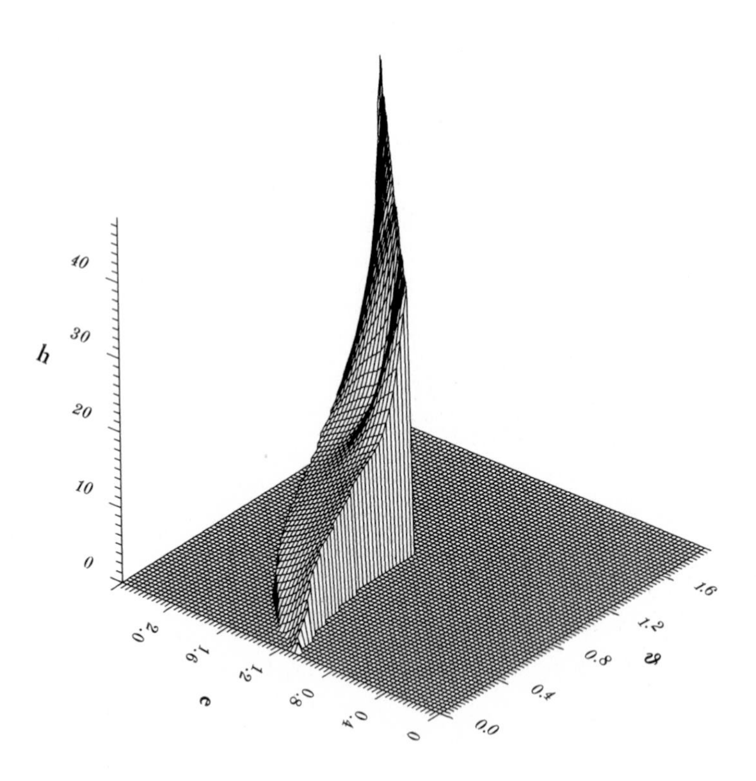

Fig.1 Graph showing experimental values of the vapour pressure h (in kPa), the void ratio e and the moisture ratio ϑ

them in more practical quantities, the following relations result

$$dV = \frac{J(V,p_I)}{J(w,h)}dp_I + \frac{J(V,w)}{J(p_I,h)}dw + \frac{J(V,h)}{J(p_I,w)}dh$$
$$dp_I = \frac{J(p_I,h)}{J(w,V)}dh + \frac{J(p_I,w)}{J(h,V)}dw + \frac{J(p_I,V)}{J(h,w)}dV \tag{51}$$

where the geometric interpretation of the Jacobians figuring in (51) may be drawn from figures 1–5; a closer quantitative study of this point will be given later in this report. If definite values be assigned to p_I, w and h variations (say) then

$$\mathscr{E}_{p_I} = \frac{J(V,p_I)}{J(w,h)}\ ;\ \mathscr{E}_w = \frac{J(V,w)}{J(p_I,h)}\ ;\ \mathscr{E}_h = \frac{J(V,h)}{J(p_I,w)}$$
$$\wp_V = \frac{J(p_I,V)}{J(w,h)}\ ;\ \wp_w = \frac{J(p_I,w)}{J(V,h)}\ ;\ \wp_h = \frac{J(p_I,h)}{J(V,w)} \tag{52}$$

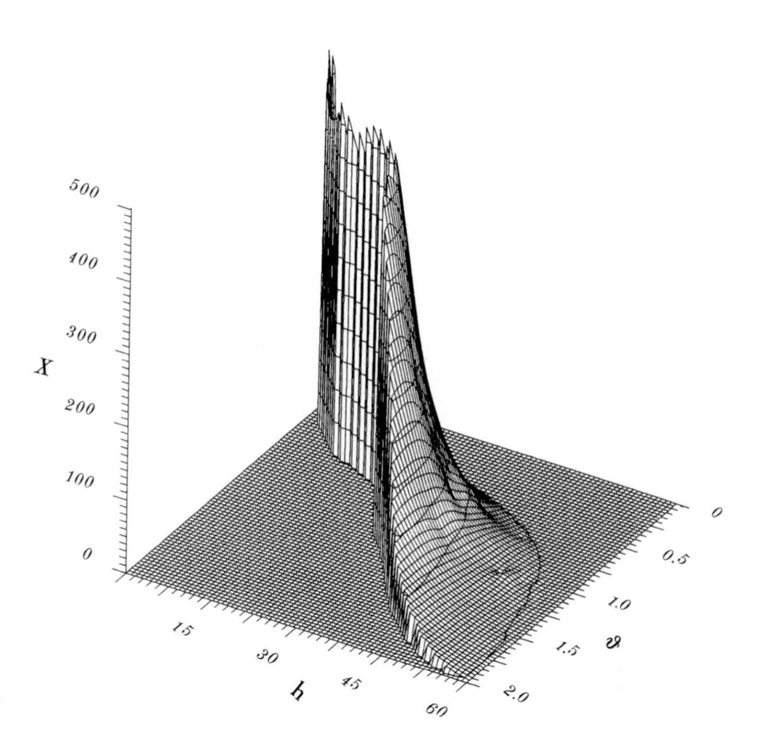

Fig.2 Graph showing experimental values of X (in kPa), h (in kPa) and ϑ

and equations (51) read

$$dV = \mathscr{E}_{p_I} dp_I + \mathscr{E}_w dw + \mathscr{E}_h dh$$
$$dp_I = \wp_V dV + \wp_w dw + \wp_h dh \tag{53}$$

where p_I, w, h, V are supposed to be restricted to values near $(p_I)_o$, w_o, h_o, V_o; representing a small portion of the tangent plane passing through $(p_I)_o$, w_o, h_o and V_o. By successively putting dp_I=0, dV=0, dh=0 and dw=0 in (51), it can be proved

$$\left(\frac{\partial p_I}{\partial V}\right)_{h,w}\left(\frac{\partial V}{\partial w}\right)_{h,p_I}\left(\frac{\partial w}{\partial p_I}\right)_{h,V}=-1$$
$$\left(\frac{\partial p_I}{\partial w}\right)_{V,h}\left(\frac{\partial w}{\partial h}\right)_{V,p_I}\left(\frac{\partial h}{\partial p_I}\right)_{V,w}=-1$$
$$\left(\frac{\partial w}{\partial h}\right)_{p_I,V}\left(\frac{\partial h}{\partial V}\right)_{p_I,w}\left(\frac{\partial V}{\partial w}\right)_{p_I,h}=-1 \tag{54}$$
$$\left(\frac{\partial h}{\partial p_I}\right)_{w,V}\left(\frac{\partial p_I}{\partial V}\right)_{w,h}\left(\frac{\partial V}{\partial h}\right)_{w,p_I}=-1$$

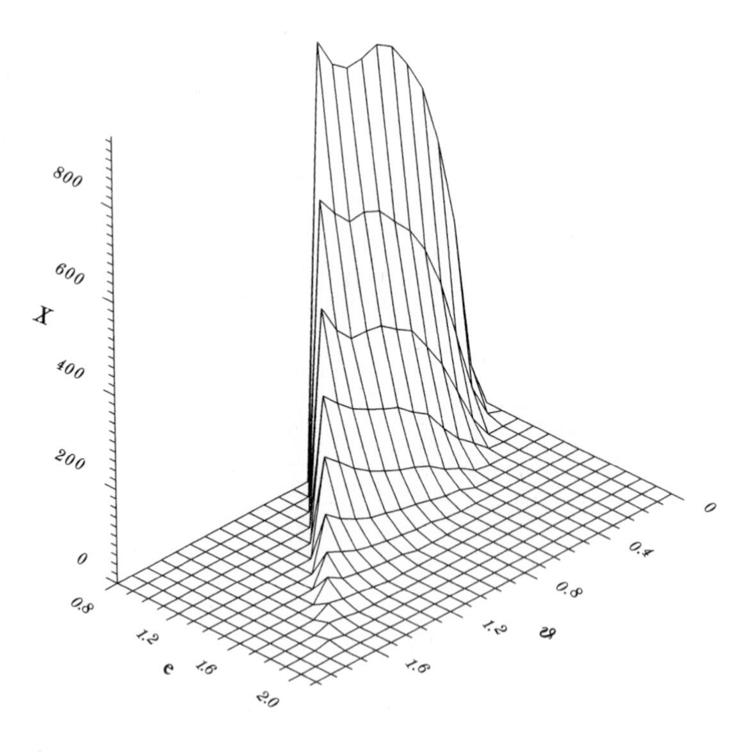

Fig.3 Graph showing experimental values of X (in kPa), e and ϑ

and expressed in terms of simple Jacobians

$$\frac{J(p_I,w)}{J(V,w)}\frac{J(V,p_I)}{J(w,p_I)}\frac{J(w,V)}{J(p_I,V)}=(-1)^3$$
$$\frac{J(p_I,h)}{J(w,h)}\frac{J(w,p_I)}{J(h,p_I)}\frac{J(h,w)}{J(p_I,w)}=(-1)^3$$
$$\frac{J(w,V)}{J(h,V)}\frac{J(h,w)}{J(V,w)}\frac{J(V,h)}{J(w,h)}=(-1)^3 \qquad (55)$$
$$\frac{J(h,V)}{J(p_I,V)}\frac{J(p_I,h)}{J(V,h)}\frac{J(V,p_I)}{J(h,p_I)}=(-1)^3$$

Equation of state. Investigating the compressibility and other mechanical properties of a non–hydrostatically non–saturated swelling soil it is essential to include in the equation of state the effective pressure driving swelling. No practical instruments exist for the

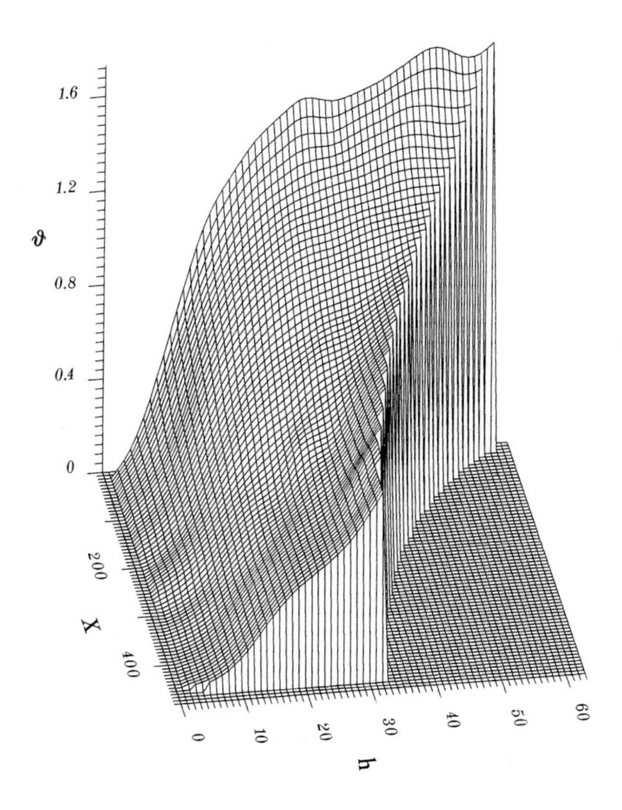

Fig.4 Graph showing experimental values of X (in kPa), h (in kPa) and ϑ

measurement and control of the effective pressure driving swelling since this term is far from simple to measure; in most treatments of porous material behaviour only the effective pressure for saturated sands is measured according to Terzaghi's equation. For swelling soils Karalis (1990b) has given a new effective swelling law in the form

$$s_x \delta X + s_y \delta Y + s_z \delta Z - \frac{\upsilon}{\upsilon^t} \delta h_v = Q' \delta w \tag{56}$$

where Q' is the effective stress driving swelling (i.e., Q' is that function of the load pressure and negative pore water pressure which controls the mechanical effects on volume change), X,Y,Z designates the normal swelling stresses, δw is the variation of the moisture density content, υ is the specific volume of the vapour phase, υ^t ($=V/m^s$) is the specific volume of the bulk soil; and s_i (i=x,y,z) is an unscaled swelling coefficient defined as the deformation due to swelling ($\delta x/x$, $\delta y/y$, $\delta z/z$) per change of the moisture density content

$$s_x = \frac{1}{x}\frac{\delta x}{\delta w} \; ; s_y = \frac{1}{y}\frac{\delta y}{\delta w} ; s_z = \frac{1}{z}\frac{\delta z}{\delta w} \tag{57}$$

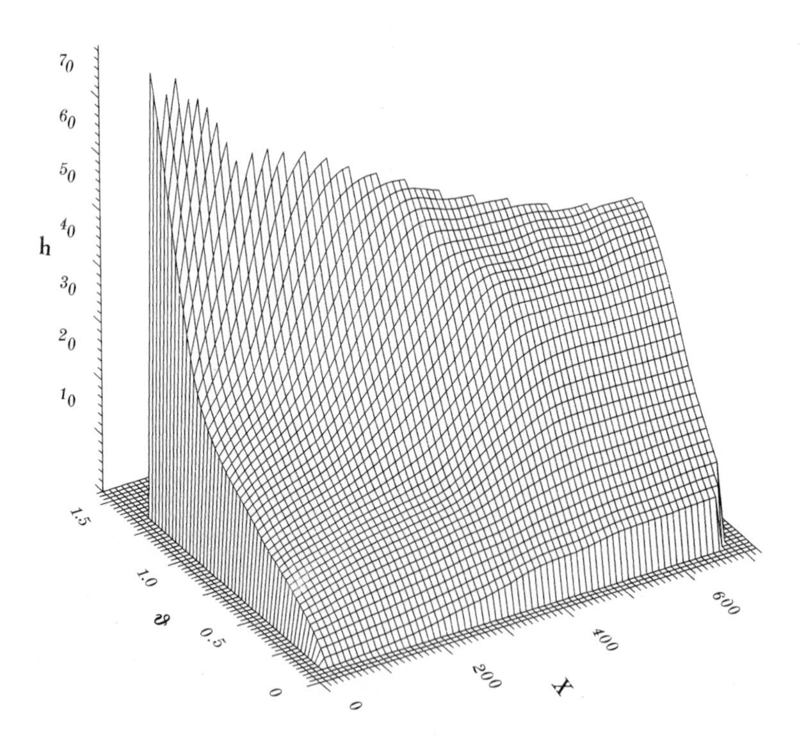

Fig.5 Graph showing experimental values of X (in *kPa*), *h* (in *kPa*) and ϑ

x,y,z designating the dimensions of the test piece volume.

Combining (56) with the component of the tensor of the stresses driving swelling expressed in terms of their isotropic and deviatoric parts

$$\frac{1}{3}\delta X=\delta p_I-\delta p_{G,xy}-\delta p_{G,zx}$$
$$\frac{1}{3}\delta Y=\delta p_I-\delta p_{G,xy}-\delta p_{G,yz} \qquad (58)$$
$$\frac{1}{3}\delta Z=\delta p_I-\delta p_{G,xy}-\delta p_{G,yz}$$

the state equation results in the form

$$\frac{\delta X}{\delta h_v}=\bar{\upsilon}\frac{\delta p_I-\delta p_{G,xy}-\delta p_{G,zx}}{s_{p_I}\delta p_I+s_G\delta p_G-Q'\delta w}$$
$$\frac{\delta Y}{\delta h_v}=\bar{\upsilon}\frac{\delta p_I-\delta p_{G,xy}-\delta p_{G,yz}}{s_{p_I}\delta p_I+s_G\delta p_G-Q'\delta w} \qquad (59)$$
$$\frac{\delta Z}{\delta h_v}=\bar{\upsilon}\frac{\delta p_I-\delta p_{G,yz}-\delta p_{G,zx}}{s_{p_I}\delta p_I+s_G\delta p_G-Q'\delta w}$$

where $\upsilon = \upsilon'/3\upsilon$. Designating with B the modulus of swelling due to changes in the load pressure components (X,Y,Z) and putting

$$B_X=\frac{1}{V}\left(\frac{\delta V}{\delta X}\right)_{hwYZ}$$
$$B_Y=\frac{1}{V}\left(\frac{\delta V}{\delta Y}\right)_{hwXZ} \qquad (60)$$
$$B_Z=\frac{1}{V}\left(\frac{\delta V}{\delta Z}\right)_{hwXY}$$

and

$$s_{p_I}=\frac{1}{V}\frac{\delta V}{\delta w}\ ;\ B_h=\frac{1}{V}\frac{\delta V}{\delta h} \qquad (61)$$

and also substituting (60) and (61) into (56) the effective pressure driving swelling can be derived from

$$Q'=\frac{s_x}{s_{p_I}}B_X+\frac{s_y}{s_{p_I}}B_Y+\frac{s_z}{s_{p_I}}B_Z-\frac{\upsilon}{\upsilon^t}B_h \qquad (62)$$

where s_x/s_{pI}, s_y/s_{pI}, s_z/s_{pI} designate the fractions of the unscaled swelling along the directions x,y,z to the unscaled volumetric swelling s_{pI}.

The effective stress driving swelling Q' from geometric considerations. The effective stress driving swelling Q' can be derived from the adsorption isotherm, $dh/d\vartheta$. From the figure 6 (viz., Karalis (1990b))

$$\frac{KL}{LG}=\frac{EF}{FG} \tag{63}$$

and since

$$FG=AG-AF;\ \frac{KL}{EF}=\frac{LG}{AG-AF} \tag{64}$$

Putting

$$KL=\frac{d(X+Y)}{2}=dp_I\ ;\ \ EF=\frac{d(X'+Y)}{2}=dp'_I \tag{65}$$

where X' designates the lateral pressure driving swelling into the oedometer, equation (63)

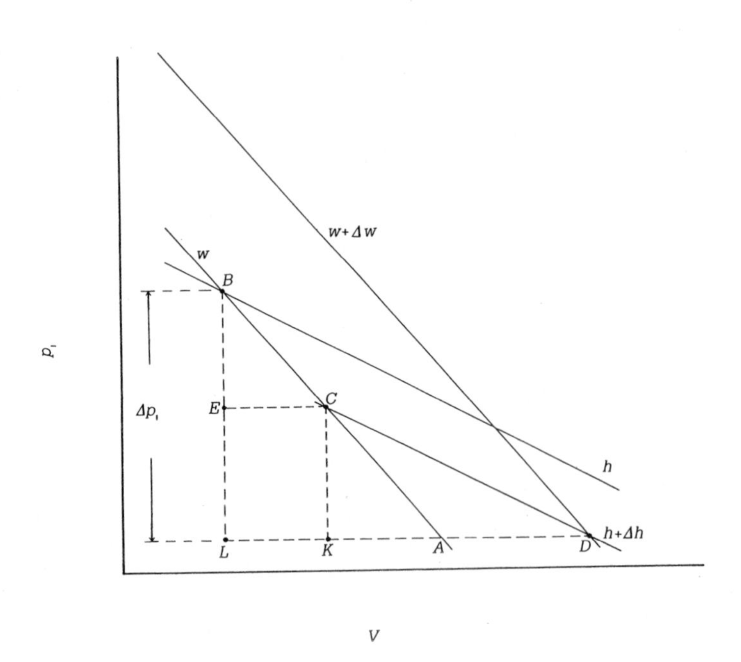

Fig.6 Graph representing w and h in the plan of p_I and V

reads

$$\left\{\frac{d(X'+Y)}{d(X+Y)}\right\}_w = \frac{AG-AF}{LG} = \frac{s_{p_I}-s'_{p_I}}{s_{p_I}} \tag{66}$$

where s'_{pI} is the unscaled swelling corresponding to the conditions that the oedometer reads. Furthermore the state equation (56) reads

$$s_{p_I} d\left(\frac{X+Y}{2}\right) = \upsilon\, dh + Q' dw \tag{67}$$

Hence,

$$\frac{d(X'+Y)}{dw} = \left(\upsilon \frac{dh}{dw} + Q'\right) \frac{s_{p_I}-s'_{p_I}}{(s_{p_I})^2} \tag{68}$$

From the compressibility equation

$$\frac{dp'_I}{dw} = \frac{B_X}{V}(s_{p'_I}-s_{p_I}) \tag{69}$$

Hence,

$$\frac{B_X}{V}(s_{p'_I}-s_{p_I}) = \left(\upsilon \frac{dh}{dw} + Q'\right) \frac{s_{p_I}-s_{p'_I}}{(s_{p_I})^2} \tag{70}$$

and the effective stress in the oedometer is given by

$$Q' = \frac{B_X}{V}(s_{p_I})^2 - \upsilon \frac{dh}{dw} \tag{71}$$

(v) Large deformability. Referring the stress and the strains due to swelling to the principal directions and taking into account the properties of symmetry of the material, the stress–strain relations reduce to

$$X = \phi\left(\frac{\delta x}{x}, w, h\right);\; Y = \chi\left(\frac{\delta y}{y}, w, h\right);\; Z = \psi\left(\frac{\delta z}{z}, w, h\right) \tag{72}$$

and the following analysis deals with the development of the expressions for finite strains due to any amount of the adsorbed liquid at a given hygrometric state. Defining the normal stresses driving swelling X, Y, Z by

$$\boldsymbol{X} = yzX;\; \boldsymbol{Y} = xzY;\; \boldsymbol{Z} = xyZ \tag{73}$$

where $\boldsymbol{X}, \boldsymbol{Y}$ and $\boldsymbol{Z}$ are the forces driving swelling along the three principal directions, it is considered that the condition $dx/x + dy/y + dz/z = 0$ (satisfied for constant volume) does not apply for a swelling soil. By differentiation of (73) the following equations result

$$\begin{aligned}
\delta \boldsymbol{X} &= yz\delta X + X(y\delta z + z\delta y) = yz\left\{\delta X + X\left(\frac{\delta z}{z} + \frac{\delta y}{y}\right)\right\} \\
\delta \boldsymbol{Y} &= xz\delta Y + Y(x\delta z + z\delta x) = xz\left\{\delta Y + Y\left(\frac{\delta z}{z} + \frac{\delta x}{x}\right)\right\} \\
\delta \boldsymbol{Z} &= xy\delta Z + Z(x\delta y + y\delta x) = xy\left\{\delta Z + Z\left(\frac{\delta y}{y} + \frac{\delta x}{x}\right)\right\}
\end{aligned} \tag{74}$$

Considering that from the state ${}^{1}(X,h,w)_{YZ}$ to the state ${}^{3}(X,h,w)_{YZ}$ the same results can be evaluated by considering first the variation of the hygrometric state when the load pressure X is held constant and thereafter varying the load pressure keeping the hygrometric state h constant, we can write

$$\delta w_{YZ} = \partial w|_{hYZ} - \partial w|_{XYZ} \tag{75}$$

where $\partial w/_{XYZ}$ is the variation of the moisture density content when the load pressure is held constant and $\partial w/_{hYZ}$ is the variation of the moisture density content when the hygrometric state is held constant. We can write for the volume variation

$$\left(\frac{\delta V}{\delta w}\right)_{YZ} = \left(\frac{\partial V}{\partial w}\right)_{hYZ} - \left(\frac{\partial V}{\partial w}\right)_{XYZ} \tag{76}$$

and the components of the unscaled swelling along the three directions are given by

$$\begin{aligned}
\frac{1}{x}\left(\frac{\delta x}{\delta w}\right)_{YZ} &= \frac{1}{x}\left(\frac{\partial x}{\partial w}\right)_{hYZ} - \frac{1}{x}\left(\frac{\partial x}{\partial w}\right)_{XYZ} \\
\frac{1}{y}\left(\frac{\delta y}{\delta w}\right)_{ZX} &= \frac{1}{y}\left(\frac{\partial y}{\partial w}\right)_{hZX} - \frac{1}{y}\left(\frac{\partial y}{\partial w}\right)_{YZX} \\
\frac{1}{z}\left(\frac{\delta z}{\delta w}\right)_{XY} &= \frac{1}{z}\left(\frac{\partial z}{\partial w}\right)_{hXY} - \frac{1}{z}\left(\frac{\partial z}{\partial w}\right)_{XZY}
\end{aligned} \tag{77}$$

Considering the load pressures X,Y,Z applied along the directions x,y,z, the moisture density content and the hygrometric state as the independent variables, we may write

$$
\begin{aligned}
&\delta x=(\frac{\partial x}{\partial w})_{hXYZ}\delta w+(\frac{\partial x}{\partial h})_{wXYZ}\delta h\\
&+(\frac{\partial x}{\partial X})_{hwYZ}\delta X+(\frac{\partial y}{\partial Y})_{hwZX}\delta Y+(\frac{\partial z}{\partial Z})_{hwXY}\delta Z\\
&\delta y=(\frac{\partial y}{\partial w})_{hXYZ}\delta w+(\frac{\partial y}{\partial h})_{wXYZ}\delta h\\
&+(\frac{\partial x}{\partial X})_{hwYZ}\delta X+(\frac{\partial y}{\partial Y})_{hwZX}\delta Y+(\frac{\partial z}{\partial Z})_{hwXY}\delta Z\\
&\delta z=(\frac{\partial z}{\partial w})_{hXYZ}\delta w+(\frac{\partial z}{\partial h})_{wXYZ}\delta h\\
&+(\frac{\partial x}{\partial X})_{hwYZ}\delta X+(\frac{\partial y}{\partial Y})_{hwZX}\delta Y+(\frac{\partial z}{\partial Z})_{hwXY}\delta Z
\end{aligned}
\tag{78}
$$

Putting

$$
\begin{aligned}
&s^{x}_{hXYZ}=\frac{1}{x}(\frac{\partial x}{\partial w})_{hXYZ}\ ;\ \beta^{x}_{wXYZ}=\frac{1}{x}(\frac{\partial x}{\partial h})_{wXYZ}\\
&s^{y}_{hXYZ}=\frac{1}{y}(\frac{\partial y}{\partial w})_{hXYZ}\ ;\ \beta^{y}_{wXYZ}=\frac{1}{y}(\frac{\partial y}{\partial h})_{wXYZ}\\
&s^{z}_{hXYZ}=\frac{1}{z}(\frac{\partial z}{\partial w})_{hXYZ}\ ;\ \beta^{z}_{wXYZ}=\frac{1}{z}(\frac{\partial z}{\partial h})_{wXYZ}
\end{aligned}
\tag{79}
$$

the following equations result

$$
\begin{aligned}
&\frac{\delta x}{x}=s^{x}_{hXYZ}\delta w+\beta^{x}_{wXYZ}\delta h\\
&+\frac{1}{x}(\frac{\partial x}{\partial X})_{hwYZ}\delta X+\frac{1}{x}(\frac{\partial y}{\partial Y})_{hwZX}\delta Y+\frac{1}{x}(\frac{\partial z}{\partial Z})_{hwXY}\delta Z\\
&\frac{\delta y}{y}=s^{y}_{hXYZ}\delta w+\beta^{y}_{wXYZ}\delta h\\
&+\frac{1}{y}(\frac{\partial x}{\partial X})_{whYZ}\delta X+\frac{1}{y}(\frac{\partial y}{\partial Y})_{whZX}\delta Y+\frac{1}{y}(\frac{\partial z}{\partial Z})_{whXY}\delta Z\\
&\frac{\delta z}{z}=s^{z}_{hXYZ}\delta w+\beta^{z}_{wXYZ}\delta h\\
&+\frac{1}{z}(\frac{\partial x}{\partial X})_{hwYZ}\delta X+\frac{1}{z}(\frac{\partial y}{\partial Y})_{hwZX}\delta Y+\frac{1}{z}(\frac{\partial z}{\partial Z})_{hwXY}\delta Z
\end{aligned}
\tag{80}
$$

Considering the identities expressing the condition that the hygrometric state remains constant

$$
\begin{aligned}
\left(\frac{\partial y}{\partial Y}\right)_{hwZX} &= -\left(\frac{\partial w}{\partial Y}\right)_{hyZX}\left(\frac{\partial y}{\partial w}\right)_{hYZX} \\
\left(\frac{\partial x}{\partial X}\right)_{hwYZ} &= -\left(\frac{\partial w}{\partial X}\right)_{hxYZ}\left(\frac{\partial x}{\partial w}\right)_{hXYZ} \\
\left(\frac{\partial z}{\partial Z}\right)_{hwXY} &= -\left(\frac{\partial w}{\partial Z}\right)_{hzXY}\left(\frac{\partial z}{\partial w}\right)_{hZXY}
\end{aligned}
\tag{81}
$$

and that each of them transliterates the fact that $\delta Y=\delta Z=0$, $\delta Z=\delta X=0$, $\delta X=\delta Y=0$ respectively, then (47) transform to

$$
\begin{aligned}
\frac{\delta x}{x} &= s^{x}_{hXYZ}\delta w+\beta^{x}_{wXYZ}\delta h-\frac{1}{x}\left(\frac{\partial w}{\partial X}\right)_{hxYZ}\left(\frac{\partial x}{\partial w}\right)_{hXYZ}\delta X \\
&-\frac{1}{x}\left(\frac{\partial w}{\partial Y}\right)_{hyZX}\left(\frac{\partial y}{\partial w}\right)_{hXYZ}\delta Y \\
&-\frac{1}{x}\left(\frac{\partial w}{\partial Z}\right)_{hzXY}\left(\frac{\partial z}{\partial w}\right)_{hZXY}\delta Z
\end{aligned}
\tag{82}
$$

. . . .

. . . .

4. THERMODYNAMIC CONSIDERATIONS

Let us suppose that the dry soil mass m^s occupying the volume V_1 under the load pressure $(p_I)_1$ $(X=Y=Z=p_I)$ at temperature T swells (when imbibed with water) to the volume V_2 while the mean pressure driving swelling goes to $(p_I)_2$; the temperature rising to T + dT. The work that might have been done is $(p_I)_2V_2-(p_I)_1V_1$, whereas for the soil swelling in a cylinder it would have been

$$
\int_1^2 p_I dV; \tag{83}
$$

the difference

$$
\{(p_I)_2V_2-(p_I)_1V_1\}-\int_1^2 p_I dV=\int_1^2 V dp_I \tag{84}
$$

being the useful work that fails to be done. Denoting with G the energy responsible for the work that fails to be done (responsible however for the increase in the mean pressure driving swelling) and with E the energy expended for the work actually done, the energy for the work that might have been done is $E + G$.

Considering also: (*i*) that G is an extensive quantity equivalent in all respects to the potential energy producing swelling and (*ii*) that the potential G can be uniquely determined by the independent variables of state chosen to describe swelling, any decrease in the thermodynamic potential G is given by

$$dG=\left(\frac{\partial G}{\partial p_I}\right)_{m^w}dp_I+\left(\frac{\partial G}{\partial m^w}\right)_{p_I}dm^w. \tag{85}$$

Besides, designating the chemical potential of the soil water with $\mu^w=(\partial G/\partial m^w)_{pI}$ and the volume by $V=(\partial G/\partial p_I)_{mw}$, furthermore taking the Maxwell relations for μ^w and V and considering $\partial\mu^w = \upsilon\partial h$, then (85) implies

$$\left(\frac{\partial V}{\partial m^w}\right)_{p_I}=\upsilon\left(\frac{\partial h}{\partial p_I}\right)_{m^w} \tag{86}$$

or choosing different sets of independent variables of state namely e, p_I, h, w for a soil of constant mass (86) reads

$$\left(\frac{\partial V}{\partial w}\right)_{p_I}=\frac{\upsilon}{\upsilon^w}\left(\frac{\partial h}{\partial p_I}\right)_{w}. \tag{87}$$

However, although the analysis for the thermodynamic description of a swelling isotropic soil is quite straightward, there are many conceptual problems entering into the formulation of the free energy components, for a soil swelling non-hydrostatically. The subsequent discussion focuses on this point of view and precisely on the description of the soil behaviour in the more general case namely along the three different principal stresses. It is herein assumed that the axes of orthogonal anisotropy in the soil structure coincide with the directions of principal stresses and the changing three principal stresses keep the order of their magnitude constant. The analysis may then be put forward in the space of thermodynamic potential, using the correspondence between extensive and intensive coordinates. We may write for the oncotic energy (Karalis, 1991b)

$$O(T,\boldsymbol{r},{}^D\boldsymbol{F})\equiv G(T,{}^D\boldsymbol{F})-F(T,\boldsymbol{r})={}^D\boldsymbol{F}\boldsymbol{r} \tag{88}$$

where O is the oncotic energy and $F(T,r)$ is the partial free energy defined by (Gibbs, 1906)

$$\delta F=T\delta T-{}^D\boldsymbol{F}\delta\boldsymbol{r}-\sum_a\mu^a\delta n^a \tag{89}$$

${}^D\boldsymbol{F}$ being the force driving swelling and $\boldsymbol{r}$ is the length changed during swelling.

Oncotic energy variation in terms of the normal stress driving swelling for different values of the moisture density and void ratio may be drawn from figures (1–5). The expressions $\delta e/\delta\vartheta$, $\delta X/\delta\vartheta$ may be drawn (for different values of the normal stress driving swelling and void ratio) from which F and G can be calculated. Then the ratio G/F may be drawn which represents the amount of energy that is actually available for doing work per work done and $G(T,{}^{D}F)$ is the partial free energy given by

$$\delta G = S\,\delta T + \boldsymbol{r}\,\delta^{D}\boldsymbol{F} - \sum_{a} \mu^{a}\delta n^{a} \quad . \tag{90}$$

Therefore the change in the free energy $G(T,{}^{D}F)$ is a directional property and can be studied further considering ΔG as the difference between the amount of heat entering ΔQ into the soil and the amount of heat $T\Delta S$ absorbed by the soil

$$\Delta G = \Delta Q - T\Delta S . \tag{91}$$

Additionally, the change in the chemical potential is also a directional property dependent on the anisotropy of the soil and can be calculated from the adsorption isotherm where an amount of water is transferred from the liquid in bulk to the soil at constant temperature and load pressure; the latter being responsible for the length change along the direction $\boldsymbol{r}$. According to (91) the associated change in the chemical potential may be written

$$\Delta\mu = \left(\frac{\partial \Delta G}{\partial n^{a}}\right)_{\cdot {}^{D}\boldsymbol{F},T} = \left(\frac{\partial \Delta Q}{\partial n^{a}}\right)_{\cdot {}^{D}\boldsymbol{F},T} - T\left(\frac{\partial \Delta S}{\partial n^{a}}\right)_{\cdot {}^{D}\boldsymbol{F},T} \quad . \tag{92}$$

Proceeding to write down the analytical expressions for the two terms figuring in the *RHS* of (91) one can state that the change in the heat of swelling occurring between two adsorption moisture densities, each corresponding to a different hygrometric state, can be obtained from thermal measurements and then the accompanying change in the chemical potential can be computed if the change in entropy is known. Considering that the system is maintained at constant temperature and that the deviator swelling stress tensor and the deviator swelling strain increment tensor are coaxial and that the soil is quasistatically compressed with an initial mean stress $(p)_{Io}$ to a final mean stress $(p)_{If}$, it can be proved that the change in entropy can be computed directly from

$$dS = \left(\frac{\partial S}{\partial^{D}\boldsymbol{F}_{i}}\right)_{T,n^{a},{}^{D}\boldsymbol{F}_{j}} d^{D}\boldsymbol{F}_{j} = -V\left(a_{x}dX + a_{y}dY + a_{z}dZ\right) \tag{93}$$

since

$$\frac{\partial S}{\partial^{D}\boldsymbol{F}} = \frac{\partial \boldsymbol{r}}{\partial T} \tag{94}$$

where a_i $(i=x,y,z)$ is the volume expansion coefficient defined by $(1/i)$ (di/dT).

Considering the change in the internal energy given by

$$\delta U = T\delta S - V\left(X\delta x + Y\delta y + Z\delta z\right) \tag{95}$$

one can easily find

$$\delta U=-TVD^{-1}[a_xE^x{}_{hwYZ}[(1-\nu_{zy}\nu_{yz})(\frac{\delta x}{x}-s^x{}_{hXYZ}\delta w-\beta^x{}_{wXYZ}\delta h)$$
$$-(\nu_{xy}-\nu_{xz}\nu_{zy})(\frac{\delta y}{y}-s^y{}_{hXYZ}\delta w-\beta^y{}_{wXYZ}\delta h)$$
$$-(\nu_{xz}-\nu_{xy}\nu_{yz})(\frac{\delta z}{z}-s^z{}_{hXYZ}\delta w-\beta^z{}_{wXYZ}\delta h)]$$
$$+a_yE^y{}_{hwXZ}[(1-\nu_{xz}\nu_{zx})(\frac{\delta y}{y}-s^y{}_{hXYZ}\delta w-\beta^y{}_{wXYZ}\delta h)$$
$$-(\nu_{yz}-\nu_{yx}\nu_{xz})(\frac{\delta z}{z}-s^z{}_{hXYZ}\delta w-\beta^z{}_{wXYZ}\delta h) \qquad (96)$$
$$-(\nu_{yx}-\nu_{yz}\nu_{zx})(\frac{\delta x}{x}-s^x{}_{hXYZ}\delta w-\beta^x{}_{wXYZ}\delta h)]$$
$$+a_zE^z{}_{hwXY}[(1-\nu_{xy}\nu_{yx})(\frac{\delta z}{z}-s^z{}_{hXYZ}\delta w-\beta^z{}_{wXYZ}\delta h)$$
$$-(\nu_{zx}-\nu_{zy}\nu_{yx})(\frac{\delta x}{x}-s^x{}_{hXYZ}\delta w-\beta^x{}_{wXYZ}\delta h)$$
$$-(\nu_{zy}-\nu_{zx}\nu_{xy})(\frac{\delta y}{y}-s^y{}_{hXYZ}\delta w-\beta^y{}_{wXYZ}\delta h)]]$$
$$-V(X\delta x+Y\delta y+Z\delta z)$$

and for the change in the chemical potential

$$d\mu=(\frac{\partial\mu}{\partial^D\boldsymbol{F}_i})_{T,n^a,{}^D\boldsymbol{F}_j}d^D\boldsymbol{F}_j=-Vs_Idp_I=-V(s_xdX+s_ydY+s_zdZ) \qquad (97)$$

since

$$\frac{\partial\mu}{\partial^D\boldsymbol{F}}=\frac{\partial\boldsymbol{r}}{\partial n^a} \qquad (98)$$

Hence, the differential expressions (93) and (97) for ΔS and $\Delta\mu$ respectively may be applied to describe the thermodynamic swelling potential of the soil by means of conventional techniques. Returning to the first order differential an interesting case arises when p_I, w and h do not vary independently but are considered to be functions of the temperature T themselves

$$dp_I=\frac{\partial p_I}{\partial T}dT\ ,\quad d\vartheta=\frac{\partial\vartheta}{\partial T}dT\ ,\quad dh=\frac{\partial h}{\partial T}dT \qquad (99)$$

Then (53a) reads

$$\frac{de}{dT}=(1+e)\left\{p_I\beta^e{}_{p_I}\chi_{p_I}+s_\vartheta\frac{\partial\vartheta}{\partial T}+h\beta^e{}_h\chi_h\right\}$$
$$=(1+e)\left\{\left(\frac{c_{p_I}\beta^e{}_{p_I}}{TVa}\right)_{p_I}+s_\vartheta\frac{\partial\vartheta}{\partial T}+\left(\frac{c_h\beta^e{}_h}{TVb}\right)_h\right\} \tag{100}$$

where

$$\chi_{p_I}=\frac{1}{p_I}\frac{\partial p_I}{\partial T}\ ;\ \chi_h=\frac{1}{h}\frac{\partial h}{\partial T}\ ;\ \frac{dS}{dp_{p_I}}=-Va\ ;\ \frac{dS}{dh}=-Vb$$
$$\beta^e{}_{p_I}=\frac{1}{1+e}\left(\frac{\partial e}{\partial p_I}\right)_{\vartheta,h};\ \beta^e{}_h=\frac{1}{1+e}\left(\frac{\partial e}{\partial h}\right)_{p_I,\vartheta}$$

and c_{pI} stands for the specific heat at constant pressure driving swelling $(=T(\partial S/\partial T)_{pI}$; representing the quasi-static heat flux per mole required to produce a unit increase in temperature when the system is maintained at constant load pressure.

5. SUMMARY AND CONCLUSIONS

In this article, the general field equations which describe the behaviour of unsaturated swelling soils are considered. Also in view of the practical importance of the problem of calculation of the change in the mechanical properties of a swelling non–saturated soil, certain aspects forming a logical sequence have been selected for the theoretical treatment from as general a point of view as possible. Formulas are derived for the stress–strain relationship, the strength of swelling, the change in entropy of a swelling rectangular element of soil, all of them in terms of the moisture density content, the hygrometric state and other measurable quantities.

6. REFERENCES

Biot, M.A. 1941 General theory of three–dimensional consolidation. *J.Appl.Phys.* **12**: pp. 155–164.

Gibbs, J.W. 1906 On the equilibrium of heterogeneous substances, in collected works of J. Willard Gibbs. *Yale Univ. Press, New Haven, Conn.*

Karalis, T.K. 1990a Compressibilite des argiles non saturees a partir des essais rheologiques. *Can. Geoth. J.* **67**: pp. 90–104.

Karalis, T.K. 1990b On the elastic deformation of non–saturated swelling soils. *Acta Mechanica* **84**: pp. 19–45.

Karalis, T.K. 1991a Flow through swelling soils (submitted for publication)

Karalis, T.K. 1991b Quelques resultats theoriques et experimentaux concernant le gonflement anisotrope. *J. Phys. II France* **1**: pp. 717–738.

McNabb, A. 1960 A mathematical treatment of one–dimensional soil consolidation Quarterly. *Appl. Math.* **17**: pp. 337–347.

Philip J.R. 1969 Hydraustatics and hydrodynamics in swelling soils, *Water Resources Res.* **5**: pp. 1070–1077.

Terzaghi, K. 1923 Reprinted from theory to practice in soil mechanics, *Wiley and Sons New York*: pp. 133–136.

OPERATIONAL ASPECTS OF THE MECHANICS OF DEFORMING POROUS MEDIA: THEORY AND APPLICATION TO EXPANSIVE SOILS

Philippe Baveye
Bradfield Hall
Cornell University
Ithaca, New York 14853 (U.S.A.).

Abstract: *The analysis of transport processes in expansive porous media generally involves the use of some form of a referential, or material, coordinate transformation. The most commonly used of these relationships involves a ratio of two macroscopic bulk densities, that associated with the configuration of the soil at a reference time (e.g., t = 0) and that associated with its configuration at a later (arbitrary) time. This coordinate transformation is traditionally derived in a conceptual framework that entirely disregards the interactions between the observer and the porous media being observed. The purpose of the present chapter is to show that an operational coordinate transformation may be derived in a conceptual framework, called relativist, that takes the measurement process explicitly into account. Even when there is no solid phase production in the system, this generalized coordinate transformation still differs from its traditional counter-part by the presence of an additional term in the numerator of the bulk density ratio. This additional term vanishes identically when the measuring instrument used has space-and time-invariant properties. Measuring techniques are considered for which this stringent condition appears to be met, in first approximation. In particular, it is argued that such is the case with the dual-energy gamma-ray probe commonly used to study the deformation of soils in laboratory columns and in situ. Another method routinely used to measure bulk densities in situ in expansive soils, based on the simultaneous use of gamma-ray and neutron probes, is analyzed from the same perspective. It is argued that insufficient information is available at present about the characteristics of its volume of influence to conclude whether or not this method is incompatible with the operational constraints inherent in the traditional coordinate transformation.*

1. Introduction

The measurement process is an integral and essential step in the application of the principles of rational continuum mechanics to the description of porous media. However, the traditional formalism of rational continuum mechanics, axiomatic by nature, does not attempt at any time to account for this process explicitly (see, *e.g.*, *Malvern*, 1969;*Truesdell*, 1977; *Raats*, *1984)*.

The principal objective of the present chapter is to show that it is possible to

NATO ASI Series, Vol. H 64
Mechanics of Swelling
Edited by T. K. Karalis

remediate this very serious deficiency and to account explicitly for measuring instruments in the description of deforming porous media.

The present chapter is organized as follows. The section immediately following this introduction presents a rapid overview of the kinematics of deforming continua, in particular as it has been applied to expansive porous media. In the subsequent section, the limitations of the "classical" (REV-based) approach are briefly outlined and a radically different conceptual viewpoint, termed relativist, is described. This viewpoint is adopted in the fourth section, where an operational coordinate transformation is derived. As a result of this derivation, the operational constraints embodied in the traditional coordinate transformation are elucidated. In the next section, several measurement techniques commonly used to evaluate the macroscopic mass density of the solid phase of soils are analyzed in detail to determine whether they satisfy the operational constraints of the traditional coordinate transformation, or whether they require the use of a more general transformation.

2. Kinematics of deforming continua

In rational continuum mechanics, continuous media (or "bodies") and material particles are primitive concepts, defined *a priori*. A body is usually defined as a set, any element of which is called a material particle or, more commonly, a body-point (*Slattery*, 1981; *Truesdell*, 1977). A one-to-one, continuous, invertible mapping, *i.e.*, a homeomorphism of this set onto a region of Euclidian three-dimensional space exists at any time and is called a configuration (*Slattery*, 1981, p.2) or a placement (*Truesdell*, 1977) of the body.

The continuum mechanical analysis of the deformation of a body is made by describing the motion of its body-points. Four types of description are in common use, all based on classical nonrelativistic kinematics (see, *e.g.*, *Truesdell*, 1952, 1966, 1977; *Malvern*, 1969; *Batchelor*, 1970). They are usually referred to as material, referential, spatial and relative. Only the second and third will be used in the following.

The ***referential*** description adopts as reference a particular configuration of the body. This configuration may be chosen arbitrarily. In elasticity theory, it is usually taken as the unstressed state, *i.e.* the configuration to which the body will return when it is unloaded (*Malvern*, 1969, p. 138). In soil mechanics, a hypothetical configuration of zero porosity is often used in the description of deforming soils (*McNabb*, 1960; *Smiles and Rosenthal*, 1968). *Euler* (1762) introduced the particular referential description, often erroneously termed "Lagrangian", in which the Cartesian coordinates of the position $\mathbf{X}$ of a given body-point at a time $t_0 = 0$ are used as a label for that body-point. In this referential description, the independent variables are the coordinates $\mathbf{X}$ and the time t.

The ***spatial*** description, which was introduced by *d'Alembert* (1752) (cf *Truesdell*, 1952, p. 139) and is termed "Eulerian" by hydrodynamicists, focuses attention on a given region of space and adopts as independent variables the time t and the positions $\mathbf{x}$ occupied by the body-points at time t.

The mathematical relation between the spatial and referential descriptions is expressed formally as (see, *e.g.*, *Malvern*, 1969, p. 140):

$$\mathbf{x} = \mathbf{x}\,(\mathbf{X},t) \tag{1}$$

This equality means that **x** is the position at time t of the body-point that occupied the position **X** in an arbitrary reference configuration, for example that at time $t_o = 0$. In eq. (1), both **x** and **X** are defined with respect to the same space-fixed coordinate frame. The components x_i and X_i of the two position vectors **x** and **X** will be referred to hereafter as spatial and referential coordinates, respectively.

At a given time t, the total differential of **x**, written d**x**, is expressed mathematically as a function of the total differential d**X** as follows (see, *e.g.*, *Truesdell and Toupin*, 1960, p. 248):

$$dx_i = \frac{\partial x_i}{\partial X_K} dX_K \equiv x_{i,K}\, dX_K \tag{2}$$

where the indicial notation is adopted, involving summation over repeated indices. The quantities $x_{i,K}$ are known in continuum mechanics as the components of the deformation gradient tensor, usually denoted by $\mathfrak{F}$.

The positive-definite Jacobian determinant, $\mathfrak{J}$, of the deformation gradient tensor satisfies the so-called material equation of continuity, first derived by *Euler* (1762):

$$\rho(\mathbf{x}(\mathbf{X},t),t)\,\mathfrak{J} = \rho(\mathbf{X},t_o) \quad \text{or} \quad \rho(\mathbf{x}(\mathbf{X},t),t) = \rho(\mathbf{X},t_o)\,\mathfrak{J}^{-1} \tag{3}$$

where $\rho(\mathbf{x}(\mathbf{X},t),t)$ and $\rho(\mathbf{X},t_o)$ correspond to the mass density in the configuration of the body at time t and in the reference configuration, respectively. The Jacobian $\mathfrak{J}$ is, by definition, the determinant of the matrix of elements $\{x_{i,K}\}$ and is, consequently, a scalar function of the components of the deformation gradient tensor $\mathfrak{F}$. However, when the motion of the body-points occurs along one axis of the reference coordinate frame, for example in the x_3 direction, the determinant of the deformation gradient tensor, written in matrix form, reduces to (see, *e.g.*, *Raats and Klute*, 1969; *Raats*, 1987):

$$\mathfrak{J} = \det\{x_{i,K}\} \equiv \det \begin{pmatrix} 1 & 0 & 0 \\ 0 & 1 & 0 \\ 0 & 0 & \dfrac{\partial x_3}{\partial X_3} \end{pmatrix} = \frac{\partial x_3}{\partial X_3} \tag{4}$$

which can be substituted for $\mathfrak{J}$ in eq. (3) to yield:

$$dx_3 = \frac{\rho(X_3,t_o)}{\rho(x_3(X_3,t),t)} dX_3 \quad (5)$$

A particular form of this equation is used to describe the one-dimensional deformation of expansive soils, along a "vertical" direction z perpendicular to the soil surface (*Raats and Klute*, 1969; *Sposito et al.*, 1976; *Baveye et al.*, 1989):

$$dz = \frac{\rho_{sm}(Z,t_0)}{\rho_{sm}(z(Z,t),t)} dZ \quad (6)$$

In this equation, ρ_{sm} stands for the macroscopic mass ("bulk") density of the solid phase, either in the reference configuration or in the configuration at time t. According to *Raats* (1984), eq. (6) implies that the ratio $\rho_{sm}(z, t_0)/\rho_{sm}(z(Z,t),t)$ at a given time t can be deduced from the value of the ratio dz/dZ, measured with markers moving along with the deforming soil. Various types of markers, for example small steel plates, have been developed in the past (see, *e.g.*, *Schothorst*, 1977; *Hallaire*, 1987).

The integrated form of eq. (6) is of broad applicability in the interpretation of both laboratory and field experiments on deforming soils, as long as the deformation is essentially one-dimensional (*Sposito et al.*, 1976). *Baveye et al.* (1989) raise a number of issues concerning the integration of the coordinate transformation of eq. (6). They suggest, in particular, that a systematic comparison between results obtained using two mathematically equivalent methods of integration of eq. (6) may reveal occurrences of lateral expansion or systematic errors in experimental data.

3. The relativist viewpoint

The above derivation assumes *a priori* the existence of a body, with which a number of macroscopic fields are associated, like the mass density ρ. This postulate is in perfect keeping with the axiomatic nature of rational continuum mechanics and is acceptable as long as one remains in the realm of pure abstraction. However, in any situation where either eq. (5) or (6) is to be put to practical use, merely postulating the existence of this body is not sufficient. A one-to-one correspondence needs to be established between the actual material under study (*e.g.*, steel, rubber, wood, plant tissue, soil,..) and the abstract body described by rational continuum mechanics.

This question has been traditionally addressed in the literature on porous media by the introduction of the concept of a Representative Elementary Volume (REV), inspired by a similar concept in fluid mechanics (see, *e.g.*, *Batchelor*, 1970, p. 4). This REV is supposed to be large enough to enclose many voids and solid grains but, at the same time, to be small enough to qualify as a mathematical neighborhood of its centroid, in the mathematical sense (*Sposito*, 1978). Macroscopic fields, like the mass densities and mass flux densities of the various phases present in a porous medium, may then be defined by volume averaging, over the REV, of the corresponding microscopic variables (see, *e.g.*, *Bear*, 1972; *Hassanizadeh and Gray*, 1979).

The REV concept has been argued to be both unnecessarily restrictive and experimentally unverifiable (*Baveye*, 1983; *Baveye and Sposito*, 1984). In addition, it imposes stringent operational constraints that are virtually impossible to satisfy in practice. This point, seldom explicitly acknowledged in studies based on the REV concept (see, *e.g.*, *Bear*, 1972; *Hassanizadeh and Gray*, 1979; however see also *Molz et al.*, 1990), is best appreciated by outlining the practical "program" that a researcher should go through when trying to compare measured and theoretically predicted values of a given macroscopic variable at different points in a field soil. In the case of theoretical predictions in REV-based theories, there is assumed to exist at any point an absolute value of the macroscopic variable, dependent only on the characteristics of the porous medium and independent of the observer. Because this absolute variable is defined by volume averaging over an REV, the only consistent way to compare its theoretically predicted value with a measured one is to use a measuring device that performs precisely the same volume averaging. This instrument should be designed so as to be equally sensitive to all portions of its sphere of influence, which has exactly the same volume, shape, and orientation as the REV of the porous medium under investigation. Since determining the geometric features of the REV is at present entirely unfeasible, even in the simplest conceivable cases, the program just outlined cannot be carried out practically.

In order to establish a correspondence between an abstract body and the actual porous medium it is meant to represent, a radically different conceptual framework may be adopted. It was first introduced and termed relativist by *Baveye* (1983), and is described in detail in *Baveye and Sposito* (1984, 1985a,b). Its origins may be traced back to the elegant work of *Massignon* (1957) who, in the statistical mechanics of fluids, introduced variables, that in French he termed "*semi-fines*", to describe instrumental responses. Examples of the use of the relativist viewpoint in other contexts than that described in the following may be found in *Cushman* (1986), *Moltyaner* (1990), *Molz et al.* (1990), *Bachu and Cuthiell* (1990) and *Maneval et al.* (1990).

Unlike the REV concept, the relativist viewpoint does not assume that a given macroscopic field has an "absolute" value at any point within a porous medium; on the contrary, it considers that the local value taken by a particular macroscopic field depends of course on the properties of the porous medium but is also, unavoidably, relative to the instrument used by the observer.

In practice, every measuring instrument is characterized by a "volume of influence", defined as the volume of porous medium that actually influences the numerical outcome of the measurement process. Within this volume of influence, some portions may have a more determinant effect than others. For instance, *Baker and Lascano* (1989) suggest that the sensitivity of the neutron probe, commonly used to measure the volumetric water content in soils, decreases with the distance away from the access tube in which the probe is lowered.

Pivotal in the relativist framework is the idea that the instrumental response may be described mathematically using a so-called instrumental weighting function. This function, denoted m, is defined such that its geometrical support (the volume where m takes non-zero values) coincides with the volume of influence of the instrument. The value of the weighting function at a particular point within its geometrical support represents the degree to which the instrument probes the soil at that point or, alternatively, the degree to which the soil at that particular location influences the output of the instrument. Selection of the origin of the coordinate frame in which this function m is defined mathematically is arbitrary. A possible choice is the centroid of the geometrical support of m.

The microscopic level of description of a porous medium is that at which the constituting phases can be looked at as continua, in the traditional sense (see, *e.g.*, *Malvern*, 1969; *Truesdell*, 1977). At the macroscopic scale, the porous medium itself may be construed as a continuum. The "macroscopic" scale therefore corresponds to the scale at which measurements are traditionally made. According to the relativist framework, the **macroscopic** mass density of the solid phase of a porous medium, $\rho_{sm}(\mathbf{x},t)$, measured at a point $\mathbf{x}$ and at a time t by a dual-energy gamma-ray probe to which is associated a weighting function m, is defined by the "instrumental transform" of ρ_s, the **microscopic** mass density of the soil solid phase (*Baveye and Sposito*, 1985):

$$\rho_{sm}(\mathbf{x},t) = \rho_s \circ \widetilde{m} \equiv \int \rho_s(\mathbf{x}+\xi,t)\, m(\xi,\mathbf{x},t)\, d^3\xi \qquad (7)$$

where the integral extends over all space, $\widetilde{m}(\xi,\mathbf{x},t) \equiv m(-\xi,\mathbf{x},t)$ and ρ_s is the microscopic mass density of the soil solid phase. The explicit dependence of the weighting function m on $\mathbf{x}$ and t allows the description of the measurement process when the properties of the instrument are affected by the environment in which it is placed. It is clear from eq. (7) that in the relativist framework, unlike in R.E.V.-based theories, there are in principle as many "macroscopic scales" as there are measuring instruments.

The instrumental transform of eq. (7) establishes a one-to-one correspondence between the solid phase of the porous medium and an abstract "body" or continuum, characterized by a number of macroscopic variables. In the context of rational continuum mechanics, this body is viewed as a collection of body-points, *i.e.* infinitesimal volumes of the body. These body-points, like the body that they constitute, are abstract beings, relative to a given weighting function, and should in no way be confused, as is sometimes the case (see, *e.g.*, *Raats*, 1984) with actual physical volumes or "parcels" of the porous medium. Indeed, while a "parcel" centered at a given location Z at time t_0 moves in a unique manner to occupy a position z at time t, the location at t of a body-point that was at Z at t_0 may be strongly dependent on the weighting function used to define the body to which this body-point belongs. In other words, unlike in the case of physical "parcels" of the solid phase of a porous medium, the relation z(Z,t) for macroscopic body points of the body associated with this phase is relative to a given weighting function.

4. Derivation of an operational coordinate transformation

The derivation, in the relativist framework, of an operational coordinate transformation for the solid phase of a porous medium starts from considerations on mass balance at the microscopic level. At this level, the spatial form of the mass balance equation for the solid phase is, in indicial notation (*Baveye and Sposito*, 1984)

$$\frac{\partial \rho_s}{\partial t} + (\rho_s v_i^s)_{,i} = \sigma_s \tag{8}$$

where σ_s corresponds to the rate of production of mass of the solid phase and v_i^s is a component of the microscopic velocity vector in the solid phase.

According to the definition (7), the macroscopic mass flux density of the solid phase is defined as follows (*Baveye and Sposito*, 1985)

$$J_i^{sm} \equiv (\rho_s v_i^s) \circ \widetilde{m}. \tag{9}$$

When the weighting function m is time- and space-dependent, the spatial and temporal derivatives of the macroscopic variables ρ_{sm} and J_i^{sm} may be expressed in terms of their microscopic counterparts using the generalized regularization theorems (*Baveye and Sposito*, 1985), which for an arbitrary macroscopic variable a_{sm} are expressed as:

$$\frac{\partial a_{sm}(x,t)}{\partial x} = \left(\left\{\frac{\partial a_s}{\partial x}\right\} + \sum_{\beta \neq s} [a_s]\, \mathbf{n}\, \delta_{s\beta}\right) \circ \widetilde{m} + a_s \circ \frac{\partial \widetilde{m}}{\partial x} \tag{10}$$

$$\frac{\partial a_{sm}(x,t)}{\partial t} = \left(\left\{\frac{\partial a_s}{\partial t}\right\} - \sum_{\beta \neq s} [a_s]\, n_i V_i\, \delta_{s\beta}\right) \circ \widetilde{m} + a_s \circ \frac{\partial \widetilde{m}}{\partial t} \tag{11}$$

The derivative inside the braces on the right side of eqs. (10) and (11) is a derivative of a_s in the usual sense, defined everywhere except on the interface $S_{s\beta}$ between the soil solid phase and the other phases β present in the soil. The jump of a_s at these interfaces is denoted by $[a_s]$, while **n** represents the normal unit vector at $S_{s\beta}$ oriented toward the solid phase, $\delta_{s\beta}$ is the Dirac distribution in three dimensional space, defined on $S_{s\beta}$,, and V_i is the ith component of the velocity of the interface $S_{s\beta}$.

The definitions (7) and (9) and the generalized regularization theorems of eqs. (10) and (11) lead to the transformation of (8) into:

$$\frac{\partial \rho_{sm}}{\partial t} + (J_i^{sm})_{,i} = \sigma_{sm} + \sum_{\beta \neq s} \left(([\rho_s v_i^s] - [\rho_s] V_i) n_i \delta_{s\beta} \right) \circ \widetilde{m} + \rho_s \circ \frac{\partial \widetilde{m}}{\partial t} + (\rho_s v_i^s) \circ \widetilde{m}_{,i} \tag{12}$$

The second term on the right side of eq. (12) vanishes identically if, at the interfaces between the solid phase and the phases β, the velocities of the two phases are identical (*Baveye and Sposito* , 1984). This is satisfied if the solid phase represents a no-slip boundary condition for the phases β and if there is no production of mass at this interface. Under these conditions, eq. (12) can be transformed into the spatial differential form of a generalized mass balance equation for the solid phase in a soil:

$$\frac{\partial \rho_{sm}(x,t)}{\partial t} + (J_i^{sm}(x,t))_{,i} = \sigma_{sm}(x,t) + (\rho_s \circ (\frac{\partial \widetilde{m}}{\partial t}))(x,t) + \left((\rho_s v_i^s) \circ \widetilde{m}_{,i} \right)(x,t) \tag{13}$$

The two terms on the right side of eq. (13) account for the time and space variability of the weighting function m, respectively. They vanish identically for a space- and time- invariant weighting function, in which case eq. (13) reduces to a conventional macroscopic mass balance equation (see, *e.g.*, *Baveye and Sposito*, 1984). In the following, we shall denote by $\mathcal{M}(x, t)$ the sum of the last two terms in eq. (13). With this simplifying notation, eq. (13) becomes:

$$\frac{\partial \rho_{sm}(x,t)}{\partial t} + (J_i^{sm}(x,t))_{,i} = \sigma_{sm}(x,t) + \mathcal{M}(x,t) \tag{14}$$

The integral of both sides of eq. (14) over V(t), after rearrangement, is given by:

$$\int_{V(t)} \left\{ \frac{\partial \rho_{sm}(x,t)}{\partial t} + (J_i^{sm}(x,t))_{,i} - \sigma_{sm}(x,t) - \mathcal{M}(x,t) \right\} dv = 0 \tag{15}$$

where dv is an infinitesimal element of the volume V(t), expressed in spatial coordinates. Application of the extended form of Reynolds' transport theorem including a sink/source term (see, *e.g.*, *Muller*, 1975; *Slattery*, 1981) transforms eq. (15) into:

$$\frac{d}{dt} \int_{V(t)} \rho_{sm}(x,t) \, dv = \int_{V(t)} \left\{ \sigma_{sm}(x,t) + \mathcal{M}(x,t) \right\} dv \tag{16}$$

Integration of both sides of eq. (16) with respect to time, from the reference time t_o to the time t, yields:

$$\int_{V(t)} \rho_{sm}(x,t) \, dv - \int_{V(t_o)} \rho_{sm}(x,t_o) \, dv_o = \int_{t_o}^{t} \int_{V(t)} \left\{ \sigma_{sm}(x,t) + \mathcal{M}(x,t) \right\} dv \, dt \tag{17}$$

The integrals over V(t) in eq. (17) can be evaluated by a change of the variables of integration from spatial to referential coordinates. In order to make this change of variables, an expression is required that relates an infinitesimal element of volume dv $= dx_1dx_2dx_3$, in spatial coordinates to one, $dv_0 = dx_1dx_2dx_3$, in referential coordinates. This relation is given by (see, *e.g.*, *Truesdell and Toupin*, 1960, p. 249; *Malvern*, 1969, p. 168):

$$dv = \mathfrak{J}\, dv_0 \tag{18}$$

where $\mathfrak{J}$ is the strictly positive Jacobian determinant of the spatial-referential coordinate transformation. $\mathfrak{J}$ is, by definition, the determinant of the matrix of elements $\{x_{i,K}\}$ and, therefore, is a scalar function of the components $x_{i,K}$ of the deformation gradient tensor. With the introduction of eqs (7) and (18), eq. (17) becomes, after rearrangement:

$$\int_{V(t_o)} \{\rho_{sm}[x(X,t),t]\, J - \rho_{sm}(X,t_o)\}\, dv_o =$$

$$\int_{t_o}^{t}\int_{V(t_o)} \{\sigma_{sm}[x(X,t),t] + \mathcal{M}[x(X,t),t]\}\, \mathfrak{J}\, dv\, dt \tag{19}$$

where the notation $\mathbf{x}(\mathbf{X},t)$ expresses the dependence of the spatial coordinates $\mathbf{x}$ on the referential coordinates $\mathbf{X}$ as well as on the time t.

Because the domain of integration $V(t_o)$ in eq. (19) is independent of time, the order of the two integrations may be reversed. Equation (19) then becomes:

$$\int_{V(t_o)} \Big\{\rho_{sm}[x(X,t),t]\, J - \rho_{sm}(X,t_o) -$$

$$\int_{t_o}^{t} \sigma_{sm}[x(X,t),t]\, \mathfrak{J}\, dt + \int_{t_o}^{t} \mathcal{M}[x(X,t),t]\, \mathfrak{J}\, dt\Big\}\, dv_o = 0 \tag{20}$$

which holds for an arbitrary initial volume $V(t_o)$. Because the integrand in eq. (20) is a continuous function of space, it must vanish identically for all $\mathbf{X}$, according to the Du Bois-Reymond Lemma (see, *e.g.*, *Segel*, 1977). The resulting equation is a macroscopic mass balance in terms of referential coordinates, which is expressed by:

$$\mathfrak{J}\cdot\rho_{sm}[x(X,t),t] \;=\; \rho_{sm}(X,t_o) + \Phi_m[x(X,t),t] + M_m[x(X,t),t] \tag{21}$$

where $\Phi_m[x(\mathbf{X},t),t]$ and $M_m[x(\mathbf{X},t),t]$ stand, respectively, for the first and second time integral in equation (20).

Because this relation involves only the determinant $\mathfrak{J}$ of the matrix of elements $\{x_{i,K}\}$ and not the elements themselves, eq. (21) is, in general, of no help to express the components $x_{i,K}$ in experimentally measurable terms. In one particular case, however, eq. (21) may be used to this effect. When the soil deformation process can be considered one-dimensional (e.g., in the x_3- or z-direction, oriented perpendicularly to the soil surface), the determinant of the macroscopic deformation gradient tensor of the solid phase, written in matrix form, reduces to dz/dZ (cf eq. (4)), which can be substituted for J in the one-dimensional form of equation (21) to give:

$$dz = \frac{\rho_{sm}(Z,t_o) + \Phi_m[z(Z,t),t] + M_m[z(Z,t),t]}{\rho_{sm}[z(Z,t),t]} dZ \qquad (22)$$

Even when there is no production of solid phase in the system, *i.e.* when $\Phi_m[x(X,t),t]$ is equal to zero, this equation still differs from eq. (6) by the presence of the term denoted by $M_m[z(Z,t),t]$. For eq. (22) to be of practical use, this term should be measurable, as is the case with the bulk densities $\rho_{sm}[Z,t_o]$ and $\rho_{sm}[z(z,t),t]$. Unfortunately, it is not so at present and, in view of the mathematical definition of $M_m[z(Z,t),t]$ (cf eq. (20)), it is unclear whether one will ever be able to evaluate it directly.

When both the time derivative and the gradient of the weighting function m are equal to zero for any point z situated on the spatial path followed by each particular body-point Z, the term $M_m[z(z,t),t]$ in eq. (22) vanishes identically and this equation then becomes formally equivalent to eq. (6). It is clear therefore that the latter equation involves very stringent constraints on the weighting function associated with the instrument used in practice to measure the bulk densities $\rho_{sm}[Z,t_o]$ and $\rho_{sm}[z(z,t),t]$. Consequently, it is important to determine whether these constraints are met by instruments traditionally used in soils *in situ* to obtain the bulk density data needed for the application of eq. (6).

5. Analysis of measurement techniques commonly used in soils

Probably the most common way to evaluate the macroscopic bulk density at a given point in a soil profile is to use an corer to obtain a sample centered on that point, to weigh this sample once its moisture has been evaporated in an oven, and to divide the weight of soil by the known volume of the sample. Perroux et al. (1974), among others, have used this method to study the relationship between bulk density and gravimetric moisture content in an expansive Thai soil. Since the corer has a fixed geometry, one expects the function m associated with this measurement technique to be essentially time- and space-invariant, if of course one neglects possible problems due to soil compaction by the corer (see, *e.g.*, *Hassan et al.*, 1983). Unfortunately, the fact that this method is destructive makes its use in connection with equations like (6) or (22) conditional on the absence of spatial variability over short distances. Indeed, repeated measurements of the bulk density at various depths and at the same location in a field are not possible for obvious reasons. In order to estimate the bulk densities appearing in eq. (6), one would then have to assume that samples taken a short distance away from the initial location would be in all respects similar to those that might have been obtained at this initial location at time t, had the soil at this initial

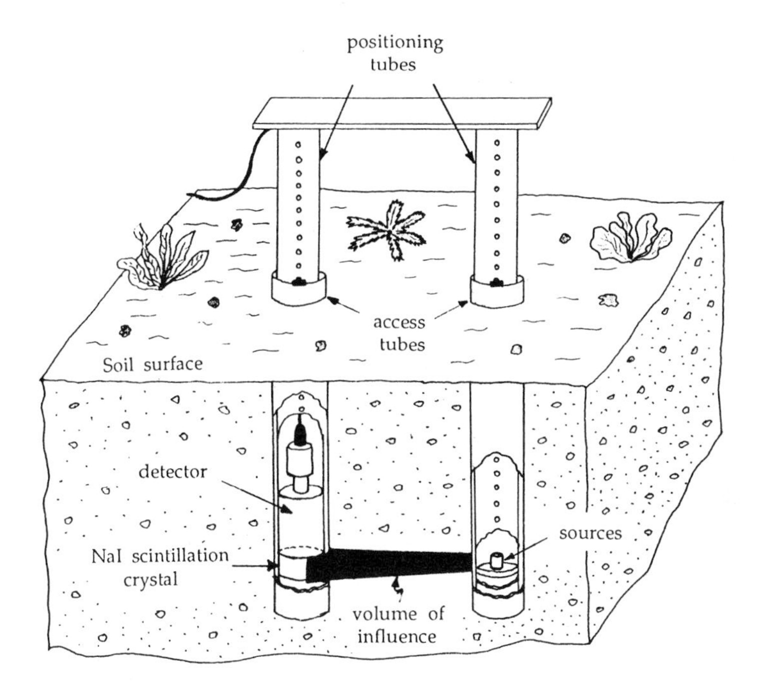

Figure 1: *Schematic illustration of the main features of a dual-energy gamma-ray probe and of its volume of influence.*

location not been removed. Lateral variability makes the validity of this assumption very questionable in many soils, especially in expansive soils with an important macroporosity (see, *e.g.*, *Wilding and Puentes*, 1988).

The dual energy gamma-ray probe (cf Figure 1) is also commonly used to assess the deformation of expansive soils under field conditions (*Reginato*, 1974a and b; *Sposito et al.*, 1976). This equipment allows the simultaneous measurement, *in situ*, of the macroscopic mass density, ρ_{sm}, and of the volumetric water content, θ_{vm}, simultaneously at a vertical set of positions in deforming soil profiles (see, *e.g.*, *Reginato*, 1974 a and b). The two gamma-ray sources, typically ^{137}Cs and ^{241}Am, are lowered in one access tube to the desired depth in the soil. A detector consisting of a NaI scintillation crystal mounted on a photomultiplier is lowered in the second access tube. During the measurement, the scintillation crystal and the dual-energy source are positioned in such a way that their centers of mass are at the same depth in the soil. In going from the sources through the soil to the scintillation crystal, a portion of the emitted photons looses energy by Compton- or photoelectric absorption or by Compton scattering, resulting in an attenuation of the gamma-ray intensity. Within the soil, the photons interact with both the liquid- and the solid phases. Mathematically, the effect of these interactions on the gamma-ray intensity corresponding to a given energy level is expressed as follows (see, *e.g.*, *Reginato*, 1974a; *Bruckler*, 1983):

$$\ln I = \ln I_o - x\,\mu_s\,\rho_{sm} - x\,\mu_w\,\rho_w\,\theta_{vm} \qquad (23)$$

where I is the gamma-ray intensity after passage through the soil, I_o is the unattenuated intensity, x is the thickness of the soil between the two access tubes, μ_s is the mass attenuation coefficient for the soil solid phase, μ_w is the mass attenuation coefficient for the liquid phase, and ρ_w is the mass density of the liquid phase (usually assumed equal to 1 Mg m^{-3}). Since eq. (23) involves two unknowns (ρ_{sm} and θ_{vm}), a single gamma-ray source does not allow the experimenter to determine the contribution of each one of the two phases to the observed attenuation, even when the attenuation characteristics μ_s and μ_w are known. However, when two sources are used, e.g. ^{137}Cs and ^{241}Am, a solvable system of two equations in two unknowns is obtained, which allows the simultaneous determination of both ρ_{sm} and θ_{vm} (*Reginato*, 1974a and b). The volume of soil that effectively influences the measured values of ρ_{sm} and θ_{vm}, i.e. the so-called "volume of influence" of the instrument, is contained in the solid angle subtended by the scintillation crystal with respect to the radioactive sources.

To decide which coordinate transformation, eq. (6) or (22), should be used in analyzing data obtained with this equipment, one needs to determine the space- and time dependence of the weighting function associated with it, i.e. establish whether M_m in eq. (22) is equal to zero within a good approximation. There are clearly three aspects to this question: the space- and time invariance of the characteristics of the measuring device itself, the invariance of its volume of influence and, finally, the sensitivity of the measuring instrument toward the actual distribution of matter within its volume of influence.

As the configuration of the probe (i.e. the type of source and detector used and the distance between them) is typically not modified during the measurement process, the only characteristic of the instrument that may possibly vary in time is the radiation intensity of its radioactive sources. This potential cause of time variation, due to decay, is negligible under most circumstances because the half-lives of the two sources of gamma-rays, respectively 30 years for ^{137}Cs and 458 years for ^{241}Am (*Corey et al.*, 1971) are much longer than the usual time frame of *in-situ* swelling experiments.

With respect to the space- and time dependence of the geometrical support of the weighting function, it is to a very large extent a matter of design of the instrument's detector. To understand this point, one could consider for example the case of a source composed of ^{241}Am, whose gamma-ray spectrum exhibits a principal pulse peak corresponding to an energy of 9.55 x 10^{-15} J (59.6 keV). If the detector is designed to be sensitive only to the photons that have an energy very close to that value, i.e. if its detector's energy window is a narrow band around 9.55 x 10^{-15} J, almost all the photons detected have gone unattenuated on a rectilinear path from the source to the detector. The larger the detector's energy window, the larger the number of photons that, after one or several Compton scattering events, are backscattered into the detector and influence the measurement (see, *e.g.*, *Vachaud et al.*, 1970). When virtually no backscattering is allowed, the theory predicts that the volume of influence of the probe corresponds to the solid angle subtended by the scintillation crystal with respect to the radioactive source, independently of the soil features. *Van Bavel* (1959) and *Van Bavel et al.* (1985) show experimentally that the vertical extent of this geometrical support (that they call the "resolution" of the instrument) is, as expected on theoretical grounds, equivalent to the detector's thickness and is independent of the medium in which the instrument is placed.

The third question that needs to be addressed concerns the space- and time invariance of the value of the weighting function at any particular point within its geometrical support. In the case of the gamma-ray attenuation technique, only the occurrence or non-occurrence of an interaction between the emitted photons and the matter placed between the source and the detector does matter. Provided the detector's energy window is properly dimensioned to screen out backscattered photons, the exact location within the "volume of influence" where the photon-matter interactions occur is of no importance. In support of this statement, *Van Bavel* (1959) concludes from laboratory experiments that the horizontal position within the instrument's "volume of influence", of an effectively infinitely tall slab of a material with a high density has no influence on the overall counting rate. A similar conclusion is also reached by *Van Bavel et al.* (1985). Therefore, it appears reasonable to assume, as a first approximation, that the weighting function associated with the dual-energy gamma-ray attenuation instrument is space- and time invariant and to use the referential coordinate transformation of eq. (6) to analyze field data on soil swelling and shrinking obtained with this instrument.

The situation is quite different in the case of another method commonly used to estimate ρ_{sm} *in situ* in expansive soils. This method (see, *e.g.*, *Gardner and Calissendorff*, 1967; *Christensen*, 1974; *Christaller and Thies*, 1983; *Jayawardane*, 1984; *Culley and McGovern*, 1990) is based on the combined use of a single-energy gamma-ray probe and of a neutron moisture meter. The latter instrument provides a value of the volumetric water content and thereby allows the equation for the attenuated gamma-ray intensity, eq. (23) to be solved for the bulk density.

During the measurement of the volumetric water content with the neutron probe, fast neutrons are emitted by the probe source radially into the soil. These fast neutrons scatter through repeated collisions with atomic nuclei, gradually reducing their speed to that characteristic of free Maxwellian particles at ambient temperature. This process is called thermalization. Use of a calibration equation permits the evaluation of θ_{vm} on the basis of the number of thermalized neutrons reaching the probe's detector. The volume of soil in which the swarm of thermalized neutrons is generated, usually called the "sphere" of influence or effective volume of measurement of the neutron probe, depends on the physical characteristics of the probe as well as on the concentration of (principally hydrogen) nuclei in the soil near the probe. In homogeneous soils with a uniform volumetric water content, the volume of influence of the neutron probe is usually considered to be spherical, with a diameter ranging from 150 mm in wet soils to 700 mm or more in dry soils (*Baker and Lascano*, 1989). In general, instrumental variability and the heterogeneity of natural soils together with the continuous modification of their properties by physicochemical process can cause the volume of influence of a particular neutron probe to depend on the position in the soil at which the probe is located and on the time at which the measurement is performed. There is unfortunately no information available on the distribution of sensitivity (or weighting) within this volume of influence. Some authors (see, *e.g.*, *Baker and Lascano*, 1989) have suggested that the relative sensitivity declines with distance from the source/detector assemblage but the validity of this statement remains to be assessed.

The use of a neutron moisture meter, instead of a second gamma-ray source, to evaluate the volumetric water content θ_{vm} in eq. (23) has two direct consequences. The first is that a significant error may be introduced in the measurement process. Indeed, the volumes of influence of the gamma-ray probe and of the neutron moisture meter have very different geometries. Probing different volumes of soil, these instruments are likely to yield numerically different values for θ_{vm}, resulting in different values for ρ_{sm}. A second consequence is that the volume of soil influencing the measured ρ_{sm} value is no longer simply that, highlighted in Fig. 1, between the two access tubes of the gamma-ray probe. It also includes the soil probed by the neutron moisture meter. In other words, the volume of influence associated with the measured ρ_{sm} value is the union, in a mathematical sense, of the (constant) volume of influence of the dual-energy gamma-ray probe and that of the neutron moisture meter. These respective volumes of influence of the two instruments may or may not overlap, depending on the location of the various access tubes.

In the relativist framework, the combined volume of influence of the two-probes system is looked at as the geometric support of a weighting function m associated via eq. (7) with the measurement of the bulk density ρ_{sm}. Because the extent of this combined volume of influence varies in space and time, one would be tempted to conclude that so does m and, therefore, that the use of eq. (6) in connection with bulk density data obtained with a two-probes system is not legitimate. However, such a conclusion cannot be reached at present because nothing is known about the shape of the weighting function associated with the two-probes system. It is entirely possible that the geometrical support of m only corresponds to a constant subset of the combined volume of influence of the two-probes system. Over this constant subset, m would take values that may or may not be space- and time-invariant. Only a detailed analysis of the dynamics of photons and neutrons in heterogeneous soils (following, *e.g.*, *Wilson and Ritchie*, 1986) would provide some insight on the form and properties of the weighting function m and would make it possible to determine whether the use of eq. (6) in connection with data obtained with the two-probes system is warranted.

As a side remark, before turning to the general conclusions of this chapter, it is worthwhile to note that the relativist framework provides new insight into the possible limitations of the use of markers to monitor the deformation of expansive soils. As we mentioned earlier, the relationship between the spatial and referential coordinates used to describe the deformation of a body is relative to the weighting function used in defining this body. Therefore, if the Jacobian dz/dZ is evaluated on the basis of measurements made with markers, the ratio $\rho_{sm}(z, t_o)/\rho_{sm}(z(Z,t),t)$ that one can deduce from dz/dZ at a given time t via eq. (6) must be relative to the weighting function associated with the particular marker used. Since this latter weighting function is likely to be very different from that associated with a typical gamma-ray probe, the mass density ratio obtained on the basis of marker displacement measurements should be expected, in general, to differ from that measured directly via gamma-ray attenuation.

6. Conclusions

The key result of the research reported here is contained in eq. (22), derived in the so-called relativist framework (*Baveye and Sposito*, 1984, 1985a,b). This new coordinate transformation, unlike that of eq. (6), allows the weighting function associated with the instrument used in practice to measure the bulk density ρ_{sm} to vary in space and time. When the measuring instrument is space- and time-invariant, the term $M_m[z(Z,t),t]$ on the right side of eq. (22) vanishes identically. When there is no production of mass in the system, eq. (22) reduces to the traditional coordinate transformation of eq. (6).

Two methods commonly used for the *in situ* measurement of bulk densities in soils are shown to have properties that, in first approximation, can be considered space- and time-invariant. The first is based on corer sampling and gravimetric determination of the bulk density. The second, the dual-energy gamma-ray probe, is non-destructive and is therefore widely used in field studies on expansive soils. An analysis of the attenuation of the photons in the (constant) volume of influence of the probe suggests that, in first approximation, the weighting function associated with this instrument is space-and time-invariant. On the other hand, it is argued that insufficient information is available at present concerning the weighting function of another method, based on the simultaneous use of gamma-ray and neutron probes, to allow us to determine whether this method is compatible with the operational constraints embodied in eq. (6). This conclusion emphasizes the need for further research in this area.

7. Acknowledgements

Sincere gratitude is expressed to Dr. Garrison Sposito (University of California) who provided helpful comments on an early draft of this paper. Dr. Charles Boast (University of Illinois) and three anonymous reviewers provided a number of editorial suggestions and constructive criticisms that were particularly helpful in writing the final version of this chapter.

8. References

Baker, J.M., and R.J. Lascano. 1989. The spatial sensitivity of time-domain reflectometry. *Soil Science* **147**: 378-384.

Batchelor, G.K. 1970. *An introduction to fluid dynamics.* Cambridge University Press, New York;.

Baveye, P. 1983. *Theory of water transport through heterogeneous soils: a relativist approach,* Unpublished M.S.E. Thesis, The Johns Hopkins University, Baltimore, Maryland.

Baveye, P., C.W. Boast, and J.V. Giráldez. 1989. Use of referential coordinates in deforming soils. *Soil Sci. Soc. Amer. J.* **53(5)**: 1338-1343.

Baveye, P. and M.B. McBride (eds.) 1992. *Clay swelling and expansive soils.* Kluwer Scientific Publishers. Amsterdam. (in press)

Baveye, P., and G. Sposito. 1984. The operational significance of the continuum hypothesis in the theory of water movement through soils and aquifers. *Water Resour. Res.* **20**: 521-530.

Baveye, P., and G. Sposito. 1985a. Macroscopic balance equations in soils and aquifers: the case of space- and time-dependent instrumental response, *Water Resour. Res.* **21**: 1116-1120.

Baveye, P., and G. Sposito. 1985b. Reply to a comment on "The operational significance of the continuum hypothesis in the theory of water movement through soils and aquifers" by B. Berkowitz and J. Bensabat, *Water Resour. Res.* **21**: 1295-1296.

Bachu, S., and D. Cuthiell. 1990. Effects of core-scale heterogeneity on steady-state and transient fluid flow in porous media: Numerical analysis. *Water Resources Research* **26**: 863-874.

Bear, J. 1972. *Dynamics of fluids in porous media,* Elsevier, New York.

Bruckler, L. 1983. Analyse théorique des erreurs de mesure de teneur en eau et de masse volumique par atténuation gamma. *Bulletin du Groupe Français d'Humidimétrie à Neutrons* **13**:7-33.

Christaller, G. and R. Thies, 1983. Two new designs of two-source soil-moisture gauges. pp 489-499. In: *Proc. 1983 Symp. on Isotope and Radiation Techniques in Soil Physics and Irrigation Studies,* International Atomic Energy Agency, Vienna.

Christensen, E.R., 1974. Use of the gamma density gauge in combination with the neutron moisture probe. pp 27-44. In: *Proc. 1973 Symp. on Isotope and Radiation Techniques in Soil Physics and Irrigation Studies,* International Atomic Energy Agency, Vienna.

Corey, J.C., S.F. Peterson, and M.A. Wakat. 1971. Measurement of attenuation of ^{137}Cs and ^{241}Am gamma rays for soil density and water content determinations. *Soil Sci. Soc. Am. Proc.* **35**: 215-219.

Culley, J.L.B. and M.A. McGovern 1990. Single and dual probe nuclear instruments for determining water contents and bulk densities of a clay loam soil. *Soil & Tillage Research* **16**: 245-256.

Cushman, J.H. 1986. On measurement, scale and scaling. *Water Resources Research* **22**: 129-134.

Cushman, J.H. (ed.) 1990. *Dynamics of fluids in hierarchical porous media.* Academic Press, London.

Euler, L. 1762. Recherches sur la propagation des ébranlements dans un milieu élastique (Lettre de M. Euler à M. de La Grange). *Opera* **(2) 10**: 255-263.

Gardner, W.H., and C. Calissendorff, 1967. Gamma-ray and neutron attenuation in measurement of soil bulk density and water content. pp. 101-113. In: *Proc. 1967 Symp. on Isotope and Radiation Techniques in Soil Physics and Irrigation Studies,* International Atomic Energy Agency, Vienna.

Hallaire, V. 1987. Retrait vertical d'un sol argileux au cours du dessèchement. Mesures de l'affaissement et conséquences structurales. *Agronomie* **7(8)**: 631-637.

Hassan, H.M., A.W. Warrick and A. Amoozegar-Fard, 1983. Sampling volume effects on determining salts in a soil profile. *Soil Sci. Soc. Amer. J.* **47**: 1265-1267.

Hassanizadeh, M., and W.G. Gray. 1979. General conservation equations for multi-phase systems, 1, Averaging procedure, *Advances in Water Resources* **2**: 131-144.

Jayawardane, N.S. 1984. Determination of the swelling characteristics of a soil using neutron and gamma density meters. *Aust. J. Soil Res.* **22**: 389-399.

Malvern, L.E. 1969. *Introduction to the mechanics of a continuous medium.* Prentice-Hall, Englewwod Cliffs, N.J.

Maneval, J.E., M.J. McCarthy, and S. Whitaker. 1990. Use of nuclear magnetic resonance as an experimental probe in multiphase systems: Determination of the instrument weight function for measurements of liquid-phase volume fractions. *Water Resources Research* **26**: 2807-2816.

Massignon, D. 1957. *Mécanique statistique des fluides,* Dunod, Paris.

McNabb, A. 1960. A mathematical treatment of one-dimensional soil consolidation. *Quarterly Appl. Math.* **17**:337-347.

Moltyaner, G.L. 1990. Field studies of dispersion: Radioactive tracer experiments on scales of 20, 40 and 260 metres. In: Cushman (1990). P. 7-36.

Molz, F.J., O. Güven, J.G. Melville and C. Cardone 1990. Hydraulic conductivity measurement at different scales and contaminant transport modelling. In: Cushman (1990). P. 37-59.

Müller, I. 1975. Thermodynamics of mixture of fluids. *J. de Mécanique* **14**: 267-303.

Perroux, K.M., U. Aromratana, and S. Boonyoi. 1974. Volume change and air-water relations of Chai Nat soil. *Thai. J. Agr. Sci.* **7**: 23-35.

Raats, P.A.C. 1965. *Development of equations describing transport of mass and momentum in porous media, with special reference to soils.* Ph.D. dissertation. Univ. of Illinois. Urbana-Champaign. Illinois.

Raats, P.A.C. 1984. Applications of the theory of mixtures in soil physics. In: Truesdell (1984). p. 327-343.

Raats, P.A.C. 1987. Applications of material coordinates in the soil and plant sciences. *Netherlands Journal of Agricultural Science* **35**: 361-370.

Raats, P.A.C. 1992. Application of material coordinates to describe the deformation of the solid phase and the movement of water in nonrigid soils, In: Baveye and McBride (in press, 1992).

Raats, P.A.C., and A. Klute. 1969. One-dimensional, simultaneous motion of the aqueous phase and the solid phase of saturated and partially saturated porous media. *Soil Sci.* **107**: 329-333.

Reginato, R.J. 1974a. Gamma radiation measurement of bulk density changes in a soil pedon following irrigation. *Soil Sci. Soc. Am. Proc.* **38**: 24-29.

Reginato, R.J. 1974b. Water content and bulk density changes in a soil pedon measured with dual energy gamma-ray transmission. *Can. J. Soil. Sci.* **54**: 325-328.

Schothorst, C.J. 1977. Subsidence of low moor peat soils in the western Netherlands. *Geoderma* **17**; 265-291.

Segel, L.A. 1977. An introduction to continuum theory. *Lectures in Applied Mathematics* 16:1-60.

Slattery, J.C. 1981. *Momentum, energy and mass transfer in continua.* Second edition. R.E. Krieger Publ. Co. Huntington, New York.

Smiles, D.E., and M.J. Rosenthal. 1968. The movement of water in swelling materials. *Aust. J. Soil Res.* **6**: 237-248.

Sposito, G. 1978. The statistical mechanical theory of water transport through unsaturated soil, 1, The conservation laws, *Water Resour. Res.* **14**: 474-478.

Sposito, G., J.V. Giráldez, and R.J. Reginato. 1976. The theoretical interpretation of field observations of soil swelling through a material coordinate transformation. *Soil Sci. Soc. Am. J.* **40**: 208-211.

Truesdell, C.A. 1952. The mechanical foundations of elasticity and fluid dynamics, *J. Ratioal Mech. Anal.* **1**:125-300.

Truesdell, C.A. 1977 *A first course in rational continuum mechanics,* Volume 1: General concepts. Academic Press, New York.

Truesdell, C.A. (ed.) 1984 *Rational thermodynamics,* 2nd ed. Springer Verlag, New York.

Truesdell, C.A., and R.A. Toupin (1960) *The classical field theories.* Handbuch der Physik. III/1. 226-793.

Vachaud, G., J. Cisler, J.L. Thony and L.W. De Backer. 1970. Utilisation de l'émission gamma de l'Américium pour la mesure des teneurs en eau d'échantillons de sols non saturés. *Proc. of the Isotope Hydrology Symposium of I.A.E.A.*: 643-661.

Van Bavel, C.H.M. 1959. Soil densitometry by gamma transmission. *Soil Science* **87**: 50-58.

Van Bavel, C.H.M., R.J. Lascano, and J.M. Baker. 1985. Calibrating two-probe, gamma-gauge densitometers. *Soil Science* **140**: 393-395.

Wilding, L.P., and R. Puentes (Editors). 1988. *Vertisols: their distribution, properties, classification and management.* Texas A&M. University Printing Center, College Station, Texas.

Wilson, D.J. and A.I.M. Ritchie. 1986. Neutron moisture meters: the dependence of their response on soil parameters. *Aust. J. Soil Res.* **24**: 11-23.

THE OSMOTIC ROLE IN THE BEHAVIOR OF SWELLING CLAY SOILS

S.L. Barbour, D.G. Fredlund and D.E. Pufahl

Department of Civil Engineering,
University of Saskatoon, Saskatchewan,
CANADA S7N OWO

ABSTRACT.– *Clay soils are commonly viewed in geotechnical engineering as having a net negatively charged surface. In order to become electroneutral, a diffuse double layer of cations and anions develops around the particle. Theoretical descriptions of this diffuse double layer have assisted engineers in describing the volume change behavior of clay soils. The salt content in the soil affects the size of the double diffuse layer. The difference in the salt concentration in the free pore fluid and that adjacent to the clay particle gives rise to the osmotic pressure concept for clay soils.*

For many problems in engineering practice, it has not been necessary to consider the osmotic pressure because it has not been significantly altered during the engineering application or because the changes in the osmotic pressure have been simulated in laboratory tests used to evaluate relevant soil parameters. In other words, the classic effective stress approach to the behavior of saturated soils has proved sufficient.

In recent years, geotechnical problems have involved situations where the pore fluid chemistry of the soil is changed. It then becomes necessary to develop a more complete set of state variables to describe the behavior of the soil.

The main objective of this paper is to present a more complete description of the state variables for soils where there is a change in the osmotic pressure of the pore fluid. The theoretical justification for the state variables is based on multiphase continuum mechanics. It is shown that for saturated soils two state variables are required to describe the stress sate. These include the classic effective stress (i.e., $(\sigma - u_w)$*) and a physico–chemical stress state variable which represents the net electrostatic repulsive stresses between the soil particles (i.e.,* $(R-A)$*). This physicochemical stress variable is controlled by the osmotic pressure of the pore fluid.*

The proposed model state variables are then used to describe volume change behavior in a saturated soil. The volume change behavior of compressible clay soils may be strongly influenced by physicochemical effects when concentrated pore fluids are introduced to the soil. A macroscopic description of the osmotic volume change behavior of a clay soil undergoing changes in pore fluid chemistry is provided.

The theoretical descriptions of two potential mechanisms of osmotic volume change (osmotic consolidation and osmotically induced consolidation) are presented. Osmotic consolidation occurs as a result of a change in the electrostatic repulsive–minus–attractive stresses, $(R-A)$ *between clay particles. Osmotically induced consolidation occurs because of fluid flow out of the clay in response to osmotic gradients. A numerical simulation is used to*

NATO ASI Series, Vol. H 64
Mechanics of Swelling
Edited by T. K. Karalis

demonstrate the characteristic behavior of a clay soil undergoing either of these volume change processes. The results of a laboratory testing program on two clay soils exposed to concentrated sodium chloride solutions are used to illustrate that the dominant mechanism of osmotic volume change in surficial clay soils is osmotic consolidation.

1.0 INTRODUCTION

The mechanical behavior of clay soils may be strongly influenced by osmotic effects resulting from the physico-chemical interactions between the clay particles and pore fluid. Conceptual models have been proposed to explain the influence of pore fluid chemistry on the behavior of clays. These models have been developed at a microscopic scale and have provided a qualitative description of interactions. In this paper the impact that strong electrolyte solutions have on the volume change behavior of clay soils is used to illustrate the influence of osmotic effects on the mechanical behavior of clay soils.

The time-dependent volume change behavior of soil is controlled by two properties. First, the compressibility of the soil, that is, its change in volume related to changes in stress state; and secondly, by the ability of the soil to conduct pore fluid. The ability of osmotic (i.e., physico-chemical) effects to influence both soil compressibility and fluid flow through soil have been demonstrated in the literature.

The presence of osmotic flow in clay aquitards has been studied by numerous researchers including Hanshaw and Zen (1965), Marine and Fritz (1981), and Neuzil (1986). The influence of osmotic flow processes on the transport of salt has been studied by Kemper and van Schaik (1966), Kemper and Rollins (1966), Elrick et al (1976), Greenberg et al (1973), and others. Deviations from Darcy's law have also been attributed to osmotic flow by investigators such as Kemper (1961), Low (1955), Bolt and Groenevelt (1969) and Olsen (1985).

Extensive studies of the compressibility and swelling of clays in response to changing pore fluid concentrations have been conducted over the last thirty years, and include the work of Bolt and Miller (1955), Bolt (1956), Warkentin et al. (1957), Aylmore and Quirk (1962), Blackmore and Miller (1962), Warkentin and Schofield (1962), Mesri and Olson (1971) and others. Bailey (1965) and Mitchell (1973a) provide reviews of this literature.

A comprehensive theoretical description of the combined influence of osmotic flow and volume change on the behavior of clay soils has been lacking up till now. The objective of this paper is to provide a theoretical description of the two primary mechanisms for osmotic volume change, referred to in this paper as osmotically induced consolidation and osmotic consolidation.

When a clay is exposed to a concentrated salt solution the volume of the clay will change as a result of the combined influence of the processes of osmotic flow and osmotic compressibility. As salt is transported into the clay, changes in the interparticle repulsive stresses between clay particles will occur. In turn, this will lead to changes in void ratio. The time dependent volume change associated with changes with the osmotic pressure of the pore fluid will be termed osmotic consolidation.

A second mechanism of volume change will also develop as a result of fluid flow within the clay in response to osmotic gradients. For example, if a clay soil is exposed to electrolyte solutions osmotic flow out of the sample will occur. These outward flows cause negative pore fluid pressures to develop within the sample which then lead to increases in the interparticle mechanical stresses and consequent volume change. The volume change that occurs is still in response to changes in effective mechanical stress (σ - u_f); however, because the changes in pore-water pressure are induced as a result of osmotic pressures, this form of volume change is called osmotically induced consolidation.

Characteristic features of these two processes are demonstrated through the use of a numerical model and the relative importance of these two independent processes are demonstrated from the results of a laboratory testing program.

Consolidation is the result of the combined influence of fluid flow and soil compressibility. Changes in pore fluid chemistry may significantly influence both volume change and fluid flow through clay. Theoretical descriptions of these "osmotic" phenomena are presented in the next two sections.

2.0 THEORETICAL DESCRIPTION OF OSMOTIC FLOW

Osmosis is the term used to describe the phenomenon by which a solvent passes from a solution of lower solute concentration through a semi-permeable membrane into a solution of higher solute concentration. A membrane is described as semi-permeable if it allows the passage of solvent but not solute. If the flow of water is restricted a pressure imbalance equal to the osmotic pressure difference between the two solutions would have to be present. The osmotic pressure can be calculated from thermodynamic principles (Robinson and Stokes, 1968). The osmotic pressure can also be approximated by the van't Hoff equation (Metten 1966):

$$P = R\,T\,C \qquad [1]$$

where:

P = osmotic pressure (kPa)
C = sum of the molar concentrations in solution (moles/litre)
R = universal gas constant = 8.32 (litre.kPa/K.mole)
T = absolute temperature (K)

The osmotic flow of water in soil can be described using a flow law similar in form to Darcy's law:

$$q_\pi = k_\pi \frac{\Delta \pi}{\Delta x} = \xi \, k_h \frac{\Delta \pi}{\Delta x} \qquad [2]$$

where:

q_π = water flux (m/s)
k_π = coefficient of osmotic permeability (m/s)
k_h = coefficient of (hydraulic) permeability (m/s)
π = osmotic pressure head = $P/\rho_f g$
ρ_f = pore fluid density (kg/m^3)
g = gravitational acceleration (m/s^2)
ξ = osmotic efficiency
x = distance (m)

If the soil behaves as a perfect semi-permeable membrane the coefficient of osmotic permeability will be equal to the coefficient of hydraulic permeability. In this case only pure water will flow in response to osmotic gradients. However, if the membrane is "leaky", the osmotic permeability will be equal to the hydraulic permeability multiplied by an osmotic efficiency. Under these conditions, water, carrying with it some dissolved salts, will flow in response to osmotic gradients. The osmotic efficiency is a measure of the degree to which the clay behaves as a perfect semi-permeable membrane.

The osmotic efficiency of clays has been studied by Kemper and Rollins (1966), Bresler (1973), Olsen (1972), and others. Results of these studies (Fig. 1) illustrate that osmotic efficiency is strongly dependent on pore fluid chemistry, pore fluid concentration, void ratio, and interparticle spacing. High osmotic efficiencies have only been observed at low pore fluid concentrations, or low void ratios.

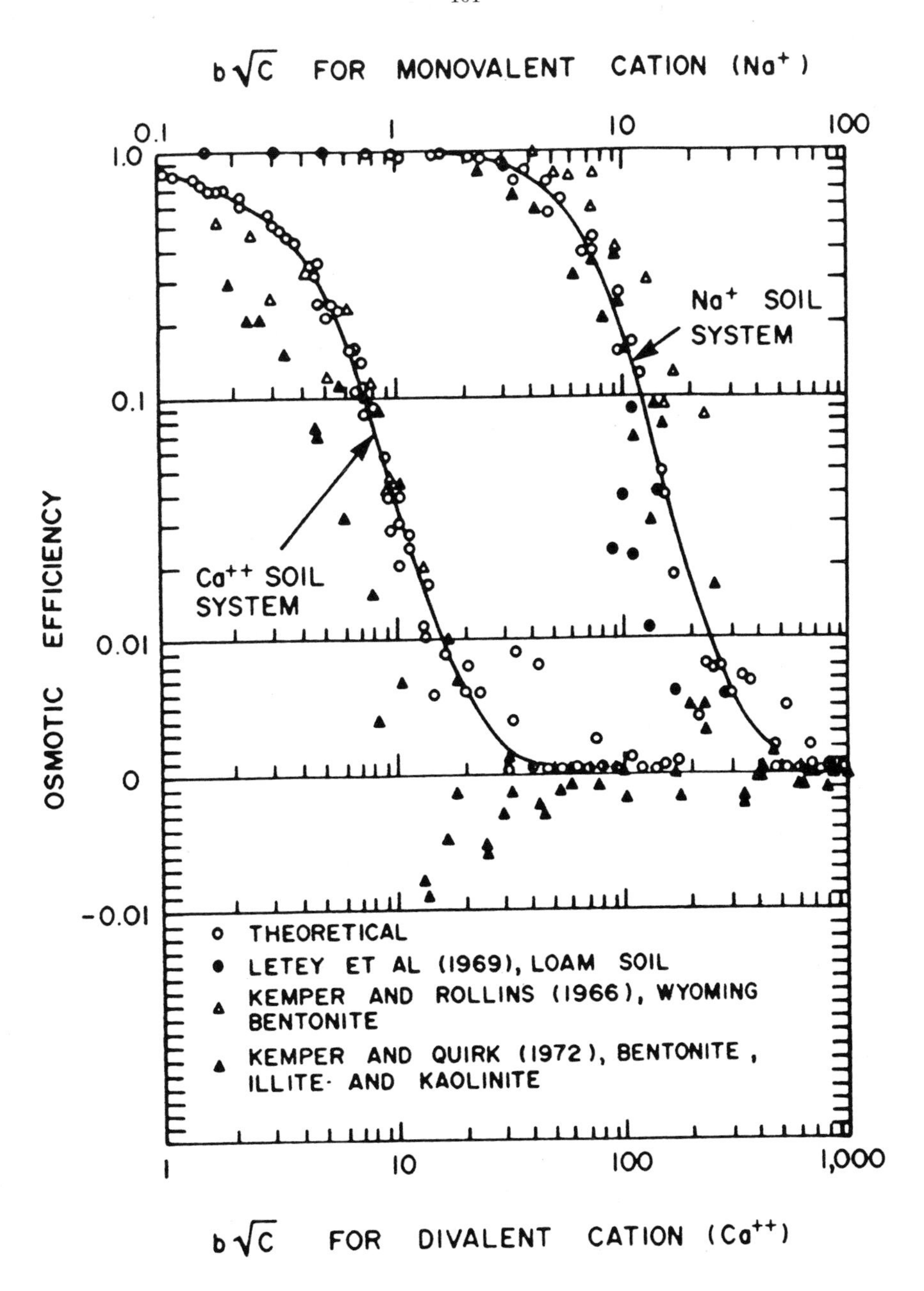

FIG. 1 Osmotic efficiency for a monovalent anion, unsaturated soil system, as a function of $b\sqrt{C}$ (C is pore fluid concentration in Normality, 2b is the film thickness in angstroms, (Bresler 1973))

Figure 1 illustrates the dependence of osmotic efficiency for an unsaturated soil on pore fluid chemistry, pore fluid concentration and fluid film thickness between soil particles. The film thickness is obtained by dividing the volumetric water content by half the soil surface area.

3.0 THEORETICAL DESCRIPTION OF OSMOTIC COMPRESSIBILITY AND SWELLING

Before the appropriate stress state and constitutive relationships for osmotic volume change can be developed, the mechanisms responsible for the sensitivity of clay soils to changes in pore fluid chemistry need to be described. In this section, a general description of the diffuse double layer and the mechanisms responsible for interparticle physico-chemical stresses are provided.

3.1 Osmotic Effects on Interparticle Stresses - Diffuse Double Layer and the Osmotic Pressure Concept

Most clay minerals have a net negative surface charge. In order to obtain electroneutrality, a diffuse double layer of cations (and anions) develops around the clay particle. The negatively charged clay surface and the charge distribution adjacent to the clay particle is termed the diffuse double layer. Theoretical descriptions of the diffuse double layer were first proposed by Gouy (1910) and Chapman (1913) and later modified by Stern (1924). Detailed developments of diffuse double layer theory are provided by Mitchell (1976) and van Olphen (1977). The distribution of charge density and electric potential in the diffuse double layer are described by an equation known as the Poisson-Boltzman equation.

The charge density and electric potential vary as a function of distance from the clay surface, surface charge density, surface potential, electrolyte concentration and valence, dielectric constant of the fluid, and temperature. The influence of these factors can be seen in the equation for the distance to the center of gravity of charge density surrounding a semi-infinite negatively charged clay particle. This distance can be viewed as being representative of the "thickness" of the diffuse double layer (Mitchell, 1976):

$$\frac{1}{K} = (DkT/8\,\pi\, n_o \varepsilon^2 \upsilon^2)^{1/2} \qquad [3]$$

where:

$1/K$ = "thickness" of the double layer (cm)

D = dielectric constant

k = Boltzmann constant (ergs/K)

n_o = bulk solution electrolyte concentration (ions/cm^3)

ε = unit electronic charge (esu)

υ = cation valence

T = absolute temperature (K)

Equation [3] illustrates that the thickness of the diffuse double layer changes in response to changes in dielectric constant, concentration and valence of the pore fluid. Changes in the behavior of clay soils due to changes in these pore fluid properties have been described by Mesri and Olson (1971).

In clay soils, long range attractive and repulsive forces develop between particles. The long range attractive force is due primarily to London van der Waals forces. The long range repulsive force develops as a result of electrostatic repulsion between two adjacent clay particles. In many cases the dominant long range force between clay particles is electrostatic repulsion (Bailey, 1965; Mitchell, 1976).

The osmotic pressure concept can be used to calculate the repulsive stress between clay particles. The clay particle system is assumed to exist as a series of parallel clay particles. The Poisson-Boltzman equation for a single particle can be integrated to obtain the mid-plane electrolyte concentration and mid-plane potential between two clay particles. The half spacing between the clay particles can be estimated from the void ratio and the specific surface of the clay particles (Mitchell 1976).

In the osmotic pressure concept, the overlapping diffuse double layers are considered to behave as a semi-permeable membrane. The difference in osmotic pressure between the bulk pore fluid and the fluid between the clay particles is then taken to be equivalent to the electrostatic repulsive pressure between the particles. Detailed developments of the osmotic pressure concept are provided by Bolt (1956) and Mitchell (1976). Researchers have attempted to use the osmotic pressure concept to predict the void ratio versus pressure relationship for clays containing low concentrations of homoionic pore fluid solutions. The osmotic pressure concept has been found, however, to provide a quantitative description of soil compressibility only for soils at high void ratios. The major limitation of this approach is the assumption of a perfectly dispersed parallel arrangement of individual clay particles, and the assumption that the dominant interparticle forces are the electrostatic repulsive force (Bailey 1965).

3.2 Theoretical Description of the Osmotic Stress State and the Constitutive Relationships for Volume Change

The development of the principle of effective stress by Terzaghi (1936) heralded the beginning of theoretical soil mechanics. This principle stated that changes in the volume or shear strength of a soil occur as the result of changes in the effective stress, that is, the difference between the total stress, σ, and the pore-water pressure, u_w. This principle is consistent with, and can be shown to be based on continuum mechanics. In this approach the "state" and "stress state" of a soil system is defined for a Representative Elementary Volume (REV) of the continuum as follows (Fredlund 1987):

(1) State: non material variables required for the characterization of a system.
(2) Stress state variable: the variables required for the characterization of the stress condition.

Changes in the stress state of the system can be linked to the deformation and the shear strength through constitutive relationships which contain material properties. It follows that the effective stress principle can be extended to assert that no change in behavior will occur unless there is a change in the stress state variable.

When clay soils are subject to changes in pore fluid chemistry, a difficulty arises with the use of the classic effective stress state variable for saturated soils (i.e., $\sigma - u_w$). Although there may be no changes in the total stress or pore fluid pressure in a specimen, changes in the strength, volume change, and permeability have been observed (Bolt 1956, Mesri and Olson 1971, Dunn and Mitchell 1984, Kenney 1967). In practice, these changes in soil behavior have usually been incorporated into the soil properties by ensuring that tests conducted in the laboratory duplicate the pore fluid chemistry present in the field. This approach, however, cannot account for transient changes in soil behavior during changes in pore fluid chemistry and may not prove to be the best approach for geotechnical practice. In modern geotechnique more and more engineers are being faced with design of waste containment systems in which natural and compacted soils routinely experience wide variations in pore fluid chemistry over their design life. Consequently, a more general approach is needed to allow changes in the "chemical" stress state to be incorporated into conventional analyses.

In this section a theoretical model of how osmotic effects can be incorporated into the definition of stress state and constitutive relationships for clay soils will be developed.

3.3 "True Effective Stress"

The work described in Section 3.1 and 3.2 dealt with the physico-chemical stresses at a microscopic level rather than the macroscopic level used in the definition of classic effective stress for a saturated soil. During the 1960's and 1970's several researchers, including Lambe (1960), Balasubramonian (1972) and Chattapadhyay (1972), attempted to extend the effective stress equation to include the net long range electrostatic stress as follows:

$$\sigma^* = (\sigma - u_w) - (R\text{-}A) \qquad [4]$$

where:

σ^* = "true" effective stress

σ = total stress

u_w = pore water pressure

(R-A) = net long range electrostatic stress

This "true effective stress" was developed from an analyses of static equilibrium of normal forces perpendicular to a wavey plane passing between soil particles. Although it provides a statement of the equivalence of stress, it is not based on an appropriate free body diagram and consequently is not an appropriate means of developing stress state variables (Fredlund, 1987).

To be an appropriate stress state variable, the components of the stress state must be shown to be equally effective in controlling soil behavior. Fredlund and Morgenstern (1977) demonstrated this concept for stress states for unsaturated soil through the use of null tests. Similar tests for "true" effective stress were conducted by Balasubramonian (1972) on clay shales. Morgenstern and Balasubramonian (1980) demonstrated that during leaching of a dense clay shale, changes in the predicted (R-A) stress were balanced by changes in the total stress required to maintain a constant specimen volume. It is important to note that the shale used was predominately montmorillonitic and consequently the assumption regarding the dominance of electrostatic repulsion is likely valid. This would not be true for clay soils in general.

4.0 PHYSICO-CHEMICAL STRESS STATE FROM CONTINUUM MECHANICS

In the continuum mechanics of multiphase mixtures, the behavior of each phase is dictated by the stresses acting on that phase. Fredlund (1973) states that the first two steps in this approach are the description of the physical multiphase element or Representative Elementary Volume (REV), and the establishment of the state variables associated with each phase.

4.1 Representative Elementary Volume (REV)

An element of saturated soil can be considered as a three phase system consisting of the soil particles, the pore fluid, and the diffuse double layer or adsorbed fluid hull surrounding each soil particle (Fig. 2). The diffuse double layer includes the surface charge along the clay particles. By definition, a phase must possess differing properties from the contiguous homogeneous phases and must have a continuous bounding surface throughout the element. The unique properties of the diffuse double layer have been documented by Mitchell (1976). The existence of a distinct boundary surface is not as well defined, but the extent of the adsorbed fluid phase is described by diffuse double layer theory.

The porosities of each phase of the elementary volume may be defined as follows:

n_p = volume of soil particles / total elemental volume
n_f = volume of bulk pore fluid / total elemental volume
n_d = volume of diffuse double layer / total elemental volume
The sum of the phase porosities is equal to unity.

4.2 Equilibrium equations for each phase

The principle of superposition of coincident equilibrium stress fields may be used as described in continuum mechanics (Truesdell, 1966). The assumption is made that an independent continuous stress field is associated with each phase of the multiphase system. The number of independent force equilibrium equations that can be written is equal to the number of cartesian coordinate directions multiplied by the number of phases constituting the continuum (Fredlund 1987).

Force equilibrium for the element is ensured by independently considering equilibrium for each phase. The elemental forces consist of surface tractions, gravity body forces and interaction forces between the phases. The interaction body forces between the phases are as follows:

F_f = interaction force between the diffuse double layer and pore fluid
F_d = interaction force between the soil particles and the diffuse double layer

Figures 3, 4 and 5 illustrate the equilibrium stress systems acting on the overall element and the diffuse double layer and bulk pore fluid phases. i) We can assume that the electrostatic interactions between particles can be described by the net repulsive minus attractive stress (R-A) acting within the diffuse double layer. This is conceptually similar to the osmotic pressure concept, in which the repulsive stress is equated to the osmotic pressure difference within the

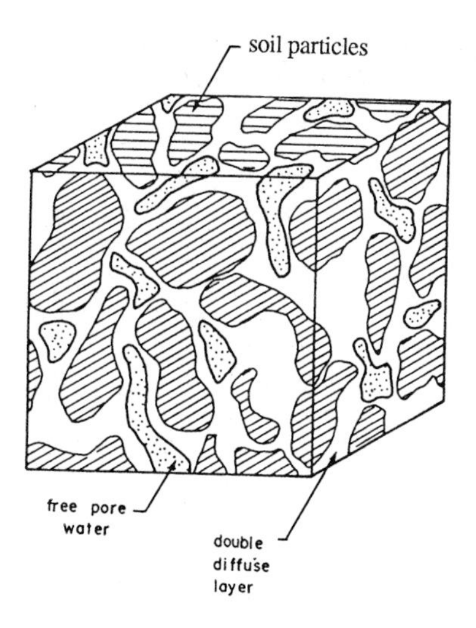

FIG. 2 Element of Saturated soil illustrating the diffuse double layer

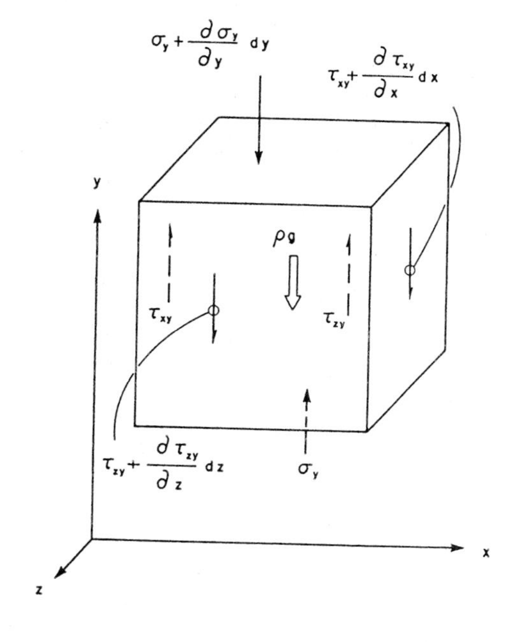

FIG. 3 Total stress equilibrium for an element of saturated soil

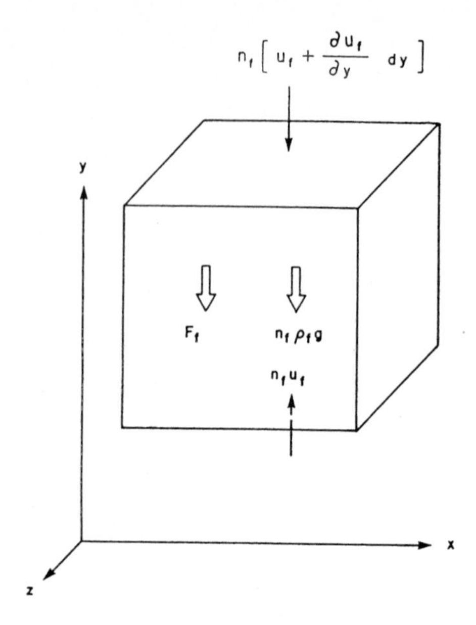

FIG. 5 Stress equilibrium for the diffuse double layer

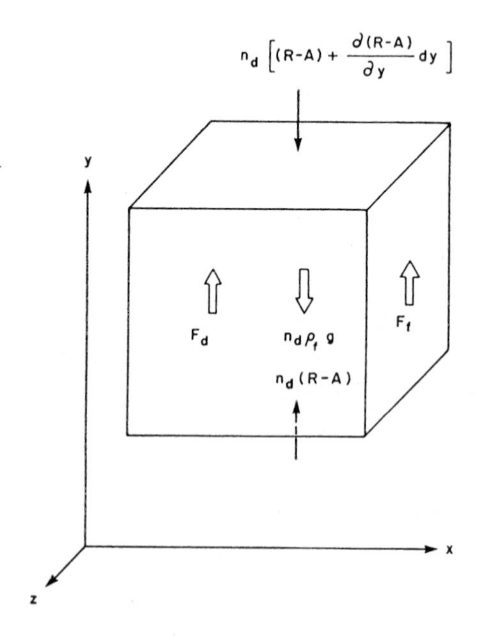

FIG. 4 Stress equilibrium for the fluid phase

pore fluid. ii) A second assumption is that when (R-A) is expressed as a stress tensor it is isotropic and does not contain shear stresses. iii) The density of the diffuse double layer is assumed to be equal to the density of the pore fluid. iv) Equilibrium equations can be written for the overall soil element as well as for each phase. A one-dimensional form of the equilibrium equations in the y-direction can be written as follows:

Overall Element:

$$\left(\frac{\partial \tau_{xy}}{\partial y} + \frac{\partial \sigma_y}{\partial y} + \frac{\partial \tau_{zy}}{\partial y} + \rho_t g\right) dx\, dy\, dz = 0 \qquad [5]$$

Pore Fluid:

$$\left(n_f \frac{\partial u_f}{\partial y} - F_f + n_f \rho_f g\right) dx\, dy\, dz = 0 \qquad [6]$$

Diffuse Double Layer:

$$\left(n_d \left(\frac{\partial (R\text{-}A)}{\partial y} - F_d - F_f + n_d \rho_f g\right) dx\, dy\, dz = 0 \qquad [7]$$

where:

σ_y = normal stress in the y-direction

τ_{xy} = shear stress on the x-plane in the y-direction

τ_{zy} = shear stress on the z-plane in the y-direction

u_f = pore fluid pressure

ρ_t = total density of the soil

ρ_f = fluid density

Equations [5] through [7] are statements of equilibrium for the overall element, the pore fluid, and the double layer. Volume change of the soil structure is controlled by the interparticle interactions between the soil particles. However these interactions cannot be measured directly. An equation for the stresses effective at the interparticle contacts can be written as the difference between the statement of equilibrium for the entire element, and those for the pore fluid and double layer phases. This equation is formulated as the algebraic sum of [5], [6] and [7].

$$\frac{\partial \tau_{xy}}{\partial y} + \frac{\partial \sigma_y}{\partial y} + \frac{\partial \tau_{zy}}{\partial y} + \rho_t g - n_f \frac{\partial u_f}{\partial y} - n_d \frac{\partial (R\text{-}A)}{\partial y} +$$

$$2F_f + F_d - \rho_f g\,(n_d + n_f) = 0 \qquad [8]$$

By adding and subtracting $n_d\,(\partial u_f/\partial_y)$ and $n_f\,\partial(R\text{-}A)/\partial_y$ and combining terms, the following equilibrium equation for the soil particles can be obtained:

$$\frac{\partial\tau_{xy}}{\partial y} + \frac{\partial(\sigma_y - u_f - (R\text{-}A))}{\partial y} + \frac{\partial\tau_{zy}}{\partial y} + \rho_t g + (1 - n_f)\frac{\partial u_f}{\partial y} +$$

$$(1 - n_d)\,\frac{\partial(R\text{-}A)}{\partial y} + 2F_f + F_d - (n_f + n_d)\,\rho_f g = 0 \qquad [9]$$

An examination of [9] reveals the presence of three sets of surface tractions (i.e., (σ_y - u_f - (R-A); u_f; (R-A))). In a similar development for the stress state variables for unsaturated soils, Fredlund and Morgenstern (1977) suggested that the (u_f) term can be eliminated if the assumption is made that the soil particles are incompressible. If the diffuse double layers in the soil are extensive and completely overlapping, then a similar assumption for the (R-A) term would also be valid. The remaining stress state variable is identical to that of true effective stress (i.e., (σ_y - u_f - (R-A))). The test results of Morgenstern and Balasubramonian (1980) provide verification for this combination of stresses as a single stress state variable for the case of a montmorillonitic clay shale.

In a more general case, it is likely that the diffuse double layers will not be extensive and consequently two stress state variables remain (i.e., (σ_y - u_f - (R-A)) and (R-A)). It is not appropriate to include (R-A) in both stress state variables. If only the term ($n_d\,\partial u_f/\partial y$) is added and subtracted from [8], the resulting equilibrium equation is as follows:

$$\frac{\partial\tau_{xy}}{\partial y} + \frac{\partial(\sigma_y - u_f)}{\partial y} + \frac{\partial\tau_{zy}}{\partial y} + \rho_t g + (1 - n_f)\frac{\partial u_f}{\partial y} - n_d\frac{\partial(R\text{-}A)}{\partial y} -$$

$$(n_f + n_d)\,\rho_f g + 2F_f + F_d = 0 \qquad [10]$$

If the u_f term is eliminated as before from this equation two stress state variables emerge (i.e., (σ_y - u_f) and (R-A)). These two stress state variables then control soil behavior; the effective stress, (σ - u_f), and the physico-chemical stress represented by (R-A). This is similar in form to the stress state variables for unsaturated soil (i.e., (σ_y - u_a) and (u_a - u_w)) proposed by Fredlund and Morgenstern (1977). Fredlund and Morgenstern (1976) used a three-dimensional constitutive surface for volume change to relate changes in void ratio to changes in matric suction and net total stress. A similar constitutive surface for the physico-chemical stress state variables drawn from the compressibility data of Mesri and Olson (1971) is illustrated in Figure 6. Fredlund and Morgenstern (1976) showed that for unsaturated soils two variables are essential because changes in matric suction are not as "efficient" in producing changes in volume change and shear strength as are changes in total stress. This has also been shown to be true for (R-A) and (σ_y - u_f). As illustrated in Figure 6, each stress state variable is linked to volume change through a separate modulus.

4.3 Practical physico-chemical stress state variables

A difficulty remains in the practicality of the physico-chemical stress state variables. Fredlund (1987) suggested that the choice of an appropriate set of stress state variables has to also consider whether:

(1) the variables can be experimentally tested,
(2) the variables can be theoretically justified,
(3) the variables are operational in practice, and
(4) the variables have characteristics acceptable within the definition of stress state variables.

4.4 R-A: Stress State Variable

(R-A) is not a stress which can be directly measured in laboratory testing. In addition, the prediction of (R-A) is possible only under ideal conditions. This is a severe restriction to the use of (R-A) as a component of the stress state variable. Because of this, it may be necessary that a deduced quantity, which is more easily measured than (R-A), be used.

The osmotic pressure concept relates the change in repulsive stress between clay particles, to the change in the osmotic pressure difference between the interparticle fluid and the bulk pore fluid. Mitchell (1962) suggested that "double layer interactions and the consequent repulsion between opposing particles are reflected by the osmotic pressure". The osmotic pressure is described as having two components; i) one associated with the free salt in the bulk solution, and ii) the other caused by the additional ions required to satisfy double layer requirements. The osmotic pressure of the bulk solution can be readily evaluated using samples of the pore fluid; however, the osmotic pressure of the inter particle fluid cannot be measured. Consequently, the difference between the osmotic pressure of the bulk fluid and that between the clay particles cannot be evaluated.

In most practical problems, the change of osmotic pressure is of more interest in regard to the process of osmotic compressibility than the absolute value of the components of osmotic pressure. In problems of osmotic compressibility, changes in the osmotic pressure of the bulk solution will be large relative to the initial osmotic pressure within the sample.

Consequently, it can be assumed that the change in the osmotic pressure of the bulk solution forms a satisfactory independent stress state variable. However, the osmotic pressure cannot be incorporated into the single stress state, $[\sigma - u_f - (R\text{-}A)]$, because osmotic pressure, π, may not be as effective as $(\sigma - u_f)$ in producing behavioral changes in the soil.

The change in the osmotic pressure of the bulk pore fluid is the driving force for osmotic flow. Consequently, the use of osmotic pressure as a stress state variable provides for consistency in the stresses used for the description of volume change as well as for the flow of the pore fluid.

4.5 Constitutive Relationship for Volume Change

A general form for the relationship linking the stress state variables and volumetric strain can be written as follows:

$$\varepsilon = -\frac{dV}{V} = -\frac{1}{V}\left[\frac{\partial V}{\partial(\sigma - u_f)} d(\sigma - u_f) + \frac{\partial V}{\partial \pi} d\pi\right] \qquad [11]$$

where:

ε = volumetric strain

V = initial soil volume

Equation [11] takes the form of a three-dimensional constitutive surface as was illustrated in Figure 6. Along a plane in which π is constant, the coefficient of volume change (compressibility), m_v, can be obtained. Along a plane in which $(\sigma - u_f)$ is constant, the slope can be used to define the osmotic coefficient of volume change (osmotic compressibility), m_π. The equation corresponding to a point on the constitutive surface can be written as:

$$d\varepsilon = -m_v d(\sigma - u_f) - m_\pi d\pi \qquad [12]$$

It should be noted that m_v and m_π are not necessarily material constants. Just as m_v must be evaluated over an appropriate stress range, so must m_π be evaluated over an appropriate change in pore fluid chemistry.

5.0 THEORETICAL DESCRIPTION OF OSMOTIC VOLUME CHANGE - OSMOTIC CONSOLIDATION AND OSMOTICALLY INDUCED CONSOLIDATION

Osmotically induced consolidation occurs as a result of the flow of water out of a soil in response to chemical or osmotic gradients. This outward flow causes negative fluid pressure to develop within the clay; and consequently, produces an increase in effective stress. The second osmotic volume change process is termed osmotic consolidation. In this process, volume change occurs due to a reduction in interparticle repulsive stresses as a result of changes in pore fluid chemistry.

Osmotically induced consolidation was addressed in a study by Greenberg (1971) and subsequently discussed by Greenberg et al (1973) and Mitchell (1973b, 1976). The theoretical formulation of volume change due to osmotically induced consolidation was originally developed by Greenberg (1971) using the theory of irreversible thermodynamics. Barbour (1987a, 1987b) formulated the problem from a phenomenological basis more familiar to geotechnical engineers. This formulation described transient volume change due to both osmotic and osmotically induced consolidation.

5.1 Formulation for Osmotic Volume Change - Osmotic Consolidation and Osmotically Induced Consolidation

In order to predict transient volume change, changes in the stress state variables with time are related to volume change through constitutive equations. Continuity is used to relate these volume changes to the divergence of the fluid flux from the element. These fluid fluxes are in turn defined by the flow laws for the fluid phase. It is assumed that only hydraulically and osmotically induced flows of the fluid occur. The potential of ion streaming to produce electrical gradients is not considered. The rate of movement of the dissolved salt is described by the advection/diffusion equation (Freeze and Cherry 1979). It is assumed that no geochemical interactions of individual ion species occurs with the soil solids; however, retardation of dissolved salts due to adsorption onto the soil solids may be accommodated in this formulation.

Table 1 summarizes the flow laws and constitutive equations required to develop a description of transient volume change of clay soils due to osmotic or osmotically induced consolidation. The development and solution of the governing differential equations describing these phenomena were developed in detail by Barbour (1987a).

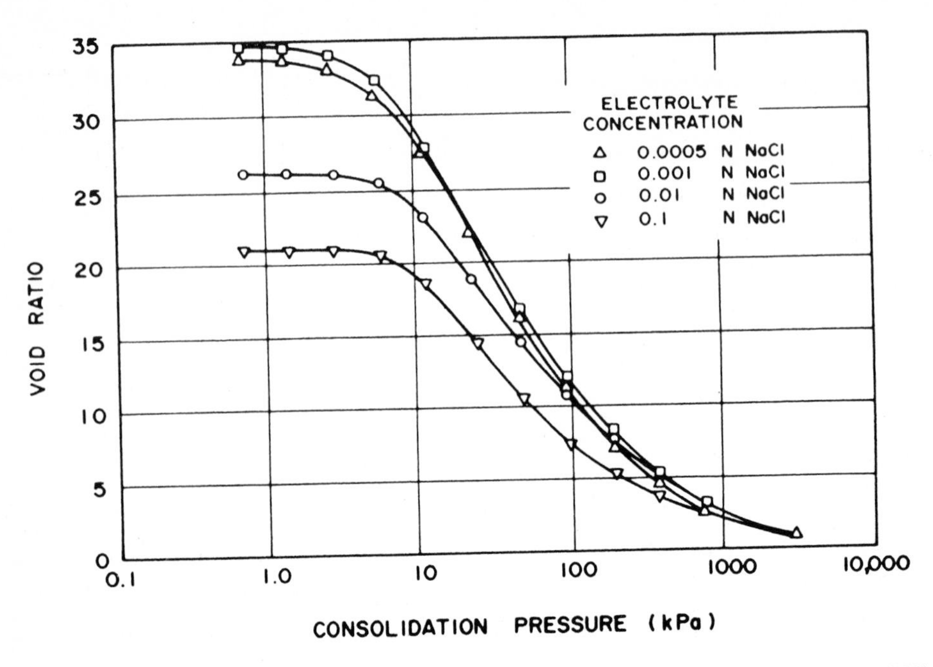

FIG. 6 a) Consolidation curves for Na^+ Montmorillonite (after Mesri and Olson 1971)

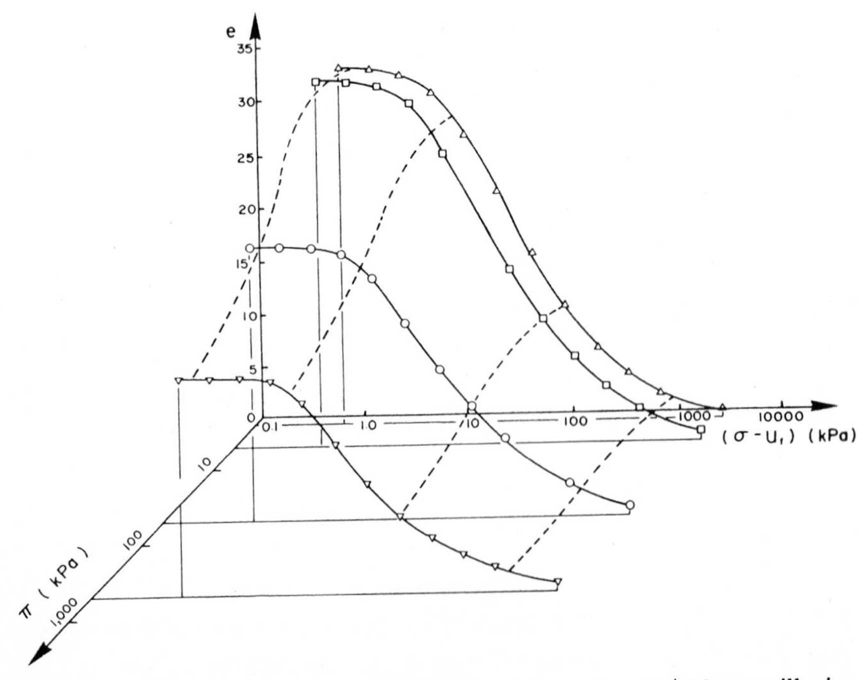

FIG. 6 b) Three-dimensional constitutive surface for Na^+ Montmorillonite

6.0 NUMERICAL SIMULATION

The equations given in Table 1, describing transient osmotic volume change were used to develop a one-dimensional numerical solution describing transient fluid flow and salt migration during osmotic volume change (Barbour 1987a). The numerical model was verified by comparing the results with the simulations of Greenberg (1971) and Mitchell (1973a, 1976) for osmotically induced consolidation, and to analytical solutions of one-dimensional conventional consolidation and contaminant transport. At present models of osmotic consolidation do not exist against which the present model can be verified. Rather this model will be used to illustrate behavior that is typical of clay specimens undergoing either osmotic or osmotically induced consolidation.

Table 1

Summary of theoretical relationships for osmotic flow and volume change

Flow laws

For fluid: $q_h = -k_h(\partial h/\partial x)$

$q_\pi = k_\pi(\partial \pi/\partial x)$

$q_f = q_h + q_\pi$

For salt: $q_s = q_f C - nD(\partial C/\partial x)$

Constitutive relationships

$\Delta e/(1 + e_o) = \Delta(\sigma - u_f)m_v$

$\Delta e/(1 + e_o) = \Delta\pi m_\pi$

Material properties

k_h = coefficient of permeability
k_π = coefficient of osmotic permeability
D = coefficient of diffusion
n = porosity
e = void ratio
m_v = coefficient of volume change
m_π = osmotic coefficient of volume change
C = concentration

The simulation was conducted for a 1 cm thick specimen of clay exposed to a 4.0 M NaCl brine along its upper surface. Two cases with different types of boundary conditions were analysed (Table 2). In Case 1, the base of the specimen was considered to be sealed so that no flow occurred across the base. In Case 2, a zero pore pressure condition was maintained at the base. Salt flux across the base was not permitted in either case.

Two sets of material properties were used for the simulations to represent conditions in which

Table 2

Numerical simulation example cases

General properties	Specific properties
$k_h = 1.0 \times 10^{-10}$ m/s	Osmotically induced consolidation
$m_v = 5.0 \times 10^{-4}$ kPa^{-1}	$k_\pi = 5.0 \times 10^{-13}$ m/s
$m_{vs} = 5.0 \times 10^{-5}$ kPa^{-1}	$m_\pi = 0$
$c_v = 2.0 \times 10^{-4}$ cm^2/s	Osmotic consolidation
$D = 5.0 \times 10^{-6}$ cm^2/s	$k_\pi = 0$
$n = 0.5$	$m_\pi = 2.5 \times 10^{-6}$ kPa^{-1}
m_{vs} = coefficient of volume change during swelling	

Geometry boundary conditions		
Sample thickness:	1 cm	
Initial equilibrium conditions:	$u = 0, C = 0$ for all y; $t \leq 0$	
Boundary conditions:	Case 1 (Base sealed)	Case 2 (Base open)
Top:flow boundary	1st type; $u = 0$; $t > 0$	1st type; $u = 0$; $t > 0$
Bottom:flow boundary	2nd type; $q_f = 0$; $t > 0$	1st type; $u = 0$; $t > 0$
Top-salt transport boundary	1st type; $C = 4M$; $t > 0$	1st type; $C = 4$ M; $t > 0$
Bottom-salt transport boundary	2nd type; $\partial C/\partial y = 0$; $t > 0$	3rd type; $qC = 0$; $t > 0$ $C = 0$; $t > 0$

osmotically induced consolidation or osmotic consolidation was the dominant mechanism of volume change. Osmotically induced consolidation will be the principal mechanism of volume change in the first example. The clay was assumed to behave as a semi-permeable membrane with an osmotic efficiency of 0.005 (i.e., k_π is equal to 0.005 k_h). In the second example, osmotic consolidation is the dominant mechanism of volume change. The clay was given an osmotic compressibility equal to 0.005 times the conventional compressibility of the sample, but with an osmotic efficiency of zero. These values of osmotic compressibility and osmotic efficiency are typical of those obtained in laboratory tests described later in the paper.

The results of the simulations are illustrated in Figs. 7 and 8. For Case 1 (Fig. 7) it is apparent that the time-deflection behavior is characteristically different in the two mechanisms. Osmotically induced consolidation occurs within the same time frame as conventional consolidation. The maximum deflection of the specimen occurs at a time factor of approximately 1. Rebound then occurs as the negative pore pressures induced by osmotic gradients dissipate as the salt diffuses into the soil specimen. Osmotic consolidation, however, develops slowly in response to the diffusion of salt into the clay.

In Case 2, (Fig. 8) in which fluid flow across the base of the specimen was permitted, the results from the two simulations are quite different. During osmotic consolidation (Fig. 8b), fluid is expelled from the base as the advancing salt front causing a reduction in volume. However, for osmotically induced consolidation (Fig. 8a), strong upward flow across the base begins at a time factor of approximately 1.0 in response to the strong osmotic gradients near the top of the specimen. The rate of flow through the base becomes nearly constant with time as the salt distribution reaches an equilibrium position. The final flow rate is approximately equal to the product of osmotic permeability and the osmotic pressure gradient across the soil specimen.

7.0 LABORATORY TESTING

Laboratory testing was conducted to establish the dominant mechanism of osmotic volume change in two clay soils. A simple modified oedometer was used to consolidate a clay slurry to various initial consolidation stresses. The specimen was then exposed to concentrated NaCl solutions along either the upper soil surface only, or along the top and bottom of the specimen simultaneously. Measurements of specimen deflection were made for both circumstances. When a NaCl solution was placed on the top surface of the sample, the amount of osmotic flow across the base of the specimen or the pressure that developed along the base of the sample when flow was prevented, was measured. The specimens tested were between 0.5 cm to 1.7 cm thick.

One natural soil and one artificial soil mixture were used in the tests. The natural soil was the minus #10 sieve fraction of air-dried samples of Regina clay. The artificial soil was a mixture of 20% Na-montmorillonite and 80% Ottawa sand. The slurries were prepared by mixing the air-dried soil to water contents 10 to 20% above the liquid limit. Tables 3 and 4 provide a summary of the physical and chemical characteristics of the two soil types. In contrast to the Na-montmorillonite/sand mixture the Regina clay is predominately a Ca-montmorillonitic soil.

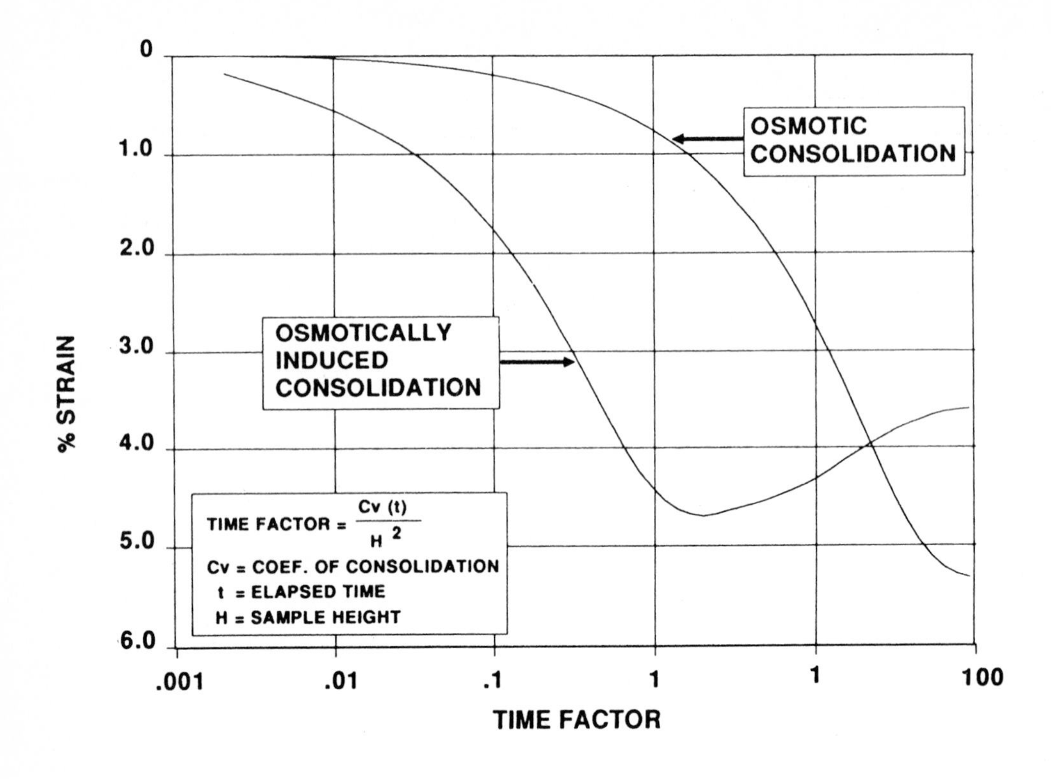

FIG. 7 Consolidation curves from numerical simulation of osmotic volume change for Case 1 (no fluid flow across the base)

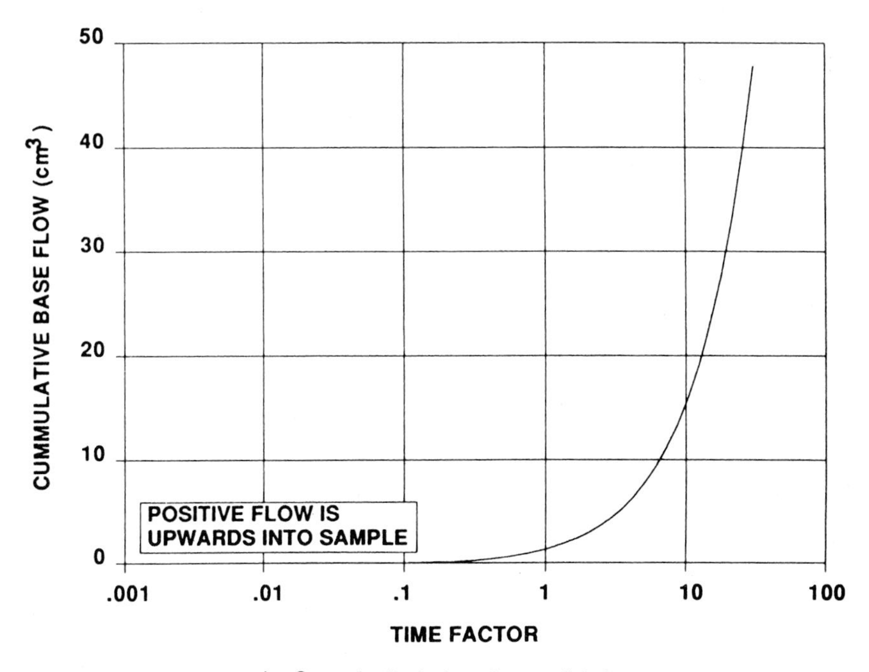

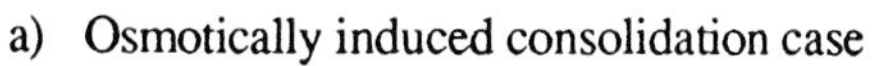
a) Osmotically induced consolidation case

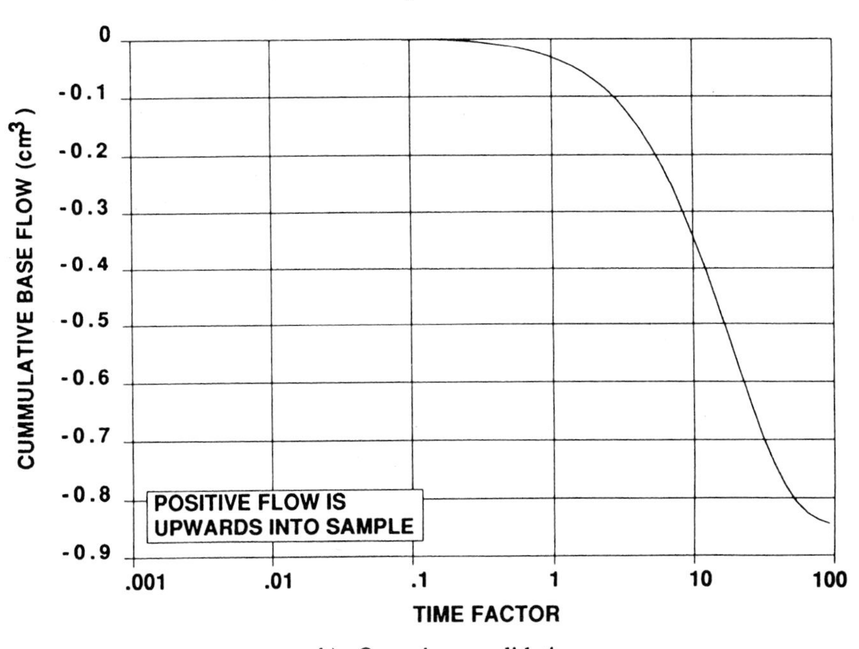

b) Osmotic consolidation case

FIG. 8 Cumulative base flow curves from numerical simulation: Case 2 - fluid flow across the base permitted.

8.0 TEST RESULTS

Figure 9 illustrates a deflection curve for a specimen of Regina clay undergoing normal effective stress consolidation from 100 to 200 kPa loading. At an elapsed time of 8000 minutes, the upper surface of the specimen is exposed to a 4.0 M NaCl solution and further consolidation of the specimen takes place. Typical curves of deflection versus time due to osmotic consolidation are illustrated in Figs. 10 and 11. These curves have been corrected for secondary consolidation as illustrated in Fig. 9. For the Regina clay, the

Table 3
Summary of classification tests

Test	Regina Clay	Na-mont-morillonite	Ottawa Sand	Mixture of sand-mont morillonite
Specific gravity	2.83	2.56	2.65	2.63
Atterberg limits				
Liquid limit	75.5%	-	-	62.1%
Plastic limit	24.3%	-	-	Nonplastic
Plasticity index	51.2%	-	-	-
Grain-size distribution				
Sand sizes	0%	0%	100%	80%
Silt sizes	34%	0%	0%	0%
Clay sizes	66%	100%	0%	20%
Mineralogical composition-less than 2 m μ[a]				
Montmorillonite	45.2%	80%[b]	-	-
Illite	27.7%	7%	-	-
Kaolinite	17.7%	7%	-	-
Trichlorite	9.4%	0%	-	-
Quartz	0.0%	8%	-	-
Gypsum	0.0%	3%	-	-
Carbonate	0.0%	2%	-	-
Specific surface (m^2/g)	53[c]	700-840[d]	-	140-168
Exchange capacity (mequiv./100g)[e]	31.7	80-150[d]	-	-
Total exchangeable bases by NH_4Ac leaching (mequiv./100g)				
Magnesium	15.3	-	-	-
Calcium	54.4	-	-	-
Potassium	0.59	-	-	-
Sodium	1.77	-	-	-

[a]Test performed by Saskatchewan Research Council;[b]Quigley (1984);
[c]Test performed by Department of Soil Science, University of Saskatchewan, by ethylene glycol sorption (Black 1965a, Part 1);[d]Mitchell (1976);[e]Fredlund (1975)

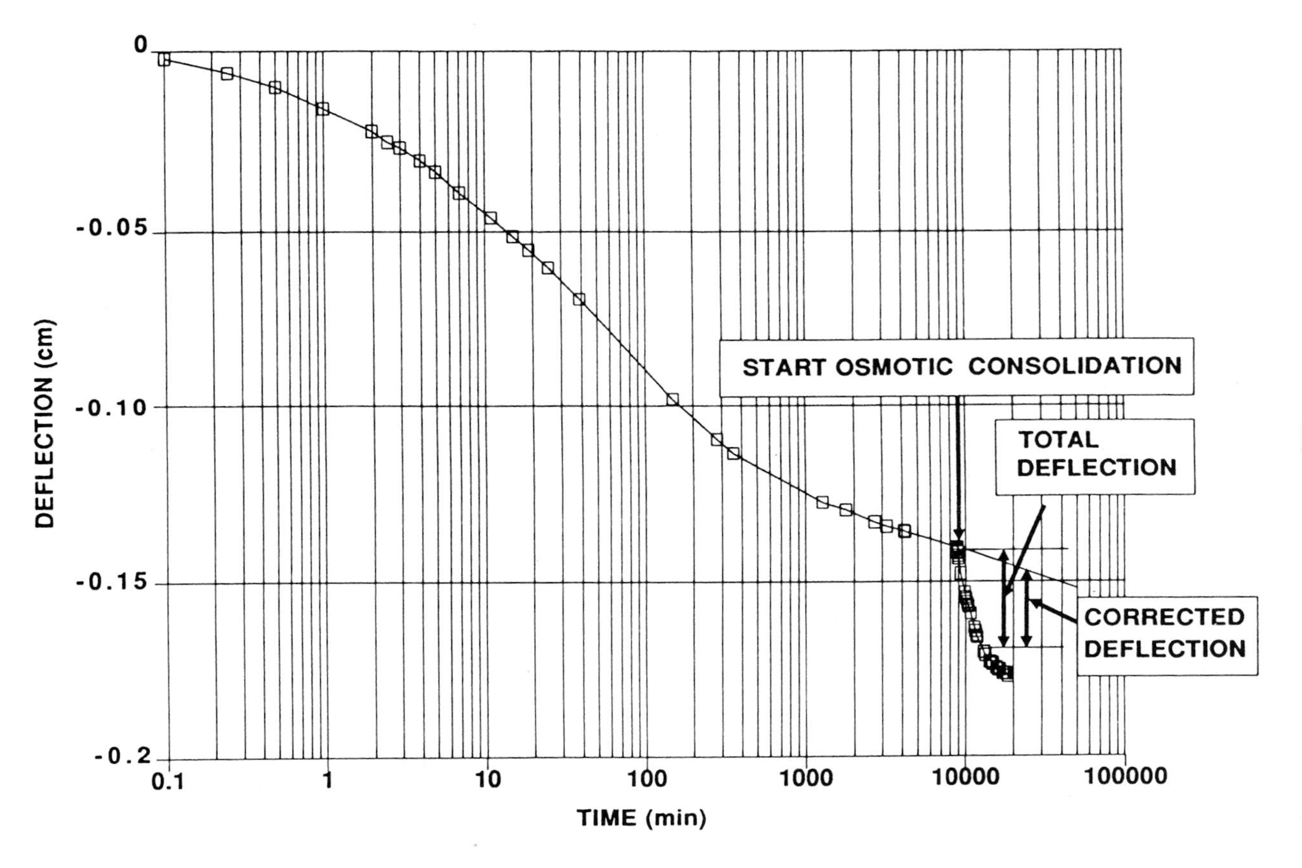

FIG. 9 Effective stress consolidation and osmotic volume change of Regina Clay

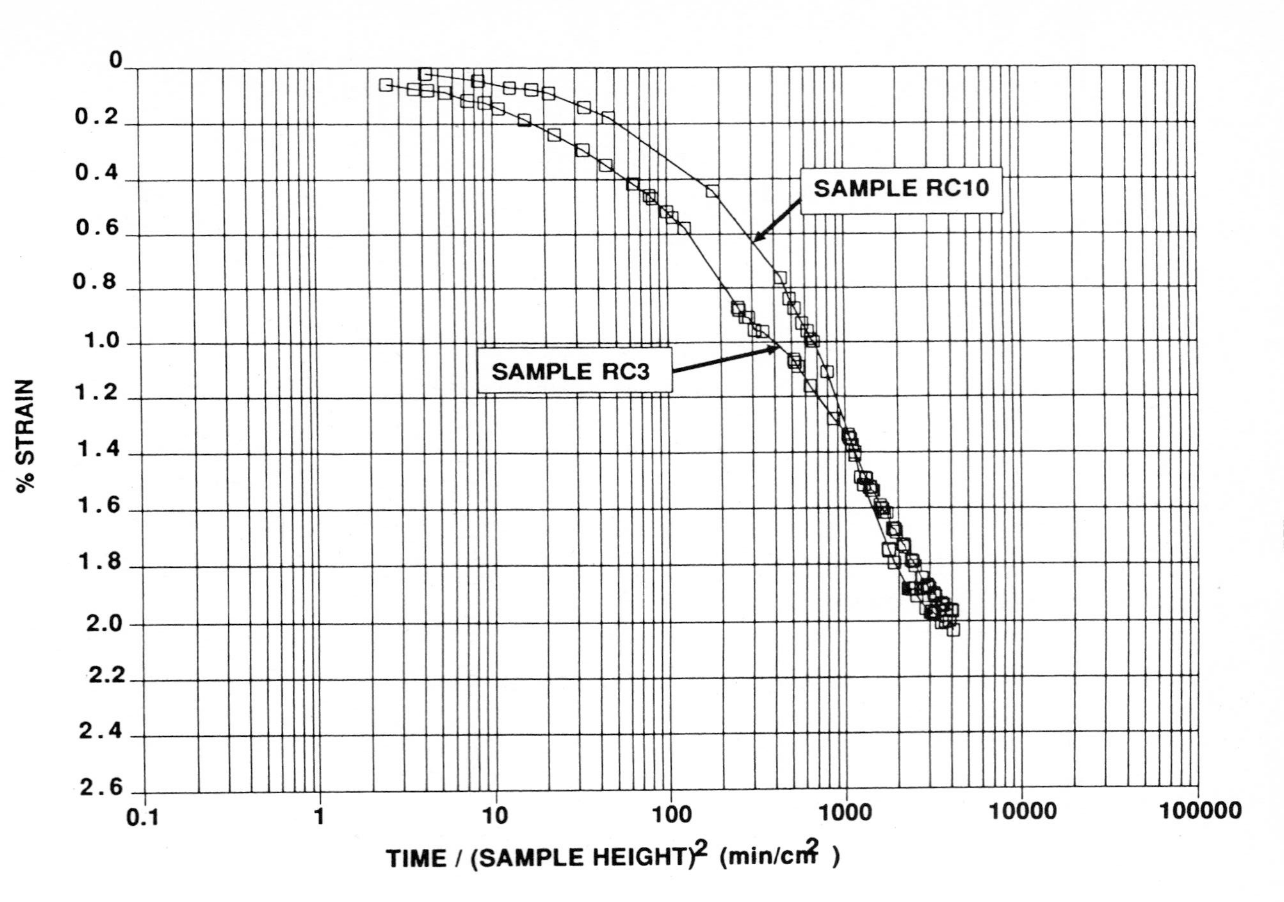

FIG. 10 Time dependent volume change of Regina Clay by 4.0 M NaCl solution (specimen preconsolidated to 200 kPa)

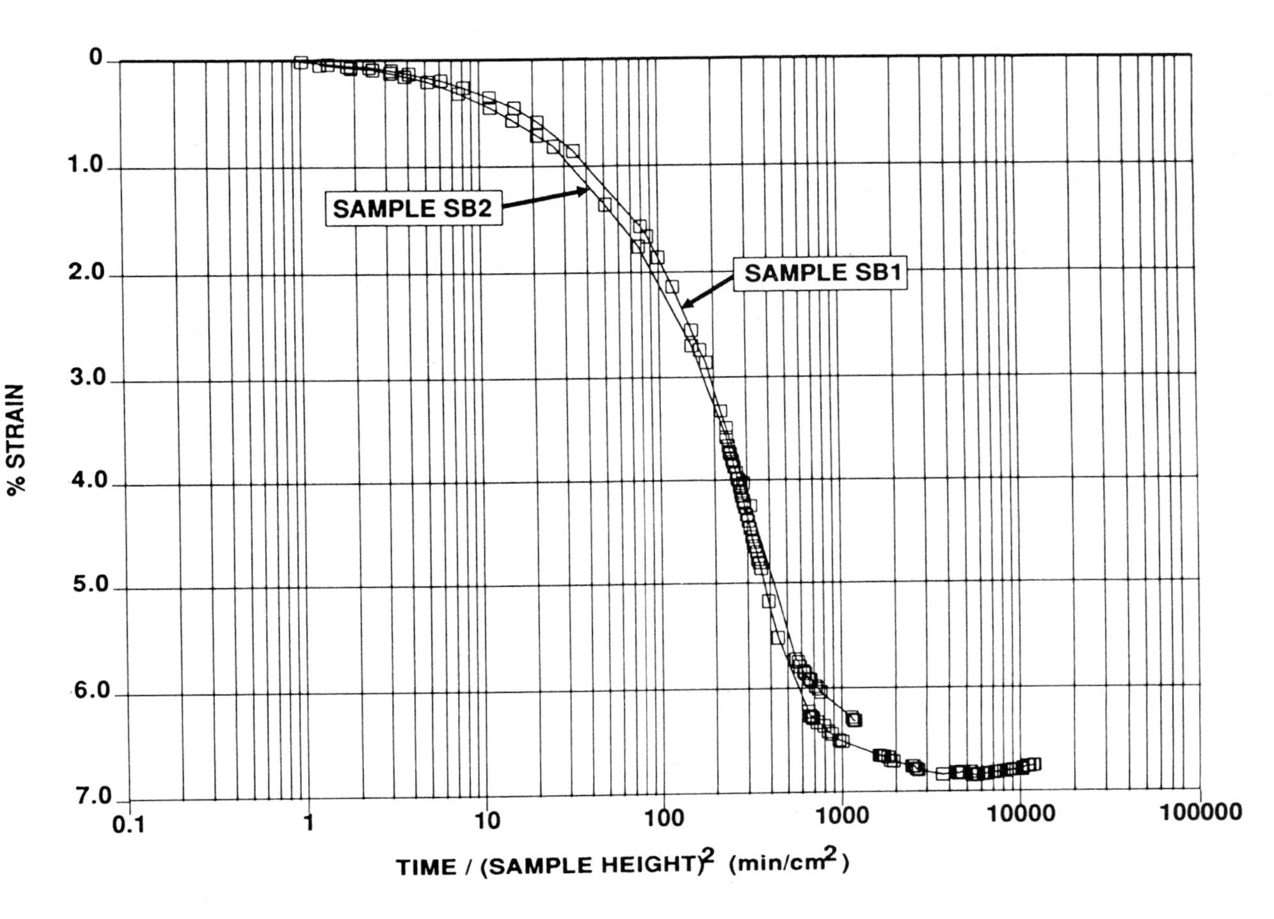

FIG. 11 Time dependent volume change of sand/montmorillonite by 4.0 M NaCl solution (specimen preconsolidated to 200 kPa)

consolidation process is extremely slow and appears to be controlled by the rate of diffusion of salts into the specimen. The specimens of Na-montmorillonite/sand appear to undergo more rapid consolidation than the Regina clay specimens. At first observation, this rapid consolidation would appear to be similar to that anticipated to occur for osmotically induced consolidation (Fig. 7).

Table 4

Chemical analysis of saturated extract[a]

Test result	Regina Clay	Na-montmorillonite
Water content		
of extract (%)	88	550
pH	7.4	8.8
Conductivity	3.7	4.0
(mS/cm)		
Ions in extract		
(mg/μL)		
Na^+	210	910
Ca_2^+	546	23
Mg_2^+	183	7
K^+	35	9
Cl^-	133	81
SO_4^{2-}	1700	1400
Sodium adsorption ratio	2.0	42.2

[a] Analyses performed by Saskatchewan Soil Testing Laboratory (after Black 1965b)

Figure 12 illustrates the effect of osmotic volume change as viewed on a conventional void ratio versus effective stress (σ - u_f) diagram. After osmotic volume change, the specimen appears to be overconsolidated to some higher stress level. Upon reloading, the specimen recompresses back towards the original virgin branch.

For the cases where only the top of the specimen was exposed to brine, measurements were made of the flow across the base of the specimen. The reservoir for this flow was maintained at the same elevation as the brine level on the top of the specimen. Figure 13 illustrates typical patterns of cumulative flow with time for the Regina clay and Na-montmorillonite/sand specimens. In the case of the Regina clay, small flows into the

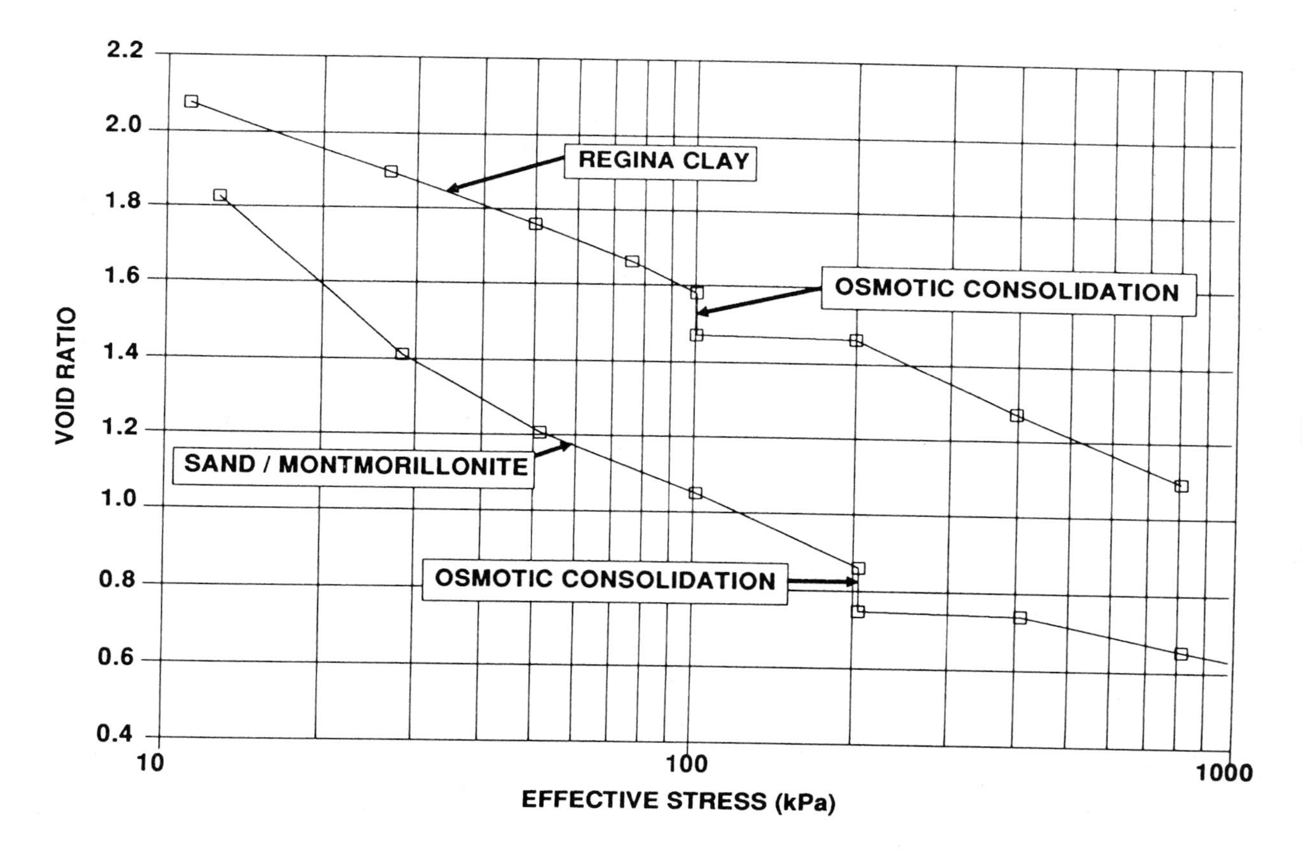

FIG. 12 Influence of osmotic consolidation on the virgin compression branch

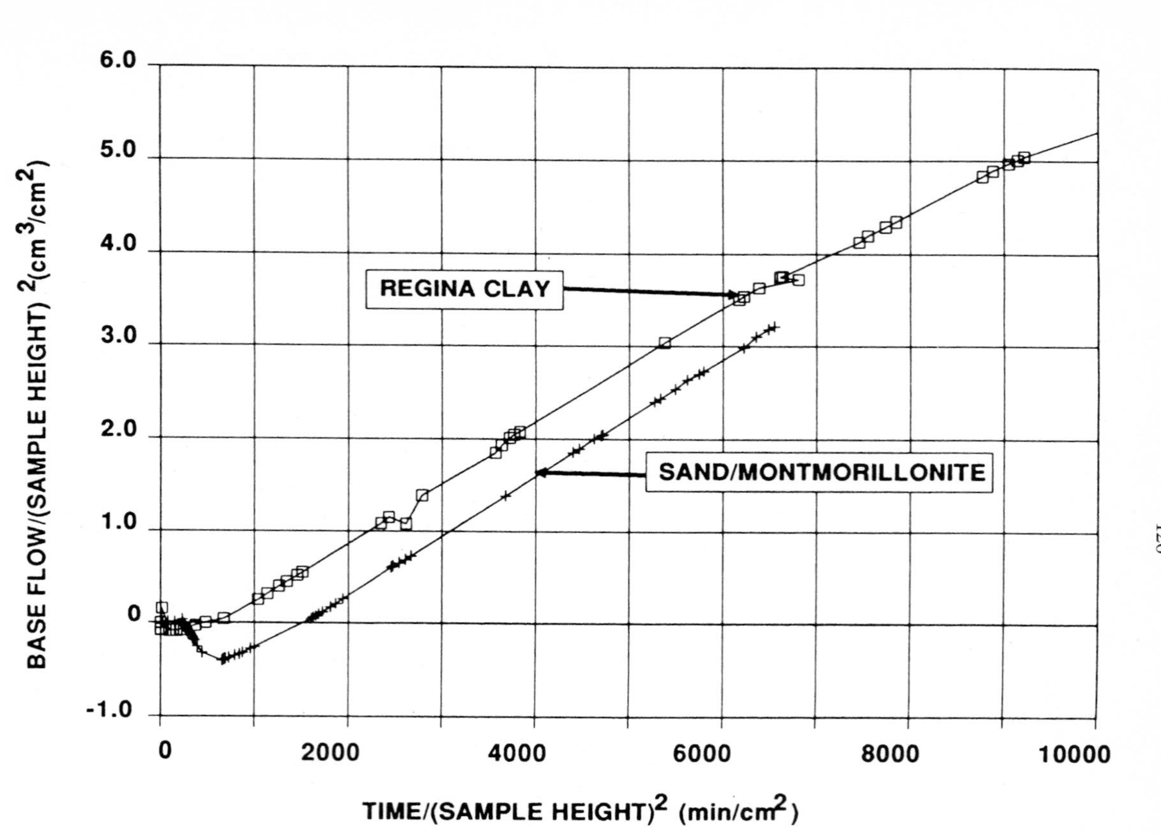

FIG. 13 Cumulative base flow for Regina Clay and sand/montmorillonite exposed to 4.0 M NaCl solution along upper surface (specimen preconsolidated to 200 kPa)

specimen begin at time factors of approximately 1.0 and build to nearly constant flow rates. For the Na-montomorillonite/sand specimens, the initial flow was out of the specimen. These flows reverse at a time factor of approximately 1.0 and begin to flow into the specimen, again increasing to essentially fairly constant flow rates.

9.0 ANALYSES OF RESULTS

The test results may be interpreted in light of the theory presented previously, in order to evaluate the role played by osmotic flow (as represented by the osmotic efficiency) and osmotic compressibility in producing volume change.

9.1 Osmotic Efficiency

Nearly constant osmotic flow rates were developed in both soil types after the completion of volume change within the specimen. The osmotic flow rates were observed during exposure to a number of NaCl solution concentrations. Using [2] the osmotic efficiency of the soils at different confining stresses and different solution concentrations were calculated. These efficiencies are shown in Figs. 14 and 15 along with osmotic efficiencies calculated from the work of Bresler (1973). The data of Bresler requires an estimate of the film thickness of the solution between two charged soil particles. This value was obtained using estimates of specific surface, and the equation for interparticle spacing between packets of particles as given by Shainberg et al. (1971). Assuming different values of specific surface shifts the lines of efficiency to the right or left on the graph.

In spite of considerable scatter, a number of significant observations can be made. First, fairly high efficiencies are obtained at low solution concentrations and low efficiencies are obtained at high solution concentrations. This is consistent with the results described by Kemper and Rollins (1966). Secondly, the induced negative fluid pressure that would have occurred due to the observed flows is small. These pressures were calculated from [2] using the osmotic pressure calculated using [1], the full specimen thickness for Δx, and the measured value of k_h. Efficiencies (ξ) of 0.01 at 0.01 M and 0.001 at 1.0 M would only provide a negative pressure response in the range of 0.5 kPa to 5 kPa. This would not be sufficient to account for the magnitude of volume change observed in Fig. 12. Finally, Fig. 14 indicates that for the Regina clay, the number of particles bound into packets decreases as higher concentrations of NaCl are used. Although there is insufficient data, the pattern in the Na-montmorillonite/sand seems to indicate that the number of particles bound up in packets remains unchanged. Regina clay is predominately a Ca-montmorillonite; consequently, the clay would be expected to form packets of clay particles bound together by adsorbed Ca^{++} during sedimentation. However, as the pore fluid

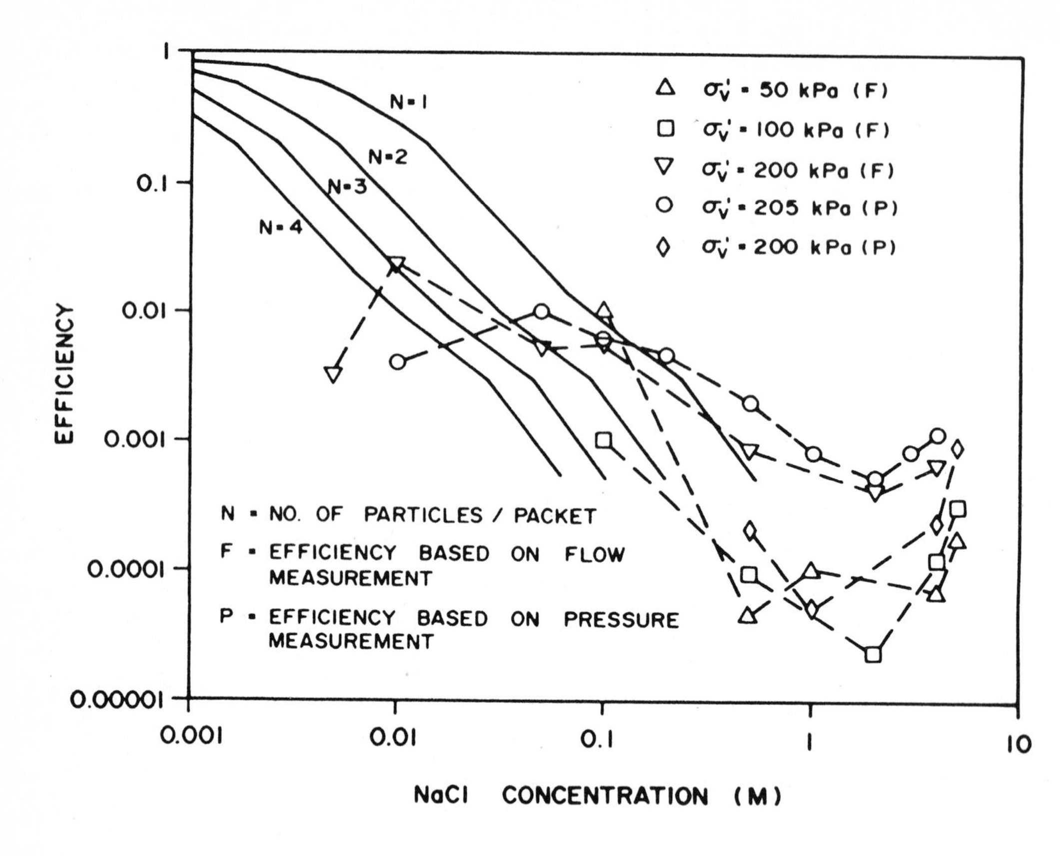

FIG. 14 Theoretical and experimental osmotic efficiencies for Regina Clay

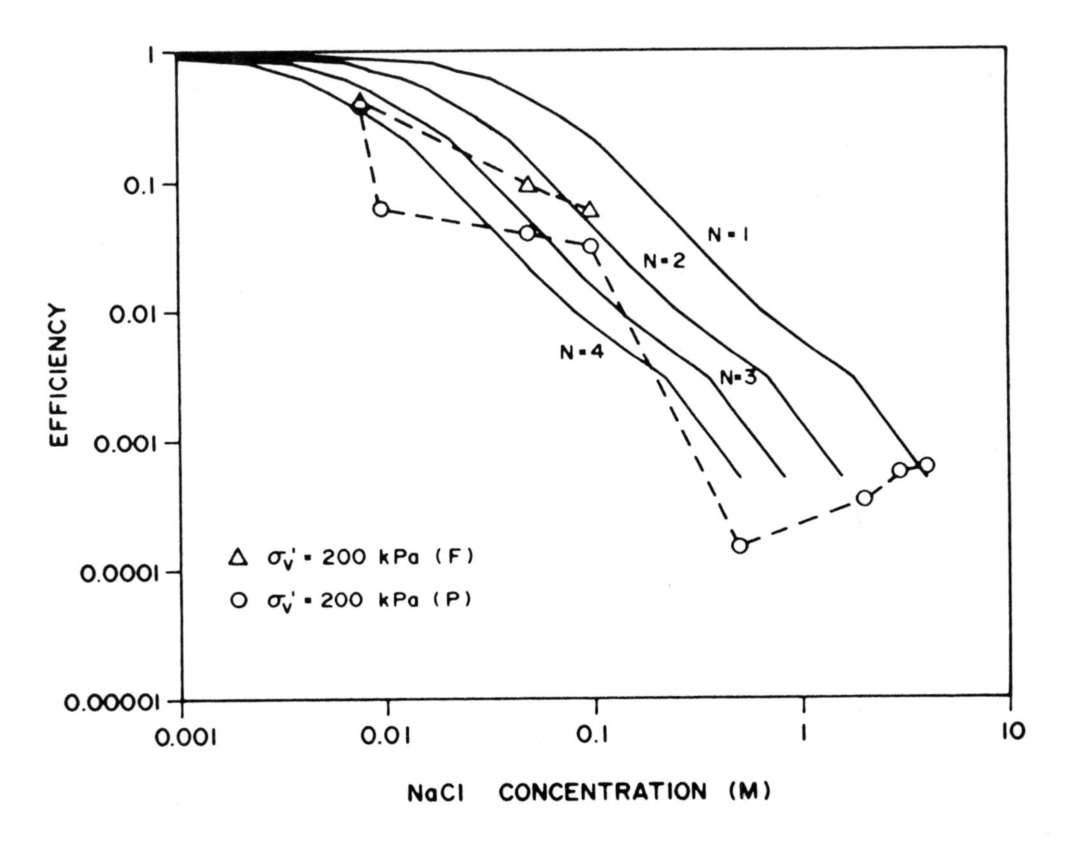

FIG. 15 Theoretical and experimental osmotic efficiencies for sand/montmorillonite

becomes concentrated, these packets would begin to open up due to the dominance of Na^+ in the double layer, resulting in fewer particles per packet.

9.2 Osmotic Compressibility

The analyses of the osmotic flow and pressure measurements indicate that the predominate mechanism of osmotic volume change for both types of soil cannot be osmotically induced consolidation. However, if osmotic consolidation is the dominant mechanism, then the rate of consolidation will be controlled by the rate of salt transport into the specimen by diffusion.

Steady state diffusion tests were conducted on a single specimen of each material. The combined coefficient of diffusion for Na^+ and Cl^- from tests on both materials was approximately 4×10^{-6} cm^2/s. The coefficient of diffusion from steady state diffusion tests does not take into account the retardation of Na^+ migration due to adsorption onto the soil solids. Since the Regina clay is a Ca^{++} dominated clay, it would be likely that the effective coefficient of diffusion during transient conditions would be lower than that measured in the steady state tests.

This lower rate of diffusion in the Regina clay samples may in part explain the more rapid changes in volume as a result of osmotic consolidation for the Na-montmorillonite/sand samples (Fig. 11) over those for the Regina clay samples (Fig. 10). The difference in the time to 100% consolidation for the two soil types is very large, however, and it is unlikely that this difference could be completely explained on the basis of the differences in the effective coefficients of diffusion.

A second explanation for the difference in the rates of osmotic volume change is that the magnitude of the osmotic compressibility of the two samples is substantially different. The osmotic compressibility for each of these materials is illustrated in Fig. 16. Because of the high osmotic compressibility of the Na-montmorillonite/sand mixture at low salt concentrations, a majority of the osmotic volume change within the specimen would have occurred rapidly, prior to full equalization of salt concentrations within the specimen.

In the case of Regina clay, the constant value of osmotic compressibility would require nearly full equalization of salt concentrations within the specimen before osmotic consolidation would be complete. Consequently, the time to the completion of osmotic consolidation of the Regina clay was much longer than that for the sand/ montmorillonite mixture.

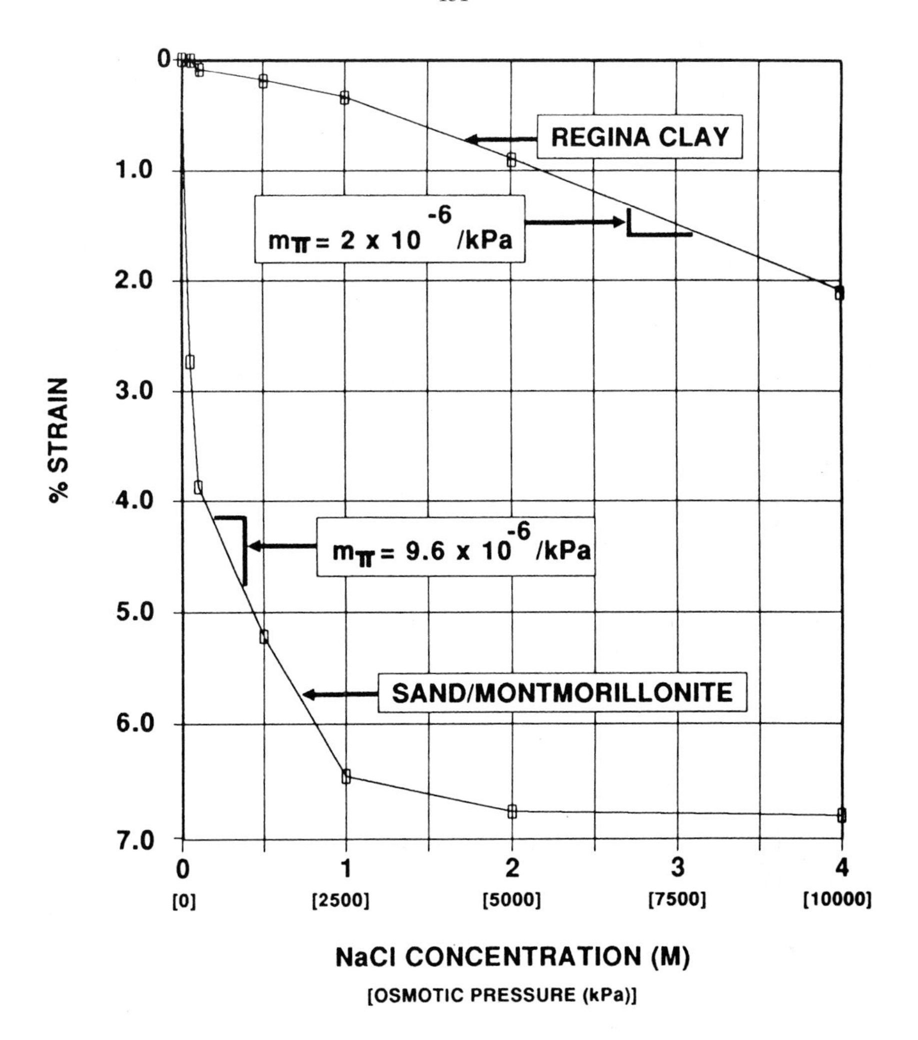

FIG. 16 Osmotic volumetric strain versus solution concentration

The nearly linear relationship between strain and concentration for the Regina clay (Fig. 16) does not appear to be consistent with he suppression of the double layer around individual clay particles. This linearity may be due to the dispersion of the clay packets within the Regina clay, as the concentration increases, as was also indicated by the osmotic efficiency data. The shape of Na-montmorillonite/sand and Regina clay curves (Fig. 12) are consistent with data described by Aylmore and Quirk (1962) for pure Na-montmorillonite and Ca-montmorillonite, respectively. These authors also attributed the anomalous behavior of the Ca-montmorillonite respectively (similar to the Regina clay) to the presence of clay packets.

9.3 Apparent (R-A) Stress

The stress path followed during consolidation, osmotic consolidation, and then reloading, as viewed on the three-dimensional constitutive surface presented previously, is shown in Fig. 17a. When this pathway is viewed on the void ratio versus (σ - u_f) plane (Fig. 17b) a path different from that presented in Fig. 12 develops. The actual observed behavior suggests that upon reloading, the osmotic pressure of the bulk pore fluid is not constant but is decreasing. In fact, the pore fluid concentration is unchanged during reloading. This behavior is consistent, however, with a view of using (R-A) as the stress state variable representing physico-chemical interactions. It would be reasonable that upon reloading, small decreases in void ratio would occur which would produce an increase in the net repulsive stress between the particles as they are forced closer together. An indirect estimate of the change in the net repulsive stress within the soil mass could be obtained from the void ratio versus stress plots. Figure 18 illustrates a typical virgin branch for the two soil types tested. A second line, below the virgin branch, was drawn by subtracting the volume changes due to osmotic consolidation under a 4.0 M NaCl solution (Fig. 16) from the virgin branch. The difference between these two virgin branches can be taken as a measure of the change in the (R-A) stress present in the specimen at a particular void ratio when the soil is exposed to a 4.0 M NaCl solution.

Figure 19 illustrates how the change in the repulsive stress within the two soils varies with confining stress. Also shown is the predicted repulsive stress based on the osmotic pressure concept for the Na-montmorillonite/sand specimens. It is of interest that in the case of the Na-montmorillonite/sand specimens, the theoretical values of (R-A) agree reasonably well with the measured values when a value of specific surface equal to 580 m^2/gm is selected. Agreement between the theoretical and measured values for (R-A) for Regina clay was not obtained. Barbour (1987a) indicated that this may because the clay particles are bound up in

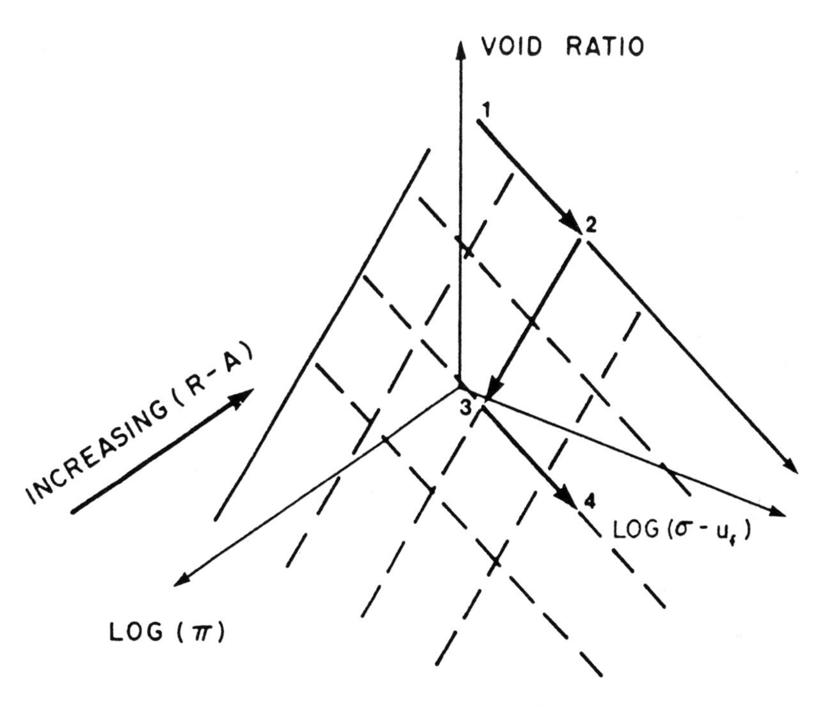

a) Three-dimensional surface

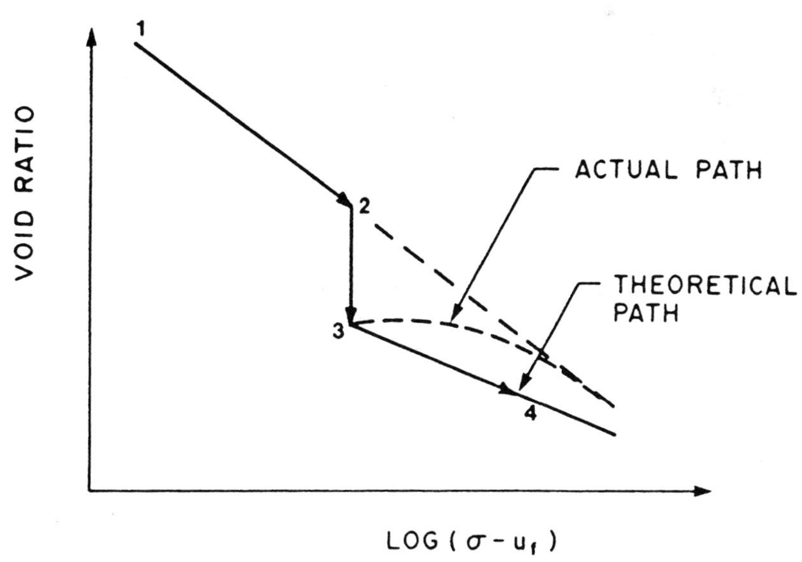

b) Projection onto effective stress axis

FIG. 17 Stress paths during effective stress and osmotic consolidation

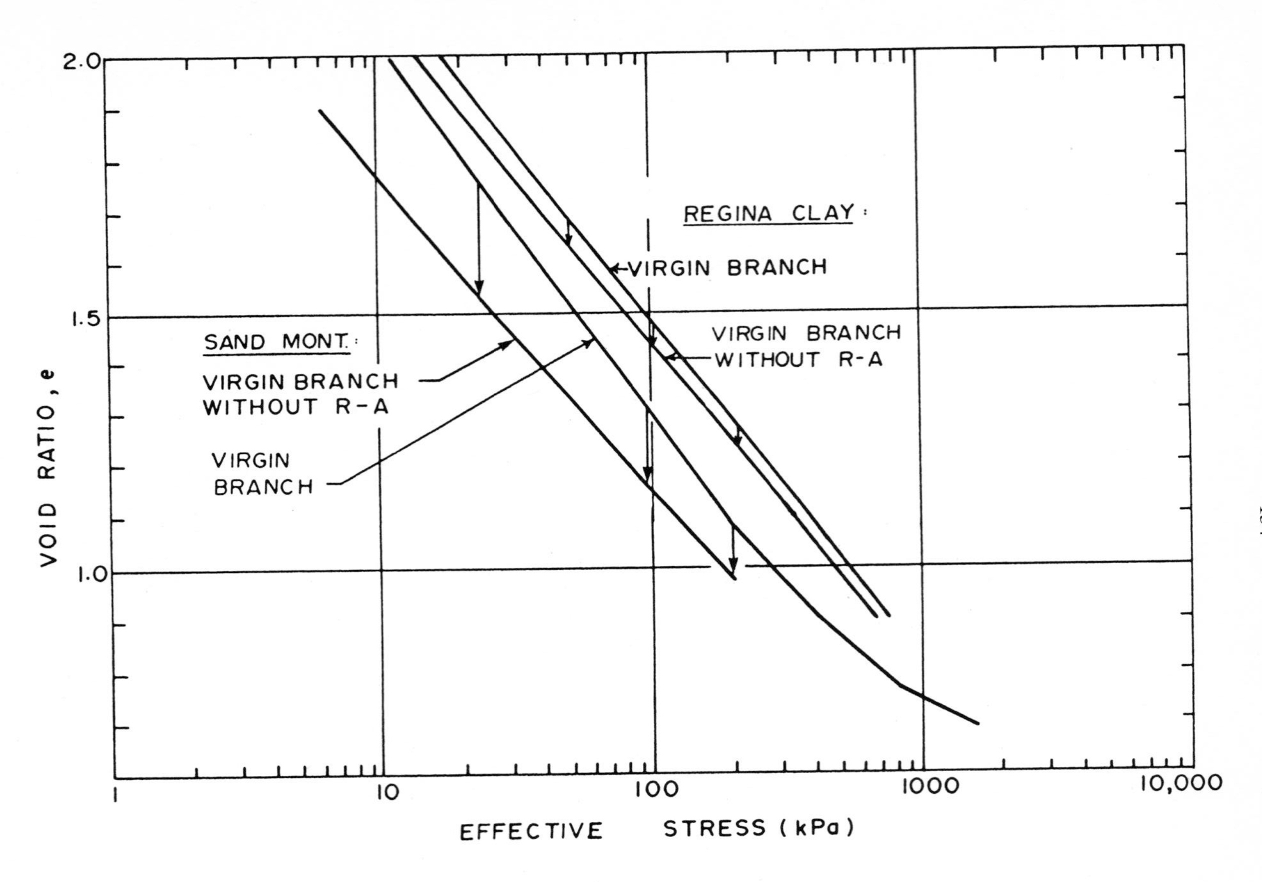

FIG. 18 Indirect evaluation of net (R-A) stresses from the virgin curve

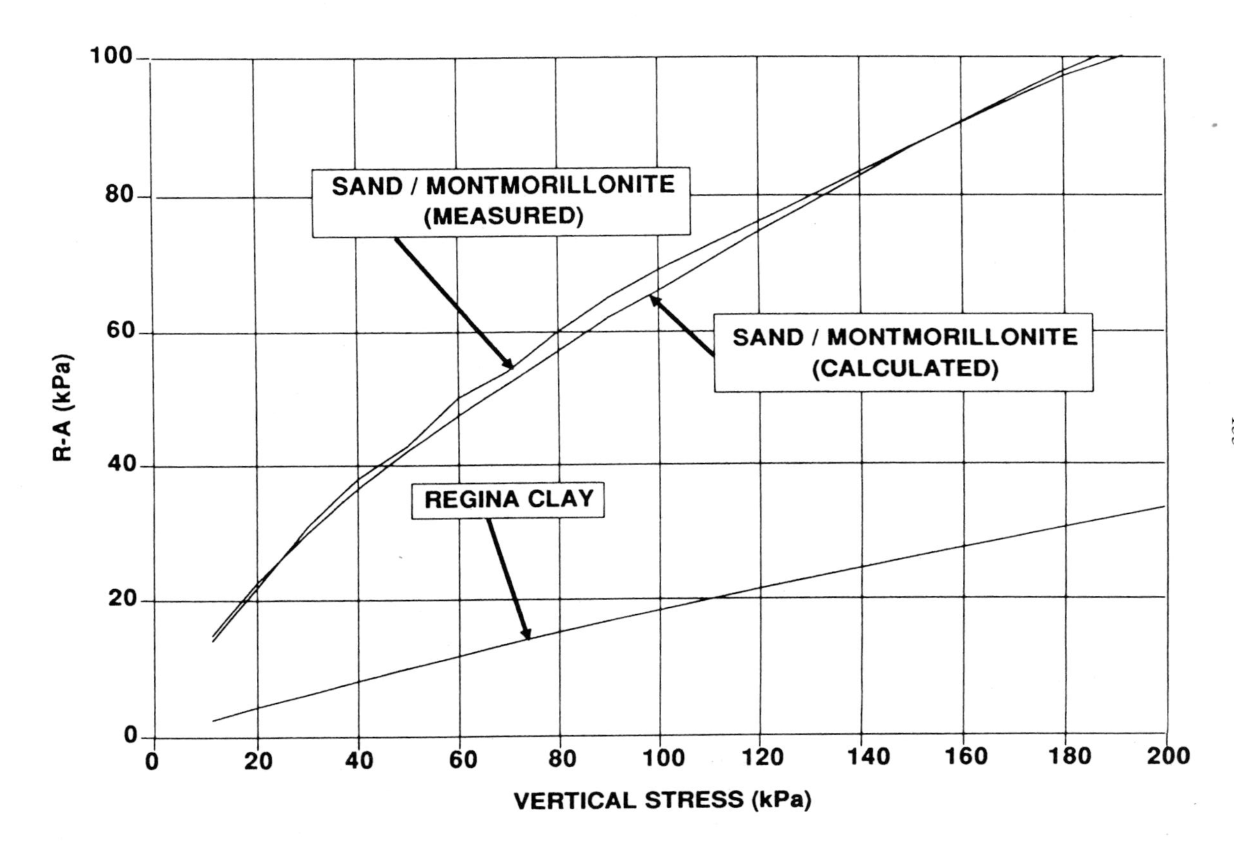

FIG. 19 (R-A) versus vertical effective stress

packets and do not exist as individual parallel particles, as is assumed with the osmotic pressure concept method of calculating (R-A).

10.0 CONCLUSIONS

The theoretical development highlighted in this paper provides for a generalized description of osmotic flow and volume change based on a continuum mechanics approach consistent with existing practice. The volume change behavior of a soil undergoing osmotically induced consolidation and osmotic consolidation can be described using the following properties: osmotic compressibility (m_π), compressibility (m_v), coefficient of permeability (k_h), coefficient of osmotic permeability (k_π), and the coefficient of diffusion (D) for the dissolved salts. Laboratory tests on modified conventional oedometers can be used to establish the osmotic compressibility and permeability with techniques similar to those used to test for conventional compressibility and permeability.

In the two soils tested, the dominant mechanism of volume change associated with brine contamination was osmotic consolidation. This process is characterized by a slow rate of volume change controlled by the rate of transport of the dissolved salt into the specimen. Osmotically induced consolidation was shown to be of little significance with respect to volume change, although osmotic flow did occur.

The results of this study provide a comprehensive approach to dealing with the volume change processes that occur during permeation of clay soils with concentrated pore fluids. This understanding can be used to predict the magnitude and rate of associated volume changes. However, a second possible use of this approach is to incorporate the effects of changing repulsive stresses within the soil mass, into conventional geotechnical analyses through the use of true effective stress and an estimate of (R-A). Barbour (1987b) illustrated the potential use of this approach in the analyses of the alteration of the permeability and shear strength of clays after permeation with brine.

The microscopic, mechanistic models such as diffuse double layer theory and the osmotic pressure concept have provided valuable insights into the role that pore fluid chemistry has on soil behavior. Building on this understanding, a macroscopic, phenomenological model provides a way for this understanding to be quantified in design. The description of physico-chemical interactions using the same phenomenological framework as conventional soil mechanics would allow existing analyses to include physico-chemical interactions in design.

REFERENCES

Aylmore, L.A.G. and Quirk, J.P., 1962. The structural status of clay systems. Clays and Clay Minerology, 9, pp. 104-130.

Bailey, W.A., 1965. The effect of salt content on the consolidation behaviour of saturated remoulded clays. USAE Waterways Experimental Station, Vicksburg, Contract Report 3-101, p. 162.

Balasubramonian, B.I., 1972. Swelling of Compaction Shale. Ph.D. Dissertation, Department of Civil Engineering, University of Alberta, Edmonton, p. 236.

Barbour, S.L., 1987a. Osmotic Flow and Volume Change in Clay Soils. Ph.D. Dissertation, Department of Civil Engineering, University of Saskatchewan, Saskatoon, p. 255.

Barbour, S.L., 1987b. The role of physico-chemical effects on the behavior of clay soils. Keynote address, 40th Canadian Geotechnical Conference, Regina, pp. 323-342.

Black, C.A., ed. 1965. Methods of Soil Analysis, Part 1 and 2. No. 9 in the Series Agronomy, American Society of Agronomy, Inc., Madison, Wisconsin.

Blackmore, A.V., and Miller, R.D. 1962. Tactoid size and osmotic swelling in calcium montmorillonite. Soil Science Society of America Proceedings, 25: 169-173.

Bolt, G.H., and Miller, R.D., 1955. Compression studies of Illite suspensions. Soil Science Society America Proceedings, 19, pp. 285-288. Bolt, G.H., 1956. Physico-chemical analyses of the compressibility of pure clay. Geotechnique, 6(2), pp. 86-93.

Bolt, G.H. 1956. Physico-chemical analyses of the compressibility of pure clay. Geotechnique, 6: 86-93.

Bolt, G.H., and Groenevelt, P.H., 1969. Coupling Phenomena as a possible cause of "non-darcian" behavior of water in soil. Bulletin International Association Scientific Hydrology, XIV(2), pp. 17-28.

Bresler, E., 1973. Anion exclusion and coupling effects in non steady transport through unsaturated soils: I. Theory. Soil Science America Proceedings, 37(5), pp. 663-669.

Chapman, D.L., 1913. A contribution to the theory of elecrtrocapillarity. Philosophical Magazine, 25(6), pp. 475-481.

Chattopadhyay, P.K., 1972. Residual Shear Strength of Some Pure Clay Minerals. Ph.D. Dissertation, Department of Civil Engineering, University of Alberta, Edmonton, p. 340.

Dunn, R.J. and Mitchell, J.K. (1984). Fluid conductivity testing of fine grained soils. Journal of the Geotechnical Engineering Division, ASCE 110 (11), 1648-1665.

Elrick, D.E., Smiles, D.E., Baumgartner, N., and Groenevelt, P.H., 1976. Coupling phenomena in saturated homo-ionic montmorillonite: I. Experimental. Soil Science Society of America Journal, 40, pp. 490-491.

Fredlund, D.G., 1975. Engineering Properties of Expansive Clays. Transportation and Geotechnical Group Publication IR-7, Department of Civil Engineering, University of Saskatchewan, p. 58.

Fredlund, D.G. (1973). Volume Change Behavior of Unsaturated Soils, Ph.D. Dissertation, University of Alberta, Edmonton, Alberta.

Fredlund, D.G. (1987). The stress state for expansive soils. Sixth International Conference on Expansive Soils, Lectures - Additional papers, Centre Board of Irrigation and Power, New Delhi, India, 1-9.

Fredlund, D.G. and Morgenstern, N.R., 1977. Stress state variables for unsaturated soils. Journal of Geotechnical Engineering Division, ASCE, 103 (GT5), pp. 447-466.

Fredlund, D.G. and Morgenstern, N.R. (1976). Constitutive relations for volume change in unsaturated soils. Canadian Geotechnical Journal 13(3), 261-276.

Freeze, R.A., and Cherry, J.A., 1979. Groundwater, Prentice-Hall, Englewood Cliffs, N.J. Gouy, G., 1910. Sur la constitution de la charge electrique a la surface d'un electrolyte. Anniue Physique (Paris), 9(4), pp. 457-468.

Gouy, G. 1910. Sur la constitution de la carge electrique a la surface d'un electrolyte. Annales de Physique (Paris), Serie 4, 9: 457-468.

Greenberg, J.A., 1971. Diffusional flow of salt and water in soils. Ph.D. Dissertation, Department of Civil Engineering, University of California, Berkeley, p. 231.

Greenberg, J.A., Mitchell, J.K. and Witherspoon, P.A., 1973. Coupled salt and water flows in a groundwater basin. Journal of Geophysical Research, 78(27), pp. 6341-6353.

Hanshaw, B.B. and Zen, E., 1965. Osmotic equilibrium and overthrust faulting. Bulletin Geological Society of America, 76, pp. 1379-1386.

Kenney, T.C. (1967). The influence of mineral composition on the residual strength of natural soils. Proceedings of the Oslo Geotechnical Conference, Vol. 1, reprinted in NGI Publ. No. 76, 37-43.

Kemper, W.D., 1961. Movement of water as affected by free energy and pressure gradients: I. Application of classic equations for viscous and diffusive movements to the liquid phase in finely porous media. Soil Science Society of America Proceedings, 24, pp. 244-260.

Kemper, W.D., and Rollins, J.B., 1966. Osmotic efficiency coefficients across compacted clays. Soil Science Society of America Proceedings, 30, pp. 529-534.

Kemper, W.D. and J.P. Quirk, 1972. Ion mobilities and electric charges of external clay surfaces infrared from potential differences and osmotic flow. Soil Science Society of America, Proceedings, 36, pp. 426-433.

Kemper, W.D., and van Schaik, J.C., 1966. Diffusion of salts in clay-water systems. Soil Science Society of America Proceedings, 30, pp. 534-540.

Lambe, T.W., 1960. A mechanistic picture of shear strength in clay. Proc. A.S.C.E. Research Conference on the Shear Strength of Cohesive Soil, Colo., pp. 555-580.

Low, P.F. 1955. The effect of osmotic pressure on diffusion rate of water, Soil Science, 80: 95-100.

Marine, I.W., and Fritz, S.J., 1981. Osmotic model to explain anomalous hydraulic heads, Water Resources Research, 17(1), pp. 73-82.

Mesri, G. and Olson, R.E. 1971. Consolidation characteristics of montmorillonite. Geotechnique, 21:341-352.

Metten, U., 1966. Desalination by Reverse Osmosis. M.I.T. Press, Cambridge, Massachusetts. Mesri, G., and Olson, R.E., 1971. Consolidation characteristics of montmorillonite. Geotechnique, 21(4), pp. 341-352.

Mitchell, J.K. 1962. Components of pore water pressure and their engineering signifcance. Clays and Clay Minerals, 10: 162-184.

Mitchell, J.K., 1973a. Recent advances in the understanding of the influences of minerology and pore solution chemistry on the swelling and stability of clays. Third International Conference on Expansive Soils, II, pp. 11-25.

Mitchell, J.K., 1973b. Chemico-Osmotic effects in fine grained soils. Journal of the Soil Mechanics and Foundation Division, ASCE, SM4, pp. 307-322.

Mitchell, J.K., 1976. Fundamentals of Soil Behavior. John Wiley and Sons, New York, p. 422.

Morgenstern, N. and Balasubramonian, B.I., 1980. Effects of pore fluid on the swelling of clay-shale. Proceedings of the 4th International Conference on Expansive Soils, Denver, pp. 190-205.

Neuzil, C.E., 1986. Groundwater flow in low-permeability environments. Water Resources Research, 22(8) pp. 1163-1195.

Olsen, H., 1972. Liquid movement through kaolinite under hydraulic, electric, and osmotic gradients. Bull. Amer. Assoc. Petr. Geol., 56(10), pp. 2022-2028.

Olsen, H., 1985. Osmosis: a cause of apparent deviations from Darcy's law. Canadian Geotechical Journal, 22(2), pp. 238-240.

Quigley, R.M., 1984. Quantitative minerology and preliminary pore-water chemistry of candidate buffer and backfill materials for a nuclear fuel waste disposal vault. Atomic Energy of Canada Ltd., Whiteshell Nuclear Research Establishment, Pinawa, Manitoba.

Robinson, R.A., and Stokes, R.H., 1968. Electrolyte solutions. Second edition, Butterworths, London, England.

Shainberg, I., Bresler, E. and Y. Klausner, 1971. Studies on Na/Ca montmorillonite systems I. The swelling pressure. Soil Science, 3(4), pp. 214-219.

Stern, O., 1924. Zur Theorie der elktrolytischen Doppelschriht. Zeitschrift Electrochem, 30, pp. 508-516. van Olphen, H., 1977. An Introduction to Clay Colloid Chemistry. John Wiley and Sons, New York.

Terzaghi, K. (1936). The Shearing resistance of saturated soils. Proceeding of the First International Conference on Soil Mechanics, Vol. 1.

Truesdell, C. (1966). The Elements of Continuum Mechanics, Springer-Verlag, Inc., New York, N.Y.

Van Olphen, H. 1977. An introduction to clay colloid chemistry. John Wiley and Sons, New York, NY.

Warkentin, B.P., Bolt, G.H., and Miller, R.D., 1957. Swelling pressure of montmorillonite. Soil Science Society of America Proceedings, 21, pp. 495-497.

Warkentin, B.P., and Schofield, R.K., 1962. Swelling pressures of Na-Montmorillonite. Soil Science, 13, p. 98.

Part 2

Plant Growth

OSMOTIC ADJUSTMENT IN PLANT CELLS EXPOSED TO DROUGHT AND TEMPERATURE STRESS: CAN A CAUSE AND EFFECT HYPOTHESIS BE FORMULATED AND TESTED?

L. Boersma,[1] Yongsheng Feng,[2] Xiaomei Li[2]

INTRODUCTION

The study of plant response to environmental stresses has developed rapidly during the past several decades. Attention has increasingly been devoted to the biochemical, biophysical, and physiological processes involved in the changes in solute concentration and content of cells which occur in response to changes in the external water potential. One aspect of these changes is osmotic adjustment.

In this report osmotic adjustment is defined as changes in the intracellular osmotic potential which occur in response to changes in the water potential outside of the cells of interest. The term osmotic adjustment originated in the salinity literature to describe a change in the osmotic potential of leaves parallel to that in the soil water (Bernstein, 1961). The decrease in the osmotic potential of plant cells was considered to be an adjustment to the decrease in potential of the extracelluar water. The term was later used to describe drought responses as well as salt responses. Osmotic adjustment is now generally considered to be the decrease in the osmotic potential of cells, resulting from the increase in osmolyte concentration in the cells rather than from a change in the cell volume.

Osmotic adjustment is thought to be an important adaptation to the environmental stress which impacts water relations because it helps to maintain turgor pressure, which in turn is thought to maintain growth. Munns (1988) noted that much of the recent interest in osmotic adjustment derives from the belief that plants which maintain a high turgor pressure will grow and photosynthesize more nearly at optimum levels when exposed to stress than those that do not.

[1] Department of Crop and Soil Science, Oregon State University, Corvallis, OR 97331, USA
[2] Department of Soil Science, University of Alberta, Edmonton, Canada

NATO ASI Series, Vol. H 64
Mechanics of Swelling
Edited by T. K. Karalis

Here we describe results of experiments which made it possible to evaluate the hypothesis that turgor pressure is necessary for growth and to provide insights regarding the mechanism of osmotic adjustment. These experiments were initiated for the purpose of measuring response of plant growth to water and temperature stress (Feng et al., 1990). Plants are highly integrated biological systems and responses cannot be explained on the basis of single observations such as correlation between turgor pressure and growth rate, or between degree of osmotic adjustment and growth rate. A theoretical basis for analysis of results was provided by Johnson and Thornley (1985). This report starts with a presentation of the theoretical basis followed by discussion of experimental procedures and results, analyses, and development of hypotheses to be evaluated in future studies.

THEORY

The rate at which biological processes proceed increases with temperature to an optimum and then falls off as the temperature increases further (Ingraham, 1958). Although this type of response has generally been recognized, at least in a qualitative manner, a general procedure which can be used to characterize such a response with physically meaningful parameters has been lacking.

Water stress also affects plant processes and often interacts with temperature (Barlow et al., 1977; Harrison et al., 1986). Little quantitative research has been reported where both water potential and temperature are involved (Barlow et al., 1977). In this case the usual statistical analysis of variance yields little information about the nature of these responses. A more theoretically based approach would be preferred.

The Arrhenius equation for the rate of a chemical reaction as a function of temperature states that

$$K' = K_a e^{-E_a/RT}, \tag{1}$$

where E_a (J Mol^{-1}) is the activation energy and K_a is a constant, which can be viewed as the maximum rate of reaction when there is no energy barrier (activation energy) between reactant and product.

The rate of an enzyme catalyzed reaction also depends on the fraction of the enzymes in the active state. The assumptions that an enzyme can exist in either an active or an inactive state and that the Boltzman distribution can be used to describe the distribution of enzymes between the two states lead to:

$$f_a = \frac{1}{1+e^{-dG/RT}}, \tag{2}$$

where f_a is the fraction of enzymes in the active state, and dG (J Mol^{-1}) is the free energy difference between inactive and active states of the enzyme (Johnson and Thornley, 1985). The total rate of reaction is then given by

$$K = f_a K'. \tag{3}$$

Combining equations (1), (2), and (3) with the relation

$$dG = -TdS + dH \tag{4}$$

results in

$$K = \frac{K_a e^{-\frac{E_a}{RT}}}{1+e^{\frac{dS}{R}-\frac{dH}{RT}}} \tag{5}$$

where dS (J Mol^{-1} K^{-1}) and dH (J Mol^{-1}) are, respectively, the entropy and enthalpy differences between the inactive and active enzyme states and K_a is the rate constant. The term "constant" can be misleading because K_a varies with water stress as will be shown later.

We simplified equation (5) by defining the constants

$$B = \frac{E_a}{R}, \quad (°K)$$
$$C = \frac{dS}{R}, \text{ (dimensionless)} \tag{6}$$
$$D = \frac{dH}{R}, \quad (°K)$$

and substituting these into (5), yielding

$$K = \frac{k_a e^{-\frac{B}{T}}}{1+e^{C-\frac{D}{T}}} \tag{7}$$

At the optimum temperature $T = T_0$, $dK/dT = 0$, so that

$$T_o = \frac{D}{C+\ln(\frac{D}{B-1})}, \tag{8}$$

or

$$C=\frac{D}{T_o}-\ln\left(\frac{D}{B-1}\right). \qquad (9)$$

From a practical point of view, the optimum temperature, T_0, is a more meaningful parameter than the entropy change between active and inactive states of enzymes, especially when the relation is applied to complex processes such as plant growth where the active and inactive states of the enzyme system are not easily defined. Substituting (9) into (7) gives

$$K=\frac{K_a e^{-\frac{B}{T}}}{1+\frac{B}{D-B}e^{D\left(\frac{1}{T_o}-\frac{1}{T}\right)}} \qquad (10)$$

Water Potential Effects

In the case where both temperature and soil water potential are involved, one can write

$$K=f(T,\psi) \qquad (11)$$

where Ψ is leaf water potential (MPa), or for any fixed leaf water potential

$$K=f(T|\psi=\text{constant}). \qquad (12)$$

The assumption that leaf water potential affects K in equation (10) implicitly by affecting its parameters leads to the relationships:

$$\begin{aligned} K_a&=K_a(\psi);\\ B&=B(\psi);\\ D&=D(\psi);\\ T_o&=T_o(\psi). \end{aligned} \qquad (13)$$

Equations (10) and (13) are then specific expressions of equations (11) and (12). Feng et al. (1990) found that K_a and B as functions of water stress are related by

$$\ln K_a(\psi)=\ln K_o+\left(\frac{\kappa}{T_o}\right)B(\psi), \qquad (14)$$

where κ is a dimensionless coefficient. Substituting equation (14) into equation (10) yields

$$K=K_o \frac{e^{B(\kappa/T_o-1/T)}}{1+\frac{B}{D-B}e^{-D(1/T-1/T_o)}} \qquad (15)$$

Values of the parameters K_0, B, D, κ, and T_0 can be obtained by fitting equation (15) to data sets consisting of measurements of growth rate at combinations of soil temperature and soil water potential. The statistical procedures for fitting equation (15) to the data set should be conducted in such a manner that the water potential dependency of the parameters given by equation (13) is also evaluated.

Summary

The parameters in the original equation [equation (5)] described by Johnson and Thornley (1985) include activation energy, the maximum rate of reaction, and entropy and enthalpy changes between the active and the inactive states of the enzyme. These parameters are well defined when applied to enzyme reactions, but strict physical meanings of the parameters are not clearly defined for complex plant processes. Plant growth is the combined result of the operation of a complex set of enzyme controlled reactions. When the Arrhenius equation derived for a single enzymatic reaction is used to describe these complex processes, the parameters represent the combined responses of a multitude of enzyme systems. However, equation (10) can be used as a basis for comparison of the sensitivity of different plant processes to water stress (Feng et al., 1990).

EXPERIMENTS AND RESULTS

Growth rates of leaf area and shoot dry mass of spring wheat (*Triticum aestivum cv. Siete cerrors*) seedlings and sudangrass (*Sorghum vulgare* var Piper) seedlings were measured at soil temperatures and soil water potentials maintained at predetermined, constant, levels. Eight pre-germinated seeds were planted in an 0.8 cm thick soil slab described by Barlow et al. (1977). A total of 100 slabs were planted. The seedlings were thinned when they were 3 cm tall, leaving 5 plants in each soil slab. The plants were first grown in a growth chamber with the light period of 14 hours and the light intensity of 250 μmol m^{-2} s^{-1}. The plants were supplied with water every other day by placing the soil slabs in a 3 cm deep, half strength, Hoagland solution (Hoagland and Arnon, 1950), so that the entire soil column was wetted by capillary rise.

After 16 days the plants were taken to a walk–in growth room for the experiments. Procedures were as described by Barlow et al. (1977). Light period and light intensity were the same as in the growth chamber. Soil water potential and root temperature treatments were imposed by replacing the removable side walls of the soil slabs with a semipermeable membrane and placing the assemblies in a polyethylene glycol (PEG) solution of desired osmotic potential. The temperature of the PEG solution was controlled with a constant temperature water bath. Mineral nutrients were provided by mixing nutrients with the PEG solution in concentrations corresponding to a half strength Hoagland solution (Hoagland and Arnon, 1950). Five soil water potentials and six root temperatures were used (Table 1).

Leaf areas were measured at day 0, 2, 3, 4, 5, and 7 after treatments were imposed. The slopes of least square regressions of the natural log of leaf area against time were taken as the relative rates of growth (Table 1). Procedures used for measurement and calculation of potentials and results were described by Li et al. (1991).

Equation (15) was fitted to the experimental observations using a least square technique. The mathematical procedures are described in detail by Feng et al. (1990) and Li et al. (1991).

DISCUSSION

General Observations

The two plant species differed with respect to osmotic adjustment (Fig. 1). The osmotic potential of spring wheat decreased linearly with time during the full light period. There was little difference in osmotic potential between the –0.03 MPa and –0.25 MPa soil water potential treatments despite the large difference in leaf water potential, indicating that osmotic adjustment was negligible in spring wheat under conditions of this experiment. As a result of this lack of osmotic adjustment, the turgor potential changed in parallel with total leaf water potential (Fig. 1).

The sudangrass showed a significant degree of osmotic adjustment. The osmotic potential decreased quickly after the light period started (Fig. 1). The difference between the osmotic potentia+ls at +–0.03 MPa and –0.25 MPa treatments was approximately the same as the difference in total leaf water potential. Turgor potentials between the –0.03 MPa and the –0.25 MPa soil water potential treatments were nearly the same as a result (Fig. 1). Osmotic adjustment in sudangrass resulted in almost complete turgor maintenance under the conditions of this experiment.

Table 1. Relative rates of increase in leaf areas of spring wheat and sudangrass at the indicated combinations of root temperatures and soil water potentials. Temperature treatments for wheat and sudangrass differed as shown.

Soil water potential	Root temperature (°C)					
	14	17	22	27	29	32
MPa			day^{-1}			
Spring wheat						
−0.03	0.108	0.119	0.126	0.138	0.111	0.095
−0.06	0.107	0.104	0.117	0.132	0.106	0.083
−0.10	0.103	0.110	0.110	0.131	0.091	0.067
−0.17	0.079	0.086	0.096	0.100	0.086	0.064
−0.25	0.058	0.075	0.089	0.092	0.064	0.051
Sudangrass	16	22	28	32	34	36
−0.03	0.150	0.201	0.222	0.242	0.230	0.215
−0.06	0.110	0.167	0.182	0.195	0.178	0.164
−0.10	0.071	0.135	0.155	0.161	0.144	0.129
−0.15	0.051	0.111	0.125	0.136	0.125	0.108
−0.25	0.035	0.080	0.095	0.117	0.095	0.078

During the afternoon, turgor potential increased slightly for both species, because osmotic potential continued to decrease while leaf water potential approached steady state values.

Values of the parameters B, D, K, T_o, and κ were obtained by fitting equation (15) to the data sets (Table 2). Explanations for the differences in response between the two species were sought by comparing values of the parameters.

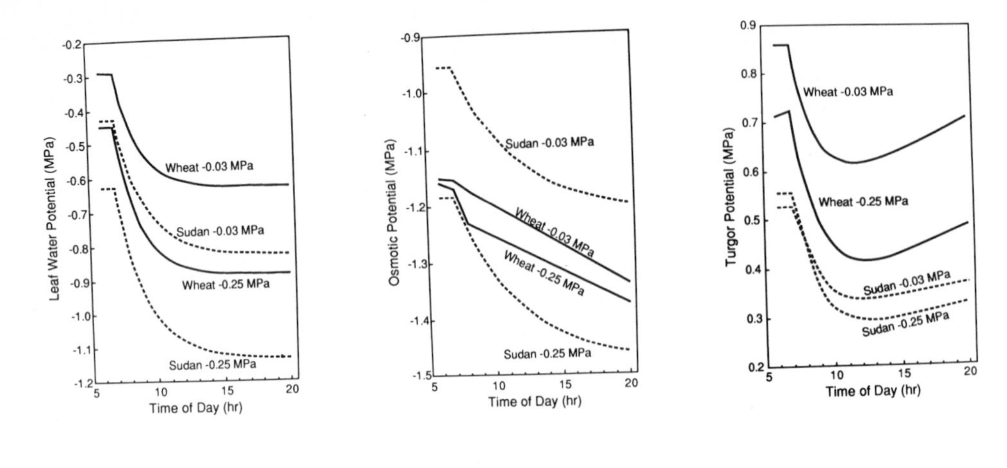

Fig. 1. Leaf water potentials, osmotic potentials and turgor potentials of wheat and sudangrass as a function of time. Lights were turned on at 7:00 hrs and remained at a constant intensity during the period for which results are shown. Lights were turned off at 20:00 hrs and water potentials recovered to 7:00 hr values during the dark period.

Sensitivity to Water Potential

The activation energy, B, of the relative growth rate of leaf area increased with decreasing leaf water potential for both spring wheat and sudangrass (negative b_1), but the rate at which B increased with decreasing water potential, given by the absolute value of b_1, differed between the two species (Table 2). Increases in activation energy, as a result of decreasing leaf water potential, resulted in decreased growth rates. The cause and effect relationship of this response is not immediately clear and may differ between plant species. Although the activation energy of sudangrass increased less than that for spring wheat as leaf water potential decreased (Table 2), the growth rate of sudangrass decreased more rapidly as a function of increasing B than that of wheat (Fig. 2). This suggests that the growth rate of sudangrass was more sensitive to increasing activation energy than that of wheat.

The sensitivity of growth rate to increasing activation energy is indicated by the partial derivative of growth rate (K) with respect to activation energy (B). Equation (15) suggests that this quantity depends on temperature. For the purpose of comparison, we calculated this quantity for wheat and sudangrass at their respective optimum temperatures. At $T = T_0$, the sensitivity of K to B is

$$S_B = \left|\frac{1}{K}\frac{\partial K}{\partial B}\right| = \frac{(1-\kappa)}{T_o} + \frac{1}{(D-B)}, \tag{16}$$

where S_B is the sensitivity of K to B. Results show that S_B decreased with increasing water stress for both species (Fig. 3) suggesting that, on a relative scale, the growth rate of sudangrass was affected more by a given decrease in leaf water potential than wheat. Equation (16) indicates that S_B increases as the value of B approaches that of D. Table 2 shows that D for wheat was 27000°K. For sudangrass, D, calculated using the appropriate π_p values, ranged from 8932 to 10962°K. Because of the higher values of D, the growth rate of wheat was less sensitive to water stress than the growth rate

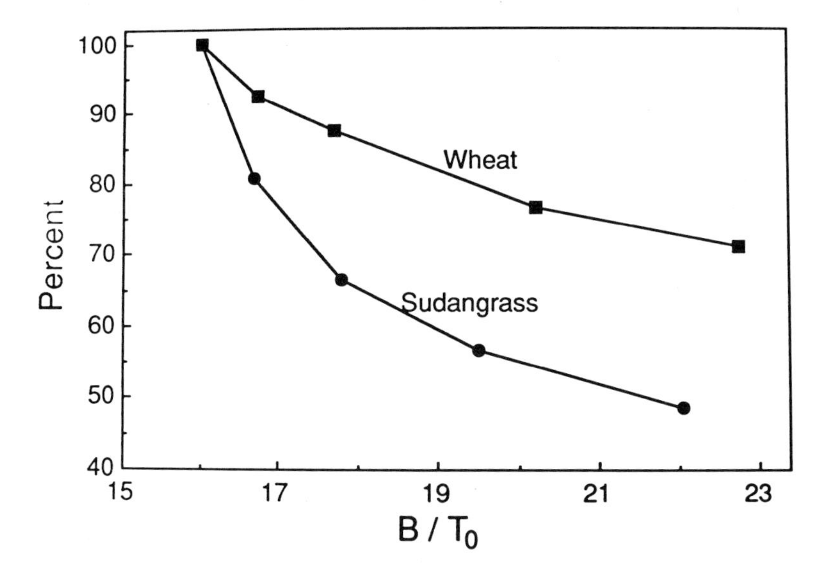

Fig. 2. Relative rate of increase of leaf area as percent of area at the −0.03 MPa soil water potential treatment, plotted as a function of activation energy. The data used for the diagram were for the optimum temperatures, 22.0°C for wheat and 32.0°C for sudangrass.

Table 2. Least square estimates with estimated standard errors of the parameters of equation (15) as functions of leaf water potential for the relative growth rate of leaf area of spring wheat and sudangrass.

Parameter	Relation with water potential	Least square estimates of parameter values		
Spring wheat				
K_0 (day^{-1}) MPa^{-1}	$= k_1\Psi_p$	k_1	$= 0.21 \pm 0.03$	day^{-1}
B (°K)	$= b_1\Psi_1$	$b_1 \times 10^{-3}$	$= -8.9 \pm 0.2$	°K MPa^{-1}
D (°K)	= constant	$D \times 10^{-4}$	$= 2.7 \pm 0.4$	°K MPa^{-1}
T_o (°K)	= constant	T_o	$= 2.98.5 \pm 0.5$	°K
κ	= constant	κ	$= 1.00$	
$R^2 = 0.90$				
Sudangrass				
K_0 (day^{-1}) MPa^{-1}	= constant	K_0	$= 1.93 \pm 0.36$	day^{-1}
B (°K)	$= b_1\Psi_1$	$b_1 \times 10^{-3}$	$= -6.9 \pm 0.4$	°K MPa^{-1}
D (°K)	$= d_1\pi$	$d_1 \times 10^{-4}$	$= -0.81 \pm 0.08$	°K MPa^{-1}
T_o (°K)	= constant	T_o	$= 305.0 \pm 0.5$	°K
κ	= constant	κ	$= 0.91 \pm 0.005$	
$R^2 = 0.92$				

of sudangrass (Fig. 2), despite the fact that its activation energy increased more rapidly with decreasing leaf water potential.

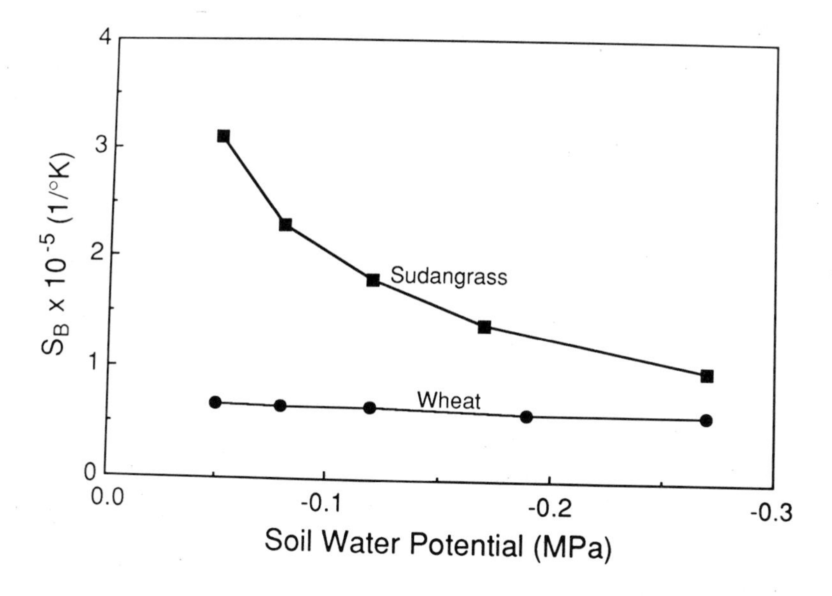

Figure 3. Sensitivity of relative leaf area growth rates of spring wheat and sudangrass to increasing activation energy as a function of leaf water potential. These calculations were carried out at the optimum temperatures of 22°C for wheat and 32°C for sudangrass.

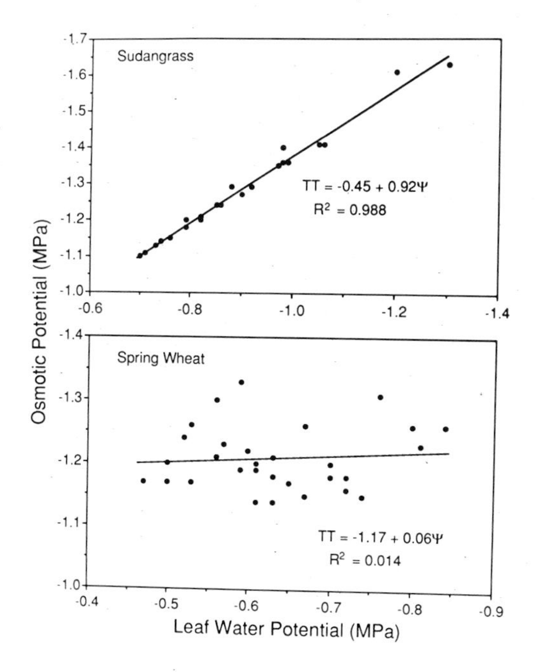

Figure 4. Daily average osmotic potential as a function of daily average total leaf water potential for all root temperature and soil water potential combinations.

Osmotic Adjustment and the Parameter D

The parameter D, which represents the enthalpy difference between active and inactive states of an enzyme, was found to increase linearly with decreasing osmotic potential for sudangrass according to $D = -8120\pi$, while it was constant for wheat. The degree of osmotic adjustment can be expressed by

$$\alpha_\pi = \frac{d\pi}{d\psi_1}, \tag{17}$$

where $d\Psi_1$ is the change in total leaf potential in response to water stress and $d\pi$ is the corresponding change in osmotic potential, referred to as osmotic adjustment. The value of α_π ranges from 0, when there is no osmotic adjustment, to 1, for complete osmotic adjustment when for each change in Ψ_1 a corresponding change in π of the same magnitude occurs. The coefficient α_π is the slope of the plot of osmotic potential as a function of total leaf water potential. Figure 4 shows that $\alpha_\pi = 0.92$ for sudangrass, indicating strong osmotic adjustment, and $\alpha_\pi = 0$ for spring wheat, indicating no osmotic adjustment.

Because spring wheat showed no osmotic adjustment, its osmotic potential may be represented by the average osmotic potential of all treatments, π_{avg}. Assuming a linear relation between D and osmotic potential, similar to the one for sudangrass, we may write for spring wheat,

$$D = d_1 * \pi = \left(\frac{D_{ls}}{\pi_{avg}}\right) * \pi \tag{18}$$

where D_{ls} is the constant, least square estimate of the parameter D for spring wheat, $d_1 = D_{ls}/\pi_{avg}$ is equivalent to the slope between D and π; and π_{avg} is the average osmotic potential for spring wheat. Since there was no osmotic adjustment for spring wheat, π was constant, so that, according to equation (18), D was constant. This approach provides a unified relation for both plant species.

A similar approach can be used for the relationship between K_0 and turgor potential. The above arguments lead to the conclusions that for both species, K_0 increases linearly with increased turgor potential, B increases linearly with decreasing total leaf water potential, and D increases linearly with decreasing osmotic potential. Mathematically this may be expressed as

$$K_o=k_1\psi_p, B=b_1\psi_1, D=d_1\pi. \tag{19}$$

Values of k_1, b_1, and d_1 are in Table 3.

Application

An important adaptation of plants to drought or salinity stress is the increase in the concentration of intracellular solutes. This phenomenon, referred to as "osmotic adjustment", is generally regarded as beneficial because turgor and cell volume are maintained, which are thought to be important for maintaining growth, or at least survival under saline conditions or when available soil water is limited. The relationship between turgor pressure and growth has often been studied (Ray et al., 1972). Munns (1988) recently questioned this view, noting that a physiological rationale for beneficial consequences is lacking and that the expected relationship between rate of cell elongation and turgor pressure is often not found. Our results seem to confirm this lack of correlation (Fig. 5). However a closer inspection of the data, based on equation (15) reveals the underlying mechanism implicit in the Lockhart equation (Lockhart, 1965).

$$\text{Rate of growth}=m(\sigma-\sigma_y) \tag{20}$$

for rate of cell growth where rate of growth is a relative rate (s^{-1}), m is cell wall extensibility (s^{-1} MPa^{-1}), σ is the total applied mechanical stress, usually equated with turgor pressure (MPa) and σ_y is the cell wall yielding stress (MPa). We now compare equation (20) with equation (15). For equation (15) write:

$$K = K_0(f)$$

where f is the function between square brackets. From experimental results:

$$K_o=k_o+k_1\sigma_p \tag{21}$$

so that:

$$K=(k_o+k_1\sigma_p)(f)$$
$$K=k_1\left(\frac{k_o}{k_1}+\sigma_p\right)(f) \tag{22}$$
$$K=k_1(f)\left(\sigma_p+\frac{k_o}{k_1}\right)$$

so that in analogy to equation (15)

$$m=m_o \frac{\exp[-B(\frac{1}{T}-\frac{\kappa}{T_o})]}{1+\frac{B}{D-B}\exp[-D(\frac{1}{T}-\frac{1}{T_o})]} \tag{23}$$

where $m_o = k_1 = 0.214$ for spring wheat and $m_o = k_1 = 5.08$ for sudangrass, and the remaining parameters are as obtained from experiments. Equation (23) shows that extensibility is affected by temperature and water stress in a manner analogous to a metabolic process, suggesting that extensibility is metabolically controlled. We found a linear relationship between K_0 and turgor pressure, suggesting that equation (23) is valid. The often observed lack of correlation between growth and turgor (Munns, 1988) is not sufficient to disprove the role of turgor in growth. Our experiments showed this lack of correlation also. However, analysis shows this lack of correlation to be result of variations in cell extensibility caused by the range of environmental conditions. A linear relation between growth and turgor emerges when the base rate, K_0, is used for the analysis.

MECHANISM OF OSMOTIC ADJUSTMENT--A HYPOTHESIS

The ability of organisms to adjust to changes in the potential of water external to the cell walls is necessary for survival and growth. Mechanisms involved in osmotic adjustment have only recently been studied. One important result of these studies has been the finding of a similarity between bacteria and plants in their response to
changes in external water potential (Yancey et al., 1982). In the preceding part of this manuscript, plant response to water and temperature stress was described by a mechanistically-based model, which allowed calculation of several thermodynamic parameters. These parameters describe the combined effects of many biochemical and biophysical reactions and processes. Plausible explanations for the behavior of the thermodynamic constants remain to be determined. The experiments from which these results were derived were initiated to gain a better understanding of mechanisms of osmotic adjustment and the importance of osmotic adjustment in determining plant growth.

An important conclusion regarding osmotic mechanisms is that the number of osmolyte systems found in water stressed organisms is very small and that these systems may be under genetic control. Specific mechanisms for genetic control of osmotic adjustment have also been identified, e.g. turgor pressure regulation of the expression of the *kdp* operon in *Esoherichia coli,* which in turn regulates potassium concentrations in the cell.

Table 3. Values of k_1, b_1, and d_1 in the unified relations between thermodynamic parameters and leaf water status.

Parameter	Spring wheat	Sudangrass
k_1 (day^{-1} MPaξ−1)	0.214	5.08
$b_1 \times 10^{-3}$ (°K MPa^{-1})	-8.9	-6.9
$d_1 \times 10^{-4}$ (°K MPa^{-1})	-2.3	-0.81

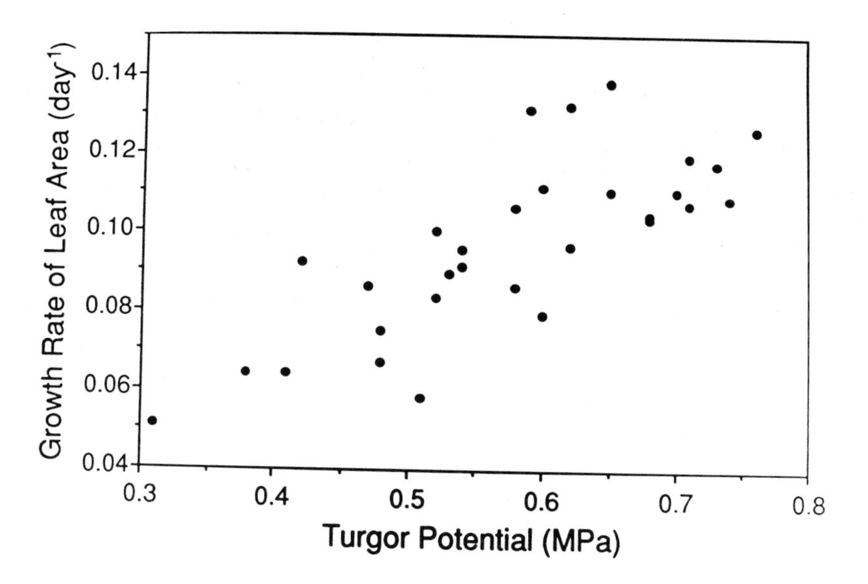

Fig. 5. Growth rate of leaf area plotted as a function of turgor potential showing results for all combinations of soil water potential and root temperature treatments.

We suggest that osmotic adjustment in plants should be considered from the perspective of plants as integrated systems. This prespective leads to the hypothesis that effects of water stress on phloem transport from source regions to sink regions could explain several of the phenomena, which have been correlated with osmotic adjustment. The hypothesis is in part based on the observation that the important compatible solutes involved in osmotic adjustment are all part of biochemical pathways. These solutes are: potassium ions, the amino acids glutamate, glutamine, proline, γ–amino butyrate, and alanine, the quaternary amines glycinebetaine, and other fully N–methylated amino acid derivatives, and the sugars sucrose, trehalose, and glucosylglycerol (Yancey et al., 1982; Csonka, 1989). These solutes are not able to cross cell membranes without the aid of a specific transport system and can be accumulated by *de novo* synthesis or by uptake from the culture medium.

Osmolyte Systems

The possibility of genetic control of osmotic adjustment was considered by Yancey et al. (1982) in a paper entitled, "Living With Water Stress: Evolution of Osmolyte Systems." They found what they referred to as "a striking convergent evolution in the properties of the organic osmotic solutes in bacteria, plants, and animals." The three types of osmolyte systems found in all water stressed organisms, except the halobacteria, were: polyhydric alcohols, free amino acids and their derivatives, and combinations of urea and methylamines. They noted that the "selective advantages of the organic osmolyte systems are, first, a compatibility with macromolecular structure and function at high or variable (or both) osmolyte concentrations, and, second, greatly reduced needs for modifying proteins to function in concentrated intercellular solution." They suggest that the nature of the low molecular weight substances, both inorganic ions and small organic molecules, which comprise the bulk of the osmolytes present in all cells has, to a large extent, been ignored.

This observation endorses all views, expressed above, that osmotic adjustment should be evaluated from the perspective of plant response as a unified system. Yancey et al. (1982) further note that the osmolytes form part of the environment for the biochemical reactions of living systems, and the solute compositions of cells may have been subject to stringent selection. This observation could be changed to state that the biochemical pathways have been subject to stringent selection. The strongest selective pressures exist for organisms that experience some form of environmental water stress due to high or fluctuating salinity, desiccation, or freezing.

Yancey et al. (1982) propose that the repeated adoption of a few classes of organic osmolytes by phylogenetically diverse organisms, encountering varied types of water stress, is a reflection of two important phenomena, namely the physical–chemical interactions between solutes, water, and macromolecules which establish which types of solutes are compatible with macromolecular structure and function and a phenomenon they refer to as "genetic simplicity." Through the use of compatible solute systems, proteins are able to work in the presence of high or variable solute concentrations, and the modification of large numbers of proteins is avoided. In this manner, a single form of a given protein can be used over a wide range of osmotic concentrations.

These observations by Yancey et al. (1982), which have been cited and elaborated upon by numerous authors, are logical when one takes as a starting point the hypothesis that a mechanism exists for regulating osmotic potential *per se*. However, it is also possible to look upon the osmolytes identified in this discussion as intermediates in chains of biochemical events leading to an end product expressed as growth. Thus one could interpret the data as selections of optimum biochemical systems for growth. When the chain of events is interrupted, the concentration of certain compounds must increase. One would expect these compounds to be mostly amino acids. Thus water stress should lead to an increase in free amino acid concentrations as was observed by Barlow et al. (1976) and can also be seen in data reported by Jones et al. (1980).

Genetic Control

Osmotic control of transcription in bacteria was reviewed by Csonka (1989), who noted that only a few genes have been identified whose transcription is regulated by osmotic potential. These include *kdp*, *pro*U, *pro*P, *omp*F, and *omp*C. The reviewer raised the question "Is there a global osmosensing regulatory protein" and concluded that at present there is not sufficient information to decide whether there is a single underlying mechanism for such transcriptional control of osmotic adjustment. Morgan (1991) presented evidence that a single gene may control differences in osmoregulation in response to water stress between wheat species. The gene is concerned with regulation of turgor pressure and water content.

Hypothesis

Evaluation of the information regarding osmotic adjustment shows that:

- much of the detailed information now available comes from studies with microorganisms
- none of the studies on osmotic adjustment considered osmotic adjustment as part of an interacting system of biochemical reactions
- important osmolytes in microorganisms are:
  proline; glycinebetaine; choline
- important osmolytes identified in plants include:
  sugars; free amino acids; proline
- plant studies have emphasized measurement of osmotic and turgor potentials but ignored identification of osmolytes
- differences in osmotic adjustment exist between wheat and sundangrass
- plant studies have considered phenomena and processes as being local phenomena and have not considered interactions between various plant parts
- the time–courses of osmotic adjustment and of the concentrations of specific osmolytes have not been measured so that cause and effect relationships cannot be identified
- activation energy increases with decreasing water potential.

Physiological processes in plants growing under optimum conditions are in balance so that throughout the plant imports are equal to exports both in terms of mass and energy. Stress disturbs this equilibrium and sets in motion regulation oriented towards restoring the original balance or achieving a new balance.

An important consequence of decreasing xylem water potential is a decrease in rate of phloem transport (Boersma et al., 1991). As a result product flows are interrupted and therefore new equilibrium concentrations of all the products and their intermediates must occur, e.g. an increase in sugars and free amino acids is expected.

A hypothesis which has not been examined to our knowledge is that the phenomena which occur in plant leaves as a consequence of decreased water potential in the external solution are the result of interrupted transport. Examine a plant which functions at optimum conditions, with dynamic processes such as photosynthesis and transpiration occurring at a steady rate. Light energy is captured and from that follows a sequence of biochemical events which ultimately results in the loading of photosynthates to the phloem followed by phloem transport to the sink regions of the plant. When water potential decreases, this steady state condition is interrupted and a series of consequences is initiated. Many of the observed phenomena can be explained on the basis of this interruption of rate of transport. These consequences are not observable as one event, all biochemical processes participate in the change.

For demonstration we describe the following consequences. When the water potential at the root surface is lowered, water potential in the xylem decreases throughout the plant, with the largeast decrease occurring in the leaves. As a result of this decrease,

Table 4. Contribution of different solutes to the decrease in osmotic potential at full turgor in sorghum leaves at two levels of stress. The osmotic potential due to the accumulation of each solute was calculated from the change in content of solute and the total symplastic water content compared with those in well–watered controls.

		Control	Mild stress	Severe stress
Predawn ψ_l (MPa)		−0.19	−0.85	−1.28
Osmotic potential		−0.90	−1.15	−1.39
$\Delta\psi_\pi$ at full turgor (MPa)			−0.25	−0.49
$\Delta\psi_\pi$ (MPa) due to increase in				
sugars	sucrose		−0.04	−0.05
	glucose		−0.04	−0.06
	fructose		−0.02	−0.03
ions	potassium		−0.07	−0.06
	chloride		−0.06	−0.06
	potassium carboxylate		ns	−0.08
total free amino acids			−0.02	−0.04
sugar phosphates			ns	−0.03
$\Delta\psi_\pi$ accounted for (%)			100	84

the cells lose turgor which has as a consequence a decrease in the pressure gradients from source to sink regions, reducing rate of transport so that concentrations of organic compounds increase, when input is not reduced. The molecules which accumulate will be those that are precursors to the final product of photosynthesis, such as free amino acids and prolines. This increase in solutes continues until a new balance is achieved; however all phenomena now occur at a higher concentration of osmolytes. There is evidence which supports this hypothesis, including the increase in organic molecules. Jones et al. (1980) allowed predawn leaf water potentials of sorghum to decline to −0.85 and −1.30 MPa in treatments designated as mild stress and severe stress, respectively (Table 4). Following rehydration, the osmotic potentials at full turgor were −0.25 MPa and −0.49 MPa less than those of well–watered controls at full turgor. Table 4 shows the contribution of sugars, inorganic ions, total free amino acids, and sugar phosphates to the decrease in the ψ_π from control to mild stress and from control to severe stress.

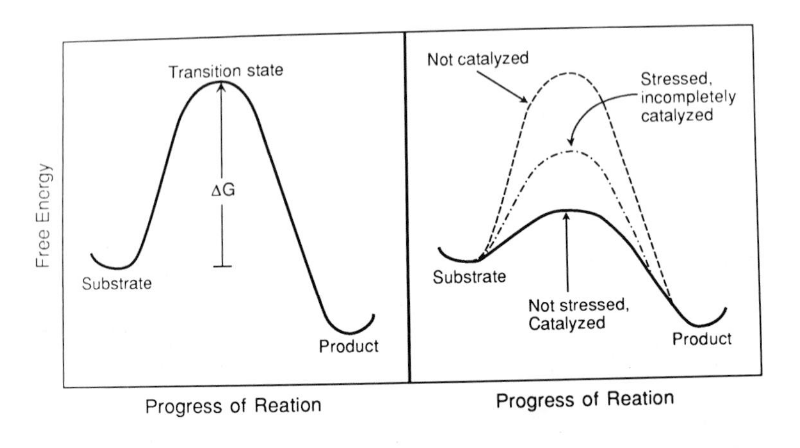

Fig. 6. Left–hand side shows definition of activation energy, right–hand side shows decreased activation energy due to enzymes (not stressed, catalyzed) and increased activation energy from catalzyed level when water potential decreases concentration of enzymes (stressed incompletely catalyzed).

The increase in activation energy with decreasing water potential could also be explained as a consequence of the interruption of transport. Enzymes reduce the activation energies of reactions catalyzed by them (Fig. 6). However when enzymes are not present in sufficient concentration, the overall activation energy of the catalyzed reaction will be increased. The decreased availability of enzymes could be a consequence of reduced rate of phloem transport.

These results can be interpreted in support of our hypothesis, but more unequivocal evaluation of the hypothesis is needed. The necessary supporting data can be obtained in experiments similar to those conducted by Jones et al. (1980) and Feng et al. (1990).

REFERENCES

Barlow EWR, Boersma L, Young JL (1977) Photosynthesis, transpiration, and leaf elongation in corn seedlings at suboptimal soil temperatures. Agron J 69:95–100

Barlow EWR, Ching TM, Boersma L (1976) Leaf growth in relation to ATP levels in water stressed corn plants. Crop Science 16:405–407

Barlow EWR, Munns RE, Scot NS, Reisner AH (1987) Water potential, growth, and polyribosome content of the stressed wheat apex. J Exp Bot 28:909–916

Bernstein L (1961) Osmotic adjustment of plants to saline media. 1 Steady state. Am J Bot 48:909–981

Boersma L, Lindstrom FT, Childs SW (1991) Model for steady state complex transport in xylem and phloem. Agron J 83:401–408

Bonnemain JL, Delrot S, Lucas WJ, Dainty J (1991) Phloem transport and assimilate compartmentation. Ouest Editions, Nantes, France, 344 p

Csonka LN (1989) Physiological and genetic responses of bacteria to osmotic stress. Microb Reviews 53:121–147

Feng Y, Li X, Boersma L (1990) The Arrhenius equations as a model for explaining plant response to temperature and water stresses. Ann Bot 66:237–244

Harrison PM, Walton DWH, Rothery P (1986) The effects of temperature and moisture on dark respiration in the foliose lichen *Umbilicaria antarctica* New Phytol 103:443–445

Hoagland DR, Arnon DI (1950) The water culture method for growing plants without soil. California Agricultural Experiment Station, Circular No. 347

Ingraham JL (1958) Growth of psychrophilic bacteria. J Bacteriol 76:75–80

Johnson IR, Thornley DHM (1985) Temperature dependence of plant and crop processes. Ann Bot 5:1–24

Jones MM, Osmond CB, Turner NC (1980) Accumulation of solutes in leaves of sorghum and sunflower in response to water deficits. Aust J Plant Physiol 7:193–205

Laimins L, Rhoads DB, Epstein W (1981) Osmotic control of *kdp* operon expression in *Escherichia coli*. Proc Natl Acad Sci USA 78:464–468

Li X, Feng Y, Boersma L (1991) Comparison of osmotic adjustment responses to water and temperature stresses between two plant species. Ann Bot (in press)

Lockhart JA (1965) An analysis of irreversible plant cell elongation. J Theor Biol 8:264–275

Morgan JM (1991) A gene controlling differences in osmoregulation in wheat. Aust J. Plant Physiol 18:249–257

Munns R (1988) Why measure osmotic adjustment? Aust J Plant Physiol 15:717–726.

Ray PM, Green PB, Cleland R (1972) Role of turgor in plant growth. Nature 229:163–164

Yancey PH, Clark ME, Hand SC, Bowlus RD, Somero GN (1982) Living with water stress: Evolution of osmolyte systems. Science 217:1214–1222

ON THE KINEMATICS AND DYNAMICS OF PLANT GROWTH

Wendy Kuhn Silk
Department of Land, Air, and Water Resources
University of California
Davis, CA 95616
U.S.A.

1. Introduction

If swelling is defined as water uptake leading to an increase in volume, then growth of plants must be considered an ecologically important swelling process. The irreversible expansion of plant cells is thought to involve an osmotically driven uptake of water and concomitant yielding of the cellulosic cell wall under turgor pressure. For my contribution to this symposium, I emphasize analysis of plant growth in continuum mechanical terms. I will review some experimental and theoretical work on kinematics of growth of maize roots and the underlying biophysics. Topics include description of the growth field, the water potential field which sustains the observed growth pattern, and the distributions (within the growing region) of osmotically active species and some rheological properties. This article is related to others in the symposium. I am relying on the presentation of L. Boersma for a review of the structure and water relations of the plant cell. Mechanical aspects of plant growth are addressed by J. Passioura who describes some intriguing empirical results showing plant elongation rate as a function of applied pressure. And J. Nakielski generalizes the one-dimensional growth analysis to a two-dimensional tensorial treatment using a natural coordinate system to describe apical growth.

2. Growth Kinematics

2.1. Reference frame for the velocity field in the maize root

The tip of the root of a maize plant is a radially symmetrical, tapering structure made of files of cells which are cemented in a contiguous network (Figure 1). Under adequate nutrition and constant environmental conditions (temperature, water availability, and soil impedance), the root elongates at a constant rate for many hours. Growth of the root is produced by the expansion of its component cells and accompanied by cell division.

Longitudinal expansion may be analyzed in terms of the displacement of cellular particles away from the origin of a one-dimensional coordinate system (Erickson and Sax, 1956). If the root apex is chosen as origin of a moving reference frame, then a point on the organ will be characterized by

NATO ASI Series, Vol. H 64
Mechanics of Swelling
Edited by T. K. Karalis

its distance from the apex, x. In the moving reference frame the velocity may be time invariant, in which case it is denoted by $v(x)$. The velocity field for a maize root (observed by time lapse photography of marked roots growing in vermiculite in plexiglass boxes at 29° C) is shown in Figure 2 (from Silk et al, 1986). Rate of displacement from the apex is shown to increase monotonically to a constant value at the base of the growth zone.

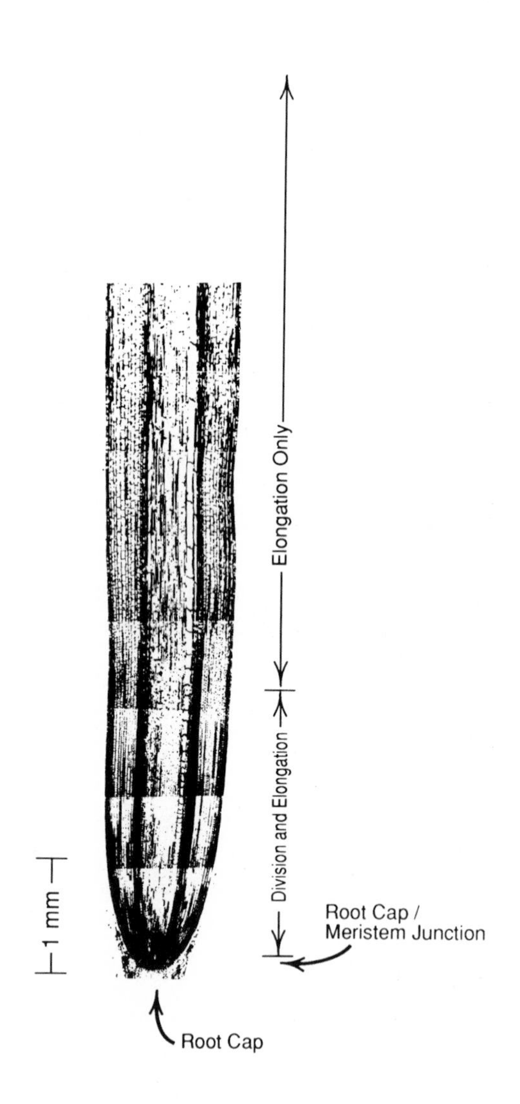

1. Composite micrograph of the growing part of a maize root. Cell division occurs in the meristem, which is found in the apical 2.5 mm of the root proper (above the "root cap/root junction"). Cell elongation occurs in the apical 12 mm of the root as described in the text.

2.2 Local growth rates.

Growth at a spatial location may be described by velocity gradients (Richards and Kavanagh, 1943; Erickson and Sax, 1956), or in more general terms by growth rate tensors (Silk and Erickson, 1979; Hejnowicz, 1984; Hejnowicz et al, 1984; Silk, 1984, 1989). In one dimensional studies the instantaneous rate of elongation of an element relative to its length is called the relative elemental growth (REG) rate, or (in fluid dynamical terms) the longitudinal strain rate. It is given by $\partial v/\partial x$ and is calculated from the velocity field. The spatial distribution of the REG rate for the maize root is shown in Figure 2. The curve rises from zero at the apex to a maximum of 0.46 h^{-1} at the location 4 mm from the tip and then falls to zero at the base of the growth zone as extension ceases (12 mm from the tip).

It is interesting to compare the growth field for the maize root to other swelling phenomena described in this workshop. Note that the maximum REG rate is a large rate of swelling in comparison to the data presented for

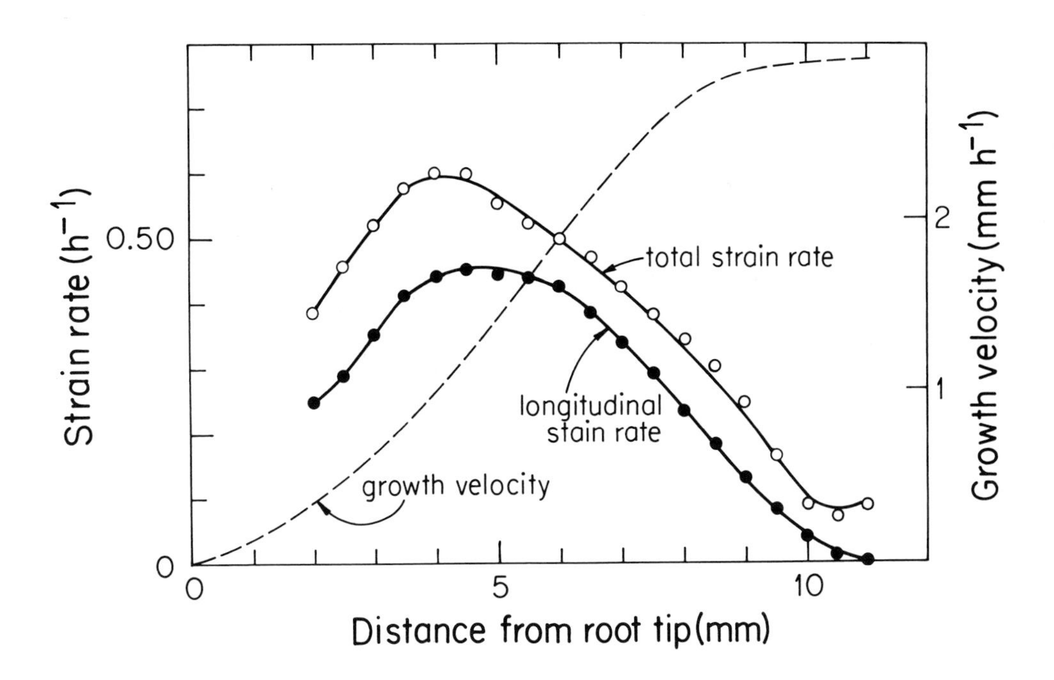

2. Spatial distribution of growth. Growth velocity (rate of displacement from root tip, dashed line) increases with distance to a maximum of 2.9 mm h^{-1} at the base of the growth zone. Derivative of growth velocity with respect to position gives REG rate (relative elemental rate of length increase, ●). Addition of a term for radial and tangential growth rate components yields the relative elemental rate of volume increase(O).

tissues such as cartilage (A. Maroudas; J.P.G. Urban). For root growth the spatial scale is on the order of millimeters, large compared to the double layer interactions of clays and biological membranes (described by S. Marcelja) but small compared to the Darcy scale movement of water in soils (described by J. Philip). Note also that the convention in plant growth analysis is to use strain rates rather than the finite strains which have been used to characterize swelling of fibrous connective tissues (described by Y. Lanir).

2.3. Spatial and material specifications of developmental variables.

Root elongation may be described either in terms of activities at fixed distances from the apex or in terms of activities associated with material elements as they are displaced from the tip during growth (Silk and Erickson, 1978, 1979; Gandar, 1983). The first approach gives a spatial or Eulerian description of growth; for example, $v(x,t)$ describes growth velocity as a function of position along the axis and of time. Alternatively, the material or Lagrangian description associates growth activities with cellular particles. Growth variables are expressed as functions of the initial position X occupied by a particle at some reference time and of current time; e.g. $V(X,t)$ denotes the velocity at time t of a particle which was initially located at point X. If growth is "steady," then the velocity field is time invariant; and the Lagrangian and Eulerian descriptions of growth coincide. Thus during steady growth, as tissue elements are displaced through the growth zone, they exhibit sequentially the velocities which are characteristic of each spatial position. From the Eulerian description of the steady growth field, we can infer that particles accelerate through the growth zone to a constant velocity equal to the rate of elongation of the organ.

2.4. Growth trajectories and space-time interrelationships

The growth trajectory, a plot of particle position versus time, shows the pathline of a particle with initial position X as it is displaced through the growth zone. The Lagrangian description of particle motion is denoted by $x = x(X,t)$ which specifies the position x, at time t, of a particle that was at X when $t = 0$. The growth trajectory can be calculated by numerical integration of particle velocity, following the particle as it is displaced through the velocity field (Gandar, 1983). During steady growth, successively formed cellular particles exhibit similar growth trajectories. Therefore during steady growth the growth trajectory may be used to infer the time course of developmental variables with known spatial distributions (Silk and Erickson, 1979).

The velocity and REG rate fields and the growth trajectory provide the basic phenomenology on which studies of the mechanics of growth are based. The following theoretical and empirical studies probe the physiology underlying the observed spatial and temporal patterns of growth.

3.Growth sustaining water potentials.

3.1. Derivation of an equation relating growth rate and local ψ.

Plant cells expand as the water entering the cell extends the surrounding wall. Thus to a first approximation, volume increase equals local water uptake. Plant physiologists have generally accepted Lockhart's model for cell elongation, in which relative growth rate is related to hydraulic conductivity, and, in a second governing equation, to turgor pressure and two wall properties, extensibility and yield threshold (Lockhart, 1965). In generalizing the water potential equation for three-dimensional tissue continua, one must consider the growth velocity, $\underset{\sim}{u}$ of the moving tissue element as well as the flux $\underset{\sim}{j}$ (volumetric flow rate per unit cross-sectional area) of the water movement, for only water which is moving faster than the growing element will cross the element boundary to enlarge the cell (8). In symbols,

$$\int_S (\underset{\sim}{j}-\underset{\sim}{u}) \cdot \underset{\sim}{n} dS = \int_S -(\underset{\approx}{C} \cdot \nabla\psi) \cdot \underset{\sim}{n} dS \tag{1}$$

Flux across surface = conductivity · driving force

where $\underset{\sim}{n}$ represents the unit normal to the surface, S, is water potential, and $\underset{\approx}{C}$ is the hydraulic conductivity tensor. Then by the divergence theorem

$$\int_V \nabla \cdot (\underset{\sim}{j}-\underset{\sim}{u})\, dV = \int_V \nabla \cdot (-\underset{\approx}{C} \cdot \nabla\psi)\, dV \tag{2}$$

Since water is largely incompressible

$$\nabla \cdot \underset{\sim}{j} \approx 0$$

And we find that the fundamental relationship between growth at a point and water potential at a point in an expanding continuum is given by

$$L = \nabla \cdot (\underset{\approx}{C} \cdot \nabla\psi) \tag{3}$$

where $L = (\nabla \cdot \underset{\sim}{u})$ is recognized as the REG rate. The implication is that if the spatial distributions of L and K are known, the growth sustaining distributions of ψ can, in theory, be calculated.

3.2. Solutions of the water potential equation for the maize root.

In practice, equation 3 is often difficult to solve. However, in the corn root certain simplifying assumptions can be made; and empirical studies have provided data on both REG rate and radial conductivity. One can assume that the tissue is cylindrical, with radius r, and growing only in the direction of its long axis. The distribution of ψ is axially symmetric. The growth pattern is steady, so that the equation may be treated as a time independent problem with coordinate origin at the root tip. It is recognized that the frame of reference is moving at constant velocity, and that one is solving for spatial values of ψ. Conductivities in the radial (K_r) and longitudinal (K_z) directions are independent so that radial flow is not modified by longitudinal flow. If these assumptions are made then equation 3 becomes

$$K_z \frac{\partial^2 \Psi}{\partial z^2} + \frac{K_r}{r} \frac{\partial}{\partial r}\left(r \frac{\partial \Psi}{\partial r}\right) + \frac{\partial K_z}{\partial z}\frac{\partial \Psi}{\partial z} + \frac{\partial K_r}{\partial r}\frac{\partial \Psi}{\partial r} = L(z) \qquad (4)$$

with $\psi = 0$. The boundary condition is to assume that the root is not transpiring and is growing in pure water or saturated air, i.e. $\psi = 0$ in the nongrowing region and outside the root.

Solutions of equation 4 displayed in Figure 3 show the spatial pattern of growth sustaining water potential. The magnitude of ψ increases with distance from the bounding water source and also parallels the strain rate to produce egg-shaped water potential shells. Individual tissue elements are displaced through the standing pattern and experience a decrease and then an increase in ψ as they approach the nongrowing region. The solutions imply a small water potential differential (~ 0.01 MPa) sustains the maximum REG rate.

The growth-sustaining water potentials inferred from solution of equation 4 and displayed in Figure 3 are derived from thermodynamic considerations. Since growth (swelling) results from local water influx, then the chemical potential energy of water in the tissue must decrease from source to the expanding region (if, as is usually assumed, water transport is passive); or else water influx must be coupled to an energy yielding process (if water transport is active). Insight into the physiology of the swelling process requires measurements of the components of the water potential.

4.Spatial distributions of solutes and solute deposition rates in the growth zone of the maize root.

In plant cells the components of water potential are the osmotic and pressure potentials. Thus understanding the physiology of growth in plants involves determination of the spatial distribution of solute concentration within the growth zone. Using freezing point depression to measure solute potential in frozen samples of root tissue sectioned into 1-mm segments, my colleagues and I found that the solute potential is rather uniform throught the growth zone of well-watered maize roots (Silk et al, 1986 Figure 4).

To quantify the deposition rates of water and of solutes, we made calculations based on the continuity equation

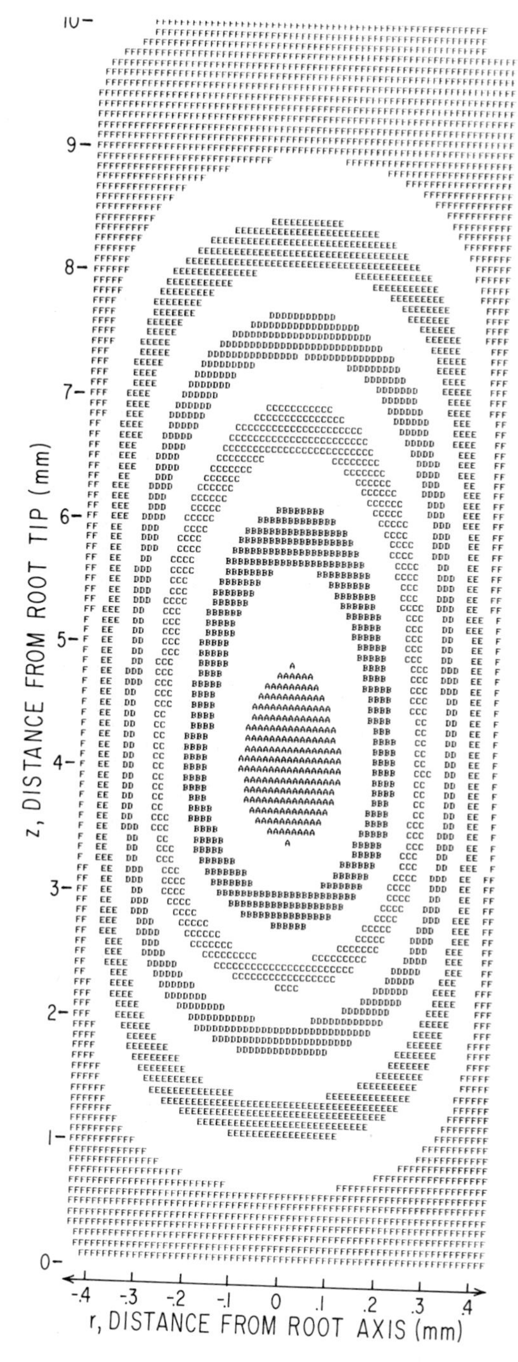

3. Growth sustaining water potentials in the maize root. Contour map shows the distribution of LΨ in a longisection of the growth zone of the maize root. The root is presumed to be growing in pure water (Ψ =0) without transpiration. Region "A" in the centre has $-L\Psi > 5 \times 10^{-8}$ cm^2s^{-1}; and water potentials become progressively less negative with distance from the root centre and with decreasing growth rate. (Redrawn from Silk and Wagner, 1980.)

$$D = \partial S/\partial t + \partial(Sv_z)/\partial z \qquad (5)$$

where D represents the local deposition rate, S is the local density of solutes or water, t is time, z is distance from the root tip, and v_z is local rate of displacement from the root apex.

The calculations based on equation 5 confirm that the deposition rates of solutes and of water parallel the REG rate (Figure 5). Thus in well watered roots, deposition of water and solutes is synchronized to maintain a uniform osmotic potential throughout the length of the growth zone.

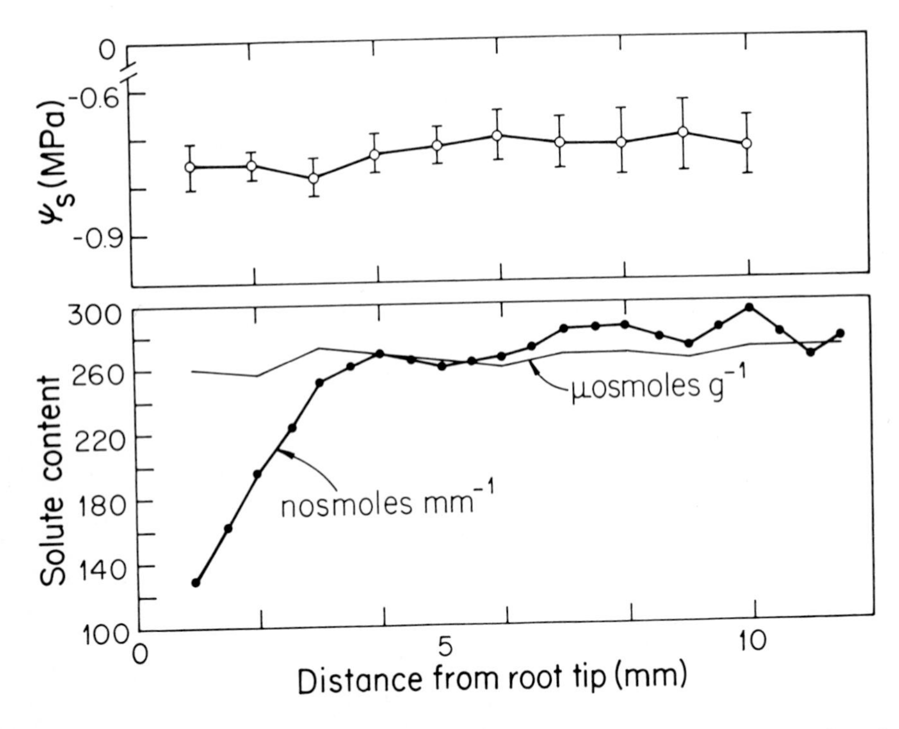

4. Spatial distribution of solutes. (Top) Spatial distribution of solute potential. (Bottom) solute density calculated from solute potential and expressed per unit root length (—●—) or per g fresh weight (—).

The implication of the uniform solute potential is that gradients of osmotic potential do not determine the pattern of the REG rate in the seedling root of maize. In other tissues the solute concentration may increase after growth has slowed (e.g. Sharp et al, 1990) or may parallel the distribution of the REG rate (Cosgrove, 1985). In general there does not seem to be a causal relationship between osmotic potential and REG rate. To find the physiological controls of the growth rate we should look for gradients in

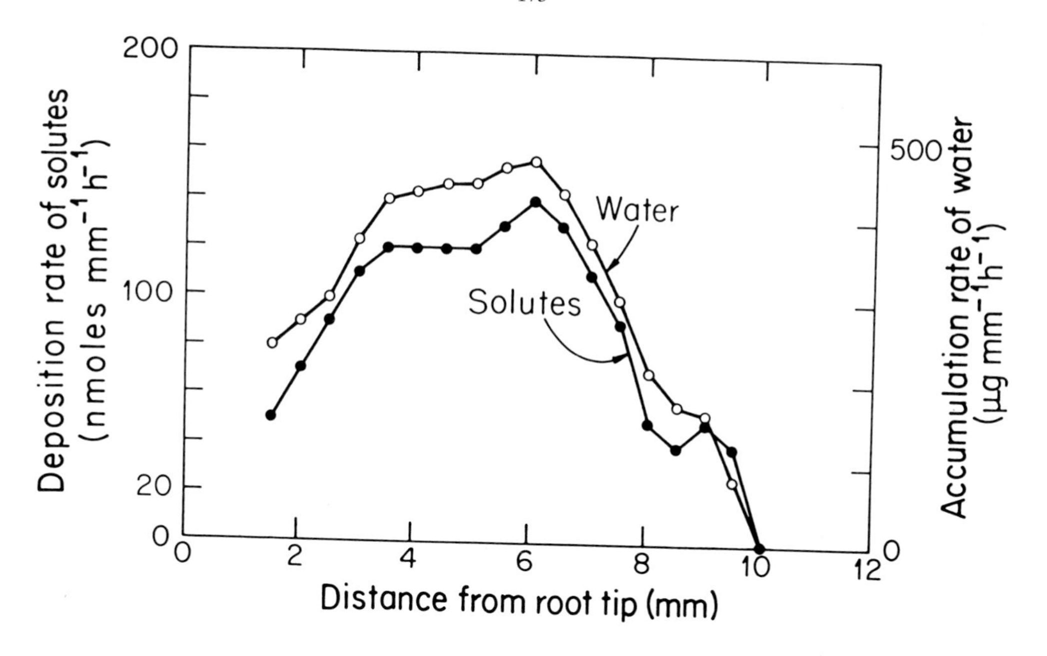

5. Spatial distributions of solute deposition rate (●) and water accumulation rate (O).

pressure potential and the related problem of cell wall rheology.

5.Spatial distribution of Young's modulus.

Growth rate for a particular turgor pressure depends on the extensibility of the cell wall, and the intracellular hydrostatic pressure is itself regulated by rheological properties of the cell wall. Thus the literature in plant physiology includes extensive studies of the mechanical properties of the cell wall in vivo and in vitro (reviewed in Taiz, 1984; Cosgrove 1986). As a step toward resolving the spatial distribution of rheological properties within the growth zone, J. Beusmans and I evaluated the local curvatures which develop in response to an imposed bending moment and used the engineering theory of bending to infer Young's modulus (Silk and Beusmans, 1988; Beusmans and Silk, 1988).

According to beam theory, a rod subjected to bending moment M assumes a curved shape. Local curvature, K, is proportional to bending moment and inversely proportional to flexural rigidity; flexural rigidity is the product of a rheological property, Young's modulus, E and a geometrical property, I, the moment of inertia:

$$K = M/(E)(I)$$

This equation inspired our experimental plan, which was to bend gently the growing part of a root, observe the force exerted, measure the spatial distribution of curvature, and compute the distribution of Young's modulus. Our experimental device is diagrammed in Figures. 6 and 7. Curvature and moment of inertia were computed from digitized photographs of the bent roots.

The force measurements confirmed that the bent root experiences stress relaxation, and that there are reversible and irreversible components of bending strain. Young's modulus is approximately 19 MPa at 3 mm, declines to approximately 5 MPa at 6 mm, then remains low until 11 or 12 mm where it rise abruptly. For comparison with the growth rate pattern, it is useful to plot the distribution of compliance, the ratio of strain to stress. Evaluated as the reciprocal of the bending modulus, compliance can be considered as a measure of extensibility. The compliance curve does not parallel the growth rate distribution (Figure 8). The discrepancy between the fields of REG rate and compliance occurs whether elastic (reversible) or plastic (irreversible) components of bending strain are used for the comparison (not shown).

For the plant physiologist, the discrepancy between the distributions of compliance and REG rate is puzzling. Perhaps bending changes wall rheology, or a time dependence of compliance is important to the mechanics of the growth process. Another puzzle is that recently published measurements of intracellular turgor pressure appear to show that the pressure potential decreases slightly with distance from the center of the growth zone (Spollen and Sharp, 1991; Pritchard et al, 1991). The pressure results together with the solute potential data shown in Figure 3, imply that growth related water flux is not in the direction dictated by the water potential gradient.

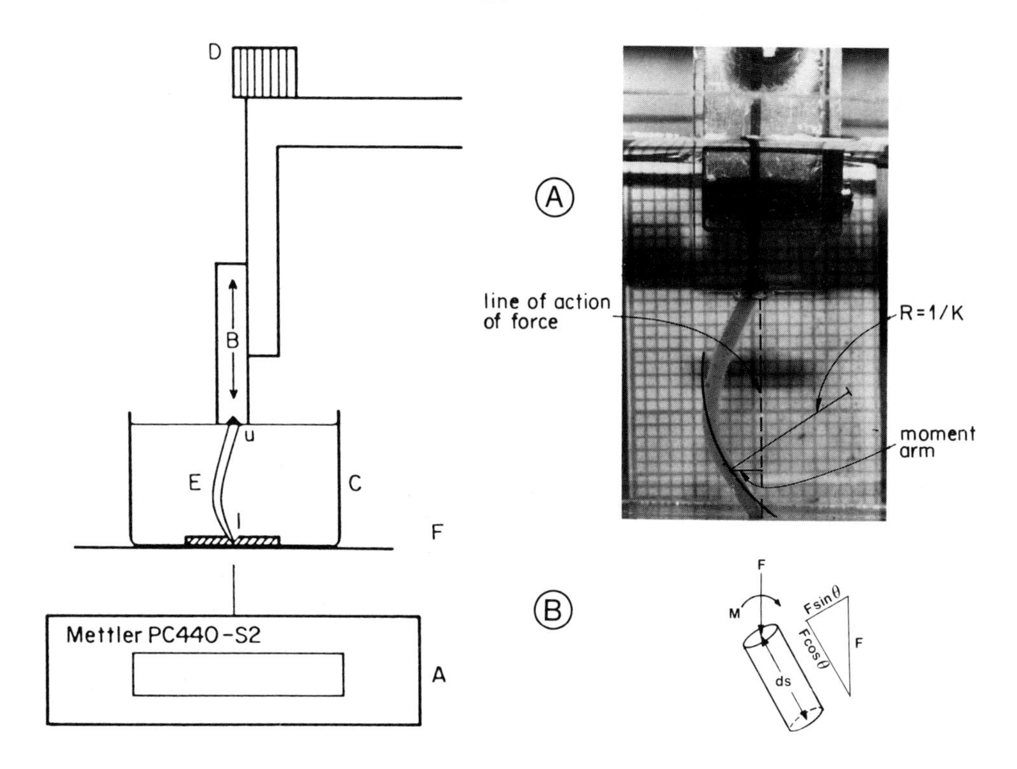

6. Diagram of the device to bend roots and observe the force exerted on an electron balance. A. Mettler PC440-S2 weighing machine. B. Bar which can be moved vertically by turning knob. C. Plexiglass container filled with mannitol solution. D. Knob to move bar b vertically. E. Root tip (20 mm long) inserted into holes O and U. F. Load pan of balance (From Silk and Beusmans, 1988).

7. Diagrams of some of the variables in the bending equation.A. Photograph of a root bent with the apparatus of Figure 1. Dashed line indicates line of action of force. The radius *R* of the circle which just fits the curve at *s=5* mm is the reciprocal of the curvature, *K*, at *s*. The line *d* between s and the line of action of the force is the moment arm. B. A root element in the' bending apparatus is subjected to bending moment, *M*, and a vertically acting force F which can be resolved into axial and shearing forces (From Silk and Beusmans, 1988).

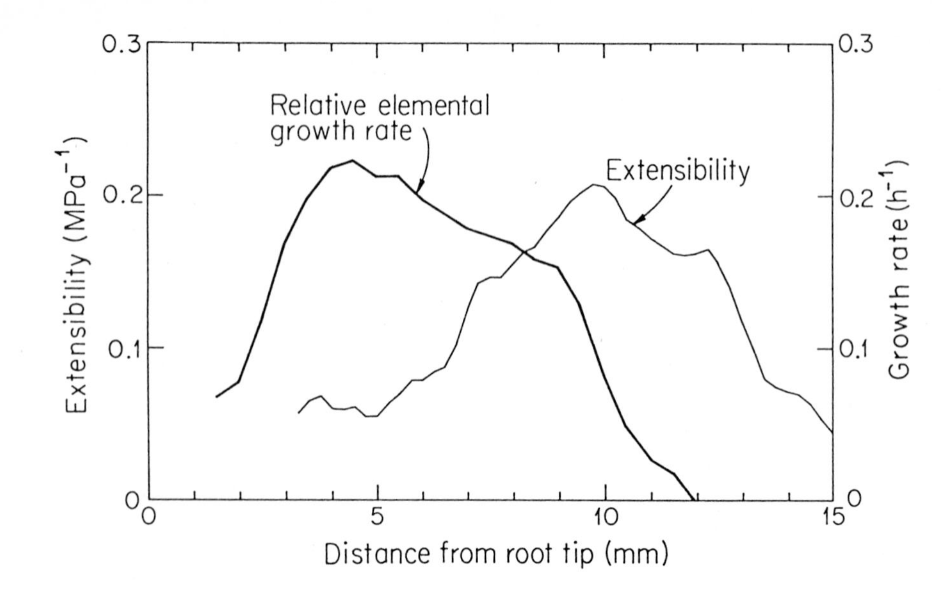

8. Comparison of compliance (light line, calculated as the reciprocal of bending modulus) and local growth rate (heavy line, denoting the REG rate) (From Beusmans and Silk, 1988.)

7.Conclusion.

We are just beginning to see physiology of growth analyzed on the spatial scale pertinent to the REG rate distribution. The current theory cannot explain the observed distribution of growth in thermodynamic or mechanical terms. Much work remains to understand the dynamics of growth.

REFERENCES

Beusmans JM, Silk WK (1988) Mechanical peoperties within the growth zone of corn roots investigated by bending experiments. II. Distributions of modulus and compliance in bending. Am J Bot 75:996-1002.

Cosgrove DJ (1985) Cell wall yield properties. Evaluation by in vivo stress relaxation. Plant Physiol 78:347-56.

Cosgrove DJ (1986) Biophysical control of plant cell growth. Ann Rev Plant Physiol 37:377-405.

Erickson RO, Sax KB (1956) Elemental growth rate of the primary root of Zea mays. Proc Am Philos Soc 100:487-98.

Gandar PW (1983) Growth in root apices. I. The kinematic description of growth. Bot Gaz 144:1-10.

Hejnowicz Z (1984) Trajectories of principal directions of growth natural coordinate system in growing plant organ. Acta Soc Bot Polon 53:29-42.

Hejnowicz Z, Nakielski J, Hejnowicz K (1984) Modeling of spatial variations of growth within apical domes by means of the growth tensor. I. Growth specified on dome axis. Acta Soc Bot Polon 53:17-28.

Lockhart JA (1965) An analysis of irreversible plant cell elongation. J Theor Bio 8:264-75.

Pritchard J, Jones RGW, and Tomos AD (1991) Turgor, growth and rheological gradients of wheat roots following osmotic stress. Exp Bot 42:1043-1049.

Richards OW, Kavanagh AJ (1943) The analysis of the relative growth gradients and changing form of growing organisms: Illustrated by the tobacco leaf. Am Nat 77:385-99.

Sharp RE, Hsiao TC, Silk WK (1990) Growth of the maize primary root at low water potentials. II. Role of growth and deposition of hexose and potasssium in osmotic adjustment Plant Physiol 93: 1337-1346.

Silk WK (1989) Growth Rate Patterns which Maintain a Helical Tissue Tube J Theor Biol 138:311-327.

Silk WK, Beusmans JM (1988) Mechanical properties within the growth zone of corn roots investigated by bending experiments. I. Preliminary observations. Am J Bot 75:990-995.

Silk WK, Erickson RO (1978) Kinematics of hypocotyl curvature. Am J Bot 65:310-319.

Silk WK, Erickson RO (1979) Kinematics of plant growth. J Theor Bio 76:481-501.

Silk WK, Wagner KK (1980) Growth sustaining water potential distributions in the primary corn root. Plant Physiol 66:859-63.

Silk WK, Hsiao TC, Diedenhofen U, Matson, C (1986) Spatial distributions of potassium, solutes, and their deposition rates in the growth zone of the primary corn root. Plant Physiol 82:853-58.

Spollen WG, Sharp RE (1991) Spatial distribution of turgor and root growth at low water potentials. Plant Physiol. 96:438-443.

Taiz L (1984) Plant cell expansion: Regulation of cell wall mechanical properties. Annu Rev Plant Physiol 35:585-657.

REGENERATION IN THE ROOT APEX; MODELLING STUDY BY MEANS OF THE GROWTH TENSOR

Jercy Nakielski

Department of Biophysics and Cell Biology
Silesian University
Jagiellonska 28
40–032 Katowice
Poland

INTRODUCTION

Plant organisms, unlike a great majority of those in animals, grow symplastically. They have meristems as the zones devoted to growth". Cells in the meristem are capable of elaborating new organs by an organized proliferation, and their proliferating activities supply the cellular material necessary for growth. Symplastic growth (Erickson 1986) means a uniform swelling by which neighbouring cells move "in an organized way" into new positions. Maintaining their mutual contact in time, they do not slide or glide over one another. Such growth is considered continuous regardless of the cellular structure of the tissue. It can be well described by tensor fields (Silk and Erickson 1979, Hejnowicz and Romberger 1984, Silk 1984).

The root apex is a meristematic, dome–like end of the root. It has at its tip a group of cells called a cap which are more or less visible in longisections depending upon the "closed" or "open" type of the root construction. In roots of the closed type, including that of the radish which is modeled in this paper, the cap is distinguished from the root proper by a distinct cell wall that forms a common boundary, whereas, it is not so in roots of the open type (Clowes 1961, Barlow 1975). Both the cap and the root proper may be regarded as being composed of columns of cells but they differ from each other in the growth organization. Initial cells of the cap columella grow downwards in the direction parallel to the axis of the root, while the orientation of cell lineages of the root is such that the hemispheral zone of mitotically inactive cells, at the pole of the root proper (except for

NATO ASI Series, Vol. H 64
Mechanics of Swelling
Edited by T. K. Karalis

roots with an apical cell and merophytes), is formed. This zone is called the quiescent centre (*QC*) (Clowes 1961). The cells in *QC* proliferate much more slowly than any of the surrounding cells of the meristem, so assuming that middle dimensions of cells are approximately constant there is a minimum of growth rate. It is postulated that such a region must exist to perpetuate the cellular pattern in the apex; therefore the cap imposes quiescence on cells at the pole of the root (Clowes 1961, Barlow 1984).

What is the growth inside the root apex? The growth field in the apex with the *QC* is shown in Fig.1. Note that there is no growth in the region referring to the *QC* even though it is relatively large around this zone. The map of the growth rate in Fig.1 was done using the growth tensor method (Hejnowicz and Romberger 1984). The growth tensor (*GT*) for root apices, introduced by Hejnowicz (1989), requires knowledge of the vector field of the displacement velocity V of material points in a growing organ. The *GT* is a covariant derivative of V (Hejnowicz and Romberger 1984). The relative rate of linear growth ($RERG_1$) in the direction es at a given point is

$$RERG_{1(s)} = \nabla(\mathbf{V} . \mathbf{e}_s) . \mathbf{e}_s$$

where ∇ denotes a gradient. The vector field V from which the tensor and the growth field in Fig.1 is obtained was postulated by Hejnowicz and Karczewski (preparation). It will be considered further.

An inherent characteristic of the growth tensor is that three mutually orthogonal directions are distinguished in each point of the *GT* field. They are indicated by the eigenvectors of a characteristic equation for a symmetric part of the *GT* field. They are indicated by eigenvectors of a characteristic equation for a symmetric part of the *GT* and called principal directions of growth, *PDG* (Hejnowicz and Romberger 1984). In these directions the $RERG_1$ attains extreme values: maximal, minimal and one of a saddle type. If the tensor field is steady, which is approximately true in the case of the root apex, the *PDGs* obtained for many points arrange into lines giving the pattern of growth trajectories (trajectories of *PGDs*, Hejnowicz 1984). This pattern is formed at the organ level. In the case of apical meristems it can be well approximated by a curvilinear orthogonal coordinate system (u,w,ϕ) called a natural coordinate system (Hejnowicz 1984). The one for the root in Fig.1, seen in the background of the figure, was obtained assuming (for ϕ=const.):

$$x=\frac{a}{\pi}\ln(\cosh^2 u-\sin^2 w) \qquad y=\frac{2a}{\pi}\,arctg(tghu.\,tgw)$$

where a is the scale constant. The V field usually defined by its tensorial components in (u,w,ϕ) is subordinated to the growth tensor in a way that the *GT* operating on V gives the required pattern of *PDG* lines.

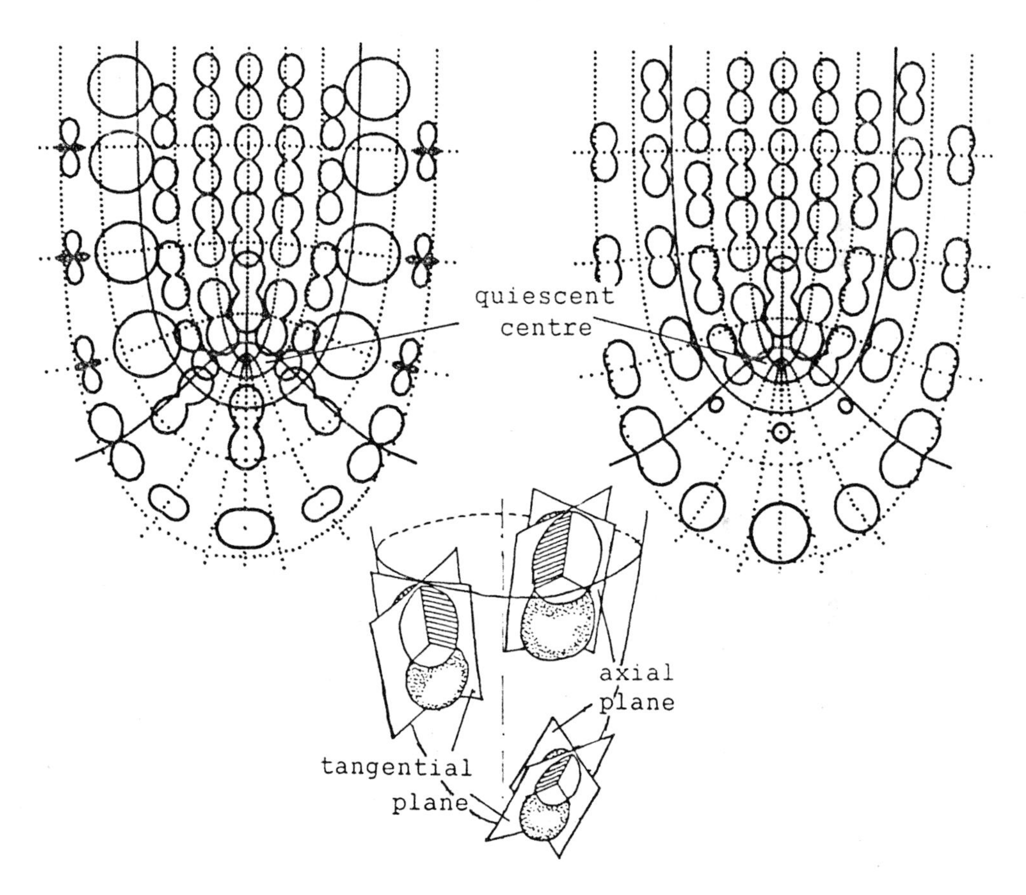

Figure 1. Spatial and directional variations of the relative elemental rate of growth in length, $RERG_1$, in the root apex in the zone of the quiescent centre. $RERG_1$ is a tensorial quantity and its graphical image at a given point is the 3–dimensional closed surface around the point. Here, values of $RERG_1$ around a number of points lying in the axial plane are displayed in the form of 2–dimensional plots showing the growth rates in the axial plane (a) and tangential planes (b); the orientation of plots and both types of planes in 3–dimension is shown in diagram in the center. In the figure (b) each plot of $RERG_1$ has changed orientation as a result of the rotation of the tangential plane by 90° around the symmetry axis. The pattern of lines of principal growth directions in the axial plane is seen in the background of each figure. It is represented by curvilinear orthogonal coordinate system defined in the text (Computer drawings).

What does the pattern of *PDG* lines mean? Firstly, the growth trajectories are seen in the cell wall network through the organ unless the growth is isotropic (Hejnowicz, 1989). There are two types of mutually orthogonal lines known as pericline and anticlines (Fig.2). They are considered smooth regardless of the zig–zag resulting from irregularities of the cell walls. Since pericline and anticlines maintain their orthogonality during growth we conclude that they represent the growth trajectories, and the same represent the *PDG* lines in the axial plane of the apex. Secondly, the alignment of the line elements in the cell wall network seems to be indicated by *PDGs*, namely, a new partition wall lies usually in a principal plane (defined by a pair of *PDGs*) giving rise to the pericline and anticlines; two mutually orthogonal elements of the cell wall network aligned initially along *PDGs* maintain their orthogonality during growth, whereas, the originally right angle between two similar elements but not lying along *PDGs* changes, it becomes acute or obtuse. The above mentioned properties are responsible for the fact that if the anatomy and the organization of growth are known we can anticipate what pattern of *PDGs* is the most appropriate for a given organ. Removing the root cap or the cap with a small apical part of the root proper can iniate regeneration in the root apex. It depends on the

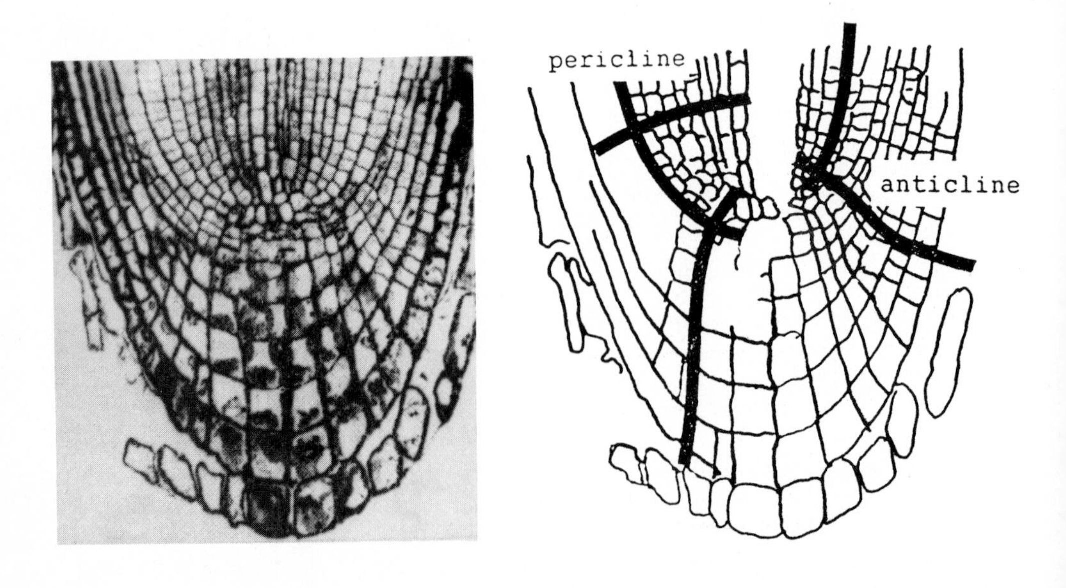

Fig.2. The axial longisection through the root apex of radish (a) and some periclines and anticlines seen in this section b): the section redrawn from Kadej (1970).

progressive reorganization of the cell architecture in the remaining part in such a way that the apex becomes complete again. This is done by cell division and differentiation which requires modification of mitotic activites and reorientation of the planes for division. All this is preceded by biomechanical changes at ther nucleous and cytoplasmic level. At present, the chronology of these changes of the regeneration itself still remains to the explained.

The question arises: can the pattern of *PDGs* control the regeneration at the organ level? I think so and the hypothesis that I make is the following. There is a steady pattern of *PDGs* in the root apex. It organizes the root growth in the sense that it governs the required periclinal and anticlinal alignment of elements in the cell wall network, therefore, they agree with *PDG* trajectories at any time. The cutting off of the cap disturbs this stability, the *PDG* pattern moves back in the direction inside the apex (Fig. 3) Since that moment it controls the growth from a new position. I postulate the displacement of the *PDG* pattern to be proportional to the number of layers of meristematically active cells which were lost by decapitation.

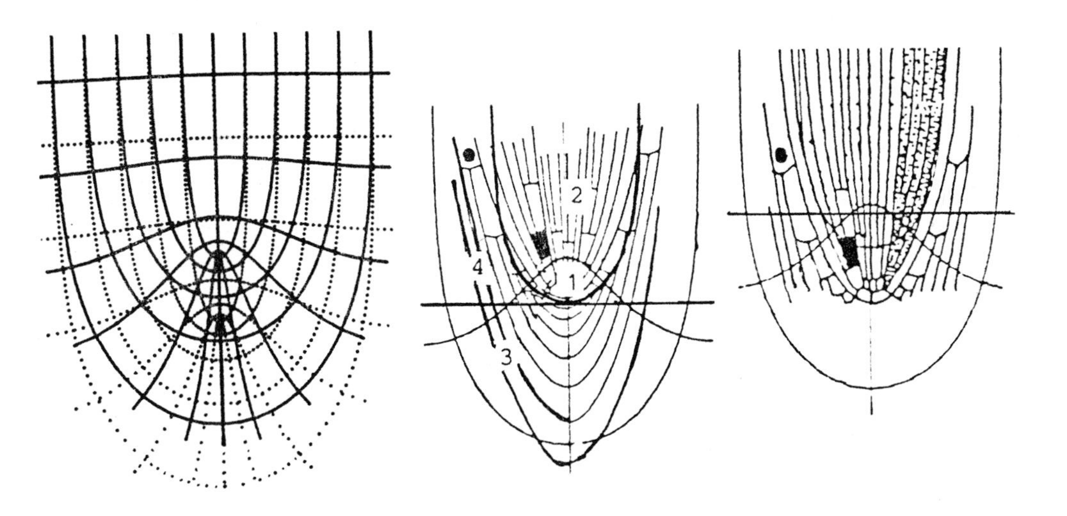

Fig. 3. The idea of moving back of the pattern of PDG trajectories after cutting off the cap: (a) lines of the pattern before (dotted lines) and after decapitation; they were moved upwards along the root axis; (b) (c) position of some lines of the pattern (four lines only) in relation to the cell wall arrangement in the apex before (b) and after (c) the excision; The cell wall arrangement in (b) comes from Kadej (1970); (1)–(4) are the zones inside the apex.

METHOD OF MODELING

For verifying the above hypothesis a computer simulation was worked out. The apex of the radish shown in Fig. 2 was used for modeling. It is assumed that the curvilinear coordinate system and the growth field from Fig. 1 well describe organ growth. The consideration was restricted to the case of cutting off the cap at the level of the root/cap junction. It was assumed that moving back the *PDG* pattern in the direction inside the root proper was about *1/3* of the cap height taking into account that not more than *4–5* layers of cells counting from the pole of the root proper are active meristematically. Positions of points in time were calculated by integration of tensorial components of *V* field that was as following (Hejnowicz and Karczewski, in preparation):

$du/dt=0$	$dw/dt=0$	for $u<u_o$ and $w<w_o$	(1)
$du/dt=c(u-u_o)$	$dw/dt=0$	for $u<u_o$ and $w>w_o$	(2)
$du/dt=0$	$dw/dt=-d\ sin4w$	for $u>u_o$ and $w<w_o$	(3)
du/dt	$dw/dt=-d\ sin4w$	for $u>u_o$ and $w>w_o$	(4)

The zones (1), (2), (3), (4) are denoted in Fig. 3. The *GT* was specified for $c=1$, $d=0.3$, $u_o=0.45$, $w_o=\pi/4$; the coordinate $w=w_o$ represents the root/cap junction, whereas, $u=u_o$ is a distal border for the *QC*. The scale factor for coordinate system was as $a=45$.

RESULTS

Temporal deformation of the cell wall organization typical of the apex of the radish is shown in Fig.4. The cell arrangements change gradually during regeneration. The zone 1 corresponding to the *QC* displaces upwards after decapitation. It lies in the place of the "old" axial cylinder now. Outside the *QC* the complexes change on the dotted areas. The complex *A*, which before decapitation is an element of the cal, grows downwards in the way typical of the columella. It is relatively wide which indicates the occurrence of periclinal divisions in its scope. Let us note that the upper edge of the complex "hooked" by its right end to the zone 1 turns clockwise by about *45°*. The complex *B* gradually becomes straight and thicker. Moreover, below the zone 1 its orientation changes into one typical; of the cell files in the cap.

Let us pay attention to the alignment of line elements, which originally were mutually perpendicular or parallel (Fig. 5) At $t=0$ lines l_1 and l_2 aligned transversally were parallel to each other. They changed both their mutual alignment and the orientation in the figure plane. This is so because neither of the lines lies in *PDGs*. The l_2 moving upwards deform nonuniformly forming discontinuity for $w_o=p_i/4$. This results from different velocities of points in zones *2* and *3*. The slower moving in the zone *3* makes the cells stay longer in the zone which can be favorable for flaking of the cal. The line l_2 moving downwards forms something like a cap. The line l_3 that was perpendicular to both previous ones takes the alignment along w=const. in the part above the zone *1* and

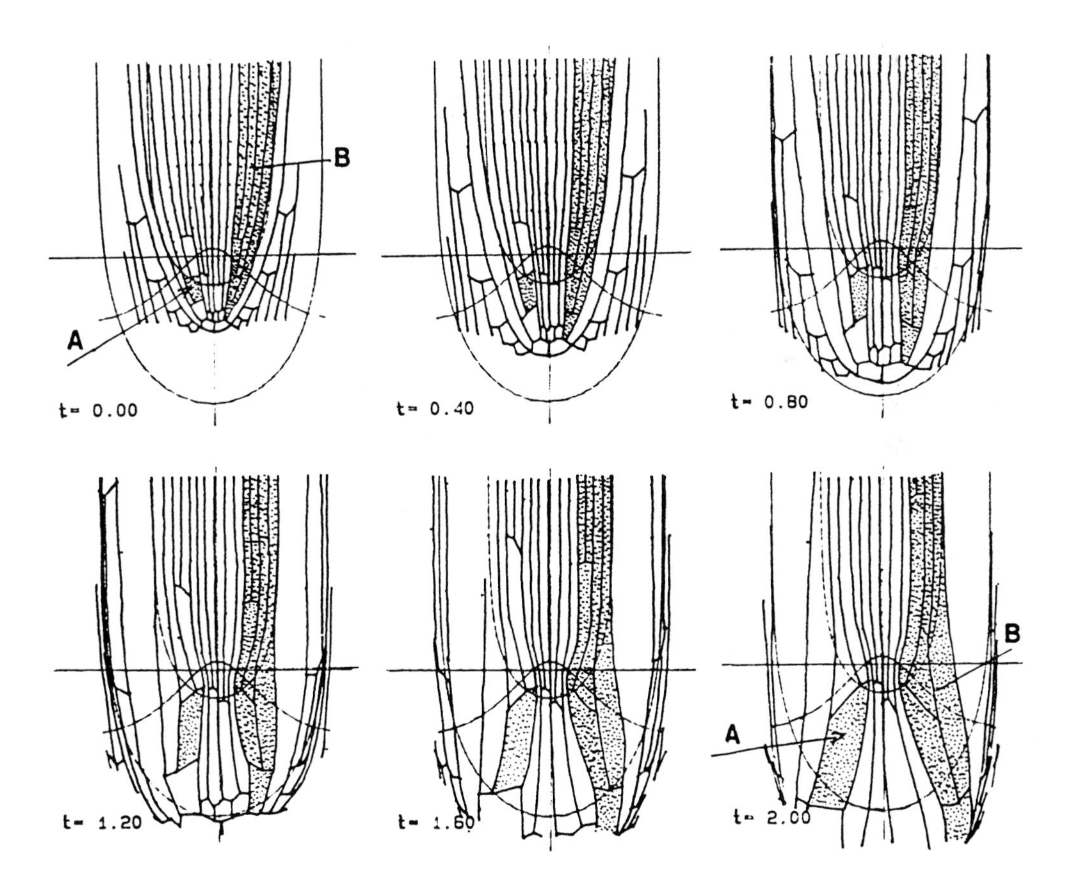

Fig. 4 The computer simulation of changes of the cell wall arrangement in the root apex of radish after cutting off the cap. The growth field from Fig.1 was used to the simulation. Two complexes: A and B are denoted.

along u=const. in the part corresponding to the cap at $t=2.8$. This means that it almost reaches the position indicated by *PDG* trajectories.

DISCUSSION

The organization of growth in the axial longisection of the root apex in radish changes during regeneration. This can be seen in Fig.6 redrawn from Kadej (1970). A substitute pattern is of the open type. A new wall architecture is formed step by step "like

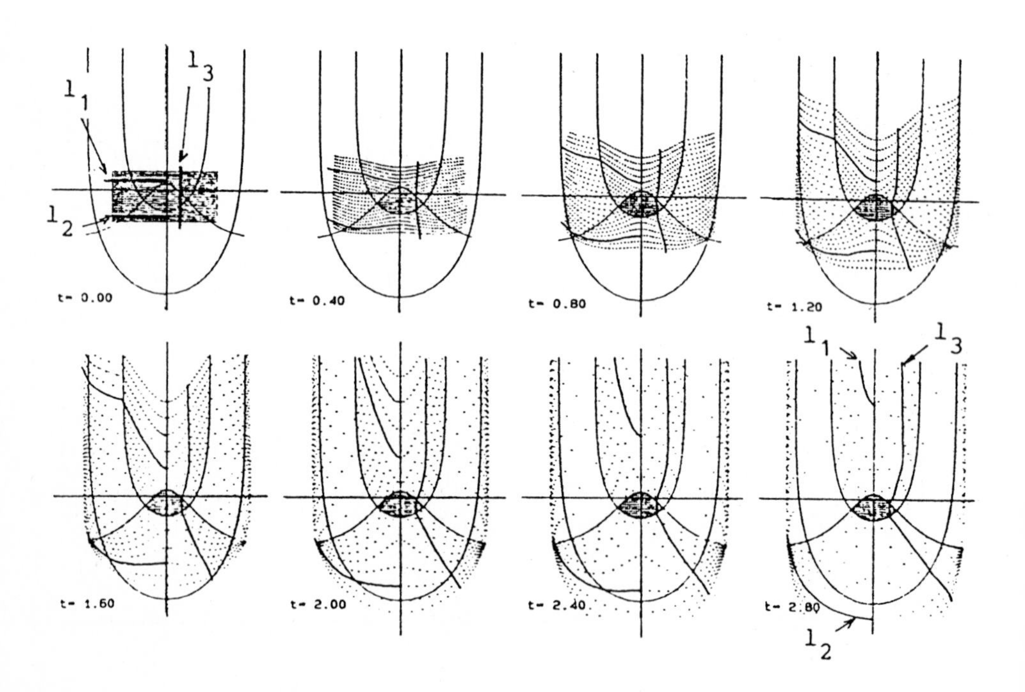

Fig. 5. Deformation of lines which originally were perpendicular or parallel using the same simulation as in Fig.4.

slowly developed film" (Kadej 1970). The reorganization was originated in a new construction centre. It is interesting to note that the centre was located asymmetrically when the excision had not been made perpendicular to the axis of the root.

In closed root meristems the cap boundary is not completely inviolate, for it can be broken by treatments that reduce the general rate of cell division in the meristem, such as X–irradiation (Clowes 1959) or recovery from cold–induced dormancy (Clowes and Steward 1967). The constraint exerted on the quiescent centre by normally actively growing cells surrounding it can cause the cells of the quiescent centre to grow, divide and burst through the cap boundary (Clowes and Wadekar 1989); the old cap of the pattern of principal growth directions governs such modifications.

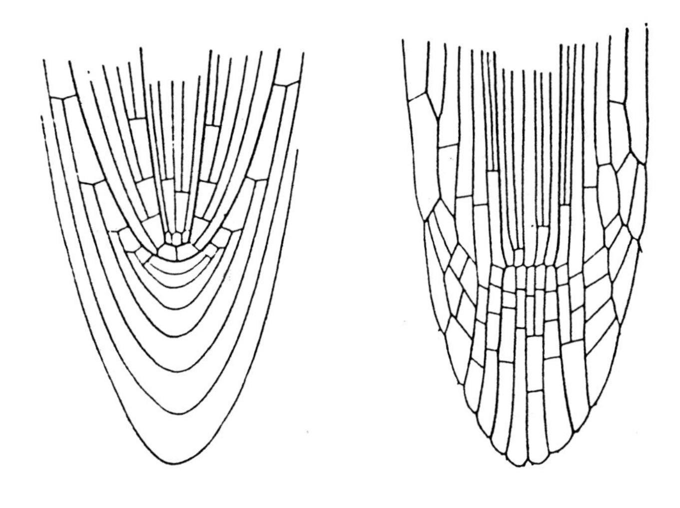

Figure 6. The organization of the growth in the root apex of radish before (a) and after (b) regeneration (redrawn from Kadej 1970).

The difference between open and closed meristems depends on what happens in the cells immediately distal to the stellar pole in the quiescent centre (Clowes and Wadekar 1989). In closed meristems they are quiescent whereas in open meristems they are "only transiently" quiescent. Before the decapitation cells of zone 1 cannot grow because their

$$\frac{du}{dt}=0;\ \frac{dw}{dt}=0$$

However, after the displacement of the PDG pattern, previously inactive cells of zone 1 "moved" to zone 3 where there is relatively large growth. Therefore, there is no condition for maintaining the common boundary between the cap and the root proper, probably until the moment when the new *QC* is fully formed. It should be noted that although in the case of the radish the reversal of the closed structure was not observed, it is generally possible and the substitute pattern seems to be transient (Barlow 1974, Clowes and Wadekar 1989).

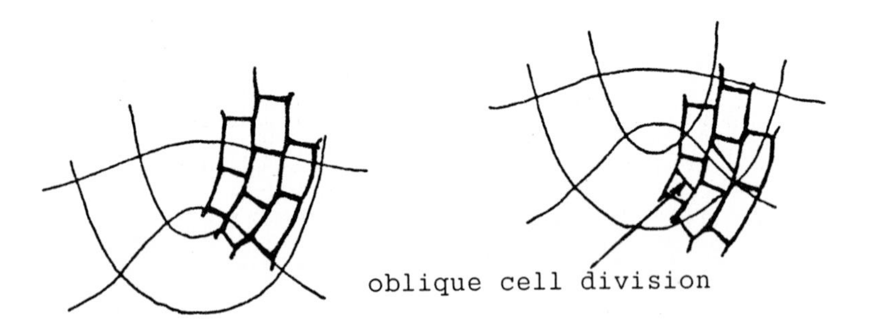

Figure 7. The formation of the oblique division: the same group and the PDG pattern (a) before and (b) after decapitation.

The oblique divisions in cells are observed during regeneration (Kadej 1970, Barlow 1974). Most of them are seen in the stele just above the *QC*. In many points of this region there is a large angle between elements of the "old" cell wall and the trajectories of *PDGs* crossing a given point (Fig.5). The old cell wall cannot change its orientation rapidly if the growth in the location of its new alignment is not large. However, when the cell divides, a new division wall is oriented just "right" i.e., in a particular plane. This means the new division wall must be oblique. It is characteristic that the number of oblique divisions in the region of their occurence is low at first, then increases to a maximum before decreasing again (Barlow 1974).

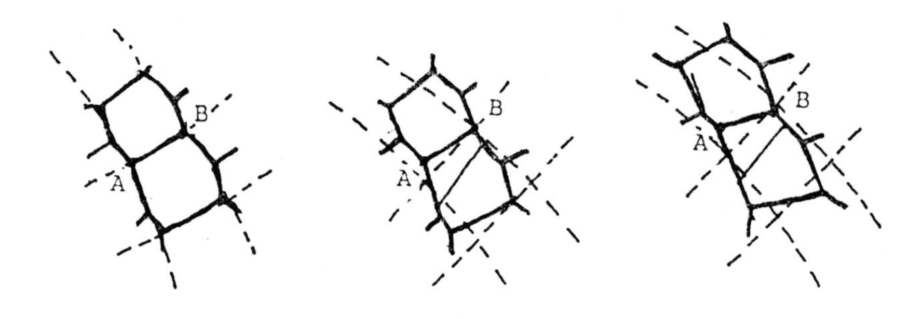

Figure 8. The formation of "sheared" cells–walls of two cells in relation to PDG trajectories: (a) before decapitation, (b) just after decapitation, and (c) some time after decapitation.

It is observed that "sheared" cells together with oblique divisions occur in some regions of regenerating roots. Let us explain their formation by means of Fig.8. In the intact apex the great majority of cell walls agree with *PDG* lines and as seen in Fig.8a points *A* and *B* lying at the same anticlinal trajectory move uniformly, having the same velocity before decapitation. After the *PDG* pattern has been displaced points A and B are not at the same orthogonal trajectory (Fig.8b). Hence, they have different velocities especially when the gradient of *V* is large. This gives the effect of "shearing" shawn in Fig.8c.

The question arises as to how it is possible that the pattern of *PDG* can shift to a position further inside the apex. The idea of a movement of the *PDG* pattern seems to be naive but it does explain some features of reformed root apices. Moreover, a steady 2growth field which "moves" together with an organ can describe the formation of new root apices from a group of cells (Hejnowicz Z and Hejnowicz K 1991). It happens that the *PDG* pattern moves away from the level of the previous base of the apex to avoid the development of a concavity at the base of the newly formed apex. Thus, the principal growth directions and their pattern at the organ level may, in part, control the growth or may take part in such control along with other factors (*e.g., stresses).* It is not yet clear though and requires further study.

REFERENCES

Barlow P. W. (1974) Regeneration of the cap of primary roots of Zea mays. New Phytol 73:937–954

Barlow P. W. (1975) The root cap. In: Torrey J.G., Clarson D.T. (eds). The development and function of roots, Academic Press, London, pp.21–54.

Barlow P. W. (1984) Positional control in root development. In: Barlow P.W., Carr D.J. (eds.). Positional control in plant development, Cambridge University Press, Cambridge, p.502

Barlow P. W., Hines E. R. (1982). Regeneration of the root cap of Zea mays L. and Pisum sativum L.: A study with the scanning electron microscope. Ann Bot. 49:521–529.

Clowes F. A. L. (1959) Reorganization of root apices after irradiation. Ann. Bot. 23: 205–210

Clowes F. A. L. (1961) Apical meristems, Blackwell, Oxford

Clowes F. A. L., Stewart H. E. (1967) Recovery from dormancy in roots. New Phytol. 66: 115–123

Clowes F. A. L, Wadekar R. (1989) Instability in the root meristem of Zea mays L. during growth. New Phytol. 111: 19–24

Erikson R. O. (1986) Symplastic growth and symplasmic transport. Plant Physiol.82:11–53.

Feldman J. L. (1976) The de movo origin of the quiescent center regenerating root apices of Zea mays. Planta 1:207–212.

Hejnowicz Z. (1984) Trajectories of principal growth direction. Natural coordinate system in plant growth. Acta Soc. Bot. Pol. 53:29–42

Hejnowicz Z. (1989) Differential growth resulting in the specification of different types of cellular types of cellular architecture in root meristems. Envir. Exper. Bot. 29:85–93

Hejnowicz Z., Hejnowicz K. (1991) Modeling the formation of root apices. Planta 184:1–7

Hejnowicz Z. and Romberger J. A. (1984) Growth tensor of plant organs. J. Theor. Biol. 110:93–114

Kadej F. (1956) Przebieg regeneracji wierzcholka korzenia Hordeum vulgare. Acta Sco.

Bot Pol 25: 682–712

Kadej F. (1970) Apical meristem regeneration in root of Raphanus sativus. Acta Soc Bot Pol 39:373–381

Nakielski J. (1991 to be published) Distribution of growth rates in different directions in root apical meristems. Acta Soc. Bot. Pol.

Silk W. K. (1984) Quantitative descriptions of development. Ann. Rev. Plant Physiol. 35:479–518

Silk W. K., Erikson R. O. (1979) Kinematic of plant growth. J. Theor Biol. 76:481–501.

DAILY VARIATIONS OF STEM AND BRANCH DIAMETER: SHORT OVERVIEW FROM A DEVELOPED EXAMPLE

Thierry Ameglio and Pierre Cruiziat

Laboratoire de Physiologie Intégrée
de l' Arbre Fruitier Centre INRA
de Clermont-Theix,
Domaine de Crouelle,
63039 Clermont-Ferrand
France.

INTRODUCTION

Reversible and irreversible changes in size of plant tissues have been observed for almost a century. Reversible changes are accounted for by changing degrees of hydration, while irreversible changes, often superposed on the previous ones, result from continuous growth of tissues through cell division and enlargement. During the last decade many attempts have been made to use the variations of stem and fruit to schedule irrigation, especially in orchards where more traditional methods are difficult to apply. The major advantage of this "micromorphometric" approach is that changes in diameter are a very sensitive indicator of the plant response to water supply conditions, more than the others usually employed (like leaf water potential), and can be recorded continuously. However much work should be done to achieve a precise interpretation of these variations of diameter. Supported by an example taken in our current research, this paper gives a rapid overview of some typical results and open questions in this domain.

I. SOME BASIC FEATURES

Water movement through the soil–plant–atmosphere system results from the existence

NATO ASI Series, Vol. H 64
Mechanics of Swelling
Edited by T. K. Karalis

of gradients of water potential which are maintained by transpiration. This loss of transpired water from the leaves, equivalent to evaporation, causes the air–water interfaces in the leaf to retreat into the finally porous spaces between cellulose fibers in the cell wall. By capillary effect tension is created within the conducting system and water is drawn from the soil through the plant roots (viz., Tyree and Sperry 1989). Due to the fact that a plant, from the hydraulic point of view, behaves like a very complex network of capacitances and resistances, absorption from the roots usually lags behind transpiration. On sunny days this lag stays as long as transpiration, which follows the solar radiation, increases.

Internal water deficits develop during this first phase. When transpiration declines with solar radiation, during the afternoon, absorption begins to exceed transpiration and the plant rehydrates. These internal water deficits are usually reduced or eliminated during late afternoon or during the night, provided that normal soil water supply exists (Fig. 1). Corresponding to these hydration changes, diurnal shrinkage and expansion have been reported for many different species and for various plant parts, including leaves, branches, stems, roots and reproductive structures like fruits and cones of conifers. Figure 2 shows two lines of evidence for a storage capacity in stem and leaves. The first is the decline in water content of the external tissues of the trunk which occurs even in irrigated trees. For these trees however, recovery is almost complete at the end of the afternoon. The second line of evidence comes from observations of the relationship between transpiration rate E and the leaf water potential ψ. Any "hysteresis" in this relationship, with lagging behind E such that ψ is lower at given values of E in the afternoon than in the morning, can be interpreted in terms of water storage within leaves.

II. DEVELOPED EXAMPLE: YOUNG WALNUT TREES

The principal of this example, which is part of the undergoing research of the first of the authors is to point out some important features concerning the daily variations of trunk and branch diameter and their interpretation. Experiments have been performed with *2* and *3* year old potted trees. Variations of diameter have been measured with linear displacement transducers and sap flux measurements with the heat balance method (Valancogne and Nasr, 1989). Variations of diameter with time can be divided into three categories (Milne et al, 1983).

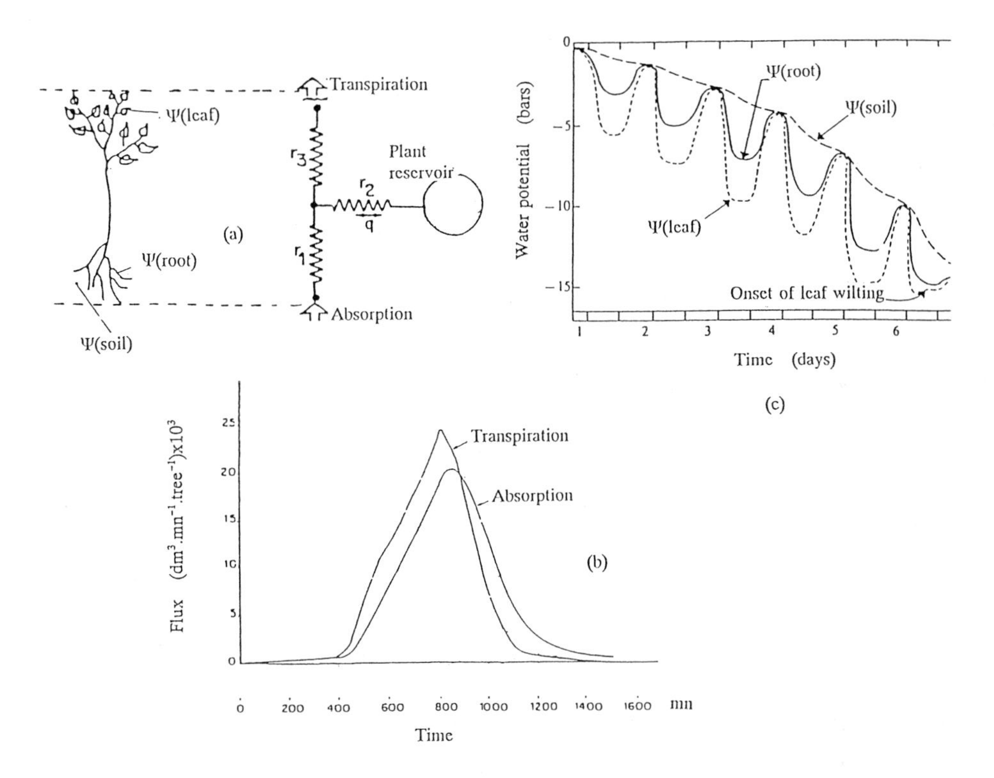

Figure 1. Illustration of some basic features of the "plant water relations": (a) the diagram in the middle illustrates the simplest electric analogue of a plant: an non–branched circuit of two or three resistances with one capacitance. A flow of water is driven from the soil to the atmosphere throughout the plant by water potential gradients. The magnitude of these gradients reflect the hydraulic resistances r_1, r_2, r_3 of the interlinked components. (b) During a normal day in summer, absorption lags transpiration and dehydration occurs leading to shrinkage processes. In afternoon the situation reverses and plant rehydrate (swelling phase– *From Cruiziat et al 1989*). (c) Schematic representation of daily changes in the water potentials in the soil, root and leaf of a plant in an initially wet soil that dries out over a one week period (*From Nobel, 1983*).

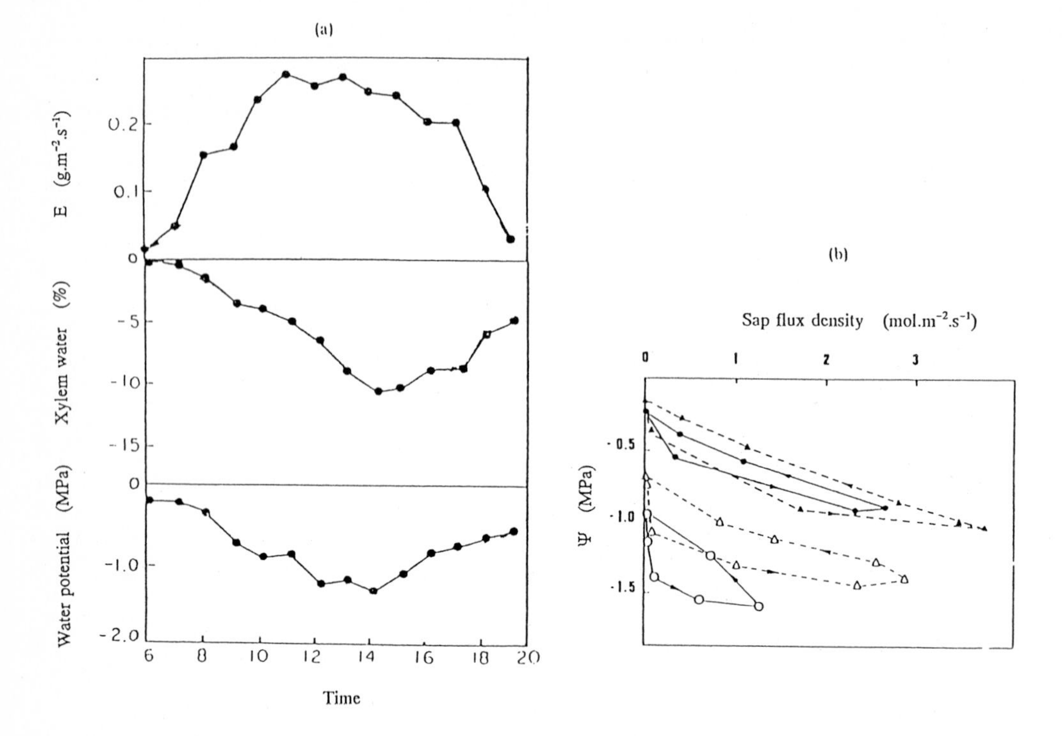

Figure 2. (a) Evidence of diurnal variation of water content (middle graph) in the trunk of an apple tree well irrigated, in july; top: time course of the transpiration; bottom: time course of leaf water potential (*From Brough et al, 1986, simplified*). (b) Examples of diurnal relationships between sap flux and shoot water potential (ψ) in coniferous trees (Granier *et al*, 1989).

- *Trend*: variations occurring over periods longer than a day.
- *Periodic*: diurnal changes, but repetitive patterns with other periods may also occur.
- *Random*: short term variations within the day not included in the other two categories. We will deal essentially with the two first categories.

a) Sap flux and stem diameter of an irrigated tree and unirrigated tree during a ten day period (Fig.3)

* For the irrigated tree, the trend expresses the growth rate of the trunk diameter. This rate is roughly *25* microns per day.
Between two plateau (maximum) values of the diameter, the periodic (daily) variations can be decomposed into three phases:

- *First phase*: it corresponds to a rapid decrease of the trunk diameter, starting soon after sunset. The trunk reservoir loses water during the rapid increase of transpiration.
- *Second phase*: a rapid increase of the trunk and branch diameter corresponding to rehydration. It starts soon after the afternoon decrease of transpiration.
- *Third phase*: a second much slower increase of diameter, which reaches a plateau value around *3* or *5* a.m., follows the second phase. This plateau is higher than the previous plateau value, one day before. This phase is interpreted as a growth phase. There is no clear limit between phases *2* and *3*.

* For the unirrigated tree the trend shows two parts:During the first four days without irrigation, there is no trend! The maximum trunk diameter value does not change. After that period, however, there is a clear decrease of the plateau value. For that tree, no growth occurs, but furthermore, its trunk diameter continuously decreases. The most probable interpretation is that, during this "dry" period without irrigation, no complete water recovery is possible owing to the regular fall of the predawn water potential (this value is the soil water potential at the end of the night, as experienced by the plant).

Concerning the periodic variations, phases *1* and *2* are more pronounced (both irrigated and unirrigated trees have the same diameter), but phase *3* is absent (no trend). During the first four days, recovery seems to be complete despite the decrease of the predawn water potential. After that period, no complete recovery is possible.

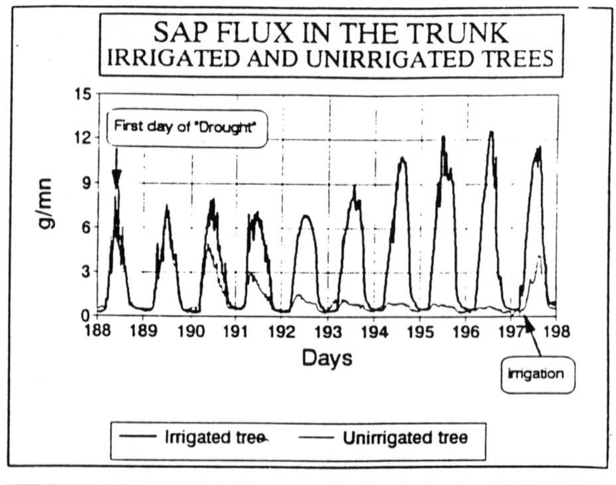

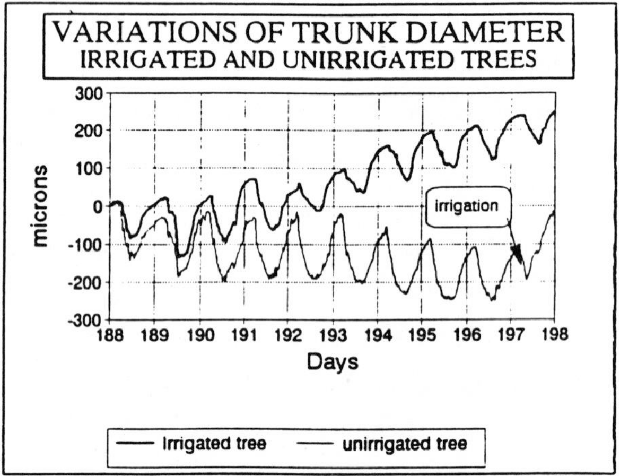

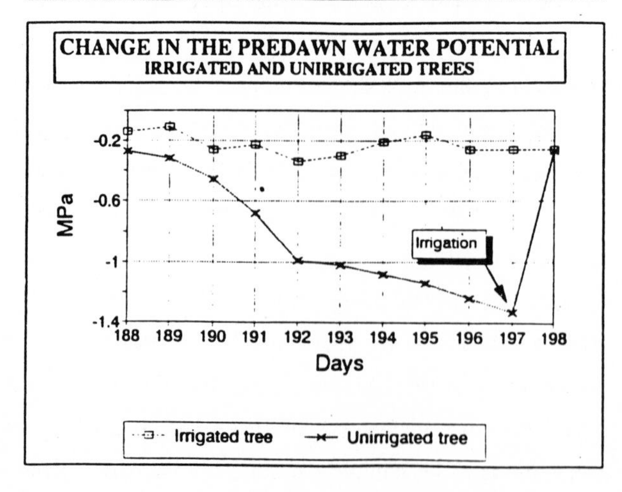

Figure 3. (a) Sap flux in the trunk. (b) Variations of trunk diameter. (c) Change in the predawn water potential irrigated and unirrigated trees

b) Sap fluxes and branch diameter of the irrigated tree during the day 198 (Fig. 4)

The main points here, are quite a good synchronisation between the different fluxes on the one hand and between the different variations of branches and trunk on the other hand.

c) Relationship between sap flux and stem diameter for the irrigated tree during three consecutive days (#190, 191, 192). (Fig. 5)

Several remarks can be drawn from this figure:

* As for leaf water potential and transpiration, the relationship between flux and stem diameter exhibits loops, as expected.
* The same value of sap flux leads to very different variations of trunk diameter (compare, for example days *191* and *192*). We have no clear explanation of this fact.
* During cloudy days, with irregular transpiration and due to the short lag between sap flux and trunk diameter, it is frequent to observe short periods where stem diameter decreases while transpiration diminishes too.

e) Relationship between sap flux and trunk diameter for the unirrigated tree during three consecutive days of water shortage (days 190, 191, 192) (Fig. 6)

The most intriguing point here is that despite a rapid fall of transpiration (viz. Fig.3) the amplitude of variations of the trunk diameter remains the same. For that fact also no clear explanation can be put forward, although the following remarks can be made.
Two possible opposite effects are involved in these stem variations. For high values of the soil water potential, the higher the transpiration the larger the trunk variations. For a given value of transpiration, the lower the soil water potential, the larger the variations of the trunk diameter.

f) Amount of water involved during trunk and branch shrinkage and swelling (Fig. 7)

As it can be deduced from Fig. 7, the quantities of water involved in the growth of the trunk of the irrigated tree are small in comparison with the daily transpiration. For the unirrigated tree the trunk contribution to the transpiration represents also some percents of that transpiration and of the quantity of water stored within the trunk. Besides it is interesting to

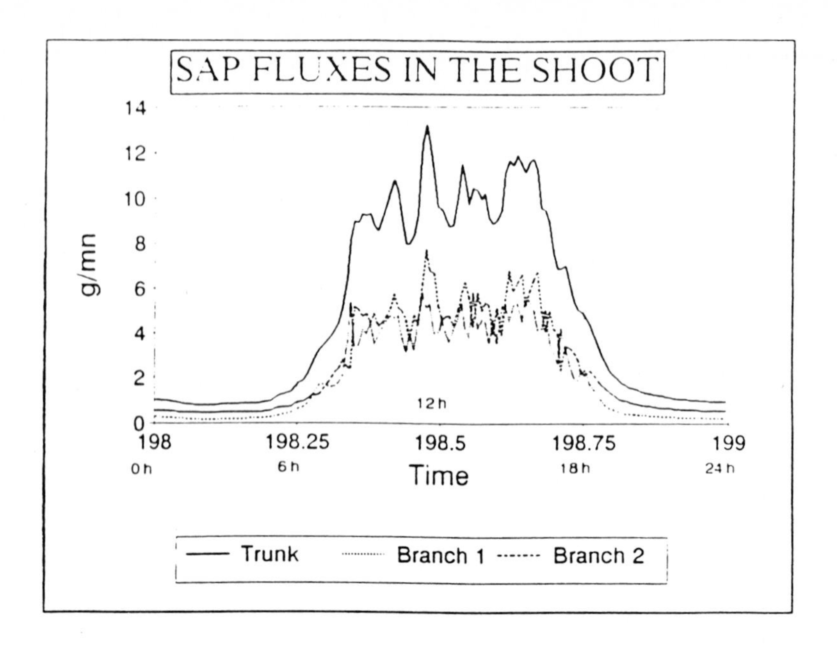

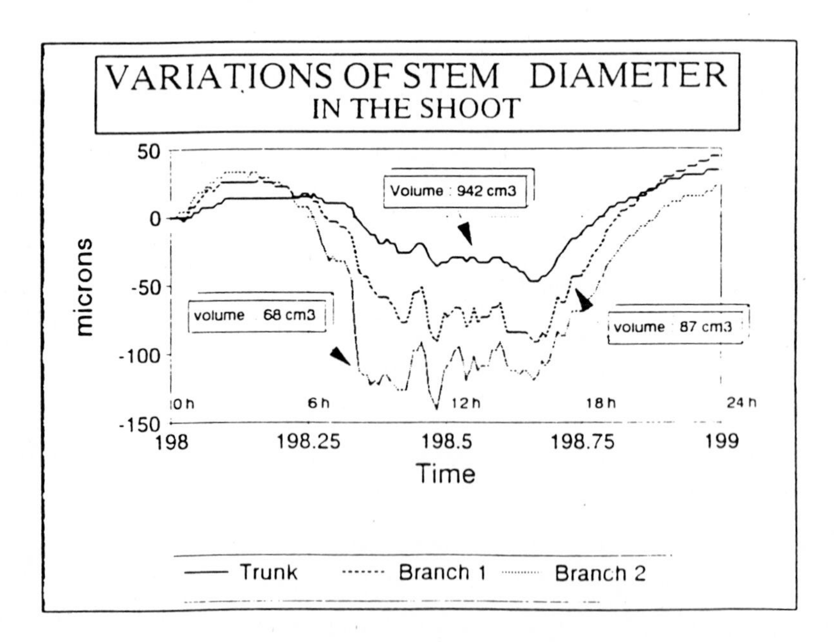

Figure 4. (a) Sap fluxes in the shoot. (b) Variations of stem diameter in the shoot

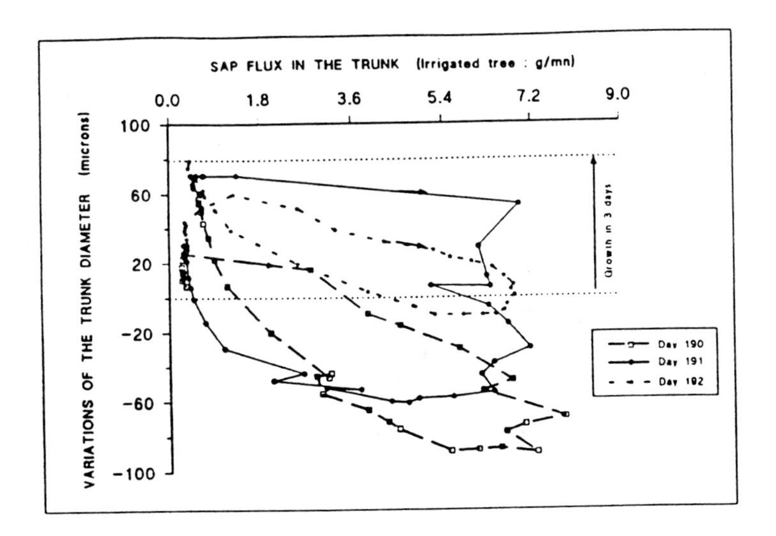

Figure 5. Sap flux in the trunk

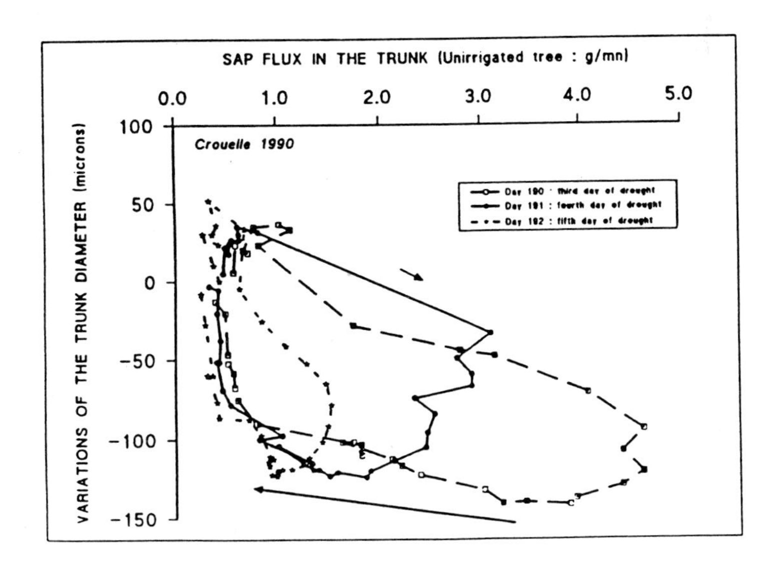

Figure 6. Sap flux in the trunk

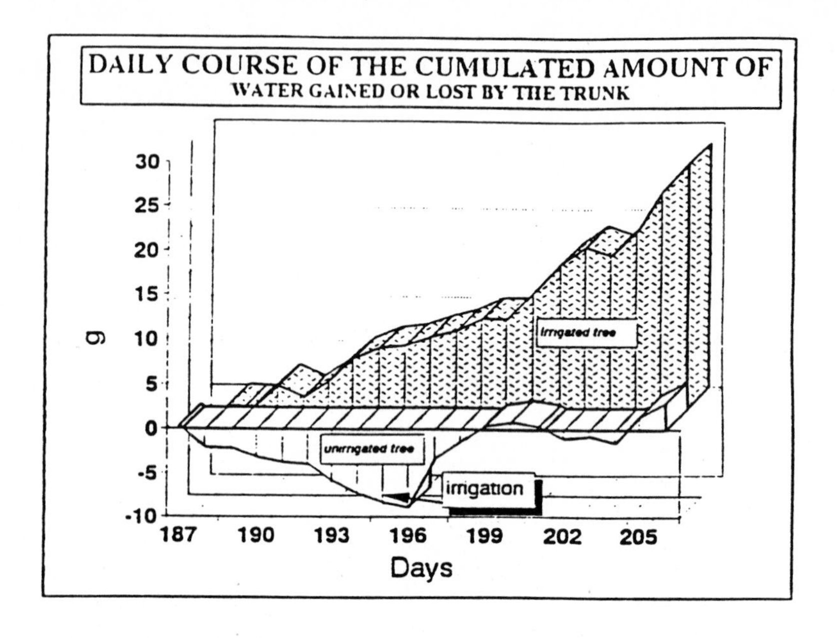

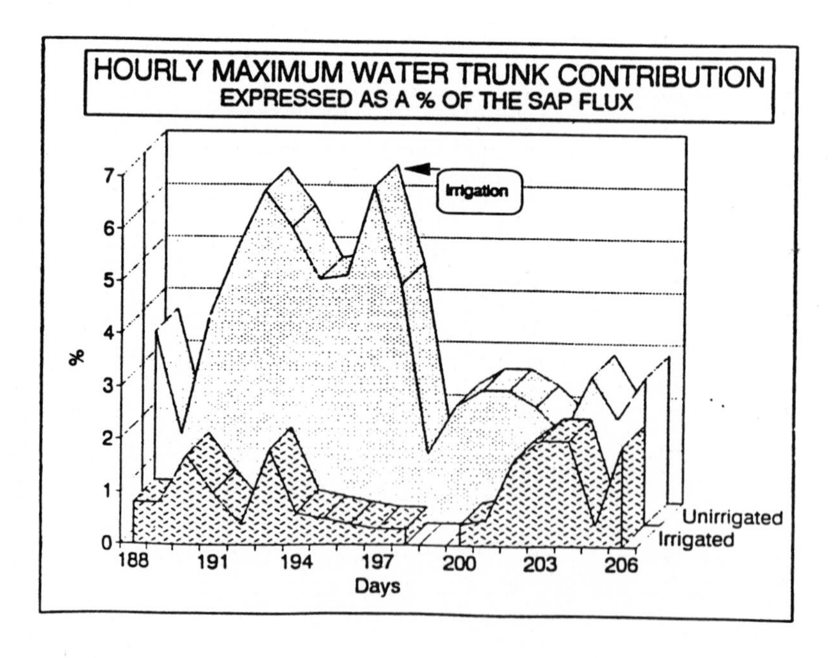

Figure 7. Water gained or lost by the trunk

notice that growth of the trunk can require more water than can be used to match the transpiration during periods without irrigation.

III. CONCLUSIONS

The previous example illustrates some of the features which should be taken into account to achieve a complete and quantitative interpretation of the shrinkage and swelling processes which occur within a stem. Accordingly the following points should be kept in mind:

a) It is very important to know the anatomy of the woody organs to determine whose tissues are involved in shrinkage and swelling processes. Although it seems quite certain that the external elastic tissues are responsible for almost the major part of reversible changes, the situation is not so clear during drought and growth. Furthermore, anatomical differences which occur during the plant development or exist between species, may play a very important role in the dynamics of these variations of dimension.

b) Water relations are, of course, the main frame for the interpretation of such variations. It is therefore of primary importance to measure, in addition to the stem variation, the different interfering variables (water potential, flux) and parameters (resistances capacitances ...). In this respect trunk and branch capacitances are very difficult to estimate precisely.

c) We have seen the importance of growth in these variations. This factor is probably the weakest point in these studies because the mechanism of growth diameter depends on many factors which are still poorly known.

IV. REFERENCES

Brough D.W., Jones H.G., Grace J. (1986) Diurnal changes in water content of the stems of apple trees, as influenced by irrigation. Plant Cell Environ., 9, (1): 1–7.

Cruiziat P., Granier A., Claustres J.P., Lachaize D. (1989) Diurnal evolution of water flow and potential in an individual spruce: experimental and theoretical study. In: Dreyer E., Aussenac G., Bonnet Masimbert M., Dizengremel P., Favre J.M., Garrec J.P., Le Tacon F.,

Martin F. (Eds), "Forest tree physiology" Proc. Intern. Symp., INRA, et Univ. Nancy, Nancy (FRA), 1988/09/25–30. Ann. Sci. for. 46, (suppl.): 353s–356s.

Granier A., Breda N., Claustres J.P., Colin F. (1989). Variation of hydraulic conductance of some adult conifers under natural conditions. In: Dreyer E., Aussenac., Bonnet Masimbert M., Dizegremel P., Favre J.M., Garrec J.P., Le tacon F., Martin F. (Eds.), "Forest tree physiology". Proc. Intern. Symp., INRA, et Univ. Nancy, Nancy (FRA), 1988/09/25–30. Ann. ASci. For., 46, (Suppl.): 357s–360s.

Milne R., Ford E.D., Deans J.D. (1983) Time lags in the water relations of Sitka spruce. For.Ecol.Manag., 5: 1–25.

Nobel P.S. (1983) Biophysical plant physiology and ecology. Freeman W.H., New York, (USA), 608 p.

Tyree M.T., Sperry J.S. (189) Vulnerability of xylem to cavitation and embolism. Annu. Rev. Plant Physiol. Plant mol. Biol., 40: 19–38.

Valancogne C., Nasr Z. (1989) Une methode de mesure du debit de sève brute dans de petits arbres par bilan de chaleur. Agronomie, 9 (6): 609–617.

PHYSICAL PRINCIPLES OF MEMBRANE DAMAGE DUE TO DEHYDRATION AND FREEZING

Joe Wolfe and Gary Bryant
School of Physics, University of New South Wales
PO Box 1, Kensington 2033, Australia

Water is the solvent for the ions, the organic solutes and many of the biochemicals necessary for active life, and the substrate for some of its reactions including photosynthesis. Water also allows relatively rapid diffusion of the reagents and products of vital reactions. Water is necessary, and it must be liquid water: ice is a much poorer solvent, and the diffusion in ice of solutes other than the hydroxyl ion is very slow.

Therefore there is almost no active life in organisms or tissues from which most of the water is removed, or in which most of the water is frozen. Any remaining unfrozen water forms highly concentrated solutions with very high viscosities, or is bound tightly to other molecules. The rates of diffusion and of biochemical reactions are very slow or zero.

As a result, the nearly dry or frozen state is, for some tissues and organisms, a state of suspended animation[1]: the processes of biochemistry and physiology slow almost to a halt, and recommence only when liquid water reappears. Seeds, for instance, survive with only a small fraction of water (5-15% by weight); so do the spores of bacteria. The tissues of resurrection plants are capable of rehydration to active life. The animals and micro-organisms of the soil such as nematodes, rotifers and tardigrades survive drying and rehydration. Similarly, some organisms, tissues and cells survive temperatures which freeze the environmental water (eg. woody tissues of plants, some insects, suspensions of some cells).

[1] The observation that seeds are a state of suspended animation must be almost as old as agriculture. Several millenia later, van Leeuwenhoek (1702) observed that some lower animals too could be resurrected to vigorous life from the almost inert, desiccated state. Three centuries after van Leeuwenhoek, anhydrobiology is now a recognized interdisciplinary research area (Leopold, 1986) which incudes physiology, cell and molecular biology, biochemistry, colloid and surface science, thermodynamics and statistical mechanics.

NATO ASI Series, Vol. H 64
Mechanics of Swelling
Edited by T. K. Karalis

Such cases, however, are the exception rather than the rule: removal of most of the liquid water is usually fatal to cells, tissues and organisms. The explanation for this involves other properties of water: its surface tension and its capacity to form hydrogen bonds. These properties are of central importance in cellular ultrastructure.

In a textbook illustration of a cell, it appears that the cell membranes hold together the largely aqueous cytoplasm and the organelles. The reverse is also true: water holds cell membranes together. The surface tension of water - or more explicitly the high free energy associated with putting CH_2 and other non-polar groups into water - provides the attractive term in the free energy of the lipid matrix of biological membranes. The surface tension of water is also a dominant effect contributing to the tertiary structure of proteins. Proteins in their native state are folded so that their hydrophobic regions (such as the α helices) are in contact with other hydrophobic groups rather than water. Hydrogen bonding both within the protein and with water may also be involved.

For this ultrastructural role, the water must be liquid. The surface tension of ice is only about half that of liquid water, and the geometry with which it forms hydrogen bonds is different. As a result, proteins denature in ice. This denaturation may be irreversible, because there are often many possible tertiary structures in water. For an enzyme, the exact molecular geometry is important for its biochemical function, and only one (or a small number) of the possible foldings performs that function. Intracellular ice formation appears to be universally fatal (the authors know of no reported case of a cell surviving intracellular freezing) and it is likely that one of the causes is this irreversible denaturation of some vital enzymes.

In those cases where cells survive sub-freezing temperatures, the plasma membrane of the cell plays the crucial role in preventing the entry of ice. In nature, and in many of the clinical situations where cell suspensions are frozen, this entails allowing most of the intracellular water to leave the cell. Therefore cells which withstand freezing temperatures must also withstand dehydration, and in many cases the damage caused to cells by freezing temperatures is the damage caused by dehydration rather than by low temperature *per se*. For this reason we shall consider dehydration and freezing together in this paper.

The hydration of biological cells

In their active state, nearly all biological cells are surrounded by a water-permeable membrane. Most are also surrounded by an external medium which is or includes a relatively dilute aqueous solution. The hydraulic permeability of a membrane L_P is defined by $J_V = L_P(\Delta P - \sigma\Delta\Pi)$ where J_V is the volume flux of water, P the pressure, Π the osmotic pressure and σ the reflection coefficient for solutes. Consider a cell (initial volume V_o, area A_o), with an ideal semipermeable membrane ($\sigma = 0$), with $P = 0$ and with an osmotic pressure Π_o. Suppose that exposure to a change in osmotic pressure $\Delta\Pi$ results in a final volume change of ΔV. In the approximation $\Delta V \cong -V_o\Delta\Pi/\Pi_o$,

$$\frac{dV}{dt} = -AL_P\Delta\Pi \cong \frac{A_oL_P\Pi_o}{V_o}\Delta V = \frac{3L_P\Pi_o}{d}\Delta V = \frac{\Delta V}{\tau} \qquad (1)$$

where $d = 3V_o/A_o$ is a characteristic length of the cell (the radius in the case of a sphere) and where $\tau = d/3L_P\Pi_o$ is the characteristic time for equilibration of water through the membrane. For an order of magnitude, take $d = 10\ \mu m$, $L_P = 0.1\ pms^{-1}Pa^{-1}$ and $\Pi_o = 1$ MPa. Thus the characteristic time for osmotic equilibration is typically tens of seconds[2]. Because changes in humidity and temperature and other relevant variables are usually slower than this, the intracellular and extracellular water are usually at or near equilibrium, at least for relatively high hydration[3].

Equilibrium between intracellular and extracellular water (subscripts i and e) requires that the chemical potential of water μ be constant:

$$\mu = \mu^o + kT \ln a_i + P_i v_W = \mu^o + kT \ln a_e + P_e v_W \qquad (2)$$

where a is the activity of water, P its pressure, v_W its molecular volume and where μ^o, the standard chemical potential, is the value of μ for pure water at atmospheric pressure. The plasma membrane of animal cells cannot resist a substantial pressure difference, so at moderate or high hydration $a_i = a_e$ (in many cases this implies that the extracellular and intracellular solutions have similar (total) solute concentrations). In contrast, plant cell plasma membranes are supported by a cell wall which resists stretching beyond several percent, and so limits the cell volume. Under normal conditions the extracellular

[2] It is assumed that diffusion is much more rapid than permeation (as is the case for small cells) or that there is rapid mixing of the internal and external solutions.
[3] At sufficiently low hydration the elevated concentration of solutes gives the intracellular solution a very high viscosity and thus reduces the rate of water diffusion, particularly at low temperature. More on this anon.

solution is much more dilute (a_e approaches 1) than the intracellular solution, so $P_i > P_e$. If P_e is not too negative, P_i is positive and this pressure excess (called turgor pressure) distends the cell wall and contributes to the rigidity of leaves. The cell wall of most plant cells resists only small negative (internal) pressures, however, so cells collapse rather than suffer substantially negative P_i (Tyree and Hammel, 1972).

There is therefore a large range of hydration over which (intact) plant and animal cells behave approximately osmotically: changes in μ change a_i but $P_i \approx 0$. The water content (and thus the volume) of the cell is determined by μ and the number of solutes in the cell. Non-osmotic behaviour becomes pronounced at low levels of hydration.

Under most conditions, the water content of cells is typically about 90% by weight, where the 10% non-aqueous content comprises varying amounts of membranes, macromolecules and a range of solutes which give an osmotic pressure of between a few hundred kPa (for animal cells) and several MPa (for plants adapted to relatively dry environments). The water content of seeds and "dried" tissues is typically several percent. The water content of intact cells can fall to several percent if the chemical potential of the external water is lowered by about kT/3. All terms in μ may be varied: the pressure and temperature may be changed; μ^o may be varied by a phase change[4]; and the activity of water depends on the solute concentration. So a reduction in μ of kT/3 can be caused by:

i) equilibration with an atmosphere of relative humidity about 80%,
ii) freezing the external water to about -25 °C,
iii) increasing the external solute concentration to several $kmol.m^{-3}$,
iv) applying a suction in the external water of about 30 MPa.

All of the above occur in nature - not to mention the laboratory[5] - with a range of severity. Equilibration with an unsaturated atmosphere confronts seeds, spores and many organisms separated from a source of water. Freezing of the external medium befalls not only the habitants of shallow ponds in

---

[4] $(\mu^o_{water} - \mu^o_{ice})$ and $(\mu^o_{vapour} - \mu^o_{water})$ are the latent heats of fusion and evaporation respectively.

[5] where scientific ingenuity has produced other methods for dehydration, such as the application of large pressures in a "pressure bomb" (Scholander *et al.*, 1964).

intemperate climes, but also the cells in any tissue exposed to sufficiently low temperatures. Hyperosmosis (low a_e) is another insult imposed by extracellular

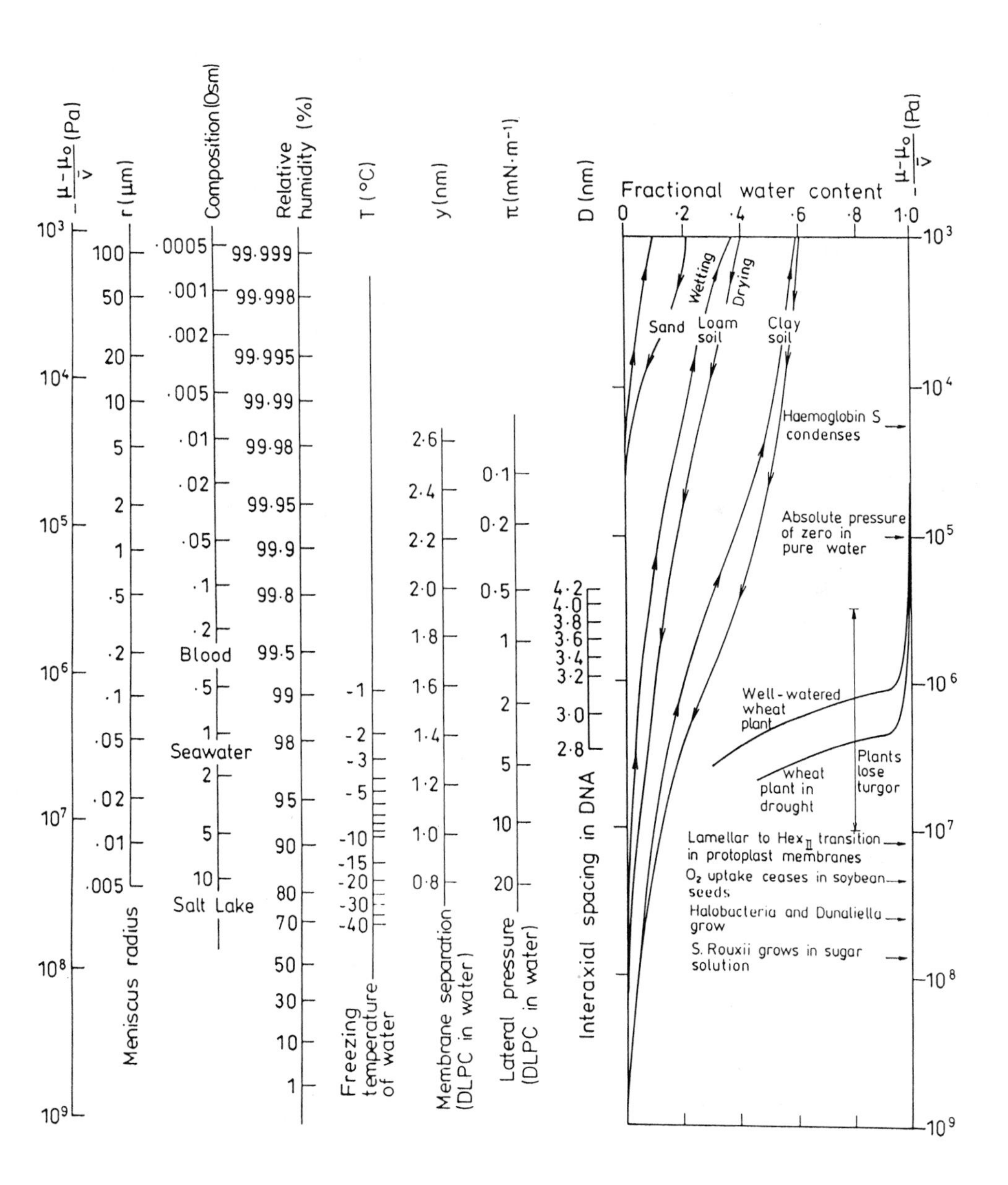

Figure 1. The equilibrium relations among several quantities relevant to low hydration. The data for dilaurylphosphatidylcholine are from Lis *et al.*, (1982) and those for DNA are from Parsegian *et al.*, (1986).

freezing, and it also strikes the residents of small, evaporating volumes of water evaporating ponds. Suctions are produced in the microscopic water volumes in soil or in plant cell walls - these suctions are applied by evaporation and are supported by highly curved menisci.

Relations among the relevant physical parameters are shown in Figure 1. Biologists use the quantity $(\mu-\mu^o)/v_w$ (Slatyer, 1967) which has dimensions of pressure and which they call the water potential Ψ. In the approximation that v_w is constant, (eqn 2) gives[6]: $\Psi = P - \Pi$. The logarithmic scale on the left is $-\Psi$, and it can be considered as the suction (the negative gauge pressure) in Pa. The next scale is the radius of a spherical meniscus of pure water which would support this suction in air at atmospheric pressure. The third scale is the solute concentration (in osmolal or moles per kg of water for an ideal solution) of a solution at atmospheric pressure in equilibrium which has the same μ. The fourth scale is the relative humidity of an atmosphere which has the same μ; the fifth is the temperature of (pure) ice which has the same μ. Thus the table shows for example that seawater can be in equilibrium with ice at about -3 °C (as is the nearly the case in arctic oceans), with an atmosphere of about 98% relative humidity, or pure water with a suction about 3 MPa, such as could be supported by a meniscus[7] with radius about 50 nm. Hydration relations for a lipid-water lamellar phase, DNA, a wheat leaf and for several soils are also shown.

Freezing of cells in aqueous suspension

Figure 2 shows several of the possible responses of cells in suspension to the freezing and thawing of the suspending medium. The extracellular medium freezes first for two reasons. First, the nucleation of freezing is a probabilistic phenomenon and, as the external solution has a much greater volume than any intracellular solution, nucleation is much more likely to occur outside. Second, it seems that many cells have a low concentration of ice nucleators.

---

[6] Another component of Ψ called the Matric Potential is sometimes identified. This quantity can always be identified with components of the pressure or osmotic pressure. Passioura (1980) explains its use, and recommends against it.

[7] The law of Young and Laplace is applicable for menisci whose radii of curvature is as small as a few nm (Fisher and Israelachvili, 1980).

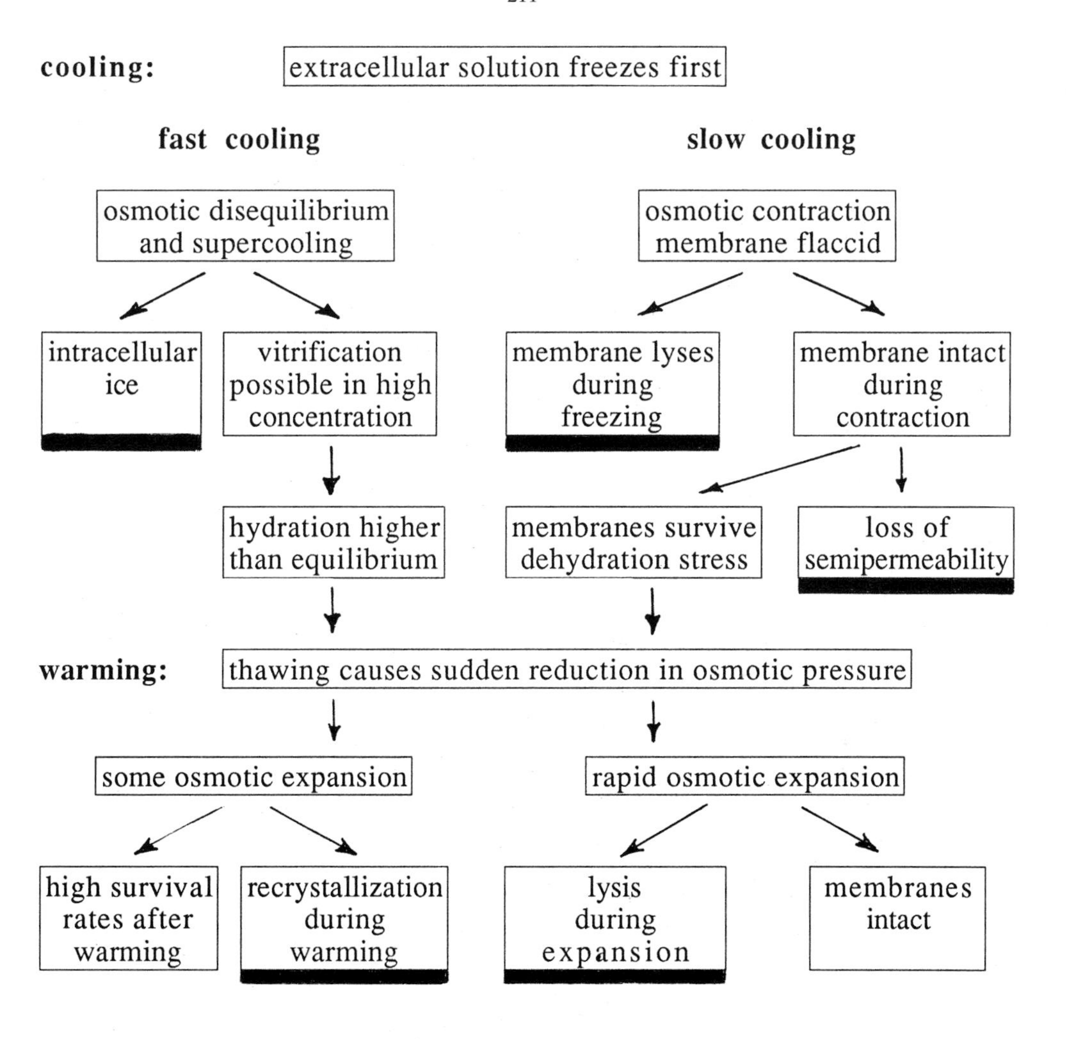

Figure 2. Pathways in a freeze-thaw cycle

Whatever the composition of the external medium (low concentration for plants, higher for animals), the solutes are much less soluble in ice than water, so external freezing produces a large fraction of nearly pure ice and a small fraction of concentrated solution whose chemical potential is determined by the temperature and pressure (see figure 1). Most of the external water freezes within a few degrees of zero so there can be a very rapid reduction in μ_e and a sudden increase in Π_e, especially if the solution supercools by a few degrees. Water diffuses out of the cell i.e. the cell contracts osmotically. In the figure "slow cooling" means that the temperature falls sufficiently slowly that water can leave to maintain $\mu_i \cong \mu_e$. "fast cooling" means that heat leaves the cell faster than water does, there is relatively little osmotic contraction, and so the internal solution becomes supercooled. Sufficient supercooling will usually cause ice nucleation in the cell at some temperature above about -40 °C, and thus cell death. (In figure 2 the horizontal lines signify cell rupture or death.)

A supercooled intracellular solution may however remain unfrozen indefinitely if it vitrifies[8], i.e. if it becomes exceedingly viscous (10^{12} to 10^{14} Pa s), due to the effects of high solute concentration and low temperature (Franks *et al.*, 1990). This is the route pursued in the cryopreservation of cell suspensions *in vitro*, where the internal concentration is often increased by adding a membrane-permeating solute called a cryoprotectant (eg. dimethylsulphoxide). Another method of provoking vitrification[9] is to allow some osmotic contraction (eg. by cooling first to a relatively warm sub-zero temperature), and then rapidly cooling. In many cases, vitrified cells may be stored in liquid nitrogen for years. The Scylla and Charibdis that face the cells during warming are recrystallization and osmotic rupture. Recrystallization, a paradoxical freezing during warming, occurs if the temperature rise causes the viscosity to fall sufficiently to allow the supercooled water molecules to orient and to form crystals. On the other hand there will be some osmotic expansion as the external medium thaws, and this may rupture the membrane. If these straits be safely passed, high rates of cellular survival are possible. Among the problems which limit the wider application of cryopreservation are these: the required concentrations of cryoprotectants may be toxic; and the cooling rates required for vitrification may not be achievable. The latter is particularly a problem for macroscopic organs[10] as distinct from cell suspensions.

---

[8] Icecream is a common example of a metastable, supercooled solution whose viscosity at very low temperatures prevents it from separating into ice and a concentrated solution of sugar with suspended fats. If the mixture is warmed and cooled without stirring, it separates into its thermodynamically stable configuration.

[9] In some dehydration tolerant species, vitrification occurs at warm temperatures due to high concentrations of internal solutes. Some solutes such as trehalose and sucrose are reputed to be effective in this role (Crowe *et al.*, 1988).

[10] To any reader who may be contemplating paying a large sum of money for the right to have his/her body or parts thereof frozen in the hope of being reanimated by a more competent future civilization, the authors suggest that the vendors of this service be asked whether the cytoplasm of the cells of vital organs will be vitrified or crystallized. If the latter, then reanimation would require rapid reconstruction of denatured proteins *in situ*. If the former, then there are many serious cryobiolgists who would like to know the details, so as to be able to preserve human organs.

The right hand side of figure 2, the response to slow cooling, is the régime that is relevant to freezing and thawing in nature. Again the external medium contains some almost pure ice and some highly concentrated solution. The osmotic contraction that follows freezing to even several degrees below freezing usually requires a severalfold increase in osmotic pressure (see figure 1). This usually produces a severalfold decrease in the volume of the cell. Even for a cell that was spherical initially, the result is an irregular shape with a flaccid membrane. Thawing will produce a sudden decrease in osmotic pressure and allow a rapid osmotic expansion. The survival of a cycle of freezing and thawing has much in common[11] with the survival of osmotic dehydration and rehydration.

The big question is: how can a cell retain its ultrastructure during such a cycle? Maintaining the conformation of the proteins is obviously important, because denaturation can be irreversible. In the case of freezing, the maintenance of the cell membrane is vital, because its rupture allows ice to nucleate into the cytoplasm, and this, so far as the authors know, is always fatal. The integrity of membranes is almost equally important in the survival of any cycle of dehydration and rehydration because of their important topological function in separating different solutions - in the case of the plasma membrane, the quick from the dead.

On the right hand side of figure 2 (slow cooling and warming) there are three dead ends for the cell:

i) lysis (i.e. rupture of the cell membrane) during freezing,
ii) lysis during osmotic expansion, and
iii) loss of osmotic response at low hydration.

Our research on the first two was conducted in collaboration with Steponkus and co-workers at Cornell University. These studies used protoplasts (single cells from which the wall has been removed) isolated from rye seedlings that

[11] There are also many differences. First the low temperature itself (rather than its osmotic effects) may be injurious: biochemical reaction rates are differentially sensitive to temperature and so metabolic imbalances and the buildup of potentially toxic intermediary products is one possibility. The direct temperature effect is however less important in the study of freezing than it is in the study of chilling (damage inflicted by cold non-freezing temperatures). This may be because at freezing temperatures reaction and diffusion rates are so slow that metabolism is nearly stopped. Other differences are that environmental freezing usually lowers the chemical potential less than does air drying (see figure 1), and that the macroscopic mechanical effects in tissues are rather different for the two cases.

had either been exposed to cool but not freezing temperatures (acclimated) or not (nonacclimated). Currently we are concentrating on (iii).

One dead end in figure 2 is lysis during freezing. This is surprising, as one would not expect large mechanical stresses to appear in the membrane during the osmotic shrinkage that accompanies freezing: the membrane in cells of ordinary size bends rather than support compressive stress. For protoplast suspensions at least, this injury is unique to freezing and cannot be simulated by osmotic manipulation. This rupture occurs in the vicinity of the ice front and we accuse the very strong local electric field near the freezing interface of rupturing the membrane electrically. The evidence is circumstantial but strong: the rupture occurs in those conditions which give rise to a large transient potential difference and does not occur when the measured transient is small; and rupture of this type occurs with higher frequency in cells which are also damaged easily by electric fields (Steponkus *et al.*, 1984, 1985).

Another dead end for cells is rupture during osmotic re-expansion. The damage caused to protoplast suspensions by freezing to warm sub-zero temperatures and then thawing can be simulated at room temperature by equivalent osmotic manipulation. This damage has been correlated with frost tissue damage in non-hardy tissues. When protoplasts from non-acclimated seedlings are osmotically contracted, they become spherical at the new volume

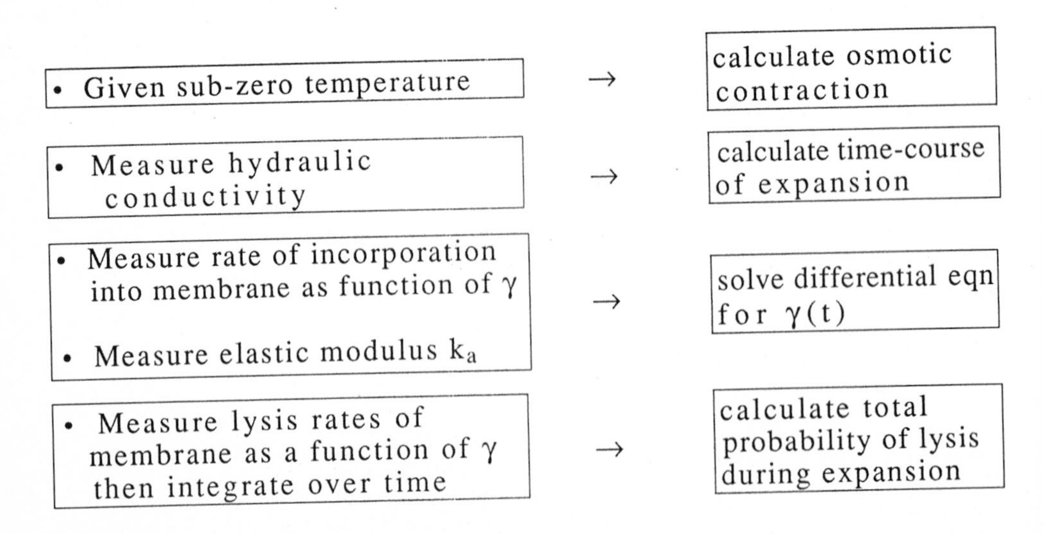
• Given sub-zero temperature
→
calculate osmotic contraction
• Measure hydraulic conductivity
→
calculate time-course of expansion
• Measure rate of incorporation into membrane as function of γ
• Measure elastic modulus k_a
→
solve differential eqn for γ(t)
• Measure lysis rates of membrane as a function of γ then integrate over time
→
calculate total probability of lysis during expansion

Figure 3. Analysis of cell lysis during thawing

by removing membrane material from the cell surface. When they are re-expanded (by dilution or thawing of the external medium), the surface tension in the membrane rises, and membrane material is re-incorporated. The chance of cell survival depends on the dynamics of this process, and on the dependence of membrane rupture on membrane tension. The relevant parameters may all be measured independently using micropipette aspiration, and the analysis of the process is set out in Figure 3. In this analysis, all parameters are directly measured, and there is good agreement with experiment (Wolfe and Steponkus, 1982; Wolfe *et al.*, 1985; Dowgert *et al.*, 1987).

Membrane damage at low hydration

Cells without walls behave osmotically in aqueous suspension over a large range of hydration: down to about 10 or 20% water. Protoplasts isolated from cold acclimated plants usually survive dehydration-rehydration or freeze-thaw cycles in the osmotic range, but suffer membrane damage while exposed to very low hydration, whether imposed osmotically or by freezing. They do not expand osmotically when the external solution is thawed or diluted.

The phenomenon is more general: many types or symptoms of damage occur in the range of water contents below about 20% by weight. This is also the range over which the mechanical stresses in membranes cease to be negligible in comparison with the relevant elastic moduli[12].

Removal of water at low hydration involves close approach of everything else in the cell, including the many internal membranes which are folded and forced together. When the thickness of solution separating the non-aqueous components of the cell is reduced to a few nanometres or less, large, repulsive hydration forces are encountered[13]. This repulsion can support a negative solution pressure or suction. Thus, at very low hydration, reductions in μ must

[12] At a final, summary session of a Rockefeller conference on anhydrobiology, C. Vertucci observed that many speakers had reported damage or unusual effects in the hydration range several-20%. (The conference was held in Bellagio, Italy. A.C. Leopold was the convenor and editor of the proceedings, 1986). Vertucci enquired whether there might be a general phenomenon involved, and Wolfe made the above observation about the magnitude of the stresses in membranes (elaborated by Wolfe, 1987).
[13] Aspects of hydration and other interfacial forces are treated by Israelachvili, Marcelja, Parsegian, Simon and Rand in other chapters in this volume.

produce reductions in pressure P as well as reductions in a_i, and this gives rise to non-osmotic behaviour[14].

Nearly all of our knowledge of hydration forces comes from simple systems. The model of biological membranes is the lipid bilayer or lamellar phase. These have been investigated by Israelachvilli, Parsegian, Rand and co-workers. Applying the same arguments to cells is a leap of faith, but a leap which is made less vertiginous by the observation of hydration repulsions among other cellular constituents including proteins and DNA (eg Parsegian *et al.*, 1986); and by the geometric similarities between various lipid water phases and the ultrastructure of dehydrated cells inferred from electron microscopy.

For membranes at very close approach, reduction of the water volume causes both reduction of intermembrane separation and contraction in the plane of the surface. Very close approach is opposed by the strong repulsion and the lateral contraction produces compressive stresses in the membrane. The stresses in the membrane interior are highly anisotropic - large and compressive in the plane of the membrane, much smaller in the normal direction. Several different strains are possible - of which the simplest is contraction in the plane of the membrane and thickening in the normal direction (Lis *et al.*, 1982). Others involve phase changes and phase separations.

In a quasi-two dimensional system such as a membrane, the lateral stress π (the compressive force per unit length in the plane of the membrane) is analogous to the bulk pressure in a three dimensional system, and it has a similar effect on the phase properties (Evans and Needham, 1987). Lipid bilayers contract in area (and increase in thickness) when they "freeze" from the fluid to the gel phase, so the temperature of this transition is lowered by lateral pressure (and raised by applied surface tension). Thus dehydration to the range which increases lateral pressure raises the transition temperature (Ulmius *et al.*, 1977; Kodama *et al.*, 1982; Seddon *et al.* 1984; Lynch and Steponkus, 1989).The area change is typically 0.2 nm^2, the latent heat of order 5×10^{-20} J per molecule, and so the change is of order 1 K for each mNm^{-1} of applied stress.

---

[14] Hydration forces between membranes may be measured directly (the Israelachvili method) or else by applying a suction hydrostatically or osmotically (the method of Rand, Parsegian and co-workers).

Hydration forces and the lateral stresses they cause may produce another effect in membranes at low hydration: fluid-fluid phase separations or demixing. This is possible because of the large range of magnitudes of the hydration force among different molecular species - among lipids this range is two orders of magnitude (Rand and Parsegian, 1989). Consider a lamellar phase with water and only two lipid components: one strongly hydrating which (on its own) gives rise to a large hydration repulsion, and another which hydrates weakly and has a small repulsion. If an homogeneous mixture of the lipids separates into one phase rich in the weakly hydrating species, and another rich in the strongly repelling species, then the former will have a smaller inter-membrane separation than the homogeneous mixture, and the latter a larger separation (see figure 4). Now because the hydration repulsion is strongly nonlinear, this configuration has a lower potential energy than does the homogeneous phase. If the difference in energy is greater than the entropy of demixing associated with the separation, then the separation is stable. An analysis of this effect is reported by Bryant and Wolfe (1989).

Such a separation has been observed in a model system. Palmitoyloleoyl-phosphatidylcholine (POPC) and palmitoyloleoylphosphatidylethanolamine (POPE) were chosen for the study because they are mixed chain unsaturated phospholipids (and thus are typical of plant lipids), and because the hydration repulsion of POPC is much greater than that of POPE. Above the thermotropic phase transition temperatures of both, these two lipids are

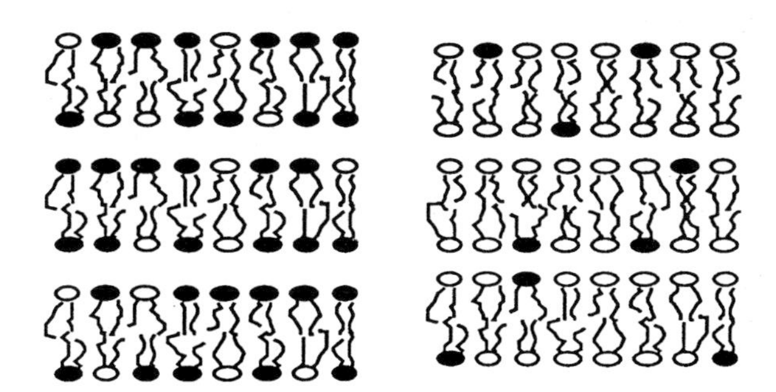

Figure 4. Lateral demixing in a lipid mixture. The lipids whose head groups are shaded in this sketch have a larger repulsion.

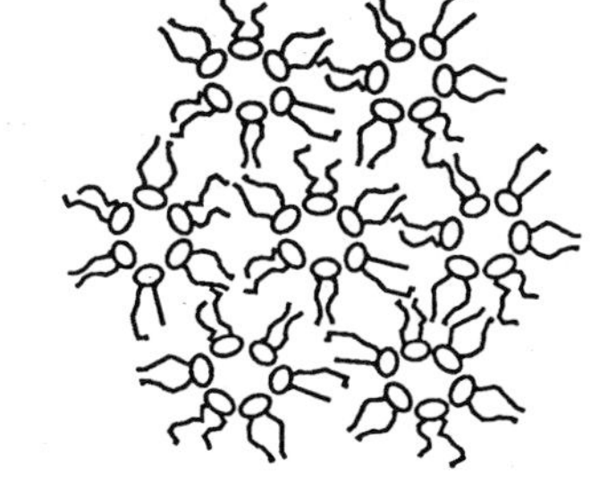

Figure 5. A section through an inverted hexagonal phase, perpendicular to the axes of the cylinders.

completely miscible in excess water. At 315 K and 10% water content, however, they separate into two fluid phases with different repeat spacings (Bryant, 1991). X-ray diffraction patterns of the mixture show two different repeat spacings consistent with fluid phases of POPE-rich and POPC-rich compositions. Further, the D_2O-NMR spectra of the sample shows two superimposed Pake patterns indicating that the water is distributed between two regions with different degrees of ordering. The ratio of hydration repulsions expected for these two lipid is within the range of the ordinary components of biological membranes, and so it is likely that similar separations occur *in vivo* in living systems[15] similarly dehydrated. The importance of such separations is that such demixing concentrates the lipids which can form the inverted hexagonal phase (discussed below), and thus may be a necessary precursor to its formation.

The transition of a lipid water mixture from the lamellar phase to the inverted hexagonal (H_{II}) phase is topologically spectacular. In the former, thin extended planes of water separate lipid bilayers; in the latter extended cylinders of water (whose axes lie on an hexagonal array, whence the name) are surrounded by the hydrophilic parts of the lipids while the hydrophobic parts fill the remaining space (Luzzatti and Husson, 1962; Luzzatti, 1968; Tardieu *et al.*, 1973; Luzzatti and Tardieu, 1974). This transition relaxes the compressive stress in the plane of the lipid-water interface and, in some cases, may also lower energy associated with the curvature moduli (see analyses by Kirk *et al*, 1984; Gruner *et al*, 1986). At a given hydration, the H_{II} phase is stable at higher temperatures than the lamellar, but at a given temperature it is stable at lower water contents. Thus the H_{II} phase may be induced by freezing cell suspensions: it is of course the osmotic dehydration and not the temperature reduction which is responsible. Gordon-Kamm and Steponkus (1984) studied electron micrographs of protoplasts that had been contracted either by freezing or by equivalent osmotic contraction at temperatures above freezing, and observed arrays of cylinders resembling the H_{II} phase. They correlated the

[15] In fact another example of such a separation is regularly reported in studies of dehydrated cells. When electron micrographs show regions of membranes rich in particles, these are interpreted as being rich in proteins, and conversely. Two effects (at least) may account for this separation: First, the proteins may be differentially soluble in two lipid phases which have demixed either due to the thermotropic transition, or due to the effect discussed above. Alternatively, the proteins may demix in a fluid membrane due to differential hydration repulsion. Membrane proteins are large and often protrude from the membrane surface, so they are expected to have a large hydration effect compared to lipids. The thermodynamics of this separation is discussed by Bryant and Wolfe (1989).

frequency of observation of these arrays with lack of osmotic response damage to protoplasts. The ultrastructural implications of this are clear: were a stack of membranes (say a folded region of the plasma membrane of a dehydrated cell, or a dehydrated mitochondrion or chloroplast) to form regions of hexagonal phase, the lumens would lose their erstwhile independence, and the osmotic response of the cell and its organelles would be lost.

The effect of cytoplasmic solutes

Plants exposed to cold but not freezing temperatures, or to low but not fatal μ_{water} may acclimate, i.e. become less susceptible to freezing or dehydration damage. Acclimated plants usually have higher cytoplasmic osmotic pressures. Dissolved sugars (trehelose in particular) are reported to contribute to the stability of membranes at low hydration (Crowe and Crowe, 1986; Crowe *et al.*, 1988; Clegg, 1986).

Cytoplasmic solutes have several effects which mitigate the damage caused by environmental dehydration: (i) they can slow or prevent the approach to equilibrium of water; (ii) they cause that equilibrium to occur at higher water content, and (iii) they reduce the strains imposed at a given water content. Let us consider these in turn:

i) Solutes are concentrated by dehydration and this may increase the viscosity to the range 10^{12} to 10^{14} Pa s where supersaturated or supercooled solutions form glasses (Franks *et al.*, 1990). In this state the diffusion of water is so slow that hydration levels above equilibrium may be sustained.
ii) High initial solute concentration (and higher osmotic pressure) at full hydration requires a larger volume of solution at any given μ_{water} due to osmotic pressure.
iii) At any given μ_{water}, higher Π implies less negative P (eqn 2). This smaller suction requires less intermembrane repulsion and thus creates less intra-membrane stress. Burke (1986) gives a review of the work on (i). We have investigated the effects (ii) and (iii) using the intensity of the narrow resonance of a D_2O-NMR signal[16] as a measure of the fraction of unfrozen water.

---

[16] The broad resonance is due to the ice, but its bandwidth is so large that there is no danger of confusing the two. This technique also yields data related to the degree of ordering of that remaining unfrozen water.

Freezing of lipid-solution mixtures or of cell suspensions allows a very simple control of the chemical potential of water[17]: $\mu - \mu_o = L(T/T^* - 1)$ where L is the latent heat and T* is the freezing temperature of pure water. Consider a lamellar phase of lipid and solution in equilibrium with ice. The pressure in the solution equals -1 times the hydration repulsion per unit area, so write

$P = - P_o e^{-y/\lambda}$, where λ is the characteristic length of the repulsion and P_o is its extrapolated value at separation y = 0, and assume that these parameters are independent[18] of solute concentration (for more detail see the chapter by Rand). Let there be a volume V_u of (unfrozen) water and n molecules of an improbably ideal solute with partial volume v. Suppose that the total (one-sided) membrane area is A so that the solution volume is $Ay = nv_s + V_u$. With the approximation that the molecular volume v_w of (liquid) water is constant, eqn 2 gives:

$$\frac{\mu - \mu_o}{v_w} = P - \Pi \cong - P_o e^{-y/\lambda} - \frac{nkT}{V_u} = - P_o e^{-(nv_s + V_u)/\lambda A} - \frac{nkT}{V_u} \qquad (3)$$

Several simple conclusions can be read from these equations. First, higher solute concentration (larger n) requires larger intermembrane separation y. Higher osmotic pressure means less negative pressures, and so smaller repulsions, which occur at larger separations. Further, solutes with larger partial volumes cause larger separations and less negative pressures. Finally, all else equal, (3) shows that solutes with larger partial volumes v_s imply a *smaller* volume of unfrozen water (though not a smaller volume of unfrozen solution). From (3):

$$\frac{dV_u}{dn} = \frac{kTA\lambda}{-PV_u} - v_s \qquad (4)$$

This function is positive for small suctions, but negative for large suctions. For typical numbers, consider membranes separated by a few nm, $\lambda \sim 0.2$ nm and v_s a few times 10^{-28} m^3. The characteristic pressure is of order MPa. The dehydration properties of dipalmitoylphosphatidylcholine (DPPC) bilayers in different D_2O solutions is shown in Figure 6. The vertical axis is the integral of the narrow resonance of the NMR spectrum (approximately proportional to the quantity of unfrozen water). For pure D_2O this falls abruptly to zero a little below the equilibrium freezing temperature (277 K). For DPPC lamellar phases, the fraction of unfrozen water decreases continuously with decreasing

[17] and thus may be useful in dehydration as well as in cryobiological studies.
[18] Increasing the solute concentration should lower the dipole concentration and the dielectric permittivity, which might lower P_o and increase λ.

temperature. At low temperatures (low chemical potentials) the quantity of unfrozen water is least for the lamellar phase with sucrose solution, greater for sorbitol solution, and greater again for lamellae plus pure D_2O. For sucrose the unhydrated partial volume is 0.52 nm^{-3} and for sorbitol it is ~0.27 nm^{-3}, so these results are qualitatively in agreement with (3) and (4).

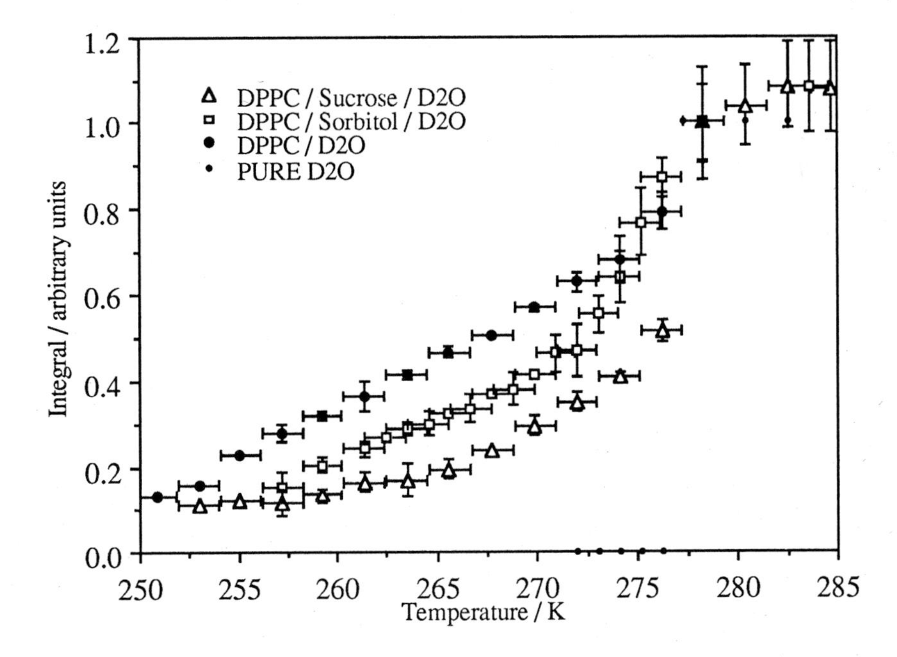

Figure 6. The intensity of the D_2O-NMR signal as a function of temperature.

The biological importance of this model is its implications for membrane damage. The lateral compressive stress π in the membrane (expressed as a lateral pressure or force per unit length) is just $\pi = |Py|$, so the presence of solutes which can enter between the membranes lowers the intra-membrane stresses at any given chemical potential of water. This effect is greater for large solute molecules than for small. Figure 7 shows the calculated π as a function of temperature and water potential for membranes with pure D_2O and with three different solutions: a hypothetical ideal solute with zero partial volume, and ideal solutes with partial volumes corresponding to sorbitol and sucrose.

This model is still highly simplified: in particular no specific chemical effects of solutes have been included. Some solutes may have a large effect on P_o due to interactions at the membrane interface, and this will influence membrane stress. Trehelose, for example, appears to have a specific interaction with the membrane and creates strains even in excess water where external stresses[19]

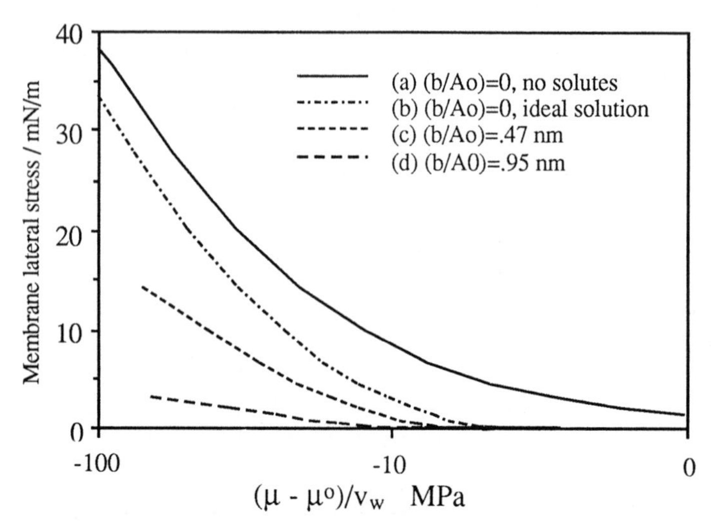

Figure 7. Calculated stresses in membranes as a function of the chemical potential of water in the presence of different solutions. The data for curves c and d are appropriate for sorbitol and sucrose respectively.

are negligible (Wistrom *et al.*, 1989). Nevertheless this simple model predicts substantial non-specific effects for large solutes on membrane freezing and dehydration behaviour. These predictions are in at least qualitative agreement with experimental results from model systems and, in preliminary experiments, from cell suspensions (Bryant, 1991). Analysis of specific solute effects on membrane stability will be a non-trivial exercise, and any such analysis must first consider the magnitude of these simple solute properties.

References

Bryant G (1991) PhD thesis. University of New South Wales, Sydney

Bryant G, Wolfe J (1989) Can hydration forces induce lateral phase separation in membranes? Eur Biophys J 16:369-374

Burke MJ (1986) The glassy state and survival of anhydrous biological

---

[19] As distinct from internal stress such as that imposed by the plane suface of a lamellar phase on an interface with an intrinsic curvature.

systems. In: Membranes, Metabolism and Dry Organisms, AC Leopold (ed) Cornell NY

Clegg JS (1986) The physical properties and metabolic status of Artemia cysts at low water contents: the water replacement hypothesis. In: Membranes, Metabolism and Dry Organisms, AC Leopold (ed) Cornell NY

Crowe JH, Crowe LM (1986) Stabilization of membranes in anhydrobiotic organisms. In: Membranes, Metabolism and Dry Organisms, AC Leopold (ed) Cornell NY

Crowe JH, Crowe LM, Carpenter JF, Rudolph AS, Wistrom CA, Spargo BJ, Anchordoguy TJ (1988) Interactions of sugars with membranes. Biochim Biophys Acta 947:367-384

Dowgert MF, Wolfe J, Steponkus, PL (1987) The mechanics of injury to isolated protoplasts following osmotic contraction and expansion. Plant Physiol 83:1001-1007

Evans EA, Needham D (1987) Physical properties of surfactant bilayer membranes: thermal transitions, elasticity, rigidity, cohesion and colloidal interactions. J Phys Chem 91:4219-4228

Fisher LR, Israelachvili JN (1980) Determination of the capillary pressure in menisci of molecular dimensions. Chem Phys Lett 76:325-328

Franks F, Mathias SF, Hatley RHM (1990) Water, temperature and life. Phil Trans R Soc Lond B 326:517-533

Gordon-Kamm WJ, Steponkus PL (1984) Lamellar-to-hexagonal$_{II}$ phase transitions in the plasma membrane of isolated protoplasts after freeze-induced dehydration. Proc Natl Acad Sci USA 81:6373-6377

Gruner SM, Parsegian VA, Rand RP (1986) Directly measured Deformation Energy of Phospholipid H_{II} Hexagonal Phases. Faraday Discuss Chem Soc 81:29-37

Kirk GL, Gruner SM, Stein DL (1984) A thermodynamic model of the lamellar to inverse hexagonal phase transition of lipid membrane-water systems. Biochem, 23:1093-1102

Kodama M, Kuwabara M, Seki S (1982) Successive phase transition phenomena and phase diagram of the phosphatidylcholine-water system as revealed by differential scanning calorimetry. Biochim Biophys Acta 689:567-570

van Leeuwenhoek A (1702) On certain animalcules found in the sediment in gutters of the roofs of houses. transl Hoole S (cited by Leopold)

Lis LJ, McAlister M, Fuller N, Rand RP, Parsegian, VA (1982) Interactions between neutral phospholipid bilayer membranes. Biophys J 37:657-666

Leopold AC (1986) Membranes, Metabolism and Dry Organisms. Cornell NY

Luzzati V, Tardieu A (1974) Lipid phases: structure and structural transitions. Ann Rev Phys Chem 25:79-94

Luzzati V, Husson F (1962) The structure of the liquid-crystalline phases of lipid-water systems. J Cell Biol 12:207-219

Luzzatti V, (1968) X-ray diffraction studies of lipid-water systems. In: Biological Membranes: Physical Fact and Function, D Chapman (ed), Academic Press

Lynch DV, Steponkus PL (1989) Lyotropic phase behaviour of unsaturated phosphatidylcholine species: relevance to the mechanism of plasma membrane destabilization and freezing injury. Biochim Biophys Acta 984:267-272

Parsegian A, Rau D, Zimmerberg J (1986) Structural transitions induced by osmotic stress. In Membranes, metabolism and dry organisms (AC Leopold, ed) Cornell NY

Passioura, JB (1980) The meaning of matric potential. J Exp Bot 123:1161-1169

Rand RP, Parsegian VA (1989) Hydration forces between phospholipid bilayers. Biochim Biophys Acta Reviews on Biomembranes 988:351-376 (1989)

Scholander PF, Hammel EA, Hemmingsen HT, Bradstreet ED (1964) Hydrostatic pressure and osmotic potential in leaves of mangroves and some other plants. Proc Nat Acad Sci USA 52:119-125

Seddon JM, Cevc G, Kaye RD, Marsh D (1984) X-ray diffraction study of the polymorphism of hydrated diacyl- and dialkyl-phosphatidylethanolamines. Biochem 23:2634-2644

Slatyer RO (1967) Plant Water Relationships. Academic, London

Steponkus PL, Stout DG, Wolfe J, Lovelace RVE (1984) Freeze-induced electrical transients and cryoinjury. Cryo-letters 5:343-348

Steponkus PL, Stout DG, Wolfe J, Lovelace RVE (1985) Possible role of transient electric fields in freezing-induced membrane destabilisation. J Memb Biol 85:191-198

Tardieu A, Luzzatti V, Reman FC (1973) Structure and polymorphism of the hydrocarbon chains of lipids. A study of lecithin-water phases. J Mol Biol 75:711-733

Tyree MT, Hammel HT (1972) The measurement of the turgor pressure and the water relations of plants by the pressure-bomb technique. J Exp Bot 23:267-282

Ulmius J, Wennerstrom H, Lindblom G, Arvidson G (1977) Deuteron nuclear magnetic resonance studies of phase equilibria in a lecithin-water System. Biochem, 16:5742-5745

Wistrom CA, Rand RP, Crowe LM, Spargo, BJ and CRowe, JH (1989) Direct transition of dioleoylphosphatidylethanolamine from lamellar gel to inverted hexagonal phase caused by trehalose. Biochim Biophys Acta 984:238-242

Wolfe J (1987) Lateral stresses in membranes at low water potential. Aust J Plant Physiol 14:311-318

Wolfe J, Dowgert MF, Steponkus PL (1985) Dynamics of incorporation of material into the plasma membrane and the lysis of protoplasts during rapid expansions in area. J Memb Biol 86:127-138

Wolfe J, Steponkus PL (1982) Mechanical properties of the plasma membrane of isolated plant protoplasts. Pl Phys 71:276-285

ION CHANNELS IN THE PLASMA MEMBRANE OF PLANT CELLS

B.R. Terry, S.D. Tyerman and G.P. Findlay
School of Biological Sciences, The Flinders University of South Australia, GPO Box 2100, Adelaide 5001, South Australia.

INTRODUCTION

Many types of ion channels have been characterised in the membranes of plant cells (Tester, 1990) but overall there is a lack of understanding of their physiological functions. The functions of plant ion channels need to be seen in context of the variable ionic environment to which plant cells are exposed. The mechanisms for acquisition of nutrients must cope with variable driving forces for uptake from low external concentrations yet simultaneously or subsequently be able to respond to excess concentrations. A proton pump energises the plasma membrane and due to a low membrane conductance and rather negative reversal potential of the pump, a large proportion of the electrochemical difference for protons ($\Delta\mu_{H+}$) is generated as an electrical potential difference (PD). Changes in the driving force due to variations in the external pH can be quickly compensated for by changes in PD. Ion channels are ideally suited to the control of the PD because large dissipative fluxes can occur through relatively few channels when they open.

Whether a particular channel modulates the PD, provides large net fluxes for osmotic adjustment, or is a primary route for nutrient uptake, there must exist comprehensive control mechanisms to ensure that the plasma membrane ion channels integrate with the activities of the proton pumps and cotransport systems.

TYPES OF ION CHANNEL IN THE PLASMA MEMBRANE

Outward rectifier (OR) cation channels: allow cation efflux (outward current) under suitable conditions, are normally activated at depolarised membrane PDs, and generally give rise to time–dependent currents in whole cell preparations (Findlay and Coleman, 1983; Bisson, 1984; Beilby, 1985; Schroeder, 1988; Ketchum et al,

NATO ASI Series, Vol. H 64
Mechanics of Swelling
Edited by T. K. Karalis

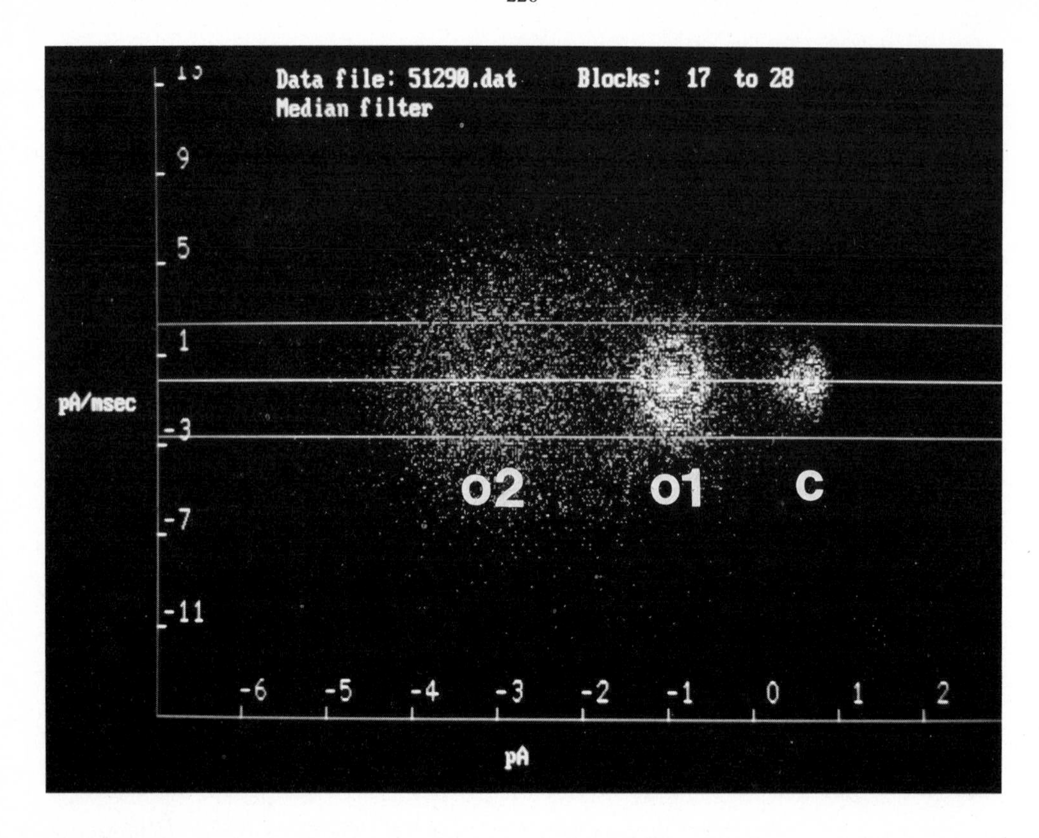

Plot of rate of change of current versus the current amplitude for a multi–conductance cl⁻ channel from the membrane of a plant cell. The nodes marked C, 01 and 02 refer to the closed state, and first and second open states respectively. The parabolic trajectory of points between the nodes are the transitions between the conductance levels.

1989). There would appear to be different types of cation OR in different plants, based on the diversity of cation selectivities that have been observed (e.g. Stoekel and Takeda, 1987). Our own work (Terry *et al.* 1991) has shown that in protoplasts from cotyledons of *Amaranthus tricolor* the cation ORs have a single channel conductance of about 35 pS (10/100 mM K^+) and a relative permeability sequence of K^+>Na^+>>Cl^- (1 : 0.16 : 0.03). The channels are active in whole–cell, outside–out and attached patches, but have not yet been observed in inside–out detached patches. Time–dependent cation ORs dominate the membrane conductance of *Amaranthus* and wheat protoplasts (Terry *et al.*, 1991; Schachtman *et al.*) although cells can switch to a mode in which no time–dependence of the currents is apparent (Schachtman *et al.*, Terry and Tyerman unpub.). We have not yet confirmed whether K^+ ions carry the non time–dependent currents, although this seems likely from the evidence we do have, nor do we know the stimulus which leads to the transition between the modes.

<u>Inward rectifier (IR) cation channels</u>: conduct cations into the cell, and are activated in a time–dependent fashion by membrane PDs negative of the channel reversal potential. Inward rectifying cation channels (cation IR) have been reported in several plant cells (Schroeder et al. 1987; Schroeder 1988; Bush et al. 1988; Ketchum et al. 1989; Moran and Satter 1989), but they are not a consistent feature of all plant cells nor is their activity predictable even in a particular type of cell. The principal differences between inward and outward rectifiers are their reversed voltage gating and the sensitivity of inward rectifiers to the level of free calcium on the cytoplasmic side; they are inactivated at >1μM $[Ca^{2+}]_{cyt}$ (Schroeder and Hagiwara 1989; Bush *et al.* 1988; Blatt *et al.* 1990). By manipulating the cytosolic concentrations of K^+ and Cl^- we have observed a cation IR in *Amaranthus* (fig. 1). The K^+/Na^+ selectivity of this channel is less than the cation OR; K^+>Na^+>>Cl^- (1 : ~0.3 : 0.03). The channel conductance is about 10 pS (10/100 mM K^+). Cation IRs are not commonly seen in any type of patch preparations of *Amaranthus tricolor.*

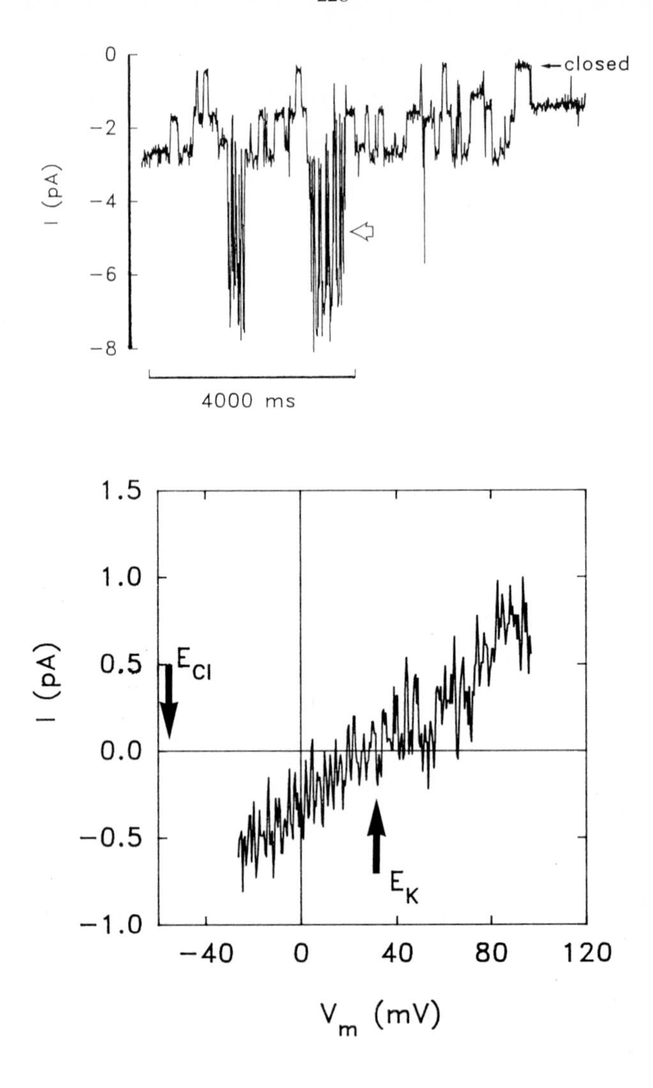

Figure 1 (a) Current trace for an outside–out patch pulsed to a membrane PD of −105 mV. Two cation IR channels are active in the patch (smaller transitions). Superimposed upon their activity are two bursts of larger current transitions which are the result of spontaneous (and uncharacteristic) openings of a cation OR channel (arrow). (b) Current–voltage relationship for a single cation IR obtained by ramping the membrane potential across an outside–out patch as the channel is in the open state (see Terry *et al.* 1991). The single channel conductance is about 10 pS. The equilibrium potentials for K^+ and Cl^- are marked E_K and E_{Cl} respectively. Solutions were (mM): Bath – 100 KCl, 10 $CaCl_2$, 10 MES, 7 KOH, 460 Sorbitol, pH6.0 Pipette – 10 KCl, 2 $MgCl_2$, 2 EGTA, $2K_2ATP$, 670 Sorbitol, 10 HEPES, 17.5 KOH, pH 7.8

Inward rectifying anion channels: conduct anions out of the cell and are usually activated only at extreme negative membrane potentials. Anion channels have been well documented in the plasma membranes of some of the giant algae by means of classical whole–cell electrophysiology and flux studies (e.g. Coster 1965; Coster 1969; Findlay and Coleman 1983; Lunevsky et al. 1983; Kataev et al. 1984; Coleman and Findlay 1985; Tyerman et al. 1986a,b; Shiina and Tazawa 1987; Zherelova 1989) but there are only a few patch–clamp studies of anion selective channels in the plasma membrane (Coleman, 1986; Schauf and Wilson, 1987; Keller et al., 1989; Schroeder and Hagiwara,1989; Boult et al, 1989; Terry et al 1991). In protoplasts of *Amaranthus tricolor* (see Terry *et al.* 1991) the anion IRs have very short open times in cell attached and whole–cell configurations, but in detached patches the channels show extremely long open times and multiple conductance levels. The single channel conductance can range from 10 to 200 pS (100/100 mM Cl^-) and the relative permeability sequence, approximately constant for all subconductance levels, is $NO_3^- > Cl^- > K^+ > Asp^-$ (2 : 1 : 0.2 : 0.04). The anion IR is progressively activated by external pH < 6.5 and effectively inactivated at external pH > 7 in outside–out detached patches. The difference in behaviour of the anion IR in intact and detached membranes suggests that its activity is under some sort of constant cellular control. $[Ca^{2+}]_{cyt}$ may be involved in this control; anion channels in several other plant cells have been shown to be activated by cytosolic Ca^{2+} (Kataev et al. 1984; Shiina and Tazawa 1987; Zherelova 1989; Keller et al., 1989; Schroeder and Hagiwara,1989). The *Amaranthus* anion IR may also be stretch–activated. A stretch sensitive anion channel has also been described by Falke *et al.* (1988) in cultured tobacco cells.

The inward rectifying anion channels usually show distinct subconductance states (Coleman, 1986; Terry et al, 1991). This may be physiologically significant since it adds another dimension to the way in which the channel can be modified (i.e mean current through channels is number of active channels x probability of opening x current, where current can vary at a constant PD for a multistate channel). The subconductance states can also give some insight about the structure of the protein, for example whether the channel is of a single or multi–barrelled form (see Fox 1987). For the *Amaranthus* channel permeability ratios of the various levels are similar but not identical; the higher conductance levels are slightly less selective for anions over cations than the lower levels. The main conductance state is also not clearly composed of equal subconductances. Therefore at present the evidence tends not to support a model consisting of identical subunits as has been proposed for some animal multistate channels (Miller, 1982; Hunter and Giebisch, 1987).

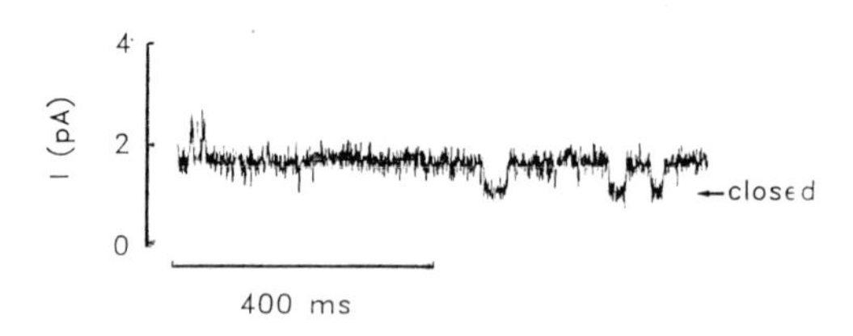

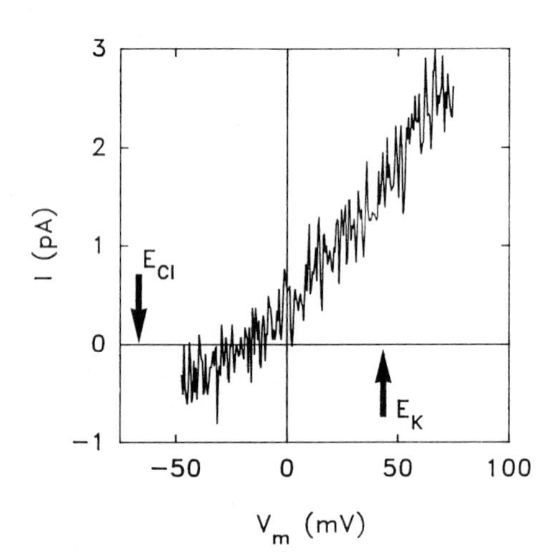

Figure 2 (a) Two portions of the current trace from an outside–out patch of membrane in which both an anion OR and cation OR are active at a membrane PD of +76 mV. The arrow indicates where a cation OR channel has opened at the same time as the anion OR is open. The baseline is marked "closed".
(b) Current–voltage curve for anion OR channel, obtained by ramping the membrane potential across an outside–out patch as the channel is in the open state (see Terry *et al.* 1991). In fact two anion OR channels were open as this ramp was recorded thus the conductance is near 25 pS rather than 12 pS. The equilibrium potentials for K^+ and Cl^- are marked E_K and E_{Cl} respectively. Solutions were (mM): Bath – 100 KCl, 10 $CaCl_2$, 10 MES, 7 KOH, 460 Sorbitol, pH6.0. Pipette – 5 KCl, 2 $MgCl_2$, 1 EGTA, $2K_2ATP$, 670 Sorbitol, 5 HEPES, 10.4 KOH, pH 7.2.

Outward rectifying anion channels: conduct anions into, and cations out of the cell. Characteristics of the anion OR from *Amaranthus are shown in fig. 2.* The voltage and time–dependent behaviour of anion OR channels in *Amaranthus* is very similar to that of the cation ORs, making it difficult to separate the two types of channel. As already indicated, the anion OR is not particularly anion selective with a relative permeability for $Cl^-:K^+$ of about 2:1. Single channel conductance is approximately 12 pS (10/100 mM K^+).

FUNCTIONS OF ION CHANNELS

Control of $\Delta\mu_{H+}$: From the studies on Charophytes it could be expected that the anion IR in the plasma membrane of higher plants should also be activated by acidic pH (Coster, 1969; Tyerman et al. 1986 a,b). The *Amaranthus* anion IR behaves in just this way (Terry and Tyerman, unpublished results). We have found inward rectifying anion channels in all other plant plasma membranes that we have patched to date (*Stylidium, Triticum, Cucumis, Hordeum,*: Terry, Tyerman, Schachtman and Double, unpublished results); such anion channels would seem to be a common feature in the plasma membrane of plant cells. If the higher plant anion IR has similar characteristics to its counterpart in *Chara* then by virtue of its pH dependence, it will help to maintain a constant $\Delta\mu_{H+}$ via its effect on the PD.

Net fluxes: For a net flux through the cation OR another conductance in the membrane must be able to move the PD positive of the reversal potential of the channel as occurs during the action potential in *Chara* (Hope and Findlay, 1964). The anion IR, once activated, would perform this function. The net result of combined activity of the two channels would result in an efflux of KCl.

There has been increasing speculation that the cation OR may be activated indirectly at depolarised PDs by a voltage dependent Ca^{2+} influx (Moran *et al*, 1990; Ketchum and Poole, 1991). If this is the case, and if the anion IR is also Ca^{2+} activated, then the situation in higher plant cells may be something like the action potential in *Chara*. The combined activation of an anion channel and the cation OR has also been used to explain the efflux of KCl during stomatal closure (Keller et al, 1989).

Selectivity and salinity tolerance: There are differences in the cation selectivities of OR channels in plants with K^+/Na^+ permeabilities ranging from 1 (Stoeckel and Takeda, 1989) to about 30 (Schachtman *et al.*). The variety found in the animal cell counterparts, the delayed rectifiers (e.g. see Hille 1984; Rudy 1988) probably indicates that with more extensive experimental investigations an increased diversity in the outward rectifier channels in plants may become apparent.

Whilst the cation OR is termed a rectifier, its rectification derives from voltage dependent kinetics, and not from some extreme non–linearity of the single channel current–voltage relationship. An influx of K^+ would occur through the cation OR if the proton pump were running and the channel remained open. In both wheat and *Amaranthus* we have observed the cation OR to continue to open in the long term at PDs more negative than the reversal potential, albeit with very different kinetics, therefore allowing some K^+ influx. The fact that it is the same channel which passes outward current is easily demonstrated using the fast voltage ramp technique (Tyerman and Findlay, 1989; Terry *et al*, 1991, Schachtman *et al*). An influx of another cation, for example Na^+, may also occur through the channel if there is an appropriate inward driving force and a significant Na^+ permeability. We have presented the argument that Na^+ entry through the K^+ outward rectifier can account for most of the measured Na^+ permeability of glycophyte roots (Schachtman *et al.*). Furthermore, an efflux of K^+ through the channel would occur in high NaCl concentrations, at least for glycophytes where the membrane potential becomes depolarised (Cakirlar and Bowling, 1981; Katsuhara and Tazawa, 1986; Kourie and Findlay, 1990). Halophytes may not show such a depolarisation (Cheeseman, Bloebaum and Wickens, 1985). The fluxes of Na^+ and K^+ that would occur through the channel could easily account for both the accumulation of Na^+ and the loss of K^+ often observed under saline conditions (e.g. Tyerman *et al.*, 1989 and Schachtman, *et al.* 1989).

Despite showing that a significant Na^+ uptake could occur through the cation OR, we were unable to discern any difference in K/Na selectivity between two varieties of wheat differing in their sodium fluxes to the shoot and their salinity tolerance (Schachtman *et al.* 1989). The anion IR could also allow a significant Na^+ influx by virtue of its high conductance and relatively low anion/cation selectivity (Terry *et al*, 1991) but this has not yet been examined in detail. The limited information that we have for the cation IR in *Amaranthus* suggests that it too has a high relative permeability for Na^+, but we do not know whether this is so for wheat cells. It seems likely that salt tolerance may derive from the way that channels are controlled in different plant varieties. At least one second messenger system responds to NaCl treatment with an increase in $[Ca^{2+}]_{cyt}$ (Lynch *et al.* 1990; Bittisnich *et al.* 1989). Phosphorylation of channels (or assoc. proteins) (Levitan, 1985) or direct modulation by ATP or G proteins (Brown and Birnbaumer, 1990) are also processes likely to be involved in the control of ion channels.

ACKNOWLEDGMENT

This work was supported by an Australian Research Council grant to S.D.T, G.P.F and D.C. Elliott.

REFERENCES

Beilby MJ (1985) Potassium channels at *Chara* plasmalemma. J Exp Bot 36:228–239

Bisson MA (1984) Calcium effects on electrogenic pump and passive permeability of the plasma membrane of *Chara corallina*. J Membrane Biol 81:59–67

Bittisnich D, Robinson D, Whitecross M (1989) Intracellular free calcium levels in the root cells of plants showing differential tolerance to salinity. In: Dainty J, De Michaelis MI, Marre E, Rasi Caldagno F (eds) Plant membrane transport: the current position. Elsevier, Amsterdam, 681–682

Blatt MR, Thiel G, Trentham DR (1990) Reversible inactivation of K^+ channels ov *Vicia* stomatal guard cells following the photolysis of caged inositol 1,–4,5–trisphosphate. Nature 346:766–769

Brown AM, Birnbaumer L (1990) Ionic channels and their regulation by G protein subunits. Ann Rev Physiol 52:197–213

Bush DS, Hedrich R, Schroeder JI, Jones RL (1988) Channel–mediated K^+ flux in barley aleurone protoplasts. Planta 176:368–377

Cakirlar H, Bowling DJF (1981) The effect of salinity on the membrane potential of sunflower roots. J Exp Bot 32:479–485

Cheeseman JM, Bloebaum PD, Wickens LK (1985) Short term $^{22}Na^+$ and $^{42}K^+$ uptake in intact, mid–vegetative *Spergularia marina* plants. Physiol Plant 65:460–465

Coleman HA (1986) Chloride currents in *Chara* – A patch–clamp study. J Membrane Biol 93:55–61.

Coleman HA, Findlay GP (1985) Ion channels in the membrane of *Chara inflata*. J Membrane Biol 83:109– 118

Coster HGL (1965) A quantitative analysis of the voltage–current relationships of fixed charge membrane and the associated property of "punch–through". Biophys J 5:669–686

Coster HGL (1969) The role of pH in the punch–through effect in the electrical characteristics of *Chara corallina*. Aust J Biol Sci 22:365–374

Falke L, Edwards KL, Pickard BG, Misler S (1987) A stretch activated anion channel in cultured tobacco cells. Biophysical Journal 51:251a

Findlay GP, Coleman HA (1983) Potassium channels in the membrane of *Hydrodictyon africanum*. J Membrane Biol 68:179–189

Fox JA (1987) Ion channel subconductance states. J Membrane Biol 97:1–8

Hille B (1984) Ionic channels of excitable membranes. Sinauer, Sunderland MA

Hope AB, Findlay GP (1964) The action potential in *Chara*. Plant Cell Physiol (Tokyo) 5:377–379

Hunter M, Giebisch G (1987) Multi–barrelled K^+ channels in renal tubules. Nature 327:522–524

Iijima T, Hagiwara S (1987) Voltage–dependent K channels in protoplasts of trap–lobe cells of *Dionaea muscipula*. J Membrane Biol 100:73–81

Kataev AA, Zherelova OM, Berestovsky GN (1984) Ca^{2+} induced activation and irreversible inactivation of chloride channels in the perfused plasmalemma of *Nitellopsis obtusa*. Gen Physiol Biophys 3:447–462

Katsuhara M, Tazawa M (1986) Salt tolerance in *Nitellopsis obtusa*. Protoplasma 135:155–161

Keller BU, Hedrich R, Raschke K (1989) Voltage–dependent anion channels in the plasma membrane of guard cells. Nature 341:450–453

Ketchum KA, Shrier A, Poole RJ (1989) Characterisation of potassium–dependent currents in protoplasts of corn suspension cells. Plant Physiol 89:1184–1192

Ketchum KA, Poole RJ (1991) Cytosolic calcium regulates a potassium current in corn (*Zea mays*) protoplasts. J Membrane Biol 119:277–288

Kourie JI, Findlay GP (1990) Ionic currents across the plasmalemma of *Chara inflata* cells II Effects of external Na^+, Ca^{2+} and Cl^- on K^+ and Cl^- currents. J Exp Bot 41:151–163

Levitan, IB (1985) Phosphorylation of ion channels. J Membrane Biol 87:177–190

Lunevsky VZ, Zherelova OM, Beretovsky GN (1983) Excitation of *Characeae* cell membranes as a result of activation of calcium and chloride channels. J Membrane Biol 72:43–58

Lynch J, Polito VS, Lauchli A (1990) Salinity stress increases cytoplasmic Ca activity in maize root protoplasts. Plant Physiol 90:1271–1274

Miller C (1982) Open–state substructure of single chloride channels from *Torpedo* electroplax. Phil Trans R Soc Lond (Ser B) 299:401–411

Moran N, Satter RL (1989) K^+ channels in plasmalemma of motor cells of *Samanea saman* In: Dainty J, De Michaelis MI, Marre E, Rasi Caldagno F (eds) Plant membrane transport: The current position. Elsevier, Amsterdam, 529–530

Moran N, Fox D, Satter RL (1990) Interaction of the depolarization–activated K^+ channel of *Samanea saman* with inorganic ions: a patch–clamp study. Plant Physiol 94:424–431

Rudy B (1988) Diversity and ubiquity of K channels. Neuroscience 25:729–749

Schachtman DP, Tyerman SD, Terry BR (to be published) K^+/Na^+ selectivity of a cation channel in the plasma membrane of root cells does not account for salinity tolerance in wheat. Plant Physiol

Schauf CL, Wilson KJ (1987) Properties of single K^+ and Cl^- channels in *Asclepias tuberosa* protoplasts. Plant Physiol 85:413–418

Schroeder JI (1988) K^+ properties of K^+ channels in the plasma membrane of *Vicia faba* guard cells. J Gen Physiol 92:667–683

Schroeder JI, Hagawari S (1989) Cytosolic calcium regulates ion channels in the plasma membrane of *Vicia faba* guard cells. Nature 338:427–430

Schroeder JI, Raschke K, Neher E (1987) Voltage dependence of K^+ channels in guard cell protoplasts. Proc Nat Acad Sci 84:4108–4112

Stoekel H, Takeda K (1989) Calcium–activated voltage–dependent non–selective cation currents in endosperm plasma memnbrane from higher plants. Proc R Soc Lond Ser B 237:213–231

Terry BR, Tyerman SD, Findlay GP (1991) Ion channels in the plasmalemma of *Amaranthus* protoplasts: one cation and one anion channel dominate the conductance. J Membrane Biol 121:223–236

Tester M (1990) Plant ion channels: whole–cell and single–channel studies. New Phytol 114:305–340

Tyerman SD, Findlay GP (1989) Current–voltage curves of single Cl^- channels which coexist with two types of K^+ channel in the tonoplast of *Chara corallina*. J Exp Bot 40:105–117

Tyerman SD, Findlay GP, Paterson GJ (1986a) Inward membrane current in *Chara inflata* I A voltage– and time–dependent Cl^- component. J Membrane Biol 89:139–52

Tyerman SD, Findlay GP, Paterson GJ (1986b) Inward membrane current in *Chara inflata*: II Effects of pH, Cl–channel blockers and NH_4^+, and significance for the hyperpolarized state. J Membrane Biol 89:153–61

Zherelova OM (1989) Activation of chloride channels in the plasmalemma of *Nitella syncarpa* by inositol 1,4,5–triphosphate. FEBS Lett 249:105–107

THE EFFECTS OF LOW O_2 CONCENTRATION AND AZIDE ON THE WATER RELATIONS OF WHEAT AND MAIZE ROOTS.

S.D. Tyerman, W.H. Zhang and B.R. Terry
School of Biological Sciences, The Flinders University of South Australia, G.P.O. Box 2100, Adelaide, South Australia 5001.

SUMMARY

The water relations parameters of wheat and maize roots were measured using pressure probe techniques. We investigated the short term effects of O_2 deficiency and the TCA cycle inhibitor sodium azide on the water relations parameters of single cells of wheat and maize, and whole roots of maize. Hydraulic conductivities (L_p) of single cortex cells were obtained by measuring either turgor pressure relaxation after turgor pressure was quickly changed, or volume flow while turgor pressure was clamped. Whole root water relations were obtained only on maize using the root pressure probe technique (Steudle and Jeschke, 1983).

Both low O_2 concentration (0.038 – 0.042 mol m^{-3}) and 1 mol m^{-3} NaN_3 substantially decrease the hydraulic conductivity of single cells of wheat and maize but have only a small effects on the whole root hydraulic conductivity of maize. The pressure probe can also be used to evaluate the apparent osmotic volume of single cells. In aerated solutions the osmotic volume of wheat root cells is about 2 times higher than their geometric volume, but in low O_2 concentration and NaN_3 solutions the osmotic cell volume is reduced and approximately equal to the geometric volume. The results for wheat indicate that the decrease in osmotic volume and hydraulic conductivity of cortex cells induced by both low O_2 concentration and NaN_3 may be partly due to the closure of plasmodesmata. Calculations based on estimated values of pore diameters and fluid viscosity within plasmodesmata and densities of plasmodesmata indicate that their occlusion may account for a proportion of the decrease in L_p. However, changes in the plasma membrane may also account for some reduction in L_p.

NATO ASI Series, Vol. H 64
Mechanics of Swelling
Edited by T. K. Karalis

INTRODUCTION

When the roots of plants are exposed to an O_2–deficient environment, as will occur in water logged soil, they exhibit many changes in physiology, morphology and biochemistry (Jackson and Drew 1984). In many plants an initial sign of oxygen stress is that of wilting of leaves (Kramer and Jackson 1954) and this is attributed to continued transpiration under the influence of a reduced hydraulic conductivity (L_p) of roots (Bradford and Hsiao 1982; Everard and Drew 1987; 1989). The change in L_p occurs in the radial pathway across the root to the xylem but the reasons for this change in root L_p have not yet been determined. It has been proposed that the decrease in L_p of roots treated anaerobically may be due to a decrease in L_p of those root cells which offer the main resistance to water flow from the external solution to the xylem (Everard and Drew 1989), but there has been no convincing experimental evidence to confirm this proposal mainly because of experimental difficulties.

It is possible that reduced L_p of roots under hypoxia and other conditions affecting their energy status may be due to reduced symplasmic flow of water via plasmodesmata (Pitman *et al.* 1981). There is some evidence that the energy status of the cell effects transport via plasmodesmata. Drake (1979) demonstrated that 1 mol m^{-3} NaN_3 an inhibitor of the TCA cycle, reduces the electrical coupling between cells in oat coleoptiles. This reduction of electrical coupling suggests an increase in plasmodesmata resistance caused by their occlusion. The occlusion of plasmodesmata would also reduce L_p of the cells, as measured with the pressure probe, because under normal circumstances water flow should occur both across the plasma membrane and through plasmodesmata when the water equilibrium is disturbed. To test the significance of plasmodesmata in accounting for changes in flow through the symplast the pressure probe has been used to measure changes in osmotic volume of single cortex cells of wheat (Zhang and Tyerman). In this paper we further describe these experiments on wheat and extend the work by examining the effect of low O_2 on L_p of cortex cells and whole roots of maize.

MATERIALS AND METHODS

Wheat (*Triticum aestivum* L. cv. Machete) and maize (*Zea mays* L. Var. XL–35) seedlings were grown in fully aerated 1/2 Hoagland solution. All the roots used in the experiments were 4–7 days old and the pressure probe measurements were made on cells 10–20 mm (wheat) and 20–30 mm (maize) from the root apex.

Single cell water relations

For pressure probe experiments on single cells, an excised root was held in a plexiglass chamber which allowed observation of the microcapillary and cortex cells under a microscope. A 2 x 5 mm opening on one side of the chamber allowed access for the microcapillary of the pressure probe to be inserted into the root. The root was about 3 mm from the surface of the water meniscus at the access port. A glass slide was sealed onto the top of the chamber with vacuum grease. The bathing solution (1/2 Hoagland) was continuously recirculated (7 $cm^3\ min^{-1}$) from a reservoir (about 250 ml), into the root chamber and back to the reservoir. Low oxygen concentration was achieved by bubbling the solution in the reservoir with O_2 –free N_2 gas. The low oxygen concentrations ranged between 0.038 – 0.042 mol m^{-3} as measured just upstream and downstream of the root chamber. Fully aerated solution had an O_2 concentration of 0.276 mol m^{-3}. The temperature was maintained at 20°C. For experiments using NaN_3, the bathing solution was changed to fully aerated 1/2 Hoagland solution with 1 mol m^{-3} NaN_3. Measurements were after 30 min of treatment.

Pressure relaxation experiments

In these experiments the meniscus between the oil and sap in the capillary was quickly changed to a new position using the remote controlled piston in the pressure probe (Hüsken *et al.*, 1978) (Fig.1). While the meniscus was maintained manually at the new position, the turgor pressure recorded on a chart recorder would relax to a new steady level as water flow proceeded across the effective cell membrane (Fig. 2). The hydraulic conductivity (L_p) was calculated using the following equation (Zimmermann and Steudle 1978):

$$L_p = \frac{V.\ln 2}{A.T_{1/2}.(\varepsilon+\pi)}, \quad (1)$$

where V and A are the cell volume and surface area, respectively; $T_{1/2}$ is the half–time for the turgor relaxation and π is the internal osmotic pressure of the cells. The volumetric elastic modulus (ε) was determined independently with the pressure probe as described by Tyerman and Steudle (1982). After measurements of ε and of several turgor relaxations in aerated solution, the oxygen content of the flowing solution was reduced. The measurements were then repeated after 30 min in the low O_2 solutions. At least two turgor relaxations (endosmotic and exosmotic volume flow) were measured in both aerated and low O_2 solutions. A similar protocol to the treatments was used as a control to check for any consistent changes in L_p with time in aerated solution. For each cell this consisted of a set of pressure relaxations, 30 min break, then another set of pressure relaxations.

The pressure relaxation experiments can also be used to determine osmotic volume as described by Malone and Tomos (1990). The osmotic volume (V_0) is given by:

$$V_0 = \frac{P_0.dV}{P_0-P_e}, \quad (2)$$

where dV is the volume change to generate a turgor relaxation and was measured from the change in the position of the oil/sap meniscus in the capillary, and (P_0–P_e) is the difference between the initial equilibrium turgor pressure (P_0) and final equilibrium turgor pressure (P_e) after the relaxation. The osmotic volume was determined in both high O_2 and low O_2 for 5 cells of wheat.

Pressure clamp experiments

The pressure clamp technique as described by Wendler and Zimmermann (1982) was used on wheat cells. The turgor pressure was changed by a certain amount (dP) from its steady state value and manually clamped at a new value using the remote controlled piston of the pressure probe to keep the output on the chart recorder constant. The volume flow required to clamp the turgor pressure was measured by a second operator from the movement of the oil/sap boundary along the microcapillary with time. The time course of the volume relaxation (V(t)) is given by:

$$V(t) = \frac{dP}{\pi} . V_0 . [\exp(-t/t_c) - 1] , \qquad (3)$$

where:

$$t_c = \frac{V_0}{A . L_p . \pi} . \qquad (4)$$

The half time of a volume relaxation under pressure clamp is longer than that for a turgor relaxation (Wendler and Zimmermann, 1982). The slower kinetics allow L_p to be accurately measured from the initial slope (S_v) of the volume relaxation, which does not require a measurement of $\in$, π and V:

$$L_p = \frac{-\ S_v}{A . dP} . \qquad (5)$$

The osmotic volume (V_0) of the cell can be estimated from the total change in cell volume at t –> ∞ (dV_∞):

$$V_0 = \frac{-dV_\infty \ . \ \pi}{dP} . \qquad (6).$$

The volume relaxations could not reach equilibrium because the changes in volume could not be accommodated by the microcapillary. Equation (2) was fitted to the data (least squares method) and extrapolated to obtain dV_∞. It was suggested by Wendler and Zimmermann (1982) that a more exact equation be used to measure V_o (their Eq. 5) since in the derivation of Eq. 6 the approximation is made that $\Delta\pi = -\pi\Delta V/V$. We found that Wendler and Zimmermann's method was experimentally more difficult so we opted for a slight loss of precision provided that any changes in osmotic volume could also be confirmed by the pressure relaxation method (Eq. 2). After several pressure clamp experiments (both endosmotic and exosmotic) had been performed in aerated solutions, the bathing solution was changed to one of low O_2 concentration or to 1 mol m^{-3} NaN_3 solution. The pressure clamp experiments were repeated on the same cells after 30 min of these treatments.

Cell dimensions

After an experiment the cell was highlighted by filling it with silicone oil from the pipette. This allows for more accurate measurement of cell dimensions. The cells were approximated to cylinders.

Whole root experiments

The whole root pressure probe technique (Steudle and Jeschke, 1983; Steudle *et al.* 1987) was used on maize roots. So far our attempts with wheat roots have been unsuccessful due to sealing problems with the smaller diameter root. The pressure probe was modified so that an excised root (length 66–95 mm) could be connected via a silicone rubber seal (Sylgard, Dow Corning). Connected between the root–seal and the pressure probe was a 0.25 mm (inside diameter) capillary. Distilled water filled the root–seal section and about half way along the capillary. Silicone oil filled the rest of the capillary and the transducer section. The interface between the silicone oil and the distilled water (the meniscus) could be viewed under a microscope so that changes in volume could be measured of the root and water filled portion of the apparatus. The root was supported horizontally in a chamber made air tight except for the port where the root is connected to the probe. Solution (1/2 Hoagland) could be pumped through the chamber from the distal end of the root to the proximal end and out through the access port. This allowed solutions with low oxygen concentration to be passed along the full length of the root. The O_2 concentration measured upstream of the root chamber was between .03 and .04 mol m^{-3}.

After root pressure developed in aerated solution pressure relaxation experiments were performed. The hydraulic conductivity was measured from the initial slope of the relaxation curve according to Steudle and Jeschke (1983):

$$L_{pr} = \frac{1}{A_r.(P_{ro}-P_{ra})} \cdot (\Delta V_s/\Delta P_r) \cdot (dP_r/dt)_{t=0} \, , \tag{7}$$

where $P_{ro}-P_{ra}$ is the difference between the equilibrium value of root pressure and the peak at the beginning of the relaxation (see Fig. 3). The initial slope of the pressure relaxation $(dP_r/dt)_{t=0}$ was obtained from fitting the first portion of the digitised relaxation curve to an exponential equation and finding the time derivative of the fitted equation at t=0. The compressibility $(\Delta V_s/\Delta P_r)$ was measured during the experiments by changing volume (movement of the meniscus) and measuring the change in pressure. The measurements of L_{pr} are on a root surface area (A_r) basis over conducting xylem. The portion of the root containing non–conducting xylem was measured by successively cutting the root from the tip until the root pressure declined.

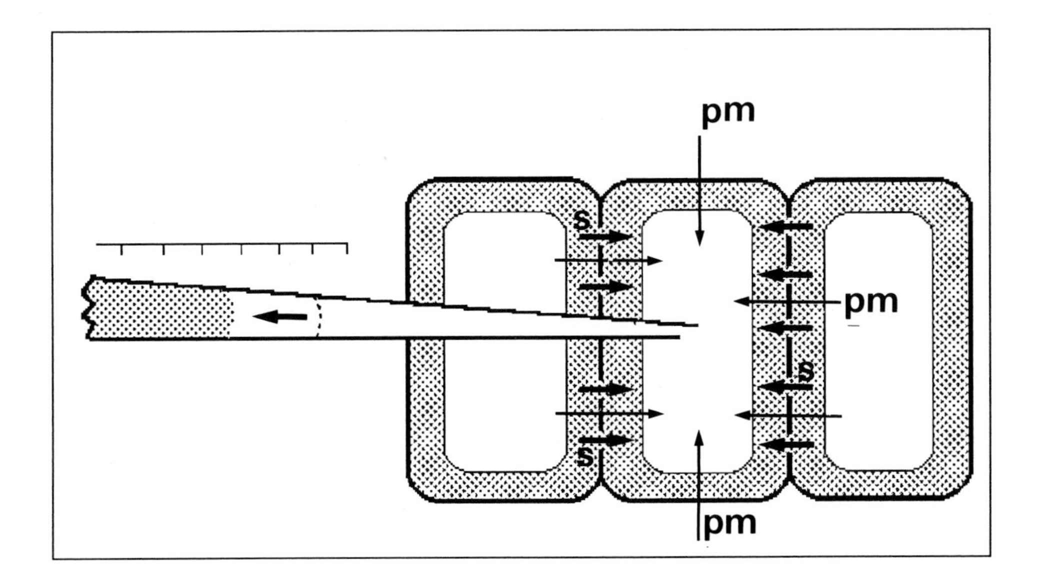

Figure 1. Diagram showing placement of the pressure probe microcapillary in a single cortex cell of a root. The meniscus in the capillary is viewed with a microscope and its position measured with an ocular micrometer. The two pathways of water flow from the penetrated cell are labelled pm (plasma membrane) and s (symplast via plasmodesmata).

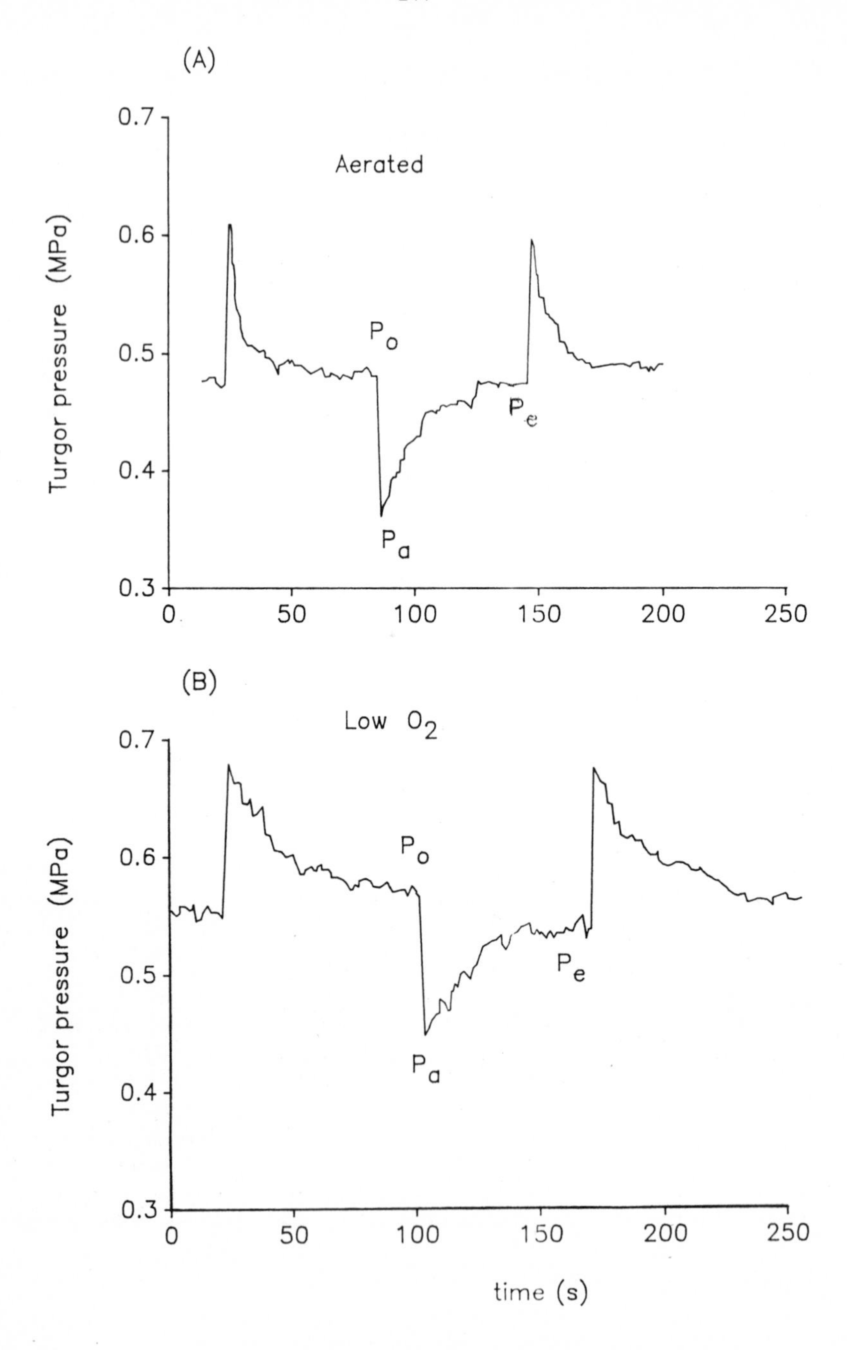

Figure 2. Tracing of a chart recording of turgor relaxation curves in high O_2 (A) and low O_2 (B) on the same maize cell and showing the increase in $T_{1/2}$ and the increase in the difference between initial equilibrium turgor (P_o) and final equilibrium turgor (P_e).

RESULTS

Turgor pressure

Turgor pressure of cortical cells of wheat and maize roots was in the range 0.4 to 0.75 MPa. The turgor pressure (P) of individual cells was monitored continuously for up to 1 hr after transferring the roots from high to low O_2 concentration. Most cells showed an initial increase in P (less than 0.1 MPa, e.g. Fig. 2) and then a gradual decrease in P. In some cells no changes in P were evident. A decline in P would indicate a loss of osmotica from the cell. Leakage of organic and amino acids, and K^+ and Cl^- ions from excised barley roots treated by anoxia has been reported (Hiatt and Lowe 1967).

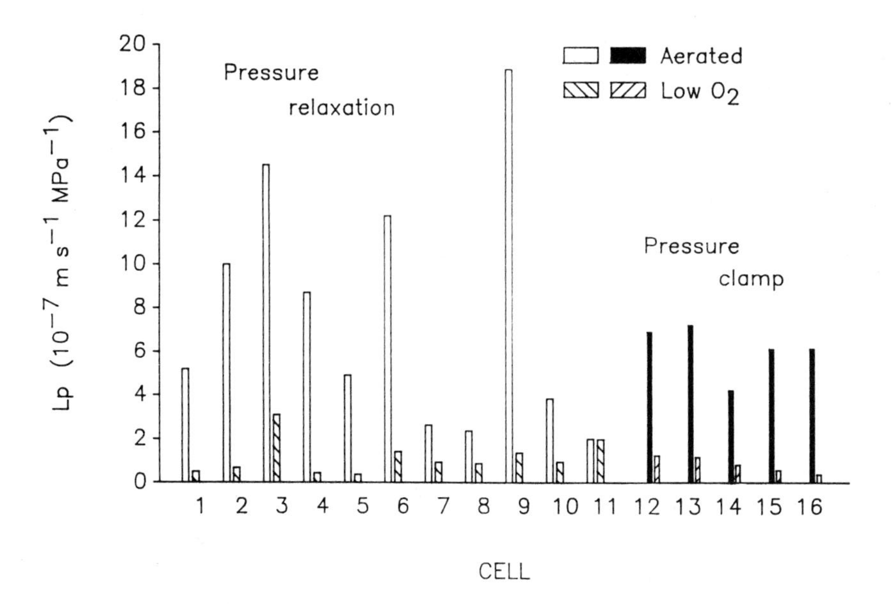

Figure 3. Effect of low O_2 on L_p recorded for individual cortex cells of wheat roots using either pressure relaxation or pressure clamp experiments.

Hydraulic conductivity of single cells

Pressure relaxation experiments show that the half time of water exchange increases significantly in low O_2. Fig. 2 shows pressure relaxation experiments performed on a maize cortical cell before and after exposure to low O_2 solution and illustrates the increase in $T_{1/2}$. This increase in $T_{1/2}$ in low O_2 indicates a marked decrease in the hydraulic conductivity of the cortical cell since ε did not change. Accordingly, the mean values for L_p were 5.574 and 2.296 x 10^{-7} m s^{-1} MPa^{-1} in high and low O_2, respectively. A more extensive set of L_p values for wheat cortex cells before and after low O_2 treatment are shown in Fig. 3 (from Zhang and Tyerman). Fig. 3 also shows that pressure clamp experiments yield the same result. The pressure clamp, however, gives significantly less variation in L_p since calculation of L_p does not require separate determinations of ε, V and π (see equation 4). Azide (1 mol m^{-3}) also significantly reduces L_p of wheat cortex cells as determined by pressure relaxation and pressure clamp experiments (Fig. 4). Control experiments under aerated conditions showed no consistent changes in L_p with time (Zhang and Tyerman). The values of L_p for cortical cells of wheat and maize roots in aerated solutions are in the same range of those reported by Jones *et al.* (1983) and Steudle *et al.* (1987) respectively.

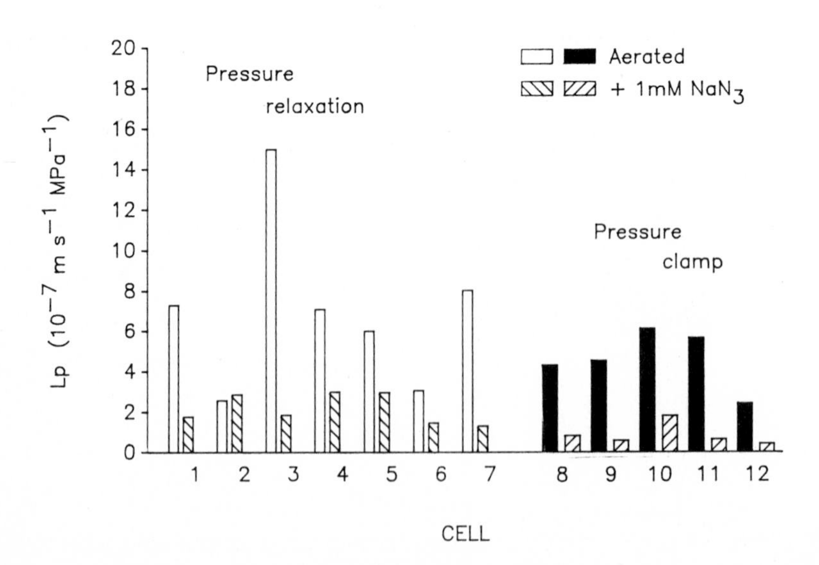

Figure 4. Effect of 1 mol m^{-3} NaN_3 on L_p recorded for individual cortex cells of wheat roots using either pressure relaxation or pressure clamp experiments.

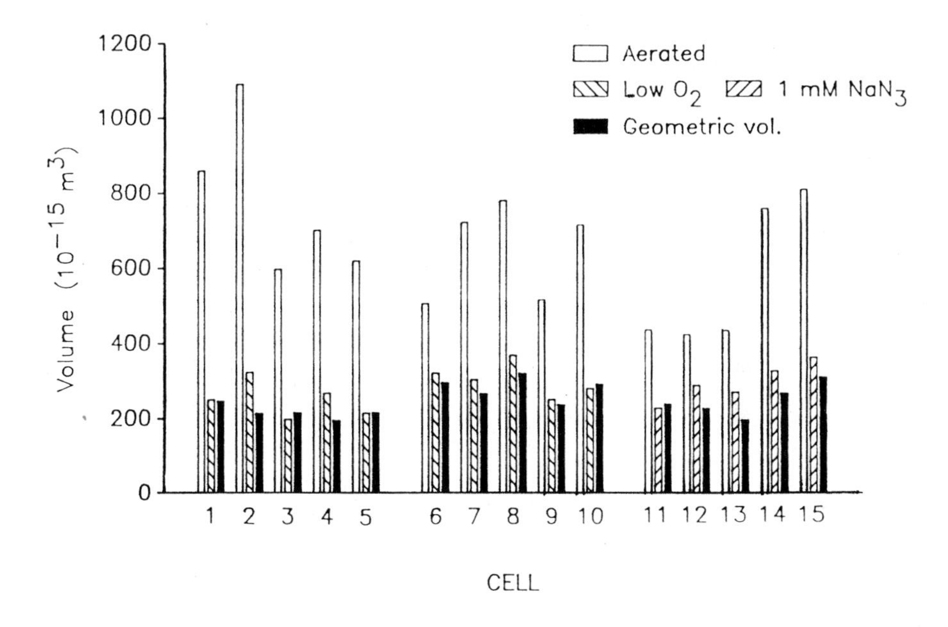

Figure 5. Comparison of geometric volume with osmotic volume in aerated, low O_2 and NaN_3 treatments. For cells 1 to 5 the osmotic volume was measured from pressure relaxation experiments (Eq. 2).

Apparent osmotic volume

The pressure clamp technique can also be used to estimate the osmotic cell volume. In high O_2 the osmotic volume of the cells is about twice the geometric cell volume (Fig. 5). Despite variations among the cells, in low O_2 and 1 mol m^{-3} azide the osmotic cell volume is reduced to near the value of the geometric cell volume. Calculations by Zhang and Tyerman show that the high osmotic volumes in aerated solution are not due to an effect of the hydraulic conductivity of the pipette. Also the values of osmotic volume measured from pressure relaxation experiments are similar to those measured by pressure clamp experiments (Fig. 5). In high O_2, the difference between initial turgor (P_o) before a relaxation and the subsequent equilibrium turgor (P_e) after a relaxation is small (see Fig. 2a for maize); this leads to high values of V_0 (Eq. 2). In low O_2, however, when the same ΔV for the turgor relaxation was used as for high O_2, there was a larger difference between P_o and P_e (Fig. 2b) indicating that the osmotic volume is reduced.

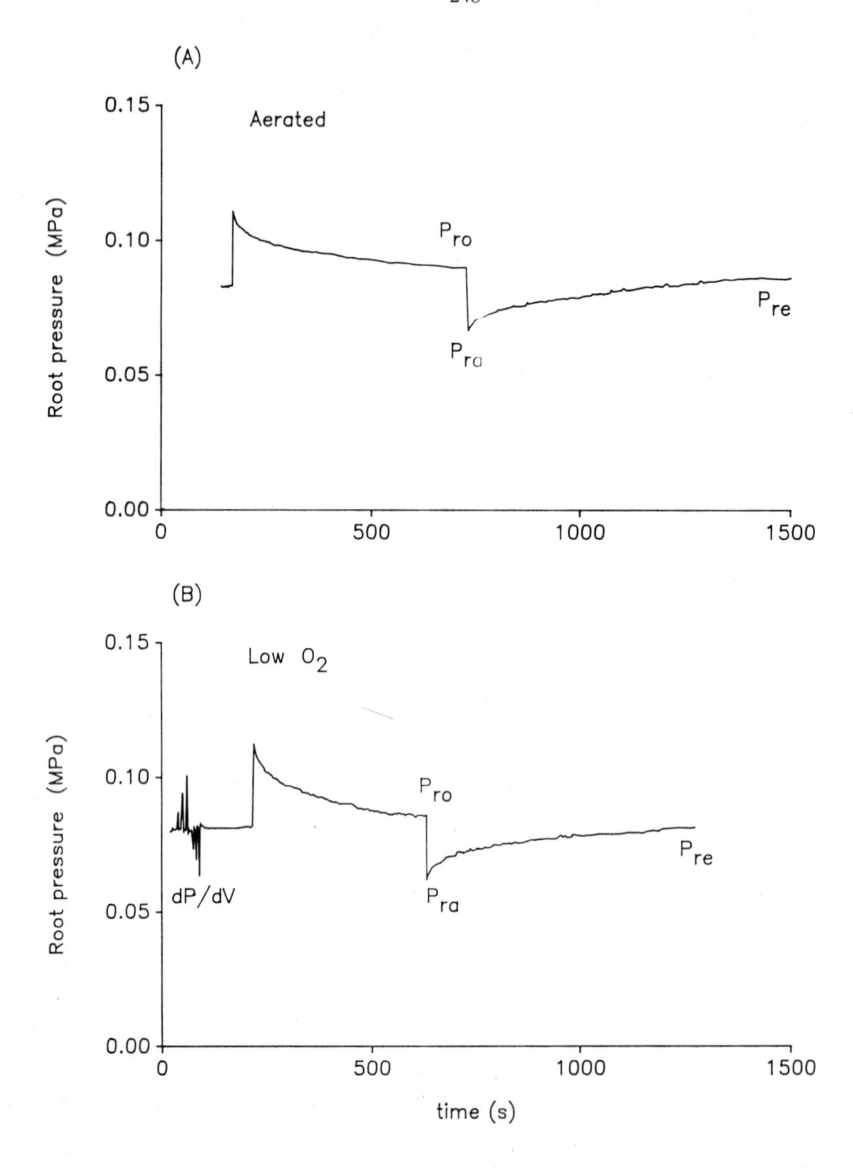

Figure 6. Root pressure probe experiment on a maize root in aerated (A) and low O_2 (B) solution. The root was treated with low O_2 for 30 min. before the second set of relaxations were performed. There was no obvious difference in L_p after another 30 min. of treatment. At the beginning of the trace in B is a determination of the compressibility (dP/dV).

Maize; whole root experiments

Figure 6 shows a root pressure probe experiment on a maize root. After connecting the root to the probe and maintaining the meniscus at a constant position the root pressure builds up over 30 to 60 minutes. The root pressure ranged from 0.08 MPa to 0.14 MPa for 5 roots (mean = 0.11 MPa). In response to both low O_2 and azide the root pressure declined by 20%. The mean L_p before treatment was 1.6×10^{-7} m s^{-1} MPa^{-1}. In low O_2 L_p was reduced by an average of 18%. Roots in azide showed a 33% reduction in L_p. The mean L_p under both treatments was 1.2×10^{-7} m s^{-1} MPa^{-1}.

DISCUSSION

Wendler and Zimmermann (1982, 1985) found that the osmotic volumes of single giant algal cells determined using the pressure clamp technique were in good agreement with the geometric cell volume. However, when the same technique is used in higher plant tissues, the cells may not behave the same, osmotically, as isolated algal cells because of exchange of solutes between cells in a tissue via plasmodesmata. An understanding of why the osmotic cell volume in aerated roots should be higher than the geometric volume can be obtained by considering what would happen in a simple pressure relaxation experiment on a cell which was connected to other cells via plasmodesmata which may allow rapid diffusion of solutes between the cells. If the turgor pressure is decreased in the test cell, water will flow into the cell across its membrane and possibly via mass flow through the plasmodesmata from adjacent cells. The flow of water into the test cell across its membrane will decrease the concentration of solutes in the cell leading to diffusion of solutes through the plasmodesmata from adjacent cells. In addition solutes will move with the mass flow through the plasmodesmata. The result will be that, for the volume flow induced into the cell, the concentration of solutes in the cell will not decrease as much as it should otherwise have if there were no plasmodesmata connected to adjacent cells. This leads to P_e being not very much less than P_o for an exosmotic pressure relaxation, as shown in Fig. 2a for example. A similar explanation applies to pressure clamp experiments. Essentially the cell behaves as if it had a very leaky membrane except that the cells are leaky with respect to one another rather than with the apoplast. The apparent osmotic volume of the cell will probably depend on a number of factors including the overall hydraulic resistance and diffusive resistance of the plasmodesmata relative to that for the membrane. If the plasmodesmata should close this would reduce the hydraulic conductivity measured on a cell surface basis and it would reduce the osmotic volume to near the geometric volume.

Another explanation of the osmotic volume results is that the pressure changes used to invoke water flow in the cell also result in rapid solute flows across the cell membrane. Such would be the result if the cell regulated turgor pressure very rapidly so that, for example during a pressure clamp experiment in which the turgor pressure is increased, solutes would flow out of the cell in order to reduce the internal osmotic pressure and turgor pressure. This would lead to an overestimation of osmotic volume. A similar argument could apply for reductions in turgor pressure. Low O_2 concentrations could stop the turgor mediated solute fluxes resulting in osmotic volumes being similar to geometric cell volumes. Although turgor mediated solute fluxes have been described in root cells of higher plants (Tyerman *et al.* 1989) this response is slow, i.e. half times of 20 minutes, not several seconds. Also during pressure relaxations in which turgor pressure is perturbed for only a short period we obtain similar osmotic volumes to those derived from pressure clamp experiments. Another difficult fact against the possibility that turgor regulation may be responsible is that *Lamprothamnium* (a known turgor regulator) had equal osmotic and geometric volume (Wendler and Zimmermann, 1985). However, *Lamprothamnium* being a giant celled alga, has a V/A ratio about an order of magnitude higher than a small higher plant cell so that turgor mediated solute fluxes may not be evident in the short term. Overall we prefer the explanation that solute flow through plasmodesmata is responsible for our large osmotic volume results particularly since this correlates with high L_p. We will lean this way through the rest of the discussion but in the words of Hank Greenway; "the jury may still be out on this issue".

Our results for osmotic volume under high O_2 contrast to the results obtained by Malone and Tomos (1990) on leaf epidermal cells of wheat, where it was found that the osmotic volume and geometric volume were about equal. There are two possible reasons for this. Firstly the turgor pressure changes used by Malone and

Tomos were quite large (larger than 0.2 MPa) which could lead to a closure of plasmodesmata. Pressure gradients can rapidly induce closure of plasmodesmata in *Chara* (Ding and Tazawa, 1989) and *Nitella* (Zawadzki and Fensom 1986). Secondly there may be very few plasmodesmata in the epidermis of wheat leaves reminiscent of the fact that in *Egeria densa* the epidermis does not seem to be connected to the cortex (Erwee and Goodwin, 1985).

Much research has been carried out on whether the plasmodesmata open or close in response to various cell perturbations. Treatments which have been reported to close plasmodesmata include azide and cyanide (Drake, 1979), osmotic shock (Drake *et al.*, 1978; Erwee and Goodwin, 1984), turgor pressure gradients (Ding and Tazawa, 1989; Zawadzki and Fensom, 1986), high intracellular Ca2+ concentration (Tucker, 1990) and microinjection of 1,4–inositol bisphosphate and 1,4,5–inositol trisphosphate (Tucker, 1988). Since azide and anoxia can induce an increase in intracellular Ca^{2+} concentration (Gilroy *et al.*, 1989), the reduction in conductivity of plasmodesmata in wheat roots could be caused by high intracellular Ca^{2+} concentration. Davies *et al.* (1987) have reported that the endoplasmic reticulum condenses into whorls when the energy charge of pea root tips is reduced below 0.6 by either hypoxia (.81% O_2) or an uncoupler of oxidative phoshorylation (FCCP). However, if most transport through a plasmodesma occurs via the cytoplasmic annulus, whorl formation of the ER is unlikely to be of significance.

The reduction of osmotic volume in low O_2 and NaN_3 solutions may be explained in terms of the plasmodesmata between the cells becoming occluded. However, possible changes in conductance of the plasmodesmata in response to low O_2 and NaN_3 may not necessarily account for all of the decrease in L_p. If all the change in L_p caused by hypoxia and NaN_3 was due to occlusion of plasmodesmata this would mean that plasmodesmata account for an L_p of about 6×10^{-7} m s^{-1} MPa^{-1}. Higher plant plasmodesmata have a tube of endoplasmic reticulum running through the pore (the desmotubule). The transport of solutes and water between cells most probably occurs in the cytoplasmic annulus between the desmotubule and plasma membrane (Overall *et al.*, 1982; Terry and Robards, 1987); quite large solutes can pass freely through the annuli (upto an effective Stokes radius of 1 nm; Terry and Robards, 1987) so that most solutes that account for the osmotic pressure of the cytoplasm could equilibrate with adjacent cells. For the purposes of calculating L_p we will assume that these pores can also allow bulk flow of water under a pressure gradient. We will assume that the conclusion of Terry and Robards (1987) applies; i.e. that the cytoplasmic annulus consists of 10 to 20 pores each with a radius of 1.5 nm. Using Eq. 12a of Tyree (1970) derived from Poiseuille's law, we can calculate L_p for plasmodesmata on a membrane surface area basis to see if it is possible for the reduction in L_p due to low O_2 and azide to be accounted for by closure of plasmodesmata. To account for an L_p of about 6×10^{-7} m s^{-1} MPa^{-1} (the difference between control and treatments for wheat) the plasmodesmatal density would need to be 2 μm^{-1} (at 20 pores per plasmodesma, 0.2 μm long) or 6 μm^{-1} (at 10 pores per plasmodesmata). These densities are in the upper range observed for roots. This calculation shows that it is possible to account for the change in L_p due to plasmodesmata becoming closed but there are some imponderable assumptions in this calculation, for example the viscosity of the pore fluid was taken to be that of pure water (.01 poise) since we do not believe that there would be a bulk flow of cytoplasm through the pores. Closure of plasmodesmata could account for some fraction of the difference in L_p between the treatments in our work. It remains to be seen what processes in the plasma membrane could also lead to a reduction in L_p.

There is evidence that transport systems in the plasma membrane are affected by hypoxia. Buwalda *et al.* (1988) showed that the membrane of wheat cortex cells depolarises from –120 mV to –80 mV and that there is a loss of K^+ from the cells. This result could be explained by the opening of K^+ channels in the plasma membrane. Both inward or outward rectifier K^+ channels are found in plasma membranes of plant cells (see Terry *et al.* this volume) and the outward rectifier K^+ channel is known to be present in the cortex cells of wheat roots (Schachtman *et al.*). However, opening of the outward K^+ rectifier channel in *Chara inflata* is not correlated with a change in L_p of the plasma membrane (Tyerman, unpublished). It is entirely possible that there other hydrophilic pores spanning the membrane which need to be metabolically maintained in some way (Wayne and Tazawa, 1990).

The values of L_p of higher plant cells so far investigated show a large variation between species. Root cortex cells are no exception to this with L_p for tobacco root cells being some 10 fold smaller than those for wheat and maize

(Tyerman *et al.* 1989). The state of the plasmodesmata might well account for this large variation. A continuous decrease in L_p when turgor relaxation experiments are repeated on the same cell has been observed in epicotyl segments of pea (Cosgrove and Steudle 1981) and in maize root cells (Steudle *et al.* 1987). Possible reasons for this decrease in L_p have been discussed by Cosgrove and Steudle (1981). Our results support their hypothesis that the decrease in L_p may be due to partial closure of plasmodesmata rather than an artefact of the technique. The decrease in L_p observed in low O_2 and NaN_3 solutions was not due to a progressive decrease in L_p with time since control experiments showed no reduction in L_p over 1 h.

Steudle *et al.* (1983) concluded that in response to hydrostatic pressure gradients the predominant water flow pathway across maize roots was in the apoplast. This is because L_p of the whole root is similar to that for individual cortex cells rather than an order of magnitude lower as occurs with barley roots. The reduction in root L_p of only 25% in response to azide and low O_2 despite a larger reduction in L_p of single cortex cells partly supports this notion. Nevertheless, a reduction in root L_p would not be expected if the flow pathway was exclusively apoplastic. Also Everard and Drew (1987) using hydrostatic pressure gradients observed somewhat larger reductions in L_p of maize roots subjected to hypoxia. This suggests that water flow must pass through at least one cell layer; most probably the endodermal cells. For wheat the picture may be somewhat different. The values of L_p for the cortical cells were reduced by a factor of seven in low O_2 compared to those in high O_2 so that water flow via a cell–to–cell pathway will inevitably encounter a greater hydraulic resistance. It is notable that L_p for whole wheat roots is considerably smaller than for maize roots (Jones *et al.* 1983), but this needs to be checked using the root pressure probe or the method of Everard and Drew (1987). The reduction in L_p of cortical cells under hypoxic conditions may have different effects on total water flow across the roots depending upon the predominant pathway. Plants with predominant apoplastic water flow should be less affected by hypoxic rooting media than those with cell–to–cell water flow.

Acknowledgments

We acknowledge the support of a Flinders University Research Grant and a

Commonwealth Overseas Postgraduate Research Scholarship to W.H. Zhang. We are grateful for Hank Greenway's general comments about our ideas.

REFERENCES

Bradford, KJ, and Hsiao, TC (1982) Stomatal behaviour and water relations of waterlogged tomato plants. Plant Physiol 70:1508–13

Buwalda, F, Thomson, CJ, Steigner, W, Barrett–Lennard, EG, Gibbs, J, and Greenway, H (1988) Hypoxia induces membrane depolarization and potassium loss from wheat roots but does not increase their permeability to sorbitol J Exp Bot 39:1169–1183

Cosgrove, D, and Steudle, E (1981) Water relations of growing pea epicotyl segments. Planta 153:343–50

Ding, D–Q and Tazawa, M (1989) Influence of cytoplasmic streaming and turgor pressure gradient on the transnodal transport of rubidium and electrical conductance of *Chara corallina*. Plant and Cell Physiol 30:739–748

Drake, GA, Carr, DK and Anderson, WP (1978) Plasmolysis, plasmodesmata and the electrical coupling of oat coleoptile cells. J Exp Bot 29:1205–14

Drake, G (1979) Electrical coupling, potentials, and resistances in oat coleoptiles: Effects of azide and cyanide. J Exp Bot 30:719–25

Erwee, MG and Goodwin, PB (1984) Characterisation of the *Egeria densa* Planch. leaf symplast: response to plasmolysis, deplasmolysis and to aromatic amino acids. Protoplasma 122:162–68

Erwee, MG and Goodwin, PB (1985) Symplast domains in extrastellar tissues of *Egeria densa* Planch. Planta 163:9–19

Everard, JD, and Drew, MC (1987) Mechanisms of inhibition of water movement in anaerobically–treated roots of *Zea mays* L. J Exp Bot 38:1154–65

Everard, JD, and Drew, MC (1989) Mechanisms controlling changes in water movement through the roots of *Helianthus annuus* L. during continuous exposure to oxygen deficiency. J Exp Bot 40:95–104

Gilroy, S, Hughes, A and Trewavas, AJ (1989) A comparison between Quin–2 and aequorin as indicators of cytoplasmic calcium levels in higher plant cell protoplasts. Plant Physiol 90:482–91

Hiatt, AJ, and Lowe, RH (1967) Loss of organic acids, amino acids, K^+ and Cl^- from barley roots treated anaerobically and with metabolic inhibitors. Plant Physiol 42:1731–36

Hüsken, D, Steudle, E, and Zimmermann, U (1978) Pressure probe technique for measuring water relations of cells in higher plants. Plant Physiol 61:158–163

Jackson, MJ, and Drew, MC (1984) Effects of flooding on growth and metabolism of herbaceous plants. In: Kozlowski T.T. (ed) Flooding and Plant Growth. Academic Press, London, p 47

Jones, H, Tomos, AD, Leigh, RA, and Wyn Jones, RG (1983) Water relations parameters of epidermal and cortical cells in the primary roots of *Triticum aestivum* L. Planta 158:230–36

Kramer, PJ, and Jackson, WT (1954) Causes of injury to flooded tomato plants. Plant Physiol 29:241–45

Malone, M and Tomos, AD (1990) A simple pressure–probe method for the determination of volume in higher–plant cells. Planta 182:199–203

Overall, RL, Wolfe, J, and Gunning, BES (1982) Intercellular communication in *Azolla* roots. I. Ulstrastructure of plasmodesmata. Protoplasma 111:134–50

Pitman, MG, Wellfare, D, and Carter, C (1981) Reduction of hydraulic conductivity during inhibition of exudation from excised maize and barley roots. Plant Physiol 61:802–8

Schachtman, DP, Tyerman, SD, and Terry, BR (to be published) The K^+/Na^+ selectivity of a cation channel in the plasma membrane of root cells does not differ in salt–tolerant and salt–sensitive wheat species. Plant Physiol

Steudle, E, and Jeschke, WD (1983) Water transport in barley roots: measurements of root pressure and hydraulic conductivity of roots in parallel with turgor and hrdraulic conductivity of root cells. Planta 158:237–48

Steudle, E, Oren, R, and Schulze, E–D (1987) Water transport in maize roots: measurementsof hydraulic conductivity, solute permeability, and of reflection coefficients of excised roots using the root pressure probe. Plant Physiol 84:1220–32

Terry, BR and Robards, AW (1987) Hydrodynamic radius alone governs the mobility of molecules through plasmodesmata. Planta 171:145–157

Tucker, EB (1988) Inositol bisphosphate and inositol trisphosphate inhibit cell–to–cell passage of carboxyfluorescein in staminal hairs of *Setcreasea purpurea*. Planta 174:358–63

Tucker, EB (1990) Calcium–loaded 1,2–bis(2–aminophenoxy)ethane–N,N,N',N'–tetraacetic acid blocks cell–to–cell diffusion of carboxyfluroscein in staminal hairs of *Setcreasea purpurea*. Planta 182:34–38

Tyerman, SD, and Steudle, E (1982) Comparision between osmotic and hydrostatic water flow in a higher plant cell: determination of hydraulic conductivity and reflection coefficients in isolated epidermis of *Tradescantia virginiana*. Aust J Plant Physiol 9:461–79

Tyerman, SD, Oats, PM, Gibbs, J, Dracup, M and Greenway H (1989) Turgor–volume regulation and cellular water relations of *Nicotiana tabacum* roots grown in high salinities. Aust J Plant Physiol 16:517–31

Tyree, MT (1970) The symplast concept. A general theory of symplastic transport according to the thermodynamics of irreversible processes. J Theor Biol 26:181–214

Wayne, R and Tazawa, M. (1990) Nature of the water channels in the internodal cells of *Nitellopsis*. J Membrane Biol 116:31–39

Wendler, S, and Zimmermann, U (1982) A new method for the determination of hydraulic conductivity and cell volume of plant cells by pressure clamp. Plant Physiol 69:998–1003

Wendler, S, and Zimmermann, U (1985) Determination of the hydraulic conductivity of *Lamprothamnium* by use of the pressure clamp. Planta 164:241–245

Zhang, WH and Tyerman, SD (to be published) The effect of low O_2 concentration and azide on hydraulic conductivity and osmotic volume of the cortical cells of wheat roots. Aust J Plant Physiol

Zawadzki, T, and Fensom, DS (1986) Transnodal transport of ^{14}C in *Nitella flexilis*. II. Tandem cells with applied pressure gradients. J Exp Bot 37:1353–63

THE EXPANSION OF PLANT TISSUES

J.B. Passioura
CSIRO, Division of Plant Industry,
GPO Box 1600, Canberra, 2601,
Australia

Plants grow by expanding newly-divided cells. These cells are enclosed in tough envelopes, called cell walls, which are comprised mainly of cellulose microfibrils tethered to each other by single-stranded polysaccharides such as xyloglucan (Fry, 1989). From a simple physical viewpoint, there are two essential features of these cells' expansion. The first is that the cell wall must extend plastically, to allow for the extra volume. The second is that water must enter the cell to occupy this extra volume.

Because expanding plant tissues are geometrically complex, and may contain hundreds to millions of simultaneously expanding cells of various shapes, it is difficult to measure the stress/strain relations of the cell walls, and so it is common to assume that plastic expansion obeys:

$$dV/dt = m(P - Y) \qquad (1)$$

where dV/dt is the rate of increase of volume, V, with time, t; m is the extensibility of the wall; P is the hydrostatic pressure (turgor pressure) within the cell (which determines the stress in the wall); and Y is a threshold value for P below which responses of V to changes in P are entirely elastic, and no plastic expansion occurs.

The water relations of the expanding cells are also difficult to measure, because attempts to measure their water potential disrupt the approximately steady flows of water within them, which unavoidably changes the water potential.

NATO ASI Series, Vol. H 64
Mechanics of Swelling
Edited by T. K. Karalis

But it is common to assume that the flow of water into an expanding cell can be described by:

$$dV/dt = L(\Delta\Psi) = L(\Delta P - \Delta\pi) \qquad (2)$$

where L is the hydraulic conductivity of the membrane, $\alpha\nu\delta\,\Delta\Psi$, ΔP, and $\Delta\pi$, are the respective differences in water potential, hydrostatic pressure, and osmotic pressure, across the membrane. One could write a similar equation for the expansion of a tissue, with, say, $\Delta\Psi$ being the difference in water potential between the xylem and the expanding cells.

Equations (1) and (2) can be combined to give:

$$dV/dt = mL(\Psi_o + \pi - Y)/(m + L) \qquad (3)$$

which is known as the Lockhart equation (Lockhart, 1965), and in which Ψ_o is the water potential in the xylem that supplies the cells with water. In practice L is usually so large that the simpler equation (1) is adequate (Cosgrove, 1986).

These equations have an elegant, but beguiling, simplicity. It looks as though they encapsulate the essential physical aspects of cell expansion, and indeed they do. But they have turned out to be of limited physiological use because the quantities m and Y, which look like parameters, are not parameters at all, but respond rapidly to changes in P, the notional main physical variable, to modulate expansion rate according to a set of rules that has yet to be discovered. A physiologist interested in growth in the sense of the unfolding of a plant through time gets little joy out of integrating equation (1) because even if he knows $P(t)$ he will be largely ignorant of $m(t)$ and $Y(t)$.

The rapid adjustment of m and/or Y to perturbations in P is illustrated in Fig.1, which shows the elongation rate, as a function of time, of a growing leaf of a barley plant whose roots were contained in a pressure chamber. When the pressure in the chamber was suddenly increased by 100 kPa, at the time denoted by the arrow, the pressure of water through-

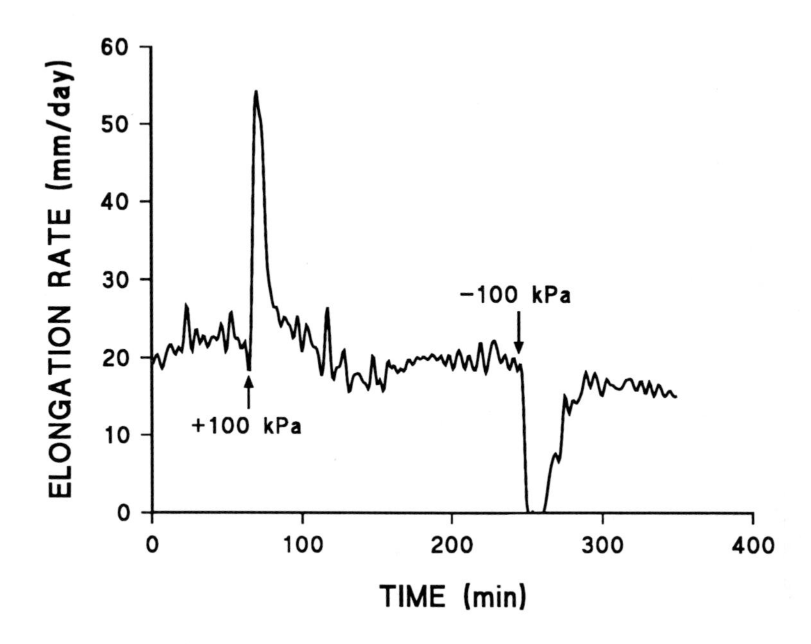

Fig.1. Elongation rate of an expanding barley leaf as a function of time. The pressure within the expanding cells was increased by 100 kPa (by applying a pressure of 100 kPa to the plant's roots, which were contained within a pressure chamber) during the period between the arrows.

out the leaves, including P, the pressure within the expanding cells, increased correspondingly. The result was a sudden spurt in elongation rate, to more than double the previous rate, consistent with equation (1). But this increase was quite transient, and was followed by a rapid exponential fall in the elongation rate, with a time constant of about 15 minutes, so that the rate was virtually back to its original value after about 30 minutes. When the pressure in the root chamber was returned to normal, the elongation rate followed a trajectory that was a vertical reflection of Fig.1 (apart from the truncation at zero elongation rate), that is, the rate fell sharply then recovered exponentially to near its original value. It is clear that the rate of leaf expansion is being controlled by processes occurring at a time scale of tens of minutes that override any perturbations in P. Similar perturbations in leaf expansion rate occur when the lights are turned on or off in growth chambers, thereby inducing step changes in leaf water status (Christ, 1978).

One could argue that the perturbations in elongation rate shown in Fig. 1 were annulled because the cells may have a turgor sensor that controls their osmotic pressure to maintain the turgor constant. But similar experiments by Shackel et al. (1987), who manipulated the water status of grape leaves by changing the transpiration rate, and followed the changes in turgor of epidermal cells of expanding leaves by using a pressure probe, also showed a transient change in expansion rate despite a sustained change in turgor. Thus m or Y must have changed, for rapid changes in L are improbable. and in any case L is likely to be too large to be influential.

From the point of view of a plant physiologist it is perhaps not surprising that effects of large increases in turgor on cell expansion are transient, for the continued expansion of the cell wall requires that new wall be synthesised, and there is no reason to assume that the rate of synthesis is unfettered. Furthermore, if expansion proceeded faster than synthesis, the wall would get thinner, with a consequent increases in stress that could lead to failure of the wall unless it were accompanied by some form of strain-hardening.

Although Fig. 1 shows that the leaf reacts to perturbations in P to maintain an approximately constant expansion rate, longer term experiments in which the elongation rates of leaves are measured by ruler twice daily show that when P is maintained artificially high (by growing the plants for many days with their roots in pressurised chambers) the leaves elongate slightly more quickly during the day, an increase that is difficult to discern in the comparatively noisy data shown in Fig.1. But this increase during the day becomes a decrease, relative to the unpressurised controls, at night (Fig.2), so that the elongation rate averaged over the whole day is essentially unaffected by the pressurising treatment. Thus, in addition to the short term control evident in Fig. 1, which modulates *m* and/or *Y* to return the elongation rate to its approximately "normal" value after a perturbation, there are controls operating at a time scale of a day that bring the daily elongation rate even closer to its "normal" value.

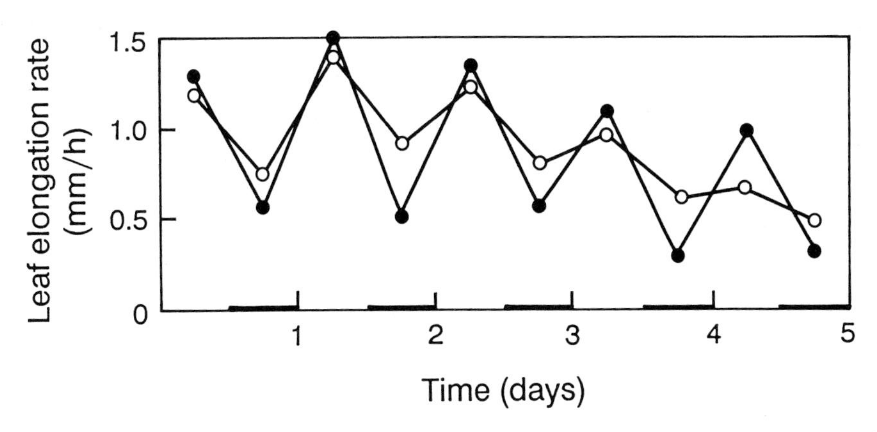

Fig.2. Diurnal and nocturnal elongation rates of expanding leaves on pressurised (O) and unpressurised (O) wheat plants. The pressurising treatment (applied to the roots) kept the pressure in the leaf cells high at all times, approximately 500 kPa higher during the day, and 100 kPa higher during the night, than in the cells of the unpressurised controls. The thick lines on the time axis denote the nights. (Reproduced from Passioura, 1988).

An interesting feature of leaves that have experienced all of their expansion while the plants have been continuously pressurised is that although they have the same areas as those of unpressurised plants, they have a different shape: they are 10% wider and 10% shorter (Passioura and Gardner, 1988). A similar change of shape with altered water status occurs in roots that are growing in drying soil. Normally, roots that are growing in such soil get fatter, even though they have a low water potential, because they find it difficult to penetrate the soil, which unless it is very sandy gets harder as it dries. But Sharp et al. (1988) found when they grew roots in quite dry vermiculite (Ψ = -1.5 MPa), which remains easy to penetrate even when dry, that the roots became not only shorter but also thinner, presumably because their turgor was much reduced. The expanding cells in both these roots and in the barley leaves of Figs.1 and 2 are cylindrical, and their expansion is almost entirely longitudinal. The cellulose microfibrils in the stress-bearing parts of the walls are arranged transversely, and so resist transverse expansion. It is remarkable that in the barley leaves the cells appear to conserve volume despite the large changes in turgor that alter their conformation; the slightly larger transverse expansion in the cells is apparently compensated by a *reduction* in longitudinal expansion despite the much greater stress in their walls.

Most of the above remarks apply to well-watered plants. When a plant is growing in drying soil, one would expect the expansion rate of its leaves to fall after a time, because the increasing dehydration of the plant must eventually reduce P to zero, when no expansion is possible. But the expansion rate falls well before P becomes zero, and will do so even if P is kept artificially high by pressurising the roots as described above. Indeed, perturbing P may have no effect on expansion rate, as shown in Fig.3. Equation (1) is seemingly irrelevant here; the expansion rate is being controlled, at a time scale of days, not by the falling water status of the expanding cells, but by an inhibitory signal that is sent to the leaves by the roots that are experiencing the drying soil. The stress/strain relations of the cells must adjust to accord with the overriding control.

Thus, the expansion of plant tissues is under the control of complex systems that increase in influence at increasing time scales. The challenge now is to elucidate the biochemical and physicochemical process that modulate the yielding of the cell wall, for it is these that control cell expansion at time scales greater than a few minutes. Fry's (1989) model of the structure and physical chemistry of the

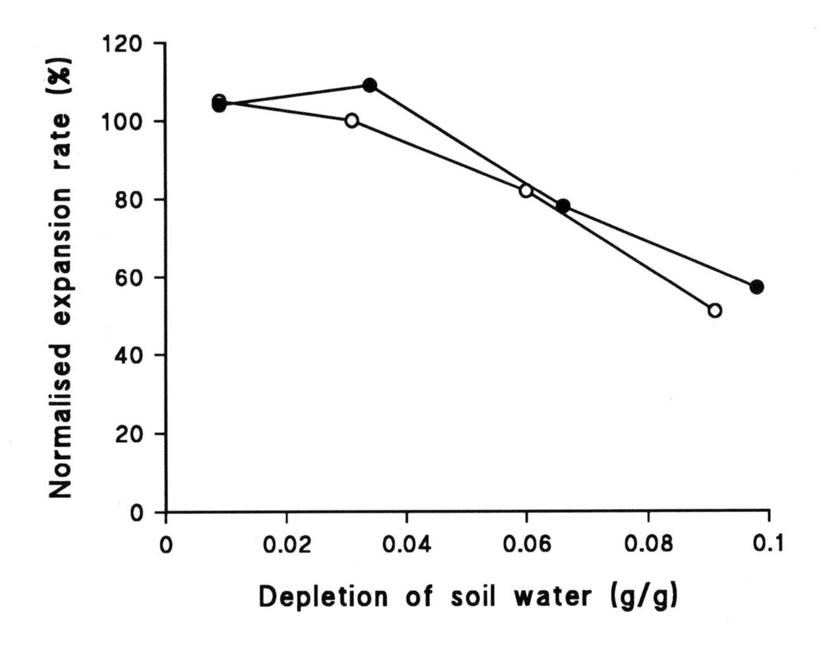

Fig.3. Expansion rate of leaves of pressurised (O) and unpressurised (O) wheat plants growing in drying soil (% of well-watered control) as a function of depletion of water from the soil. The leaves of the pressurised plants remained fully hydrated despite the drying of the soil. (Adapted from Passioura, 1988).

expanding cell wall is a useful starting point. He envisages that the cellulose microfibrils, which comprise several strands of cellulose combined together into crystals of about 3 nm diameter, are tied together by strands of xyloglucan, some of which are taut and some loose. Enzymes that cleave the taut strands allow the microfibrils to separate until some of the previously loose strands tighten. Cellulase is an enzyme that can cut the tethers, but endotransglycosylase, a recently-discovered enzyme (Smith and Fry, 1991), is able both to cut the xyloglucan tethers and to unite previously cut tethers that are attached at only one of their ends. It is conceivable that this enzyme may control the rate of cell expansion, although it is not immediately obvious how it could overcome sudden changes in stress to produce the steady rate of strain evident in Fig.1.

References

Christ RA (1978) The elongation rate of wheat leaves. II. Effect of sudden light change on the elongation rate. J Exp Bot 29:611-618

Cosgrove D (1986) Biophysical control of plant cell growth. Ann Rev Plant Physiol 37:377-405

Fry SC (1989) Cellulases, hemicelluloses and auxin-stimulated growth: a possible relationship. Physiol Plantarum 75:532-536

Lockhart JA (1965) An analysis of irreversible plant cell elongation. J Theor Biol 8:264-275

Passioura JB (1988) Root signals control leaf expansion in wheat seedlings growing in drying soil. Aust J Plant Physiol 15:687-693

Passioura JB, Gardner PA (1990) Control of leaf expansion in wheat seedlings growing in drying soil. Aust J Plant Physiol 17:149-57

Shackel KA, Matthews MA, Morrison JC (1987) Dynamic relation between expansion and cellular turgor in growing grape (*Vitis vinifera* L.) leaves. Plant Physiol 84:1166-1171

Sharp RE, Silk WK, Hsiao TC (1988) Growth of the maize primary root at low water potential. I. Spatial distribution of expansive growth. Plant Physiol 87:50-57

Smith RC, Fry SC (1991) Endotransglycosylation of xyloglucans in plant cell-suspension cultures. Biochem J (in press)

Part 3

Cells Morphology, Function and Deformation

MDCK CELLS UNDER SEVERELY HYPOOSMOTIC CONDITIONS

James S. Clegg
Bodega Marine Laboratory
University of California (Davis)
Bodega Bay, California 94923
USA

INTRODUCTION

The literature on animal cells exposed to anisosmotic solutions is truly massive as one may appreciate from a few selected recent reviews on the subject (Gilles et al. 1987; Macknight, 1988; Chamberlin and Strange, 1989; Kleinzeller and Ziyadeh, 1990; Beyenbach, 1990). No attempt will be made here to review this research, but a few introductory comments on hypoosmotic exposure are in order. The extent to which osmolarity of the surrounding medium has been reduced varies widely, according to the intentions of the investigators, but commonly amounts to reductions of 50 percent or less. A huge effort has gone into the means by which such cells restore their volumes (if they do so). Occasional reference has also been made to the remarkable ability of certain animal cells to survive exposure to, "essentially", distilled water (see, for example, Kleinzeller and Ziyadeh, 1990; Macknight, 1987). However, to my knowledge these reports are anecdotal, and no detailed account of this extreme situation seems to have been published. Study of the behavior of cells in such dilute solutions might provide insight into the potential involvement of cell ultrastructure in cell swelling. In addition, some interesting questions arise about their metabolic status. For these reasons we carried out a series of experiments using Madin-Darby canine kidney cells (MDCK). This cell type was selected because it is a stable, well-established epithelial line, and widely used in studies involving volume changes (see Roy and Sauvé 1987; Helig et al. 1990; Mills, 1987). Moreover, the cytomatrix of MDCK cells has been described in reasonable detail (see Fey and Penman, 1986; Mitchell, 1990). As we shall see, their response to extremely dilute solutions is impressive.

NATO ASI Series, Vol. H 64
Mechanics of Swelling
Edited by T. K. Karalis

MATERIALS AND METHODS

MDCK cells were obtained from the American Type Tissue Culture Collection and cultured in sealed 25cm^2 plastic flasks using Medium 199 (M199) containing 10% fetal calf serum (cs), from Flow Laboratories, at 37°C.

Unless otherwise noted the standard protocol for hypoosmotic incubation involved washing each flask twice with Ca^{+2} and Mg^{+2}-free Dulbecco's phosphate buffered saline (DPS), CMF, and allowing the flasks to drain upright for 2 minutes to remove the maximum amount of CMF by aspiration. Five ml of M199cs diluted 1 to 100 with distilled water was then added to each flask while it was upside-down to avoid contact with the cells. To start the incubation the flasks were placed cells-downward and the medium moved throughout the cell layer. Control experiments using flasks without cells showed that the osmotic pressure of the combined trapped CMF and 5ml of 1:100 diluted M199cs ranged between 10-13 mOsm/kg (measured with a Precision Systems "μOsmette"). The cell contents, even when fully permeabilized by exposure to this solution, contributed negligibly to the osmotic pressure of the 5ml of medium as the following calculation shows. Each flask contained 3-4$\times$10^6 cells. Using a cell water content of 2μl/10^6 cells (Roy and Sauvé, 1987) the total <u>cell water</u> per flask was 6-8μl; therefore, the total diffusible cell solutes (supposed to be about 285mM) would be diluted by a factor of 600-800, and well below the limits of osmotic pressure measurements carried out in this work.

After incubation in 1:100 diluted M199cs (hereafter referred to as 1:100) the medium was removed and replaced by 5ml of the appropriate solution, usually M199cs. Observations on cells were made with an inverted phase microscope and survival was determined by counting cells per standard field from photographs.

Plasma membrane permeability was determined, in part, by incubation with ethidium bromide (12μg/ml) which fluoresces when intercalated with nuclear DNA but not otherwise. Cytoskeleton perturbing compounds (Sigma Chemical Corporation) and ethidium bromide were dissolved in dimethylsulfoxide (DMSO) as stock solutions; controls were run on DMSO but in no case produced an observable effect.

Incubations with ^{14}C-U-glucose (310 mCi/mmole) and ^{14}C-U-leucine (328 mCi/mmole), both from Amersham, were carried out at final concentrations of 1μCi/ml incubation medium. Aliquots of cell extracts and incubation media were analyzed for radioactivity using a scintillation spectrometer. The confluent cells were rinsed with DPS (2×30ml) to remove the medium, and the cells extracted with 72% ethanol (2ml/flask) to obtain low molecular weight metabolites. The remaining insoluble cell fraction (still attached to the flask), consisting of macromolecules, was dissolved with 2ml of 88% formic acid for scintillation counting.

Production of $^{14}CO_2$ from ^{14}C-glucose was evaluated by acidifying the flask contents and trapping in 200μl of hyamine hydroxide (Sigma) contained in small cups, followed by scintillation counting. Chemiluminescence of hyamine hydroxide was removed by addition of 20μl of glacial acetic acid to the scintillation cocktail (5ml). Protein in cell extracts was estimated using the Pierce assay, ATP was measured using a cycling-enzyme procedure based on NAD/NADH, and lactate was determined by Sigma Chemical Corporation procedure no. 826-UV; details of these have been published (Clegg and Jackson, 1988).

Analysis of the distribution of ^{14}C-labeled metabolites in ethanol-soluble extracts was carried out by thin layer chromatography on 100μm thick cellulose sheets (Eastman) using (v/v) isopropanol (7): water (2): 88% formic acid (1) as developing solvent. Chromatograms were exposed to X-ray film for autoradiography. Standard SDS-PAGE analysis of proteins in media and cell extracts utilized a Mini-Protean Electrophoresis apparatus (BioRad); gels were stained with Coomassie blue and/or evaluated by autoradiography after drying.

RESULTS

I begin with a general description of the light-microscope morphology of cells exposed to 1:100 media, followed by their return to normal growth medium (M199cs). That account also provides the usual protocol to be followed in subsequent studies.

Swelling, Shrinking and Recovery

Confluent MDCK monolayers exposed to 1:100 media for 30 min undergo obvious swelling with characteristic structural changes, most notable in the nuclei whose contents seem to "dissolve" (Fig. 1). Cytoplasmic granules in most cells do not undergo Brownian

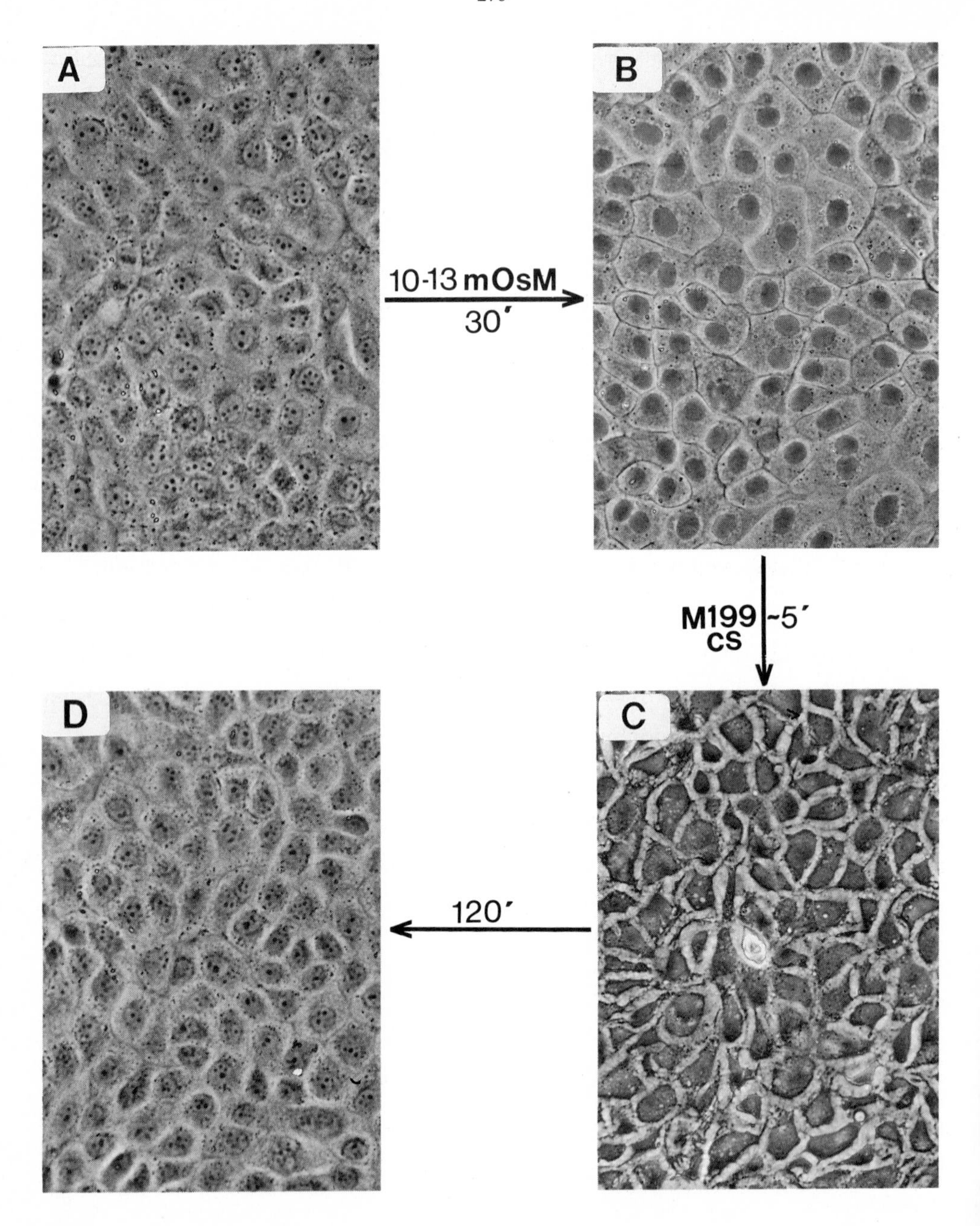

Figure 1. Control MDCK cells (A) exposed to 1:100 diluted M199cs for 30 minutes (B) and then to M199cs (regular growth medium) for 5 minutes (C). Panel D Shows the morphology of such cells after 2 hours recovery.

motion under these conditions, suggesting that even this osmotic insult does not release the particles from whatever restrains them under physiological conditions. When normal growth medium is restored the cells undergo a severe contraction which, however, does not begin immediately but requires several minutes (Fig. 1C). Normal morphology is restored after about 2 hours of recovery (Fig. 1D) during which time the cells exhibit a cycling of swelling and shrinking phases not readily described by separate photographs, but clearly evident in videos. Figure 2 shows the appearance of such recovering cells at different times following replacement of 1:100 medium with M199cs.

As can be expected, the percentage of cells that survive this treatment varies, but is always impressively high for healthy cultures. By counting cells in standard fields the survival rate has been determined to range from 75-90%, scored at 4 hours after hypoosmotic exposure. The surviving cells go on to divide normally, and no selection of cells adapted for surviving hypoosmotic exposure is evident from single exposure; that is, cultures of cells recovering from one such exposure do not exhibit appreciably higher survival when exposed to another hypoosmotic episode.

Figure 3 illustrates a typical time course of the swelling and shrinking phases depicted in Figures 1 and 2, in this case for non-confluent cultures. Several features are worth comment. Dissolution of the nuclear contents is apparent; however, the nuclei do not exhibit an obvious change in volume (compared to controls) based on estimates of nuclear dimensions, a result that was also observed for nuclei in confluent cells (Fig. 1B). Although cell swelling does not occur immediately upon exposure to an osmolarity about 30 times lower than physiological, it seems to take place faster than for cells in confluent monolayers (Fig. 1). Thus, both swelling and shrinking behavior exhibit lag times.

It is very difficult to quantify changes in thickness and volume of cells attached to culture dishes and no attempt has been made here, although heroic effects have been published (see Roy and Sauvé, 1987; Litniewski and Bereiter-Hahn, 1990). The cell in the upper right part of each segment of Figure 3 is judged to have just completed telophase but has not begun to spread and flatten. In any event, this cell is clearly attached and, appearing to be spherical, its change in volume can be approximated by diameter measurements using a calibrated ocular micrometer. Relative to the volume of this cell before 1:100 exposure (top left, Fig. 3) an increase of 3.3 times that volume has

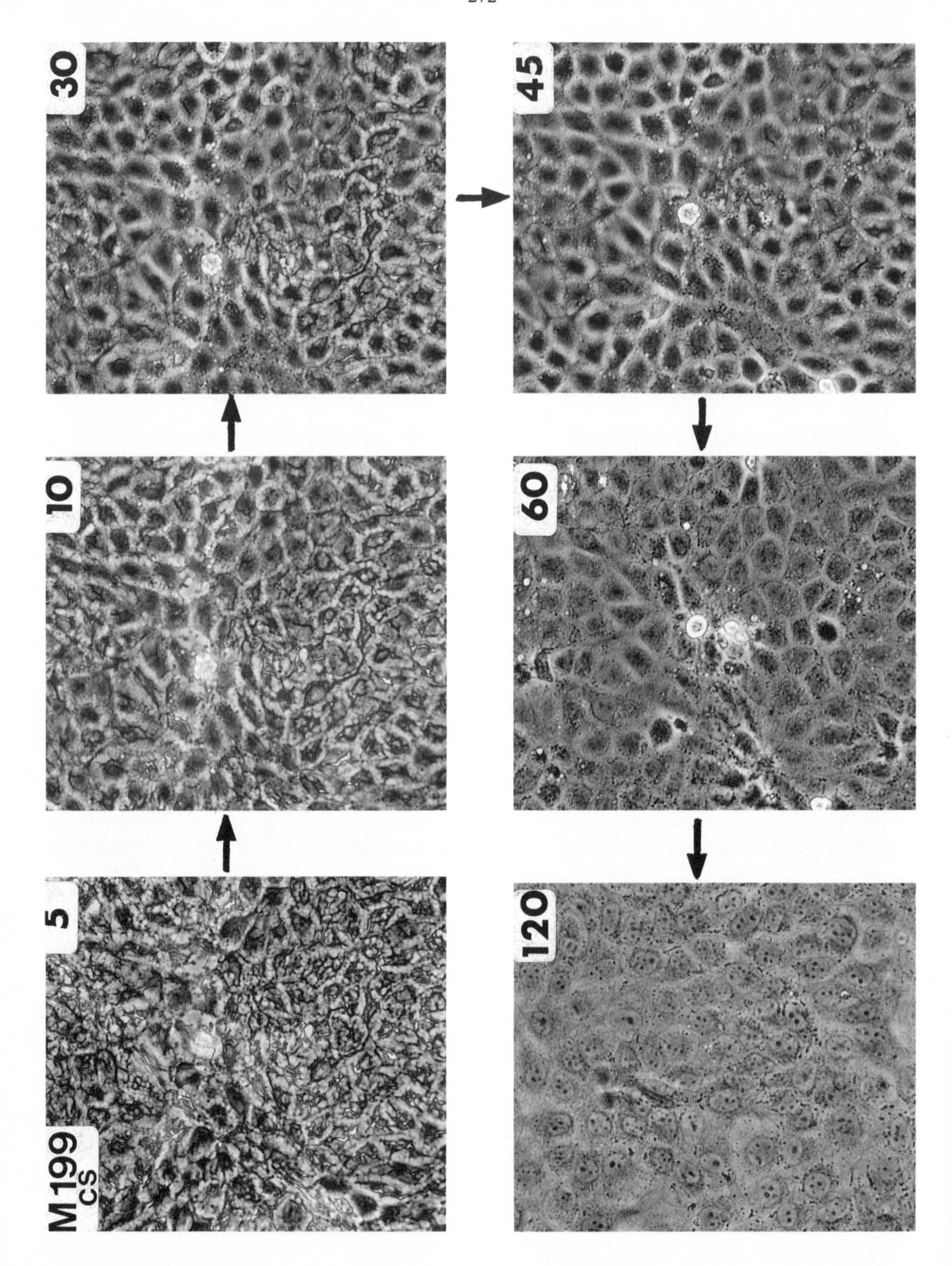

Figure 2. (above) Snapshots of cells exposed to 1:100 diluted medium and then incubated in M199cs. Numbers in the right top refer to minutes following addition of M199cs. The same field is shown for this time period.

Figure 3. (opposite page) Subconfluent cells exposed to 1:100 medium and then placed in normal growth medium (M199cs). The numbers in the lower right of each panel refer to minutes of incubation. Volume estimates for the cell in the upper right are given in the text.

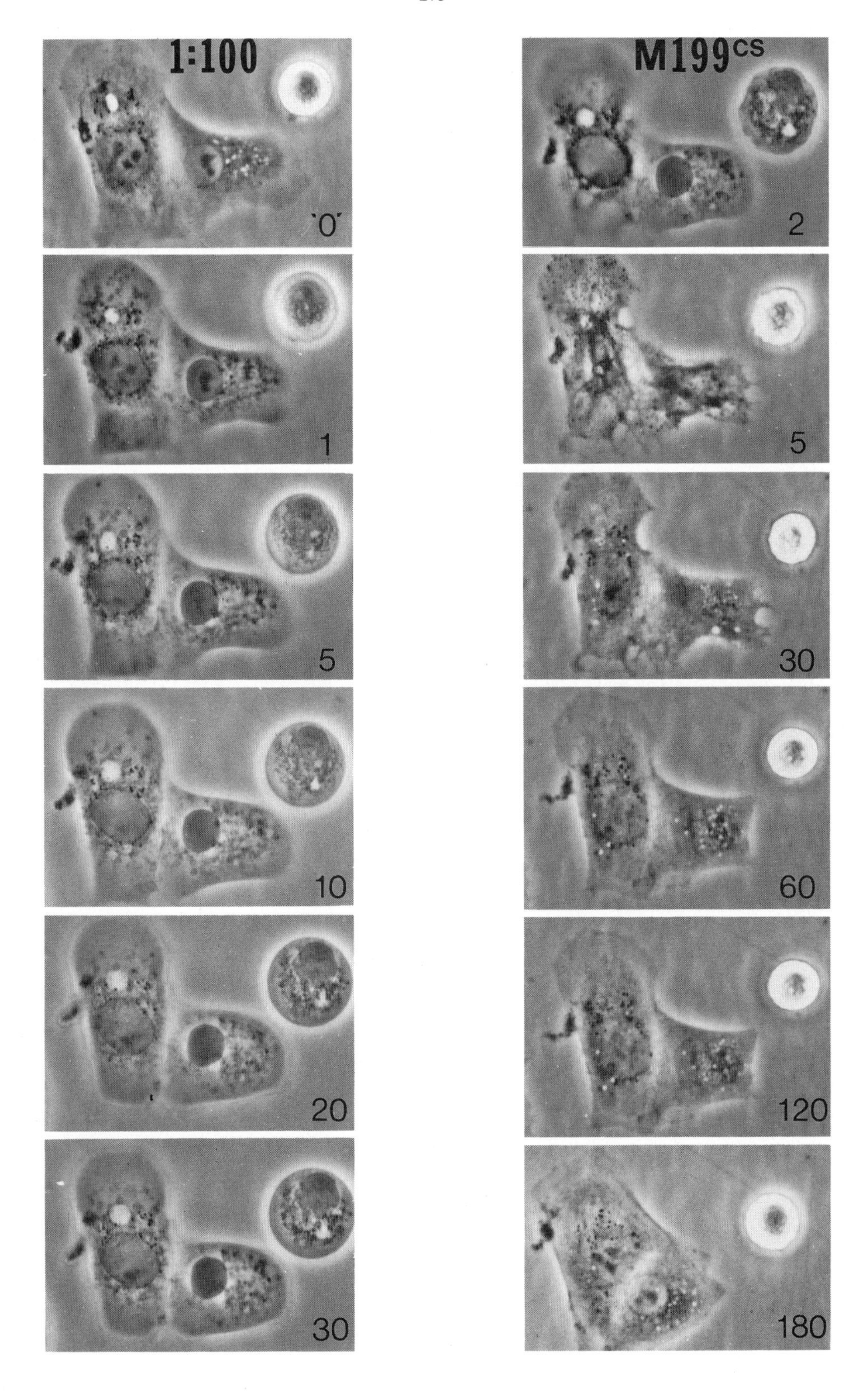
1:100
'0'
1
5
10
20
30
M199cs
2
5
30
60
120
180

taken place after 20 minutes of exposure, and no measurable change takes place after 30 minutes, suggesting that some restraining force is operating, or that this cell has reached swelling equilibrium. A series of such measurements has been carried out with the following result: for control cells the diameter was 15.9±1.7μm (SE, n=15) while cells exposed to 1:100 media had a diameter of 25.5±2.2μm (n=15) after 30 minutes. The resulting volumes are 2100μm^3 for controls, and 8700μm^3 for 1:100 media exposure, amounting to a relative volume increase of about 4 fold. While no quantitative data are available for confluent cells, the impression one gets (Figs. 1 and 4) is that these cells exhibit comparable increases in volume.

Figure 3 also shows that the extensive contraction and recovery that occur when swollen confluent cells are returned to normal growth medium also occurs in "isolated" cells. Note that the cell in the upper right of each panel has achieved, after 2 hours, a volume very similar to that before it was exposed to 1:100 medium.

A question of general interest concerns the permeability of the plasma membranes of these cells as they swell and shrink. That subject is examined next.

Plasma Membrane Integrity of Cells Exposed to Hypoosmotic Media (1:100) and Subsequent Recovery

Our first approach was to follow the uptake of ethidium bromide during exposure to 1:100 media. Figure 4 shows a typical outcome. Fluorescence of nuclei was not easily observed until the cells had experienced this condition for about 2 minutes (Fig. 4A), becoming more evident as incubation proceeded (Fig. 4B). Confluent cells exhibit essentially the same behavior (data not shown). This lag is due chiefly to the time required for the ethidium bromide to intercalate the nuclear DNA, and not to its inability to cross the plasma membrane. That is so because ethidium bromide added after the cells were first exposed to 1:100 media for 15 minutes, and are clearly permeable, still required about 2 minutes before nuclear fluorescence was detected (data not shown). Thus, it seems that the plasma membranes of cells in 1:100 media become permeable to solutes of at least this mass (about one kDa) within a minute after exposure. Figure 4 also provides further description of the morphology of swelling in subconfluent cultures.

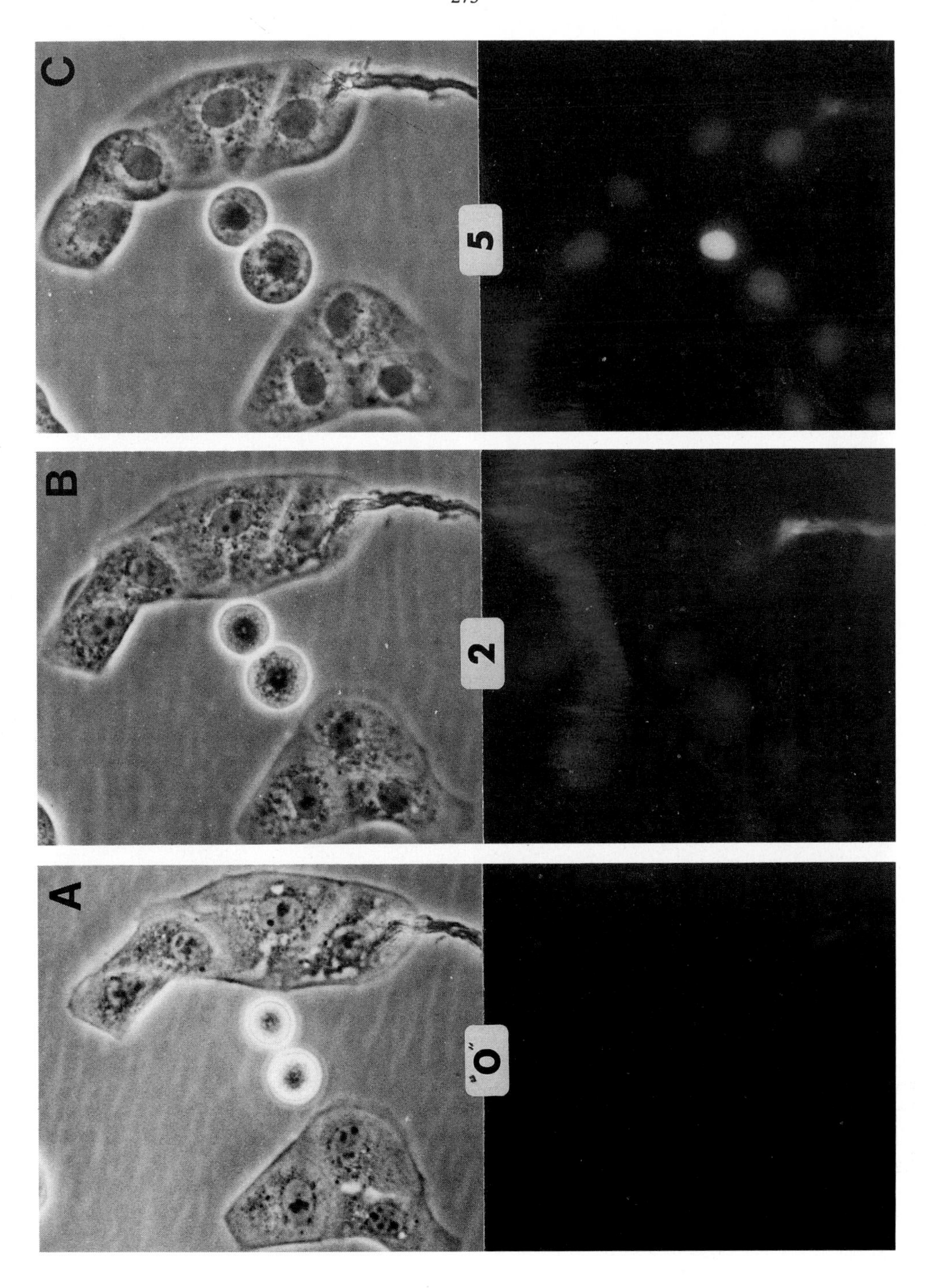

C
5
B
2
A
"0"

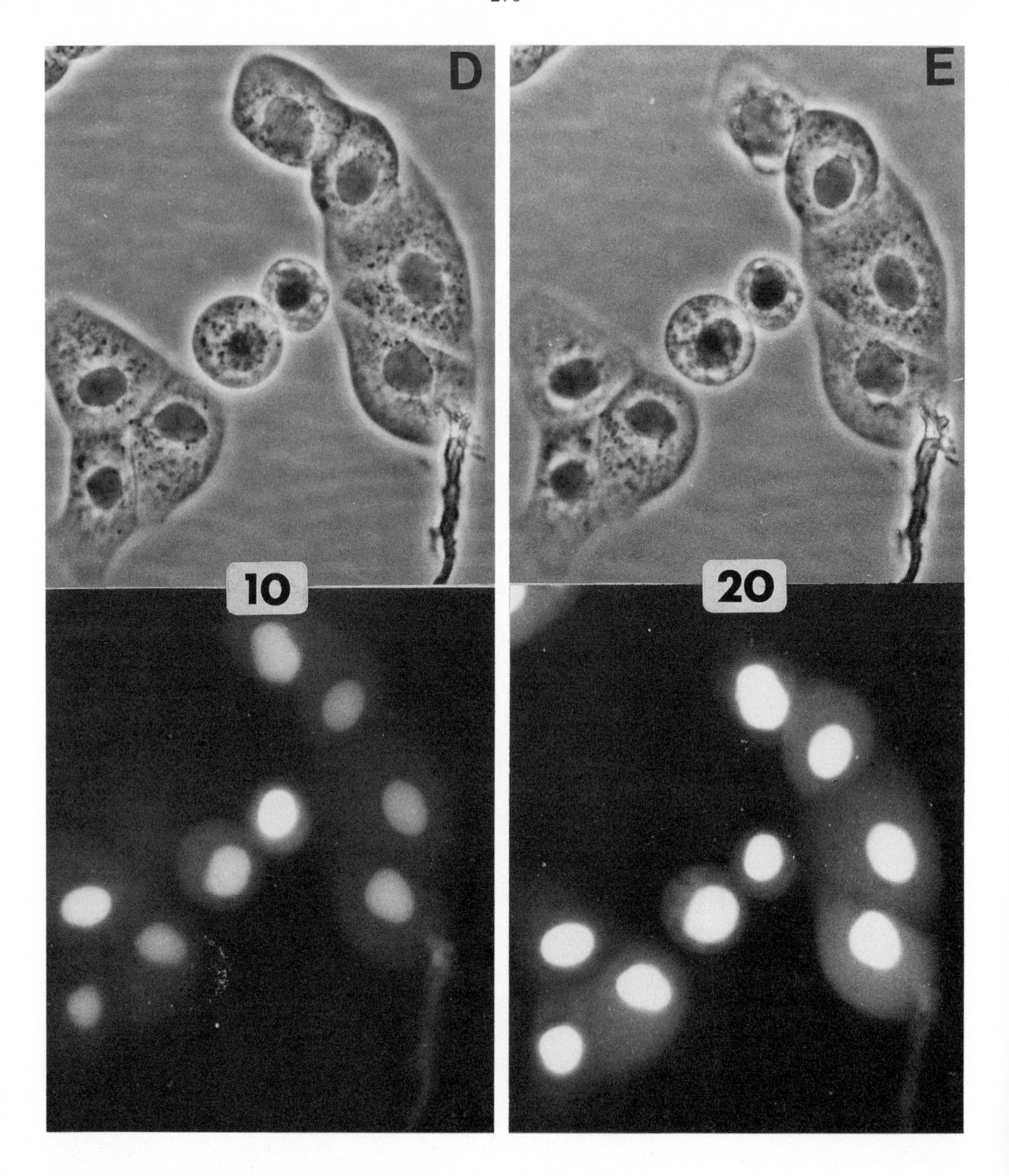

Figure 4. (4A opposite page, 4B top of this page). Subconfluent cells exposed to 1:100 medium containing ethidium bromide, the exposure times (minutes) being given in the middle of each panel. The bottom part of each panel illustrates epifluorescence images of the same cells viewed by ordinary phase microscopy (top part of each panel).

Effects of Cytoskeleton-Perturbing Compounds on Swelling in Hypotonic Media and Subsequent Recovery

The participation of cytoplasmic structure in the extraordinary response of MDCK cells to severe hypoosmotic stress seems certain in view of the results presented thus far. Therefore, a wide variety of compounds known to perturb cytoskeletal structure was examined next. Table 1 summarizes the outcome of these studies. While it is recognized that different exposure times and concentrations might yield different results, the data obtained clearly point to the importance of microfilament involvement in the response of MDCK cells to severe hypoosmotic treatment. Thus, pre-treatment with cytochalasin-D at 10μM concentration for 20 minutes results in the death of all cells when subsequently exposed to 1:100 media. Noteworthy in this regard is that the microfilament stabilizing agent, phalloidin, actually increases the survivorship of cells exposed to 1:100 media, compared to controls, and also apparently saves some cells subsequently treated with cytochalasin-D (Table 1). There is no indication from these data that microtubules play an obvious role in surviving hypoosmotic exposure: while colchicine results in cell death following exposure to 1:100 media, control cells (not exposed to 1:100 media) also eventually die after such exposure, which does not provide information on a possible connection between hypoosmotic treatment, cell death, and microtubules. Finally, it should be pointed out that, except for cytochalasin-D, none of these compounds altered the pattern of cell swelling in 1:100 medium.

The effects of cytochalasin-D are dramatic as illustrated in Figure 5. Unlike controls, these cells seem to lose cell-cell contact and swell as individuals (Fig. 5B) rather than as a confluent sheet (Fig. 1) when exposed to 1:100 media. That observation suggests that microfilaments are also somehow involved in connections between cells, as well as intracellularly. When normal growth medium is restored to cells pre-treated with cytochalasin-D and exposed to 1:100, they contract vigorously, never to recover their original morphology, and soon die (Figure 5). The case will be made in the Discussion section that microfilaments play a major role in the response to hypoosmotic exposure, as well as subsequent recovery.

Table 1. Effects of pre-treatment by cytoskeleton perturbing compounds on subsequent swelling and recovery from hypoosmotic exposure.

Compound	mM in M199	swelling, contraction, recovery	% survivors 2 hrs	24 hrs
Control (none)	--	usual response	84	confl.
Griseofulvin	0.10	" "	85	"
Nocodazole	0.10	" "	85	"
Colchicine	0.02	usual, but altered morphology	79	0
Vinblastine	0.05	usual, " " "	67	42
Phalloidin	1.00	usual response	93	confl.
Cytochalasin-D	0.01	swelling unusual, stay contracted	0	0
Phall.+Cyto-D	1.0+0.01	usual response, some stay contracted	31	confl.

Confluent monolayers were first exposed to these compounds in M199 for 20 minutes, then washed with 2×30ml CMF, and incubated in M199cs, diluted 1:100 with distilled water, for 30 minutes. The monolayers were drained and 5ml M199cs added. The "control" was incubated in 1% DMSO in M199 for 20 minutes before subsequent hypoosmotic treatment as described. Viabilities were scored after incubation in M199cs. "Usual response" refers to swelling, contraction and recovery behavior exhibited by controls incubated without DMSO.

Intracellular Events Associated With Exposure to Hypoosmotic Conditions and Subsequent Recovery

It is clear that MDCK cells exposed to 1:100 media quickly become permeable to solutes whose molecular weights are less than 1kDa (Fig. 4). Thus, we may suppose that all freely diffusing ions and small metabolites rapidly escape from cells treated this way. In contrast, we have found that very little of the total cell protein content of monolayers is released into the medium, even after 30 minutes of incubation in 1:100 medium (Fig. 6) and that is reduced in less-diluted media. It seems most likely that proteins found in the medium originate from cells that detach and do not survive hypoosmotic exposure, rather than escaping equally from all the cells, although no direct proof of that is on hand.

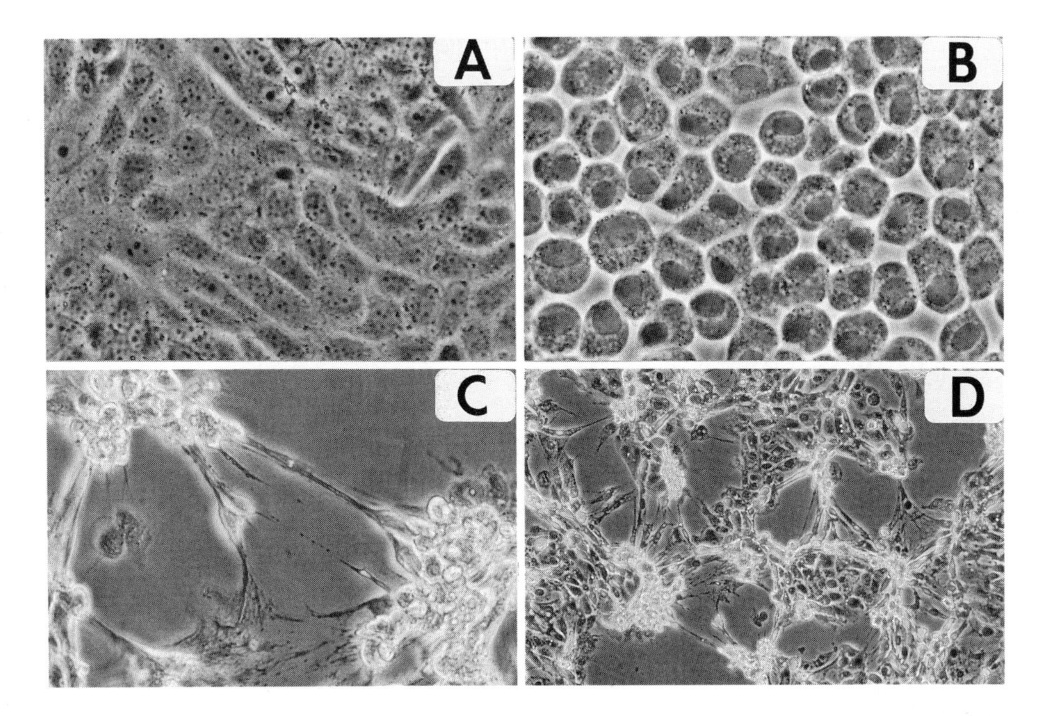

Figure 5. Morphology of confluent cells (A) after exposure to cytochalasin-D followed by swelling in 1:100 medium for 30 minutes (B) and then incubation in M199cs for 1 hour (C,D). Panels A-C are the same magnification, whereas panel D shows a larger field (lower magnification) of panel C.

The properties of hypoosmotic cells were explored further by pre-labeling with ^{14}C-glucose or ^{14}C-leucine for 33 hours in M199cs prior to incubation in 1:100 media (Fig. 7). As described in the methods section, these pre-labeled cells were fractionated into ethanol-soluble (S) and insoluble (I) components (metabolites and macromolecules, respectively) before (B) and after (A) a 30 minute exposure to 1:100 media. The incubation medium (M) was also examined for radioactivity (Fig. 7). Results were similar for glucose and leucine pre-labeling: release of isotope into the medium was accompanied by a decrease in the macromolecular fraction (I) and an increase in small metabolites (S), the total sum of all three remaining about the same. This study has been replicated twice with similar outcome.

Further study was made of the metabolite fraction (S) using autoradiography of thin layer chromatograms (Fig. 8). Although the distribution is complex, the general impression one gets is that most compounds present in pre-labeled cells prior to 1:100

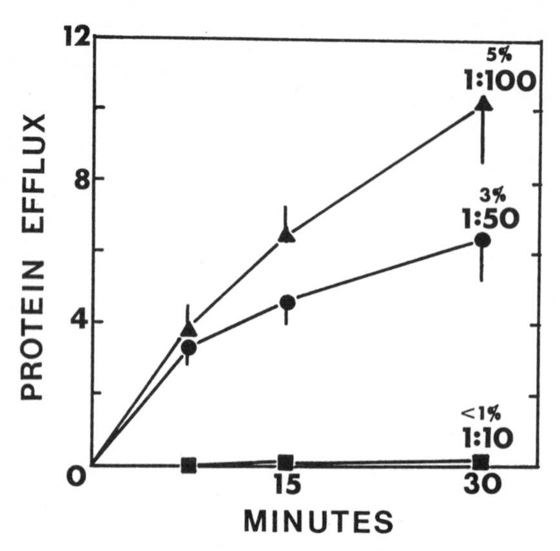

Figure 6. Protein efflux (μg protein/10^6 cells) from cells exposed to different dilutions of M199cs given above the data points for the 30 minute measurements. The percentage values given here refer to the amount of total cell protein found in the incubation medium after 30 minutes of incubation, corrected for protein added in the diluted M199cs. Bars refer to standard errors for n=4 (replicates).

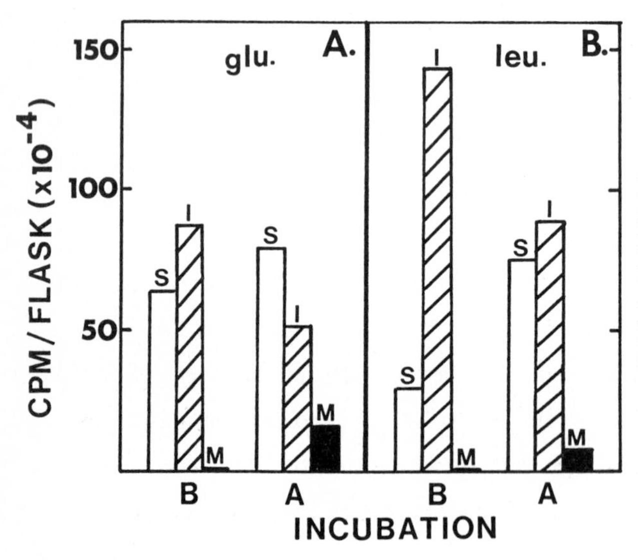

Figure 7. Distribution of radioactivity in the macromolecular fraction (I) metabolite fraction (S) and incubation medium (M) of cells pre-labeled with ^{14}C-labeled glucose (glu.) and leucine (leu.) for 33 hours. B (incubation) refers to control cells, before exposure to 1:100 medium for 30 minutes (A). Each flash contained 3.7×10^6 cells (±10%).

exposure are released into the medium during subsequent incubation. This study has been replicated twice with comparable results.

The data in Figs. 7 and 8 suggest that hypoosmotic incubation leads to marked hydrolysis of cellular macromolecules, the products of this breakdown being released from the cells which we know are freely permeable to small molecules (Fig. 4).

Attempts to detect the existence of an ongoing energy metabolism or the occurrence of biosynthetic pathways in cells exposed to 1:100 media have been negative. Thus, no measurable release of $^{14}CO_2$ or production of ^{14}C-lactate from ^{14}C-glucose has been observed for cells in 1:100 medium, nor has incorporation of ^{14}C-leucine into protein been detected. It seems that cells exposed to 1:100 medium may essentially be ametabolic, except for the hydrolytic reactions they carry out (Figs. 7, 8). That possibility was supported by measurements on ATP. No ATP was detected in cells incubated for 30 minutes in 1:100 media (limit of detection <3 nmoles ATP/10^6 cells); cells before such incubation (controls) contained about 22 nmoles ATP/10^6 cells. Moreover, no ATP was detected in the medium of 1:100 incubated cells (limit of detection $<4\mu M$). That limit can be compared to the concentration of ATP that would occur if control cells simply released their ATP when incubated in 1:100 media: $19\mu M$. A reasonable interpretation of all these data is that cells in 1:100 media utilize their ATP but do not replenish it at levels that can be detected.

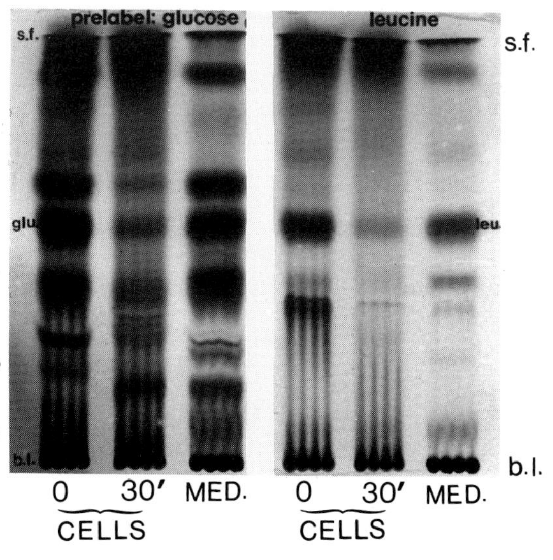

Figure 8. Autoradiography of thin-layer chromatograms of the alcohol-soluble metabolites from fractions described in Figure 7(S). The segment to the left of each chromatogram (0) describes labeled compounds in the cells prior to 30 minutes incubation in 1:100 medium, and the middle segment describes the same fraction after this incubation (30′). MED. refers to the labeled compounds in the incubation medium after this incubation. b.l. refers to base line, and s.f. to the solvent front.

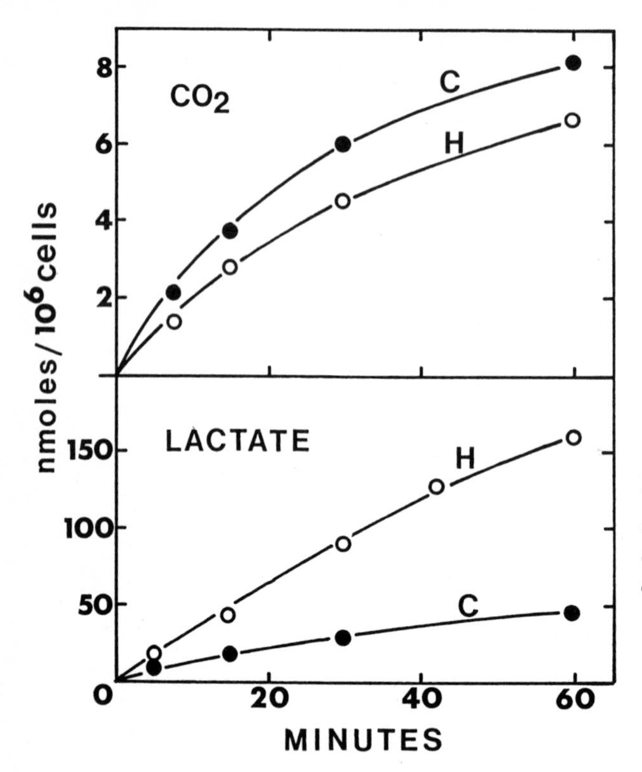

Figure 9. Conversion of glucose into CO_2 and lactate by cells (H) experiencing 30 minutes of exposure to 1:100 medium followed by incubation in M199cs (zero time), compared to controls (C) which had not experienced hypoosmotic conditions.

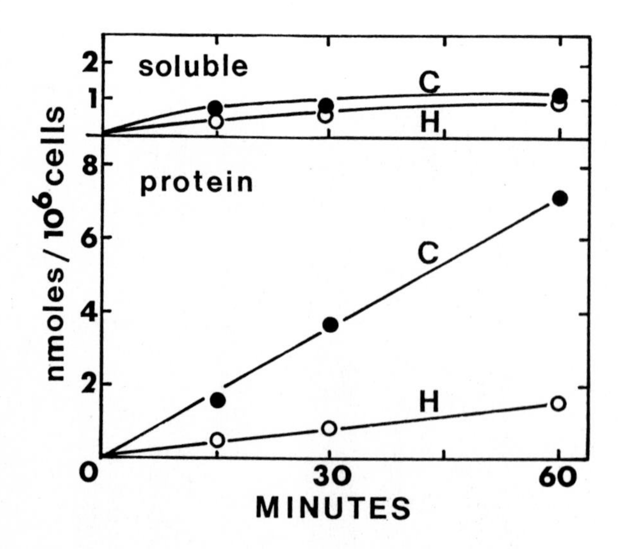

Figure 10. ^{14}C-leucine incorporation into protein by cells (H) previously exposed to 1:100 medium for 30 minutes, then incubated in M199cs (zero time) compared to controls (C) which had not experienced hypoosmotic conditions. The "soluble" fraction refers to ^{14}C-leucine in the "free amino acid pool" of C and H cells.

Attention was next turned to the metabolism of cells recovering from exposure to severe hypoosmotic conditions, after addition of M199cs.

Figure 9 shows the production of CO_2 and lactate by cells first exposed to 1:100 media, and then restored to normal growth media (H). CO_2 production was not remarkably different from controls (C) suggesting that oxidative pathway(s) of glucose metabolism are rapidly resumed. Moreover, the glycolytic pathway of cells previously exposed to hypoosmotic conditions was much faster than controls, presumably due to the need to repair damage done during the osmotic insult. It is remarkable that the recovery of energy metabolism is so rapid in view of the apparent loss of metabolic intermediates, coenzymes, etc. during hypoosmotic exposure.

Protein synthesis in cells previously exposed to 1:100 media was also examined during recovery (Fig. 10). Incubations with ^{14}C-leucine produced comparable levels in the free amino acid pool of these cells. However, markedly reduced rates of incorporation of ^{14}C-leucine into cell protein are shown by cells previously exposed to 1:100 media (H) compared to untreated cells (C). Clearly, the recovery of net protein synthesis in cells exposed to hypoosmotic conditions and then incubated in M199cs is much slower than the recovery of energy metabolism (Fig. 9). Of course, these studies do not evaluate protein breakdown, which may be extensive in cells recovering from hypoosmotic treatment (recall Figs. 7, 8).

DISCUSSION

Cytoplasmic Organization

The first point to be made is that cytoplasmic structure must be involved in the response of MDCK cells to extremely hypoosmotic conditions (Figs. 1-4). Thus, it seems worthwhile to briefly summarize current views on the organization of cytoplasm before turning to the results presented here. Until a decade or two ago the orthodox view considered the cytoplasm as a crowded chaotic solution (the so-called "cytosol") within which the various organelles were suspended and, for the most part, structurally independent of each other. Even the discovery of the three major cytoskeletal systems of microfilaments (MF), microtubules (MT) and intermediate filaments (IF) did not immediately change that description; indeed, these three systems were initially considered not to interact in any major way with each other, or with other cell structures in non-muscle cells. Now, however, the evidence for interaction of cytoplasmic structures is so abundant that it seems a paradigm-shift has occurred, moving from one involving cytoplasmic chaos to one dominated by the theme of organization, both structural and functional (see Welch and Clegg, 1986; Srere, 1987; Jones, 1988; Luby-Phelps et al. 1988; Negendank and Edelmann, 1988; Klymkowsky et al. 1989; Albrecht-Buchler, 1990; Carmo-Fonseca and David-Ferreira, 1990; Srere et al. 1990; Welch, 1991). It is interesting to realize that cytologists of the late 19th and early 20th century considered cells to be highly organized (see Porter, 1984, for history). Somehow those views, consensus for the time, were replaced by the "crowded chaotic cytoplasm", probably a result of the advent of the reductionist approach of modern biochemistry from the 1940s onward which captured the attention of those who worked in this area (see Clegg 1988). It seems we have come full circle,—at least most of us.

Nevertheless, the extent to which cytoplasm is organized and the nature and regulation of that organization, have not yet "gelled" into a consensus view. I believe that Keith Porter and his associates have provided us with a good approximation of what the global cytoplasmic structure might be like (see Porter 1986 for review). Applying high voltage electron microscopy (HVEM) to whole, critical-point-dried cells in culture they observed a highly branched network (termed the microtrabecular lattice, or MTL) that apparently made contact with (or enmeshed) virtually all organelles, and other ultrastructurally described structures. Porter has kindly provided me with some HVEM images for this paper (Fig. 11) which illustrate the MTL for those not familiar with this work. Porter has championed the "unit structure" nature of the MTL, and considers the aqueous volume surrounding it to be very dilute. That is a vital point since, if correct, it means that most (all?) enzymes of intermediary metabolism previously believed to be "soluble" or "cytosolic" are instead intimately associated with structure, the potential metabolic importance of which is evident and far-reaching. There are a few outspoken critics who consider the MTL to be an artifact and, probably worse, many others simply ignore the MTL when interpreting results on cytoplasmic structure. In my judgment, the "artifact-objections" have been answered satisfactorily (see Porter, 1986). Moreover, their is now abundant independent evidence, from the study of living cells that provides strong support for the existence of such a global structure or something very much like it.

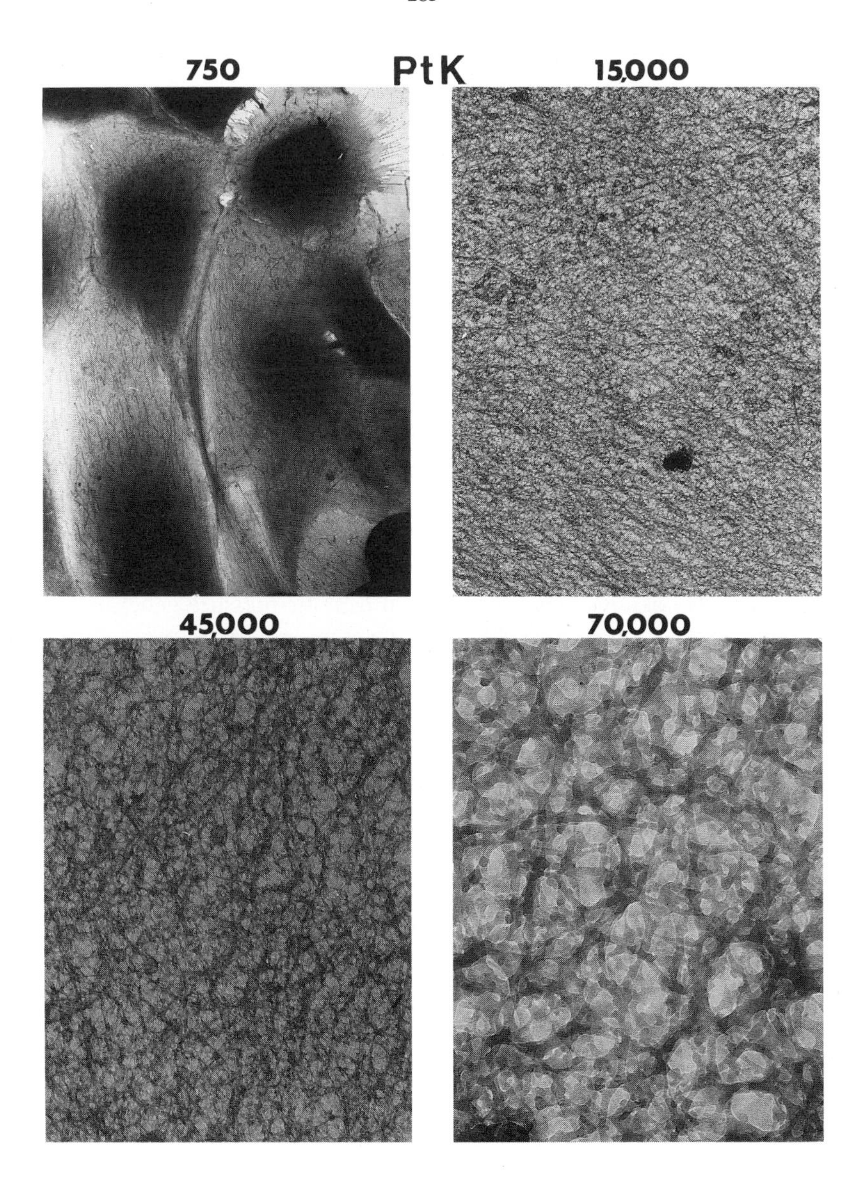

Figure 11. High voltage electron photomicrographs of critical-point-dried. PtK cells in culture generously provided by K.R. Porter. Whole cells are shown in the upper left panel, while increasing magnifications of the cytoplasm are depicted in the other panels. The numbers refer to magnification; thus, the lower right panel describes an area of cytoplasm magnified about 100 times compared to the whole cell. Detailed accounts of these images, methods employed, etc. are given by Porter (1986).

Needless to say, the existence of such a global structure should dominate the responses of cells to osmotic perturbation, and it should be noted that Porter (1986) has previously shown that the MTL shrinks or expands when mammalian cells are exposed to media of modestly altered osmolarity (150-600mOsm/kg). Against this brief background we may turn to the results of the present paper.

Cell Swelling in Extremely Hypoosmotic Media

Figures 1, 3, and 4 show clearly that MDCK cells do not continue to increase in volume, but swell to some limiting level, approximated to be 2-4 times the control volume (see text). While these cells eventually do rupture, that requires several hours (unpublished results). As pointed out previously by others (see Kleinzeller and Ziyadeh, 1990, and Lechene, 1985) one expects cells exposed to osmotic gradients of the kind studied here to rupture quickly, in the absence of some restraining mechanism. Thus, the MTL provides a basis for interpreting these data since it could be involved in that restraint.

It is reasonably assumed that the internal osmotic pressure generated by diffusible ions and metabolites is relieved since the plasma membranes quickly become freely permeable upon hypoosmotic exposure (Figs. 4, 7, 8) allowing their escape into what is, essentially, an infinite sink (see Materials & Methods section). Nevertheless, the cells should continue to swell due to the internal colloid osmotic pressure. All the factors that determine the limit to cell swelling are unknown, but the possibility clearly exists that these do not involve the continuous provision of free energy (ATP), a point to which I will return. Thus, it may be that swelling is limited by cell structure, not dynamics. Finally, it should be noted that, although the quantitative data are limited, it appears that single cells, small groups of cells and cells in close confluence swell to similar extent. Thus, cell-cell contact does not seem to affect swelling in any obvious way, although that may be a factor in survival when cells are returned to normal growth medium.

Effects of Cytoskeletal-Perturbing Agents

Of the six compounds tested cytochalasin-D (CD) produces the most obvious effect (Table 1) clearly implicating MF and actin as important components of the swelling response (Fig. 5), an interpretation in agreement with the studies by Mills (1987), Cornet et al. (1988) and Kajstura and Reiss (1989). Most notable is the apparent release of contacts between the confluent cells (Fig. 5B); however, the extent of swelling is similar to confluent cells whose contacts are maintained (Fig. 1). Fey and Penman

(1986) have presented ultrastructural evidence that confluent MDCK cells maintain contact through intermediate filaments (IF) and not microfilaments (MF). Because the action of CD is complex (see Cooper, 1987) and since it is quite likely that it could lead to a cascade of events involving IF, MF, and the plasma membrane (at the least) further discussion here seems pointless in view of our limited data. I will not consider the massive literature on actin and microfilaments except to provide access through selected reviews (Stossel, 1989; Pollard, 1990; Jamney, 1991; Matsudaira, 1991).

The effects of CD,3 applied and removed prior to swelling, are dramatic when 1:100 treated cells are returned to normal growth medium (M199cs) as shown in Figure 5C and D. The cells seem to undergo contractive seizures which result in death of all the cells if the concentration of CD used is 10μM or greater. It is as if a global collapse of the cytoplasm occurs. Whatever the mechanism involved, it is clear that CD treatment prevents 1:100-exposed cells from undergoing the processes that eventually restore normal cell structure and function, the topic taken up next.

Recovery After Hypoosmotic Exposure: Swell-Shrink Cycling

Observations on cells exposed to 1:100 media then returned to M199cs reveal a peculiar behavior not evident from the snapshots shown here (Figs. 1-3). The plasma membranes of swollen cells appear to remain permeable to small solutes for a few minutes following addition of M199cs based on ethidium bromide studies; in any event the cells do not reduce in volume during that time, and probably equilibrate with the constituents of M199. Rather suddenly the cells shrink. That response does not seem to be a simple osmotic one since it may be assumed that the cells would have previously taken up all the small molecules in M199cs and approached osmotic equilibrium. Although no direct proof is available, this shrinkage appears to be an active contraction, perhaps involving the cytomatrix. Following the initial shrinkage the cells do not then undergo a slow continuous return to their normal volume. Rather, they undergo readily observed cycles of swelling and shrinking. Because the cells are not synchronized (some are swelling, while others are shrinking) I would guess that these average out when bulk measurements are made on volume changes for huge numbers of cells. In any event, this cycling is a dramatic scene to witness and, to my knowledge, one not previously described.

Although we have not examined the exogenous requirements for recovery in any detail, a complex medium does not seem to be required to achieve normal morphology such as shown in Figures 1-3. Thus, 1:100 cells exposed to simple buffered salt solutions (CMF and DBS for example) do almost as well as cells in complex normal growth media, at least over the 2 hour recovery period examined; no exogenous energy source is required. That is truly remarkable since the cells must have lost all their complement of freely diffusing metabolites, coenzymes adenine nucleotides, and so on. This observation suggests strongly that 1:100-treated cells can regenerate all metabolites necessary to achieve "recovery", as defined here, and raises some interesting questions about their metabolism during hypoosmotic exposure as well as recovery. Next to be considered are our limited current results on this subject.

The Metabolic Status of Cells During Hypoosmotic Exposure

No indication has been obtained that 1:100 treated cells carry on an energy metabolism based on glucose utilization or reflected by lactate production and occurring at rates detectable during an hour incubation. The question arises: how do these cells maintain their integrity over 30 minutes (and, for some cells, more than 4 times that long)? This issue presents interesting prospects for further study. It may be that the gel properties of these cells allow them to reach an equilibrium swelling volume, much like any inanimate highly entangled and/or cross-linked gel (see Ross-Murphy, 1991; Bereiter-Hahn and Strohmeier, 1987; Lechene, 1985; Tanaka, 1981), and that a constant supply of free energy (ATP) is not required to maintain that structure for long periods of time.

There is, however, evidence of enzymatic activity in hypoosmotic cells (Figs. 7 and 8), but strictly of the hydrolytic kind, essentially converting polymers into monomers. Since these cells were exposed to ^{14}C-glucose and leucine for at least one doubling time before exposure it seems likely that they are "uniformly labeled" and the results should reflect total cell composition. Thus, hypoosmotic cells are degrading their polymers, but in a reversible fashion for survivors, over the 30 minute period studied.

In contrast to the quiet metabolism of swollen cells in 1:100, those returned to normal growth media exhibit a surprisingly vigorous energy metabolism, resumed at impressive rates compared to controls (Fig. 9). Because these cells (H) are initially permeable to low molecular weight compounds it may be assumed that they rapidly equilibrate with M199, thus providing a source of ions and metabolites required for the restoration of metabolism. It will be worth repeating these studies using simple salt solutions (instead of M199cs) to evaluate the extent to which exogenous metabolites are required. In any event, the glycolytic pathway clearly dominates recovery energy metabolism compared to oxidative pathways (Fig. 9).

There are many possible reasons why the rate of amino acid incorporation (reflected by leucine) into protein by H-cells is much lower in recovering cells compared to controls (C). The levels of leucine in the "free amino acid pools" of H and C-cells are similar, so that does not seem to be a limiting factor. Since cells in 1:100 media actively degrade proteins (Figs. 7, 8) it is possible that this continues in cells returned to normal growth medium resulting in lower net incorporation. Another likely possibility involves the well-documented association of polysomes with the cytomatrix (See Fey and Penman, 1986) which may be disrupted to some extent during cell swelling.

CONCLUDING REMARKS

The ability of MDCK cells to survive severely hypoosmotic conditions is not only dramatic but tells us something important about their structure. In order to maintain structure in a reversible state during swelling there must exist extensive entanglement and/or cross-linking of the cytomatrix, internally as well as with the plasma membrane. Thus, the absence of blebbing (see Fig. 3) suggests that unsupported portions of the plasma membrane do not exceed a spacing of about 40nm (see Jacobson, 1983; Mills 1987). However, it also seems very likely that the gel properties of cytoplasm are critical (Lechene, 1985; Wiggins and van Ryn, 1990); indeed, the swelling limit we've observed here might be due solely to these properties (also see Ross-Murphy, 1991). I have not considered the issue of cell water and its possible involvement in the osmotic response; but that could also be of major importance, as pointed out in an interesting series of papers by Wiggins and her colleagues (see Wiggins et al. 1991). That is a topic of such complexity that it would require a separate treatment.

Observations on Brownian motion of the numerous granules in these cells are also relevant. Most cells in 1:100 medium do not exhibit Brownian motion of these granules (whose diameters are 1μm or less) in spite of an increase in cell volume of 2-4 times. In the absence of a restraining structure these granules would certainly diffuse. Of course, not all cells survive these hypoosmotic insults and these cells also provide insight into the swelling process. Although not documented here, we have observed that cells that do not survive always exhibit Brownian motion of their contents, a result easily obtained by tracking individual cells in video segments. However, the presence of Brownian motion in a given cell does not always serve as a reliable prediction of its death, since some of these cells do survive, suggesting restoration of restraining mechanisms. We may suppose that the cytomatrix, whatever its nature, eventually reaches a critical point of global and irreversible disruption and collapse when cells are exposed to prolonged severe hypoosmotic conditions, and that this critical point varies amongst the individual cells of a population. That suggestion is in agreement with numerous studies on the patterns of protein efflux from cells opened with detergents (see Schliwa et al. 1981; Kellermayer et al. 1986; Ridsdale and Clegg, 1991) which also indicate the occurrence of a sudden and global collapse of cytoplasmic structure.

Brief comment should be made about the potential involvement of the cytomatrix in establishing and maintaining cell volume. To my knowledge, we still lack a complete understanding of the mechanism(s) that determine the volume of a given cell. There can be no doubt that movements of inorganic ions and the production and release of "osmolytes" play a critical role and these have been given massive study. However, the participation of the cytomatrix is generally neglected, in spite of the chance that it could also be a vital player in generating cell volume, as others have previously pointed out (see Lechene, 1985; Bereiter-Hahn and Strohmeier, 1987; Mills, 1987; Cornet et al. 1988; Kleinzeller and Ziyadeh, 1990). If the cytomatrix can perform the Herculean osmotic feats observed in this study it certainly can be involved in "setting" cell volume under physiological conditions, as well as responding to changes in external osmolarity.

I have tended to fit the results presented here into a "procrustean bed" of interpretation in which the cytomatrix reigns supreme and dominates mechanism. That is not because I believe the cytomatrix is the whole story, explaining everything about the physical properties of cells. However, it seems to me that its exclusion from descriptions of the physiology of cells will lead us astray, and provide a seriously incomplete picture. For that reason I have taken a somewhat myopic stance in this paper in order to focus attention on the remarkable structure of cytoplasm.

ACKNOWLEDGMENTS

I am very grateful to Diane Cosgrove for her skill and patience in the preparation of this camera-ready manuscript. Previously unpublished research presented here was supported by a grant from the US National Science Foundation (DCB 88-20347). The HVEM photographs in Figure 11 were generously provided by K.R. Porter.

REFERENCES

Albrecht-Buehler G (1990) In defense of "nonmolecular" cell biology. Int Rev Cytol 120:191-241

Bereiter-Hahn J, Strohmeier R (1987) Hydrostatic pressure in metazoan cells in culture: its involvement in locomotion and shape generation. In: Bereiter-Hahn J, Anderson OR, Reif W-E (eds) Cytomechanics. Springer-Verlag, Berlin, pp 261-272

Beyenbach KW (ed) (1990) Cell volume regulation. S. Karger, Basel

Clegg JS (1988) The aqueous intracellular compartments. In: Jones DP (ed) Microcompartmentation. CRC Press, Boca Raton, pp 1-16

Clegg JS, Jackson SA (1988) Glycolysis in permeabilized L929 cells. Biochem J 255:335-344

Carmo-Fonseca M, David-Ferreira JF (1990) Interactions of intermediate filaments with cell structures. Electron Microsc Rev 3:115-141

Chamberlin ME, Strange K (1989) Anisomotic cell volume regulation: a comparative view. Am J Physiol 257:C159-C173

Cooper JA (1987) A review of the effects of cytochalasin and phalloidin. J Cell Biol 105:1473-1478

Cornet M, Delpire E, Gilles R (1988) Relations between cell volume control, microfilaments and microtubule networks in T2 and PC12 cultured cells. J Physiol Paris 83:43-49

Fey EG, Penman S (1986) New views of cell and tissue cytoarchitecture: embedment-free electron microscopy and biochemical analysis. In: Welch GR, Clegg JS (eds) Organization of cell metabolism. Plenum, New York

Gilles R, Kleinzeller A, Bolis L (eds) (1987) Cell volume control: fundamental and applied aspects. Academic Press, New York

Helig CW, Brenner RM, Yu ASL, Kone BC, Gullans SR (1990) Modulation of osmolytes in MDCK cells by solutes, inhibitors, and vasopressin. Am J Physiol 259:F653-F659

Jamney PA (1991) Mechanical properties of cytoskeletal polymers. Curr Opin Cell Biol 2:4-11

Jones DP (ed) (1988) Microcompartmentation. CRC Press, Boca Raton

Kajstura J, Reiss K (1989) F-actin organization influences the osmotic reactions of animal cells. Folia Histochem Cytobiol 27:201-208

Kellermayer M, Ludany A, Jobst K, Szucs G, Trombitas K, Hazlewood CF (1986) Cocompartmentation of proteins and K^+ within the living cell. Proc Nat Acad Sci USA 83:1011-1015

Kleinzeller A, Ziyadeh FN (1990) Cell volume in epithelia—with emphasis on the role of osmolytes and the cytoskeleton. In: Beyenbach KW (ed) Cell volume regulation. S. Karger, Basel, pp 59-86

Klymbowsky M, Bachant JB, Domingo A (1989) Functions of intermediate filaments. Cell Motil Cytoskeleton 14:309-331
Lechene C (1985) Cellular volume and cytoplasmic gel. Biol Cell 55:177-180
Litniewski J, Bereiter-Hahn J (1990) Measurements of cells in culture by scanning acoustic microscopy. J Microsc 158-95-107
Luby-Phelps K, Lanni F, Taylor DL (1988) The submicroscopic properties of cytoplasm as a determinant of cellular function. Annu Rev Biophys Biophys Chem 17:369-396
Macknight ADC (1987) Volume maintenance in isosmotic conditions. Cur Top Memb Transp 30:3-43
Macknight ADC (1988) Principles of cell volume regulation. Renal Physiol 3:114-141
Matsudaira P (1991) Modular organization of actin crosslinking proteins. Trends Biochem Sci 16:87-92
Mills JW (1987) The cell cytoskeleton: possible role in volume control. Curr Top Membr Transp 30:75-101
Mitchell JJ, Low RB, Woodcock-Mitchell JL (1990) Cytomatrix synthesis in MDCK epithelial cells. J Cell Physiol 143:501-511
Negendank W, Edelmann L (eds) (1988) The state of water in the cell. Scanning Microscopy International, Chicago
Pollard TD (1990) Actin. Curr Opin Cell Biol 1:33-40
Porter KR (1984) The cytomatrix: a short history of its study. J Cell Biol 99:3s-12s
Porter KR (1986) Structural organization of the cytomatrix. In: Welch GR, Clegg JS (eds) Organization of cell metabolism. Plenum, New York, pp 9-26
Ridsdale JA, Clegg JS (1991, to be published) Evidence for cooperativity of protein dissolution in Brij-58 permeabilized L929 cells. J Cell Physiol
Ross-Murphy SB (1991) Physical gelation of synthetic and biological macromolecules. In: DeRossi (ed) Polymer gels. Plenum, New York, pp 21-40
Roy G, Sauvé R (1987) Effect of anisotonic media on volume, ion and amino acid content and membrane potential of kidney cells in culture. J Membr Biol 100:83-96
Schliwa M, van Blerkom J, Porter KR (1981) Stabilization of the cytoplasmic ground substance in detergent-opened cells and a structural and biochemical analysis of its composition. Proc Nat Acad Sci USA 78:4329-4333
Srere P (1987) Complexes of sequential metabolic reactions. Annu Rev Biochem 56:21-62
Srere PA, Jones ME, Matthews CK (eds) (1990) Structural and organizational aspects of metabolic regulation. Wiley-Liss, New York
Stossel TP (1989) From signal to pseudopod: how cells control cytoplasmic actin assembly. J Biol Chem 264-18261-18264
Tonaka T (1981) Gels. Sci Am 244:124-138
Welch GR (ed) (1991, to be published) Metabolic organization and cellular structure. Curr Top Cell Regul 33
Welch GR, Clegg JS (eds) (1986) Organization of cell metabolism. Plenum, New York
Wiggins PM, van Ryn RT (1990) Changes in ionic selectivity with changes in density of water in gels and cells. Biophys J 58:585-596
Wiggins PM, van Ryn RT, Ormrod DGC (1991, to be published) The Dounan membrane equilibrium is not directly applicable to distributions of ions and water in gels or cells. Biophys J

OSMOTIC SWELLING-PRESSURIZATION-RUPTURE OF ISOLATED CELLS AND DISJOINING OF CELL AGGREGATES IN SOFT TISSUES

Evan Evans
Departments of Pathology and Physics
University of British Columbia
Vancouver, B.C. Canada V6T 2B5

Introduction

All biocellular organisms and tissues are subjected to some variation in osmotic environment. Dilution of the osmotic environment can lead to swelling of cells and to pressurization when the cell structure rigidly opposes increases in volume. For a sufficiently-large osmotic pressurization, mechanical failure of the cell structure may ensue. This failure can result in immediate rupture of the cell permeability barrier (lysis) and loss of cytosolic contents or less-catastrophic chemical/mechanical degradation of the cytoarchitecture. Hence, the sequence of swelling, pressurization, and rupture depends on cell "design" as well as the intrinsic strength of cell "materials". Nature has employed different engineering designs - appropriate to the osmotic environment - for each class of organism (prokaryotes and eukaryotes).

Simply viewed, prokaryotes (e.g. bacteria) are stiff capsular structures which respond rigidly to osmotic stresses. Most likely, the strong cell wall supports the pressure differential since some organisms survive extreme variations in osmotic environment. The large pressures would easily create wall stresses greater than the strength of the surfactant bilayer(s) used as the permeability barrier to small solutes. Here, Nature has taken advantage of the "law of Laplace" to diminish wall stresses by keeping the organisms small. [Note: the wall stress σ scales as the pressure x radius of curvature / wall thickness.] Based on the osmotic stress (~ 10 Atm) implicit in the environments of some organisms and their sizes (~ 10^{-4} cm), the strength requirement for a cell-wall material of 100 Å thickness is estimated to be $\hat{\sigma}$ ~ 10^{3} Atm which is high strength. This requirement may have guided early (evolutionary) developments of these organisms. At first glance, wall strength seems to be an obvious design criterion; but there are hidden subtleties and

NATO ASI Series, Vol. H 64
Mechanics of Swelling
Edited by T. K. Karalis

frustrations associated with compatibility of the "soft" and "hard" material components of prokaryotes which have also been accommodated in the design (e.g. thermal expansivity of the "soft" bilayer >> expansivity of the "hard" cell wall; compression of the "soft" bilayer is compressed against the "hard" cell wall).

By comparison, eukaryotes (e.g. animal and plant cells) appear to have developed according to very different design guidelines. Here, two tactics have been employed by Nature to deal with chemical and physical stresses in the environment: (1) cells are organized into structurally-reinforced aggregates (tissues) so that stresses are distributed over larger distances; (2) homeostatic control (physiological chemistry) is used to narrow the variation of local osmotic stress to which cells are exposed (e.g. osmotic stresses experienced by cells in most animal tissues are reduced to a small fraction of an atmosphere). For plants, strong cellulose "containers" provide structural shelter for cells, a concept analogous to the cell wall of prokaryotes. Likewise, there are interesting and important chemical mechanisms involved in protecting plants against osmotic "injury". A different situation exists for the soft-cellular constituents of animal tissues. Consistent with large compliance, cells of soft tissues are exquisitely sensitive to changes to osmotic environment, i.e. pressures on the order of 10^{-1} Atm usually lead to lysis! Thus, Nature has developed "tight" permeability designs and chemical "pumping" to regulate the osmotic activity in tissues. As witness to this protection, we do not "bloat" when bathing in salt-free water nor "wither" when swimming in the ocean! However, if the permeability barrier is violated or the chemical pumps are poisoned, cells readily swell or shrink in response to changes in osmotic environment which can result in cell destruction. Further, since tissues are made-up of adhesively-bonded capsules, osmotic swelling can cause disjoining of the cellular aggregate as membrane tensions increase under pressurization. Ultimate failure of a tissue may result from membrane lysis, disruption of intercellular bonding, or rendering of cytoskeletal structure. In the discussion to follow, the focus will be on the general features of (eukaryotic) cell structure and interfacial properties in soft-animal tissues; these factors determine deformation, stress, and failure response to osmotic pressurization.

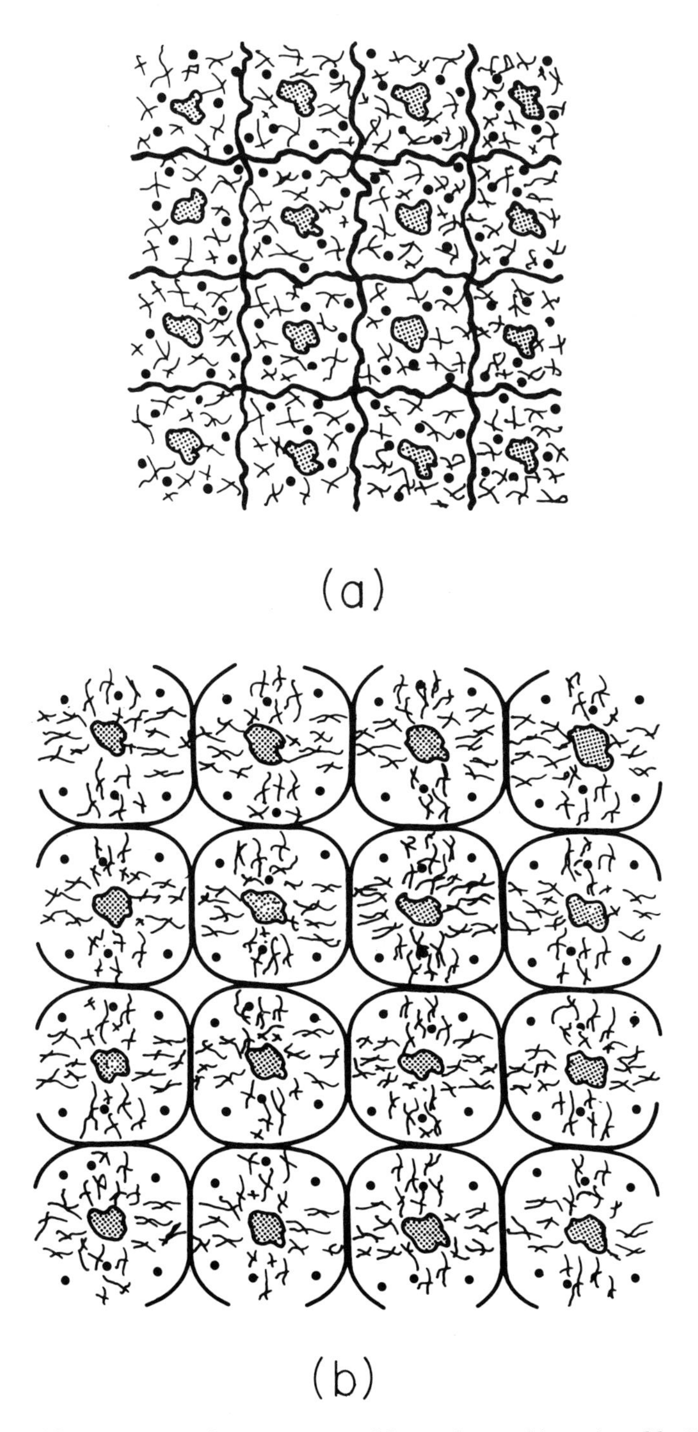

1. Schematic illustration of osmotic swelling (a to b) and cell disjoining in "soft" tissue aggregates of cells.

Mechanics of Swelling Deformations in Soft Tissues: Qualitative Features

As mentioned above, animal tissues are not usually exposed to significant osmotic stress because of restricted permeability and chemically-driven "pumps". These important life-preserving features will be ignored here in order to examine the mechanical response of cells and tissues when made vulnerable to extreme hydration. Figure 1 is the conceptual starting point and illustrates the qualitative behavior expected for swelling-shrinking of a multicellular aggregate (tissue). Although there are significant variations in detailed architecture amongst different tissue cells, the principal structural components are represented schematically in Fig. 1: i.e. a redundant - wrinkled and gathered - plasma membrane envelope; a cytoskeletal scaffolding or network; nonstructural elements like nucleus and small organelli (endoplasmic reticulum, golgi, mitochondria, granules, etc.). Functionally, the cytoskeletal network - often composed of actin filaments - is usually the dominant "load-bearing" element. In other words, when the cell is deformed, external forces are transmitted to the cytoskeletal network through interfacially-bonded contacts between cells . Although not precisely known, the network can most likely sustain stresses up to 1 Atm, maybe higher. By comparison, the plasma membrane envelope is a condensed-fluid "insulator" composed of lipids and other cosurfactants which isolates the cytoplasmic chemistry from the external environment (i.e. the permeability barrier). The plasma membrane envelope freely flows to assume new shapes with little resistance to displacement provided that the area of the membrane is not required to dilate (i.e. before the cell becomes pressurized into a spherical shape). If the plasma membrane is forced to dilate, it will fail at tensions $\hat{\tau}$ in the range of 2-10 dyn/cm which will be produced by hydrostatic pressures in the cell on the order of only 0.1 Atm. Hence, cells of tissues possess large redundancy of plasma membrane to allow for major deformations without causing stress on the membrane. As shown in Fig. 1b, tissue hydration and swelling is accommodated by the excess plasma membrane until the "wrinkles" are pulled smooth where-upon regions of the surface approach spherical contours. Also, swelling causes the cytoskeletal structure to be drawn into patches at the cell-cell contacts as the cells disjoin to form interstitial spaces. It is expected that further increases in cell pressure (due to hydration) will rupture the cell membrane when the pressure reaches $2\hat{\tau}/R_c$ even though the labyrinth of cytoskeletal networks remain linked together.

There are three important scenarios involved in stress build-up and composite rearrangements: (1) First, cytoskeletal-interfacial connections may be sparse and easily dragged through the membrane to produce the condition shown in Fig. 1b. For this case, the total curvature $\bar{c}$ = "$2/R_c$" of the membrane boundary (which borders the open-interstitial space) will be limited by the size of the intercellular contacts (patches) and the residual stiffness of the network elements within the cell. (2) Second, the cytoskeletal connections between cells may be dense and immobile. As such, the network would distribute the hydrostatic pressure Π amongst fibres in the cell to give forces of approximately $f \sim \Pi \bar{a}$ (where $\bar{a}$ is the area per fibre over the intercellular contact region). When this force exceeds the rupture strength of the network, then the network would locally render and portions would coalesce along the interface to again produce the situation depicted in Fig. 1b. (3) Finally, the swelling properties of cells in the aggregate may not be uniform. This would result in local network displacements that are incompatible with the geometric boundaries of cell envelopes. Hence, the network would be <u>pulled through</u> the surface as the cell became pressurized. From these scenarios, it is clear that excessive hydration can cause tissue destruction by either rendering of the cytoskeletal network or lysis of the plasma membrane envelope. The criterion for which process of failure will occur is based on the comparison between the level of force $\hat{f}$ required to rupture a fibre (or intercellular connection) x the number of intercellular connections per unit area of interface $1/\bar{a}$ vis a vis the tension $\hat{\tau}$ necessary to lyse the plasma membrane x the characteristic curvature $\bar{c}$ of an unsupported region of membrane boundary; failure will proceed according to which factor ($\hat{f}/\bar{a}$ or $\hat{\tau}\cdot\bar{c}$) is reached first.

Based on typical surface receptor densities ($1/\bar{a} \simeq 10^{11}$-$10^{12}/cm^2$) and the macroscopic dimensions of cells (which imply $\bar{c} \sim 10^4$ cm^{-1}), fibre strength would only have to exceed 10^{-6} dyn in order for plasma membrane lysis to be the primary mechanism of failure. The plasma membrane appears to be the most vulnerable material since even fibres held together by hydrogen bonds are probably an order of magnitude stronger than 10^{-6} dyn. It is interesting to note that when a plasma membrane is lysed, the hydrostatic pressure is rapidly released accompanied by ejection of small-cytoplasmic solutes; but the membrane <u>reseals</u>. Thus, the failure seems to primarily involve a short violation of the permeability barrier which results in loss of small solutes but from which the cell can recover. However, more subtle membrane lysis processes may also occur which would indeed cause permanent damage to the cell. For example, small organelli (e.g. mitochondria) or enzyme-containing granules may lyse which would be potentially catastrophic for the cell. These microfailure processes probably accompany osmotic swelling and perhaps precede ultimate cell rupture in a multicellular aggregate. It is clear that the anatomy and arrangement of plasma membrane, cytoskeletal network, and organelli in different tissues will result in widely varying responses to hydration. Ignoring these tissue-specific features in the sections to follow, swelling of single cells and small-bonded cell aggregates will be discussed to emphasize more general mechanical behavior.

Single Cells as Osmometers: Volume Increase-Pressurization-and Rupture

To minimize complexity, synthetic membrane vesicles and circulating blood cells will be used to illustrate osmotic swelling-pressurization-and rupture properties. From a physical perspective, the structure of blood cells differs from tissue-fabric cells (e.g. fibroblasts, endothelial cells, smooth muscles cells, etc.) primarily in the architecture of the cytoskeletal scaffolding. For tissue-fabric cells, the cytoskeleton is distributed throughout the cytoplasm in ways peculiar to each cell type. By comparison, the cytoskeletal structure in blood cells is localized to the cortex of the cytoplasm adjacent to the membrane. When isolated in suspension, both tissue-fabric and blood cells are spherical in form (except red cells) and exhibit similar osmotic responses. However, only synthetic lipid bilayer vesicles behave as "perfect" osmometers!

For comparative reference, examine the swelling and rupture behavior of a synthetic-lipid bilayer vesicle. Figure 2a shows a macroscopic-size (~ 20μm) lipid bilayer vesicle held by a suction micropipet in 0.16 M solution of glucose. With the tension in the bilayer fixed by the pipet suction pressure P, the vesicle was transferred to a higher osmotic strength solution (~ 0.2 M glucose) where the vesicle was partially dehydrated and reached the shape shown in Fig. 2b. Because the vesicle

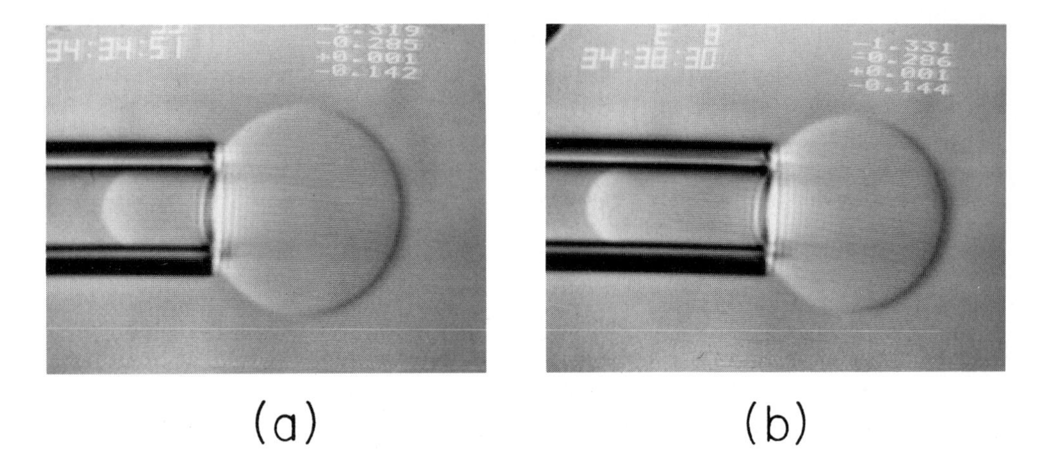

2. "Perfect" osmometer: videomicrographs of a lipid bilayer vesicle (~ 20μm diameter) held by micropipet suction (~ 10^{-3} Atm), first in 0.16 M glucose solution (a) and then transferred to 0.2 M glucose (b).

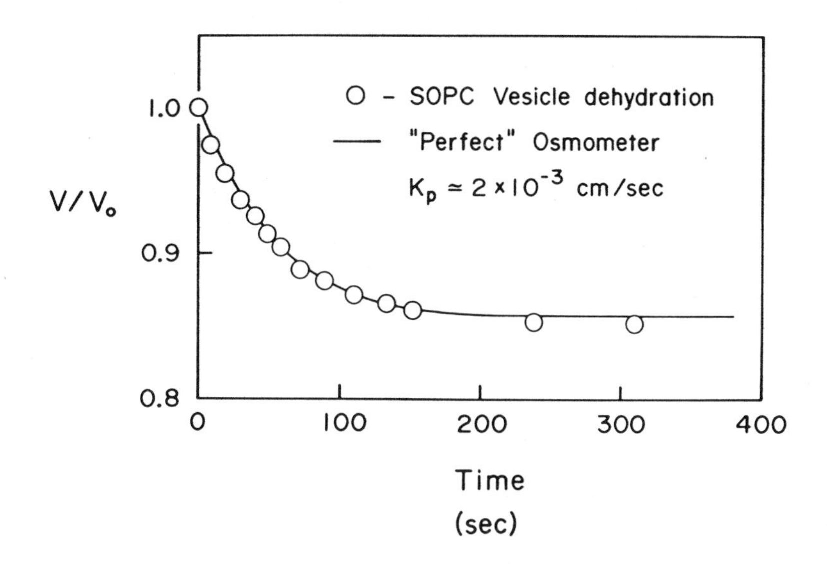

3. Time course of osmotic dehydration for a single lipid bilayer vesicle (cf. Fig. 2). The solid curve is the classical prediction for volume change assuming a uniform-constant permeability k_p.

tension was held constant (~ 0.1 dyn/cm), the vesicle area did not change throughout the dehydration process; hence, the increase in projection length inside the pipet was proportional to the decrease in vesicle volume. The time course of the vesicle dehydration is plotted in Fig. 3 along with the analytical prediction for a "perfect" osmometer assuming a constant membrane permeability. For the pure lecithin (stearoyl-oleoyl-phosphatidylcholine SOPC) vesicle studied here, the permeability was measured to be ~ 2 x 10^{-3} cm/sec. Studies with other bilayer compositions have shown that permeability to water can be as low as 10^{-5} cm/sec (1:1 sphingomyelin-cholesterol) and as high as 6 x 10^{-3} cm/sec (e.g. dimyristoyl-phosphatidylcholine DMPC). Clearly, composition is an extremely important determinant for permeability!

When a vesicle is swollen close to a spherical state, then tension increases sharply as the area is required to expand. Figure 4 presents data for tension versus area expansion which demonstrate the strong increase in tension for small area dilations up to the point of lysis which always occurs in the range of 2-4%. The slope of the data specify the elastic dilation modulus K_a for the membrane. Like permeability, the peak tension $\hat{\tau}$ at lysis (as well as the elastic modulus) varies significantly with composition of the bilayer. For example, tensions of 35 dyn/cm have been obtained for lysis of sphingomyelin-cholesterol vesicles but only tensions of ~ 2-3 dyn/cm are sufficient to lyse DMPC bilayer vesicles. Further, in plasma membranes of cells, there are many integral proteins with bilayer-spanning sequences which modulate lipid bilayer permeability (e.g. red cell membrane permeability to water is an order of magnitude larger than found for vesicles made from extracts of red cell lipids). However, lysis of cells appears to be almost entirely regulated by the strength of the lipid bilayer component (e.g. red cells lyse at tension levels in the range of 6-12 dyn/cm and vesicles made from extracts of red cell lipids lyse in the range of 10-12 dyn/cm).

Similar to synthetic-bilayer vesicles, isolated cells in suspension easily swell in response to dilution of the osmotic environment until the plasma membrane envelope is forced to dilate. Membrane expansion only becomes significant when the cell is pressurized to a spherical form. Figure 5 shows examples of red and white (polymorphonuclear) cells which have been transferred from an "isotonic" (~ 0.15 M) NaCl buffer to low salt solutions. For the red cell, the cell volume increased by about 70% before it lysed in the 0.05 M salt solution - again with no readily apparent

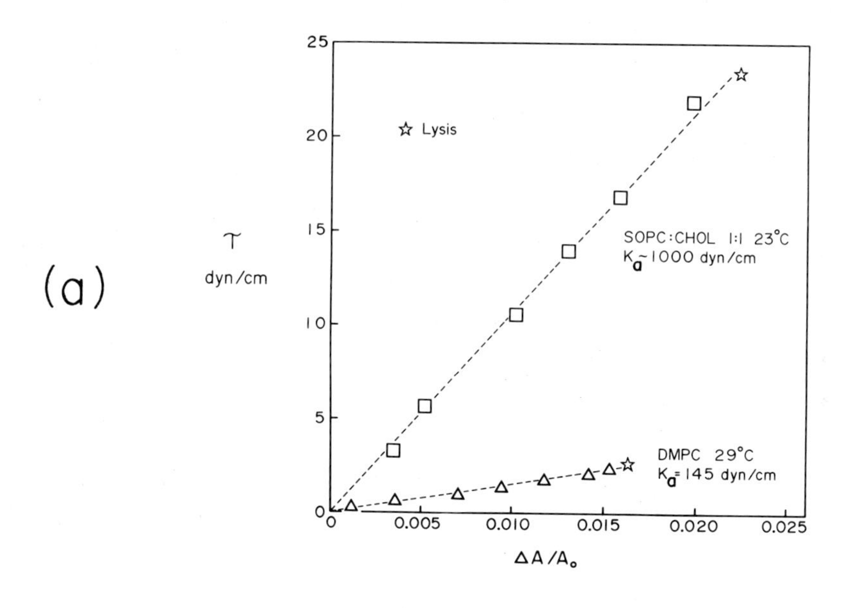

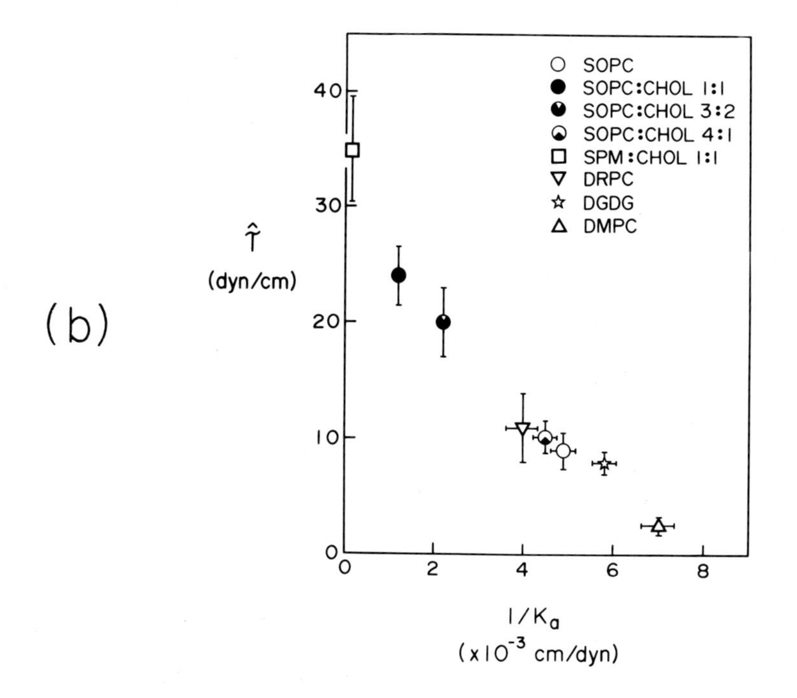

4. (a) Area dilation α and bilayer tension τ produced by pressurization of a lipid bilayer vesicle with pipet suction. The slope of the linear portion of the tension versus dilation provides the elastic area compressibility modulus of the lipid bilayer. (b) Tensions required to lyse bilayer vesicles versus elastic area compressibility which demonstrate the major effect of lipid composition.

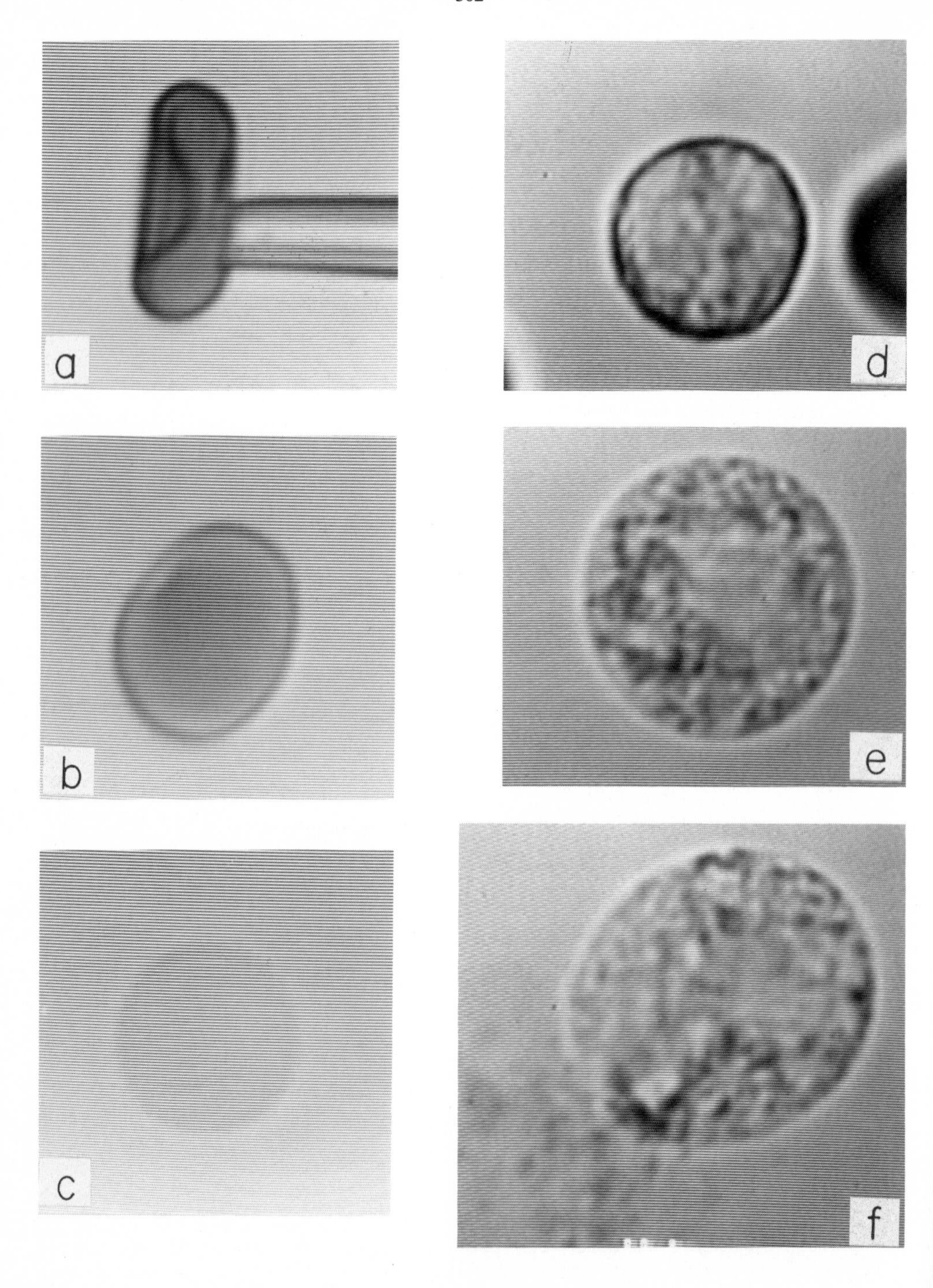

5. Videomicrographs of osmotic swelling and rupture of a red blood cell (a-c) and a polymorphonuclear (white) blood cell (d-f), both with maximum dimensions of ~ 8 μm.

change in surface area. By comparison, the white cell was swollen by 200%! without lysis in 0.025 M NaCl but rapid swelling in 0.01 M NaCl ruptured the cell as shown in Fig. 5f. The exceptional increase in volume for the white cell and the cell size at lysis are consistent with measurements of the excess area of the "wrinkled" plasma membrane surface which was found to be about 110% greater than that required to enclose the normal volume. A subtle feature of white cell swelling (different from red cells) is that they do not remain swollen in low salt buffers. With no apparent rupture in 0.025 M NaCl, the cell eventually reduces its volume as illustrated in Fig. 6. If the osmotic shock is larger (i.e. the concentration of the buffer is lowered to 0.01 M NaCl), the white cell swells at the same rate but ruptures when it reaches the critical area (Fig. 6).

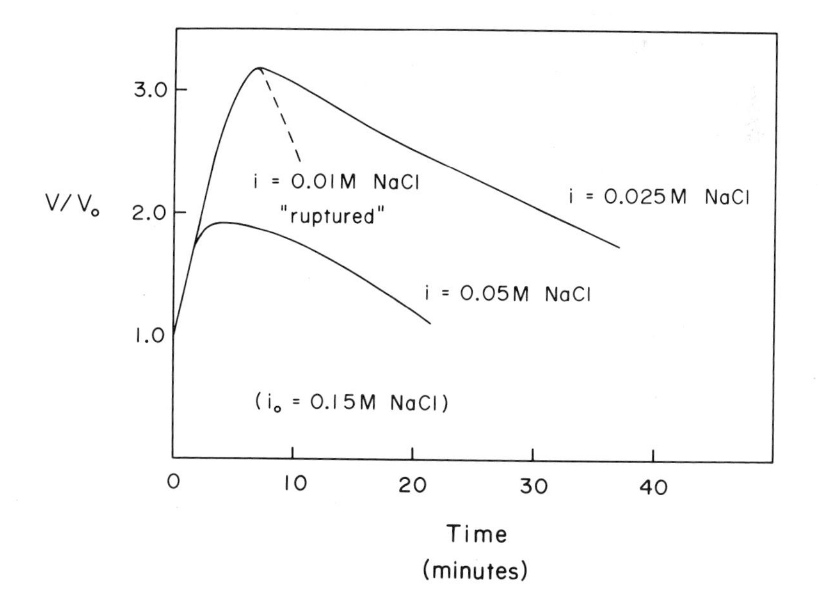

6. Time course of osmotic swelling of white blood cells in different ionic strength buffers: Rupture (cf. Fig. 5f) occurred when the buffer concentration was lowered below 0.025 M NaCl; but without rupture, the cell "regulated" volume downward and became stiff - often extending pseudopods.

Even though the plasma membrane bilayer is the vulnerable material component for lysis of cells, cells do not behave as "perfect" osmometers when they swell up to the rupture condition. Comparing the changes in swelling environment and volume for red and white blood cells, it is obvious that the cell volumes were much lower than predicted by the reduction in osmotic strength of the environment (e.g. the volume of the red cell should have increased by nearly 200%! and the white blood cell by 500%! based on the ratio of concentrations for the salt buffers). The simplest-phenomenological rationalization for deviation from ideality is to assume that some fraction (compartment) of the cell contents is osmotically inactive but the remainder of the cell interior behaves as an ideal solution. In this case, the volume $\tilde{V} = V/V_o$ (relative to a reference volume V_o at a buffer concentration where the cell can be considered to be "unstressed") is expressed in terms of the ratio $\tilde{i} = i/i_o$ for buffer concentrations in the environment by,

$$\tilde{V} = \phi / \left[\tilde{i} + 2\tau/\Pi_o R_c\right] + (1 - \phi) \tag{1}$$

$$\tau \simeq K_a \left[\left(\tilde{V}/\tilde{V}_s\right)^{2/3} - 1\right] \qquad R_c \equiv (3V/4\pi)^{1/3}$$

where ϕ is the volume fraction of "active" solution in the reference state; τ is the membrane tension; Π_o is the osmotic strength of the reference state; and R_c is the radius of curvature of the pressurized cell. [The validity of equation (1) is demonstrated in Fig. 7 by comparison with osmotic swelling of blood cells.] For biological cells with radii of a few microns, tension builds-up rapidly to the lysis point only after reaching a spherical state (represented by $\tilde{V}_s$ above) again similar to synthetic bilayer vesicles. A small reduction in buffer concentration in this state will readily achieve lysis provided that the permeability barrier is not violated (as shown by the magnitude of $2\hat{\tau}/\Pi_o R_c \simeq 10^{-2}$, equivalent to ~ 2×10^{-3} M NaCl when $i_o = 0.15$ M NaCl). In other cells with cytoskeletal networks distributed throughout the cytoplasm, the term $2\tau/\Pi_o R_c$ is replaced by σ_n/Π_o which is the ratio of network stress produced by the swelling deformation to the osmotic reference stress. Assuming the cytoskeletal network is elastic and the deformation is nearly isotropic, then the volume response can be related to the elastic-dilatation modulus K_v of the network through the following equations:

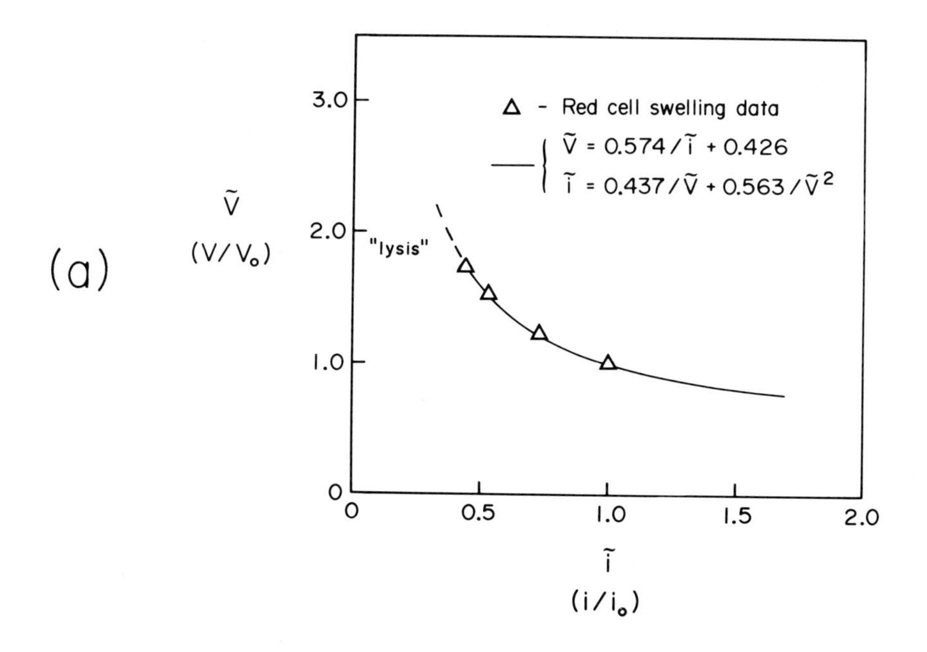

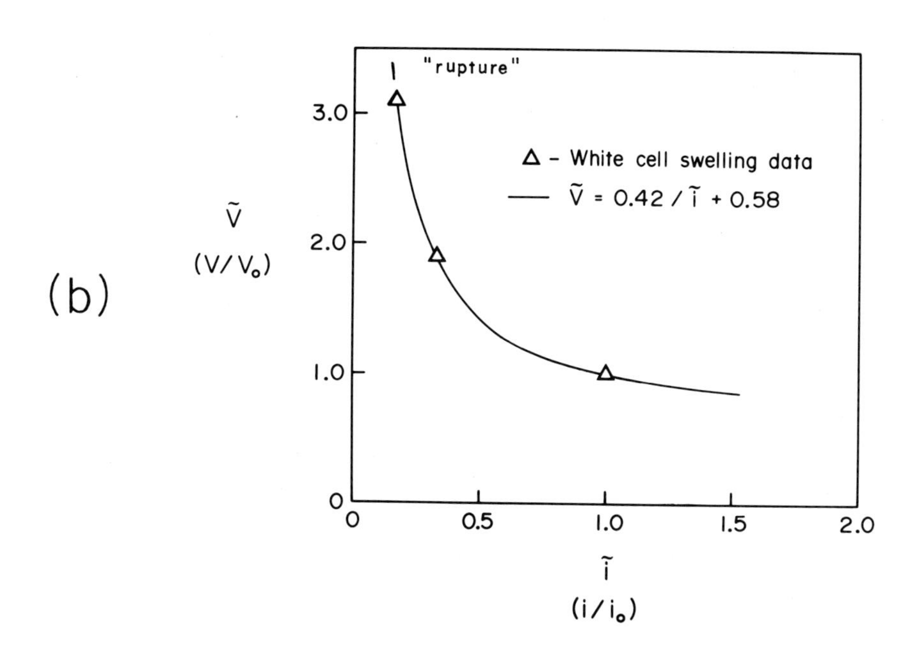

7. Blood cell volumes versus tonicity (ratio of buffer concentration i to the isotonic concentration i_o). The solid curve is the fit to the data for either an "imperfect" osmometer relation or a virial expansion.

$$\tilde{V} = \phi / \left[\tilde{i} + \sigma_n/\Pi_\circ\right] + (1 - \phi)$$

$$\sigma_n \simeq K_v \left(\tilde{V} - 1\right) \qquad (2)$$

The important feature is that build-up of network stress (either localized to the cortex of the cell or distributed through the cell cytoplasm) acts to diminish the sharp increase in volume in low salt environments. Keeping in mind the subtle aspects of cell structure and other caveats described in the previous section, elastic-network stresses could (in principle) stabilize finite volumes in pure water. The volumes are predicted by the roots to equations (2) when the buffer concentration is reduced to zero, i.e.,

$$\tilde{V} \simeq 1 + \Pi_\circ/K_v \qquad , \qquad \{\phi K_v >> \Pi_\circ\}$$

for the case of very stiff networks and,

$$\tilde{V} \simeq 1 + \sqrt{\phi \Pi_\circ / K_v} - \phi/2 \qquad , \qquad \{\phi K_v << \Pi_\circ\}$$

for soft-compliant network structures.

Another approach for expressing the nonideal-swelling behavior of cells is to expand the chemical potential of water (i.e. the osmotic pressure of the environmental solution Π x the volume per mole of water molecules) in a virial expansion, i.e.

$$\Pi = \Pi_\circ \tilde{i} = a_1/\tilde{V} + a_2/\tilde{V}^2 + \cdots$$

clearly, only the first term in this expansion is required for a "perfect" osmometer. The coefficient of the second term in the expansion above is recognized as the "excluded volume" parameter commonly used to characterize the osmotic properties of macromolecular (e.g. polymer) solutions. Correlations of equations (1) and (3) with blood cell swelling data are indistinguishable over the range plotted in Fig. 7.

Disjoining of Adherent Cells by Osmotic Swelling

As illustrated in Fig. 1, swelling of cells in aggregates is frustrated by network structural rigidity and intersurface bonding. Noted previously, the characteristics of intercellular-network connections at interfaces (e.g. sparse versus dense, mobile versus immobile) plus the network properties (e.g. crosslinking, stiffness, and strength) strongly affect the progression of tissue swelling and cell disjoining. Again, the cell-specific cytoplasmic structure will be ignored to focus on the interfacial disjoining of adherent cells driven by osmotic pressurization. It will be assumed that swelling leads to the scenario depicted in Fig. 1b where intercellular network connections are sparse and dragged to the contact zone when cells are separated. Similar accumulation of intercellular bonds commonly occurs when agglutinated blood cells are "peeled" apart - Fig. 8. The mechanics of separation are depicted in

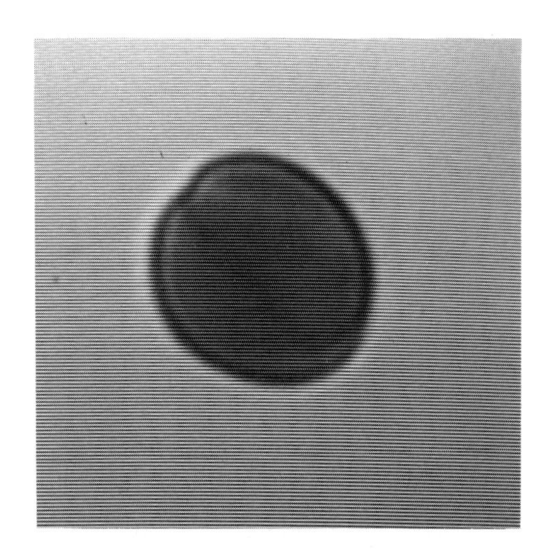

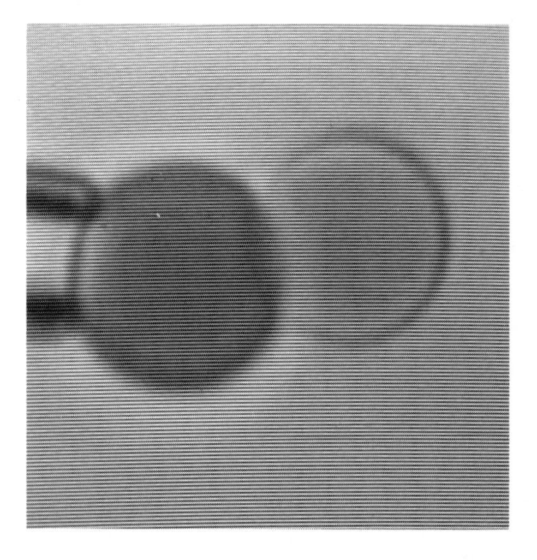

8. Videomicrographs of osmotic swelling and disjoining of an agglutinated pair of red blood cells: (a) in 0.15 M NaCl before swelling; (b) after swelling in 0.05 M NaCl where one of the cells lysed.

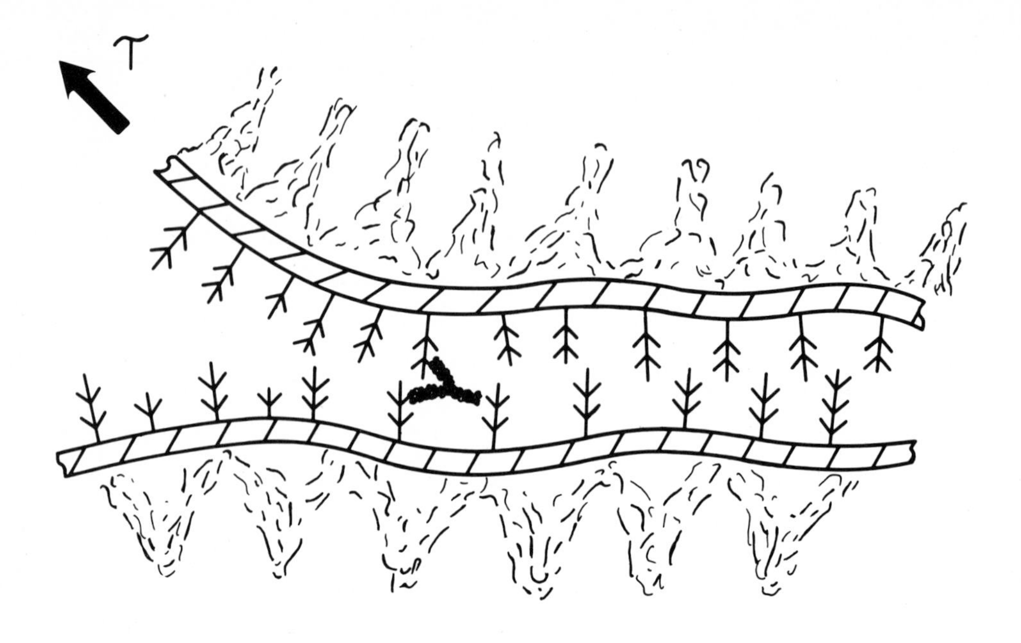

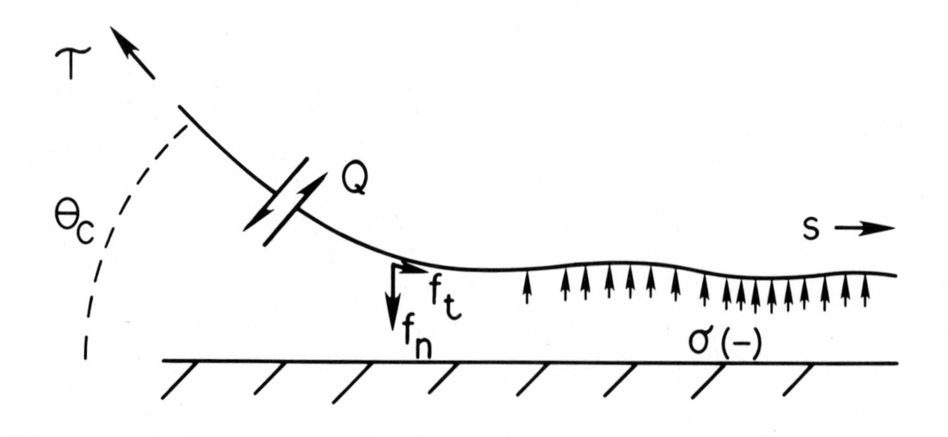

9. Schematic illustration for "fracture" of membranes bonded together by focal-molecular attachments (symbolized by the "Y" in the upper drawing).

Fig. 9. Because of the paucity of intercellular bonds, the tension force τ (i.e. "peeling" force) applied to the membrane at some macroscopic-apparent contact angle θ_c is almost entirely supported by bonds along the perimeter of the contact zone. Since the membrane is almost tangent to the substrate at the bond, the macroscopic tension force is transmitted to the microscopic bond by the transverse bending stress Q in the membrane. The bending stress is derived from the gradient of curvature along the surface contour, i.e. $Q \simeq k_c(d\bar{c}/ds)$, where k_c is the elastic-bending stiffness of the membrane ($\sim 10^{-12}$ dyn/cm); $\bar{c}$ is the local value of total curvature (sum of principal curvatures); and s is the curvilinear coordinate along the surface contour. The extensional force f_n on the intercellular bond is approximately Q which, in turn, is essentially given by $\tau \sin \theta_c$. Thus, the level of tension required to rupture bonds by extension would be given by,

$$\tau \sim \hat{f}_n / \sqrt{\bar{a}} \sin \theta_c \tag{4}$$

for immobile bonds (with surface area $\bar{a}$ for each bond). This "fracture" relation differs from the ideal adhesion concept where a uniform-reversible energy density $\tilde{w}$ ("adhesion energy"/area) represents both formation and separation of adhesive contacts. For an energy-based "fracture" process, the classical Young-Dupre equation predicts the level of tension required to disjoin the surfaces, i.e.

$$\tau = \tilde{w}\,(1 - \cos \theta_c) \tag{5}$$

The main difference between these two "fracture" criteria is that the former (force based) scales with the square root of surface density whereas the latter (energy based) scales directly with surface density. Because the actual-microscopic contact angle θ^* at the bond is not zero (i.e. the membrane is not exactly parallel to the substrate), there is also a lateral force f_t on the bond which acts to drag the bond along the surface. To do this, the lateral force must shear membrane receptor-cytoskeletal network connections. From simple geometry, the lateral force is approximately $f_t \sim Q \cdot \theta^* \simeq (\tau \sin \theta_c) \cdot \theta^*$. The microscopic angle at the bond can be represented by a length l_b for the membrane separation from the substrate at the bond and the local membrane curvature c_m, i.e. $\theta^* \simeq (2 l_b c_m)^{1/2}$. The mechanics of bending local to the bond is used to relate the curvature to membrane tension, i.e. $c_m^2 \simeq 2\tau(1 - \cos\theta_c)/k_c$. Hence when the contact angle $\theta_c < 30^o$, the lateral force on the bond is expected to scale as the tension and macroscopic contact angle to fractional powers, i.e.

$$f_t \simeq (2l_b\bar{a})^{1/2}\, \tau^{5/4}\, {\theta_c}^{3/2}/{k_c}^{1/4} \tag{6}$$

As demonstrated by this equation, the dependence on membrane bending stiffness is extremely weak. Again for contact angles $\leq 30^o$, the ratio of lateral force to extensional force on the bond is predicted to increase weakly with tension and contact angle,

$$f_t/f_n \sim (2l_b)^{1/2}\, \tau^{1/4}\, {\theta_c}^{1/2}/{k_c}^{1/4}$$

Along with the extensional strength $\hat{f}_n$ of a bond, the shear strength $\hat{f}_t$ of the surface receptor-network connections will be another major determinant for disjoining of cells in aggregates. If the ratio $\hat{f}_t/\hat{f}_n$ is large, disjoining will be governed by the "fracture" relation given in equation (4). On the other hand, if this ratio is small, then intercellular bonds will be dragged along the surface and condensed in the contact zone when the lateral force given by equation (6) reaches $\hat{f}_t$.

When adherent cells swell and pressurize to form smooth interfaces as illustrated in Fig. 1b, the contact area and angle θ_c diminish. In this state, membrane area dilation (hence tension), interfacial curvature, and contact angle become unique functions of volume and size of the contact regions. These functions are easily obtained from simple geometric relations using the unstressed area of the membrane envelope to define a characteristic volume of an unstressed sphere. The balance of cell forces leads to an equation which specifies the total curvature $\bar{c}$ for the unsupported interfaces as tension becomes uniform under high pressure, i.e.

$$\bar{c} = \sigma_n \;/\; \tau + \underbrace{2\sin\theta_c/r_c}_{(\equiv 2/R_s \text{ for sphere})} \tag{7}$$

For aggregates with nearly periodic arrangements of cells, the total curvature and surface area:volume ratio imply a lattice of "minimal" surfaces which intersect at the contact regions with prescribed values of radius r_c and angle θ_c(the angle between membranes bordering the interstitial space). The membrane tension τ and network stress σ_n will be determined by area and volume increases relative to the unstressed values.

To demonstrate these geometric features, consider pressurization of a single cell bonded to a flat substrate. Expressing the area dilation and volume increase relative to an unstressed sphere of volume V_s, the geometry of a spherical segment in contact with a flat surface prescribes the following relationship between membrane tension and contact angle as a function of the cell volume $\tilde{V} = V/V_s$ when the size of the contact r_c is less than half the size of the sphere:

$$\theta_c \simeq r_c / R_s \tilde{V}^{1/3}$$

$$(1 + \tau / K_a) \cong \tilde{V}^{2/3} \left(1 + \theta_c{}^4 / 16\right) \qquad \{\theta_c < 0.5\}$$

As shown by these equations, a nonzero contact angle (the angle between the flat surface and the spherical contour) can result from either volume reduction or dilation of the membrane surface. When the size of the contact is stabilized by intercellular-network connections, increases in volume slowly diminish the contact angle but greatly increase the membrane tension. Figure 8 shows an example of agglutinated red blood cells swollen to nearly spherical shapes where upon one of the cells was forced to lyse. Thus, the constitutive relations for stresses and geometric relations for contact angle along with the appropriate failure mechanism (equations 4-6) predict the course of disjoining of adherent cells under osmotic swelling.

Summary

Cell aggregates in tissues and even isolated cells are composite-hierarchical structures with a wide range of mechanical properties. Viewed simply, cells are capsules enclosed by a plasma membrane envelope of lipids and other cosurfactants to which a cytoskeletal network is tethered. When cells are specifically bonded together to form tissue aggregates, integral membrane proteins (the tethering sites for the network) transmit stresses from one cell cytoskeleton to the other. Because the plasma membrane is a nearly unexpandable-surface fluid, compatibility of membrane and network deformations is provided by a large redundancy of plasma membrane (in wrinkles and folds) through which the tethering-integral proteins of the network easily move. Hence, osmotic swelling initially stretches the network and unwrinkles the plasma membrane. Once the membrane has become smooth, pressurization produces sharp increases in membrane tension which lead to lysis of the cell at pressures of about 0.1 Atm. Throughout swelling, cells do not behave as "perfect" osmometers and some even "regulate" their osmotic activity to diminish swelling and relieve pressure. In cell aggregates, the build-up of membrane tension and sphering of cells act to disjoin the interfacial contacts between cells leaving interstitial gaps of solution. "Peeling" stress applied to the membrane produces lateral stress on intercellular-network bonds which may render the network and coalesce the connection sites within the interfacial contact region. The ultimate cause of cell death under hypoosmotic shock can result from rendering of network or lysis of plasma membrane. Most likely, the cell can rebuild network and may reseal its outer-plasma membrane envelope if its cytoplasmic chemistry remains patent. Thus, the crucial determinant of cell survival will be preservation of cytoplasmic chemistry which rests on the viability of subcellular organelli (e.g. mitochondria and enzyme-containing granules) and the reproductive apparatus of the cell (endoplasmic reticulum, chromosomes, centrioles, microtubules, etc.).

Acknowledgement

This work was supported by grants HL45099, HL31579 from the US National Institutes of Health, and grant MT7477 from the Medical Research Council of Canada.

Related References

Bloom M, Evans E, Mouritsen OG (1991) Quart. Rev. Biophys. in press

Cosgrove DJ (1987) Planta (Berl) 171:266-278

Evans E, Fung YC (1971) Microvascular Res. 4:335-347

Evans E, Skalak R (1980) Mechanics and thermodynamics of biomembranes, C.R.C. Press, Florida

Evans E (1985) Biophys. J. 48:175-183, 184-192

Evans E, Needham D (1987) J. Phys. Chem. 91:4219-4228

Evans E (1990) Colloids and Surfaces 43:327-347

Evans E, Berk D, Leung A, Mohandas N (1991) Biophys. J. 59:838-848, 849-860

Koch AL, Pinette MFS (1987) J. Bacteriol. 169:3654-3663

ENDOCRINE PITUITARY CELL CULTURES: CELLULAR MORPHOLOGY, PROTEIN SECRETION, AND SUSCEPTIBILITY TO WEAK BASES AND IONOPHORES

Athanassios Sambanis
School of Chemical Engineering
Georgia Institute of Technology
Atlanta GA 30332-0100
U.S.A.

INTRODUCTION

Animal cells are of significant use in bioprocessing technology, tissue engineering, and gene therapy applications. In bioprocess technology, cells cultured in bioreactors produce specific recombinant or, occasionally, endogenous proteins. Compared to bacteria and yeast, animal cells have the disadvantages of slow growth to relatively low densities, and of complex and expensive culture media requirements. On the other hand, they offer the distinct advantage of possessing the enzymatic machinery necessary for performing post-translational modifications on recombinant proteins. Such modifications are essential for biological activity of a protein product and may not be feasible in yeast or, even more so, in bacteria. Animal cells are thus indispensable for production of pharmaceuticals such as tissue plasminogen activator (tPA), factor VIII, or other complex proteins.

In tissue engineering, cells cultured *in vitro* and usually associated with biological polymers are implanted in animals or human beings for long-term restoration of function (Nerem, 1991). Examples are the development of artificial skin (Bell *et al*, 1981), artificial blood vessels for use in bypass and replacement of diseased arteries (Weinberg and Bell, 1986), and artificial organs by immobilization of endocrine cells (Colton *et al*, 1988; Galletti and Aebischer, 1988). For organ development, cells can be macroencapsulated by membranes in hollow fiber units or microimmobilized in beads or capsules of biocompatible polymers. In either case, the pores of the immobilizing agent should be large enough to allow passage of nutrients and metabolites, especially polypeptide hormones, but small enough to exclude antibodies or immune cells of the host, thus immunoprotecting the implant. Calcium alginate/ polylysine copolymers and

NATO ASI Series, Vol. H 64
Mechanics of Swelling
Edited by T. K. Karalis

ultrafiltration membranes of proper molecular weight cutoff have been used with success for micro- and macroimmobilization, respectively. Other important issues in bioartificial organ design are the long-term survival and function of immobilized cells. Cells should retain for prolonged periods of time differentiated properties, such as synthesis and secretion of specific proteins or metabolic or secretory responses to biochemical stimuli. For example, pancreatic implants should secrete correct amounts of insulin when blood glucose levels raise above certain values, and they should retain this ability for at least several months so that implantation procedures are not needed often.

Finally, in gene therapy, cells genetically engineered to express certain property at a high level are introduced into a host for long-term restoration or replacement of function (Nerem, 1991). For example, cells genetically modified for enhanced secretion of the thrombolytic agent tissue plasminogen activator (tPA) are implanted into the body for treatment of cardiovascular disease.

Cell viability and metabolic activity, and protein synthesis, correct processing and secretion, are all essential for the success of any of the above applications. Processes related to protein trafficking events may be particularly susceptible to stressful conditions caused by nutrient depletion or metabolite accumulation. Such stresses occur in mass culture due to the metabolic activities of cells, and in implant devices due to diffusional resistances imposed by immobilization or cell aggregation. In this paper, we focus on the industrial and medical applications of specialized cells derived from the endocrine system, and we describe how their function is correlated to cellular morphology and affected by chemical stresses. Particular emphasis is given to inhibitory processes which occur coordinately with osmotic swelling of subcellular, vesicular compartments involved in protein trafficking events.

ENDOCRINE CELLS

Cells of limited life span or immortal cell lines derived from endocrine glands, such as the pancreas or the pituitary, can be used for production of enzymes and polypeptide hormones in culture, as well as implants in bioartifical organ development. These applications are strongly dependent on certain unique characteristics of protein trafficking in these cells, so a description of these intracellular processes should be offered first. A discussion of the relevant bioprocessing and tissue engineering applications is then presented.

Secretory protein trafficking

Figure 1 shows the main secretory protein trafficking events in endocrine cells. Although this diagram has been proposed for the mouse pituitary cell line AtT-20, it is believed that similar processes take place in other endocrine cells too, such as insulin-producing β pancreatic cells.

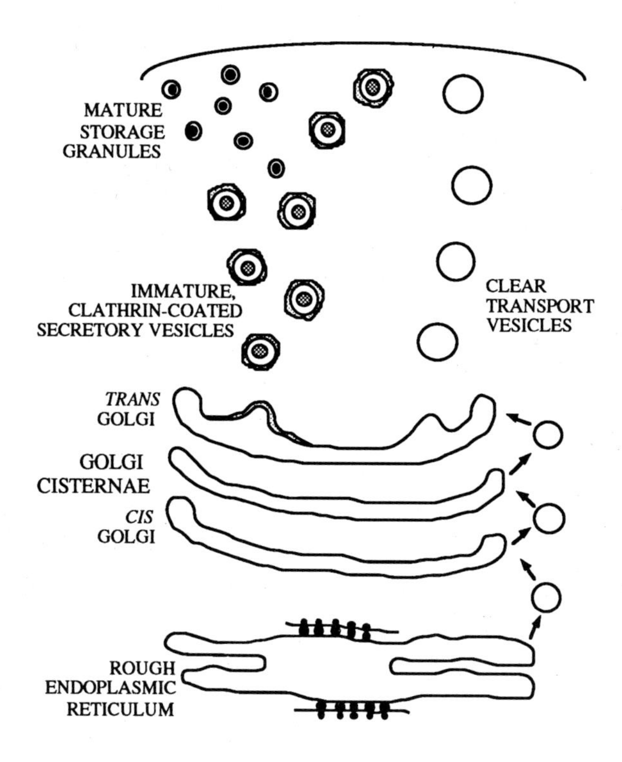

Figure 1 (from Sambanis *et al*, 1991) **Schematic of secretory protein trafficking in mouse pituitary AtT-20 cells.**

The synthesis of secretory protein begins with DNA transcription to form messenger RNA (mRNA) in the nucleus. The original mRNA formed is modified enzymatically by the removal of noncoding introns and the splicing of coding exons to form mature mRNA. This mRNA is then transported through pores in the nuclear membrane into the cytoplasm, where it becomes associated with several ribosomes to form a polyribosomal complex. Cytosolic enzymes, cytoskeletal proteins, and proteins to be incorporated in the membranes or lumen of specific organelles, such as mitochondria or peroxisomes, are translated from polyribosomal complexes with ribosomes unbound to membranes (Darnell *et al*, 1990). Secretory, lysosomal, and plasma membrane proteins, on the other hand, code for a leading signal sequence which binds to a signal recognition particle (SRP) immediately after it emerges from the ribosome. The signal sequence-SRP complex binds to an SRP receptor located on the membrane of the endoplasmic reticulum (ER), so that ribosomes appear clustered on the cytosolic side of the ER giving it a rough appearance (rough ER, RER). As the protein is translated, it passes through the membrane of the RER into the lumen of the organelle, where the signal sequence is cleaved, disulfide bonds are formed, and glycosylation and protein folding occur. Sugar groups are added to the amide nitrogen of asparagine residues, so glycosylation in the RER is of the N-linked type. For secretory proteins, this is the only membrane-crossing step that occurs; all subsequent transport and final secretion take place through budding and fusion of vesicles with organelles or the plasma membrane (Darnell *et al*, 1990; Wickner *et al*, 1985).

From the RER, secretory proteins are transported to the *cis* (closest to the nucleus) phase of the Golgi apparatus, apparently by a bulk flow mechanism. In the Golgi, additional post-translational modifications take place, which include o-glycosylation (addition of sugars to the hydroxyl-group oxygen of serine, threonine or hydroxylysine), oligosaccharide processing, proteolysis, phosphorylation and sulfation (Darnell *et al*, 1990; Mains *et al*, 1987). Routing of proteins to specific vesicles for further processing and transport, referred to as protein sorting, occurs also at the Golgi. Lysosomal proteins are sorted to lysosomes at *medial* Golgi after addition of a recognition marker while the protein is at the *cis* Golgi cisternae. Secretory proteins, on the other hand, pass through *medial* to *trans* (farthest from nucleus) Golgi, where they enter vesicles that take them quickly or after some additional events to the plasma membrane for exocytosis (Darnell *et al*, 1990; Kelly, 1985).

In most animal cells (fibroblasts, hepatomas, hybridomas), secretory proteins enter at the *trans* Golgi transport vesicles that take them constitutively, i.e., continuously and at the rate at which they are synthesized, to the plasma membrane. Transport vesicles appear clear under the electron microscope, and pulse-chase experiments reveal that they have half residence times in the cytoplasm of approximately 10 minutes. In specialized animal cells of the endocrine and exocrine systems, on the other hand, a *regulated* pathway of secretion exists in addition to the generic, constitutive one described above. Specific secretory proteins are actively sorted, apparently by receptor-like molecules, into secretory vesicles coated by the protein clathrin, which bud off from clathrin-coated regions of the *trans* Golgi. Under the electron microscope, secretory vesicles appear to have an electron-dense core surrounded by a clear halo. The core consists of protein condensed at the *trans*-most Golgi cisternae and in the vesicles themselves. The clathrin-coated vesicles undergo a maturation process to become storage vesicles or granules which accumulate in the cytoplasm and fuse with the plasma membrane at a significant rate only when cells are triggered with proper secretion agonists (secretagogues) (Burgess and Kelly, 1987; Darnell *et al*, 1990; Kelly, 1985; Tooze and Tooze, 1986). The maturation process involves shedding of the clathrin coat, acidification of the vesicular contents through action of ATP-dependent proton pumps, a further condensation of the protein core, and possibly enzymatic processing (including proteolysis, amidation or acetylation) of secretory proteins (Mains *et al*, 1987; Orci *et al*, 1985, 1987). Secretagogues act by elevating, directly or indirectly, cytosolic levels of Ca^{++} or cAMP. They can be hormones, calcium ionophores, membrane depolarizing agents (causing opening of voltage-gated calcium channels), membrane permeable cAMP analogs, activators of the enzyme adenylate cyclase which catalyzes conversion of ATP to cAMP, or inhibitors of phosphodiesterase which catalyzes hydrolysis of cAMP to AMP (Darnell *et al*, 1990; Kelly, 1985).

Current understanding is that secretory proteins are directed into the regulated pathway of secretion by specific signal sequences, whereas no such sequences are needed for constitutive secretion. Although no sorting sequences have been identified yet, experimental work has shown that considerable homology exists among endocrine, or even endocrine and exocrine cell lines. In particular, endocrine and exocrine proteins secreted via the regulated pathway in their cells of origin were also secreted in a regulated fashion when expressed as recombinant products in pituitary AtT-20 cells (Chung *et al*, 1989; Moore, 1986).

The AtT-20 cell line

The mouse cell line AtT-20, derived from an anterior pituitary tumor, has been used in numerous studies on the characterization of protein secretion pathways, as well as in evaluations of production and biomedical applications. Recombinant clones have been established expressing human proinsulin (Moore *et al*, 1983), human growth hormone (hGH) (Moore and Kelly, 1985), or rat anionic trypsinogen (Burgess *et al*, 1985).

AtT-20 cells can be cultured both on surfaces and as tumor spheroids in suspension. When offered with surfaces proper for attachment, cells seeded at low densities form monolayers by attaching, spreading and extending processes (Figure 2A). Experiments with cells stained for immunofluorescence with anti-ACTH antibody indicated that secretory granules tend to accumulate at the tips of processes (Kelly, 1985). Surface cultures exposed to growth-promoting medium become confluent, forming locally multilayers of spherical cells termed foci (Figure 2B). Upon continuous exposure to growth medium, cells in foci outgrow attached cells resulting in a change of the prevailing cell morphology in culture. Cells seeded in vessels containing no surfaces proper for attachment aggregate spontaneously forming tumor spheroids (Figure 2C).

AtT-20 cells synthesize endogenous proopiomelanocortin (POMC), which becomes proteolytically cleaved to smaller molecular weight polypeptides, the smallest of which is mature adrenocorticotropic hormone (ACTH). The final processing steps take place in maturing secretory vesicles, and ACTH accumulates in storage granules. ACTH is thus secreted efficiently only when cells are exposed to secretagogues. Part of unprocessed POMC becomes missorted into the constitutive pathway of secretion, from which it is secreted in the absence of stimulus (Gumbiner and Kelly, 1982).

Recombinant proinsulin is recognized by the sorting apparatus of AtT-20 cells and thus targeted into clathrin-coated secretory vesicles. Endogenous pituitary proteases

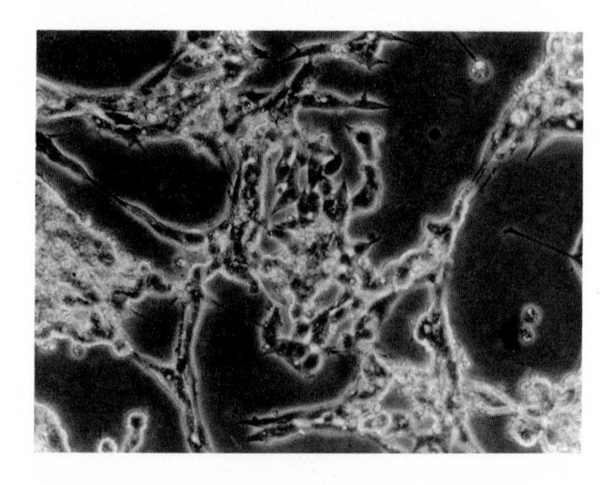

2A

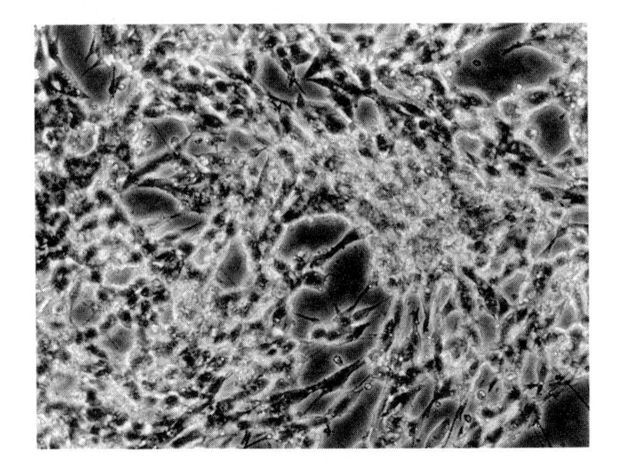

2B

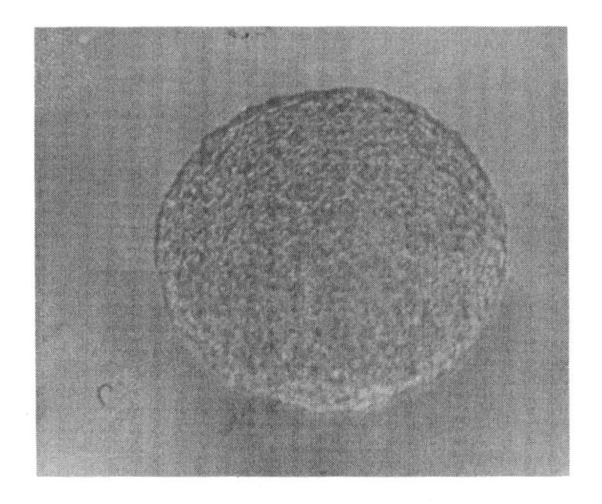

2C

Figure 2 Cultures of recombinant AtT-20 cells secreting human growth hormone.

A. Subconfluent surface culture. Cells are attached to the substrate, and they are flat, spread, and extend processes. Secretory granules tend to accumulate at the tips of processes.

B. Confluent surface culture. Cells are clustered together forming locally multiple layers, and only cells next to gaps appear to extend processes. Multilayered aggregates in surface cultures are termed foci. Cultures containing many foci exhibit inferior induced secretory response compared to cultures containing only flat, spread cells.

C. Spheroid in suspension. Spheroid size can range from 50 to more than 800 μm in diameter.

residing in these vesicles act on proinsulin and cleave it to a polypeptide not yet proved identical to, but showing many of the features of, authentic human insulin. For the sake of brevity, in what follows, this polypeptide will be referred to as insulin. Insulin is secreted effectively only in the presence of secretagogue. Some unprocessed proinsulin becomes missorted into the constitutive pathway, so it is secreted by uninduced cells (Moore *et al*, 1983). Recombinant hGH does not appear to undergo processing in maturing secretory vesicles. Some hGH is also missorted into the constitutive pathway, so this protein is secreted in the absence of stimulus, but larger amounts are secreted by cells exposed to secretagogue (Moore and Kelly, 1985).

Quantitative secretion results obtained with hGH-producing AtT-20 cells are shown in Figure 3. Cells in T flasks exposed until time zero to complete growth medium (C-DMEM) were switched to serum-free medium either without or with 5 mM of the secretagogue 8 BrcAMP. The amounts of hGH secreted in the media or remaining

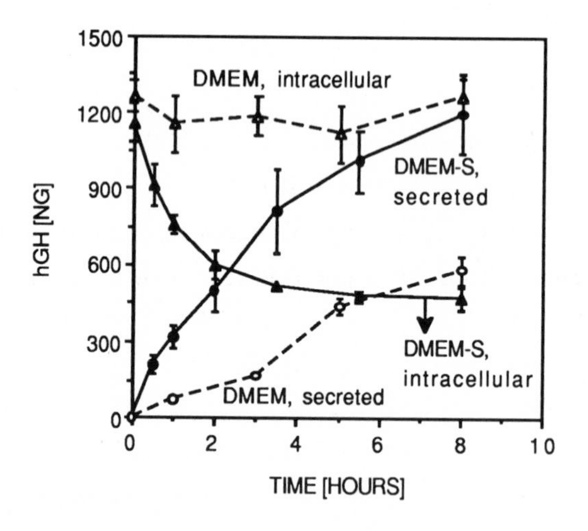

Figure 3 (from Sambanis *et al*, 1990a) **Basal and induced secretion of recombinant human growth hormone (hGH) from AtT-20 cells.**

Cells in identical 25 cm² T flasks were exposed until time zero to complete growth medium [C-DMEM: Dulbecco's modified Eagle's medium (DMEM) with 4.5 g/L glucose, supplemented with 10% fetal bovine serum and additional glutamine to a final concentration of 6 mM] and at that time were switched to serum-free DMEM without or with (DMEM-S) 5 mM of the secretagogue 8 BrcAMP. At various times, media were collected and cells extracted with a detergent-containing buffer. hGH in media and extracts was assayed by liquid-phase radioimmunoassay (RIA) using a commercially available kit. The average number of cells in each culture was 2.4×10^7. Bars represent standard deviations. Most of the variability came from RIAs, not culture duplication.

intracellularly were followed with time. Cells exposed to secretagogue-free medium (DMEM) secreted hGH at a constant, basal rate of roughly 0.32 ng/10^5 cells-hour; intracellular hGH remained constant at about 5 ng/10^5 cells. Cells exposed to medium supplemented with 8 BrcAMP (DMEM-S) secreted hGH at an initial rate of 1.7 ng/10^5 cells-hour; the secretion rate declined gradually, and after 6 hours, cells returned to basal secretion. Intracellular hGH declined from 5 ng/10^5 cells to 2 ng/10^5 cells during the period of induced secretion. The residual amount of 2 ng/10^5 cells is protein that cannot be retrieved through induction with 8 BrcAMP and presumably represents the amount present in the endoplasmic reticulum and Golgi apparatus of cells. Similar secretory patterns have been obtained with insulin-producing recombinant AtT-20 cells (G.E. Grampp, personal communication).

Protein production by cultured endocrine cells

The property of regulated secretion makes endocrine cells particularly attractive for production of endogenous or recombinant proteins in culture. The regulated pathway decouples protein synthesis from secretion, and this allows the realization of integrated production schemes in which cell culturing is conducted in such way as to facilitate downstream processing. In particular, cells can be exposed in a cycling fashion first to medium that promotes protein synthesis and intracellular accumulation, and then to a minimal volume of protein-free secretion medium containing a proper secretagogue or secretagogue combination. Stored protein is thus retrieved in the latter medium at relatively high concentration due to the small solution volume, without contamination from exogenous proteins, such as serum albumin, but only from other secreted cellular proteins. The secretion medium cannot support cells for prolonged periods of time, but cells can easily withstand it during the one to two hours needed to obtain most of the induced secretory response (Sambanis *et al*, 1990a,b).

The results of two cycling experiments with insulin and hGH-producing AtT-20 cells are are shown in Figures 4A and B, respectively. Experimental cells were exposed alternatively for prolonged time periods to complete growth medium (C-DMEM) and for short time periods to serum-free medium supplemented with 5 mM 8 BrcAMP (DMEM-S). Control cells were treated in an identical fashion, except that the serum-free medium

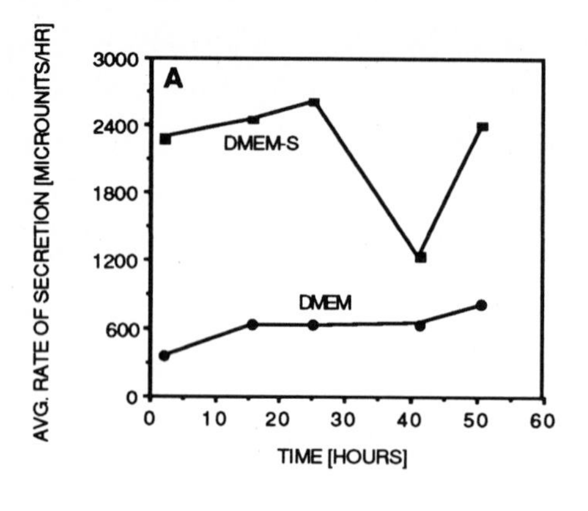

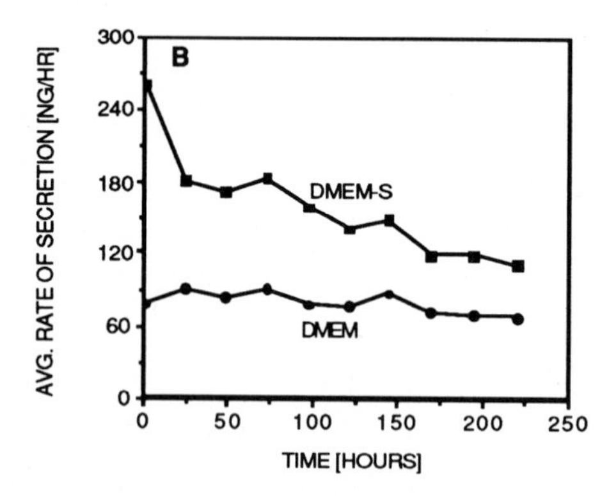

Figure 4 Cycling secretion experiments with recombinant AtT-20 cells.

A (from Sambanis *et al*, 1990b). Insulin-producing AtT-20 cells in 25 cm^2 T flasks were exposed for periods of 7.5-14.1 hours to secretagogue-free, serum-supplemented, complete growth medium (C-DMEM) and for periods of 2.0-2.1 hours to serum-free medium supplemented with 5 mM of 8 BrcAMP (DMEM-S). Fresh C-DMEM and DMEM-S (7 and 2.5 ml, respectively) were added at each medium change. A control experiment was conducted in an identical fashion, except that the serum-free medium was not supplemented with a secretagogue (DMEM). Secreted immunoreactive peptides, referred to as insulin-related peptides (IRP), were assayed by RIA. Plotted are the rates of secretion in DMEM-S and DMEM by experimental and control cells, respectively, averaged over each 2.0-2.1 hour period. Cells started to detach from tissue culture plastic after the third cycle (25 hours), and at the sixth cycle (64.1-66.1 hours, not shown) detachment was complete.

B (from Sambanis *et al*, 1990a). hGH-producing AtT-20 cells in 25 cm^2 T flasks were exposed for periods of 20.6-24.2 hours to C-DMEM and for periods of 2.0 hours each to DMEM-S. Fresh C-DMEM (8-12 ml) and DMEM-S (2.5 ml) were added at each medium change. A control experiment was conducted as in A. Secreted hGH was assayed by RIA. Plotted are the rates of secretion in DMEM-S and DMEM averaged over each 2.0 hour period. The prevailing cell morphology changed during the experiment from flat, spread cells to round cells in foci.

was not supplemented with a secretagogue (DMEM). Secreted polypeptides were assayed by liquid-phase radioimmunoassay (RIA) employing competitive binding for antibody sites of sample and ^{125}I-labelled protein. The antibody used for insulin was against the mature form of the hormone, but it also recognized proinsulin with a lower affinity. Insulin and proinsulin assayed with this technique are referred to as insulin-related peptides (IRP).

The graphs show the average rates of protein secretion in serum-free DMEM or DMEM-S by control and experimental cells, respectively, vs. time in the cycling experiment. With insulin-producing cells, cycling worked well the first three times, in the sense that induced secretion remained at least 4-fold higher than basal secretion by uninduced cells. After that, cells periodically induced to secrete started to detach from tissue culture plastic, and by the sixth cycle (not shown) detachment was complete (Sambanis *et al*, 1990b).

Induced secretion in DMEM-S by cycled hGH-producing cells was initially 3.5 times higher than uninduced secretion in DMEM by controls, but declined steadily with time and by the end of the experiment became only 1.6-1.7 times higher than uninduced secretion. A significant amount of protein (not shown in the Figure) was also secreted during the intervals of exposure of cells to growth medium. Detachment did not occur in these cultures, but the deterioration of induced secretory response correlated with a change in the prevailing cellular morphology from the spread phenotype to spherical cells in foci. The morphological change resulted from cell growth during the prolonged periods of incubation in growth medium.

In more recent cycling experiments performed with the pancreatic cell line βTC3 neither of the above problems (detachment or change of prevailing phenotype) occurred. Furthermore, secretion during exposure to growth medium was effectively suppressed without compromise in protein synthesis rate, so the time intervals needed for cells to refill their storage compartments were drastically reduced (Grampp *et al*, 1991). With AtT-20 cells, detachment of the insulin-producing clone may be prevented by coating the culture surface with adhesion polymers, but the uncontrolled growth of hGH-producing cells into foci has, so far, proved intractable.

EFFECT OF WEAK BASES AND CARBOXYLIC IONOPHORES ON INTRACELLULAR PROTEIN TRAFFICKING

Essential to the function of certain intracellular organelles is the establishment and maintenance of an acidic intravesicular pH. These organelles are the lysosomes (the most acidic, pH of 4.5 to 5.0), the endosomes, and the secretory vesicles of the regulated secretion pathway. Elements of the Golgi complex may also be mildly acidic (Mellman *et al*, 1986). The pH gradient between the cytosol and the vesicular lumen is established and maintained by ATP-dependent proton pumps (proton ATP-ases), which exhibit remarkable homology among the various organelles. An acidic intralysosomal pH is necessary for the operation of the relevant proteolytic enzymes which have acidic pH optima. Similarly, an acidic endosomal lumen is essential for ligand-receptor dissociation and recycling of receptors to the cell surface (Mellman *et al*, 1986). Acidification of secretory vesicles of the regulated secretion pathway appears to be necessary for dissociation of secretory proteins from receptor-like sorting proteins which need to be recycled to the Golgi, as well as for functioning of specific processing enzymes localized in the vesicles. There is accumulating evidence that at least proteolytic enzymes in secretory vesicles have acidic pH optima, and that conversion of prohormones to mature, biologically active hormones occurs coordinately with vesicular acidification (Docherty *et al*, 1982; Mellman *et al*, 1986; Orci *et al*, 1987).

Weak bases having a lipophilic unprotonated form and a hydrophilic protonated one, such as chloroquine, methylamine, or ammonia, may disrupt the function of acidic vesicular compartments in cells. A common mechanism that may be proposed to explain the upsetting effect of such compounds is as follows. The unprotonated form of the base may pass relatively freely through the membranes and enter the acidic vesicles. There, it becomes protonated, and since membranes are essentially impermeable to the protonated form, it becomes trapped. The transient increase in vesicular pH disturbs the function of enzymes with acidic pH optima and inhibits the dissociation of receptor-ligand complexes. The increase in vacuolar pH can amount to 1-2 pH units and may occur within a minute or two after addition of weak bases to the medium. Proton ATPases localized in the vesicular membranes try to restore the acidic pH by importing more protons, leading to osmotic swelling. The resulting dilution of enzymes and substrates reduces the rates of enzymatic reactions and thus potentiates the negative effect of alkalinization. The base itself may be also an enzyme inhibitor; for example, there is evidence that chloroquine inhibits the lysosomal protease cathepsin B_1 (Wibo and Poole, 1974).

Specific ionophores, such as the carboxylic ionophore monensin, disrupt transmembrane proton gradients by exchanging protons for other monovalent cations and have similar negative effects. However, these agents do not affect vesicular pH as rapidly as weak bases (Mellman *et al*, 1986).

Cytoplasmic vacuolation due to osmotic swelling of lysosomes has been observed in a number of studies. In experiments with mouse peritoneal macrophages, Ohkuma and Poole (1981) found distinct lysosomal dilation in cells exposed to weak bases, which they attributed to osmotic swelling. The experiments verified that it is the concentration of the free form of the weak base which determines the extent of uptake into the cells and the extent of vacuolation. In particular, uptake and vacuolation increased as the amount of base increased or as the pH was raised. The only exception was chloroquine, for which uptake and vacuolation went through a maximum for the concentration and pH ranges examined. Similarly, cultured rat fibroblasts became vacuolated in the presence of chloroquine due to lysosomal swelling (Wibo and Poole, 1974). The authors presented evidence that in the presence of the weak base, lysosomes are transformed into swollen vacuoles filled with dissolved chloroquine.

Ammonia, chloroquine and monensin inhibit receptor-mediated endocytosis of a variety of ligands, such as asialoglycoproteins (Berg *et al*, 1983; Tolleshaug and Berg, 1979), mannose glycoproteins (Tietze *et al*, 1980; Wileman *et al*, 1984), low density lipoproteins (Basu *et al*, 1981), enveloped viruses (Helenius *et al*, 1980; Marsh *et al*, 1982) and ferritin (Stenseth and Thyberg, 1989). The main difference between ammonia and chloroquine seems to be that the effects of ammonium chloride are rapidly reversed, while the effects of chloroquine are only partially alleviated upon removal of the drug (Tietze *et al*, 1980). The effect of weak bases and ionophores is attributed primarily to inhibition of receptor-ligand dissociation in alkalinized endosomes, and a consequent block in receptor recirculation to the plasma membrane leading to a depletion of surface receptor molecules. However, since unoccupied receptors may accumulate intracellularly, membrane trafficking per se may also be disrupted. Morphological studies have revealed that monensin and chloroquine cause a distinct dilation of endosomes, similar to that observed with lysosomes (Stenseth and Thyberg, 1989). In the same context, receptors which are normally carried to lysosomes for downregulation, such as epidermal growth factor receptors, accumulate in alkalinized endosomes, resulting in inhibition of degradation (Merion and Sly, 1983; Stoscheck and Carpenter, 1984). Carboxylic ionophores raise the pH of endocytic vesicles in a similar fashion (Maxfield, 1982).

Weak bases and ionophores also affect regulated secretion of proteins from endocrine cells. In cultures of parental AtT-20 cells, chloroquine at 200 μM inhibited accumulation in storage granules and induced secretion of newly synthesized, mature ACTH, whereas it did not block -and in fact, it enhanced- constitutive secretion of ACTH precursors (Moore *et al*, 1983). Similarly, ammonium chloride inhibited the packaging of secretogranins in regulated secretory granules of pheochromocytoma PC12 cells *in vivo* (Huttner *et al*, 1990). In rat β pancreatic cells, monensin caused swelling of clathrin-coated Golgi cisternae with or without condensing secretory protein and of clathrin-coated secretory vesicles. Monensin also inhibited conversion of proinsulin to insulin, which occurs in maturing vesicles of the regulated secretion pathway (Orci *et al*, 1984).

From an engineering viewpoint, it of particular significance to determine the NH_4Cl concentrations at which regulated secretion is inhibited, and whether at such concentrations other cellular processes are affected too. Since ammonium is a product of glutamine metabolism and thermal decomposition, it is generally present in spent cell culture media or nutrient solutions. Quantitative information on ammonium inhibition is essential in proposing feeding strategies for bioreactors, designing encapsulated implants of endocrine cells, or perhaps discovering indicators of proper function of the secretory apparatus of cells that are easier to monitor than secretion itself.

In a series of experiments conducted in our laboratory (Dyken *et al*, 1991a, b; also unpublished), we evaluated the effect of ammonium on basal and induced secretion of recombinant IRP from AtT-20 cells. Ammonium chloride added at 6 or 12 mM either had no effect or slightly enhanced basal secretion by uninduced cells. Induced secretion was, however, inhibited by either concentration of NH_4Cl, especially when cells were preincubated for four hours in medium with added ammonium. Interestingly, NH_4Cl at 6 mM did not appear to inhibit cell growth. Cellular vacuolation did not become noticeable in these experiments.

It is unknown which one of the two, alkalinization or osmotic swelling, is more disruptive to the function of subcellular, vesicular organelles. It is also unknown how the total cell volume changes when dilation of subcellular organelles occurs. These questions should be addressed through studies involving thorough microscopic observations and mathematical modeling approaches.

SUMMARY

Cells of the endocrine system have potential bioprocessing applications, in production of polypeptide hormones or enzymes, and tissue engineering applications, in bioartificial organ development. The property of regulated secretion, i.e., of intracellular protein storage and secretion upon stimulation with secretagogues, is essential in realizing effective production operations and biomedical life support systems. Besides lysosomes and endosomes, which are acidic vesicular compartments commonly found in mammalian cells, the regulated secretory pathway of endocrine cells consists of vesicles that become acidified as they undergo a maturation process. Acidification of secretory vesicles is essential for their function. Weak bases and carboxylic ionophores raise the pH of acidic vesicles, cause their osmotic swelling and a vacuolated appearance of cells, and disrupt vesicular function. With regard to regulated secretion, alkalinization of maturing secretory vesicles inhibits enzymes having acidic pH optima, and perhaps causes a shortage of sorting proteins at the *trans* Golgi. Dilution of substrates and enzymes due to swelling may reduce indiscriminantly the rates of all intravesicular biochemical reactions.

Acknowledgements

This work was partially supported from the National Science Foundation through the Massachusetts Institute of Technology Biotechnology Process Engineering Center, the Biotechnology Research Grant Committee of the Georgia Institute of Technology, and DuPont and Amoco Fellowships. This financial support is gratefully acknowledged.

REFERENCES

Basu SK, Goldstein JL, Anderson RGW, Brow MS (1981) Monensin interrupts the recycling of low density lipoprotein receptors in human fibroblasts. Cell 24:493-502

Bell E, Ehrlich H, Buttle D, Nakatsuji T (1981) Living tissue formed *in vitro* and accepted as skin-equivalent tissue of full thickness. Science 211:1052-1054

Berg T, Blomhoff R, Naess L, Tolleshaug H, Drevon CA (1983) Monensin inhibits receptor-mediated endocytosis in rat hepatocytes. Exp Cell Res 148:319-330

Burgess TL, Craik CS, Kelly RB (1985) The exocrine protein trypsinogen is targeted into the secretory granules of an endocrine cell line: studies by gene transfer. J Cell Biol 101:639-645

Burgess TL, Kelly RB (1987) Constitutive and regulated secretion of proteins. Ann Rev Cell Biol 3:243-293

Chung K-N, Walter P, Aponte GW, Moore H-PH (1989) Molecular sorting in the secretory pathway. Science 243:192-197

Colton CK, Dionne KE, Yarmush ML (1988) Oxygen effects on pancreatic islet insulin secretion in hybrid artificial pancreas. In: Skalak R, Fox CF (eds) Tissue Engineering, Alan R. Liss, New York, p 217

Docherty K, Carroll RJ, Steiner DF (1982) Conversion of proinsulin to insulin: involvement of a 31,500 molecular weight thiol-protease. Proc Natl Acad Sci USA 79:4613-4617

Darnell J, Lodish H, Baltimore D (1990) Molecular Cell Biology, 2nd edn. American Scientific Books, WH Freeman and Company, New York

Dyken JJ, Sambanis A (1991a) Protein secretion from endocrine animal cells: effects of chemical and physical stresses. In: Proceedings of 1991 Graduate Student Symposium, Georgia Institute of Technology, Atlanta, p 41

Dyken JJ, Vachtsevanos J, Sambanis A (1991b) Protein secretion from endocrine animal cells: effects of chemical and physical stresses. Presented at ACS National Meeting, Atlanta, Georgia, April 1991

Galletti PM, Aebischer P (1988) Bioartificial organs. In: Skalak R, Fox CF (eds) Tissue Engineering, Alan R. Liss, New York, p 211

Grampp GE, Stephanopoulos G (1991) Controlled protein secretion in a pancreatic islet-derived cell line. Presented at ACS National Meeting, Atlanta, Georgia, USA, April 1991

Gumbiner B, Kelly RB (1982) Two distinct intracellular pathways transport secretory and membrane glycoproteins to the surface of pituitary tumor cells. Cell 28:51-59

Helenius A, Kartenbeck J, Simons K, Fries E (1980) On the entry of Semliki Forest virus into BHK-21 cells. J Cell Biol 84:404-420

Huttner WB, Gerdes H-H, Rosa P, Tooze SA (1990) Biogenesis of secretory granules in vivo and in vitro. J Cell Biol Suppl 14C:9

Kelly RB (1985) Pathways of protein secretion in eucaryotes. Science 230:25-32

Mains RE, Cullen EI, May V, Eipper BA (1987) The role of secretory granules in peptide biosynthesis. Ann NY Acad Sci 493:278-291

Marsh M, Wellstead J, Kern H, Harms E, Helenius A (1982) Monensin inhibits Semliki Forest virus penetration into culture cells. Proc Natl Acad Sci USA 79:5297-5301

Maxfield FR (1982) Weak bases and ionophores rapidly and reversibly raise the pH of endocytic vesicles in cultured mouse fibroblasts. J Cell Biol 95:676-681

Mellman I, Fuchs R, Helenius A (1986) Acidification of the endocytic and exocytic pathways. Annu Rev Biochem 55:663-700

Merion MW, Sly S (1983) The role of intermediate vesicles in the adsorptive endocytosis and transport of ligand to lysosomes by human fibroblasts. J Cell Biol 96:644-650

Moore HP (1986) Factors controlling packaging of peptide hormones into secretory granules. Ann NY Acad Sci 493:50-61

Moore HP, Gumbiner B, Kelly RB (1983) Chloroquine diverts ACTH from a regulated to a constitutive secretory pathway in AtT-20 cells. Nature 302:434-436

Moore HP, Kelly RB (1985) Secretory protein targeting in a pituitary cell line: differential transport of foreign secretory proteins to distinct secretory pathways. J Cell Biol 101:1773-1781

Moore HP, Walker MD, Lee F, Kelly RB (1983) Expressing a human proinsulin cDNA in a mouse ACTH-secreting cell. Intracellular storage, proteolytic processing, and secretion on stimulation. Cell 35:531-538

Nerem RM (1991) Cellular Engineering. Ann Biomed Engng, in press

Ohkuma S, Poole B (1981) Cytoplasmic vacuolation of mouse peritoneal macrophages and the uptake into lysosomes of weakly basic substances. J Cell Biol 90:656-664

Orci L, Halban P, Amherdt M, Ravazzola M, Vassalli J-D, Perrelet A (1984) A clathrin-coated, Golgi-related compartment of the insulin secreting cell accumulates proinsulin in the presence of monensin. Cell 39:39-47

Orci L, Ravazzola M, Amherdt M, Madsen O, Vassalli J-D, Perrelet A (1985) Direct identification of prohormone conversion site in insulin-secreting cells. Cell 42:671-681

Orci L, Ravazzola M, Storch M-J, Anderson RGW, Vassalli J-D, Perrelet A (1987) Proteolytic maturation of insulin is a post-Golgi event which occurs in acidifying clathrin-coated secretory vesicles. Cell 49:865-868

Sambanis A, Lodish HF, Stephanopoulos G (1991) A model of secretory protein trafficking in recombinant AtT-20 cells. Biotechnol Bioeng 38:280-295

Sambanis A, Stephanopoulos G, Lodish HF (1990) Multiple episodes of induced secretion of human growth hormone from recombinant AtT-20 cells. Cytotechnology 4:111-119

Sambanis A, Stephanopoulos G, Sinskey AJ, Lodish HF (1990) Use of regulated secretion in protein production from animal cells: an evaluation with the AtT-20 model cell line. Biotechnol Bioeng 35:771-780

Stenseth K, Thyberg J (1989) Monensin and chloroquine inhibit transfer to lysosomes of endocytosed macromolecules in cultured mouse peritoneal macrophages. Eur J Cell Biol 49:326-333

Stoscheck CM, Carpenter G (1984) Down regulation of epidermal growth factor receptors: direct demonstration of receptor degradation in human fibroblasts. J Cell Biol 98:1048-1053

Tietze C, Schlesinger P, Stahl P (1980) Chloroquine and ammonium ion inhibit receptor-mediated endocytosis of mannose-glycoconjugates by macrophages: apparent inhibition of receptor recycling. Biochem Biophys Res Comm 93:1-8

Tolleshaug H, Berg T (1979) Chloroquine reduces the number of asialoglycoprotein receptors in the hepatocyte plasma membrane. Biochem Pharmacol 28:2919-2922

Tooze J, Tooze SA (1986) Clathrin-coated vesicular transport of secretory proteins during the formation of ACTH-containing secretory granules in AtT-20 cells. J Cell Biol 103:839-850

Weinberg CB, Bell E (1986) A blood vessel model constructed from collagen and cultured vascular cells. Science 231:397-399

Wibo M, Poole B (1974) Protein degradation in cultured cells. II. The uptake of chloroquine by rat fibroblasts and the inhibition of cellular protein degradation and cathepsin B_1. J Cell Biol 63:430-440

Wickner WT, Lodish HF (1985) Multiple mechanisms of protein insertion into and across membranes. Science 230: 400-407.

Wileman T, Boshans RL, Schlesinger P, Stahl P (1984) Monensin inhibits recycling of macrophage mannose-glycoprotein receptors and ligand delivery to lysosomes. Biochem J 220:665-67

CYTOSKELETAL NETWORKS AND OSMOTIC PRESSURE IN RELATION TO CELL STRUCTURE AND MOTILITY

Paul A. Janmey, C. Casey Cunningham, George F. Oster*, & Thomas P. Stossel
Department of Medicine
Harvard Medical School
Boston MA 02114 USA

Introduction

The motility of many cell types proceeds sporadically, by a sequence of propulsive and contractile movements. These movements appear to be controlled by proteins acting upon the actin cytoskeletal network which induce gel-sol transformations in specific regions of the cytoplasm. However, the forces that actually drive these cytoplasmic motions remain obscure; indeed there may be several force generating systems which dominate different types of motile events. For example, directed locomotion may involve the same direct mechanochemical coupling as occurs within muscles, in conjunction with other forces such as membrane bending, gel swelling and elasticity, or osmotic and hydrostatic pressures. In this chapter we will review the elastic properties of cytoskeletal protein networks. It is these networks that provide elastic resistance to cell deformation, and whose rearrangement may allow directed motion in response to externally or internally generated forces.

The morphology of living cells appears to depend on a balance between osmotic swelling pressures and elastic resistance arising within the cytoskeletal network of protein filaments in the cell periphery. Cell motions would arise when this balance is perturbed locally, by changing the osmotic pressure or by remodelling the cortical cytoskeleton in response to specific signals received at the cell membrane. Most cells contain three types of protein filaments comprised

* Departments of Molecular & Cellular Biology, and Entomology, University of California, Berkeley, CA USA

NATO ASI Series, Vol. H 64
Mechanics of Swelling
Edited by T. K. Karalis

of actin, tubulin or intermediate filament proteins such as vimentin. Recent evidence suggests that actin filament networks exhibit the greatest resistance to deformation [Janmey et al. 1991]. The cell periphery is rich in filamentous F-actin, a linear polymer composed of the globular protein G-actin. F-actin filaments can be several microns long and arranged in various ways, ranging from tightly packed bundles to isotropic networks. The form of the actin fibers depends on the activity of specific actin-binding proteins that control such parameters as filament length, stiffness, and the number and geometry of inter-filament crosslinks. Activation of actin-filament severing proteins can reduce the elastic resistance of the cell to swelling, and this could drive certain kinds of protrusive events such as pseudopod extension [Oster and Perleson, 1987]. Crosslinking of actin by the protein filamin (also called ABP) has the opposite effect of increasing the resistance to cytoplasmic deformation. Actin filament networks also impede water flow, an effect that depends on filament length and crosslinking, as demonstrated by Ito et al. [Ito, et al., 1987].

The viscoelastic properties of the three types of cytoskeletal protein networks are compared in Figure 1. Actin filament networks can withstand more than ten times as much shear stress for a given degree of deformation (strain) than can either microtubule or intermediate filament networks. Moreover, after the initial elastic deformation, F-actin networks creep very little and recover their unstrained state when the deforming force is removed. The viscoelastic differences among the three types of cytoskeletal filaments, together with the fact that most cells contain larger concentrations of actin than of the other filamentous proteins, strongly suggest that actin networks dominate the rheology of the cytoplasm.

Under shear flow, where no volume change occurs, the measured elastic moduli can be related to the bulk or elongation moduli, providing one accounts for the compressibility of the material, which in the case of protein networks depends on the flow of solvent (water) through the polymer network. *In vitro*, the shear modulus of actin networks is proportional to the square of the protein concentration, and lies in the range of 100 - 1000 Pa for typical cytoplasmic concentrations. The elastic properties of the protein network suggests that actin networks can partially resist deformation in response to physiologically significant osmotic pressure differences. For example, the osmotic pressure, Π, due to a concentration difference, Δc, of 1 mOsm is approximated by van't Hoff's law:

$$\Pi = RT\Delta c \approx 2400 \text{ Pa}$$

at physiologic temperatures. Since the cortical actin network may have a shear modulus of several thousand Pa, it could resist such a deforming stress. Therefore, local perturbation of the cortical network by selective activation of actin-binding proteins could produce site-specific motions.

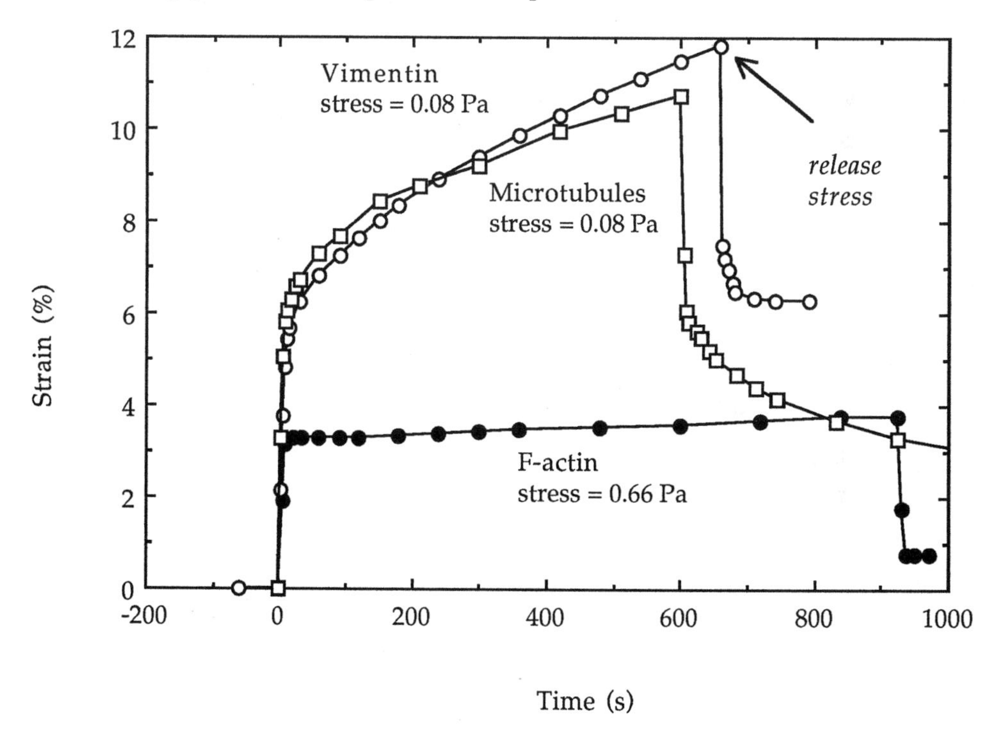

Figure 1. Comparison of the deformations induced when equal weight concentrations of networks formed by the three major classes of cytoskeletal filaments are subjected to shear stresses.

Some cells experience much larger osmotic stresses (e.g. renal epithelia). The cytoskeleton may not be able to resist these dilational forces since the elastic moduli of the three cytoskeletal polymers, and the measured elasticity of intact cells [Elson, 1988], are much smaller than the osmotic pressures in these systems. However, providing the cytoskeletal networks remain intact in their deformed state, their elasticity may guide the return to the unperturbed cell structure during volume regulation. Moreover, since the resistance of actin networks to water flow depends strongly on the length of actin filaments and their degree of network formation, hydraulic resistance provides an additional mechanism by which the cytoskeleton may resist volume changes [Ito, et al., 1987, Suzuki, et al., 1989].

Regulation of actin network viscoelasticity

Actin filaments in the cell have an average length and concentration in the so-called "semi-dilute" range, wherein the rotational motion of polymer strands is greatly restricted, but translational diffusion is not entirely suppressed [de Gennes, 1976; Doi and Edwards, 1986; Janmey, et al., 1986]. In this range, subtle changes in the length or degree of crosslinking of filaments can strongly alter their viscoelasticity [Janmey, et al., 1988; 1990]. Therefore, the rheology of the cytoskeleton *in vivo* can be regulated by activating or inhibiting specific actin-binding proteins which alter fiber length. Many actin-binding proteins have been described [c.f. reviews by Hartwig and Kwiatkowski, 1991; Pollard and Cooper, 1986; Stossel, et al., 1985]. Two proteins in particular affect actin assembly and structure in ways that alter the network's viscoelasticity. Gelsolin severs actin filaments into shorter strands, due to its ability to disrupt the non-covalent bonds holding together the subunits of the filament. A small decrease in filament length greatly reduces both elasticity and viscosity, allowing nearly unimpeded flow of an actin solution. Since gelsolin is controlled by the same intracellular signals that are generated when cells move [Yin, 1987], it is poised to initiate solation of the actin cytoskeleton necessary for locomotion. This hypothesis is supported by the finding that fibroblasts which have their gelsolin content increased by gene transfection move more rapidly than do control cells under the same conditions [Cunningham, et al., 1991a].

On the other hand, actin-binding protein (ABP) increases the viscosity and elasticity of the actin network because of its ability to crosslink individual actin filaments at near orthogonal angles [Hartwig and Stossel, 1981]. Figure 2 illustrates the opposing effects of gelsolin and ABP on the deformation (creep) of actin networks in response to a constant shear stress. Addition of gelsolin at a low molar ratio to actin greatly increases its shear deformation and decreases its elasticity, as reflected in the large degree of strain recovered when the stress is removed. Conversely, ABP reduces the degree of deformation under the same stress and increases the elastic recovery of the network when the stress is removed. The differences are greater under larger stresses and shear rates (Figure 2b) where the network of short uncrosslinked filaments flows with a viscosity close to that of the solvent while the crosslinked network retains its elastic resistance to the imposed stress. Since gelation of the cytoplasm is also important to cell motility, we expect that ABP is an essential component of locomotion. This is supported by the observation that cell lines which are deficient in ABP expression have poor locomotion. When ABP expression is restored by gene transfection, migratory ability dramatically increases, although at higher ranges of ABP expression the increase in motility begins to decrease [Cunningham, et al., 1991b]. This suggests that efficient cell locomotion depends on an optimal balance between solation and gelation. Clearly, a better understanding of the factors which alter the rheology of the actin network is necessary for understanding cell motility.

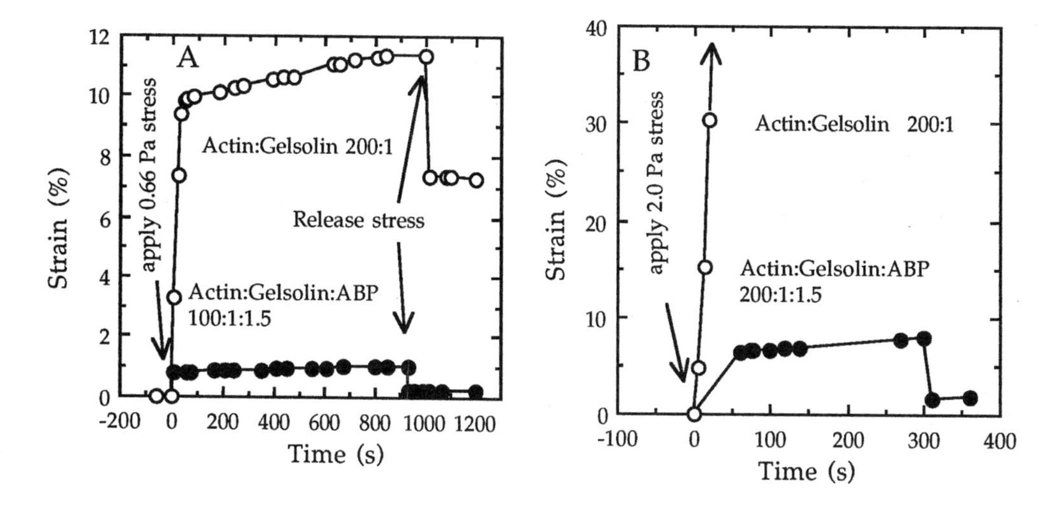

Figure 2. Shear compliance of 2 mg/ml F-actin with or without gelsolin and ABP.

Role of cytoskeletal networks in resisting propulsive motions.

The viscoelasticity of actin and other cytoskeletal networks allows them to resist deformation when mechanical forces are applied to the cell. Local alterations in the gel structure, for example by activation of gelsolin to solate the network, would lead to local shape changes. Oster and Perleson proposed a model for lamellipodial extension in which local solation of the actin gel allows the network to expand osmotically [Oster and Perleson, 1987; Oster, 1988]. Additional evidence that weakening of the actin gel can lead to cell protrusion is provided by the studies of cells deficient in actin-binding proteins.

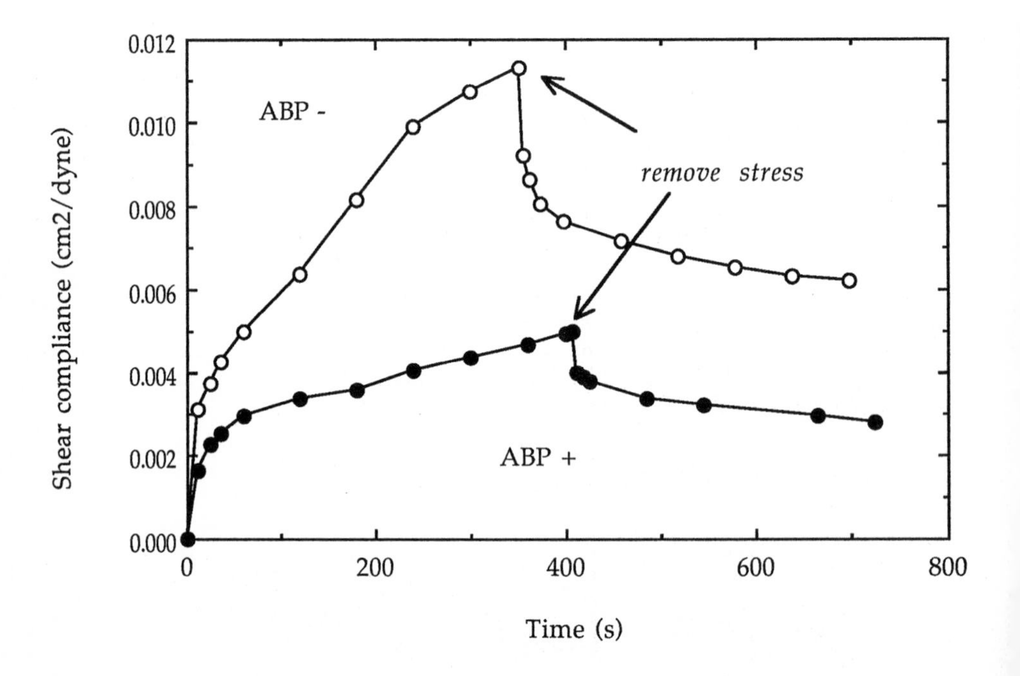

Figure 3. Shear compliance of closely-packed melanoma cells following sedimentation at 5000 x g to obtain a dense pellet that could be placed between the plates of a torsion pendulum [Janmey, 1991].

Melanoma cell lines have been isolated which do not express detectable amounts of the actin crosslinking protein ABP, and the cytoskeleton of these cells is poorly crosslinked. The effect on whole cell viscoelasticity of this lack of ABP is shown in Figure 3. Cell pellets formed by centrifugation of suspensions of melanoma cells which express or lack the ABP protein were subjected to shear stresses under identical conditions. Cells lacking ABP deform much more than normal cells subjected to the same stress, and they do not recover their undeformed state when the deforming stress is removed. This demonstrates that there is significant viscous flow during the period of strain, presumably due to the lack of crosslinks holding the actin filament network together. Optical and microscopic measurements before and after rheologic measurements confirmed that the number of cells were approximately equal in the two samples, and that they were not damaged by either the sample preparation or the measurement itself.

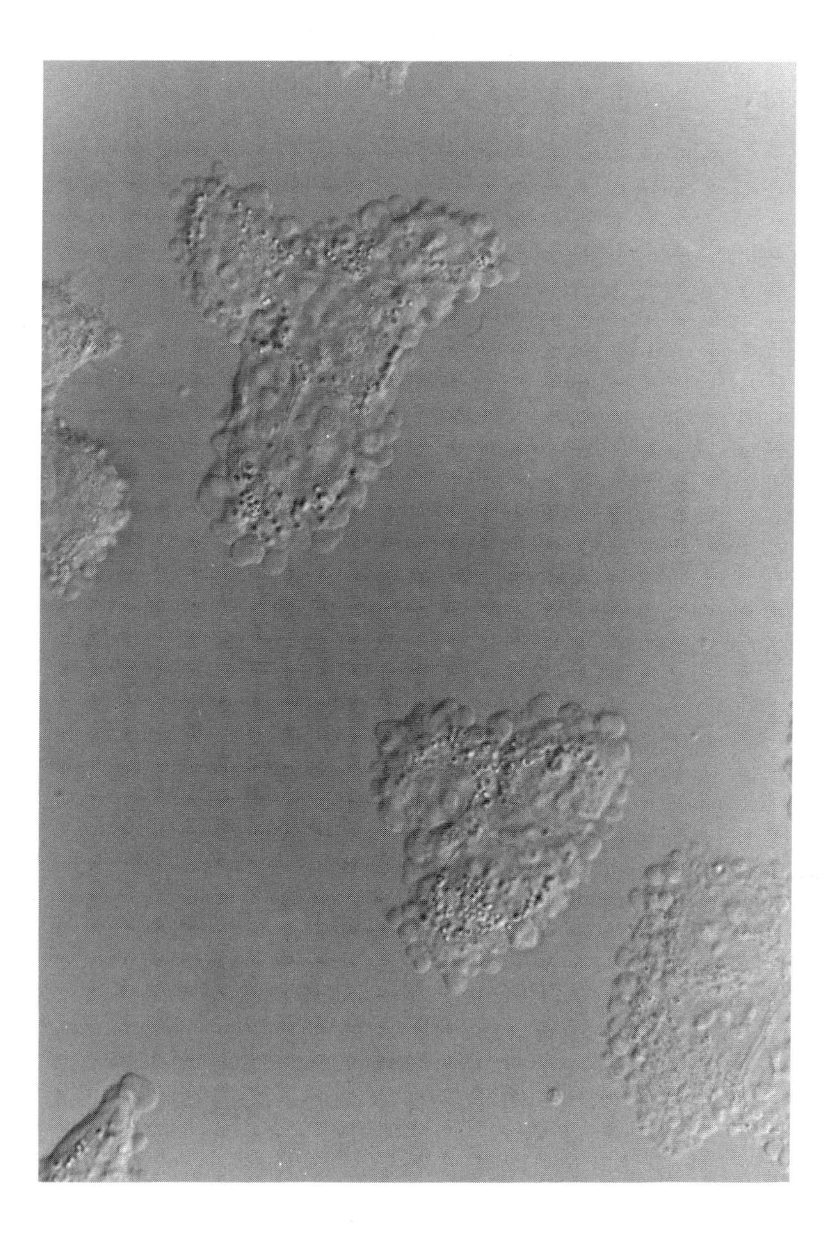

Figure 4. ABP-negative melanoma cells stimulated to bleb over their entire surface by addition of serum-containing medium.

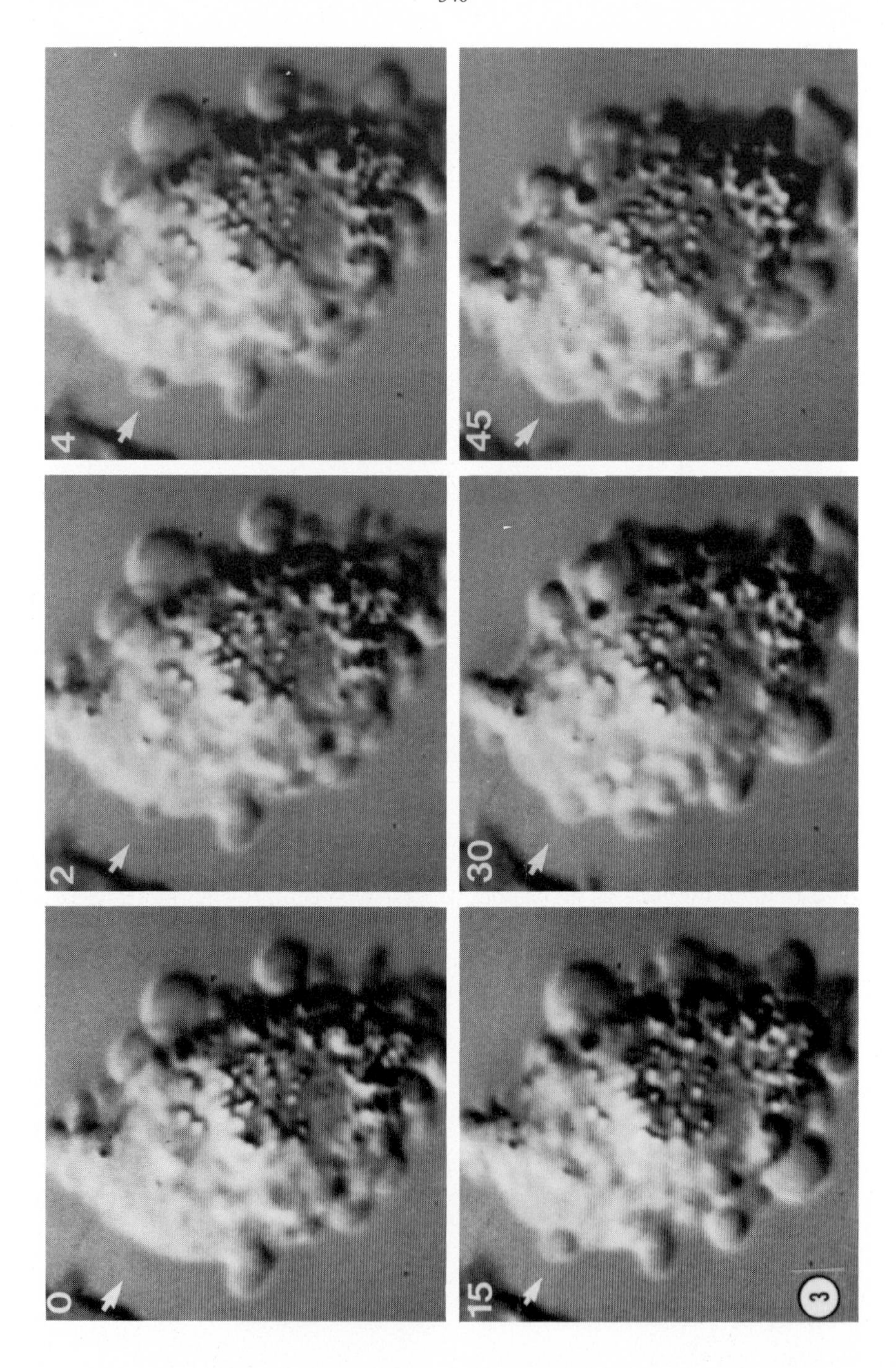

Figure 5. Individual video frames taken 4 seconds apart during the protrusion of a bleb from a melanoma cell null for the protein ABP.

When subjected to appropriate extracellular stimuli, cells lacking ABP produce large spherical blebs over their entire exposed membrane surface (Fig. 4). This blebbing responds to the same conditions which produce movement in normal (ABP+) cells. Protrusion of blebs several microns in diameter occurs within 1 - 3 seconds; the bleb remains extended for 10 - 15 seconds, and then slowly begins to retract over 30 - 60 seconds (Figure 5).

These experiments reveal several aspects of the mechanism of protrusive activity. First, protrusion itself does not require crosslinking; indeed loss of crosslinking enhances blebbing. This suggests that, in blebs, actin gel elasticity resists, rather than promotes, protrusion because of increased swelling pressures analogous to that of other crosslinked gels [Kokufata, et al., 1991][1]. Second, the rate of volume increase in a bleb is too rapid to be accounted for by water flow across the plasma membrane in response to an osmotic gradient. In fact, when the cells are permeabilized with Streptolysin-O, protrusion of the blebs still occurs, although the size of the blebs does increase under these conditions. This indicates that, while membrane osmotic forces are not necessary for protrusion itself, they may influence its extent.

Role of actin polymerization in promoting protrusive motility.

Polymerization of actin monomers into filaments and networks is involved in such motions as acrosomal process extension [Oster, et al., 1982], and the formation of filopodia and lamellae [Felder and Elson, 1990]. The idea that actin polymerization promotes outward motion is supported by morphologic and biochemical studies that identify actin polymerization as being localized at the leading edge of a moving cell, but the physics by which such polymerization could do mechanical work is not obvious [Oster and Perleson, 1987]. Nevertheless, the idea that actin polymerization can produce morphologic changes is consistent with studies of Cortese et al. [Cortese, et al., 1989] who showed a deformation of actin-loaded valinomycin-permeabilized phospholipid vesicles when the actin was polymerized by 100 mM KCl.

Figure 6 shows further evidence that actin filament assembly can produce morphologic changes superficially resembling filopod formation in the absence

---

[1] The swelling pressure of a gel is defined as the difference between the osmotic pressure and the elastic pressure: $P_{swelling} \cong \Pi_{osmotic} - P_{elastic}$.

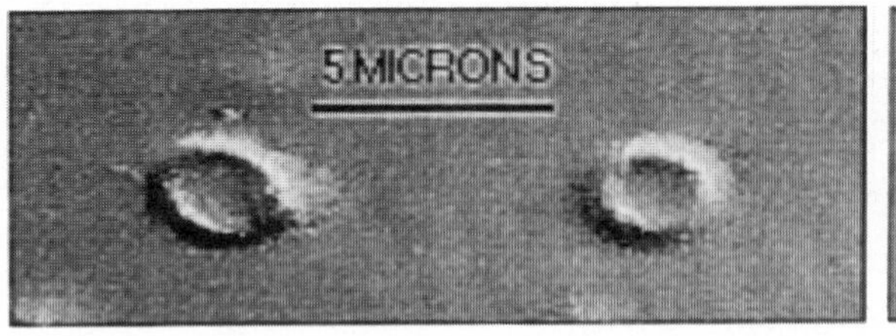

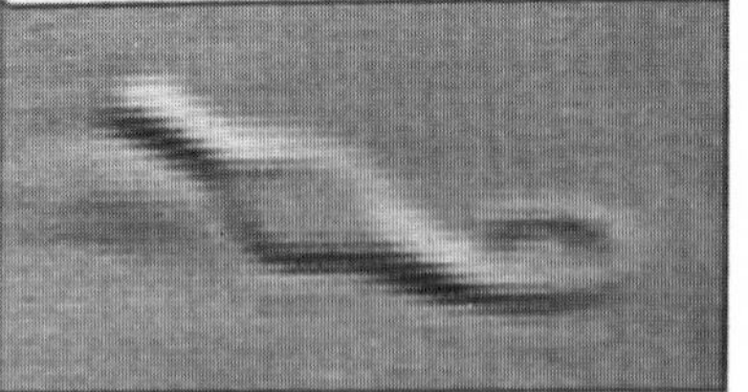

Figure 6. Morphology of large actin-loaded vesicles before (left) and after (right) actin polymerization.

of motor molecules such as myosin. Monomeric actin was incorporated within phosphatidylcholine (PC) vesicles and polymerized by permeabilizing the lipid bilayer to divalent cations by the addition of 0.1 μM ionophore A23187 and then adding 5 mM $CaCl_2$, or $MgCl_2$, a concentration sufficient to cause nearly total polymerization of actin *in vitro*. The structure of these vesicles before and after actin polymerization is shown in Figure 6. The hydrodynamic diameter of the particles was calculated from their diffusion constants as measured by dynamic (quasielastic) light scattering. Addition of ionophore had no effect on the hydrodynamic diameter of the vesicles, but subsequent addition of $MgCl_2$ caused a time-dependent increase in vesicle size within 5-10 minutes (Figure 7A).

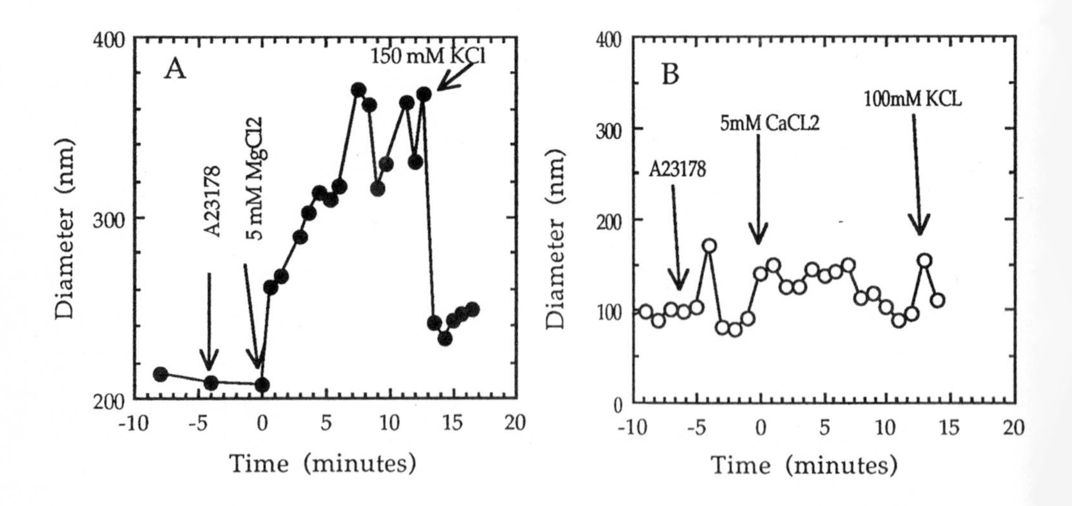

Figure 7. Average hydrodynamic radii of actin-loaded vesicles measured by dynamic light scattering before and after permeabilization of the membrane and addition of $MgCl_2$ to phosphatidylcholine vesicles containing 2 mg/ml actin.

Control experiments in which actin was omitted from the interior of vesicles (Figure 7B) confirmed that the effect of ionophore and Ca^{++} was not due to their interaction with the bilayer itself (Figure 7B). Figures 6 and 7 demonstrate that the protrusions that form in a vesicle when actin polymerizes inside proceed at a rate consistent with that of actin polymerization.

Several alternate models for polymerization-driven membrane deformation can be imagined, and further experiments are required to discriminate between them. One mechanism for the shape change involves alterations in membrane surface stress. In this model the force to bend the membrane arises from the stored elastic energy in the membrane itself [Oster, et al., 1989]. That is, a lipid bilayer is in mechanical equilibrium when the interfacial tensions and pressures on each leaflet are equal. Several effects associated with actin polymerization can upset this balance by either lowering the surface pressure, or raising the surface tension in the leaflet abutting the polymer tip. For example, the surface pressure of a lipid monolayer containing F-actin has been reported to be lower than one containing only G-actin [Llerenas and Cid, 1985]. A second mechanism depends on osmotic changes coupled to polymerization. Associated with actin polymerization are a number of effects which increase the local osmotic pressure, such as release of water molecules and counterions surrounding the free monomer and the filament. These local osmotic effects might be sufficient to drive membrane deformation, as has been shown for acrosomal extension in *Thyone* [Oster, et al., 1982]. On the other hand, the polymerization of actin occurs in vesicles that are permeabilized to divalent cations, and so osmotic pressures generated by ionic interactions may rapidly be dissipated.

A third mechanism derives from a model for random membrane deformation coupled to actin polymerization. Membrane fluctuations are rectified by the polymerizing fiber that prohibits the reverse fluctuation–the extending polymer (Odell et al., 1991). When a fluctuation larger than one monomer diameter occurs, a polymerization event take place, and the extended polymer prevents the reverse fluctuation. In addition, structural changes within the filament might increase fluctuations above the level expected from random thermal motions, and the elastic recoil of compressed filaments might help push against the membrane. Alternatively, thermally excited long wavelength instabilities can be triggered in a lipid bilayer, and these larger amplitude fluctuations can be "rectified" by the extending polymer [Gallez and Coakley,

1986, Oster, et al., 1989]. The possibility that cell protrusions can be stabilized by a polymerizing protein is supported by the observation of long extensions in erythrocytes accompanying polymerization of sickle-cell hemoglobin [Mozzarelli, et al., 1987]. In the latter case, direct force generation by the polymerizing hemoglobin seems unlikely, and the driving force for extending the membrane may likely derive from motions of the membrane itself. The polymerization of actin, however, is more complex since actin polymerization consumes ATP, and conformational changes due to hydrolysis of F-actin-bound ATP may also contribute to the protrusion of the membrane.

Conclusion

Multiple lines of evidence suggest that cell movement involves an interplay between the viscoelasticity of the cytoplasm—dominated by its cytoskeleton—and internally generated forces, of which osmotic pressure appears to be an important component. However, the degree to which osmotic pressures contribute to cell motility and the role of actin polymerization in different aspects of motility remain to be defined. For example, some forms of motility, notably the elongation of acrosomal processes and pseudopod formation by leukocytes, are strongly inhibited by 10-20 mOsm excess of neutral solutes such as sucrose in the extracellular medium. In contrast, equivalent hyperosmotic conditions do not inhibit either filopod or lamellipod formation in cultured neuronal cells [Bray, et al., 1991]. Likewise, either assembly or disassembly of actin filaments and networks can promote different aspects of motility. On the one hand local dissolution of the cortical actin network in response to receptor activation may allow directed motion in the face of pre-existing isotropic forces such as osmotic or hydrostatic pressures [Oster and Perleson, 1987]. Alternatively, local polymerization of actin may be able to harness random thermal fluctuations in the absence of an osmotic gradient. The different possible mechanisms of force generation and viscoelastic resistance to deformation may be combined in numerous ways to produce the variety of cell movements observed *in vivo*.

References

Bray D, Money N, Harold F and Bamburg J (1991) Responses of growth cones to changes in osmolality of the surrounding medium. J Cell Sci **98**:507-515

Cortese JD, Schwab B 3rd, Frieden C and Elson EL (1989) Actin polymerization induces a shape change in actin-containing vesicles. Proc Natl Acad Sci U S A **86**:5773-7

Cunningham C, Stossel T and Kwiatkowski D (1991a) Enhanced motility in NIH 3T3 fibroblasts that overexpress gelsolin. Science **251**:1233-1236

Cunningham, CC, et al. (1991b) manuscript submitted

deGennes PG (1976) Dynamics of entangled polymer solutions. I. The Rouse model. Macromolecules **9**:587-593

Elson EL (1988) Cellular mechanics as an indicator of cytoskeletal structure and function. Annu Rev Biophys Biophys Chem **17**:397-430

Felder S and Elson E (1990) Mechanics of fibroblast locomotion: Quantitative analysis of forces and motions at the leading lamellas of fibroblasts. J Cell Biol **111**:2513-2526

Gallez D and Coakley W (1986) Interfacial instability at cell membranes. Prog Biophys Molec Biol **48**:155-199

Hartwig J and Kwiatkowski D (1991) Actin-binding proteins. Curr Op Cell Biol **3**:87-97

Hartwig JH and Stossel TP (1981) The structure of actin-binding protein molecules in solution and interaction with actin filaments. J Mol Biol **145**:5630581

Ito T, Zaner KS and Stossel TP (1987) Nonideality of volume flows and phase transitions of F-actin solutions in response to osmotic stress. Biophys J **51**:745-53

Janmey P, Euteneuer U, Traub P and Schliwa M (1991) Viscoelastic properties of vimentin compared with other filamentous biopolymer networks. J Cell Biol **113**:155-160

Janmey P (1991) A torsion pendulum for measurement of the viscoelastic properties of biopolymers and its application to actin networks. J Biochem Biophys Meth **22**:41-53

Janmey PA, Hvidt S, Lamb J and Stossel TP (1990) ABP-actin gels resemble covalently crosslinked networks. Nature **345**:89-92

Janmey PA, Hvidt S, Peetermans J, Lamb J, Ferry JD and Stossel TP (1988) Viscoelasticity of F-actin and F-actin/gelsolin complexes. Biochemistry **27**:8218-27

Janmey PA, Peetermans J, Zaner KS, Stossel TP and Tanaka T (1986) Structure and mobility of actin filaments as measured by quasielastic light scattering, viscometry, and electron microscopy. J Biol Chem **261**:8357-8362

Kokufata E, Zhang Y-Q and Tanaka T (1991) Saccharide-sensitive phase transition of a lectin-loaded gel. Nature **351**:302-304

Llerenas E and Cid M (1985) The molecular interaction between F-actin and lecithin in a phospholipid monolayer system. Bol Estud Med Biol Mex **33**:33-39

Mozzarelli A, Hofrichter J and Eaton W (1987) Delay time of hemoglobin S polymerization prevents most cells from sickling *in vivo*. Science **237**:500-506

Odell, G. P. Janmey, G. Oster (1991). Actin polymerization and filopodial protrusion. (to appear)

Oster G (1988) Biophysics of the leading lamella. Cell Motil. Cytoskel. **10**:164-171

Oster, G., L. Cheng, H.-P.H. Moore A. Perelson (1989). Vesicle formation in the Golgi apparatus. J. Theo. Biol. **141**:463-504.

Oster G, Perelson A and Tilney L (1982) A mechanical model for acrosomal extension in Thyone. J Math Biol **15**:259-265

Oster G and Perleson A (1987) The physics of cell motility. J Cell Sci Suppl **8**:35-54

Pollard TD and Cooper JA (1986) Actin and actin-binding proteins. Ann Rev Biochem **55**:987-1035

Stossel TP, Chaponnier C, Ezzell RM, Hartwig JH, Janmey PA, Kwiatkowski DJ, Lind SE, Smith DB, Southwick FS, Yin HL and Zaner KS (1985) Non-muscle actin-binding proteins. Ann Rev Cell Biol **1**:353-402

Suzuki A, Yamazaki M and Ito T (1989) Osmoelastic coupling in biological structures: formation of parallel bundles of actin filaments in a crystalline-like structure caused by osmotic stress. Biochemistry **28**:6513-8

Yin HL (1987) Gelsolin: calcium- and polyphosphoinositide-regulated actin-modulating protein. Bioessays **7**:176-9

CONTROL OF WATER PERMEABILITY BY DIVALENT CATIONS

C A Pasternak
Department of Cellular and Molecular Sciences
St George's Hospital Medical School
Cranmer Terrace
London SW17 ORE
UK

SUMMARY

Leakage induced across the plasma membrane of cells by pore-forming agents is prevented by divalent cations. The same is true of leakage from liposomes or across a planar lipid bilayer. Divalent cation sensitivity appears not to depend on binding either to agent or to lipid. This paradoxical result may reflect the occurrence of a phenomenon described as 'surface conductance'.

INTRODUCTION

Swelling in cells and tissues is one of the earliest manifestations of almost all forms of injury (Robbins and Cotran 1979) and is often the result of a breach in the permeability barrier of the plasma membrane. Under normal conditions, unrestricted movement of ions like Na^+ and K^+ across the plasma membrane is limited [despite a large capacity for electroneutral processes such as Na^+ - K^+ exchange (Bashford 1990)]. The ATP-driven Na^+ pump extrudes Na^+ (and causes accumulation of K^+), thereby maintaining cells at non-equilibrium conditions. If the plasma membrane now becomes much more permeable as a result of membrane damage, or if the Na^+ pump is inhibited, - directly or through lack of ATP -, excess Na^+ will enter cells, as will Cl^- and hence water: cell swelling results. Although such swelling in erythrocytes leads to lysis (i.e. heamolysis), in most other cell types the plasma membrane is sufficiently

NATO ASI Series, Vol. H 64
Mechanics of Swelling
Edited by T. K. Karalis

dynamic to allow an expansion of cell volume to take place and thus avoid lysis. Surface microvilli, for example, can 'unfold' and thereby increase the surface area considerably (Knutton et al 1976).

We have studied the action of agents that cause cell swelling by increasing the permeability of the plasma membrane to low molecular weight compounds and ions. These agents include haemolytic viruses, bacterial and animal toxins and synthetic compounds like polylysine or detergents at sub-lytic concentration. What we found is that in every case divalent cations prevent the increase in permeability (Bashford et al 1984, 1986). Protons have the same effect (Bashford et al 1986, 1988a), though in cases where low pH is required to induce the membrane lesion in the first place (Maeda and Ohnishi 1980; Kagan et al 1981; Lenard and Miller 1981; Sandvig and Olsnes protective effect of H^+ may be masked (e.g. Patel and Pasternak 1985; Alder et al 1990).

Increased permeability resulting from physical damage to the cell membrane is also prevented by divalent cations. As long ago as 1914 it was shown that the loss of intracellular ions from plant cells by cutting roots with a knife could be prevented by Ca^{2+} (True 1914), and since 1957 every electrophysiologist who inserts electrodes into cells knows that the presence of Ca^{2+} in the extracellular medium is essential if good seals are to be obtained (Frankenhauser and Hodgkin 1957).

What is the mechanism by which divalent cations (and protons) act to minimize such permeability increases?

CHARACTERISTICS OF DIVALENT CATION - MEDIATED EFFECTS

First, it should be said that the relative efficacy of divalent cations and protons at reducing agent-induced leakage from cells is the same, irrespective of the nature of the agent: in every case the potency is $H^+ > Zn^{2+} > Ca^{2+} >$

Mg^{2+} (Bashford et al 1986, 1988a); the same is probably true of endogenous leakage from intact or physically-damaged cells (unpublished). The absolute efficacy (very approx $10^{-4}M$ Zn^{2+}, $10^{-3}M$ Ca^{2+}, $10^{-2}M$ Mg^{2+}) is such as to rule out non-specific effects like ion screening of surface charge (McLaughlin et al 1971); this is supported by the observation that dimethonium, a divalent cation that has the same effect on screening of surface charges as Mg^{2+}, Ca^{2+} or Zn^{2+} but that does not bind to anionic sites (McLaughlin et al 1983), also does not inhibit agent-induced permeability changes (Bashford et al 1988a).

Second, divalent cations are much more effective at <u>preventing</u> leakage that at <u>reversing</u> leakage; this may be partly due to the fact that when divalent cations are added to permeabilized cells, some divalent cation enters the cell before the lesion is closed or repaired, and a high concentration of intracellular Ca^{2+} (or Zn^{2+}) is known to be deleterious to cell function (Schanne et al 1979); [a <u>slight</u> increase in intracellular Ca^{2+}, on the other hand, is itself beneficial for eliminating certain lesions from the plasma membrane (Morgan and Campbell 1985)]. This, of course, assumes that where a high concentration of Ca^{2+} is effective at ameliorating leakage, it acts at the extracellular side of the plasma membrane, and this has been shown to be the case, for H^+ and Zn^{2+} as well as for Ca^{2+} (Bashford et al 1989).

If divalent cations act at the extracellular surface of cells, an obvious explanation for their action is that they compete with agent for binding to the cell surface; this would certainly be compatible with the observation that divalent cations are more effective at preventing leakage than at reducing leakage from affected cells, since displacement of bound agent (followed by repair of any remaining lesion) might not occur as readily as inhibition of binding in the first place. The fact that agents such as <u>Staphylococcus aureus</u> α toxin, melittin or polylysine are positively-charged is suggestive of such a mechanism of action. However this proved not to be the case.

DIVALENT CATIONS DO NOT ACT BY DISPLACING AGENT FROM THE CELL MEMBRANE

When binding to cells of a haemolytic virus (Sendai virus; Wyke et al 1980), melittin (Pasternak et al 1985) or activated complement (Micklem et al 1988) - all of which induce Ca^{2+}- preventable leakage - was measured by direct assay, no effect of Ca^{2+} was observed. On the contrary, in the case of the cytolysin (perforin) of cytotoxic lymphocytes, leakage is dependent on the presence of Ca^{2+}; yet high amounts of Ca^{2+} nevertheless prevent leakage (Bashford et al 1988b). Moreover if cells are incubated with agent in the presence of inhibitory concentrations of Ca^{2+} or Zn^{2+} and are then centrifuged to remove any unbound agent plus Ca^{2+} or Zn^{2+}, leakage is induced when the cells are reincubated in buffer alone; a similar result is obtained if the inhibitory Ca^{2+} or Zn^{2+} is removed by chelation with EGTA. This type of experiment has been carried out in the case of Sendai virus (Micklem et al 1985), activated complement (Micklem et al 1988), cytolysin of cytotoxic lymphocytes (Bashford et al 1988b), *S aureus* α toxin (Blomqvist and Thelestam 1986), *Clostridium perfringens* θ toxin, *E coli* haemolysin (Menestrina et al 1990), melittin and triton X100 (Alder et al 1991).

It is clear that the inhibitory action of divalent cations on agent-induced leakage cannot, - at least in the cases mentioned, - be ascribed to prevention of binding of agent to cells, or to the displacement of agent from treated cells. It must therefore be concluded that divalent cations inhibit the induction, or function, of pores themselves. What are the ligands to which they bind? In order to answer this question, we have studied pore formation in as simple a membrane system as possible, namely one consisting of pure lipids only. We have accordingly measured (a) leakage from small, unilamellar liposomes, using entrapped calcein as marker (Menestrina 1988), and (b) conductivity across planar lipid bilayers, using the Montal-Mueller (Montal and Mueller 1972) technique.

DIVALENT CATIONS CLOSE PORES INDUCED IN PURELY LIPIDIC SYSTEMS

Experiments with liposomes show that leakage induced by S aureus α toxin, C perfringens θ toxin, E coli haemolysin (Menestrina et al 1990) melittin or triton (Alder et al 1991) can be prevented by divalent cations and protons. The order of efficacy is again $H^+ > Zn^{2+} > Ca^{2+} > Mg^{2+}$, though the absolute concentrations required are somewhat higher than with cells. This is not surprising, since calcein has a mol. wt. of 630, whereas leakage from cells is generally measured with compounds having molecular weights < 300 and it is known that the larger the molecule whose leakage is being assessed, the higher a concentration of divalent cation is required to prevent it (e.g. Bashford et al 1985, Micklem et al 1988).

A better system for assessing leakage of small molecules is that of planar lipid bilayers. In this instance electrical conductivity is measured; hence the molecules need to be electrolytes, such as KCl. Experiments with a number of different agents show that in some cases, e.g. S aureus α toxin (Menestrina 1986), there is a selectivity for movement of anions (Cl^-) over cations (K^+), whereas in other cases, e.g. pneumolysin (Korchev et al 1991) or triton (T K Rostovtseva and A A Lev, unpublished experiments), selectivity is for cations over anions, and $K^+ > Na^+$. Note that from reconstituted mitochondrial lipid membranes treated with triton (Van Zutphen et al 1972), the selectivity is also $K^+ > Na^+$. What is true for every agent so far studied, is that single channels, that flicker between an open and a closed state just like endogenous ion channels, are induced. The single channel conductances vary widely, from <10 pS to >1nS. Where the effect of divalent cations and protons has been studied, several consequences have been described: an increased induction of channels, a decreased induction of channels, and increase in single channel conductance, a decrease in single channel conductance; the one effect that appears to be common to all agents is a shift towards the probability of channels being in the closed, rather than the open state. This, of course, is compatible with the prevention of leakage from liposomes and cells. The fact that closure can in some cases be reversed by voltage alone (Menestrina 1986), again emphasizes the fact that divalent cations do not act by displacing agent from the membrane. In most cases, just as in liposomes and cells, the efficacy of action of divalent cations is $Zn^{2+} > Ca^{2+} > Mg^{2+}$.

In order to probe the contribution of the lipid towards divalent cation sensitivity, channels induced by S aureus α toxin were taken as an example. First, the charged lipid normally used (a phosphatidylcholine) was replaced by an uncharged lipid (glycerol monoolein): similar-sized channels were induced, that were equally sensitive to closure by divalent cations and protons. Next, the carbonyl group of each fatty acid chain was replaced by an ether bond: again there was no effect on channel size or sensitivity to divalent cations or protons. From such experiments, and similar ones on S. aureus α toxin - induced leakage from liposomes composed of an uncharged lipid (galactosyl cerebroside) instead of a phospholipid, it must be concluded that divalent cation and proton sensitivity does not depend on binding to a charged head group or to the carbonyl group of the fatty acid side chains of the lipids present in liposomes or bilayers.

If binding of divalent cations and protons is not to lipids, it must be to the agent itself. In the case of proteins such as bacterial toxins this is possible. Indeed, channels induced across lipid bilayers by S aureus α toxin in which some of the lysine residues are modified by reaction with diethylpyrocarbonate have an altered sensitivity to closure by protons. But what of channels induced by agents such as triton X100 that lack any charged residue and have not even a free hydroxyl group? For channels induced by triton across bilayers composed of phospholipid or glycerol monoolein are closed by divalent cations and protons. Here there is clearly a paradox which we are unable to explain at this present time. A possible solution involves a phenomenon termed 'surface conductance'.

'SURFACE CONDUCTANCE' IS INHIBITED BY DIVALENT CATIONS AND PROTONS

If one constructs an apparatus such that it is possible to move a teflon septum close the underside of an air-water interface, and if one then coats with the teflon septum and the air-water interface with a single layer of amphipathic lipid, one is able to measure electrical conductance between the polar head groups of two apposed lipid monolayers (Lev, 1990). When the two monolayers are far apart, the conductance is that expected from a bulk solution of, for example, KCl. As the two monolayers are brought closer and closer together, three things are observed. First, the current is no longer continuous but is broken into a series of openings and closings: such single channels may have conductances as low as 10 pS. Second, a degree of selectivity of current is induced: cations are conducted in preference over anions. And third, current is now sensitive to inhibition by divalent cations (Y E Korchev, V Osipov and A A Lev, unpublished experiments).

This observation does not clarify what ligands within bilayers, liposomes and cells are responsible for sensitivity to divalent cations and protons. It does however indicate that no agent is necessary at all and that it is the phenomenon of 'surface conductance', - through the bilayer in the case of agent-induced pores, or along the surface in the case of the experiment described above, - that is somehow sensitive to divalent cations and protons. In each case the effect is to dehydrate the lipid surface, in a manner perhaps analogous to the substitution of water by Ca^{2+} at the head -group region of phospholipid 'crystals' (Hauser et al 1977; Hauser and Shipley 1984). For it is the entry of water into the lipid bilayer that is the crucial step in the induction of pores by the agents mentioned in this report. Since no pore-forming agent is necessary to observe the phenomenon, it may have a bearing on the prevention of leakage from damaged cells by Ca^{2+} that was referred to at the beginning of this article.

Since swelling in clays is the result of water movement through just such narrow 'cracks' as display the phenomenon of surface conductances, the experiments here reported on the effects of divalent cations and protons may have relevance beyond that of biological membranes.

ACKNOWLEDGEMENTS

I am grateful to many colleagues for informal discussion and for permission to cite unpublished material, - in particular to Drs C L Bashford, D T Edmonds, Y E Korchev, A A Lev, G Menestrina, C Pederzolli, T K Rostovtseva and R J P Williams. The work reported herein was supported by the Cell Surface Research Fund.

REFERENCES

Alder, G. M., Bashford, C. L., & Pasternak, C.A. (1990) Action of diphtheria toxin does not depend on the induction of large, stable pores across biological membranes. J. Membr. Biol. **113**, 67-74

Alder, G. M., Arnold, W. M., Bashford, C. L., Drake, A. F., Pasternak, C. A. & Zimmerman, U. (1991) Divalent cation -sensitive pores formed by natural and synthetic melittin and by triton X-100. Biochim. Biophys. Acta **1061** 111-120

Bashford, C. L., Alder, G. M., Patel, K. & Pasternak, C. A. (1984) Common action of certain viruses, toxins, and activated complement: pore formation and its prevention by extracellular Ca^{2+}. Biosci. Rep. **4**, 797-805

Bashford, C. L., Micklem, K. J., & Pasternak, C. A. (1985) Sequential onset of permeability changes in mouse ascites cells induced by Sendai virus. Biochim. Biophys. Acta **814** 247-255

Bashford, C. L., Alder, G. M., Menestrina, G., Micklem, K. J., Murphy, J. J., & Pasternak, C. A. (1986) Membrane damage by hemolytic viruses, toxins, complement and other cytotoxic agents. J. Biol. Chem. **261** 9300 -9308

Bashford, C. L., Alder, G. M., Graham, J. M., Menestrina, G. & Pasternak, C. A. (1988a) Ion modulation of membrane permeability: effect of cations on intact cells and on cells and phospholipid bilayers treated with pore -forming agents. J. Membr. Biol. **103** 79-94

Bashford, C. L., Menestrina, G. Henkart, P. A., & Pasternak, C. A. (1988b) Cell damage by cytolysin. Spontaneous recovery and reversible inhibition by divalent cations. J. Immunol. **141** 3965-3974

Bashford, C. L., Rodrigues, L. & Pasternak, C. A. (1989) Protection of cells against membrane damage by haemolytic agents: divalent cations and protons act at the extracellular side of the plasma membrane. Biochim. Biophys Acta. **983** 56-64.

Bashford, C. L. (1990) Electoneutral transport and exchange. In : Pasternak, C. A. (ed) Monovalent cations in biological systems. CRC Press, Boca Raton FL, p 103

Blomqvist, L. & Thelestam, M. (1986) Early events in the action of staphylococcal alpha-toxin on the plasma membrane of adrenocortical Y1 tumor cells. Infect. Immun. **53** 636-640

Frankenhauser, F. B., Hodgkin, A. L. (1957) The action of calcium on the electrical properties of sqid axons. J. Physiol. **137** 218-244

Hauser, H. Finer, E. G., Darke, A. (1977) Crystalline anhydrous Ca-phosphatidylserine bilayer. Biochem. Biophys. Res. Commun. **76** 267-274

Hauser, H. Shipley, G. G. (1984) Interactions of divalent cations with phosphatidylserine bilayer membranes. Biochemistry **23** 34-41

Knutton, S. Jackson, D. Graham, J. M., Micklem K. J., & Pasternak, C. A. (1976) Microvilli and cell swelling. Nature. **262** 52-54

Kagan, B. L., Finkelstein, A. & Colombini, M. (1981) Diphtheria toxin fragments form large pores in phospholipid bilayer membrane. Proc. Nat. Acad. Sci. USA. **78** 4950-4954

Lenard, J, Miller, D. K. (1981) pH-dependent hemolysis by influenza, Semliki Forest virus, and Sendai virus. Virology **110** 479-482

Lev, A. A. (1990) Sterol-dependent inactivation of gramicidin-A induced ionic channels in the cell and artificial lipid bilayer membranes In: Tenth School on Biophysics of Membrane Transport, Wroclaw p 231

Maeda, T. Ohnishi, S-I. (1980) Activation of influenza virus by acidic media causes hemolysis and fusion of erythrocytes. FEBS Lett **122** 283-287

McLaughlin, S. G., Szabo. G. Eisenman, G. (1971) Divalent ions and the surface potential of charged phospholipid membranes. J. Gen. Physiol. **58** 667-687

McLaughlin, A. Eng. W, K., Vaio, G. Wilson, T. McLoughlin, S. (1983) Dimethonium, a divalent cation that exerts only a screening effect on the electrostatic potential adjacent to negatively charged phospholipid bilayer membranes. J. Membr. Biol. **76** 183-193

Menstrina, G., Bashford, C. L., & Pasternak, C. A. (1990) Pore-forming toxins: experiments with s.aureus toxin, C. perfingens toxin and E.coli haemolysin in lipid bilayers, liposomes and intact cells. Toxicon. **28** 477 -491

Menestrina, G. (1988) Escherichia coli haemolysin permeabilizes small unilamellar vesicles loaded with calcein by a single hit mechananism. FEBS Lett **232** 217 -220

Micklem, K. J., Nyaruwe, A. & Pasternak, C. A. (1985) Permeability changes resulting from virus-cell fusion: temperature-dependence of the contributing processes. Mol. Cell. Biochem. **66** 163-173

Micklem, K. J., Alder, G. M., Buckley, G. D., Murphy, J. & Pasternak, C. A. (1988) Protection against complement -mediated cell damage by Ca^{2+} and Zn^{2+}. Complement 5 141-152

Montal, M., Mueller, P. (1972) Formation of bimolecular membranes from lipid monolayers and a study of their electricial properties. Proc Acad Nat Sci USA **69** 3561 -3566

Morgan, B. P., Campbell, A. K. (1985) The recovery of human polymorphomuclear leucocytes from sub-lytic complement attack is mediated by changes in intracellular free calcium. Biochem. J. **231** 205-208

Pasternak, C. A., Bashford, C. L. & Micklem, K., J. (1985) Ca^{2+} and the interaction of pore-formers with membranes. Proc. Int. Symp. Struct. Interactions, Suppl. J. Biosci. **8** 273-291

Pasternak, C. A., & Micklem, K. J. (1973) Permeability changes during cell fusion. J. Membr. Biol. **14** 293-303

Patel, K. & Pasternak, C. A. (1985) Permeability changes elicited by influenza and Sendai virus: separation of fusion and leakage by pH jump experiments. J. gen. Virol. **66** 767-775

Robbins, S. L., Cotran, R. S. (1979) Pathologic basis of disease. 2nd edn. Saunders, Philadelphia PA. p35

Sandvig, K. Olsnes, S. (1981) Rapid entry of nicked diphtheria toxin into cells at low pH. J. Biol. Chem. **256** 9068-9076

True, R. H. (1914) The harmful action of distilled water. Am. J. Bot. **1** 255-273

Schanne, F. A. X., Kane, A. B., Young, E. E., Farber, J. L. (1979) Calcium dependence of toxic cell death: a final common pathway. Science **206** 700-702

White, J. Matlin, K. Helenius, A. (1981) Cell fusion by Semliki Forest, influenza, and vesicular stomatitis viruses. J. Cell Biol. **89** 674-679

Wyke, A. M., Impraim, C. C., Knutton, S. & Pasternak, C. A. (1980) Components involved in virally-mediated membrane fusion and permeability changes. Biochem. J. **190** 625 -638

Van Zutphen, H. Merola, A. J., Brierley, G. P., Cornwell, D. G. (1972) The interaction of nonionic detergents with lipid bilayer membranes. Arch Biochem Biophys. **152** 755-766

MITOCHONDRIAL VOLUME HOMEOSTASIS: REGULATION OF CATION TRANSPORT SYSTEMS

Paolo Bernardi, Mario Zoratti and Giovanni Felice Azzone
CNR Unit for the Study of Physiology of Mitochondria
Institute of General Pathology, University of Padova
Via Trieste 75
I-35121 Padova
Italy

A. Mitochondrial swelling: a foreword

Since the first studies in biological energy transduction, it was established that mitochondrial swelling was accompanied by (i) loss of the matrix content of nucleotides and (ii) loss of capacity for ATP synthesis (Hunter and Ford, 1955). However, the relation between volume changes and energy conservation remained mysterious for quite a long time. During this early period, it was not at all clear whether the changes in optical density and in volume were osmotic or conformational in nature; and whether the various types of volume changes, in particular those defined as "large amplitude" or "low amplitude" based on the extent of swelling, were occurring through similar or different mechanisms.

It was only with the chemiosmotic hypothesis (Mitchell, 1966) that the question of the (im)permeability of the inner mitochondrial membrane began to assume a crucial role in the problem of energy conservation. In this hypothesis, a low membrane permeability is required for maintenance of the proton electrochemical gradient, and the uncoupling of oxidative phosphorylation observed in swollen mitochondria is directly linked to the increased permeability of the inner membrane to ions. The clarification of the nature of the various types of volume changes then took place in parallel to that of the relationship between ion transport and molecular mechanism of energy transduction.

NATO ASI Series, Vol. H 64
Mechanics of Swelling
Edited by T. K. Karalis

B. The two types of mitochondrial swelling

Since the early fifties it became customary to divide mitochondrial volume changes (measured as optical density changes) on the basis of their extent into two essentially similar processes, denoted as "large amplitude" and "low amplitude" swelling. Our laboratory provided evidence suggesting that while both the large and low amplitude swelling were osmotic in nature, quite different mechanisms were underlying the two processes. In fact, while in the case of the large amplitude swelling mitochondria could shrink and recover the original volume only after supply of energy, in the case of the low amplitude swelling the original volume could rather be recovered after addition of uncouplers, i.e. after energy dissipation (Azzone and Azzi, 1965). Due to these opposite requirements for the process of osmotic shrinkage, it was suggested that at the end of the swelling process mitochondria could be assigned to two different states: a *low energy state* in the case of the large amplitude swelling; and a *high energy state* in the case of the low amplitude swelling. It is now clear that the large amplitude and low amplitude swelling (low energy and high energy swelling, respectively, in the terminology of Azzone and Azzi, 1965) reflect indeed two completely different processes, both with respect to the energetic requirements and to the molecular pathways for solute movements across the inner membrane.

The *low amplitude (high energy) swelling* reflects the energy-linked transport of ions at the expense of either respiration or ATP hydrolysis, i.e., cation redistribution between the two sides of the inner membrane. Cations move along their electrochemical gradient a major portion of which is the voltage gradient generated by respiration or ATP hydrolysis (see Section C for details). If the membrane permeability remains low, energy dissipation is kept to a minimum, and most of the energy spent for cation transport across the membrane can be stored as an ion gradient. A striking example of this phenomenon is that respiration-inhibited mitochondria can synthesize ATP utilizing a gradient of K^+ ions (Rossi and Azzone, 1970). This type of swelling process requires that ion transport occurs through specific pathways, and has been instrumental in a large number of studies of mitochondrial ion transport both through natural carriers and through specific antibiotic ionophores.

The *high amplitude (low energy) swelling* involves movement of ions and low molecular weight solutes along their electrochemical gradients. Although osmotic in nature, this type of swelling is entirely passive, and does not imply conversion of one form of energy (respiration or ATP hydrolysis) into another (e.g., ion gradients). Rather than by specific transport pathways, the process is mediated by a large and relatively unspecific increase of the membrane permeability to solutes, and leads to complete uncoupling due to the presence of pores with diameters of at least 2.8 nm (Massari and Azzone, 1972; see review by Gunter and Pfeiffer, 1990). Studies of this type of swelling, that can be induced by a large number of compounds in a Ca^{2+}-dependent way, have been instrumental for understanding the factors affecting membrane integrity, and have therefore contributed to establish the role of the inner mitochondrial membrane as an integral part of the machinery converting electron flow to ATP synthesis. New interest in this inducible pathway for solute transport has arisen recently, and its potential role in mitochondrial volume (dis)regulation will be discussed later.

C. Mitochondrial membrane potential and the equilibrium distribution of cations

It is now firmly established that electron transport along the inner mitochondrial membrane is coupled to H^+ ejection on the redox H^+ pumps. Since the passive permeability to H^+ (the H^+ leak) and to anions is low, H^+ ejection results in the establishment of a H^+ electrochemical gradient ($\Delta\tilde{\mu}H$) that can be utilized for ATP synthesis *via* the F1F0 ATPase and for ion and metabolite transport *via* specific transport systems (Mitchell, 1966). The $\Delta\tilde{\mu}H$ can be written as

$$(1) \qquad \Delta\tilde{\mu}H = zF\Delta\Psi + RT \ln [H^+]_i/[H^+]_o$$

where $\Delta\Psi$ denotes the membrane potential difference ($\Psi_{in} - \Psi_{out}$) in mV. The magnitude of the proton electrochemical gradient is about -200-220 mV, and under physiological conditions most of the gradient is in the form of a membrane potential (see the review by Azzone et al., 1984). The existence of such a large membrane potential, negative inside, poses the major problem of cation equilibrium distribution.

Consider the case of K^+, whose electrochemical gradient across the inner mitochondrial membrane is given by

(2) $$\Delta\tilde{\mu}K = zF\Delta\psi + RT \ln [K^+]_i/[K^+]_o$$

The equilibrium condition ($\Delta\tilde{\mu}K = 0$) is given by

(3) $$-\Delta\psi = 60 \log [K^+]_i/[K^+]_o$$

Since $[K^+]_o$ is about 150 mM, with a membrane potential of -180 mV the equilibrium $[K^+]_i$ should be 150 M. Thus, the K^+ electrochemical gradient favors continuous K^+ accumulation. A similar problem exists for Na^+ ions (equilibrium $[Na^+]_i$ = 5 M for a cytosolic $[Na^+]$ of 5 mM). Since the inner membrane is highly permeable to water, K^+ (Na^+) uptake would lead to irreversible osmotic swelling of the matrix, with osmotic burst of the organelle. To survive and produce ATP in a high K^+ environment mitochondria must therefore (i) extrude K^+ against the K^+ electrochemical gradient to prevent osmotic burst; and (ii) keep electrophoretic K^+ influx to a low level to prevent excessive energy drain in futile K^+ cycling.

A similar problem exists for Ca^{2+} ions. In this case, since Ca^{2+} is transported as the fully charged species (Scarpa and Azzone, 1970), z = 2 and the equilibrium condition ($\Delta\tilde{\mu}Ca = 0$) is given by

(4) $$-\Delta\psi = 30 \log [Ca^{2+}]_i/[Ca^{2+}]_o$$

Since $[Ca^{2+}]_o$ is 0.1-1 µM, equilibrium $[Ca^{2+}]_i$ should be 0.1-1 M. The buildup of a large Ca^{2+} concentration gradient may be prevented to some extent by Ca^{2+} precipitation in the matrix, but there is little doubt that Ca^{2+} uptake can lead to increase of the osmotic pressure and therefore to matrix swelling. Besides direct osmotic effects, excessive Ca^{2+} accumulation poses additional hazards: (i) the induction of a Ca^{2+}-stimulated pore today called the *permeability transition pore* (PTP) (Haworth and Hunter, 1979b), leading to solute diffusion and membrane depolarization; and (ii) the risk of mitochondrial calcification. Again, mitochondrial survival in the cell demands (i) Ca^{2+} extrusion against the Ca^{2+} electrochemical gradient to prevent pore induction and/or mitochondrial calcification; and (ii) low electrophoretic Ca^{2+} flux to prevent energy drain in futile Ca^{2+} cycling.

In his formulation of the chemiosmotic hypothesis Mitchell (1966) was fully aware of these problems and postulated that (i) the inner mitochondrial membrane is endowed with specific electroneutral H^+-cation antiporters allowing cation extrusion without collapse of the membrane potential and (ii) the coupling membrane has a low permeability to protons and cations. Since K^+ is the main intracellular cation, most research on volume homeostasis has focused on the regulation of K^+ transport, but it is interesting to note that Mitchell (1966) did not specifically refer to monovalent cations. The recent reevaluation of the Ca^{2+}-activated PTP (Crompton et al., 1987; Broekemeier et al., 1989) and its potential role in the pathophysiology of mitochondrial volume regulation lead us to present here recent advance on the regulation of PTP activity at the single channel level, after briefly reviewing the pathways for both monovalent and divalent cation transport in mitochondria.

D. Pathways for electroneutral transport of monovalent cations: the H^+-Cat^+ antiporters

Initial studies of H^+-Cat^+ antiport indicated that mitochondria have a very active H^+-Na^+ antiport activity, while H^+-K^+ exchange was found to be very low at best (Mitchell and Moyle, 1969; Douglas and Cockrell, 1974; Brierley et al., 1978). The exchange activity has been widely attributed to a common H^+-Cat^+ antiporter with selectivity $Na^+ > Li^+ >> K^+, Rb^+, Cs^+$ (Brierley et al., 1978), which is the opposite of what would be expected on physiological grounds. The very existence of a H^+-K^+ antiporter (and therefore of chemiosmotic energy coupling) has long been questioned until it became clear that overall electroneutral H^+-K^+ exchange is low in native mitochondria, but can be induced by changes of matrix Mg^{2+} by a variety of treatments (Duszynski and Wojtczak, 1977; Azzone et al., 1978; Garlid, 1978, 1979, 1980). The demonstration that matrix Mg^{2+} indeed controls the activity of the endogenous, electroneutral H^+-K^+ antiporter was obtained by Garlid's group (Dordick et al., 1980). Shortly after, it was found that quinine inhibits the H^+-K^+ antiporter while H^+-Na^+ exchange activity is not affected by the drug (Nakashima and Garlid, 1982). Largely because of these studies, it is now generally accepted that mitochondria possess two H^+-alkali cation antiporters: the Na^+-selective H^+-Na^+ antiporter, which is constitutively active in isolated mitochondria, and was responsible for the process first reported by Mitchell and Moyle (1969); the non-selective H^+-Na^+ (K^+) antiporter, which is largely inhibited in isolated mitochondria but can be activated by Mg^{2+} depletion and by osmotic matrix swelling. The properties of these transport systems can be summarized as follows:

(i) Na^+-selective H^+-Na^+ antiporter: (i) selectivity for Na^+ and Li^+ with a pH optimum of 7.2 (Brierley et al., 1978; Nakashima and Garlid, 1982); (ii) not inhibited by Mg^{2+}, quinine or DCCD (Nakashima and Garlid, 1982; Martin et al., 1984); (iii) partially purified (Garlid et al., 1991);

(ii) non-selective H^+-Na^+ (K^+) antiporter: (i) equal rate of transport for all alkali cations with a V_{max} of about 300 nmol x mg protein^{-1} x min^{-1} (Garlid, 1988); (ii) pH optimum > 8.5 (Nakashima and Garlid, 1982; Bernardi and Azzone, 1983a); (iii) inhibited by Mg^{2+}, H^+, quinine and DCCD (Garlid, 1978, 1979, 1980; Nakashima and Garlid, 1982; Martin et al., 1984); (iv) activated by swelling *per se* (Bernardi and Azzone, 1983a); activity is mediated by a 82 kD protein (Li et al., 1990).

Both antiporters are electroneutral, and the direction of net cation flux depends on the sum of the chemical gradients for H^+ and for cations. Thus, under energized conditions both porters catalyze net K^+ efflux, protecting the mitochondrion from the hazards of K^+ and Na^+-dependent matrix swelling.

E. Pathways for electrophoretic transport of monovalent cations: the Cat^+ uniporters

In summarizing the four basic postulates of the chemiosmotic hypothesis Mitchell stated that the membrane-located ATPase, oxido-reduction chain, and electroneutral exchangers are (fourth postulate) "...localized in a specialised coupling membrane which has a low permeability to protons and to anions and cations generally" (Mitchell, 1966). With the notable exception of Brierley and his coworkers (Brierley, 1976), this statement has been widely considered to imply that the inner membrane does not possess specific pathways for electrophoretic Cat^+ transport, and that Cat^+ influx occurs only through leak pathways that are not subjected to regulation (e.g. Garlid, 1988). The idea that the very existence of a K^+ uniport cannot be reconciled with chemiosmotic energy coupling is well documented (e.g. Diwan and Tedeschi, 1975). Contrary to this assumption, we have recently shown that mitochondria possess regulated uniports for Na^+ and K^+ (Bernardi et al., 1989, 1990; Nicolli et al., 1991). Although uniport isolation has not been achieved at present, recent kinetic and thermodynamic characterization by our laboratory strongly supports the contention that electrophoretic Cat^+ fluxes are mediated by two specific systems: (i) a Na^+ (Li^+)-selective uniporter, regulated by surface Mg^{2+}; and (ii) a non-selective Cat^+ uniporter, regulated by matrix Mg^{2+}. Their properties can be summarized as follows:

(i) Na^+ (Li^+)-selective uniporter: selectivity for Na^+ and Li^+ over K^+ (Settlemire et al., 1968; Bernardi et al., 1990); pH optimum pH 7.5, activated by extramitochondrial ATP, competitively inhibited by extramitochondrial Mg^{2+} (Bernardi et al., 1990) and by ruthenium red (Kapus et al., 1990);

(ii) non-selective Cat^+ uniporter: transport of Na^+, Li^+ and K^+ at similar rates (Nicolli et al., 1991), and inhibition by matrix Mg^{2+} in the submicromolar range (Bernardi et al., 1989; Nicolli et al., 1991) and by ruthenium red (Kapus et al., 1990); K^+ conductance 0.15 nmol K^+ x mg protein^{-1} x min^{-1} x mV^{-1} x mM^{-1}, with maximal flux rates > 600 nmol K^+ x mg protein^{-1} x min^{-1} (Nicolli et al., 1991).

Both uniporters are electrophoretic, and the direction of net cation flux depends on the cation electrochemical gradient. Thus, under energized conditions both porters catalyze net Na^+ (K^+) uptake, resulting in matrix swelling.

Taken together, these findings lend support to the idea that mitochondrial K^+ and Na^+ cycling, and therefore volume homeostasis, is modulated through an interplay of regulated Cat^+ uniports and H^+-Cat^+ antiports, as first proposed by Brierley (Brierley, 1976).

F. Pathways for calcium transport

Isolated mitochondria can accumulate large amounts of Ca^{2+} *via* an energy-dependent mechanism (DeLuca and Engstrom, 1961). The problem of the disequilibrium between the Ca^{2+} accumulation ratio and the membrane potential (see Section B) was defined in our laboratory in the mid Seventies (Azzone et al., 1977; Pozzan et al., 1977) on the evidence that the steady state mitochondrial Ca^{2+} distribution is governed by kinetic rather than thermodynamic parameters (Azzone et al., 1977; Pozzan et al., 1977). That is, the rate of *net* mitochondrial Ca^{2+} uptake goes to zero when V_{influx} = V_{efflux}. Ca^{2+} distribution therefore reflects a kinetic steady state rather than a thermodynamic equilibrium (Pozzan et al., 1977). This is due to the existence of independent pathways for Ca^{2+} efflux (Vasington et al., 1972; Crompton et al., 1976; Nicholls, 1978).

Today we know that mammalian mitochondria possess three pathways for Ca^{2+} transport: (i) a Ca^{2+} uniporter; (ii) a Na^{+}-Ca^{2+} exchanger; and (iii) an active extrusion mechanism operating only under energized conditions. Since this topic has been reviewed recently, here we shall provide only a selective account while the reader is referred to the thorough review of Gunter and Pfeiffer (1990) for a detailed discussion of this field.

(i) The Ca^{2+} uniporter: this pathway catalyzes the electrophoretic transport of Ca^{2+} in the direction determined by the Ca^{2+} electrochemical gradient. This point was proven by the finding that Ca^{2+} uptake in deenergized mitochondria can be driven by the K^{+} diffusion potential set up by the ionophore valinomycin with a charge stoichiometry of 2 K^{+}/Ca^{2+} (Scarpa and Azzone, 1970) or by the diffusion potential of SCN^{-} (Selwyn et al., 1970). Thus, in energized mitochondria the uniporter catalyzes net Ca^{2+} uptake while upon deenergization (i.e., membrane depolarization) it mediates Ca^{2+} efflux (Bernardi et al., 1984). When care is taken to measure Ca^{2+} transport under rate-limiting conditions (Bragadin et al., 1979) the V_{max} is as high as 1200 nmol Ca^{2+} x mg protein^{-1} x min^{-1} and the apparent Km between 1 and 10 µM Ca^{2+}. Although Mg^{2+} increases the apparent Km and decreases the V_{max} of the uniporter (e.g., Bragadin et al., 1979), these findings flatly contradict Mitchell's fourth postulate (see section E), since the membrane permeability to Ca^{2+} is very high indeed. No matter how low cytosolic [Ca^{2+}]

is kept, and therefore how slow Ca^{2+} uptake on the uniporter is, mitochondria would eventually die of Ca^{2+} overload. Again, not only mitochondria must restrict electrophoretic Ca^{2+} flux on the uniporter (Pozzan et al., 1977): they also demand the existence of independent pathways for Ca^{2+} efflux, as proposed in the Seventies both on experimental and theoretical grounds (Vasington et al., 1972; Crompton et al., 1976; Puskin et al., 1976; Nicholls, 1978).

(ii) The Na^{+}-Ca^{2+} exchanger: this systems exchanges nNa^{+}-Ca^{2+}. It has been suggested (Crompton et al., 1976) that n = 3, but the exact stoichiometry is not established. This carrier has a V_{max} in the range 4-12 nmol Ca^{2+} x mg $protein^{-1}$ x min^{-1} and apparent Km for Na^{+} between 7 and 12 mM, depending on the source of mitochondria (Crompton et al., 1978). The exchanger requires K^{+} for optimal activity (Km 18 mM) even if K^{+} itself is not transported (Crompton et al., 1980), and is inhibited by trifluoperazine (Hayat and Crompton, 1985), diltiazem (Rizzuto et al., 1987; Chiesi et al., 1987), and amiloride (Jurkowitz et al., 1983). Independently of the exact stoichiometry, both the Ca^{2+} and Na^{+} chemical gradients favor net Ca^{2+} efflux, since the creation of a relevant Na^{+} concentration gradient is prevented by Na^{+} efflux on the Na^{+}-selective H^{+}-Na^{+} exchanger (see Section D). It must be noted, however, that the V_{max} of this system is three orders of magnitude lower than the V_{max} of the Ca^{2+} uniporter. Any increase in cytosolic $[Ca^{2+}]$ resulting in rates of uptake faster than 12 nmol Ca^{2+} x mg $protein^{-1}$ x min^{-1} will thus inevitably lead to Ca^{2+} accumulation.

(iii) The active extrusion mechanism: the third pathway for Ca^{2+} transport is independent of Na^{+}. It appears that regulation of this pathways is rather complex: it is inactive in deenergized mitochondria (Bernardi and Azzone, 1979), inhibited by uncouplers (Puskin et al., 1976; Bernardi and Azzone, 1982) and operates only when the membrane potential is higher than about 140 mV (Bernardi and Azzone, 1983b). The rate of Ca^{2+} flux through this pathway is extremely slow, and the V_{max} may not exceed 1.2 nmol Ca^{2+} x mg $protein^{-1}$ x min^{-1} (Gunter and Pfeiffer, 1990).

G. Kinetic control of mitochondrial cation distribution

From the data discussed so far, a picture emerges where the general problem of cation distribution across the inner mitochondrial membrane can be rationalized in the following terms:

(i) in energized mitochondria cation uptake takes place on regulated cation uniports; the uptake process is determined by the cation electrochemical gradient, and the rate of cation entry largely depends on the activity of the specific uniport involved;

(ii) in energized mitochondria cation extrusion takes place on independent pathways; the process prevents cation equilibration with the electrochemical gradient; thus, the cation accumulation ratio reflects a kinetic steady state rather than a thermodynamic equilibrium;

(iii) the energy drain linked to futile cation cycling is kept at a low level by restricting the rate of electrophoretic cation uptake; in the case of monovalent cations, this is achieved by a tight regulation (inhibition) of uniport activity by Mg^{2+}; in the case of Ca^{2+}, the rate of uptake is maintained at very low levels because cytosolic [Ca^{2+}] is well below the uniporter's Km;

(iv) increased electrophoretic influx of monovalent cations can be well compensated by increased efflux on the antiporters, since their V_{max} is very high; on the other hand, increased electrophoretic influx of Ca^{2+} easily leads to Ca^{2+} overload, since Ca^{2+} extrusion pathways have a very low V_{max}.

Thus, mitochondria live under the continuous hazard of Ca^{2+} overload. The possible consequences of this event include the activation of the permeability transition pore (PTP): the final part of this paper will be devoted to our current approach to this aspect of mitochondrial volume (dis)regulation.

As mentioned in Sections A and B, conditions leading to a Ca^{2+}-dependent permeability increase of the mitochondrial inner membrane have been described since the beginning of work on isolated mitochondria (Hunter and Ford, 1955). As evidence in favour of Mitchell's chemiosmotic hypothesis grew, the permeability increase has been widely considered as a "damaging" effect that may be more relevant to mitochondrial pathology than to mitochondrial physiology (but see Haworth and Hunter, 1979a, 1979b, 1979c). Our perspective on this problem has been changed by the recent finding that the permeability increase is inhibited by extremely low concentrations of cyclosporin A, a cyclic endecapeptide widely used as an immunosuppressant (Fournier et al., 1987; Crompton et al., 1988; Broekemeier et al., 1989). This finding strongly supports the idea that the permeability increase is mediated by a pore (channel), the PTP, as suggested earlier by Haworth and Hunter (1979a; 1979b; 1979c).

H. Properties of the mitochondrial megachannel (MMC)

(i) The patch-clamp tecnique - In 1987, the powerful electrophysiological technique known as "patch-clamp" (Hamill et al., 1981) was introduced in studies of mitochondrial physiology (Sorgato et al. 1987; Petronilli et al., 1989; Kinnally et al., 1989). Only a very schematic description of this technique can be given here. For details the reader is referred to the books by Sakmann and Neher (1983) and by Hille (1984). A glass pipette with a fine (normal diameter 0.5-2μm), polished tip, containing an electrode and an electrolyte solution, is pressed against the membrane of the target cell or organelle. By incompletely understood interactions, a high-resistence (Gohms) seal often forms between the glass and the membrane. Under these conditions, the conductance across the pipette mouth is essentially determined by the conductance of the membrane itself, which is extremely low unless ion-conducting channels in the membrane patch happen to be open. Thus, if a potential difference is imposed between the electrode inside the pipette and an electrode grounding the bath were the cell is immersed, an appreciable (pA) current will flow only when one or more channels are open. Transitions of the individual channels between the open and closed configuration can be easily detected. Experiments can be performed in three ways: (i) in the configuration described above *(cell attached configuration)*, which minimizes the interferences with the process being studied; (ii) after excising the membrane patch from the cell or organelle *(excised patch configuration)*; this procedure allows precise control over the solutions

bathing both sides of the patch membrane; (iii) after disrupting the patch, thus establishing a communication between the pipette and the cell interior *(whole cell configuration);* in this case, the "patch" is represented by the whole cellular membrane.

The relevance of this approach to mitochondria is obvious, since the inner mitochondrial membrane contains both specific conductance pathways for cations (see Sections E and F) and anions (Azzi and Azzone, 1967; Garlid and Beavis, 1986), and unspecific pathways (the PTP) who await molecular characterization. Indeed, the membrane density, conductance, size, kinetic behaviour, voltage-dependence and selectivity of channels can be determined with higher confidence and detail than possible with classical approaches. Thus, this technique is potentially suited to provide information on the components of the volume regulation machinery at the molecular level.

To date, patch-clamp experiments on the inner mitochondrial membrane have led to the observation of several conductances of different sizes (see review by Zoratti and Szabo', 1991). Two channels have been characterized so far: (i) a slightly anion-selective 107 pS channel (Sorgato et al., 1987, 1989); and (ii) a 1.3 nS megachannel (150 mM KCl), the MMC (Petronilli et al., 1989; Kinnally et al., 1989). The latter is of particular relevance here.

(ii) The mitochondrial megachannel - In typical patch-clamp experiments, the presence of the megachannel is revealed by a characteristic multitude of conductance steps of different sizes, appearing mostly as brief decreases from a dominant high conductance level (Petronilli et al., 1989). The latter represents the maximal conductance level of the channel(s) active in the patch, while the transitions to lower levels denote "visits" of the channel to sub-maximal conductance states. The maximal conductance varies from experiment to experiment (Szabo' and Zoratti, 1991), values of 1 and 1.3 nS being the most frequently observed. Among the substates, conductance levels close to one-half of the maximum (i.e., 500-650 pS) predominate. During transitions from the fully closed state to the maximal conductance state, and viceversa, the channel often pauses at this intermediate level. Occasionally, in multi-channel patches, single-step conductance changes as

high as 5 nS are observed. These originate from the cooperative "gating" (i.e. coordinated closure or opening) of a few channels.

Figure 1 presents, as an example, a current record suggesting a binary structure for the MMC. A "half-amplitude" conductance (approx 400 pS in this case) opens first (at A), and undergoes frequent transitions to the closed state, evident as a characteristic "flickering" of the current (at B). When both "halves" are open (full conductance: approx 800 pS in this example), they reciprocally stabilize the open state, and closures are much less frequent. Closure also takes place via the "half-amplitude" state (C). Data in reconstituted systems (Thieffry et. al, 1988; Henry et al., 1989; Moran et al., 1990), as well as the very rare observation of a "half-channel" operating without a partner in the mitoplast membrane (Zoratti et al., unpublished) suggest that the "half-channel" might actually be considered as the fundamental structure.

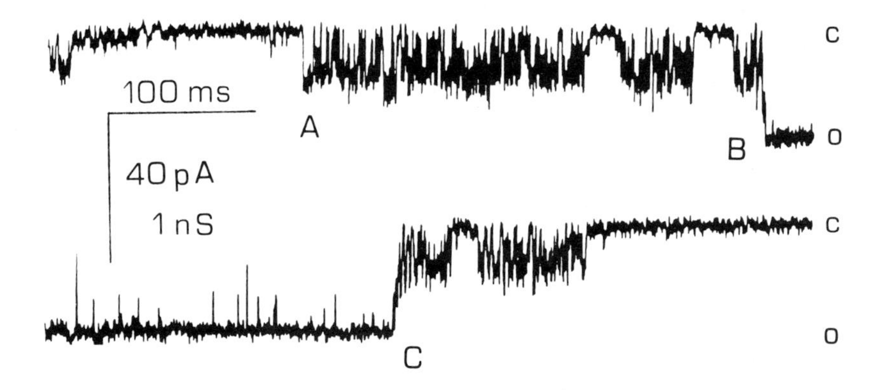

Figure 1 - Evidence for a binary structure of the MMC. A 1.54-sec current record segment obtained from a mitoplast in symmetrical 150 mM KCl, 5 mM Tris-Cl, 0.3 mM $CaCl_2$ (pH 7.2). Voltage: -40 mV. Filter corner frequency: 5 KHz. The data were stored in digitized form (22 KHz sampling frequency) and sampled for computer elaboration at 5 KHz. c, closed; o, open.

The channel is mildly voltage-dependent: increasing unphysiological potentials favor its operation in subconductance levels, i.e. its progressive closure (Petronilli et al, 1989). It would be of interest to know whether membrane stretch has any influence on its activity. This is not the case in mitoplasts, but it cannot be excluded that the channels are stretch-sensitive in the native membrane, losing this property because of the harsh shock to which the mitochondria are subjected. Such a loss of stretch sensitivity occurs in bacterial channels (Szabo' et al., 1990), to which the mitochondrial megachannel resembles in some respects (high conductance, composite nature, cooperative behavior).

I. The permeability transition pore (PTP) and the mitochondrial megachannel (MMC) are the same molecular species

The ensemble of our observations has led us to envision a channel complex composed of a number of subunits, capable of independent existence in the membrane, which can combine with a variable (within limits) stoichiometry to produce a binary structure. The channel behavior arises from the cooperative activity of the components. Conformational changes in (or possibly the disaggregation of) some of them results in lower-than-maximal conductances. Note that the permeability transition pore has also been reported to exhibit strong cooperativity (Crompton and Costi, 1988) and that a binary structure is in agreement with the presence of two cooperating binding sites for both Ca^{2+} (Haworth and Hunter, 1979b) and ADP (Hunter and Haworth, 1980) in the permeability transition pore.

Assuming a membrane thickness of 7 nm, the minimum diameter of a 1.3 nS single channel is 2.8 nm, which is the same value determined by Massari and Azzone (1972) from the Staverman coefficient for various solutes in whole mitochondria after opening of the PTP. Furthermore, recent data demonstrate that the MMC is inhibited by cyclosporin A at concentrations not far from those blocking PTP activity in whole mitochondria (Szabo' and Zoratti, 1991). Thus, available data suggest that the MMC might be the pore involved in the permeability transition of mitochondria. We have therefore carried out a detailed study where the effects of various agents on the PTP and on MMC activity were compared. The results from these experiments are presented in Table I.

Table I - A comparison of the properties of the mitochondrial megachannel (MMC, patch-clamp experiments) and of the permeability transition pore (PTP, swelling experiments).

MMC	PTP
Activation by sub-mM Ca^{2+} at a matrix site[a,b]	Activation by Ca^{2+} binding at an inner site (Km 10μM[c])
Ca^{2+}-competitive inhibition by Mg^{2+}, Mn^{2+}, Sr^{2+} and Ba^{2+} at the Ca^{2+}-binding site[a]	Ca^{2+}-competitive inhibition with similar parameters by the same divalent cations[c]
Ca^{2+}-competitive inhibition by cyclosporin A, sub-μM range[a,d]	Ca^{2+}-competitive inhibition by cyclosporin A (Ki 10nM[c])
Ca^{2+}-reversible inhibition by H^+ (matrix side site) in the physiological range[a]	Inhibition by H^+ at an inner face site other than the Me^{2+} site[c]
Inhibition by ADP, sub-mM range[b]	Inhibition by ADP, sub-mM range[e]
Inhibition by Amiodarone in the μM range[f]	Inhibition by Amiodarone in the μM range[g]
Ca^{2+}-binding at a cytoplasmic site necessary for channel activity[a]	An outward-facing Ca^{2+} site must be occupied for the pore to be open[h]
Data consistent with subunits organized into two "halves" exhibiting cooperativity	Ca^{2+} and ADP have two sites per channel[e,h]

[a]Szabo' et al., submitted; [b]Szabo' and Zoratti, submitted; [c]Bernardi et al., submitted; [d]Szabo' and Zoratti, 1991; [e]Hunter and Haworth, 1980; [f]Antonenko et al., 1991;[g]Bernardi et al., unpublished; [h]Hunter et al., 1976.

All the tested agents affect the MMC and the PTP in the same way and in approximately the same concentration range. The perfect match leaves little doubt that the mitochondrial megachannel observed in patch-clamp experiments is the permeability transition pore. It is thus possible to use the ability, afforded by the patch-clamp technique, to study phenomena at the single-channel level to gain information on the molecular structure and properties of the pore. Much progress has already been made in this direction.

J. Future perspectives

Our findings on the identity of the PTP-MMC open a new perspective in the isolation of the components of this structure. Specific mitochondrial proteins can now be tested for their activity as a high conductance, cyclosporin A-sensitive channel modulated by divalent cations and protons in reconstituted systems. Besides the PTP-MMC, other channels remain to be characterized at the molecular and electrophysiological levels (e.g. the Ca^{2+}, Na^{+} and Cat^{+} uniporters): we hope that current efforts in our laboratories will contribute to a better understanding of mitochondrial volume control through these systems.

Acknowledgments

We wish to thank all our colleagues who, through many years, contributed to the development of the concepts expressed in this review.

List of references

Antonenko YN, Kinnally KW, Perini S, Tedeschi H (1991) Selective effects of inhibitors of inner mitochondrial membrane channels. FEBS Lett in press

Azzi A, Azzone GF (1967) Swelling and Shrinkage phenomena in liver mitochondria. VI. Metabolism-independent swelling coupled to ion movements. Biochim Biophys Acta 131:468-478

Azzone GF, Azzi A (1965) Volume changes in liver mitochondria Proc Natl Acad Sci USA 53:1084-1089

Azzone GF, Bortolotto F, Zanotti A (1978) Induction of electroneutral exchanges of H^{+} with K^{+} in rat liver mitochondria. FEBS Lett 96:135-140

Azzone GF, Bragadin M, Pozzan T, Dell'Antone P (1976) Proton electrochemical potential in steady state rat liver mitochondria. Biochim Biophys Acta 459:96-109

Azzone GF, Pietrobon D, Zoratti M (1984) Determination of the proton electrochemical gradient across biological membranes. Curr Topics in Bioenergetics 13:1-77

Azzone GF, Pozzan T, Massari S, Bragadin M, Dell'Antone P (1977) H^{+}/site ratio and steady state distribution of divalent cations in mitochondria. FEBS Lett 78:21-24

Bernardi P, Angrilli A, Ambrosin V, Azzone GF (1989) Activation of latent K^+ uniport in mitochondria treated with the ionophore A23187. J Biol Chem 264:18902-18906

Bernardi P, Angrilli A, Azzone GF (1990) A gated pathway for electrophoretic Na^+ fluxes in rat liver mitochondria. Regulation by surface Mg^{2+}. Eur J Biochem 188:91-97

Bernardi P, Azzone GF (1979) ΔpH-induced Ca^{2+} fluxes in rat liver mitochondria. Eur J Biochem 102:555-562

Bernardi P, Azzone GF (1982) A membrane potential-modulated pathway for Ca^{2+} efflux in rat liver mitochondria. FEBS Lett 139:13-16

Bernardi P, Azzone GF (1983a) Electroneutral H^+-K^+ exchange in rat liver mitochondria. Regulation by membrane potential. Biochim Biophys Acta 724:212-223

Bernardi P, Azzone GF (1983b) Regulation of Ca^{2+} efflux in rat liver mitochondria. Role of membrane potential. Eur J Biochem 134:377-383

Bernardi P, Paradisi V, Pozzan T, Azzone GF (1984) Pathway for uncoupler-induced calcium efflux in rat liver mitochondria: inhibition by ruthenium red. Biochemistry 23:1645-1651

Bragadin M, Pozzan T, Azzone GF (1979) Kinetics of Ca^{2+} carrier in rat liver mitochondria. Biochemistry 18:5972-5978

Brierley GP (1976) The uptake and extrusion of monovalent cations by isolated heart mitochondria. Mol Cell Biochem 10:41-62

Brierley GP, Jurkowitz M, Jung DW (1978) Osmotic swelling of heart mitochondria in acetate and chloride salts. Evidence for two pathways for cation uptake. Arch Biochem Biophys 190:181-192

Broekemeier K, Dempsey ME, Pfeiffer DR (1989) Cyclosporin A is a potent inhibitor of the inner membrane permeability transition in liver mitochondria. J Biol Chem 264:7826-7830

Chiesi M, Rogg H, Eichenberger K, Gazzotti P, Carafoli E (1987) Stereospecific action of diltiazem on the mitochondrial Na-Ca exchange system and on sarcolemmal Ca-channels. Biochem Pharmacol 36:2735-2740

Costa G, Kinnally KW, Diwan JJ (1991) Patch clamp analysis of a partially purified ion channel from rat liver mitochondria. Biochem Biophys Res Comm 175:305-310

Crompton M, Capano M, Carafoli E (1976) The sodium-induced efflux of calcium from heart mitochondria. A possible mechanism for the regulation of mitochondrial calcium. Eur J Biochem 69:453-462

Crompton M, Costi A (1988) Kinetic evidence for a heart mitochondrial pore activated by Ca^{2+}, inorganic phosphate and oxidative stress. A potential mechanism for mitochondrial dysfunction during cellular Ca^{2+} overload. Eur J Biochem 178:488-501

Crompton M, Costi A, Hayat L (1987) Evidence for the presence of a reversible Ca^{2+}-dependent pore activated by oxidative stress in heart mitochondria. Biochem J 245:915-918

Crompton M, Ellinger H, Costi A (1988) Inhibition by cyclosporin A of a Ca^{2+}-dependent pore in heart mitochondria activated by inorganic phosphate and oxidative stress. Biochem J 255:357-360

Crompton M, Heid I, Carafoli E (1980) The activation by potassium of the sodium-calcium carrier of cardiac mitochondria. FEBS Lett 115:257-259

Crompton M, Moser R, Lüdi H, Carafoli E (1978) The interrelations between the transport of sodium and calcium in mitochondria of various mammalian tissues. Eur J Biochem 82:25-31

DeLuca HF, Engstrom GW (1961) Calcium uptake by rat kidney mitochondria. Proc Natl Acad Sci USA 47:1744-1754

Diwan JJ, Tedeschi H (1975) K^+ fluxes and the mitochondrial membrane potential. FEBS Lett 60:176-179

Dordick RS, Brierley GP, Garlid KD (1980) On the mechanism of A23187-induced potassium efflux in rat liver mitochondria. J Biol Chem 255:10299-10305

Douglas MG, Cockrell, RS (1974) Mitochondrial cation-hydrogen ion exchange. Sodium selective transport by mitochondria and submitochondrial particles. J Biol Chem 249:5464-5471

Duszynski J, Wojtczak L (1977) Effect of Mg^{2+} depletion in mitochondria on their permeability to K^+: the mechanism by which ionophore A23187 increases K^+ permeability. Biochem Biophys Res Commun 74:417-424

Fournier N, Ducet G, Crevat A (1987) Action of cyclosporine on mitochondrial calcium fluxes. J Bioenerg Biomembr 19:297-303

Garlid KD (1978) Unmasking the mitochondrial K/H exchanger: swelling-induced K^+ loss. Biochem Biophys Res Commun 83:1450-1455

Garlid KD (1979) Unmasking the mitochondrial K/H exchanger: tetraethylammonium-induced K^+ loss. Biochem Biophys Res Commun 87:842-847

Garlid KD (1980) On the mechanism of regulation of the mitochondrial K/H exchanger. J Biol Chem 255:11273-11279

Garlid KD (1988) Mitochondrial volume control. In: Lemasters JJ, Hackenbrock CR, Thurman RG, Westerhoff HV (eds) Integration of mitochondrial function. Plenum, New York

Garlid KD, Beavis AD (1986) Evidence for the existence of an inner membrane anion channel in mitochondria. Biochim Biophys Acta 853:187-204

Garlid KD, Shariat-Madar Z, Nath S, Jezek P (1991) Reconstitution and partial purification of the Na^+-selective Na^+/H^+ antiporter of beef heart mitochondria. J Biol Chem 266:6518-6523

Gunter TE, Pfeiffer DR (1990) Mechanisms by which mitochondria transport calcium. Am J Physiol 258:C755-C786

Hamill OP, Marty A, Neher E, Sakmann B, Sigworth FJ (1981) Improved patch-clamp techniques for high-resolution current recording from cells and cell-free membrane patches. Pfluegers Arch 391:85-100

Haworth RA, Hunter DR (1979a) The Ca^{2+}-induced membrane transition in mitochondria. I. The protective mechanisms. Arch Biochem Biophys 195:453-459

Haworth RA, Hunter DR (1979b) The Ca^{2+}-induced membrane transition in mitochondria. II. Nature of the Ca^{2+} trigger site. Arch Biochem Biophys 195:460-467

Haworth RA, Hunter DR (1979c) The Ca^{2+}-induced membrane transition in mitochondria. III. Transitional Ca^{2+} release. Arch Biochem Biophys 195:468-477

Haworth RA, Hunter DR (1980) Allosteric inhibition of the Ca^{2+}-activated hydrophilic channel of the mitochondrial inner membrane by nucleotides. J Membr Biol 54:231-236

Hayat LH, Crompton M (1985) Ca^{2+}-dependent inhibition by trifluoperazine of the Na^{+}-Ca^{2+} carrier in mitoplasts derived from rat heart mitochondria. FEBS Lett 182:281-286

Henry J-P, Chich J-F, Goldschmidt D, Thieffry M (1989) Blockade of a mitochondrial cationic channel by an addressing peptide: an electrophysiological study. J Membr Biol 112:139-147

Hille B (1984) Ionic channels of excitable membranes. Sinauer Associates, Sunderland

Hunter DR, Haworth RA (1980) Allosteric Inhibition of the Ca^{2+}-activated hydrophylic channel of the mitochondrial inner membrane by nucleotides. J Membr Biol 54:231-236

Hunter DR, Haworth RA, Southard, JH (1976) Relationship between configuration, function and permeability in calcium-treated mitochondria. J Biol Chem 251:5069-5077+

Hunter EF Jr, Ford L (1955) Inactivation of oxidative and phosphorylative systems in mitochondria by preincubation with phosphate and other ions. J Biol Chem 216:357-369

Jurkowitz MS, Altschuld RA, Brierley GP, Cragoe, E Jr (1983) Inhibition of Na^{+}-dependent Ca^{2+} efflux from heart mitochondria by amiloride analogues. FEBS Lett 162:262-265

Kapus A, Szaszi K, Kaldi K, Ligeti E, Fonyo A (1990) Ruthenium red inhibits mitochondrial Na^{+} and K^{+} uniports induced by magnesium removal. J Biol Chem 265:18063-18066

Kinnally KW, Campo ML, Tedeschi H (1989) Mitochondrial channel activity studied by patch-clamping mitoplasts. J Bioenerg Biomembr 21:497-506

Li X, Hegazy MG, Mahdi F, Jezek P, Lane RD, Garlid KD (1990) Purification of a reconstitutively active K^+/H^+ antiporter from rat liver mitochondria. J Biol Chem 265:15316-15322

Martin WH, Beavis AD, Garlid KD (1984) Identification of an 82,000-dalton protein responsible for K^+/H^+ antiport in rat liver mitochondria. J Biol Chem 259:2062-2065

Massari S, Azzone GF (1972) The equivalent pore radius of intact and damaged mitochondria and the mechanism of active shrinkage. Biochim Biophys Acta 283:23-29

Mitchell P (1966) Chemiosmotic coupling in oxidative and photosynthetic phosphorylation. Glynn Research Ltd, Bodmin

Mitchell P, Moyle J (1969) Translocation of some anions cations and acids in rat liver mitochondria. Eur J Biochem 9:149-155

Moran O, Sandri G, Panfili E, Stühmer W, Sorgato MC (1990) Electrophysiological characterization of contact sites in brain mitochondria. J Biol Chem 265:908-913

Nakashima RA, Garlid KD (1982) Quinine inhibition of Na^+ and K^+ transport provides evidence for two cation/H^+ exchangers in rat liver mitochondria. J Biol Chem 257:9252-9254

Nicholls D (1978) The regulation of extramitochondrial free calcium ion concentration by rat liver mitochondria. Biochem J 176:463-474

Nicolli A, Redetti A, Bernardi P (1991) The K^+ conductance of the inner mitochondrial membrane. A study of the inducible uniport for monovalent cations. J Biol Chem 266:9465-9470

Petronilli V, Szabo' I and Zoratti M (1989) The inner mitochondrial membrane contains ion-conducting channels similar to those found in bacteria. FEBS Lett 259:137-143

Pozzan T, Bragadin M, Azzone GF (1977) Disequilibrium between steady-state Ca^{2+} accumulation ratio and membrane potential in mitochondria. Biochemistry 16:5618-5625

Puskin JS, Gunter TE, Gunter, KK, Russel, PR (1976) Evidence for more than one Ca^{2+} transport mechanism in mitochondria. Biochemistry 15:3834-3842

Rizzuto R, Bernardi P, Favaron M, Azzone GF (1987) Pathways for Ca^{2+} efflux in heart and liver mitochondria. Biochem J 246:271-277

Rossi E, Azzone GF (1970) The mechanism of ion translocation in mitochondria 3. Coupling of K^+ efflux with ATP synthesis. Eur J Biochem 12:319-327

Sakmann B and Neher E (1983) Single-channel recording. Plenum Press, New York London

Scarpa A, Azzone, GF (1970) The mechanism of ion translocation in mitochondria. 4. Coupling of K^+ efflux with Ca^{2+} uptake. Eur J Biochem 12:328-335

Selwyn MJ, Dawson AP, Dunnett SJ (1970) Calcium transport in mitochondria. FEBS Lett 10:1-5

Settlemire CT, Hunter GR, Brierley GP (1968) Ion transport in heart mitochondria XIII. The effect of ethylenediamine-tetraacetate on monovalent ion uptake. Biochim Biophys Acta 162:487-499

Sorgato MC, Keller BU, Stühmer W (1987) Patch-clamping of the inner mitochondrial membrane reveals a voltage-dependent channel. Nature (London) 330:498-500

Sorgato MC, Moran O, DePinto V, Keller BU and Stühmer W (1989) Further investigation on the high-conductance ion channel of the inner membrane of mitochondria. J Bioenerg Biomembr 21:485-496

Szabo' I, Zoratti M (1991) The giant channel of the inner mitochondrial membrane is inhibited by Cyclosporin A. J Biol Chem 266:3376-3379

Thieffry M, Chich J-F, Goldschmidt D, Henry J-P (1988) Incorporation in lipid bilayers of a large conductance cationic channel from mitochondrial membranes. EMBO J 7:1449-1454

Vasington FD, Gazzotti P, Tiozzo R, Carafoli E (1972) The effect of ruthenium red on Ca^{2+} transport and respiration in rat liver mitochondria. Biochim Biophys Acta 256:43-54

Zoratti M, Petronilli V, Szabo' I (1990) Stretch-activated composite ion channels in *Bacillus subtilis*. Biochem Biophys Res Comm 168:443-450

Zoratti M, Szabo' I (1991) Channels and currents of the inner mitochondrial membrane. In: Menon J (ed) Trends in biomembranes and bioenergetics. Compilers International, Trivandrum, in press.

MECHANISMS INVOLVED IN THE CONTROL OF MITOCHONDRIAL VOLUME AND THEIR ROLE IN THE REGULATION OF MITOCHONDRIAL FUNCTION.

Andrew P. Halestrap
Department of Biochemistry
School of Medical Sciences
University of Bristol
Bristol BS8 1TD.
U.K.

Introduction

Mitochondria are organelles within the cell which, in most tissues, are primarily concerned with the oxidation of respiratory fuels to produce useful energy in the form of ATP. This is then used to drive energy requiring processes such as biosynthesis, mechanical work and solute transport. Mitochondria vary in size and shape depending on their source, but in general terms they are about 1-2 μM in diameter. They are surrounded by two phospholipid membranes, the inner one of which is elaborately folded into cristae. The outer membrane is permeable to most small molecular weight solutes, whilst the inner membrane is impermeable to the majority of solutes except those for which specific permeability pathways exist. Such transport is catalysed by proteins which span the membrane and are called carriers or transporters. The impermeability of the inner membrane to small solutes such as sucrose allow mitochondria to act as perfect osmometers; increasing the extramitochondrial osmolality causes the mitochondria to shrink in order to maintain the osmotic balance, whilst decreasing the osmolality causes swelling and unfolding of the cristae. These responses can be followed by measuring the light scattering of a mitochondrial suspension. Light scattering decreases as the mitochondrial volume increases (see Halestrap, 1989).

Osmoregulation of Mitochondria

Within the cell the osmolality of the cytosol is usually maintained at about 300mOsmolal, and thus the mitochondrial volume might be expected to remain constant. However this is not necessarily the case. Mitochondria maintain an internal pH some 0.4 pH units higher than the cytosol and a membrane potential across the inner membrane of about -180mV. Specific transport processes exist within the inner mitochondrial membrane to allow K^+ ions to enter electrogenically or to be pumped out in exchange for a proton. Movements of anions such as phosphate can occur by means of other carriers allowing charge and pH compensation to occur. The resulting net changes in osmotic balance induce changes in the matrix volume through rapid movement of water. Thus regulation of these K^+ permeability pathways will lead to

NATO ASI Series, Vol. H 64
Mechanics of Swelling
Edited by T. K. Karalis

changes in the mitochondrial matrix volume even at constant cytosolic osmolality (see Garlid, 1988; Halestrap, 1989). When isolated mitochondria are challenged with an osmotic insult, they endeavour to restore their original matrix volume by osmoregulation. A shrinking of mitochondria caused by hyperosmotic challenge leads to a decrease in the activity of the transport mechanism that pumps K^+ out of the matrix. K^+ entry continues on a separate carrier and leads to an increase in the osmolality of the matrix and consequent re-uptake of water. The opposite process occurs during a hypo-osmotic challenge. Regulation of the K^+ antiport may be achieved by $[Mg^{2+}]$, since the matrix concentration of this cation, which inhibits the pump, increases and decreases on shrinking and swelling respectively. These regulatory mechanisms are summarised in Fig. 1 and are discussed in more detail elsewhere (Garlid,1988; Halestrap, 1989).

The mitochondrial volume can regulate mitochondrial function in vitro and in vivo

Work from this and other laboratories (reviewed in Halestrap, 1989 and Halestrap *et al.*, 1990a) has shown that a large number of mitochondrial processes are activated as the mitochondrial volume is increased over quite a narrow range. These data are summarised in Fig. 2. The protocol used in these experiments was to vary the

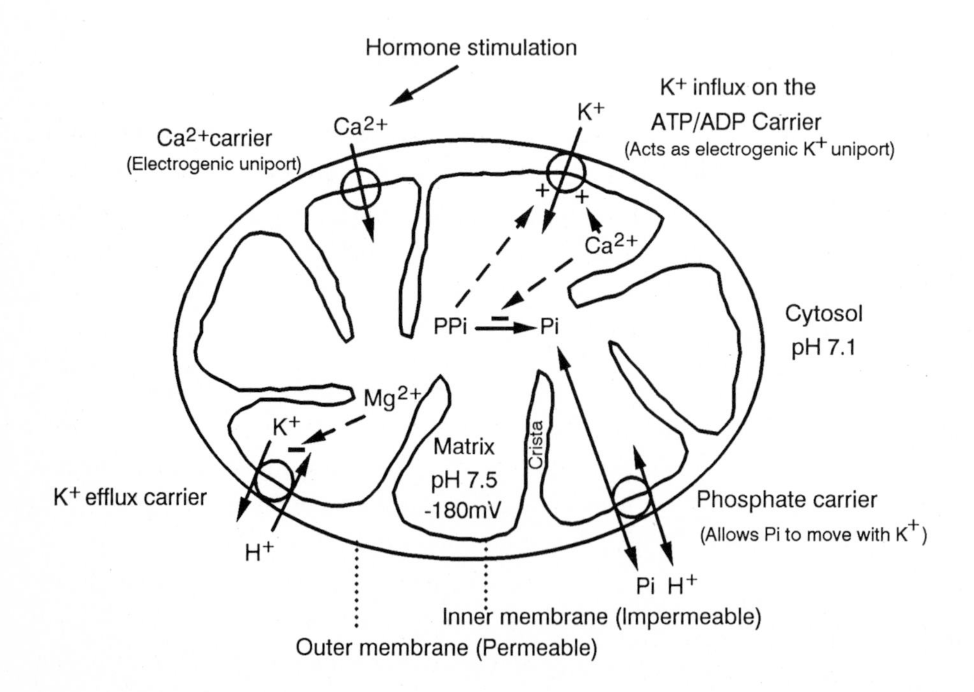

Fig. 1 Mechanisms involved in the regulation of mitochondrial matrix volume

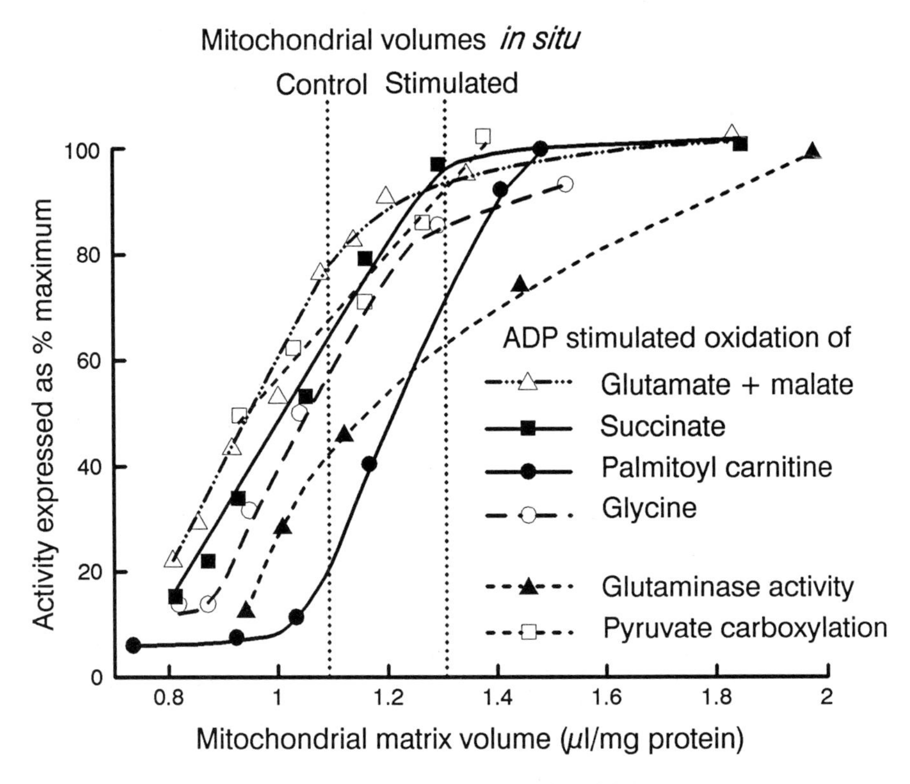

Fig. 2 The effects of matrix volume on various mitochondrial processes.

volume of the mitochondrial matrix by changing the osmolality of the the mitochondrial incubation medium. Whilst this procedure is clearly not physiological, the results would appear to be independent of the method used since low concentrations of valinomycin, a K^+ ionophore, elicit a similar response. Thus it is likely that physiological changes in mitochondrial volume (see below) would have the same effects. The mitochondrial processes that are sensitive to the matrix volume fall into three categories. Firstly electron flow through the respiratory chain into the ubiquinone pool which causes stimulation of respiration and thus ATP production. Secondly processes dependent on that increased ATP production such as pyruvate carboxylation and citrulline synthesis and which are probably regulated by the ATP/ADP ratio. Thirdly other enzymes such as glutaminase and the glycine cleavage enzyme which are involved in amino acid metabolism and are associated with the inner mitochondrial membrane (Halestrap *et al.*, 1990a).

The mechanism by which a change in matrix volume can regulate the respiratory chain and other membrane associated enzymes remains unclear and requires future research. Two mechanisms can be suggested for further investigation (Halestrap, 1989; Halestrap *et al.*, 1990a). Firstly there might be a change in protein/protein interaction as the cristae unfold during swelling, or secondly a change in membrane fluidity might occur. In cell membranes there are stretch receptors which detect the stresses on the plasma membrane produced by very modest cell swelling (Christensen, 1987; Bear, 1990), but it must be remembered that the cell membrane does not have the potential to unfold cristae that the mitochondrial inner membrane possesses. Furthermore the plasma membrane may be constrained by cytoskeletal interactions absent in mitochondria.

Whatever the mechanisms behind volume mediated regulation of mitochondrial function it is clear that it has the potential to be of physiological importance (Halestrap, 1989; Halestrap *et al.*, 1990a). Thus measurement of the liver mitochondrial matrix volume *in situ* using radioactive extramitochondrial markers and ^3H_2O has shown that it lies in the region where mitochondrial function shows its steepest dependence on mitochondrial volume (see Fig. 2). In liver cells we have shown that the mitochondrial volume can be increased by those hormones which stimulate respiration, fatty acid oxidation, amino acid degradation and glucose and urea synthesis. Since these processes require increased mitochondrial activity of those functions known to be stimulated by an increase in mitochondrial volume, it seems probable that such volume changes have an important physiological role. Furthermore, low concentrations of the potassium ionophore valinomycin can mimic many of the effects of the hormones. Recently it has been shown that some swelling of liver cells accompanies uptake of amino acids and that this has profound effects on cellular metabolism similar to those that would be predicted for an increase in mitochondrial volume (Häussinger *et al.,* 1990). The effects may well be secondary to a rise in cytosolic [Ca^{2+}] (Baquet *et al.*, 1991).

Liver mitochondria are something of a special case amongst mitochondria in that they act in more capacities than just ATP generation, being involved in the metabolic pathways listed above. The hormonal stimulation by Ca^{2+} of both NADH production to fuel the respiratory chain, and of the respiratory chain through the increase in matrix volume play an important role in making this possible as is summarised in Fig. 3. Together they allow respiration and thus ATP production to be stimulated without an increase in either the NADH/NAD^+ or ADP/ATP ratio (McCormack *et al.*, 1990). Increasing these ratios has usually been regarded as the only way of stimulating respiration, but in the liver an increase in either of them could inhibit the very metabolic pathways that need to be stimulated (Halestrap & Owen, 1991). At present we have only indirect evidence that increases in the mitochondrial volume are important in other tissues (Halestrap, 1989. However it would seem likely that most tissues would benefit from being able to stimulate respiration in response to a stimulus without the need for changes in the ATP/ADP and NADH/NAD^+ ratios (McCormack *et al.*, 1990).

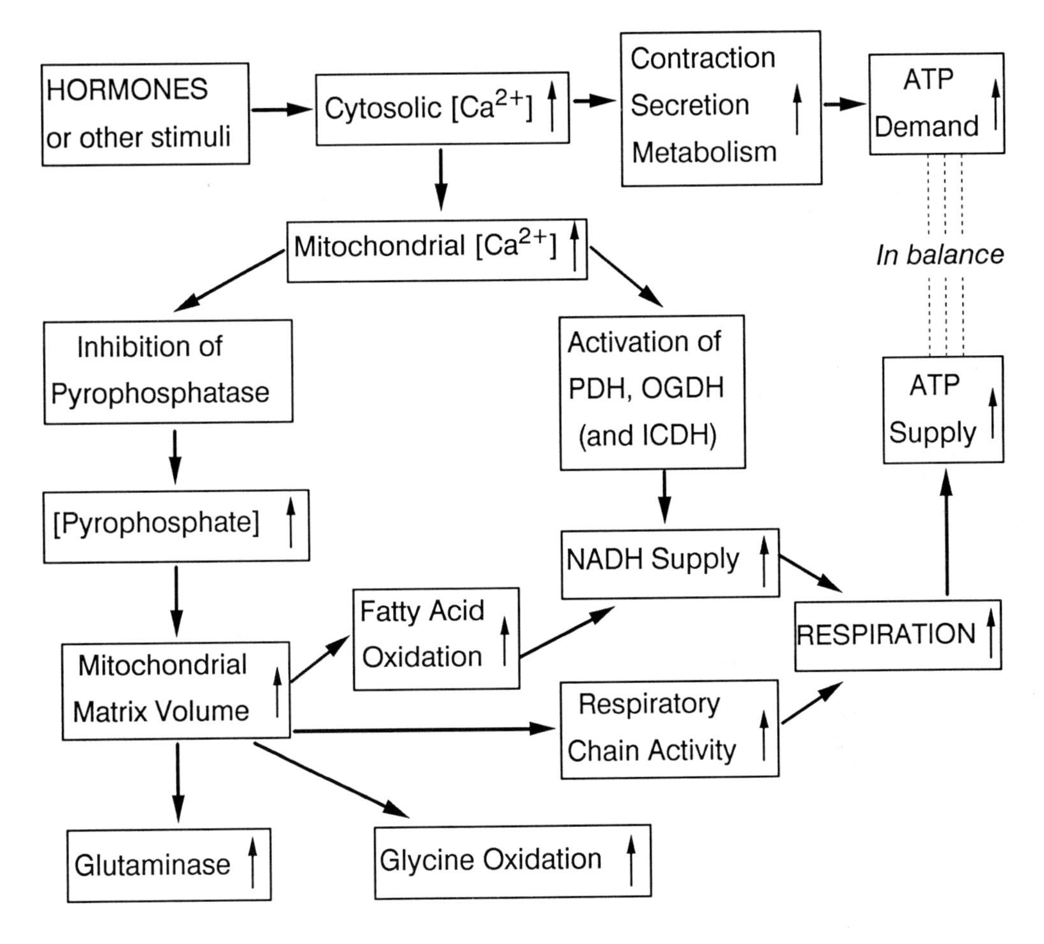

Fig. 3 Mechanisms involved in the regulation of mitochondrial metabolism in a stimulated cell.

Mechanisms involved in the regulation of liver mitochondrial volume by hormones

We have proposed above that an increase in the mitochondrial matrix volume of some 20-40% plays an important part in the hormonal regulation of liver metabolism. We have measured such changes in volume *in situ* both by the use of radioactive markers or by the use of light scattering (Quinlan *et al*, 1983). The diameter of the hepatocyte is such that changes in its volume cause only small light scattering changes. However the mitochondria within the cell have a profound effect on light scattering and the time course of their increase in volume can be followed by this technique. Use of electron microscopy to measure changes in matrix volume is difficult. Even a 30% increase in mitochondrial volume would only be associated with a 9% increase in diameter which would be hard to detect. It is also difficult to discriminate between the true matrix volume and the intermembrane and intercristae spaces. Nevertheless, although such techniques are inappropriate to the liver, there are cell types where an increase in mitochondrial volume can be seen by electron microscopy following hormonal or other stimulation of the cell (Halestrap, 1989).

Using the techniques outlined above we have demonstrated that hormonally induced increases in mitochondrial volume are mediated through increases in cytosolic [Ca^{2+}] which occur upon hormone stimulation. The response requires the presence of extracellular Ca^{2+} which must enter the cell. The resulting increase in cytosolic [Ca^{2+}] is relayed into the mitochondrial matrix where it acts to increase the permeability of the inner membrane to K^+ and so cause swelling (Halestrap *et al.*, 1986). The process can be mimicked using isolated mitochondria *in vitro* by adding sub-micromolar [Ca^{2+}] to energised mitochondria incubated in KCl (but not sucrose) media. In both hepatocytes and isolated mitochondria the increase in volume was shown to be correlated in time and magnitude with an increase in matrix pyrophosphate (PPi) concentration (Davidson & Halestrap, 1987,1988). An increase in matrix volume *in vitro* and *in situ* could also be induced by butyrate, which stimulates PPi production in mitochondria independently of [Ca^{2+}]. The rise in matrix [PPi] induced by Ca^{2+} is almost certainly the result of inhibition of the mitochondrial pyrophosphatase. We and others (Davidson & Halestrap, 1989; Baykov *et al.*, 1989) have shown that this enzyme is sensitive to inhibition by Ca^{2+} in the micromolar range. This inhibition is mediated through the formation of CaPPi which is a very potent competitive inhibitor with respect to the normal substrate MgPPi. The source of mitochondrial PPi remains unclear but it appears to be synthesised only slowly and may be a by-product of normal "house-keeping" metabolism by mitochondria (Halestrap, 1989). In contrast when butyrate is metabolised by mitochondria it produces large quantities of PPi, and when both butyrate and Ca^{2+} are added together to either isolated liver cells or mitochondria the matrix PPi rises some twentyfold (Davidson & Halestrap, 1987,1988).

The association of matrix [PPi] with mitochondrial volume changes, and the inhibitory effect of ATP on Ca^{2+}-induced swelling have led us to propose that the adenine nucleotide translocase of the inner mitochondrial membrane is involved in mediating the increase in K^+-permeability (Halestrap *et al.*, 1986; Davidson & Halestrap, 1990; Halestrap & Davidson 1989,1990). PPi is a substrate for the translocase and there is evidence from other laboratories that this carrier may be capable of providing a channel through the inner membrane for small molecules (Panov *et al.*, 1980; Dierks *et al.*, 1990). The adenine nucleotide carrier has also been implicated in the damage to mitochondria caused by Ca^{2+}-overload as discussed below and a hypothetical model incorporating these ideas is presented in Fig. 5.

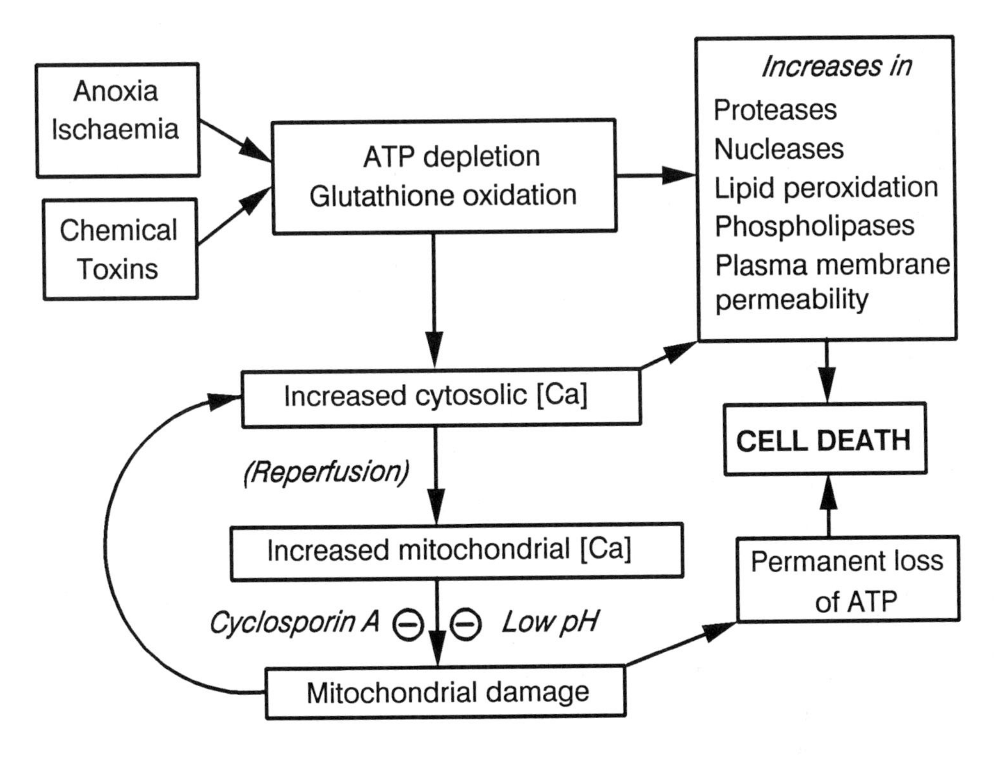

Fig. 4 The effects of chemical or hypoxic damage to cells that lead to cell death

The mitochondrial swelling and damage that is associated with cellular injury and death is mediated by a Ca^{2+}-induced non-specific pore.

When cells are insulted with hypoxia or chemicals that deplete the cell of ATP, cellular [Ca^{2+}] increases as the ATP-dependent pumps that normally expel excess Ca^{2+} from the cell are unable to operate. At these higher [Ca^{2+}] concentrations mitochondria become overloaded with Ca^{2+}, especially if they are energised as may occur under conditions of reperfusion following a period of ischaemia. This leads to massive swelling and irreversible damage to mitochondria which is a prelude to cell death (Orrenius *et al.*, 1989; Crompton, 1990; Herman *et al.*, 1990). The massive swelling is a consequence of the opening of a non-specific pore for small molecular weight solutes which become freely permeant (Crompton, 1990). The protein remaining trapped in the matrix exerts a colloidal osmotic pressure which causes water to enter and swell the mitochondria. Loss of essential cofactors contributes to make the mitochondria incompetent. A summary scheme showing events associated with cell death is shown in Fig. 4.

The nature of the non-specific pore that is opened in response to Ca^{2+} overload of mitochondria remains to be established, but several features of it are widely accepted

(Le Quoc & Le Quoc, 1988; Crompton, 1990; Gunter & Pfeiffer, 1990; Halestrap & Davidson, 1990). It accepts molecules of upto about 1500 molecular weight, requires the presence of Ca^{2+} to open, is blocked by the presence of the immunosuppressant cyclosporin A and is associated in some way with both matrix peptidyl-prolyl cis-trans isomerase (PPIase) and the adenine nucleotide translocase. The latter conclusion is based on several observations. Pore opening is greatly enhanced under conditions of adenine nucleotide depletion and by the presence of phosphate which can catalyse adenine nucleotide loss from mitochondria. Carboxyatractyloside and fatty acyl-CoA increase pore opening and these agents stabilise the carrier in the "c" conformation whilst ADP and bongkrekic acid, which stabilise the carrier in the "m" conformation inhibit pore opening. Furthermore Ca^{2+} can interact with the carrier in the "c" conformation to produce a conformational change (Halestrap & Davidson, 1990). The involvement of mitochondrial PPIase has been demonstrated by showing that there are the same number of sites of the PPIase per mg of mitochondrial protein as there are sites for the inhibition of swelling by cyclosporin (Halestrap & Davidson, 1990). Furthermore the Ki values for inhibition by a variety of cyclosporin analogues for the mitochondrial PPIase correlate well with their ability to block pore opening (Griffiths & Halestrap, 1991). These observations have led us to present a hypothetical model of the mechanism of pore opening which is summarised in Fig. 5.

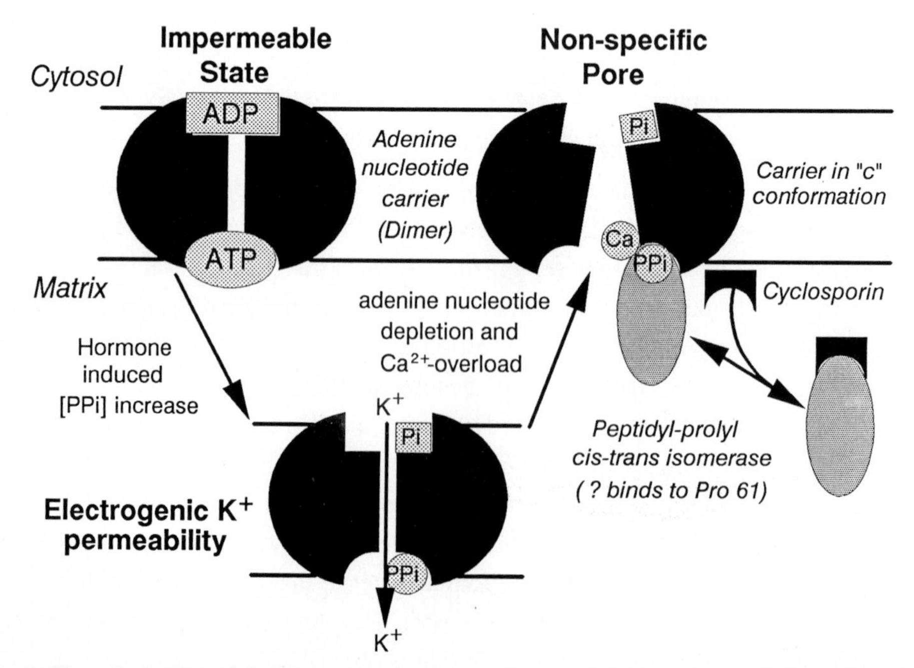

Fig. 5 Hypothetical model of how the adenine nucleotide translocase may function physiologically as a K^+ channel and also as a non-specific pore under conditions of Ca^{2+} overload.

We propose that in the presence of Ca^{2+} and under conditions of adenine nucleotide depletion the adenine nucleotide translocase takes up the "c" conformation and binds the isomerase to a proline residue exposed on the matrix face of the carrier. The resulting conformational change causes the opening of the pore. Work is in progress to test this hypothesis directly by using purified adenine nucleotide translocase and PPIase. We have purified the latter to homogeneity and performed an N-terminal amino acid sequence. In liver mitochondria there appear to be 50-60pmol PPIase per mg mitochondrial protein with a subunit molecular weight of 17.5kDa. The enzyme has strong homologies with the rat cytosolic enzyme but has a 14 amino acid N-terminal presequence (unpublished data of C.P. Connern & A.P. Halestrap).

There is some evidence to suggest that cyclosporin protects cells from ischaemia followed by reperfusion (see Goto *et al.*, 1990) which would be expected if mitochondrial Ca^{2+}-overload and consequent pore opening was an important factor in the resulting tissue damage. However a much better protective regime is lowering of the perfusion pH which also enhances recovery during re-oxygenation (see Herman *et al.*, 1990). This has now been demonstrated in heart, liver, kidney and tumour cells. Low pH tends to accompany hypoxia because of the accumulation of lactic acid within the cell.

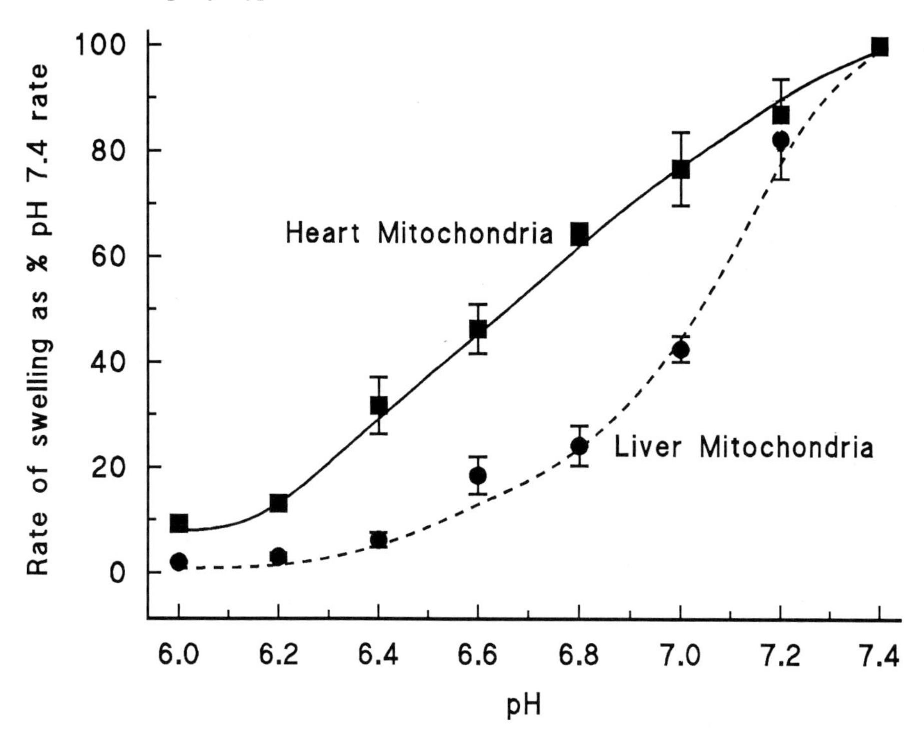

Fig. 6 The effects of pH on Ca^{2+}-induced pore opening of rat heart and liver mitochondria. The experimental protocol was as described by Halestrap & Davidson (1990)

However upon reperfusion this lactic acid will be transported out of the cell on the monocarboxylic acid transporter with a resultant alkalinisation of the cytosol (Halestrap *et al.*, 1990b). I have demonstrated that the opening of the Ca-sensitive pore within the inner membrane of both rat heart and liver mitochondria is inhibited greatly as the pH drops below 7 (Halestrap, 1991) as is shown in Fig. 6. It would seem likely that this may play an important role in the protective effect of low pH against chemical and hypoxic tissue damage

A physiological role for pore opening

Under normal physiological conditions the opening of the non-specific pore must be extremely limited, otherwise massive swelling and damage to mitochondria would result. Nevertheless there is evidence that it does occur. Firstly hormonally induced light scattering changes in hepatocytes are modulated by cyclosporin suggesting that at the cytosolic [Ca^{2+}] present in stimulated cells some pore opening is occurring (Davidson & Halestrap, 1990). Secondly [^{14}C]-sucrose introduced into hepatocytes by electro-permeabilisation enters the mitochondria over a period of hours (Tolleshaug & Seglen, 1985). Thirdly cells stained with Rhodamine G, a dye which is concentrated within mitochondria in response to their membrane potential, show oscillations in the Rhodamine fluorescence, indicating that the mitochondria may undergo cycles of energisation and de-energisation (Bereiter-Hahn *et al,* 1983). Fourthly contact sites between inner and outer membranes are enriched in the adenine nucleotide translocase (Brdiczka *et al.*, 1990; Bucheler *et al.,* 1991) and are involved in protein import which requires pore formation (Pfanner *et al.* 1990). In addition it may be significant that prolonged cyclosporin therapy can lead to severe liver and kidney damage. This is associated with changes in mitochondrial morphology and is less severe with cyclosporin G than cyclosporin A, whilst FK-506 does not suffer from these side effects. This correlates with the potency of these immunosuppressants as inhibitors of pore formation (Griffiths & Halestrap, 1991). Thus it is possible that occasional pore opening is a normal part of mitochondrial function, perhaps for protein import or for removal of waste products for which specific transporters do not exist. Total blockage would then be damaging to the mitochondria and lead to the observed side effects.

Future research

Future research will concentrate on three major areas. Firstly the molecular identification and characterisation of the K^+ transport pathways of mitochondria, and further elucidation of the mechanisms by which they are controlled. Secondly the mechanisms by which very small changes in mitochondrial matrix volume can induce large changes in the rates of mitochondrial processes. Thirdly further elucidation of the mechanisms of non-specific pore opening, with the possibility of protecting cells from ischaemia/reperfusion injury.

Acknowledgements This work was supported by the Medical Research Council

References

Baquet A, Meijer AJ, Hue L (1991) Hepatocyte swelling increases inositol 1,4,5-trisphosphate, calcium and cyclic AMP concentration but antagonizes phosphorylase activation by Ca^{2+}-dependent hormones. FEBS Lett 278:103-106

Baykov AA, Volk SE, Unguryte A (1989) Inhibition of inorganic pyrophosphatase of animal mitochondria by calcium. Arch Biochem Biophys 273:287-291

Bear CE (1990) A nonselective cation channel in rat liver cells is activated by membrane stretch. Am J Physiol 258:C421-C428

Bereiter-Hahn J, Seipel K-H, Voth M, Ploem JS (1983) Fluorimetry of mitochondria in cells vitally stained with DASPMI or Rhodamine 6 GO. Cell Biochem Function 1:147-155

Brdiczka D, Bucheler K, Kottke M, Adams V, Nalam VK (1990) Characterization and metabolic function of mitochondrial contact sites. Biochim Biophys Acta 1018:234-238

Bucheler K, Adams V, Brdiczka D (1990) Localization of the ATP/ADP translocator in the inner membrane and regulation of contact sites between mitochondrial envelope membranes by ADP. A study on freeze-fractured isolated liver mitochondria. Biochim Biophys Acta 1056:233-242

Christensen O (1987) Mediation of cell volume regulation by Ca^{2+} influx through stretch activated channels. Nature 330:66-68

Crompton M (1990) The role of Ca^{2+} in the function and dysfunction of heart mitochondria. In: Langer GA (ed.) Calcium and the Heart. Raven Press Ltd., New York, pp 167-198

Davidson AM, Halestrap AP (1987) Liver mitochondrial pyrophosphate concentration is increased by Ca^{2+} and regulates the intramitochondrial volume and adenine nucleotide content. Biochem J 246:715-723

Davidson AM, Halestrap AP (1988) Inorganic pyrophosphate is located primarily in the mitochondria of the hepatocyte and increases in parallel with the decrease in light-scattering induced by gluconeogenic hormones, butyrate and ionophore A23187. Biochem J 254:379-384

Davidson AM, Halestrap AP (1989) Inhibition of mitochondrial matrix inorganic pyrophosphatase by physiological [Ca^{2+}], and its role in the hormonal regulation of mitochondrial matrix volume. Biochem J 258:817-821

Davidson AM, Halestrap AP (1990) Partial inhibition by cyclosporin A of the swelling of liver mitochondria *in vivo* and *in vitro* induced by sub-micromolar [Ca^{2+}] but not by butyrate. Evidence for two distinct swelling mechanisms. Biochem J 268:147-152

Dierks T, Salentin A, Heberger C, Krämer R (1990) The mitochondrial Aspartate/Glutamate and ADP/ATP carrier switch from obligate counter-exchange to unidirectional transport after modification by SH-reagents. Biochim Biophys Acta 1028:268-280

Garlid KD (1988) Mitochondrial volume control. In: Lemasters JJ, Hackenbrock CR, Thurman RG, Westerhoff HV (eds) Integration of Mitochondrial Function. Plenum, New York, pp 257-276

Goto S, Kim YI, Kamada N, Kawano K, Kobayashi M (1990) The beneficial effects of pretransplant cyclosporine therapy on recipient rats grafted with a 12-hour cold-stored liver. Transplantation 49:1003-1005

Griffiths EJ, Halestrap AP (1991) Further evidence that cyclosporin-A protects mitochondria from calcium overload by inhibiting a matrix peptidyl-prolyl cis-trans isomerase - Implications for the immunosuppressive and toxic effects of cyclosporin. Biochem J 274:611-614

Gunter TE, Pfeiffer DR (1990) Mechanisms by which mitochondria transport calcium. Am J Physiol 258:C755-C786

Halestrap AP (1989) The regulation of the matrix volume of mammalian mitochondria *in vivo* and *in vitro*, and its role in the control of mitochondrial metabolism. Biochim Biophys Acta 973:355-382

Halestrap AP (1991) Calcium dependent opening of a non-specific pore in the mitochondrial inner membrane is inhibited by pHs below 7. Biochem J In Press

Halestrap AP, Davidson AM (1989) The role of the adenine nucleotide transporter in the regulation of intramitochondrial volume. In: Azzi A, Fonyo A, Nalecz MJ, Vignais PV, Wojtcak L (eds) Anion carriers of mitochondrial membranes Springer Verlag, Berlin, pp337-348

Halestrap AP, Davidson AM (1990) Inhibition of Ca^{2+}-induced large amplitude swelling of liver and heart mitochondria by Cyclosporin A is probably caused by the inhibitor binding to mitochondrial matrix peptidyl-prolyl cis-trans isomerase and preventing it interacting with the adenine nucleotide translocase. Biochem J 268:154-160

Halestrap AP, Owen MR (1991) Hormonal control of respiration in the liver. In Grunnet N, Quistorff B (eds) Regulation of Hepatic Function. Metabolic and Structural Interactions. Munksgaard, Copenhagen, pp198-213

Halestrap AP, Quinlan PT, Whipps DE, Armston AE (1986) Regulation of the mitochondrial matrix volume *in vivo* and *in vitro*. The role of calcium. Biochem J 236:779-787

Halestrap AP, Davidson AM, Potter WD (1990a) Mechanisms involved in the hormonal regulation of mitochondrial function through changes in the matrix volume. Biochim Biophys Acta 1018:278-281

Halestrap AP, Poole RC, Cranmer SL (1990b) Mechanisms and regulation of lactate, pyruvate and ketone body transport across the plasma membrane of mammalian cells and their metabolic consequences. Biochem Soc Trans 18:1132-1135

Häussinger D, Lang F, Bauers K, Gerok W (1990) Control of hepatic nitrogen metabolism and glutathione release by cell volume regulatory mechanisms. Eur J Biochem 193:891-898

Herman B, Gores GJ, Nieminen AL, Kawanishi T, Harman A, Lemasters JJ (1990) Calcium and pH in anoxic and toxic injury. Critical Rev Toxicol 21:127-148

Le Quoc K, Le Quoc D (1988) Involvement of the ADP/ATP carrier in calcium-induced perturbations of the mitochondrial inner membrane permeability: importance of the orientation of the nucleotide binding site. Arch Biochem Biophys 265:249-257

McCormack JG, Halestrap AP, Denton RM (1990) The role of calcium ions in the regulation of mammalian intramitochondrial metabolism. Physiol Rev 70:391-425

Orrenius S, McConkey DJ, Bellomo G, Nicotera P (1989) Role of Ca^{2+} in toxic cell killing. Trends Pharmacol Sci 10:281-285

Panov A, Filippova S, Lyakhovich V (1980) Adenine nucleotide translocase as a site of regulation by ADP of the rat liver mitochondrial permeability to H^+ and K^+ ions. Arch Biochem Biophys 199:420-426

Pfanner N, Rassow J, Wienhues U, Hergersberg C, Sollner T, Becker K, Neupert W (1990) Contact sites between inner and outer membranes - Structure and role in protein translocation into the mitochondria. Biochim Biophys Acta 1018:239-242

Quinlan PT, Thomas AP, Armston AE, Halestrap AP (1983) Measurement of the intramitochondrial volume in hepatocytes without cell disruption and its elevation by hormones and valinomycin. Biochem J 214:395-404

Tolleshaug H, Seglen PO (1985) Autophagic-lysosomal and mitochondrial sequestration of [^{14}C]-sucrose. Density gradient distribution of sequestered radioactivity. Eur J Biochem 153:223-229

THE DEFORMATION OF A SPHERICAL CELL SHEET; A MECHANICAL MODEL OF SEA-URCHIN GASTRULATION

Daniel Zinemanas* and Avinoam Nir**
Departments of Biomedical* and Chemical Engineering**
Technion, Israel Institute of Technology
Haifa 32000, ISRAEL

Introduction

An important aspect in the study and understanding of the biochemical and biophysical mechanism that control morphogenetic processes, e.g. embryo gastrulation, is the experimental as well as theoretical search for the forces that cause the observed morphological changes. This is of valuable help since the appearance of active forces is a direct indication of regions of particular or enhanced cellular activity.

Gastrulation is the process of cell rearrangement during which the ectoderm, endoderm and mesoderm are formed. Morphologically this process is characterized by the appearance and development of an invagination which originates at the vegetal pole and advances toward the animal pole. This invagination proceeds in two separate stages: primary and secondary invaginations (Gustafson and Kinnander 1956). During primary invagination, the archenteron is formed reaching an inward length of 1/5 to 1/2 of the vegetal to animal pole distance depending on the sea urchin species. After a pause, secondary invaginations follow which are characterized by the elongation and thinning of the archenteron until it reaches the blastocoel roof. At the onset of this second stage filopodia extend from secondary mesenchymal cells, located around the tip of the archenteron, to the inner blastocoel wall and supposedly contract in order to pull the archenteron

Although much is known on the role and distribution of the particular cells actively involved in both primary and second invaginations much less is known regarding the origin

NATO ASI Series, Vol. H 64
Mechanics of Swelling
Edited by T. K. Karalis

and nature of the forces that induce the epithelial deformations. Regarding primary invaginations, the fact that bisected embryos, deprived of their animal pole, show similar ability to deform (Moore and Burt 1939) indicates that the forces producing this primary invagination are confined to the epithelial layer at the vegetal pole of the blastula. Odell et al. (1981) have successfully simulated primary invaginations using a model that describes the active forces originating from the presence of an array of microfilament bundles in the apical surface of the epithelial cells. The rest of the cell is modelled as a viscoelastic body of constant volume and the contractions are triggered by an increase of the apical circumference. In fact this asymmetry of the location of tension elements implies that the macroscopic ensemble of filament tension is manifested as surface moments, the resutls of which may be seen in the experiments of Moore and Burt (1939).

Classically, it was accepted that the sole source of forces producing secondary invaginations lies with the filopodia pulling (Gilbert 1988). Hardin and Cheng (1986) noted the evidence that chemical and mechanical treatment which disrupt filopodia formation, interfere with gastrulation. However, such treatments may affect other processes occurring during secondary invagination. Therefore it was also noted that, although it is clear that the filopodia generate tension (Gustafson 1964), it was not obvious that filopodial contraction is a sufficient mechanism for the elongation of the gut rudiment. Hardin and Cheng (1986) have undertaken an experimental and theoretical study to investigate this hypothesis. They concluded that filopodia pulling cannot explain the elongation of the archenteron but that active forces which cause rearrangement of endodermal cells within the archenteron are of major importance. Their conclusion is based on mechanical simulations showing correct shape deformations only when the archenteron is much less stiff than the rest of the epithelial layer of the blastula. The relative importance of a pulling mechanism is also supported by the experimental findings showing that when exogastrulation is induced *in vitro* by chemical treatment of the embryo the outward archenteron elongation is achieved in the absence of filopodia. Further support to their ideas is provided by the fact that cells in the archenteron are repacked and not stretched as would be expected in a purely filopodial pulling mechanism. Filopodial role was thus suggested by Hardin and Cheng (1986) to be only of a guidance nature leading the archenteron to the site of stomadeum formation. Keller (1978) also suggested that the thinning of the wall of the archenteron could result from active cell spreading, and not necessarily produced by stretching.

The theoretical calculations of Hardin and Cheng (1986) were performed using a mechanical model for the epithelial layer. The solid mechanics model was solved using a procedure (Cheng 1987) which takes into account only surface tractions while the dissipation in the inner and outer media is neglected. This type of calculations provide only static shapes for intermediate stages and the equations cannot provide the deformation dynamics. It must also be noted that results of their simulation were shown only in a qualitative graphic way. The limitation can be overcome by using an hydrodynamic description of the system and by solving the corresponding mass and momentum balances together with the surface tractions (Zinemanas and Nir 1988) . In this communication, we present such an hydrodynamic model of gastrulation and focus our study on the distribution and characteristics of the macroscopic forces required to cause deformations of the type encountered in both normal and exogastrulations.

We examine the possible role of surface forces and moments in primary invagination and the possibility that filopodial pulling can be a valid mechanims inducing the stage of secondary invagination. In the following sections we give a brief description of the hydrodynamic model and the numerical procedure. We describe the possible forces and the dynamics which may be involved in exogastrulation. The role of filopodial pulling during normal gastrulation is investigated and compared with inward deformations due to active cell spreading.

Hydrodynamical Model and Numerical Procedure

The hydrodynamic model of gastrulation presented here is basically an extension of the one used by Zinemanas and Nir (1988,1989) in the simulations of cell cleavage, i.e., the embryo is assumed to be modelled by an incompressible viscous body surrounded by a thin enveloping layer and suspended in an ambient viscous fluid. This simplification is justified on the basis that although the inner fluid may have a particular rheological behaviour different than that of a purely viscous fluid, it is found that in slow processes the role of the internal fluid is primarily of a dissipative nature. For example, during cell cleavage substitution of the internal cytoplasm by a physiological solution (Hiramoto 1956) does not alter the phenomenology of the cytokinetic process. This is mainly due to the fact that the active forces inducing the deformations are solely confined to the surface layer, thus the particular rheological properties of the internal fluid are of secondary importance. A similar situation is encountered also in the embryo gastrulation process where the active forces originate or act on the surface regardless of the different hypothesis concerning the filopodial pulling role. Secondly, since the epithelial layer of the blastula is thin compared to the dimensions of the blastula modelling it may be well approximated by a two dimensional layer and can be described by thin shell membrane theories.

In the absence of inertial effects the slow evolution of embryo deformations are described, both in the inner and outer regions, by the equations of creepng motion together with the continuity equations and the appropriate boundary conditions which stem for the continuity of velocity and the jump of surface tractions across the interfacial layer. The set of equations and their respective boundary conditions can be reformulated as a boundary integral equation for the surface velocity, v, which assumes a simple form when equal inner and outer fluid viscosities, μ, are considered and when the outer medium is at rest far away from the body (Rallison and Acrivos 1978)

$$\mathbf{v}(\mathbf{x},t) = -\frac{1}{8\pi\mu}\int_{\partial\beta} \mathbf{J}(\mathbf{x}-\mathbf{y})\cdot\mathbf{f}(\mathbf{y},t)\; dS_y \tag{1}$$

Here **J** is the single layer potential that represents the velocity field at **x** due to a unit force applied at **y** and **f** denotes the tractions on the embryo surface, ∂B.

The dynamic evolution of the *a priori* unknown free surface location is provided by the kinematic condition

$$\frac{D\mathbf{R}}{Dt} = \mathbf{v} \tag{2}$$

where $\frac{D}{Dt}$ denotes a material derivative and R is a position vector that specifies the location of the interfacial layer.

To complete the mathematical formulation of the model, a description of the surface forces is necessary. The general expression for these forces, in the absence of inertial effects and moments normal to the surface, is given by (Waxman 1984)

$$\mathbf{f}(\mathbf{x},t) = (\gamma^{\alpha\beta}\big|_{\beta} - b^{\alpha}_{\delta} M^{\beta\delta}\,\big|_{\beta})\,\mathbf{R},_{\alpha} + (b_{\alpha\beta}\,\gamma^{\alpha\beta} - M^{\beta\alpha}\,\big|_{\alpha\beta})\,\mathbf{n} \tag{3}$$

Here $\gamma^{\alpha\beta}$ represents the surface stress tensor, $M^{\alpha\beta}$ the moment tensor and $b_{\alpha\beta}$ the second fundamental form of the surface. A comma and a bar indicate a common and a covariant derivative, respectively,. and $R,_{\alpha}$ and $\mathbf{n}$ are the basis vectors tangent and normal to the surface.

As stated previously (Zinemanas and Nir 1988,1989) the surface tractions may generally be of an active or passive nature so that both the surface tension and moment tensors are functions of the surface biochemical composition and structure as well as the interface rheological properties. The latter may be expressed through the constitutive equation that relates the stresses to the strain, rate of strain and curvature changes for a particular rheological behaviour.

Since the nature and origin of the forces acting during gastrulation are still not well understood, specific active force distributions were used in the simulations presented below. The characteristics of those distributions are discussed later. The surface tractions are approximated by nonhomogeneous surface tensions forces. This assumption is supported by the experimental findings showing a high degree of reorganization and active spreading of the cells that compose the archenteron during the secondary invagination (Hardin and Cheng 1986). Such degree of mobility is characterisitc of a fluid interface and therefore such an approximation should be valid.

Given the force distributions, the numerical solution of the equation of motion (1) and the kinematic condition (2) is calculated using the following scheme:

(a) Surface forces and shape are assumed and the interfacial velocity is calculated from equation (1) using a finite difference scheme by dividing the surface into N axisymmetric ring intervals.

(b) Using this new velocity at each surface collocation point, the shape is up-dated to its new position according to a Lagrangian scheme to integrate equation (2)

$$\mathbf{R}(\mathbf{x},t+\Delta t) = \mathbf{R}(\mathbf{x},t) + \frac{D\mathbf{R}(\mathbf{x},t)}{Dt}\Delta t \qquad (4)$$

where Δt is a time increment. A similar Newtonian integration scheme is used in the calculation of the dynamic evolution of all other variables such as surface strains and concentrations.

(c) Given the new shape, surface forces are recalculated and the next time increment is calculated (steps a and b).

A detailed description of the numerical procedure and calculation of the velocity from equation (1) may be found in Zinemanas and Nir (1988). The additional equations required to study the effects of surface rheological properties are given by Zinemanas and Nir (1989).

Exogastrulation

Exogastrulation can be induced *in vitro* by chemically treating the blastula. The outward extension of the archenteron clearly indicates that this deformation proceeds in the absence of a filopodial pullling mechanism. Thus a mechanism based on surface forces and moments should be involved. Odell et al. (1981) have modelled surface epithelial invagination using preferred and ordered apical contractions and stretching which can lead to inward and outward protrusions. In the case of exogastrulation, however their model does not agree well with the experimental findings and finally leads to the formation of an inward invagination. This may suggest that the chemical treatment that induces exogastrulation changes the characteristic of the surface active forces. In this section we present results of simulations of exogastrulae (Zinemanas and Nir 1991) which are produced by using a tension profile (Figure 1b) based on the static profile required to maintain a fully elongated archenteron (Figure 1a). The main particularity of this profile resides in the anisotropic tension distribution that shows a high level of tension in the circumferential direction at the archenteron base. It is also noted that the level of both meridional and circumferential tensions along the archenteron are low. This agrees with the fact that the cells in the archenteron may freely or actively reorganize. The shapes of the exogastrula obtained with the profile shown in Figure 1b are depicted in Figure 2 and agree well with experimental observations.

Normal gastrulation

Simulation of normal gastrulation were initially performed by neglecting the effects of filopodial pulling as in the simulation of exogastrulation and thus examining the ability of active cell spreading to fully elongate the archenteron during the secondary invagination.

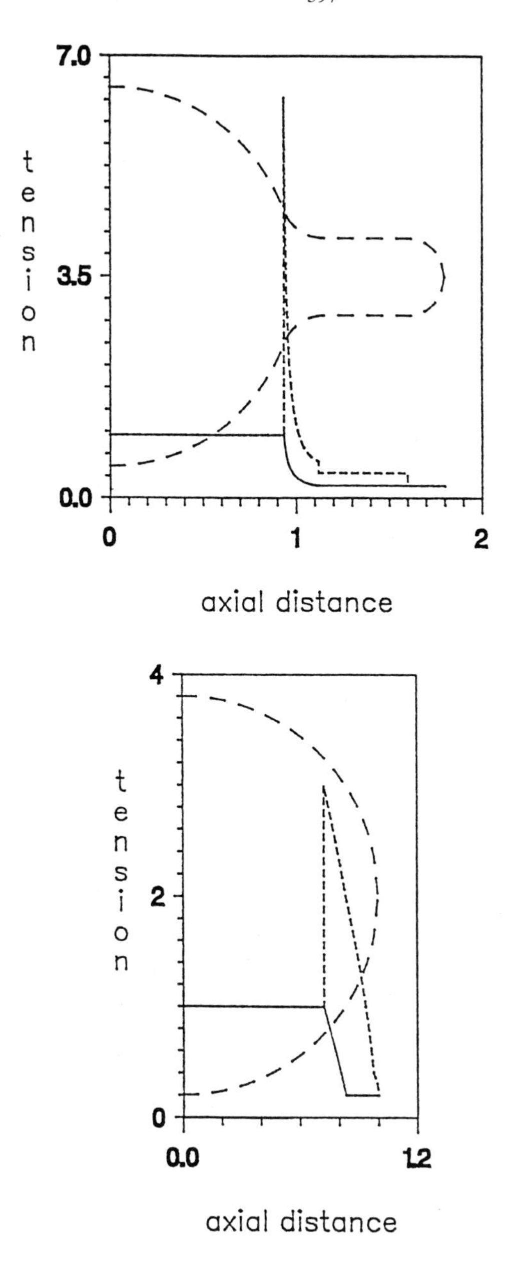

Figure 1. (a) Surface tensions required to maintain a fully developed protrusion after exogastrulation. (b) Surface tensions distribution at onset of dynamic simulation. Meridional tension _______, circumferential tension _ _ _ _ .

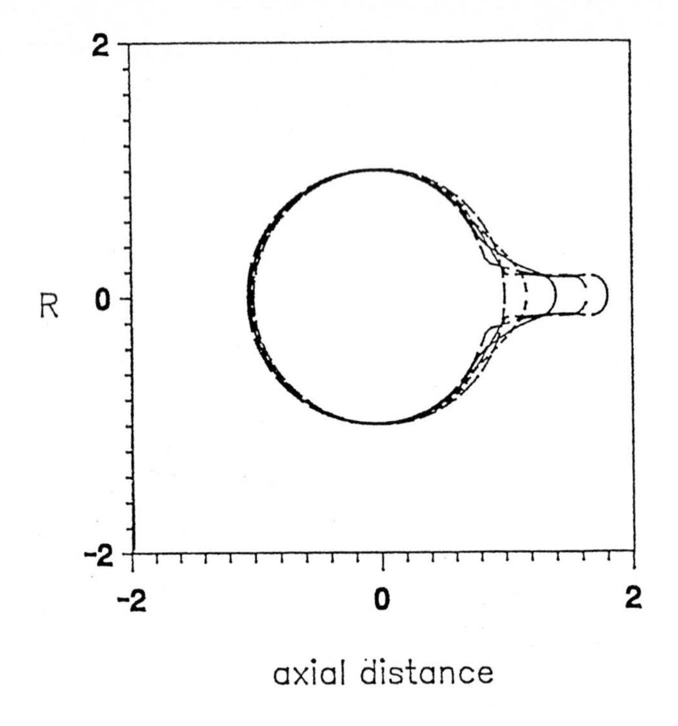

Figure 2. Calculated shapes of dynamic embryo deformation in exogastrulation.

First, a profile of active moments was assumed in order to produce the primary invagination. This profile is shown in Figure 3 together with the corresponding shape obtained which agrees well with the shapes observed experimentally and with the theoretical predictions of Odell et al. (1981). It must be noted that contrary to the case of exogastrulation in which the evagination may be produced by tensions only, the primary invagination cannot proceed without the presence of active moments. Once the archenteron is formed, a simulation of active cell spreading can be attempted by assuming negative tensions in the invagination which would also be in agreement with the tensions required to maintain a full invagination. However, such tensions were unable to produce the expected archenteron elongation. Although the archenteron attains the expected circumferential dimension, defined by the surface tension, it does not advance much in the inward direction. This behaviour is due to the fact that the meridional tension gradient tends to pull

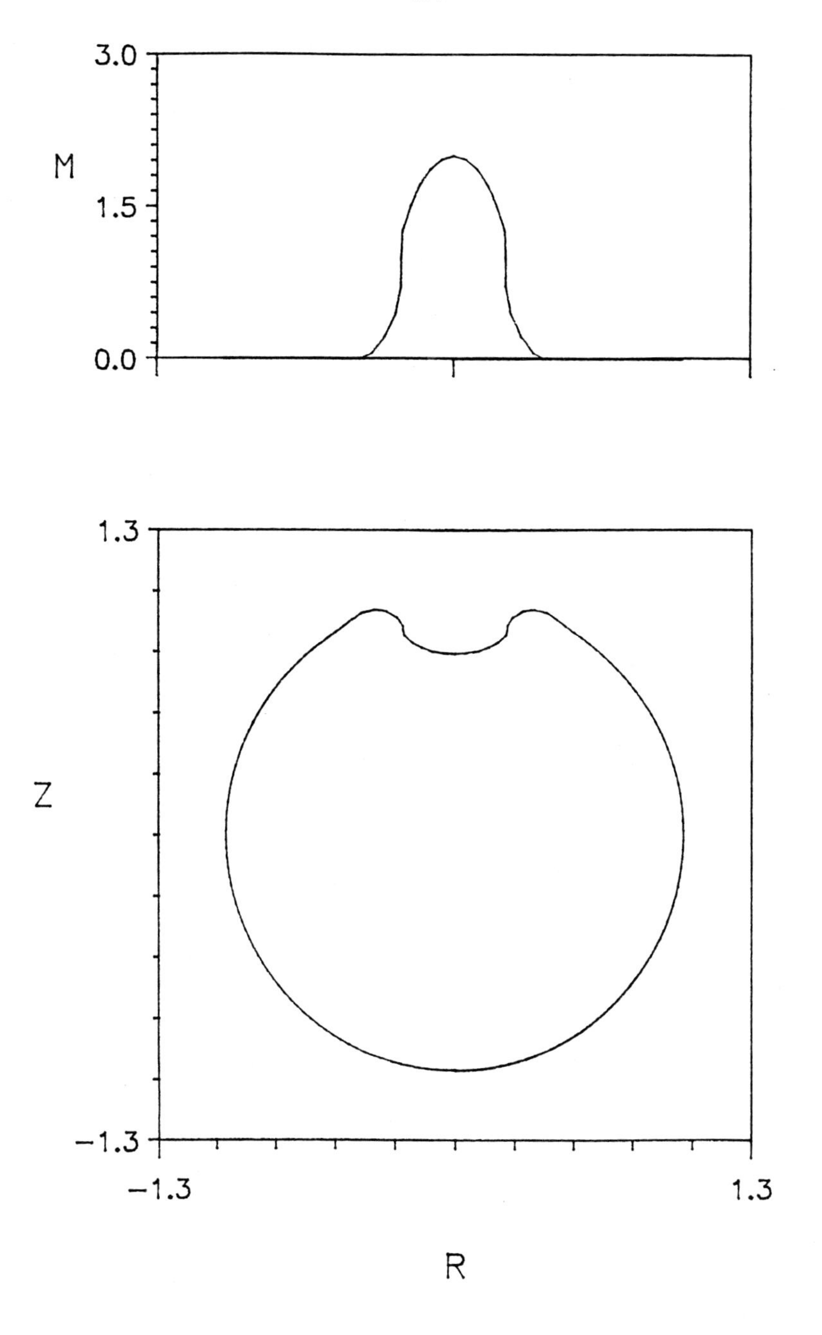

Figure 3. Moment distribution and calculated deformation during normal primary invagination

back to the base of the archenteron. A second consideration that must be taken into account is that regions with negative tensions, as those that can result from active spreading, are unstable and will tend easily to fold, behaviour that has not been observed experimentally.

Failing to obtain the correct morphological behaviour of the second invagination assuming active cell spreading the hypothesis of filopodial pulling was reconsidered. At the onset of the second stage four axisymmetrical filopodia were extended between the tip of the archenteron and the blastocoel roof. The tension in each one of the filopodia was assumed to increase linearly from the axis of symmetry outward. The initial configuration is shown in Figure 4. Since pulling will result in stretching of the cells of the archenteron and the experimental observations indicate that these are repacked into a thinner tube (Hardin and Cheng 1986), tensions in the archenteron were assumed very small relatively to filopodial tractions. This repacking behaviour, as stated above, means that interactions between cells are loose and the layer may thus be compared to a fluid interface where tension does not depend on surface deformation.

The evolution of the archenteron elongation obtained is shown in Figure 5 where the resemblance to normal gastrulation is evident. The different types of archenteron geometries related to evolution in various sea urchin species could be obtained using a different distribution of filopodia and tensions.

Discussion

In spite of the experimental and theoretical efforts invested in the study of gastrulation and although much advance has been done in the understanding of the elements taking an active role during primary and second invaginations the source and nature of the forces involved in these processes are still not well understood. Furlthermore, although during exogastrulation a mechanism of filopodial pulling is absent, it is not yet clear what is the role of filopodia during normal gastrulation since various experimental findings indicate that disruption of filopodia inhibits archenteron elongation. On the other hand Hardin and Cheng (1986) sustain that filopodial role is only secondary and that the active forces inducing the secondary invagination arise from active repacking of the cells in the archenteron.

Our simulations indicate that in fact exogastrulation may be induced by surface tensions alone without filopodia being actively involved in the process. We have used a

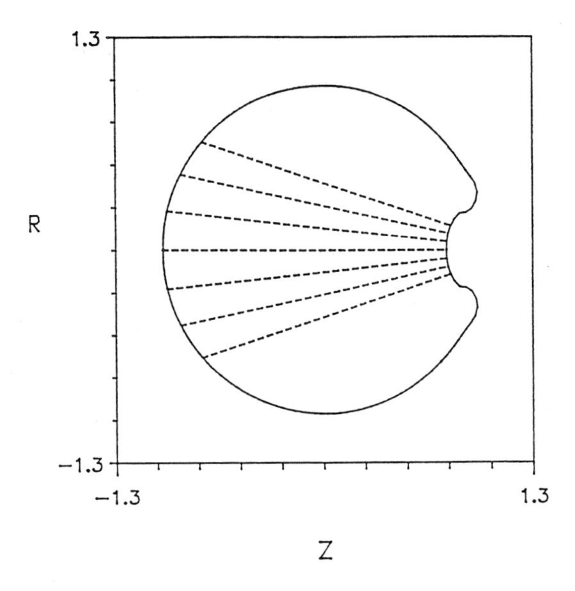

Figure 4. Initial configuration of filopodia at the onset of secondary invagination. Filopodial tension increases from 0.01 at the axis to 0.1 outward.

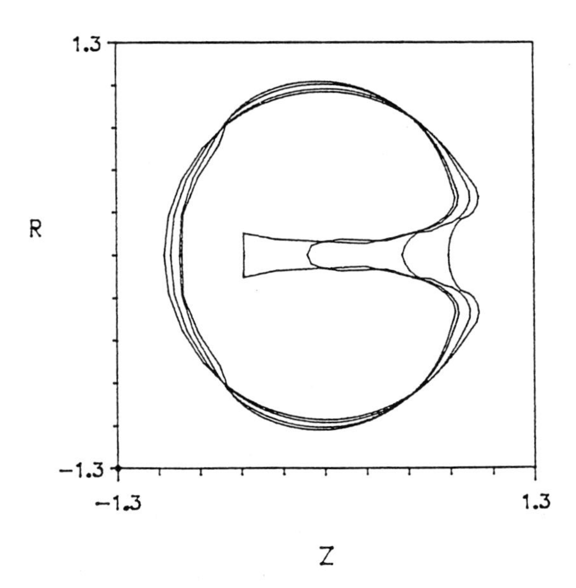

Figure 5. Simulation of the dynamics of archenteron elongation in normal gastrulation.

surface tension profile based on the tensions required to maintain a fully developed exogastrula. The tensions used are characterized then by a high level of circumferential tension at the base of the archenteron. The rest of the meridional and circumferential tensions are relatively low and the outward motion is induced by the meridional surface tension gradient.

Regarding normal gastrulation, as stated above, the question of which are the mechanisms and the biophysical and biochemical processes that produce the forces that cause the archenteron to elongate remains still unanswered. We have tried active forces and moments correponding to both hypothetical mechanism of active cell spreading and filopodial pulling. Despite an exhaustive search for surface profiles which could induce secondary invagination in absence of filopodia, archenteron elongation could not be simulated by an active cell spreading mechanism.

In a second group of simulations, when filopodial tractions were included and surface forces along the archenteron were assumed low, good agreemnet with second invagination deformations is obtained. Our results, therefore, suggest that filopodial pulling is necessary during normal gastrulation while the repacking is a passive process which may be due to the fact that the cells of the archenteron are loosely packed. This low adhesion is, in fact, manifested by the observation that cells in the archenteron do not stretch but repack. Also, as mentioned above, circumferential tensions at the base of the archenteron could have an important role and may be responsible for the narrowing of the blastopore as the archenteron elongates.

Our simulation suggests the relative importance of mechanisms by inducing embryo gastrulation using hypothetical macroscopic surface force and moment distributions and simulated filopodia. A complete model requires further studies which will yield a description that relates the macroscopic forces and the observed deformation to the ultrastructure and activity of the epithelial cells.

References

Cheng, L. Y. (1987) Deformation analysis in cell and developmental biology. Part 1 - formal methodology, Trans. ASME **109**, 10-17.

Gilbert, S.F. (1988) Developmental Biology, 2nd ed., Sinauer Associates Inc. Publishers. Gustafson, T. (1964) The role and activities of pseudopodia during morphogenesis of the sea urchin larvae. In: Primitive motile systems, Allen, R.D. and Kamiya, N., eds., 333-350, Academic Press, Orlando, Fla.

Gustafson, T. and Kinnander, II. (1956) Microaquaria for time-lapse cinematographic studies of morphogenesis in swimming laravae and observations on gastrulation, Exp. Cell Res. **11**, 36-57.

Hardin, J.D. and Cheng, L.Y. (1986) The mechanisms and mechanics of archenteron elongation during sea urchin gastrulation, Develop. Biol. **115**, 490-501.

Hiramoto, Y. (1956) Cell division without mitotic apparatus in sea urchin eggs, Exp. Cell Res. **68**, 630-636.

Keller, R.E. (1978) Time-lapse cinematographic analysis of superficial cell behaviour during the prior to gastrulation in Xenopus lacvis, J. Morphol. **157**, 223-248.

Moore, A.R. and Burt, a.s. (1939) On the locus and nature of the forces causing gastrulatin in the embroys of Dendrasterexcentricus, J:. Exp. Zool. **82**, 159-171.

Odell, G.M., Oster, G. Alberch, P. and Burnside, B. (1981) The mechanical basis of morphogenesis, Develop. Biol. **85**, 446-462.

Rallison, J.M. and Acrivos, A. (1978) A numerical study of the deformation and burst of a viscous drop in an extensional flow, J. Fluid Mech. **89**, 191-200.

Waxman, A.M. (1984) dynamics of a couple-stress fluid membrane, Stud. Appl. MAth. **70**, 63-86.

Zinemanas, D. and Nir, A. (1988) On the viscous deformation of biological cells under anisotropic surface tension, J. Fluid Mech. **193**, 217-241

Zinemanas, D. Nir,A. (1989) Surface visco-elastic effects in cell cleavage, J. Biomech. **23**, 417-424.

Zinemanas, D. and Nir, A. (1991) A fluid mechanical model of deformation during embryo exo-gastrulation, J. Biomech. (in press).

EPITHELIAL CELL VOLUME REGULATION

Kenneth R. Spring
National Institutes of Health
National Heart, Lung, and Blood Institute
Laboratory of Kidney and Electrolyte Metabolism
Building 10, Room 6N307
Bethesda, MD 20892

The cell membranes of almost all cells exhibit a sufficiently large water permeability that cell volume is determined by the content of osmotically active solutes as well as by the osmolality of the extracellular fluid. The swelling pressure created by intracellular macromolecules is constantly opposed by pumps which extrude ions. A constant cell volume is maintained both by adjustment of ion transport rates and by rapid changes of the membrane leaks. Variations in the rates of ion transport and leak constitute the primary homeostatic mechanisms available to all cells for housekeeping purposes. In addition to the ability to maintain a constant cell volume in the absence of any disturbances in the rate of solute entry or exit, most cells have some capability to counteract volume perturbations by volume regulatory processes. The tendency of osmotically swollen cells to shrink towards control volume by loss of solute and concomitant cell water is denoted regulatory volume decrease (RVD). Some shrunken cells can return towards their original volume by net

NATO ASI Series, Vol. H 64
Mechanics of Swelling
Edited by T. K. Karalis

MAJOR MODES OF EPITHELIAL CELL VOLUME REGULATION

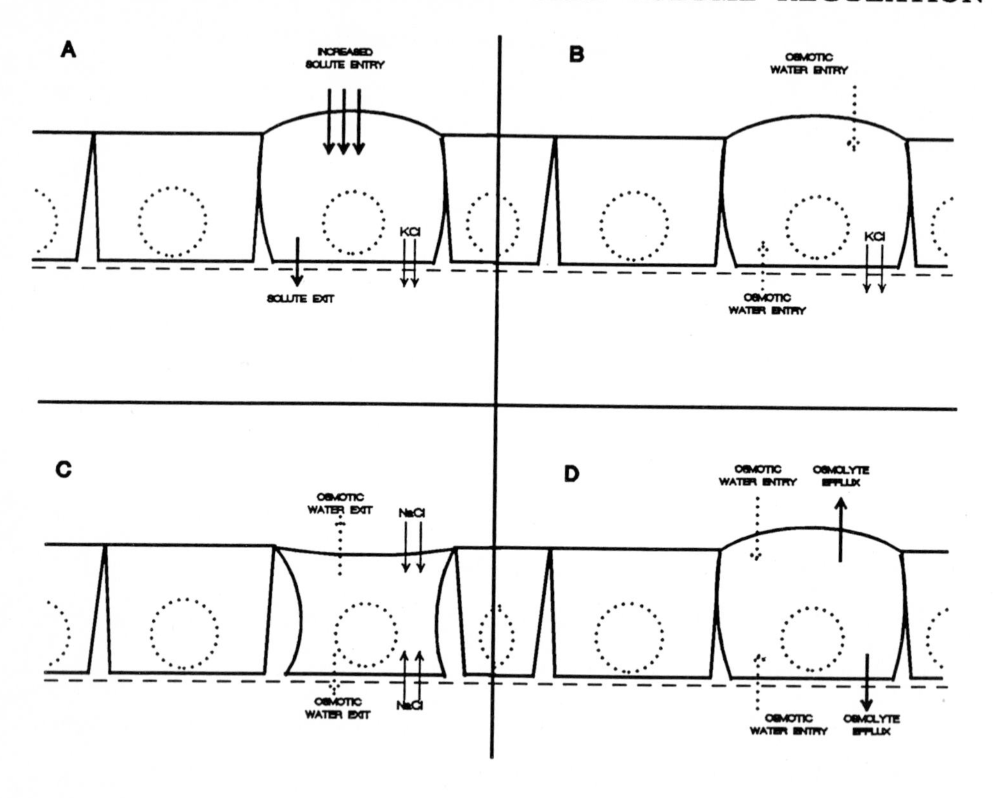

An increased rate of entry of transportable solutes leads to cell swelling and the efflux of KCl (panel A). Osmotically-induced swelling also results in KCl efflux but by pathways which may differ from those associated with swelling subsequent to solute entry (panel B). Osmotically-induced shrinkage (panel C) leads to increased NaCl uptake. Some cells respond to swelling by releasing organic osmolytes instead of KCl (panel D).

uptake of solute and water (regulatory volume increase, RVI). The ion transport systems involved in volume regulation and in the maintenance of a constant cell volume are often quiescent or operating only at a low level in the resting cell, but are activated by perturbations of the cell volume or by various external stimuli.

The large volumes of fluid which move across epithelial cell membranes would result in significant cytoplasmic dilution or concentration if fluid entry and exit were not balanced. As an example, the rates of fluid movement across renal proximal tubule cells are sufficiently great that the entire cellular volume is replaced approximately every 20 seconds. If entry or exit rates change as a result of altered solute movements, cell volume should change dramatically. Stimulating solute entry, e.g. by the addition of a sugar or amino acid to the apical bathing solution, leads to cell swelling [Furlong and Spring, 1990; Tune and Burg, 1971]. Interfering with solute exit also results in cell swelling because solute entry continues in the absence of solute exit [Jensen, Fisher, Spring, 1984; Larson and Spring, 1983; Spring and Ericson, 1982]. When solute entry into epithelial cells is blocked by removal of the entering solute or addition of an inhibitor of entry, shrinkage of the cell occurs [Guggino, 1986; Spring and Ericson, 1982; Spring and Hope, 1979]. The shrinkage is primarily due to the continued operation of the solute exit mechanism, e.g. the Na,K-ATPase [Jensen et al, 1984; Spring and Ericson, 1982; Spring and Hope, 1979].

When cells are subjected to a sudden change in medium osmolality, water rapidly enters the cells across both membranes in proportion to their relative water permeability, surface area and the chemical gradients for water. This swelling or shrinkage alters intracellular ionic and solute concentrations leading to alterations in the flux of these substances across the cell membranes as well as to the activation of otherwise quiescent transport systems. The rapid (seconds to minutes) response of most cells to osmotic swelling is RVD due to the efflux of KCl by: conductive K and Cl channels, K,Cl cotransport or functionally coupled K/H and Cl/HCO_3 exchange [Hoffmann and Simonsen, 1989]. Rapid RVD by epithelia is accomplished by activation of conductive transport pathways, primarily present in the basolateral membrane, for K and Cl or for K and HCO_3. Renal medullary cells also respond to swelling by rapidly releasing organic molecules, "organic osmolytes," via transport pathways located in either the apical or the basolateral cell membrane. These organic osmolytes (sugar alcohols, amino acids, and trimethyl amines) diffuse out of renal cells in accordance with their concentration gradients as a result of increased membrane permeability activated by the changes in cell volume [Garcia-Perez and Burg, in press].

The short term response of some osmotically shrunken single cells is to return toward their original size by undergoing RVI involving the uptake of NaCl by Na,K,Cl cotransport, Na,Cl cotransport, or coupled Na/H and Cl/HCO_3 exchange [Hoffmann and Simonsen, 1989]. Only a few epithelia are capable of rapid RVI which they accomplish by the uptake of NaCl across the apical or basolateral membranes as a result of coupled Na/H and Cl/HCO_3 exchange.

When the osmolality change is long-lived (hours-days) most cells rely on organic molecules for volume regulation. In a hypertonic environment, long term regulation by organisms from bacteria to mammals involves the accumulation of organic solutes which serve both to increase intracellular osmolality and counteract the adverse effects of high salt concentrations on protein structure and function [Garcia-Perez and Burg, in press]. Cells from the renal medulla and papilla exist in an environment of widely fluctuating osmolality. These cells cope with high extra-cellular osmolality (as high as 6000 mOsm in desert mammals) by synthesizing or transporting organic osmolytes. Four compounds, sorbitol, myo-inositol, betaine, and glycerophosphorylcholine constitute the major osmolytes [Garcia-Perez and Burg, in press]. Some other organic compounds, such as taurine, aspar-tate, butyrate, and citrate may make minor contributions. The accumulation of osmolytes is triggered by increasing osmolality, and the release is activated by decreasing osmolality. In many renal cells osmolyte changes are the major mechanism of volume regulation. Osmolytes play an important role in many non-renal tissues, such as the brain and lens of the eye.

Gradual rather than sudden alterations of medium osmolality can result in very different volume regulatory responses. The abil-ity of epithelia to maintain a constant volume during gradual wide swings of osmolality has been denoted "isovolumetric regulation" [Lohr and Grantham, 1986]. In response to a slowly changing bathing solution osmolality (1.5 mOsm/min), which mimics pathologic states in which plasma osmolality may vary widely from the normal value of ~300 mOsm, the volume of collapsed rabbit renal proximal tubules remained constant between bath osmolalities of 167 and 361 mOsm. Isovolumetric regulatory decrease was observed even when medium osmolality was reduced at rates as high as 27 mOsm/min, indicating that RVD was quite potent in these tubules. A sudden osmolality decrease causes collapsed tubules to exhibit only incomplete RVD by the conductive efflux of K and Cl. Isovolumetric regulatory increase is less potent and is not observed if the medium osmolality is increased at a rate greater than 3 mOsm/min. However, these tubules do not exhibit any RVI in response to a sudden osmolality increase [Lohr and Grantham, 1986].

The gradual accumulation or loss of organic osmolytes constitutes a form of isovolumetric regulation. The extent of involvement of organic osmolytes in the isolated proximal tubule experiments described above is not known, however, it is established that organic osmolytes play a key role in the homeostasis of renal medullary cells. The balance between the movements of electrolytes and osmolytes in isovolumetric regulation of proximal tubules as well as medullary tissues is the subject of considerable research effort at present.

The interaction of transepithelial solute transport and cell volume regulation by renal cells is not well characterized. Simply defining the transporters involved in the processes has occupied most of the research effort. The complex regulatory and feedback controls of both events must be thoroughly understood before the value of volume regulatory events can be fully appreciated.

REFERENCES

Furlong TJ, Spring KR (1990) Mechanisms underlying volume regulatory decrease by Necturus gallbladder epithelium. Am J Physiol 258 (Cell Physiol 27):C1016-C1024

Garcia-Perez A, Burg MB (to be published) Renal medullary organic osmolytes. Physiol Rev

Guggino WB (1986) Functional heterogeneity in early distal tubule of Amphiuma kidney: evidence for two modes of Cl and K transport across the basolateral cell membrane. Am J Physiol 250 (Renal Fluid Electrolyte Physiol 19):F430-F440

Hoffmann EK, Simonsen LO (1989) Membrane mechanisms in volume and pH regulation in vertebrate cells. Physiol Rev 69:315-382

Jensen PK, Fisher RS, Spring KR (1984) Feedback inhibition of NaCl entry in Necturus gallbladder epithelial cells. J Membrane Biol 82:95-104

Larson M and KR Spring (1983) Bumetanide inhibition of NaCl transport by Necturus gallbladder. J Membrane Biol 74:123-129

Larson M and KR Spring (1984) Volume regulation by Necturus gallbladder:basolateral KCl exit. J Membrane Biol 81:219-233

Lohr JW and Grantham JJ (1986) Isovolumetric regulation of isolated S_2 proximal tubules in anisotonic media. J Clin Invest 78:1165-1172

Spring KR and Ericson A-C (1982) Epithelial cell volume modulation and regulation. J Membrane Biol 69:167-176

Spring KR and Hope A (1979) Fluid transport and the dimensions of cells and interspaces of living Necturus gallbladder. J Gen Physiol 73:287-305

Tune BM and Burg MB (1971) Glucose transport by proximal renal tubules. Am J Physiol 221:580-585

INTERACTIONS BETWEEN LIVER CELL VOLUME AND LIVER CELL FUNCTION

Dieter Häussinger and Florian Lang

Medizinische Universitätsklinik

Hugstetterstrasse 55

D-7800 Freiburg

Germany

INTRODUCTION

To serve their functions, cells have to accumulate a number of osmotically active substances, which tend to swell the cells. Thus, cells had to develop volume regulatory mechanisms to defend the constancy of their volume. A great variety of volume regulatory mechanisms has been disclosed and found under appropriate conditions in virtually every cell under study. In the past three years, however, it became increasingly clear that alterations of cell volume markedly influence a variety of metabolic pathways in liver. It appears that cell swelling and cell shrinkage lead to certain opposite patterns of cellular metabolic function. Apparently, hormones and amino acids can trigger those patterns simply

NATO ASI Series, Vol. H 64
Mechanics of Swelling
Edited by T. K. Karalis

by altering cell volume, pointing to a new, not yet recognized principle of metabolic control.

CHALLENGES FOR CELL VOLUME IN LIVER

Cell volume alterations occur when the effective osmotic gradient across the cell membrane changes. During intestinal absorption portal venous blood may become slightly hypo- or hypertonic. As a matter of fact, liver swelling has been shown during intestinal absorption of water, presumably due to accumulation of solute free water. The breakdown of glycogen and proteins leads to generation of glucose-phosphates and amino acids, respectively, with the respective increase in intracellular osmolarity. Although it is at present impossible to clearly quantify the influence of given metabolic reactions on cell volume, it is beyond doubt that cellular metabolism represents a significant challenge for cell volume constancy.

Apart from hormones (see below), one of the most important challenge for cell volume homeostasis in liver is the cumulative uptake of osmotically active substances, such as amino acids [Kristensen 1980, Kristensen & Folke 1984, Bakker-Grunwald 1983, Wettstein et al 1990, Häussinger et al 1990a,b, Häussinger and Lang 1990a, Hallbrucker et al 1991a]. Na^+-dependent amino acid transporters in liver plasma membrane build up intra/extracellular amino acid concentration gradients of up to 20. Na^+ entering the hepatocyte together with the amino acid is extruded in exchange for K^+ by the electrogenic Na^+/K^+ ATPase. The accumulation of amino acids and K^+ into the cells leads to hepatocyte swelling, which in turn triggers volume regulatory K^+ efflux. Liver cell swelling and volume-regulatory K^+ efflux are observed upon addition of glutamine, alanine, proline, serine, glycine, aminoisobutyrate, phenylalanine and hydroxyproline, but not upon addition of glutamate, glucose or leucine; i.e. compounds which are not concentrated by liver cells. It recently became clear that amino acid-induced cell swelling and volume-regulatory responses occur upon exposure to amino acids in the physiological concentration range. Glutamine-induced cell swelling is halfmaximal at a concentration of 0.6-0.8 mM, i.e. the concentration normally found in portal venous blood, and is maximal at 2 mM [Wettstein et al 1990]. Accordingly, physiological fluctuations of the portal amino acid concentration are accompanied by parallel alterations of liver cell volume [Wettstein et al 1990, Hallbrucker et al 1991a]. The degree of amino acid-induced cell swelling seems largely to be related to the steady state intra/extracellular amino acid concentration gradient. This gradient and accordingly the degree of cell swelling can be modified by hormones and the nutritional state in a complex way. Their action not only involves effects on the expression of plasma membrane transport systems and the membrane potential as driving force for Na^+ coupled transport, but also alterations of intracellular amino acid metabolism. For example, starvation increases the swelling potency of glycine about 4-fold [Hallbrucker et al 1991a] largely due to adaptive up-regulation of the glycine-transporting system A [Hayes and McGivan 1982].

MODULATION OF LIVER CELL VOLUME BY HORMONES

Hormones are known to modify the activities of ion-pumps and -channels in the plasma membrane, to affect the cell membrane potential and to modulate Na^+-driven substrate transport [LeCam and Freychet 1978, Fehlmann et al 1981, Fehlmann and Freychet 1981, Jakubowski and Jakob 1990, Kilberg et al 1985, Moule & McGivan 1990]. All these actions are expected to modify cell volume.

Insulin was shown to stimulate the Na^+/H^+ exchange, the NaCl-KCl-cotransport and the Na^+/K^+ ATPase [Fehlmann and Freychet 1981, Grinstein & Dixon 1989, Jakubowski and Jakob 1990, Panet et al. 1986a, Hallbrucker et al 1991b,c]; these pumps are turned on for cell volume regulatory increase in a variety of tissues. The concerted action of these transporters should lead to cellular accumulation of potassium. Glucagon, on the other hand, has been shown to release cellular potassium. The Na^+/H^+ exchanger primarily increases intracellular sodium, which, however, is exchanged for potassium by the Na^+/K^+ ATPase [Fehlmann and Freychet 1981, Jakubowski and Jabob 1990]. In the absence of extracellular bicarbonate, the stimulation of the Na^+/H^+ exchanger leads to cellular alkalinization, in the presence of extracellular bicarbonate, however, intracellular pH is not modified by growth factors, probably due to parallel activation of HCO_3^-/Cl^- exchange [Ganz et al. 1990]. Thus, the stimulation of Na^+/H^+ exchange may serve to increase cell volume and/or to increase cellular K^+ content rather than to alkalinize the cell interior. As a matter of fact, we could show in both, perfused livers and isolated hepatocytes that insulin increases and glucagon decreases cellular volume and that these changes are dependent on the activation of the respective K^+ transport systems [Häussinger et al 1991, Hallbrucker et al 1991b,c, Häussinger & Lang 1991b]: insulin activates both, loop diuretic sensitive NaCl-KCl-cotransport and amiloride sensitive Na^+/H^+ exchange, whereas glucagon stimulates cellular potassium release through barium and quinidine sensitive K^+ channels [Hallbrucker 1991 b,c]. Accordingly, insulin-induced cell swelling is counteracted by further addition of glucagon or cAMP [Hallbrucker 1991c, Häussinger et al 1991]. Hormone-induced cell volume changes (table 1) were recently recognized to play a crucial role in mediating some metabolic responses to the hormones [Häussinger et al 1991, Hallbrucker et al 1991b,c, Häussinger & Lang 1991a]: the antagonism between insulin and glucagon regarding hepatic proteolysis is at least in part due to opposite effects of these hormones on cell volume (see below).

table 1: Effect of hormones on cell volume in isolated perfused rat liver

The intracellular water space ("liver cell volume") was determined in the perfused rat liver using a [^{3}H]inulin/[^{14}C]urea washout technique. Without effector addition, the intracellular water space was 546$\pm$9 µl/g (n=49), the data are given as % change of intracellular water space. Negative values indicate cell shrinkage, positive values cell swelling.

	change of intracellular water space (%)
insulin (35 nM)	+12$\pm$1 (15)
glucagon (100 nM)	-14$\pm$3 (6)
dibutyryl-cAMP (50 µM)	- 9$\pm$1 (4)
adenosine (50 µM)	- 6$\pm$2 (8)
vasopressin (15 nM)	- 2$\pm$2 (4)
phenylephrine (5 µM)	+ 8$\pm$1 (7)
vasopressin (15 nM) + glucagon (100 nM)	-23$\pm$3 (4)
vasopressin (35 nM) + cAMP (50µM)	-16$\pm$1 (6)
adenosine (50 µM) + phenylephrine (5 µM)	- 1$\pm$3 (3)
insulin (35 nM) + glucagon (100 nM)	+ 3$\pm$3 (4)
insulin (35 nM) + phenylephrine (5 µM)	+20$\pm$1 (4)

CELL VOLUME REGULATION

Most cells suddenly exposed to hypotonic media initially swell like more or less perfect osmometers but within minutes retain (almost) their original cell volume. This behaviour has been labelled regulatory cell volume decrease (RVD). If the cells are suddenly exposed to hypertonic media, they initially shrink but within minutes retain (almost) their original cell volume. This behaviour has been labelled regulatory cell volume increase (RVI).

RVD is achieved by reduction of intracellular osmotic activity due to extrusion of osmotically active substances (for review see [Chamberlain & Strange 1989, Lang et al 1990, Graf et al 1988]). Exposure of perfused rat liver to hypoosmotic perfusate leads to volume-regulatory release of cellular K^+, Cl^- and HCO_3^- [Graf et al. 1988, Haddad et al. 1989, Lang et al. 1989, Häussinger et al. 1990c]. The release of potassium is inhibited by barium, quinidine and SITS [Haddad et al. 1989, Häussinger et al. 1990c], pointing to activation of potassium channels in parallel to anion channels. The activation of the K^+ channels leads to the respective hyperpolarization of the cell membrane [Graf et al. 1988]. It is not yet clear to what extent organic osmolytes contribute to RVD in liver.

RVI is at least in part accomplished by uptake of ions across the cell membrane. In perfused liver, a sudden increase of extracellular perfusate osmolarity stimulates amiloride- and ouabain sensitive K^+ uptake [Graf et al. 1988, Lang et al. 1989, Häussinger et al. 1990c], which eventually leads to RVI. Loop diuretics do not appreciably modify the cell volume regulatory K^+ uptake. Accordingly, RVI in liver is apparently accomplished by activation of Na^+/H^+ exchange with subsequent extrusion of sodium in exchange with potassium. NaCl-KCl-cotransport appears not to appreciably participate in hepatic RVI [Häussinger et al 1991c], even though the carrier exists in the hepatic cell membrane and its activation by insulin is followed by increase of cell volume [Häussinger et al 1991, Häussinger & Lang 1991b, Hallbrucker et al 1991c]. RVI is similar in perfused livers, if perfusate osmolarity is increased from isotonic to hypertonic or from hypotonic to isotonic levels.

MODULATION OF HEPATIC METABOLISM BY CELL VOLUME

One consistent feature of volume regulation in most cell types studied so far is that volume-regulatory mechanisms never achieve a complete recovery of cell volume; instead, the cells are left in either a slightly swollen or shrunken state. Current evidence suggests that the extent of this remaining volume deviation somehow influences cell function. Accordingly, it was suggested that cell volume alterations may represent a new principle of metabolic control [Häussinger and Lang 1991a,b], which becomes even more interesting in view of a potent cell volume modulation by hormones (table 1). Anisoosmotic cell volume changes are accompanied by a variety of alterations of metabolic cell function (table 1) [Graf et al 1988, Lang et al 1989, Häussinger 1990a,b,d, Baquet et al 1990, Häussinger & Lang 1990, 1991,1991a, Häussinger et al 1991a, Hallbrucker et al 1991a,b, vom Dahl et al 1991], which persist throughout anisotonic exposure and are maintained even after completion of volume-regulatory ion fluxes [Graf et al 1988, Lang et al 1989, Häussinger 1990a,b,d, Häussinger & Lang 1990, 1991a,b, Hallbrucker et al 1991a,b, vom Dahl 1991, Häussinger et al 1991]. Several long-known, but mechanistically poorly understood effects of amino acids, which could not be related to their metabolism, such as stimulation of glycogen synthesis [Katz et al 1979, Lavoinne et al 1987] or inhibition of proteolysis [Schworer & Mortimore 1979, Mortimore & Pösö 1987, Seglen et al 1980, Pösö et al 1982] can simply be mimicked by hypoosmotic swelling of the cells just to the extent as the amino acids do [Häussinger et al 1990b, Baquet et al 1990, Hallbrucker et al 1991a, vom Dahl et al 1991a, Häussinger et al 1991a]. Thus, the above-mentioned amino acid effects are probably mediated by amino acid-induced cell swelling. Because liver cell swelling was shown to inhibit glycogenolysis, glycolysis [Graf et al 1988, Lang et al 1989] and proteolysis [Häussinger et al 1990b, 1991, Hallbrucker 1991a] and to stimulate glycogen synthesis [Baquet et al 1990], amino acid uptake [Häussinger et al 1990a,d, Häussinger & Lang 1990], ureogenesis from amino acids [Häussinger et al 1990d, Häussinger & Lang 1990] and also protein synthesis [own unpublished result], it seems that cell swelling acts like an "anabolic signal" [Häussinger and Lang 1991a,b] (table 2). In line with this, the "anabolic hormone" insulin induces hepatocyte swelling, whereas glucagon induces cell shrinkage [Hallbrucker et al 1991b,c, Häussinger & Lang 1991b, Häussinger et al 1991]. Because the activity of Na^+-dependent amino acid transport systems in the plasma membrane is now recognized to exert marked effects on cell volume, amino acid transporters may be seen not only as amino acid-translocating systems, but also as transmembrane signalling systems. In the following, we will focus on the regulation of proteolysis by cell volume, only.

table 2: Metabolic effects of liver cell swelling

CELL SWELLING	
stimulates	**inhibits**
protein synthesis	proteolysis
glycogen synthesis	glycogenolysis
lactate uptake	glycolysis
amino acid uptake	
glutaminase	glutamine synthesis
glycine oxidation	
ketoisocaproate oxidation	
urea synthesis from amino acids	urea synthesis from NH_4^+
glutathione efflux	

CELL VOLUME AND PROTEOLYSIS

Hepatic proteolysis is under the control of amino acids and hormones, such as insulin and glucagon; but the underlying mechanisms remained obscure (for reviews see [Mortimore and Pösö 1987]). It recently became clear that hypoosmotic cell swelling inhibits proteolysis in liver, whereas conversely, hyperosmotic cell shrinkage stimulates protein breakdown under conditions when the proteolytic pathway is not already fully activated [Häussinger et al 1990b,1991a, Hallbrucker et al 1991a]. The known antiproteolytic effect of several amino acids can at least in part be ascribed to amino acid-induced cell swelling. A close linear relationship is observed between the amino acid-induced degree of cell swelling and the inhibition of proteolysis [Häussinger et al 1990b,1991a, Häussinger & Lang 1991a, Hallbrucker et al 1991a]. The antiproteolytic effect of glutamine and glycine can fully be explained by the amino acid-induced cell swelling, since inhibition of proteolysis by these amino acids is quantitatively mimicked by hypoosmotic cell swelling when induced to the same extent as the above-mentioned amino acids do [Häussinger et al 1990b,1991a, Hallbrucker et al 1991a]. Inhibition of proteolysis by glutamine and glycine is additive to the same extent as additivity is observed with respect to cell swelling [Hallbrucker et al 1991a]. In livers from fed rats, the antiproteolytic effect of glycine is only about one fourth compared to that found in livers from starved rats. This is explained by an about 3-fold swelling potency of glycine during starvation, as a consequence of a higher activity of glycine-transporting amino acid transport system A in starvation [Hayes and McGivan 1982]. Alanine, proline and serine also lead to cell swelling, but the inhibition of proteolysis induced by these amino acids is about twice as strong as one would predict from anisotonic swelling experiments producing the same degree of cell swelling as these amino acids do [Hallbrucker et al 1991a]. Accordingly, about 50% of the antiproteolytic effect of alanine, proline and serine can be explained on the basis of cell volume changes. With phenylalanine and or a complete amino acid mixture the contribution of cell volume to the antiproteolytic action is less than 50%, even though a linear relationship between cell volume and the extent of proteolysis-inhibition is still maintained [Hallbrucker 1991a]. Thus, modulation of proteolysis by phenylalanine and also leucine (which does not lead to significant cell swelling) probably involves mechanisms distinct from cell volume.

ROLE OF CELL VOLUME IN HORMONE ACTION

The importance of hormone-induced alterations of cell volume (see table 1) has recently been studied in detail with respect to the long-known, but mechanistically unclear, hormonal control of hepatic proteolysis. In amino acid free liver perfusions, the net amount of K^+ accumulated or released by insulin or glucagon is linearly related to their effects on proteolysis [Hallbrucker et al 1991b]. This relationship is also maintained when the insulin-induced net K^+ accumulation inside the liver cell is modified by inhibitors of NaCl-KCl-cotransport, anisotonic cell swelling, the nutritional state or glucagon. The antagonism between insulin and glucagon regarding proteolysis can quantitatively be explained on the basis of opposing net K^+ shifts induced by these hormones [Hallbrucker et al 1991b]. A decisive role of the hormone-induced intracellular K^+ accumulation for the antiproteolytic action of insulin is also suggested by the finding that intracellular K^+ accumulation under the influence of the K^+ channel blocker Ba^{++} inhibits hepatic proteolysis just to the extent as observed when a similar amount of K^+ accumulates under the influence of insulin (fig. 1).

Whereas the antiproteolytic effect of insulin and Ba^{++} is parallelled by cellular K^+ accumulation, inhibition of proteolysis following hypoosmotic exposure of perfused liver is accompanied by a decrease of cellular K^+. This suggested that alterations of the intracellular K^+ concentration cannot explain the antiproteolytic effect of both, insulin and hypoosmotic exposure. It recently became clear that the strict relationship between proteolysis and the hormone- or Ba^{++} induced changes of cellular K^+ balance [Hallbrucker et al 1991b] is explained by hormone-

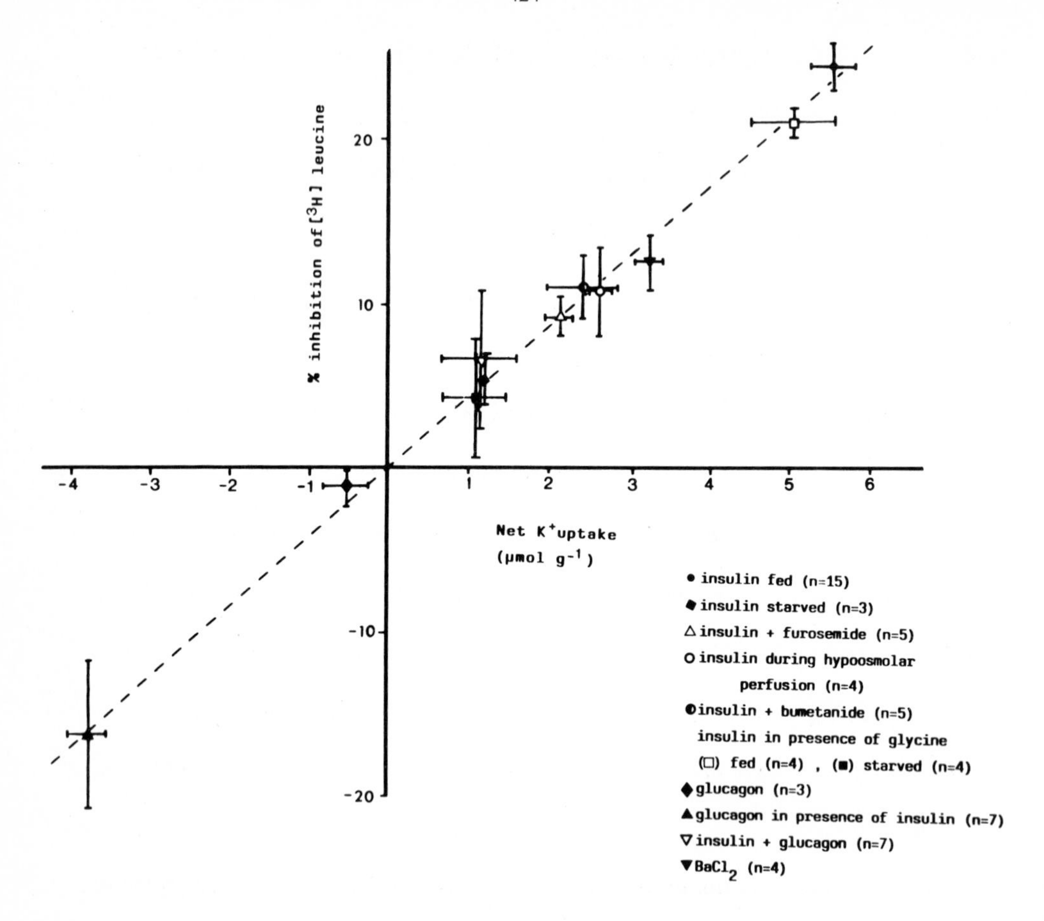

Fig. 1: Relationship between cellular K^+ balance and proteolysis

Cellular K^+ balance was modulated with insulin, glucagon and Ba^{++}, furosemide and bumetanide under a variety of experimental conditions. From [Hall-Proteolysis was assessed as [^{3}H]leucine release from perfused livers from rats, which were prelabelled in vivo by intraperitoneal injection of [^{3}H]leucine 16h prior to the perfusion experiment. From [Hallbrucker et al 1991b].

induced cell volume changes [Häussinger et al 1991a, Häussinger & Lang 1991b, Hallbrucker et al 1991c]. As shown in fig. 2, there is a close relationship between the proteolytic activity in perfused liver and the alterations of cell volume: proteolysis is inhibited during cell swelling and stimulated following cell shrinkage. This relationship is maintained regardless of whether cell volume is modified by insulin, glucagon, cAMP, inhibitors of NaCl-KCl-cotransport, glutamine, glycine, Ba^{++} or hypoosmotic exposure (fig. 2). Apparently, the extent of cell volume change produced by these effectors is the common denominator for their effects on proteolysis (fig. 2). These findings suggest that hormone-induced cell volume changes act like a "second messenger", which mediates at least some hormone effects on metabolism [Häussinger & Lang 1991b, Häussinger et al 1991a]. This view is further augmented by the finding that insulin and glucagon effects on cell volume are already observed at hormone concentrations found in portal venous blood in vivo.

Glucagon not only stimulates proteolysis, but also inhibits protein synthesis in rat liver [Woodside et al 1974]. Glucagon-induced inhibition of protein synthesis can at least in part be ascribed to glucagon-induced cell shrinkage, because hyperosmotic cell shrinkage also decreases protein synthesis in isolated rat hepatocytes (own unpublished observation).

Hyperosmotic, liver cell shrinkage stimulates glycogenolysis and glycolysis [Graf et al 1988, Lang et al 1989] and inhibits glycogen synthesis [Baquet et al 1990]. This is also observed with glucagon or cAMP and both agents lead to marked cell shrinkage [Hallbrucker et al 1991c, Häussinger et al 1991a, Häussinger & Lang 1991a]. Accordingly, one is tempted to speculate that the hormone-induced alterations of cell volume may contribute to the known effects of glucagon and insulin on hepatic carbohydrate metabolism.

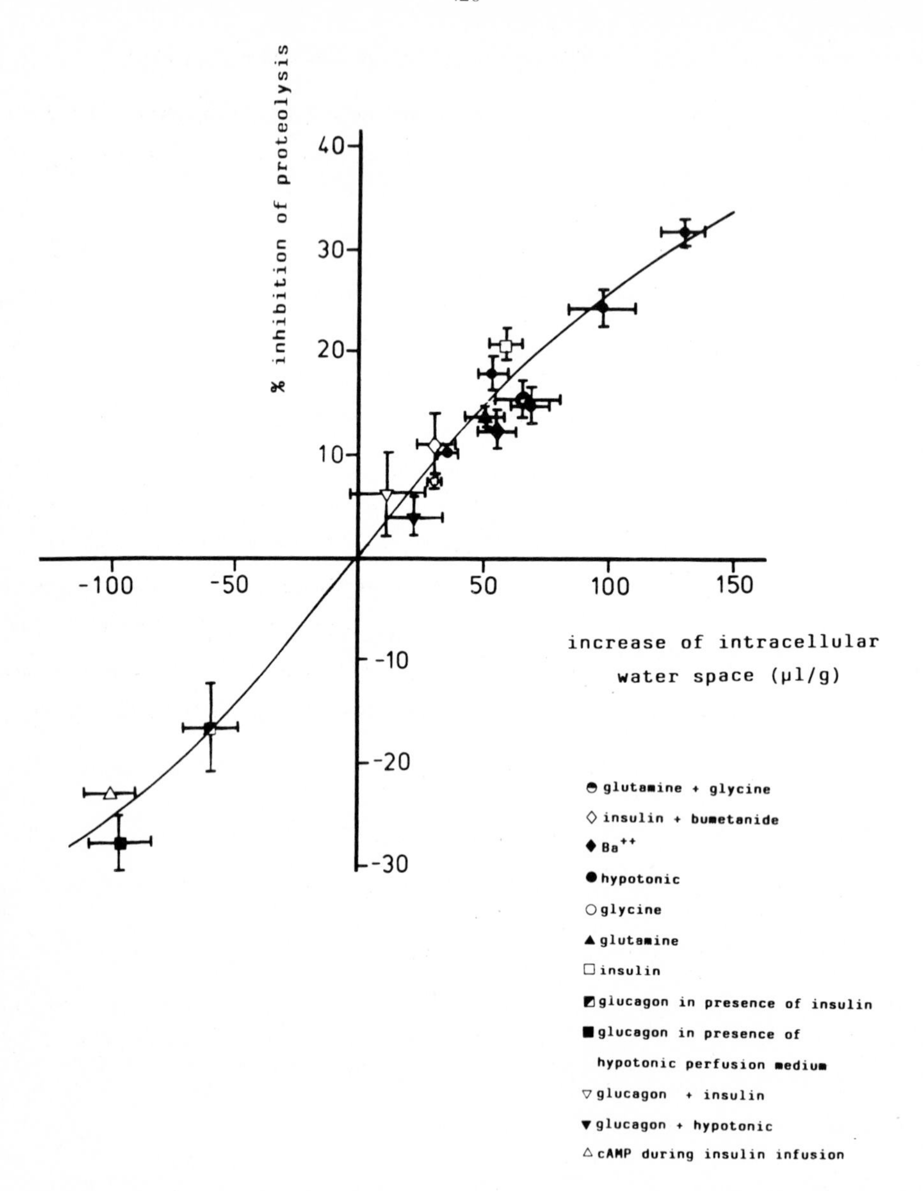

Fig. 2: Relationship between cell volume and proteolysis in liver.

Cell volume in perfused liver was determined as intracellular water space and proteolysis was assessed as [^{3}H]leucine release in effluent perfusate from perfused livers from rats, which were prelabelled in vivo by intraperitoneal injection of [^{3}H]leucine 16h prior to the perfusion experiment. Cell shrinkage stimulates proteolysis, whereas cell swelling inhibits. Cell volume changes were induced by insulin, cAMP, glucagon, amino acids, Ba^{++} or anisoosmotic exposure. From [Häussinger et al 1991].

Hormones do not only alter cell volume, but also the volume of intracellular organelles. Although there is good evidence for a glucagon-induced cell shrinkage, this hormone, like cAMP, actually swells the mitochondria (for review see [Halestrap 1989]). Mitochondrial swelling also occurs under the influence of phenylephrine or vasopressin (for review see [Halestrap 1989, Halestrap et al 1990]), although phenylephrine leads to cell swelling and vasopressin to slight cell shrinkage (table 1). Mechanisms and consequences of mitochondrial matrix volume have been reviewed recently [Halestrap 1989, Halestrap et al 1990].

SIGNALS INVOLVED IN CELL VOLUME REGULATION AND THE CONTROL OF CELL FUNCTION BY VOLUME

The question arises on the structures sensing cell volume changes and on the signals which couple cell volume changes to volume-regulatory ion fluxes and persistent alterations of cellular function. This issue is not yet settled, however, several possible mechanisms have been suggested. Cell swelling is accompanied by membrane stretching and in a variety of cells stretch-activated K^+ channels have been described (for review see [Chamberlain and Strange 1989]) which could in part mediate volume regulatory K^+ efflux following cell swelling. In addition participation of the cytoskeleton in cell volume regulation has been suggested in the liver [Rossum et al 1987]. Alanine-induced cell swelling was shown to open Ca^+-activated K^+ channels in isolated hepatocytes [Bear & Petersen 1987] and more recently a transient rise of intracellular Ca^{++} following hypotonic liver cell swelling was demonstrated in hepatocytes [Corasanti et al 1990a], in addition to slight decrease of intracellular pH [Gleeson et al 1990] and hyperpolarization of the plasma membrane [Graf et al 1988]. In liver, this Ca^{++} transient is probably due to Ca^{++} release from intracellular stores and the hypotonicity-induced stimulation of K^+ efflux from perfused livers is not blunted in the absence of extracellular Ca^{++} [vom Dahl et al 1991]. Hypoosmotic swelling of the liver [vom Dahl et al 1991] stimulates the formation of inositol 1,4,5-trisphosphate, which might trigger a rise of the intracellular Ca^{++} concentration. In perfused rat liver, after in vivo prelabeling of phospholipids with [^{3}H]myo-inositol, cell swelling by lowering the perfusate osmolarity from 305 to 225 mosm l^{-1} led to a about 3-fold stimulation of Li^+-sensitive [^{3}H]inositol release [vom Dahl et al 1991]. The maximum of hypotonicity-induced [^{3}H]inositol release preceeded maximal K^+ efflux by some 10-20s and occurred after the maximum of cell swelling. The stimulation of [^{3}H]inositol release, however, was not observed when cell swelling was performed starting from a preshrunken state (i.e. by decreasing the extracellular osmolarity from 385 to 305 mosm l^{-1}, although the volume-regulatory K^+ efflux observed under these conditions was virtually indistinguishable from that found when the osmolarity was lowered from 305 to 225 mosm l^{-1} [vom Dahl et al 1991]. This might indicate that volume-regulatory K^+ efflux is not neccessarily linked to a swelling-induced alterations of phosphoinositide turnover.

Recent studies on volume regulation in Ehrlich Ascites tumor cells have indicated an involvement of eicosanoids [Lambert 1987, Lambert et al 1987, Hoffmann et al 1988]. Although leukotrienes D_4 and C_4 and prostaglandins were shown to elicit complex K^+ fluxes in perfused rat liver, cell volume-regulatory K^+ efflux following anisotonic liver exposure was not affected in presence of inhibitors of cyclooxygenase, phospholipase A_2 and lipoxygenase or leukotriene receptor antagonists [vom Dahl et al 1991]. No effect of cell swelling on hepatic prostaglandin D_2 release was detectable. These data suggest that eicosanoids do probably not play a crucial role in liver cell volume regulation and marked differences may exist between different cell types with respect to a role of eicosanoids in triggering cell volume regulation.

Even more obscure are the mechanisms responsible for the modulation of metabolic cell function during cell volume alterations. Although liver cell swelling increases $[Ca^{++}]_i$ [Corasanti et al 1990], decreases pH_i [Gleeson et al 1990], hyperpolarizes the cell membrane [Graf et al 1988], stimulates inositol-1,4,5-trisphosphate formation [vom Dahl et al 1991], lowers significantly the cAMP/cGMP ratio [own unpublished result], the significance of these alterations for the observed effects on hepatic metabolism has not been elucidated.

PERSPECTIVE

Current evidence suggests that cell volume alterations represent a new, not yet fully recognized principle of metabolic control; they act as a "second messenger" mediating hormone and amino acid effects. Accordingly, concentrative amino acid transport systems and hormone-modulated ion transporters in the plasma membrane can be viewed as a part of transmembrane signalling systems. The major open questions refer to the mechanisms linking cell volume changes to the alterations of metabolic cell function.

REFERENCES

Bakker-Grunwald T (1983) Potassium permeability and volume control in isolated rat hepatocytes. Biochim Biophys Acta 731: 239-242

Baquet A, Hue L, Meijer AJ, van Woerkom GM, Plomp PJAM (1990) Swelling of rat hepatocytes stimulates glycogen synthesis. J Biol Chem 265: 955-959

Bear CE, Petersen OH (1987) L-alanine evokes opening of single Ca^{++}-activated K^{+} channels in rat liver cells. Pflüger's Arch 410: 342-344

Chamberlin ME, Strange K (1989) Anisosmotic cell volume regulation: a comparative view. Am J Physiol 257: C159-173

Corasanti JG, Gleeson D, Gautam A, Boyer JL (1990) Effects of osmotic stresses on isolated rat hepatocytes: ionic mechanisms of cell volume regulation and modulation of Na^{+}/H^{+} exchange. Renal Physiol Biochem 13: 164

Fehlmann M, Morin O, Kitabgi P, Freychat P (1981) Insulin and glucagon receptors of isolated rat hepatocytes: comparison between hormone binding and amino acid transport stimulation. Endocrinology 109:253-261

Fehlmann M, Freychat P (1981) Insulin and glucagon stimulation of Na/K-ATPase transport activity in isolated rat hepatocytes. J Biol Chem 256: 7449-7453

Gleeson D, Corasanti JG, Boyer JL (1990) Effects of osmotic stresses on isolated rat hepatocytes. Modulation of intracellular pH. Am J Physiol 258:G299-307

Graf J, Haddad P, Häussinger D, Lang F (1988) Cell volume regulation in liver. Renal Physiol Biochem 11: 202-220

Grinstein S, Dixon SJ (1989) Ion transport, membrane potential and cytoplasmic pH in lymphocytes: changes during activation. Physiol Rev 69: 417-481

Haddad P, Thalhammer T, Graf J (1989) Effect of hypertonic stress on liver cell volume, bile flow and volume-regulatory K^{+} fluxes. Am J Physiol 256:G563-G569

Häussinger D, Lang F, Bauers K, Gerok W (1990a) Interactions between glutamine metabolism and cell volume regulation in perfused rat liver. Eur J Biochem 188: 689-695

Häussinger D, Hallbrucker C, vom Dahl S, Lang F, Gerok W (1990b) Cell swelling inhibits proteolysis in perfused rat liver. Biochem J 272: 239-242

Häussinger D, Stehle T, Lang F (1990c) Volume regulation in liver: further characterization by inhibitors and ionic substitutions. Hepatology 11: 243-254

Häussinger D, Lang F, Bauers K, Gerok W (1990d) Control of hepatic nitrogen metabolism and glutathione release by cell volume regulatory mechanisms. Eur J Biochem 193: 891-898

Häussinger D, Lang F (1990) Exposure of perfused liver to hypotonic conditions modifies cellular nitrogen metabolism. J Cell Biochem 43: 355-361

Häussinger D, Lang F (1991a) The mutual interaction between cell volume and cell function: a new principle of metabolic regulation. Biochem Cell Biol 43: 1-4

Häussinger D, Lang F (1991b) Cell volume - a "second messenger" in the regulation of metabolism by amino acids and hormones. Cell Physiol Biochem in press

Häussinger D, Halllbrucker C, vom Dahl S, Decker S, Schweizer U, Lang F, Gerok W (1991) Cell volume is a major determinant of hepatic proteolysis. FEBS Lett in press

Halestrap AP (1989) The regulation of the matrix volume of mammalian mitochondria in vivo and in vitro and its role in the control of mitochondrial metabolism. Biochim Biophys Acta 973: 355-382

Halestrap AP, Davidson AM, Potter WD (1990) Mechanisms involved in the hormonal regulation of mitochondrial function through changes in the matrix volume. Biochim Biophys Acta 1018, 278-281

Hallbrucker C, vom Dahl S, Lang F, Häussinger D (1991a) Control of hepatic proteolysis by amino acids: the role of cell volume. Eur J Biochem 197:717-724

Hallbrucker C, vom Dahl S, Lang F, Gerok W, Häussinger D (1991b) Inhibition of hepatic proteolysis by insulin: role of hormone-induced alterations of cellular K^+ balance. Eur J Biochem in press

Hallbrucker C, vom Dahl S, Lang F, Gerok W, Häussinger D (1991c) Modification of liver cell volume by insulin and glucagon. Pflügers Arch Physiol in press

Hayes MR, McGivan JD (1982) Differential effects of starvation on alanine and glutamine transport in isolated rat hepatocytes. Biochem. J. 204: 365-368

Hoffmann EK, Lambert JH, Simonsen LO (1988) Mechanisms in volume regulation in Ehrlich Ascites tumour cells. Renal Physiol Biochem 11: 221-247

Jakubowski J, Jakob A (1990) Vasopressin, insulin, and peroxide of vanadate (pervanadate) influence Na^+ transport mediated by (Na^+/K^+)-ATPase or Na^+/H^+-exchanger of rat liver plasma membrane vesicles. Eur J Biochem 193: 541-549

Katz J, Golden S, Wals PA (1979) Glycogen synthesis by rat hepatocytes. Biochem J. 180: 389-402

Kilberg MS, Barber EF, Handlogten ME (1985) Characteristics and hormonal regulation of amino acids transport in isolated rat hepatocytes. Curr Top Cell Reg 25:113-163

Kristensen LO (1980) Energization of alanine transport in isolated rat hepatocytes. J Biol Chem 255: 5236-5243

Kristensen LO, Folke M (1984) Volume-regulatory K^+ efflux during concentrative uptake of alanine in isolated rat hepatocytes. Biochem J 221: 265-268

Lambert IH (1987) Effect of arachidonic acid, fatty acids, prostaglndins and leukotrienes on volume regulation in Ehrlich Ascites tumour cells. J Membr Biol 98: 207-221

Lambert IH, Hoffmann EK, Christensen P (1987) Role of prostaglandins and leukotrienes in volume regulation by Ehrlich Ascites tumour cells. J Membr Biol 98: 247–256

Lang F, Stehle T, Häussinger D (1989) Water, K^+, H^+, lactate and glucose fluxes during cell volume regulation in perfused rat liver. Pflügers Arch 413: 209–216

Lang F, Völkl H, Häussinger D (1990) General principles in cell volume regulation. Comp Physiol 4: 1–25

Lavoinne A, Baquet A, Hue L (1987) Stimulation of glycogen synthesis and lipogenesis by glutamine in isolated rat hepatocytes. Biochem J 248: 429–437

Le Cam A, Freychat P (1978) Effect of insulin on amino acid transport in isolated rat hepatocytes. Diabetologia 15:117–123

Mortimore GE, Mondon CE (1970) Inhibition by insulin of valine turnover. J Biol Chem 245:2375–2383

Mortimore GE, Pösö AR (1987) Intracellular protein catabolism and its control during nutrient deprivation and supply. Ann Rev Nutr 7: 539–564

Moule SK, McGivan JD (1990) Regulation of the plasma membrane potential in hepatocytes – mechanism and physiological significance. Biochim Biophys Acta 1031: 383–397

Panet R, Amir I, Atlan H (1986) Fibroblast growth factor induces a transient net K^+ influx carried by the bumetanide-sensitive transporter in quiescent BALB/c 3T3 fibroblasts. Biochim Biophys Acta 859: 117–121

Pösö AR, Schworer CM, Mortimore GE (1982) Acceleration of proteolysis in perfused rat liver by deletion of glucogenic amino acids: regulatory role of glutamine. Biochem Biophys Res Commun 107: 1433–1439

Rossum GDV van, Russo MA, Schisselbauer JC (1987) Role of cytoplasmic vesicles in volume maintenance. Curr Top Membr Transp 30:45–74

Schworer CM, Mortimore GE (1979) Glucagon-induced autophagy and proteolysis in rat liver: mediation by selective deprivation of intracellular amino acids. Proc Natl Acad Sci 76: 3169–3173

Seglen PO, Gordon PB, Poli A (1980) Amino acid inhibition of the autophagic/lysosomal pathway of protein degradation in isolated rat hepatocytes. Biochim. Biophys. Acta 630, 103–118

vom Dahl S, Hallbrucker C, Lang F, Häussinger D (1991) Role of eicosanoids, inositol phosphates and extracellular Ca^{++} in cell volume regulation of rat liver. Eur J Biochem 198: 73–78

Wettstein M, vom Dahl S, Lang F, Gerok W, Häussinger D (1990) Cell volume regulatory responses of isolated perfused rat liver. The effect of amino acids. Biol Chem Hoppe-Seyler 371: 493–501

Woodside KH, Ward WF, Mortimore GE (1974) Effects of glucagon on general protein degradation and synthesis in perfused rat liver J Biol Chem 249:5458–5463

FROM GLYCOGEN METABOLISM TO CELL SWELLING

Louis HUE, Arnaud BAQUET, Alain LAVOINNE[1] and Alfred J. MEIJER[2]

Hormone and Metabolic Research Unit, University of Louvain Medical School, and International Institute of Cellular and Molecular Pathology, Brussels, Belgium

INTRODUCTION

Our interest in the mechanics of swelling is recent and has arisen quite unexpectedly from the study of the control of glycogen metabolism in liver. Serendipity brought us to this promising field which we have explored and are still discovering with the zeal of a novice. This chapter summarizes our path from glycogen metabolism to cell swelling.

The liver is known as an organ that can store relatively large amounts of glycogen. In isolated preparations, such as perfused liver or isolated hepatocytes, glucose is a poor substrate for glycogen synthesis when given at physiological concentrations, i.e. at 5-15 mM (Seglen 1974, Hue et al. 1975, Katz et al. 1975). Addition of gluconeogenic precursors to such preparations stimulates glycogen synthesis (Hems et al. 1972, Walli et al. 1974, Whitton & Hems 1975, Howard & Widder 1976, Geelen et al. 1977). Furthermore amino acids, such as glutamine, alanine, asparagine, serine and proline, have been found to stimulate glycogen synthesis from physiological concentrations of glucose, and from fructose, lactate or dihydroxyacetone (Katz et al. 1976).

The mechanism of stimulation of glycogen synthesis by these amino acids has been extensively studied (Katz et al. 1976, Golden et al. 1979, Okajima & Katz 1979, Boyd et al. 1981, Solanki et al. 1982, Chen & Lardy 1985, Rognstad 1985, 1986, Lavoinne et al. 1987, Baquet et al. 1990, 1991a, 1991b, Carabaza et al. 1990, Plomp et al. 1990). The amino acids do not act by providing carbon atoms via gluconeogenesis, since 3-mercaptopicolinate, a known inhibitor of phosphoenolpyruvate carboxykinase, did not inhibit glycogen synthesis; on the contrary it even stimulated the process (Okajima & Katz 1979).

---

[1]Groupe de Biochimie et de Physiopathologie Digestive et Nutritionnelle, U.F.R. Médecine-Pharmacie, Rouen, France;

[2]E.C. Slater Institute for Biochemical Research, University of Amsterdam, Academic Medical Center, Amsterdam, The Netherlands

NATO ASI Series, Vol. H 64
Mechanics of Swelling
Edited by T. K. Karalis

Various compounds have been proposed as possible candidates involved in the mechanism of stimulation of glycogen synthesis by amino acids. These range from the added amino acids themselves (Rognstad 1986, Lavoinne et al. 1987), some unknown catabolites of glutamine, alanine or asparagine (Katz et al. 1976), carbamoyl phosphate (Rognstad 1985), putative purine derivatives (Solanki et al. 1982, Lavoinne et al. 1987, Carabaza et al. 1990), AMP (Carabaza et al. 1990), amino sugars or ionic changes resulting from the Na^+-dependent transport of amino acids (Lavoinne et al. 1987). The fact that aminoisobutyrate (AIB), a nonmetabolizable analogue, stimulates glycogen synthesis (Rognstad 1986) is in agreement with the hypothesis that the Na^+-dependent transport and entry of amino acids is triggering glycogen synthesis. A consequence of the accumulation of AIB, as well as of other amino acids, such as alanine, is an osmotic swelling of the cell (Bakker-Grunwald 1983, Kristensen & Folke 1984). We review in this chapter the experimental evidence, which suggests that stimulation of glycogen synthesis by glutamine and other amino acids is mediated by an increase in cell volume. We will also describe the consequences of the amino acids-induced swelling on the metabolism of lipids and we propose that swelling activates both glycogen synthase and acetyl-CoA carboxylase, the rate limiting enzymes for glycogen synthesis and lipogenesis respectively.

GLUTAMINE STIMULATES GLYCOGEN SYNTHESIS BY ACTIVATING GLYCOGEN SYNTHASE (Lavoinne et al. 1987)

Glutamine stimulates glycogen synthesis from glucose in hepatocytes prepared from fasted rats. The effect, which is half-maximal with about 3 mM-glutamine, depends on glucose concentration and is maximal at physiological concentrations of this sugar. Under these conditons, a single and linear relationship between the rate of glycogen synthesis and the proportion of glycogen synthase in the active form is obtained. This strongly suggests that the stimulation of glycogen synthesis by glutamine depends solely on the activation of synthase. The latter is not due to a change in total synthase, nor is it caused by a faster inactivation of glycogen phosphorylase as is the case after glucose (Hers 1976; Stalmans 1976). It can result from a stimulation of synthase phosphatase, since after the addition of glucagon or vasopressin, glutamine does not interfere with the inactivation of synthase, but promotes its subsequent re-activation.

STIMULATION OF GLYCOGEN SYNTHESIS BY AMINO ACIDS IS PROPORTIONAL TO SWELLING

Incubation of hepatocytes with 10 mM glutamine, proline or alanine activates glycogen synthase, stimulates glycogen synthesis and increases the cell volume (Baquet et al. 1990). On the other hand, neither D-alanine, which is not transported, nor L-leucine and L-valine, which are transported independently of Na^+ (Mc Givan & Bradford 1977), affect glycogen synthesis and cell volume. Amino acid analogues, such as AIB, methylaminoisobutyrate, hydroxyproline or β-alanine, which are poorly or not metabolized, but whose transport is Na^+-dependent (Shotwell et al. 1983), stimulate glycogen synthesis and increase cell volume. In all cases, the same direct relationship can be drawn between cell volume and glycogen synthesis, except for proline, which stimulates glycogen synthesis more than can be accounted for by the increase in cell volume (Baquet et al. 1990).

If cell swelling is essential for stimulation of glycogen synthesis, incubation of hepatocytes in hypertonic media should prevent both cell swelling and stimulation of glycogen synthesis (and activation of glycogen synthase) by amino acids. In agreement with this interpretation it was found that in the presence of media made hypertonic by the addition of 80 mM sucrose or raffinose, both the increase in cell volume and the activation of glycogen synthase induced by glutamine were inhibited (Baquet et al. 1990).

Although amino acid metabolism is not essential for increasing cell volume, the extent of swelling may be increased when amino acid catabolites accumulate inside the cells. This explains why amino acids like alanine and glutamine are more effective in causing cell swelling than nonmetabolizable amino acids. Additional evidence in support of the view that glycogen synthesis is, indeed, a function of an increase in the intracellular osmolarity, caused by accumulating amino acids, has been obtained by Plomp et al. (1990). In the experiments carried out by these authors, hepatocytes were incubated under several conditions meant to cover a large range of intracellular amino acid content, and the results indicated that one single relationship was obtained between glycogen synthesis and the sum of intracellular glutamate, aspartate, proline, asparagine, alanine, serine, glycine and citrulline concentration. Among the accumulating amino acids, glutamate, aspartate, alanine and the added amino acids were quantitatively most important. The maximal total intracellular amino acid content was about 110 µmol/g dry weight, i.e. an intracellular concentration of about 50 mM, assuming a cytosolic volume of 2.2 ml/g dry wt. (Tischler et al. 1977).

CELL SWELLING STIMULATES GLYCOGEN SYNTHESIS

The possibility that cell swelling per se could stimulate glycogen synthesis was tested. Cells were incubated without amino acids, but in media which were made hypotonic by decreasing the Na^+-content. Under these conditions, glycogen synthesis was stimulated, and, as with amino acids, cell swelling was directly related to stimulation of glycogen synthesis. If isotonicity was restored by adding sucrose instead of Na^+-ions, the stimulation of glycogen synthesis was lost and, as expected, the change in cell volume was greatly decreased (Baquet et al. 1990). This ruled out the possibility that Na^+-depletion was responsible for the stimulation of glycogen synthesis. The slight stimulation of glycogen synthesis observed in the presence of mercaptopicolinate alone (Okajima & Katz 1979) can be accounted for by cell swelling (Baquet et al. 1990) due to an increase in endogenous glutamate and aspartate (Plomp et al. 1990). In the presence of pyruvate and ammonia, mercaptopicolinate further increased both glycogen synthesis and the intracellular concentration of glutamate, aspartate and alanine (Plomp et al. 1990).

In all experimental conditions, the same linear relationship between glycogen synthesis and cell weight was obtained (Baquet et al. 1990). This suggests that the same mechanism, i.e. cell swelling, is involved in the control of glycogen synthesis.

STIMULATION OF LIPOGENESIS BY AMINO ACIDS

Besides its effect on glycogen synthesis, glutamine is also able to stimulate lipogenesis and to inhibit ketogenesis (Boyd et al. 1981, Lavoinne et al. 1987). Since other amino acids were found to stimulate glycogen synthesis (see above), their effect on lipogenesis was therefore also studied and compared (Baquet et al. 1991a). Among the other amino acids which were tested, only proline and, to a lesser extent, alanine were able to mimic the effect of glutamine on lipogenesis (Baquet et al. 1991a).

PARALLEL ACTIVATION OF GLYCOGEN SYNTHASE AND ACETYL-CoA CARBOXYLASE BY SWELLING

Lipogenesis and ketogenesis can be regulated at the level of acetyl-CoA carboxylase, which controls the concentration of malonyl-CoA, a strong inhibitor of the carntine-dependent entry and oxidation of long-chain fatty acids in mitochondria (McGarry & Foster 1980). Incubation of hepatocytes with glutamine increased malonyl-CoA concentration (Lavoinne et al. 1987), suggesting that acetyl-CoA carboxylase was activated. This enzyme is regulated by ligands, such as citrate, and by covalent modification. Like glycogen synthase, acetyl-CoA carboxylase exists in an active dephosphorylated form and an inactive phosphorylated form; furthermore, both enzymes are regulated by multi-site phosphorylation (Kim et al. 1989; Roach 1990; Cohen & Hardie 1991).

We have now shown that incubation of hepatocytes with glutamine and proline resulted in the parallel activation of glycogen synthase and acetyl-CoA carboxylase (Baquet et al. 1991c). The striking similarity between the overall effects of proline and glutamine on glycogen synthase and acetyl-CoA carboxylase activation strongly suggests the involvement of a common regulatory mechanism. That swelling is the common regulatory event triggering the activation of both glycogen synthase and acetyl-CoA carboxylase is supported by the following arguments: (i) incubation of hepatocytes in hypotonic media induced an activation of both enzymes; and (ii) the activation of both enzymes by glutamine was inhibited by the addition of raffinose to prevent cell swelling. Other results argue against swelling as a common factor for the regulation of *lipogenesis*. AIB increases cell volume, stimulates glycogen synthesis but fails to stimulate lipogenesis. In addition, cell swelling induced by hypotonic media cause much less stimulation of lipogenesis than glutamine, although changes in volume and stimulation of glycogen synthesis are quite comparable (Baquet et al. 1991a). However, under these conditions, the activation of both synthase and acetyl-CoA carboxylase were comparable indicating that, in Na^+-depleted media, lipogenesis is slightly stimulated despite an activation of acetyl-CoA carboxylase. In other words, the contradiction between the effect of swelling induced by glutamine and Na^+-depletion is only apparent and it disappears if one considers the effect of swelling on the activation of both enzymes rather than on lipogenesis (Baquet et al. 1991c).

HOW DOES CELL SWELLING ACTIVATE GLYCOGEN SYNTHASE AND ACETYL-CoA CARBOXYLASE ?

Regarding control of glycogen synthesis, changes in the intracellular concentrations of cations like K^+ and Ca^{++} are possible candidates. However, a mechanism involving K^+ is unlikely, since K^+-ions and amino acids seem to act differently despite their similar effect on swelling, glycogen synthesis and synthase activation. Indeed, K^+-ions, but not amino acids, promote phosphorylase inactivation (Hue et al. 1975) and stimulate glucose phosphorylation (Bontemps et al. 1978). Moreover, amino acids and hypotonic media tend to decrease K^+ concentration (Hoffmann & Simonsen 1989, Lang et al. 1989). A change in intracellular pH is another possible mechanism by which an increase in cell volume can result in stimulation of glycogen synthesis, and indeed intracellular pH decreases under hypotonic conditions (Lang et al. 1989). However, our preliminary results suggest that modulating the extracellular pH between 7.2 and 7.6 has little effect on glycogen synthesis in isolated hepatocytes.

Concerning possible changes in Ca^{++}, a decrease in free cytosolic Ca^{++} concentration might be responsible for an activation of glycogen synthase. It should also inactivate phosphorylase, a change which has however never been observed with glutamine. Likewise, stimulation of glycogen synthesis by amino acids persisted when hepatocytes were incubated in the presence of EGTA (Plomp et al. 1990). In addition, an increase, rather than a decrease in Ca^{++} concentration is actually found in cells submitted to a hypotonic shock (Watson 1989, Hazama & Okada 1990). Finally, recent work indicates that swelling increases the concentration of inositol 1,4,5-trisphosphate, Ca^{++} and cyclic AMP without activating glycogen phosphorylase (Baquet et al. 1991b). Such an increase in inositol 1,4,5-trisphosphate has also been found in liver perfused with hypotonic media (vom Dahl et al. 1991). In hepatocytes, the activation of phosphorylase by suboptimal concentrations of vasopressin or angiotensin II was partly antagonized by amino acids or hypotonicity despite the increase in Ca^{++} concentration (Baquet et al. 1991b). This suggests that cell swelling induces the formation of a compound which is able to overcome, at least in part, the effect of Ca^{++} and cyclic AMP on phosphorylase activation. One may speculate that this swelling-induced compound should be able to stimulate the phosphatase(s) acting on synthase and acetyl-CoA carboxylase. One may also wonder whether the other metabolic effects of swelling on protein metabolism, namely an inhibition of proteolysis (Häussinger et al. 1990a, 1990b, 1990c, Hallbrucker et al. 1991)), are also mediated by such a compound.

REFERENCES

Bakker-Grunwald, T. (1983) Potassium permeability and volume control in isolated rat hepatocytes. Biochim. Biophys. Acta, 731, 239-242

Baquet, A., Hue, L., Meijer, A.J., van Woerkom, G.M. & Plomp, P.J.A.M. (1990) Swelling of rat hepatocytes stimulates glycogen synthesis. J. Biol. Chem. 265, 955-959.

Baquet, A., Lavoinne, A. & Hue, L. (1991a) Comparison of the effects of various amino acids on glycogen synthesis, lipogenesis and ketogenesis in isolated rat hepatocytes. Biochem. J., 273, 57-62.

Baquet, A., Meijer, A.J. & Hue, L. (1991b) Hepatocytes swelling increases inositol 1,4,5-trisphosphate, calcium and cyclic AMP concentration but antagonizes phosphorylase activation by Ca^{++}-dependent hormones. FEBS Lett., 278, 103-106.

Baquet, A., Maisin, L. & Hue, L. (1991c) Swelling of rat hepatocytes activates acetyl-CoA carboxylase in parallel to glycogen synthase. Biochem. J., in press.

Bontemps, F., Hue, L. & Hers, H.G. (1978) Phosphorylation of glucose in isolated rat hepatocytes. Sigmoidal kinetics explained by the activity of glucokinase alone. Biochem. J. 174, 603-611.

Boyd, M.E., Albright, E.B., Foster, D.W. & McGarry, J.D. (1981) In vitro reversal of the fasting state of liver metabolism in the rat. J. Clin. Invest. 68, 142-152.

Carabaza, A., Ricart, M.D., Mor, A., Guinovart, J.J. & Ciudad, C.J. (1990) Role of AMP on the activation of glycogen synthase and phosphorylase by adenosine, fructose, and glutamine in rat hepatocytes. J. Biol. Chem. 265, 2724-2732.

Chen, K.S. & Lardy, H.A. (1985) Multiple requirements for glycogen synthesis by hepatocytes isolated from fasted rats. J. Biol. Chem. 260, 14683-14688.

Cohen, P. & Hardie, D.G. (1991) The actions of cyclic AMP on biosynthetic processes are mediated indirectly by cyclic AMP-dependent protein kinase. Biochim. Biophys. Acta, in press.

Geelen, M.J.H., Pruden, E.L. & Gibson, D.M. (1977) Restoration of glycogenesis in hepatocytes from starved rats. Life Sci. 20, 1027-1034.

Golden, S., Wals, P.A., Okajima, F. & Katz, J. (1979) Glycogen synthesis by hepatocytes from diabetic rats. Biochem. J. 182, 727-734.

Hallbrucker, C., vom Dahl, S., Lang, F. & Häussinger, D. (1991) Control of hepatic proteolysis by amino acids. Eur. J. Biochem. 197, 717-724.

Häussinger, D., Lang, F., Bauers, K. & Gerok, W. (1990a) Interactions between glutamine metabolism and cell-volume regulation in perfused rat liver. Eur. J. Biochem. 188, 689-695.

Häussinger, D., Lang, F., Bauers, K. & Gerok, W. (1990b) Control of hepatic nitrogen metabolism and glutathione release by cell volume regulatory mechanisms. Eur. J. Biochem. 193, 891-898.

Häussinger, D., Hallbrucker, C., vom Dahl, S., Lang, F. & Gerok, W. (1990c) Cell swelling inhibits proteolysis in perfused rat liver. Biochem. J. 272, 239-242.

Hazama, A. & Okada, Y. (1990) Involvement of Ca^{2+}-induced Ca^{2+} release in the volume regulation of human epithelial cells exposed to a hypotonic medium. Biochem. Biophys. Res. Commun. 167, 287-293.

Hems, D.A., Whitton, P.D., & Taylor, E.A. (1972) Glycogen synthesis in the perfused liver of the starved rat. Biochem. J. 129, 529-538.

Hers, H.G. (1976) The control of glycogen metabolism in the liver. Annu. Rev. Biochem. 45, 167-189.

Hoffmann, E.K. & Simonsen, L.O. (1989) Membrane mechanisms in volume and pH regulation in vertebrate cells. Physiol. Rev. 69, 315-382.

Howard, R.B. & Widder, D.J. (1976) Substrate control of glycogen levels in isolated hepatocytes from fed rats. Biochem. Biophys. Res. Commun. 68, 262-269.

Hue, L., Bontemps, F. & Hers, H.G. (1975) The effect of glucose and of potassium ions on the interconversion of the two forms of glycogen phosphorylase and of glycogen synthetase in isolated rat liver preparations. Biochem. J. 152, 105-114.

Katz, J., Wals, P.A., Golden, S. & Rognstad, R. (1975) Recycling of glucose by rat hepatocytes. Eur. J. Biochem. 60, 91-101.

Katz, J., Golden, S. & Wals, P.A. (1976) Stimulation of hepatic glycogen synthesis by amino acids. Proc. Natl. Acad. Sci. U.S.A. 73, 3433-3437.

Kim, K.H., Lopez-Casillas, F., Bai, D.H., Luo, X. & Pape, M.E. (1989) Role of reversible phosphorylation of acetyl-CoA carboxylase in long-chain fatty acid synthesis. FASEB J. **3**, 2250-2256.

Kristensen, L Ø. & Folke, M. (1984) Volume-regulatory K^+ efflux during concentrative uptake of alanine in isolated rat hepatocytes. Biochem. J. 221, 265-268.

Lang, F., Stehle, T. & Häussinger, D. (1989) Water, K^+, H^+, lactate and glucose fluxes during cell volume regulation in perfused rat liver. Pflügers Arch. 413, 209-216.

Lavoinne, A., Baquet, A. & Hue, L. (1987) Stimulation of glycogen synthesis and lipogenesis by glutamine in isolated rat hepatocytes. Biochem. J. 248, 429-437.

McGarry, J.D. & Foster, D.W. (1980) Regulation of hepatic fatty acid oxidation and ketone body production. Annu. Rev. Biochem. 49, 395-420.

McGivan, J.D. & Bradford, N.M. (1977) The transport of branched-chain amino acids into isolated rat liver cells. FEBS Lett. 80, 380-384.

Okajima, F. & Katz, J. (1979) Effect of mercaptopicolinic acid and of transaminase inhibitors on glycogen synthesis by rat hepatocytes. Biochem. Biophys. Res. Commun. 87, 155-162.

Plomp, P.J.A.M., Boon, L., Caro, L.H.P., van Woerkom, G.M. & Meijer, A.J. (1990) Stimulation of glycogen synthesis in hepatocytes by added amino acids is related to the total intracellular content of amino acids. Eur. J. Biochem 191, 237-243.

Roach, P.J. (1990) Control of glycogen synthase by hierarchal protein phosphorylation. FASEB J. 4, 2961-2968.

Rognstad, R. (1985) Possible role for carbamyl phosphate in the control of liver glycogen synthesis. Biochem. Biophys. Res. Commun. 130, 229-233.
Rognstad, R. (1986) Effects of amino acid analogs and amino acid mixture on glycogen synthesis in rat hepatocytes. Biochem. Arch. 2, 185-190.
Seglen, P.O. (1974) Autoregulation of glycolysis, respiration, gluconeogenesis and glycogen synthesis in isolated parenchymal rat liver cells under aerobic and anaerobic conditions. Biochim. Biophys. Acta, 338, 317-336.
Shotwell, M.A., Kilberg, M.S. & Oxender, D.L. (1983) The regulation of neutral amino acid transport in mammalian cells. Biochim. Biophys. Acta, 737, 267-284.
Solanki, K., Moser, U., Nyfeler, F. & Walter, P. (1982) Possible correlation between the stimulation of glycogen synthesis by some amino acids and the synthesis of purines in hepatocytes from starved rats. Experientia, 38, 732.
Stalmans, W. (1976) The role of the liver in the homeostasis of blood glucose. Curr. Top. Cell. Regul. 11, 51-97.
Tischler, M.E., Hecht, P. & Williamson, J.R. (1977) Determination of mitochondrial/cytosolic metabolite gradients in isolated rat liver cells by cell disruption. Arch. Biochem. Biophys. 181, 278-292.
vom Dahl, S., Hallbrucker, C., Lang, F. & Häussinger, D. (1991) Role of eicosanoids, inositol phosphates and extracellular Ca^{++} in cell-volume regulation of rat liver. Eur. J. Biochem. 198, 73-83.
Walli, A.K., Siebler, G., Zepf, E. & Schimassek, H. (1974) Glycogen metabolism in isolated perfused rat liver. Hoppe-Seyler's Z. Physiol. Chem. 355, 353-362.
Watson, P.A. (1989) Accumulation of cAMP and calcium in S49 mouse lymphoma cells following hyposmotic swelling. J. Biol. Chem. 264, 14735-14740.
Whitton, P.D. & Hems, D.A. (1975) Glycogen synthesis in the perfused liver of streptozotocin-diabetic rats. Biochem J. 150, 153-165.

ACKNOWLEDGEMENTS

This work was supported by the Belgian Fonds de la Recherche Scientifique Médicale, and by the Belgian State Prime Minister's Office Science Policy Programming (Incentive Program in Life Sciences 88/93-122, grant n° 20).

CELL SPREADING AND INTRACELLULAR *PH* IN MAMMALIAN CELLS

Leonid B. Margolis

Belozersky Institute of Physico–Chemical Biology
Moscow State University
Moscow 119899
USSR

1. INTRODUCTION

The geometry of cells in multicellular organisms is not constant. Different cell shapes correspond to particular cellular functions. The cell can either alter its entire shape or locally deform its plasma membrane by protruding and retracting various processes: microvilli, filopodia, lamellae, *etc*. Mechanisms of cell deformation remain obscure. According to different hypotheses it could be caused by cell swelling, by cytoplasmic streaming and/or by rearrangement of the cytoskeleton elements (Vasiliev 1985, Trinkaus 1985, Oster 1989).

Although the mechanisms by which cell deformation affects intracellular reactions remain obscure, one of the main factors causing cells' deformation is their adhesion to other cells and to non–cellular substrates. The ability to alter the shape due to cell–cell and cell–substrate contacts and triggering intracellular signalling system is retained by mammalian cells in vitro. Normal mammalian cell adhesion and spreading upon the solid substrate is obligatory for initiating *DNA* synthesis and for proliferation (Folkman and Greenspan 1975; Folkman and Moscona 1978). Only the cells which have gone extremely far in the course of tumour progression (*e.g.,* ascites cells) are able to proliferate substrate–independently.

Cell adhesion is a complex biological phenomena which includes the initial attachment of round cells to the substrate by short processes, development of longer filopodia and lamellae and finally spreading of cells on the substrate. Spreading seems to be the primary signal for the cell to enter the proliferation pool (Folkman and Moscona 1978, Vasiliev 1985).

NATO ASI Series, Vol. H 64
Mechanics of Swelling
Edited by T. K. Karalis

Unfortunately many important questions about: (*i*) the role of various plasma membrane molecules mediating cell attachment to the substrate, (*ii*) the intracellular forces which generate cellular processes and lamellae and (*iii*) the nature and pathways of the signals which are transmitted into the cytoplasm of the spread cell, remain to be answered. In an attempt to do this we have developed experimental techniques to modulate the force which generates cell processes, to bind covalently new adhesion molecules to the cell surface and studied the key elements of adhesion–mediated intracellular signalling; in particular the shift of intracellular *pH* [*pH*(*i*)].

2. ADHESION OF NON–ADHESIVE CELLS

To study cell adhesion and spreading and the role of these processes in cell stimulation we decided to choose the non–adhesive Ehrlich ascites cells and tried to modify them in order to make them adhesive. In general, for the cell to be spread on the solid surface two factors are required: the force to deform the membrane and the "adhesive molecules" to keep the cell on the solid substrate. What is lacking in Ehrlich ascites cells?

In collaboration with S. Popov a model is developed based on the system of dielectrophoresis which allowed us to apply local mechanical force directly to the cell membrane (Popov and Margolis 1988). Briefly, cells are placed on the coverslip in a non–ionic media in the gap between two planar electrodes. When the high–frequency electric field (*1 MHz*) is switched on, a force is generated applied only to the cell membrane, stretching it in the direction of the nearest electrode (Popov and Margolis 1988, Gass *et al* 1991).

Applying this system to various fibroblastic and epithelioid cells normally capable of adhesion and spreading, it is found that various protrusions (filopodia–like and lamellae–like) are generated from the cell body; the protrusions looking similar to those cells produced in physiological media. The protrusions arise on the cells which are subjected to this force, even though the active reorganization of the cytoskeleton or/and energy–dependent reactions are inhibited. We have concluded that the shape of the protrusion is determined by the structure of the membrane. Thus the force applied directly to the membrane (in principle it could be any cytoplasmic force pushing the membrane out *e.g.,* actin polymerization, cytoplasmic streaming, local swelling *etc.*) deforms it in the way pre–determined by the membrane structure. Using this system we tried to make ascitic cells spread on the substrate. However even with an additional force the cells did not spread. Only small thin protrusions were generated in a few cells which attached to the substrate (similar to those in Fig. 1A). Thus it is not the lack of the sufficient force that makes the cell deficient in ability to adhere and to spread.

To investigate the possible role of active adhesion sites in cell spreading, in collaboration with V. Torchilin and A. Bogdanov a method of reconstitution of plasma membrane is developed with artificial molecules to make the non–adhesive ascitic cells adhesive (Bogdanov *et al* 1991). We have modified Ehrlich ascites carcinoma cells by water – soluble membrane – impermeant activated biotin derivative *N*–[hydroxysulfosuccinimidyl (*6* – biotinylamido) hexanoate]. The coverslips were modified with avidin. The presence of biotin residues on the cell surface was checked by staining of modified cells with *FITC*–labelled avidin. All biotinylated cells exhibit fluorescent staining. In the presence of free biotin no fluorescent avidin was found.

Plasma membrane modification dramatically increases the number of cells attached to avidin–coated substrate. This is what one could easily predict taking into account high–affinity binding of biotin molecules to avidin (*KD* of avidin–biotin complex has been reported as *10–15 M* (Green 1963)). What does not seem trivial is that cell surface modification also promotes cell spreading (Fig 2). The spread cells attain epithelioid–like morphology with the formation of wide thin lamellae, focal contacts with substrate and circular actin bundles, as revealed by scanning electron–, interference reflection–and immunofluorescent microscopy, correspondingly.

Involvement of avidin–biotin interactions in ascites cell spreading was proved in control experiments, where non–modified cells or substrate were used (Fig.1 A) or when biotin–binding sites of immobilized avidin were blocked with the excess of free biotin. Alternatively, cells were modified with *N*–hydroxysulfosuccinimidyl–*2*– (biotinamido) ethyl–*1,3*– dithiopropionate, containing disulphide bridges. Whem the modified cells were attached to avidin–coated glass, cysteine was added to reduce disulphide bonds. The cells ceased to spread and began to round up.

On the basis of experiments described above we concluded that cell spreading requires high–affinity interaction of plasma membrane components with the substrate. The lack of the ability to spread in Ehrlich ascites cells may be explained by the lack of functionally active adhesion competent cell–surface receptors, rather than to the defects in intracellular function or organization. Intracellular machinery of cell spreading including generating of intracellular signals is preserved in these ascites cells and could be turned on by cell attachment to the substrate via artificial adhesive sites incorporated into the plasma membrane.

For physiological adhesion it was postulated a complex relation between the outer residues of integral adhesive molecules and intracellular cytoskeletal proteins which rearrange upon cell spreading. In our experiments rearrangement of the cytoskeleton is initiated by binding to avidinylated substrate of some undefined biotynilated cell surface molecules. How is it possible that these molecules can so successfully play the role of integral receptors?

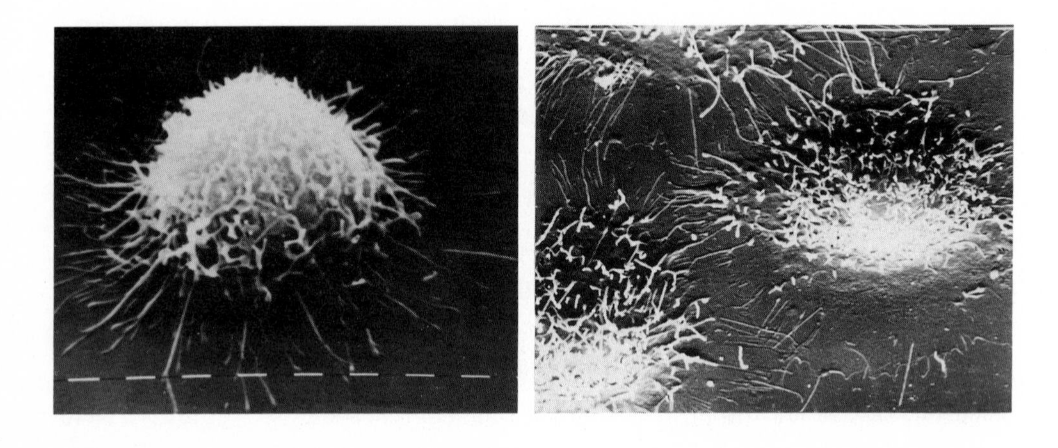

Fig.1 Adhesion of biotinylated Ehrlich ascites carcinoma cells A:Interaction with non–modified substrate. B: Spreading of biotinylated ascites cells on avidinylated substrate. Scanning electron microscopy, Bars=10um.

One of the possibilities is that biotinylation restores the activity of defective adhesion molecules which remain in the plasma membrane of ascitic cells. However it seems that other plasma membrane components can also play the role of adhesion molecules. We have developed an alternative method of plasma membrane modification with biotin residues. Suspended ascites carcinoma cells were incubated with biotinylated lipid, *N* – [(*6* – biotinylamido) hexanoyl] – phosphatidylethanolamine dioleoyl. These cells acquire the ability to attach to the substrate, to develop well – defined epithelioid morphology, and to become fully adhesive. Thus independently of the specific molecular mechanism of cell – substrate adhesion the entire intracellular machinery is turned on resulting in cell spreading and triggering intracellular signalling systems, in particular *pH(i)*–shift.

3. CELL ADHESION AND INTRACELLULAR *PH*

During the last years several works on signal transmission after cell stimulation by soluble growth factors were established. One of them is activation of Na^+/H^+ – antiporter and the resulting increase of *pH(i)* by *0.2–0.3* units (Grinstein *et al* 1984; Pouyssegur *et al.* 1984,1985, Moolenaar 1986). By measuring *pH(i)* on a single–cell basis using *pH* – sensitive dye *2′,7′*,bis(carboxyethyl) – carboxyfluorescein (*BCECF*), we have found that similar to the soluble growth factor – mediated cell stimulation, Na^+/H^+ – antiporter is also activated when cells (fibroblasts, epithelia, neutrophils) adhere to solid substrate (Margolis *et al* 1988, Schwartz *et al* 1990). When the suspended cells adhere to the solid substrate in serum – supplemented media or salt buffered solutions an increase of *pH(i)* by *0.2– 0.3* units is observed (Fig.2). Adhesion – initiated *pH(i)* – shift is reversible. The detachment of the substrate – attached cells either by trypsin or *EDTA* results in the decrease of *pH(i)* back to its initial value. Since we have measured *pH(i)* in the individual cells, we were able to establish correlation between the stage of adhesion and *pH(i)*. It seems that *pH(i)* – shift occurs in the early stages of cell spreading after the filopodia are formed, keeping the spherical cell on the substrate. Merely, physical contact with the solid substrate is not enough to initiate the *pH(i)* – shift. When non – adhesive phospholipid substrate was used, no increase of *pH(i)* was observed upon cell contact with such a surface. Together with S. Galkina we have shown that *pH(i)* in neutrophils is also increased by *0.2 – 0.3* when these cells adhere to the solid substrate. When poorly adhesive siliconized coverslips were used as substrate, *pH(i)* was either not increased or the increase was smaller than in the control.

Both in neutrophils and fibroblasts the increase in *pH(i)* was mediated by Na^+/H^+ – antiporter. Its inhibitor amiloride prevents *pH(i)* – shift and affects cell adhesion. The cells do attach to the substrate, but neither spread, nor develop lamellapodia. In Na^+–free medium, *pH(i)* of fibroblasts also remains low for at least an hour. The attached cells do not spread. However when the medium was changed for that containing *Na* the *pH(i)* immediately increased by *0.2 + 0.04* units.

Thus the *pH(i)* – shift plays an important role in cell adhesion. Since it was shown for soluble growth factors that when *pH(i)*–shift is prevented the cells do not enter *S*–phase (Pouyssegur *et al* 1984), one can suppose that adhesion– and spreading–initiated *pH(i)*–shift is one of the key elements in substrate dependence of cell proliferation. However spreading is not sufficient, since the presence of growth factors is also necessary for cell proliferation. The effects of adhesion and cell binding of soluble growth factors on *pH(i)* were additive.

Addition of Concanavalin *A* or serum growth factors to adhered cells increased *pH(i)* by another *0.2 – 0.3* units.

Why can Ehrlich ascites cells proliferate substrate–independently? These cells proliferate in a low–*pH* environment of ascitic fluid (*pH=6.8* or less). Ehrlich ascites carcinoma cells synthesize *DNA* in such a medium because synthesizing machinery of these cells is adapted to low *pH*. Our *pH(i)* measurements (Margolis *et al* 1989) demonstrate that *DNA* synthesis normally proceeds in these cells even when *pH(i)* is *6.5* and is arrested only at *pH(i) = 6.3*. Thus these cells are relatively *pH(i)* – independent. Since one of the roles of the cell spreading is to increase *pH(i)*, *pH(i)* – independence could partially explain why Ehrlich ascites cells do not require the solid substrate for proliferation. However if the cells are reconstituted with the new adhesive molecules, as described above, they become *pH(i)*–dependent: when the biotinylated cells spread on the avidinylated substrate, *pH(i)* is shifted upward by *0.2 – 0.3* units.

It seems that *pH(i)* – shift is initiated by occupation of cell surface sites involved in adhesion. *pH(i)* – shift was observed when cells contact each other in the sparse cultures. In terms of *pH(i)* – shift fibroblast adhesion can be simulated by incubation of cells with *RGD* – peptide which binds to integrins. Cell – cell adhesion can be simulated also by cell interaction with model membranes (liposomes). In that case *pH(i)* is also shifted (Galkina *et al* 1991).

Small *pH(i)* – shifts control also specialized cell – cell contacts of gap junction type (Budunova *et al* 1991). Gap junction–mediated cell – cell communication is regarded as an important factor controlling the development and differentiation of various tissues. Tumour promoters, including 12 – O – tetradecanoylphorbol – 13 – acetate, *TPA* inhibit gap junction permeability (Trosco *et al* 1982; Enomoto *et al* 1984) and simultaneously increase *pH(i)* by *0.1 – 0.3* units (Burns and Rozengurt 1983, Moolenaar *et al* 1984) (both effects are presumably mediated by activation of proteinkinase C (*PKC*)). To find out whether the increase of *pH(i)* is important for inhibition of cell – cell contact permeability we developed an experimental system where *pH(i)* can be shifted without *TPA* and *TPA* – induced *pH(i)* – shift can be prevented.

We experimentally stabilized *pH(i)* of cell cultures at the desired level using the exogenous K^+/H^+ – antiporter nigericin. In high – K^+ medium where the concentration of K^+ inside the cell is equal to that outside, nigericin equilibrates the concentration of protons bringing the *pH(i)* equal to the *pH* of the outer medium (Thomas *et al* 1979). *TPA* has no effect on gap junction permeability if by incubation in high – K^+ medium with nigericin *pH(i)* was fixed at a certain level between *7.0 and 7.5*. The *pH(i)* – shift alone (achieved by

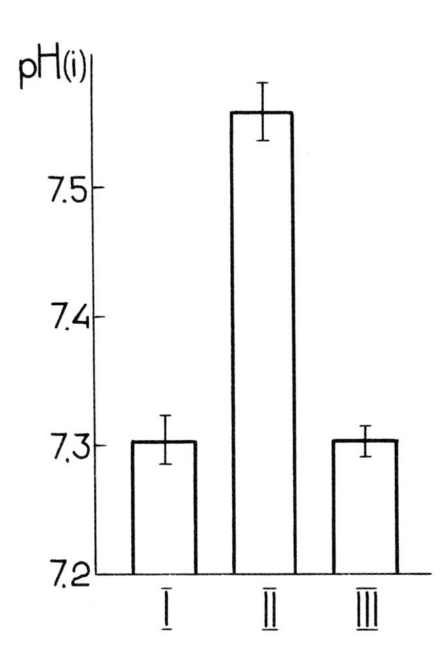

Fig.2 pH(i)–shift in fibroblasts adhered to the solid substrate. I–suspended cells; II–10 min after attachment; III–40 min on non–adhesive substrate. pH(i) of suspended and attached cells was measured on single–cell basis using microspectrophotometer–equipped microscope.

transferring the cultures from one nigericin – containing high – K^+ medium at one *pH* level to a similar medium but at higher *pH*) also had no effect on cell – cell communications without *TPA* (Fig.3). Cell – cell communication was inhibited only if *pH(i)* was shifted up by *0.1 – 0.3* units and the cells were treated with *TPA* (Fig.3). Thus an upward *pH(i)* – shift is necessary for *TPA* – induced inhibition of gap junction permeability.

That was true independently of the initial and final levels of *pH(i)*. We shifted *pH(i)* of *TPA*–treated cells upward by *0.1 – 0.5* units beginning from different initial *pH(i)* between *6.9 – 7.3*. All of these shifts allowed inhibition of gap junction permeability by *TPA* (Fig.3).

Thus it is the change in *pH(i)* rather than an upward shift to a particular level of *pH(i)* that is required for *TPA* – mediated inhibition of gap junction permeability. *pH(i)*–shifts, rather than the actual level of *pH(i)*, may be essential for other cellular functions where the involvement of small changes in *pH(i)* have been reported. Our results on *pH(i)* – shift during cell adhesion to solid substrate also confirm the conclusion about the importance of *pH(i)* – shifts rather than the absolute *pH(i)* – levels themselves.

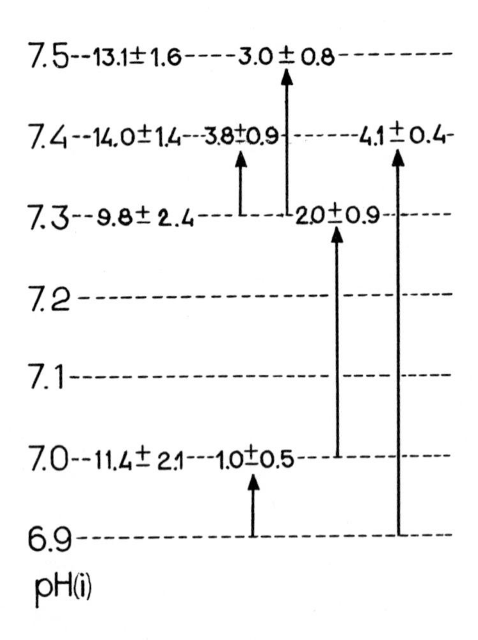

Fig.3 Inhibition of gap junction permeability by TPA depends on pH(i)– shift. Vertical axis: pH(i) levels. Figures in the chart correspond to gap junction permeability at given level of pH(i). Cells were treated with TPA 2x10 M. Using nigericin and high–K^+, pH(i) was shifted from one level to another as indicated by the arrows. The figures in the first column (without arrows) correspond to the cultures were pH(i) was not shifted. Gap junction permeability was evaluated as the number of neighbouring cells stained by Lucifer yellow when the dye was injected into one of them.

4. CONCLUSIONS

Adhesion and spreading upon solid substrate is one of the basic cellular functions requiring dramatic local deformation of the plasma membrane. It is thought that the main driving force of this process is the reorganization of the cytoskeleton. However from our experiments, it seems that by applying an external force to the cell membrane any intracellular force (*e.g.,* osmotic pressure or cytoplasmic streaming) would produce similar kinds of local deformations even in the absence of active rearrangements of the cytoskeleton elements.

Cell adhesion and spreading requires binding of the membrane to the substrate "adhesion molecules". Few types of specific adhesion molecules have been isolated and few works on their membrane localization has been discussed. Our results indicate that the actual molecular organization for such membrane molecules does not seem to be crucial: simple biotin residues covalently bound to the cell surface can fully play this role. They restore the adhesive and spreading potentials of non–adhesive ascitic cells and they trigger the intracellular signalling system. During the cell spreading the signals to enter the proliferation pool are generated. One of the key elements in the signalling system is activation of Na^+/H^+–antiporter and the increase of *pH(i)*. *pH(i)*–shifts, rather than the absolute *pH(i)* values play an important role in regulation of cell–cell and cell–substrate contact interactions.

5. REFERENCES

Bogdanov A.A., Gordeeva L.V., Baibakov B.A., Margolis L.B., Torchilin V.P. (1991). Restoration of adhesive potentials of Ehrlich ascites carcinoma cells by modification of plasma membrane. J Cell Phys 147:182–190

Budunova I., Mittelman L.A., Margolis L.B. (1991). An upward shift of intracellular *pH* rather than the final absolute *pH* value is critical for controlling gap junction permeability in tumor promotor–treated cells. FEBS Lett 279:52–54

Burns C.P.F., Rozengurt E. (1983). Serum, platelet–derived growth factor, vasopressin and phorbol esters increase intracellular *pH* in Swiss 3T3 cells. Biochim Biophys Acta 116:931–938

Enomoto T., Yamasaki H. (1985). Phorbol ester–mediated inhibition of intercellular communication in BALB/c 3T3 cells: relationship to enhancement of cell transformation. Cancer Res 45:2681–2688

Folkman J., Greenspan H.P. (1975). Influence of geometry on control of cell growth. Biochim Biophys Acta 417:211–245

Folkman J., Moscona A. (1978) Role of cell shape in growth control. Nature 273:345–349

Galkina S.I., Sudjina G.A., Margolis L.B. (1991). The effects of neutrophil and fibroblast adhesion on activation of Na^+/H^+–antiporter (in Russian). Biol Membranes 8:621–627

Gass G.V., Chernomordik L.V., Margolis L.B. (1991). (*To be published)* Local deformation of human red blood cells in high frequency electric field. FEBS Lett

Green N.M. (1963). The use of biotin C^{14} for kinetic studies and for assay. Biochem J 89:585–591.

Grinstein S., Cohen S., Rothstein A. (1984). Cytoplasmic *pH* regulation in thymic lymphocytes by an amiloride–sensitive Na^{+}/H^{+}–antiport. J Gen Physiol. 83:341–368

Margolis L.B., Rozovskaja I.A., Cragoe J. (1988). Intracellular *pH* and adhesion to solid substrate. FEBS Lett 234:449–450

Margolis L.B., Novikova I.Y.U., Rozovskaya I.A., Skulachev V.P. (1989). K^{+}/H^{+}–antiporter nigericin arrests DNA synthesis in Ehrlich ascites carcinona cells. Proc Natl Acad Sci USA 86:6626–6629

Moolenaar W.H. (1986). Effects of growth factors on intracellular *pH* regulation. Annu Rev Physiol.48:363–376

Moolenaar W.H., Tertoolen L.G.J., de Laat S.W. (1984). Phorbol ester and diacilglycerol mimic growth factors in raising cytoplasmic *pH*. Nature 312:371–374

Oster G. (1989). Cell motility and tissue morphogenesis. In: Stein MF, Bronner F (eds) Cell Shape Determinants. Regulation and Regulatory Role. Academic Press, San Diego, p.33

Pouyssegur J., Sardet C., Franchi A., L'Allemain G., Paris S. (1984). A specific mutation abolishes Na^{+}/H^{+}–antiport in hamster fibroblasts precludes growth at neutral and acidic *pH*. Proc Natl Acad Sci USA 81: 4833–4837

Pouyssegur J., Franchi A., L'Allemain G., Paris S. (1985). Cytoplasmic *pH,* a key determinant of growth factor–induced *DNA* synthesis in quiscent fibroblasts *FEBS* Lett 190:115–119

Popov S.V., Margolis L.B. (1988). Formation of cell outgrowth by external force: a model study. J Cell Sci 90:379–389

Schwartz M.A., Cragoe J.E., Lechene C. (1990). *pH* regulation in spread cells and round cells. J. Biochem 265: 1327–1332

Thomas J.A. Buchsbaum R.N., Zimnjak A., Racker E. (1979). Intracellular *pH* measurements in Ehrlich ascites tumor cells using spectroscopic probes generated in situ. Biochemistry 18: 2210–2218

Trinkaus J.B. (1985). Protrusive activity of the cell surface and the initiation of cell movement during morphogenesis. Exp Biol Med. 10:130–173

Trosco J.E., Yotti L.P., Warren, S.T., Tsushimoto G., Chang C.C. (1982). Inhibition of cell–cell communication by tumor promotors. In:Hecker E, Fuseni NE Kunz W, Marks F, Thielman H.W. (eds) Cancerogenesis, Raven Press, N.Y. p.556

Vasiliev Yu.M.(1985). Spreading of transformed and non–transformed cells. Biochim Biophys Acta 780:21–65

Part 4

Function and Deformation of Tissues

The Role of Tissue Swelling in Modelling of Microvascular Exchange

D.G. Taylor
Department of Chemical Engineering
University of Ottawa
161 Louis Pasteur
Ottawa, Ontario
K1N 6N5

1 Introduction

Connective tissues, such as skin, skeletal muscle, lung, and heart, comprise a large fraction of the human body. The extra-vascular, extra-cellular fluid filled space of the tissue (i.e., the interstitium) forms a continuum between the body's two circulatory systems: namely, the blood circulation and the lymphatic system. Fluid and solutes, including the plasma proteins, pass between the blood and the tissue spaces via the exchange vessels of the blood microcirculation (predominantly the capillaries). The initial lymphatic vessels, meanwhile, drain interstitial fluid, which contains cellular metabolites. Together, the interstitium and the exchange vessels of the blood circulation and the lymphatic network constitute the microvascular exchange system. Disturbances in the distribution of fluid and solutes within the microvascular exchange system can have

---

J.L. Bert
Department of Chemical Engineering
University of British Columbia
2216 Main Mall
Vancouver, B.C. V6T 1W5

R.K. Reed
Department of Physiology
University of Bergen
Arstadveien 19
Bergen, Norway

NATO ASI Series, Vol. H 64
Mechanics of Swelling
Edited by T. K. Karalis

life-threatening consequences, as in the case of severe burn injuries.

Figure (1) is a schematic diagram showing the organization of a typical connective tissue interstitium. The interstitial space contains collagenous and elastin fibers, high molecular weight polymers such as hyaluronan and other glycosaminoglycans and their aggregates, and plasma proteins, all within a fluid environment. Together, these components impart to the interstitium its physicochemical properties.

Figure (1). A schematic diagram of the interstitium containing hyaluronan and glycosaminoglycans, proteoglycans, elastin, collagen, interstitial plasma proteins and tissue cells (modified after Bert and Pearce, 1984).

Depending on the relative rates of flow into and out of the interstitium and its physicochemical characteristics, the interstitial space will either expand, dehydrate, or maintain a constant volume. One important property in this regard is the interstitium's swelling characteristics. (Further aspects of the swelling characteristics of connective tissue are discussed in the manuscript by Reed et al. appearing in this publication). In some tissues the interstitium may swell several times its normal volume when soaked in a fluid bath. *In vivo*, a normal state of interstital hydration and fluid composition is maintained through the dynamic and complex set of processes governing fluid and macromolecular exchange within the microcirculation.

2 Mathematical Models of Microvascular Exchange

To assess conditions within the interstitium, it is useful to formulate mathematical models of the microvascular exchange system. These models are often the only means of studying the behaviour of the overall system, as experimental measurements may be impractical or impossible. They therefore serve as useful tools for studying the interactions between components of the system, for assessing the relative influence of various properties of the microvascular exchange system on its behaviour, and for suggesting future experiments. Mathematical models can also be used as educational tools and as aids to the management of patients suffering from disturbances to the microvascular exchange system.

We have developed separately two types of models to study the microvascular exchange system. The first of these, the compartmental models, treat the interstitium as a well-mixed compartment (i.e., its properties are considered to be spatially invariant). The second class of models, namely the spatially distributed models, allow for conditions within the interstitium to vary with position. We will report briefly on our efforts with both models, noting that, while we have tested the effect of swelling characteristics on microvascular exchange using the compartmental models, we are still in the developmental stage with respect to the distributed model.

2.1 The Compartmental Models

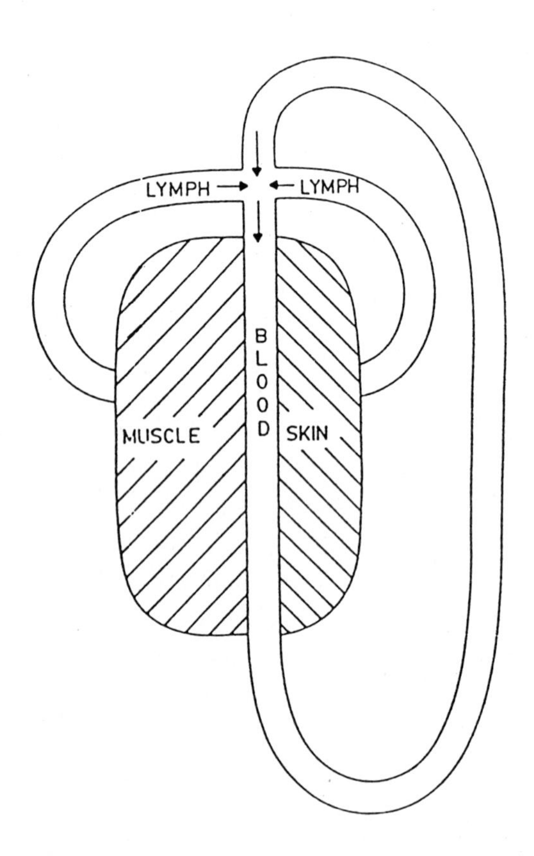

Figure (2). A schematic diagram of the rat microcirculation (from Bert et al, 1988: permission requested).

Figure (2) shows a schematic diagram of a simplified rat microvascular exchange system. Only the two largest tissue compartments that are expected to have the greatest effect on the whole animal (i.e., muscle and skin) are considered in this model. The mathematical formulation is based on mass balances for both fluid and plasma proteins within the two interstitial compartments and within the circulatory compartment. Assuming a closed system, total mass is conserved. However, the balance equations can be altered to consider the addition or removal of material to or from the system (as might occur, for example, in fluid resuscitation).

Fluid exchange between the blood and tissue compartments is described using a fluid filtration relationship such as Starling's Hypothesis, while macromolecular exchange is described by equations that consider both diffusive and convective transport across the capillary wall. To completely describe the system, many phenomenological parameters and constitutive relationships must be quantified. Their selection and/or estimation are detailed in the original works (Bert et al, 1988; Reed et al, 1989).

Simulations of Interstitial Swelling following Fluid Infusion in the Rat

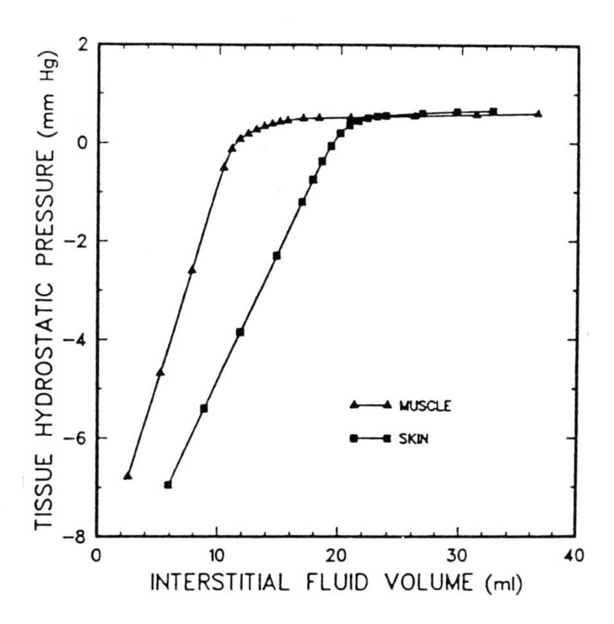

Figure (3). The compliance (tissue pressure versus interstitial fluid volume) curve for muscle and for skin in the rat (from Bert et al, 1988: permission requested).

We will focus here on the swelling characteristics of the interstitium. Figure (3) shows examples of interstitial volume versus hydrostatic pressure relationships, termed compliance curves. Typical features of the compliance curve are a linear but rapidly increasing pressure with increasing volume at low interstitial volumes (i.e., the dehydrated state), a region in which pressure increases linearly but only slightly with increasing volume at high interstitial volumes (i.e., the overhydrated state), and an intermediate region where the pressure-volume relationship is nonlinear. At extreme overhydration, one can expect the influence of boundaries to alter the linear behaviour of the curve.

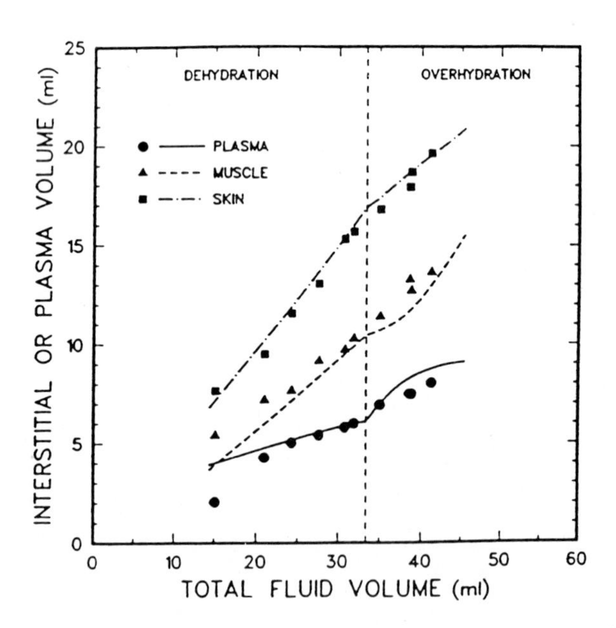

Figure (4). Experimental data and model predictions of compartmental volumes following fluid addition in the rat (from Bert et al 1988: permission requested).

The compliance characteristics proved to be of critical importance in the simulation of experimental data such as that shown in Figure (4). In this case the model was used to predict the volumes of the three compartments following the addition of fluid to the animal. Of interest is the observation that, at a high degree of overhydration, the volume of the plasma compartment is much better regulated than that of the tissue compartments, due to a loss of buffering capability against edema within the interstitium at those levels of hydration. Based on these and other simulations, we were able to conclude that the system required much longer times than previously thought to reach steady-state (approximately 30 hours following some perturbations). The model was also used to investigate the effects that different parts of the system had on the distribution and transport of fluid and plasma proteins following various perturbations, both during the transient periods and subsequent steady-states. Further, the auto-regulatory property of the compliance relationship (i.e., the ability of the tissue's swelling property to influence homeostasis with respect to interstitial fluid volume) was examined.

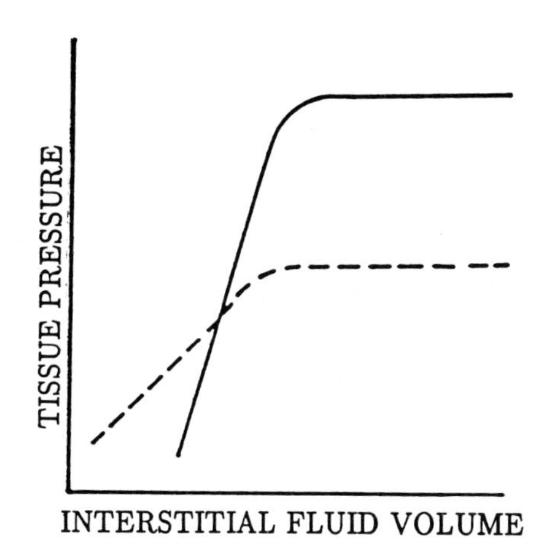

Figure (5). Modified compliance curves used in simulations of microvascular exchange in the rat. The solid line corresponds to the compliance curve of Figure (4). The dashed line corresponds to a slower rate of increase in tissue pressure with increasing interstitial fluid volume at lower states of hydration, and lower values of tissue pressure in the normal and overhydrated regions (modified after Reed et al, 1989).

The particular choice of the compliance relationships is a source of considerable controversy. To test the appropriateness of the compliance relationship used in these simulations and its influence on the model predictions, we replaced the curves shown in Figure (3) with relationships of the shape shown in Figure (5), as suggested by some investigators. The simulations using the second set of curves yielded results that were less in keeping with experimental data, compared to the results obtained with the original set of curves. From this we concluded that the compliance relationships of Figure (3) more closely represent the tissue pressure-volume characteristics of muscle and skin in the rat.

Simulation of Interstitial Swelling following Burn Injury in the Rat

An example of the dramatic effect that tissue swelling characteristics have on microvascular exchange can be found in the case of thermal injuries. The physicochemical changes to skin that result from a burn can produce transient, negative pressures of considerable magnitude (see the article by Reed et al in this publication). These changes have a profound effect on microvascular exchange following the injury. To simulate the tissue following thermal injury, the skin compartment was divided into burned and unburned subcompartments. Normal compliance relationships were used in the intact skin compartment, while experimentally measured tissue pressures served as model inputs for the injured skin compartment. Both the lower swelling pressure and altered capillary wall transport characteristics of this latter compartment contributed to the post-burn dynamics as modelled and partially validated in some of our studies.

Of particular clinical interest are the fluid management practices for burn victims. Figure (6) shows the swelling pressure in the injured skin of the rat versus time for three different regimens of fluid treatment (no fluid resuscitation, resuscitation with Ringer's solution, and resuscitation with rat plasma). The exact mechanisms responsible for the highly negative tissue pressure in the injured skin and the resolution of the injury towards a normal condition remain unexplained. However, simulations have been performed using the data of Figure (6) to compare the different fluid therapies outlined above. Some results are shown in Figure (7). With an understanding of the swelling phenomena following burn injury and through the use of the models, the effects of increased driving forces for fluid exchange within the injured tissue and changes in the capillary wall resistance to fluid flow can be uncoupled and assessed.

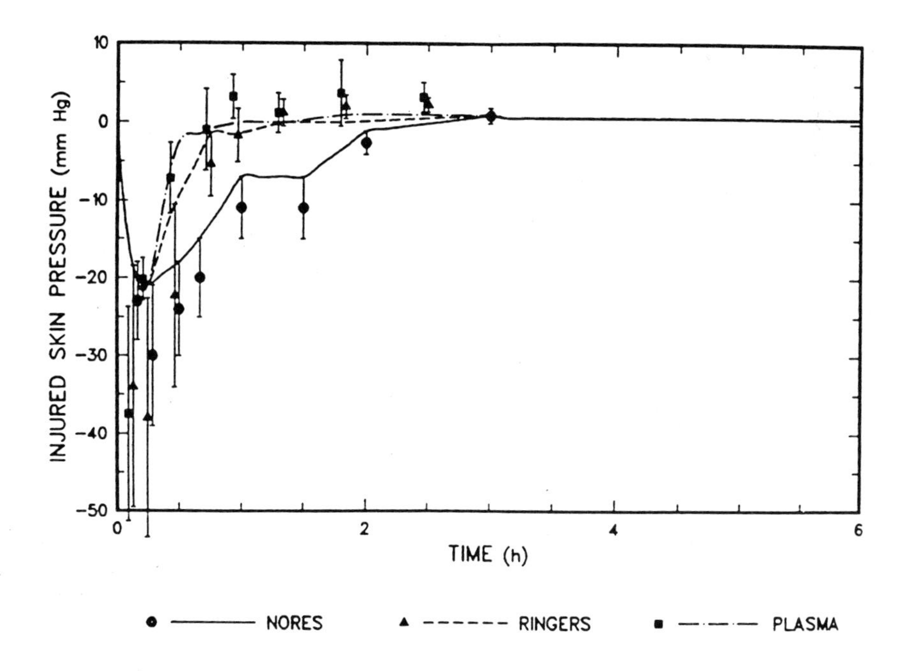

Figure (6). Transient swelling pressure following thermal injury in the rat. The three curves correspond to three resuscitation schemes: no fluid resuscitation (NORES), fluid resuscitation with Ringer's solution (RINGERS), and fluid resuscitation with rat plasma (PLASMA) (from Bert et al, In Press: permission requested).

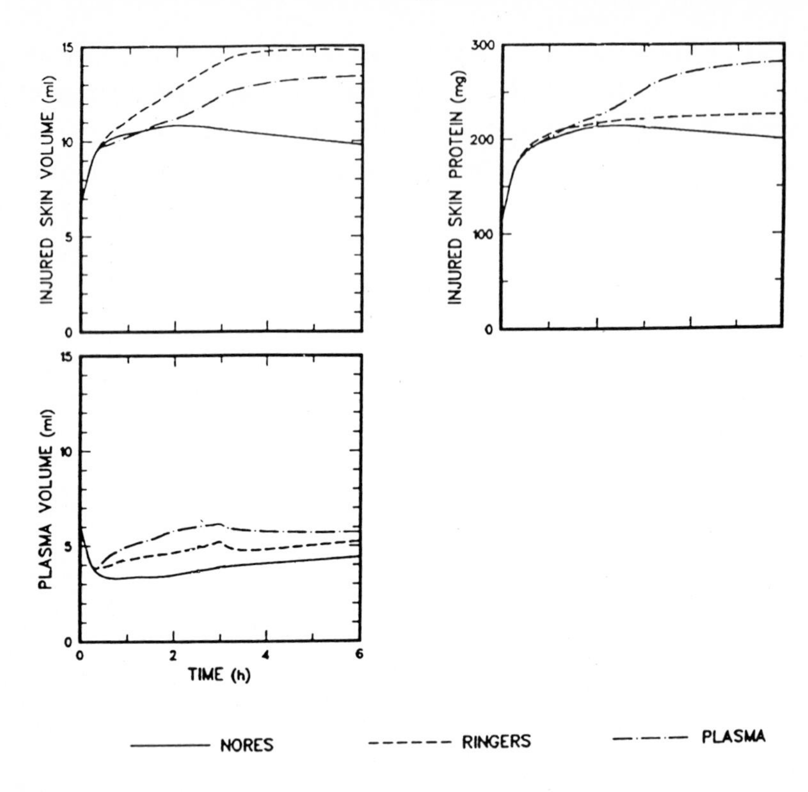

Figure (7). Transient fluid volume and protein content (injured skin) as a function of fluid resuscitation in the rat. The solid line corresponds to no resuscitation, the dashed line represents resuscitation with Ringer's, and the chain-dotted line corresponds to resuscitation with rat plasma (modified after Bert et al, In Press).

Simulation of Interstitial Swelling in Humans

Very little is known about the compliance characteristics of human tissue. Hence, an apparent weakness in the development of a model of human microvascular exchange is the absence of a pressure-volume relationship for any tissue of interest. From the available data concerning the few isolated measurements of subcutis swelling pressure and the degree of hydration (see the review by Wiig, 1990), and by assuming behaviour and properties similar to other mammalian tissues, a hypothetical compliance curve for human skin was constructed. What we termed the "most likely" compliance characteristics for human interstitial tissue are shown in Figure (8). Unfortunately, the greatest degree of uncertainty in these curves occurs in the overhydration range, which also corresponds to the region of greatest clinical importance. Therefore, during our model studies we varied the slope of the overhydration region between three different values.

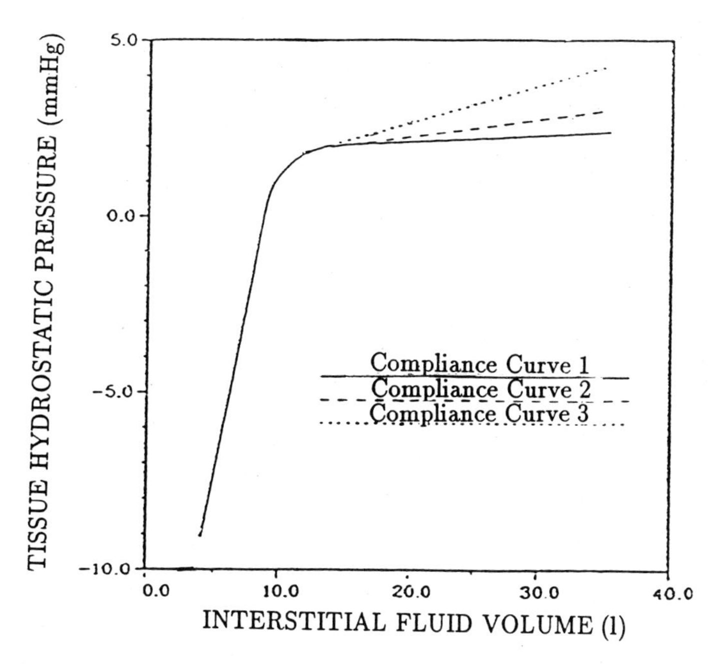

Figure (8). Compliance relationships in the human (from Chapple, 1990).

Our search for experimental data against which to test the model indicated a paucity of information concerning interstitial fluid volumes and/or plasma protein distribution in human pathological states. Results from studies on nephrotic patients served as the primary data source for parameter estimation. A statistical analysis of the results of our parameter estimation study revealed that the compliance curve with the largest slope in the overhydration region consistently yielded the best fits of the data. Further, of all the fitted parameters and relationships studied, the compliance curve proved to be the most sensitive. The data on nephrotic patients, together with the model predictions using the high compliance characteristic curve, are shown in Figure (9). Clearly, a better understanding of the swelling characteristics of human tissues and reliable experimental or clinical data are required to improve our understanding of, and ability to simulate, microvascular exchange in humans.

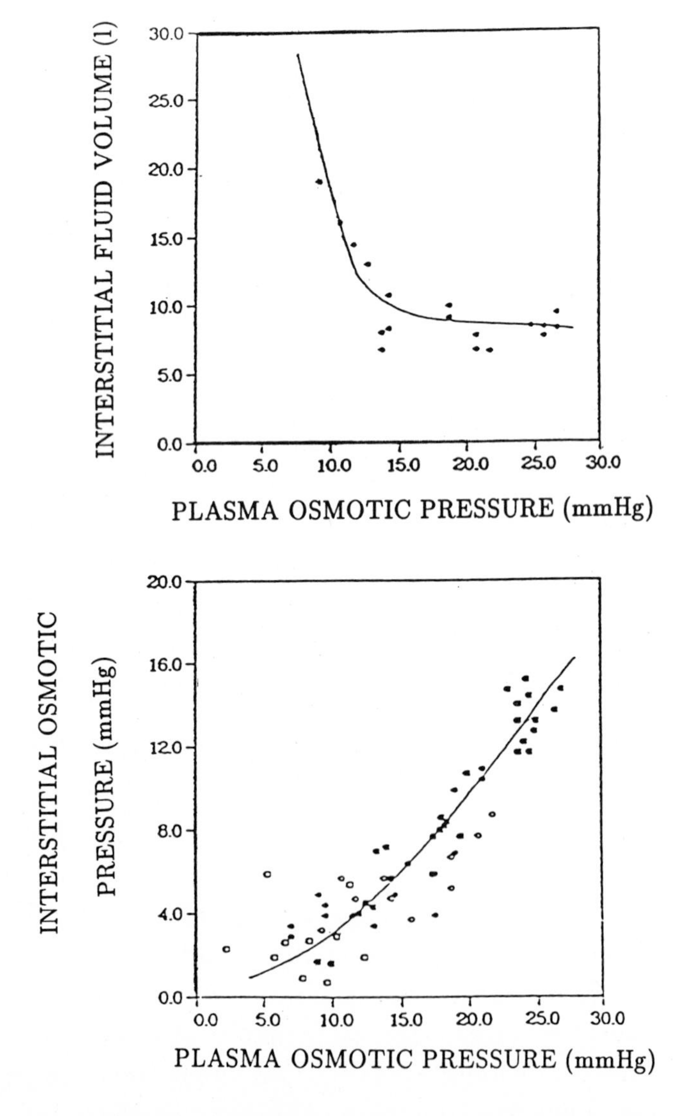

Figure (9). Model prediction versus experimental data in nephrotic humans (from Chapple, 1990).

2.2 A Distributed Model of the Interstitium

In the distributed model (Taylor et al, 1990) we treat the interstitium as a multiphase system which is under local thermodynamic equilibrium. Since fluid flow within the interstitium is of a creeping nature, it is described in the model by a form of Darcy's Law in which the driving force is the gradient in fluid chemical potential. In this way both hydrostatic and colloid osmotic pressure effects are included. Solute transport, meanwhile, is assumed to occur by hindered convection, restricted molecular diffusion, and convective dispersion. Tissue swelling is accounted for through a compliance relationship and by assuming that local deformation at any point within the interstitium is a function of the local fluid hydrostatic pressure. The analysis is based on classical elastic theory, using Terzaghi's principle of effective stress (Terzaghi, 1943), and so is limited to small deformations only.

To formulate mathematical expressions describing the distribution of fluid and macromolecules within the interstitium, we must first develop a continuum representation of the space. We begin by applying a volume averaging procedure to a representative volume of interstitial space, as illustrated in Figure (10). Volume A in Figure (10) represents a representative portion of interstitium, centered about some point P and containing the various components comprising the tissue space, such as collagen, the glycosaminoglycans and their aggregates, fluid, and interstitial plasma proteins. Through the averaging procedure, each phase within the volume is replaced by a hypothetical continuum, as shown in volume B of Figure (10). The properties of these continua represent spatial averages of the properties and processes occurring locally in the physical system, and are associated with the point P about which the representative volume is centered. In other words, the properties associated with a point within the continuum represent an average of the properties of a small volume of interstitium surrounding that point. By repeating this averaging procedure at each point within the interstitium, the original system is transformed into a multiphase continuum amenable to analysis using differential calculus.

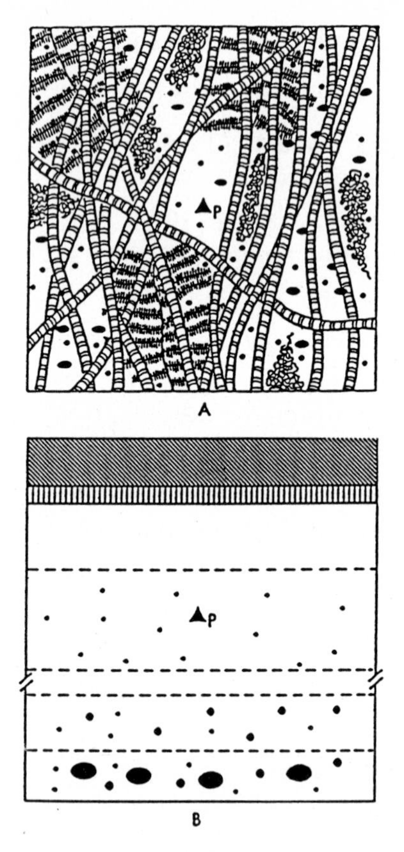

Figure (10). A schematic diagram of an elemental volume of interstitium A) before and B) after volume averaging (from Taylor et al, 1990).

Following this procedure, and assuming that m plasma protein species are present within the volume, then m+3 distinct volume fractions arise from the averaging procedure:

1. a total mobile fluid volume fraction;
2. m volume fractions corresponding to the distribution volume fractions of the m protein species;
3. a solid phase consisting of the structural elements (e.g., the

collagen, elastin, glycosaminoglycans, proteoglycans, and their aggregates); and

4. an immobile fluid volume fraction representing trapped or bound water.

We assume further that the plasma proteins are uniformly distributed throughout their respective distribution volumes. Each plasma protein species also displays its own excluded volume fraction (i.e., that fraction of the total fluid volume that is devoid of that protein species), where the excluded volume fraction of a solute species depends on the amount of solid phase present. Likewise, the immobile fluid volume fraction is some function of the solid phase volume fraction.

In considering swelling, we have assumed that the tissue behaves as an elastic porous medium. This implies that any viscoelastic behavior is attributed to fluid movement within the interstitial space, and that the solid skeleton of the interstitium behaves as an elastic body. A rigourous analysis of tissue swelling would then require a complete description of the stress distribution within the interstitium, coupled with the appropriate elastic stress-strain relationships. We have adopted a simpler approach, based in part on Terzaghi's analysis of soil deformation (Terzaghi, 1943) and employing the interstitial compliance relationship. In this way the local volumetric dilation of the interstitium is related to the average local hydrostatic pressure. Differential material balances are performed on the mobile fluid, the solid constituents (including the immobile fluid), and the macromolecular species. These yield a set of coupled partial differential equations describing the transient distribution of fluid and macromolecules within the interstitium. The governing equations must then be combined with appropriate boundary and initial conditions to complete the mathematical description.

3 Summary

The microvascular exchange system is the site for the transfer of fluid, nutrients, and metabolic wastes between the blood stream and the body tissues. This process of exchange involves three components: the blood stream, the interstitium, and the lymphatics. A proper distribution of fluid between these compartments is critical for maintaining the health and well-being of the individual. However in many types of systemic disturbances, considerable amounts of fluid can shift from the blood compartment to the interstitium, resulting in a loss of blood volume and extensive tissue swelling.

We have discussed here the use of compartmental and distributed models of the interstitium to investigate the processes governing microvascular exchange under normal and pathological conditions, based on our own work in this area and with an emphasis on tissue swelling characteristics. Much more effort is needed however, both experimentally and theoretically, to improve these models. In particular, the tissue compliance characteristics have been shown to be a critical property of the overall exchange system. Forthcoming quantitative information related to tissue swelling properties and an improved understanding of the swelling mechanisms will no doubt aid in our efforts to simulate the distribution and transport of fluid and plasma proteins within the microvascular exchange system.

References

Bert JL, Bowen BD, Reed RK (1988) Microvascular exchange and interstitial volume regulation in the rat: model validation. Am J. Physiol 254:H384-H399

Bert JL, Bowen BD, Reed RK, Onarheim H. (In Press) Microvascular exchange during burn injury IV: fluid resuscitation model. Circ Shock

Bert JL, Pearce RH (1984) The interstitium and microvascular exchange. In: Renkin E, Michel C (eds) Handbook of physiology. The cardiovascular system. Microcirculation. American Physiological Society, Bethesda, pp 521-547

Chapple C (1990) A compartmental model of microvascular exchange in humans. MASc thesis. University of British Columbia, Vancouver

Reed RK, Bowen BD, Bert JL (1989) Microvascular exchange and interstitial volume regulation in the rat: implications of the model. Am J Physiol 257: H2081-H2091

Taylor DG, Bert JL, Bowen BD (1990) A mathematical model of interstitial transport I. theory. Microvasc Res 39: 253-278
Terzaghi K (1943) Theoretical soil mechanics. Wiley, New York
Wiig H (1990) Evaluation of methodologies for measurement of interstitial fluid pressure (Pi): physiological implications of recent Pi data. CRC Crit Rev Biomed Eng 18: 27-54

INTERSTITIAL FLUID PRESSURE IN CONTROL OF INTERSTITIAL FLUID VOLUME DURING NORMAL CONDITIONS, INJURY AND INFLAMMATION

R. K. Reed, H. Wiig, T. Lund,
S. Å. Rodt, M.-E. Koller and G. Østgaard
Department of Physiology
University of Bergen
Årstadveien 19
N-5009 Bergen
Norway

Introduction.

The interstitial space is the intercellular and extravascular compartment which is present in all tissues and which comprise from a few percent of the total tissue weight in brain to forty percent of the tissue weight in skin (Aukland & Nicolaysen 1981). The fluid contained in this compartment is an ultrafiltrate of plasma and contains proteins and electrolytes. The fluid volume of the interstitial compartment is governed by the influx of

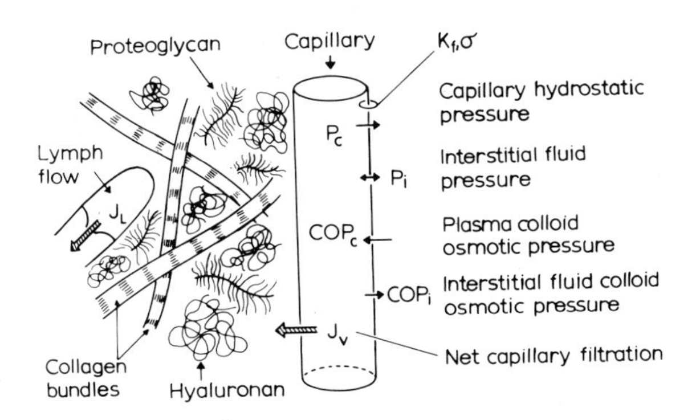

Figure 1. Schematic drawing of the interstitium (not to scale) and the pressures determining the transcapillary fluid flux. $K_{f,c}$ is the capillary filtration coefficient. (From Wiig 1990 with permission).

NATO ASI Series, Vol. H 64
Mechanics of Swelling
Edited by T. K. Karalis

fluid across the capillary wall and the subsequent drainage of fluid via the lymphatics. The general outline of this exchange system is shown in Fig. 1 where the fluid flux across the capillaries (J_V) is generated as a result of the transcapillary net filtration pressure, i.e. the net pressure gradient created by the transcapillary colloid osmotic (COP) and hydrostatic (P) pressures. The magnitude of the transcapillary fluid flux is the product of the transcapillary net filtration pressure and the capillary filtration coefficient ($K_{f,c}$).

The main focus of the present review is on the role of the interstitial fluid pressure (P_i) and the interstitial compliance in the control of the interstitial fluid volume in skin during normal conditions as well as during injury and inflammation. P_i is the pressure in the interstitial fluid acting across the capillary wall, participating in the transcapillary fluid exchange and also the pressure that serves as filling pressure for the initial lymphatics. P_i is usually measured with "fluid equlibration techniques" (Guyton, Granger & Taylor 1971), i.e. with a transducer which is connected to the interstitium via a fluid-filled column. The size of the fluid-filled column varies from a sharpened glasscapillary with tip diameter of a few μm to needles with the size of up to 1-2 mm (see Wiig 1990). The pressures obtained with these techniques in peripheral tissue like skin is generally reported to be -1 to -2 mmHg and somewhat closer to ambient pressure in skeletal muscle (Wiig 1990). The previous controversy of the magnitude of the negativity of P_i seems to be resolved since simultaneous measurement with different techniques in the same tissue yields similar pressures (see Wiig, Reed & Aukland 1987, Wiig 1990).

Role of P_i in the normal turnover of interstitial fluid volume.

Normally the transcapillary pressure gradient is about 0.5 mmHg in skin (Aukland & Nicolaysen 1981, Reed & Rodt 1991). The interstitial fluid volume in skin is 0.4 ml per g wet tissue weight (Wiig & Reed 1981) and the transcapillary net filtration pressure is sufficient to filter a volume equal to the interstitial fluid volume every 12 to 24 hours (Reed, Johansen and Noddeland 1985). The slightly negative interstitial fluid pressure will favour filtration of fluid out of the capillaries, but at the same time any excess filtration will increase the interstitial fluid volume and thereby P_i to directly oppose further filtration. The opposite will be the situation during dehydration. Then, P_i is reduced through a reduction in interstitial fluid volume, and again P_i will oppose the perturbation in capillary filtration.

It should be noted that the balance between capillary filtration and lymph flow is what normally keeps the interstitium in a slightly dehydrated state. When a biopsy is taken from normally hydrated skin or umbilical cord and soaked in saline, the tissue specimen will imbibe water to increase its water content two to three times above that seen during the normally hydrated tissue (Brace & Guyton 1979, Meyer 1983). When the same amount of swelling occurs in the tissue, the capillary fluid filtration greatly exceeds the normal lymphatic removal of fluid and a new steady state can be reached where interstitial fluid volume is greatly increased and edema (i.e. visible of swelling of the tissue) has resulted (See below).

Compliance of skin and skeletal muscle.

Results from direct and simultaneous measurement of interstitial fluid volume (extravascular ^{51}Cr-EDTA space after nephrectomy) and P_i in dog skin is shown in Fig. 2 (Wiig & Reed 1987). Compliance was measured in dog skin and skeletal muscle which together contains about 2/3 of the extracellular fluid of the body. The change in P_i that serves as a counterpressure against changes in net capillary filtration pressure is obtained from the interstitial compliance which is the derivative of the interstitial volume-pressure relationship with respect to pressure ($C=(\Delta\text{Interstitial fluid volume})/(\Delta\text{Interstitial fluid pressure})$). The interstitial volume-pressure relationship is linear, i.e. compliance is constant, in dehydration and in the initial stages of overhydration as illustrated in Fig. 2. Studies on skin and skeletal muscle of rats, cats and dogs show that compliance in these tissues and species are similar, and corresponding to 10-15% change in interstitial fluid volume per mmHg change in P_i (Reed & Wiig 1981, Wiig & Reed 1981, Reed & Wiig 1985, 1987). The compliance increases in overhydration and

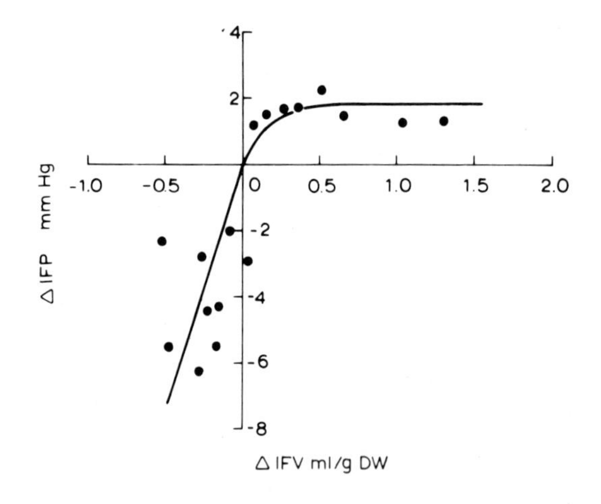

Figure 2. Interstitial volume-pressure relationship of skin (Reed & Wiig 1987). Compliance in dehydration and initial part of overhydration corresponds to 10-15% change in interstitial fluid volume (IFV) per mmHg change in interstitial fluid pressure (IFP). The maximal counterpressure towards increased filtration is 2-3 mmHg. (From Wiig & Reed 1987 with permission.)

approaches infinity when interstitial fluid volume has increased more than 50 to 100% above control. The increase in P_i that can be achieved is maximally 2-3 mmHg from control to maximal overhydration. It has been reported that compliance decreases again at excessive overhydration (Guyton, Granger & Taylor 1971).

The compliance value of 10-15% change in interstitial fluid volume per mmHg change in P_i implies that in the normal range of changes in interstitial fluid volume (±5% change from control) an increased capillary filtration pressure will be opposed by a rise in P_i and a reduction of interstitial colloid osmotic pressure (due to dilution of the interstitial proteins) that are of the similar magnitude (Wiig & Reed 1981). The total capacity of interstitial fluid pressure to counteract increased capillary net filtration pressure (i.e. the increase from control P_i to that achievable during maximal overhydration) is 2-3 mmHg in skin and skeletal muscle (Reed & Wiig 1981, Wiig & Reed 1981, Wiig & Reed 1985, Wiig & Reed 1987). This is in contrast to the capacity of the interstitial colloid osmotic pressure to counteract more than 10 mmHg increase in capillary net filtration pressure by lowering interstitial fluid colloid osmotic pressure through a dilution and reduction the interstitial protein mass (See below).

Edema-prevention.

Edema means "swelling" and describes a situation where the capillary transport of fluid and protein greatly exceeds the normal lymphatic removal. In skin the presence of edema implies that interstitial fluid volume is increased by at least 50 to 100%. The most common causes for the occurence of edema is reduction of the colloid osmotic pressure in plasma or increased capillary hydrostatic pressure (Cf. Fig. 1). The consequence of an increased capillary filtration is that the capillary filtrate becomes more dilute (Taylor & Granger 1984) which in turn will have two effects: First, the increase in filtered volume will increase the interstitial fluid volume and thereby P_i which then acts directly across the capillaries to prevent further fluid filtration. However, as seen from the interstitial volume-pressure relationship (Fig. 2), the total buffering capacity of P_i is limited to a rise of 2-3 mmHg. Second, the increased interstitial fluid volume will also cause a dilution of interstitial proteins and thereby a reduction in interstitial colloid osmotic pressure. A reduction of interstitial protein mass by lymphatic washout is a slower phenomenon, but will cause a net reduction of the interstitial protein mass (Aukland & Nicolaysen 1981). The capacity to counteract increased net filtration pressure through reduction of interstitial colloid osmotic pressure is considerably larger than the rise in P_i since interstitial colloid osmotic pressure can be reduced from a normal value of 10 to 12 mmHg to 1 or 2 mmHg at increased capillary filtration.

So far interstitial fluid pressure has been considered a pressure that counteracts alterations in capillary filtration pressure, despite the tissue's ability to imbibe fluid in vitro when soaked in saline. The reason for the seemingly discrepancy between these two observations is that in vivo the fluid flux into the tissue is controlled by a balance between P_i and COP_i that will counteract perturbations in the capillary fluid filtration: When transcapillary filtration increases, the increase in interstitial fluid volume causes P_i to increase and COP_i to decrease. Both changes will limit further fluid filtration. Furthermore, P_i is normally considered the main driving force for lymph formation. Increased P_i will then also limit increase in interstitial fluid volume by increasing lymph flow. The situation when a skin specimen is given free access to fluid in vitro is different. The plasma proteins in the tissue diffuses into the soaking solution, and there is no lymph flow. Thus there is free access to fluid and no forces limiting the swelling else than the amount of fluid that the tissue specimen can imbibe until ambient pressure is reached (i.e. the hydrostatic pressure in the soaking solution).

To summarize, under normal conditions P_i and the swelling characterics of skin and skeletal muscle are only two of several determinants of the interstitial fluid volume.

Burn injuries.

The most striking feature of burn injuries to the skin is the massive and very rapid edema-formation. Thus total tisse water has been reported to increase by 75% (corresponding to about 100% increase in interstitial fluid volume) in 10 minutes (Leape 1970). This implies that the fluid filtration is increased to at least 200 times normal since the time normally required to filter a volume equal to interstitial fluid volume is 12 to 24 hours. The fluid required to create this edema is derived from the circulating plasma volume and transported across the capillaries. Burn injuries have been found to increase the capillary filtration coefficient ($K_{f,c}$ which equals the product of hydraulic conductivity and available surface area) by a factor of two to three above normal and it has been estimated that a driving force of about 200 mmHg was required to account for the rapid capillary fluid filtration (Arturson & Mellander 1964). This pressure is far beyond what can possibly be obtained by increasing capillary pressure. The authors observed an increased osmolarity of small solutes and ascribed the increased filtration pressure to this phenomenon (Arturson & Mellander 1964).

The observation of Arturson & Mellander (1965) did not result in further studies until we recently observed that burn injury to the skin decreases P_i to the same magnitude that Arturssson & Mellander (1964) observed for the filtration pressure, i.e. from a control level of -1 mmHg to as low as -100 to -200 mmHg.

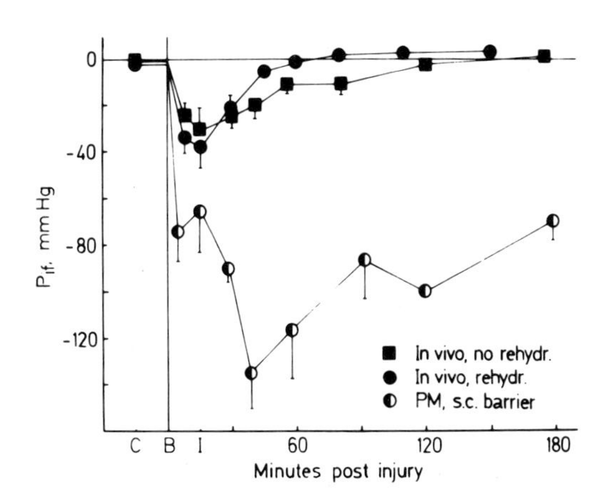

Figure 3. Interstitial fluid pressure (P_{if}) measured with micropipette following burn injury to the skin. Mean ± 1SE. (From Lund et al. 1989 with permission.)

P_i was measured with micropuncture technique (Lund et al. 1988, 1989). The lowest levels of P_i were observed after circulatory arrest had been induced prior to the injury and a diffusion barrier had been placed under the skin to prevent fluid movement from underlying skeletal muscle (Fig. 3). Furthermore, with intact circulation, the increased negativity in P_i returned to control level more rapidly since fluid was imbibed into the burned skin (Fig. 3). An even more rapid return of P_i to control was seen when intravenous fluid was given to compensate for the depletion of plasma volume that occur as fluid accumulated in the tissues and edema was formed (Fig. 3). The observations of the dramatically increased negativity of P_i (i.e. increased imbibition pressure) has later been confirmed by measurement of P_i using tissue osmometers with membranes that are freely permeable to macromolecules (Lund et al. 1989). The exact mechanisms that are underlying the observed reduction in P_i are as yet not fully understood, but the burn injury caused denaturation of collagen to water soluble gelatin, suggesting that at least part of the increase in the imbibition pressure could be caused by the collagen denaturation. Thus, as compared to the normal state, P_i or the imbibition pressure seems to be a major determinant of the rate by which edema can form following a burn injury to the skin.

Inflammation.

The observation of dramatically increased negativity of P_i in burn injury made us look closer at other conditions characterized by rapid edema formation to see if a similar reduction in P_i participated in generating the edema also under these circumstances. In the first experiments we studied P_i following xylene application to the skin (Rodt, Wiig & Reed 1990). This is a classical model of inflammation which induces extravasation of fluid at a rate corresponding to doubling of interstitial fluid volume in 10 to 15 minutes (Aschheim & Zweifach 1962). Following circulatory arrest and xylene application P_i was reduced to about -10 mmHg (Rodt, Wiig & Reed 1990). As a further extension of these studies, P_i was also measured during the rapidly forming edema on the paw of rats during the anaphylactic reaction to dextran. The measurements were performed in several different groups to outline a potential importance of P_i in the generation of the edema. In rats with intact circulation, there was an initial reduction of P_i to about -10 mmHg and then an increase to positive values as edema developed (Fig. 4) (Reed & Rodt 1991). When circulatory arrest was induced 1 min after intravenous

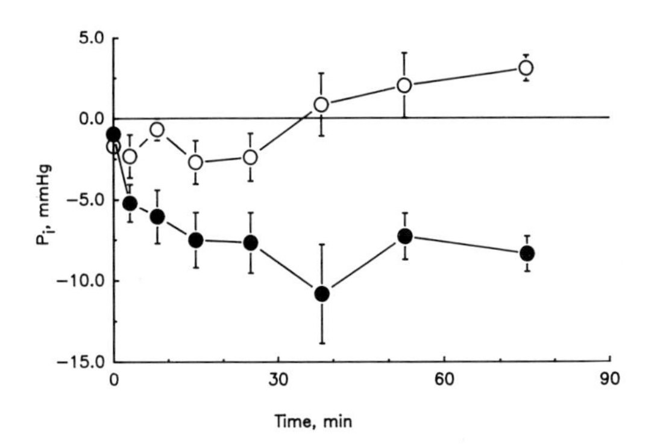

Fig. 4. Interstitial fluid pressure (P_i) in rat dermis following intravenous injection of dextran with intact circulation (open symbols) and following circulatory arrest (closed symbols). Mean ± 1SE. (From Reed & Rodt 1991 with permission.)

administration of dextran, the reduction in P_i continued until P_i had reached -10 mmHg and then remained at this level throughout the observation period (Fig. 4) (Reed & Rodt 1991). The dextran anaphylaxis is associated with degranulation of mast cells with release of histamine and serotonin (Schwartz & Austen 1984). Subsequent injection of histamine resulted in a reduction in P_i similar to that seen during dextran anaphylaxis and circulatory arrest. Injection of serotonin did change P_i (Reed & Rodt 1991). A similar reduction in P_i has also been observed during the edema formation induced in trachea as part of the anaphylactic reaction to dextran (Koller & Reed 1990).

Increased negativity of P_i has also been observed during the inflammatory edema developing after subdermal injection of carrageenan (Rodt & Reed 1990) and the complement activator zymosan (Østgaard personal communication).

The major importance of these observations is that increased imbibition pressure, i.e. reduction of P_i, in acute inflammation is a major determinant of the driving forces that create the rapidly forming edema. These studies might suggest that altered imbibition pressure could be a way to exert control of interstitial fluid balance. The mechanims behind and potential importance of these observations are, however, at present not fully resolved.

Summary and conclusions.

The interstitial fluid pressure, P_i, is normally of importance for the fluid balance of the interstitium through opposing changes in capillary fluid filtration. This is achieved by automatic adjustment of P_i through altered interstitial fluid volume. The altered interstitial fluid volume will in turn change P_i to counteract further filtration. During burn injury to the skin and in several occasions of acute inflammation, P_i will decrease, i.e. there will be an increased imbibition pressure that will be the active force in creating fluid movement from plasma into the tissue.

References.

Arturson G, Mellander S (1964) Acute changes in capillary filtration and diffusion in experimental burn injury. Acta Physiol Scand 62: 457-463

Aschheim E, Zweifach BW (1962) Quantitative studies of protein and water shifts during inflammation. Am J Physiol 202: 554-558

Aukland K, Nicolaysen G (1981) Interstitial fluid volume: Local regulatory mechanisms. Physiol Rev 61: 556-643

Brace RA, Guyton AC (1979). Interstitial fluid pressure: Capsule, free fluid, gel fluid and gel absorption pressure in subcutaneous tissue. Microvasc Res 18: 217-228

Guyton AC, Granger HJ, Taylor AE (1971) Interstitial fluid pressure. Physiol Rev 51: 527-563

Koller M-E, Reed RK (1990) Increased negative interstitial fluid pressure generates airway mucosal edema during dextran-anaphylaxis in the rat. Physiologist 33: A100

Leape LL (1970) Initial changes in burns: tissue changes in burned and unburned skin of rhesus monkeys. J Trauma 10: 488-492

Lund T, Wiig H, Reed RK (1988) Acute postburn edema: role of strongly negagative interstitial fluid pressure. Am J Physiol 255 (Heart Circ Physiol 24): H1069-H1074

Lund T, Onarheim H, Wiig H, Reed RK (1989) Mechanisms behind increased dermal imbibition pressure in acute burn edema. Am J Physiol 256 (Heart Circ Physiol 25): H940-H948, 1989

Meyer FA (1983) Macromolecular basis of globular protein exclusion and of swelling pressure in loose connective tissue (umbilical cord). Biochim Biophys Acta 755: 388-399

Reed RK, Johansen S, Noddeland H (1985). Turnover rate of interstitial albumin in rat skin and skeletal muscle. Acta Physiol Scand 125: 711-718

Reed RK, Rodt SÅ (1991) Increased negativity of interstitial fluid pressure during the onset stage of inflammatory edema in rat skin. Am J Physiol 260 (Heart Circ 29): H1985-H1991

Reed RK, Wiig H (1981) Compliance of the interstitial space in rats. I. Studies on hindlimb skeletal muscle. Acta Physiol Scand 113: 297-305

Rodt SÅ, Reed RK (1990) Increased negativity of interstitial fluid pressure (IFP) in skin during development of edema following subdermal injection of carageenan. Int J Microcirc Clin Exper 9 (Suppl 1): 159

Rodt SÅ, Wiig H, Reed RK (1990) Increased negativity of interstitial fluid pressure contributes to development of oedema in rat skin following application of xylene. Acta Physiol Scand 140: 581-586

Schwartz LB, Austen KF (1984) Structure and function of the chemical mediators of mast cells. In: Progress in Allergy, Mast cell activation and Mediator release. Ed. Ishizaka K. Karger Basel. vol 34. pp. 271-321

Taylor AE, Granger DN (1984) Exchange of macromolecules across the microcirculation. In: Handbook of Physiology, Section 2: The cardiovascular System, Volume IV, Microcirculation, Part 1. Eds: Renkin EM and Michel. CC. Bethesda, Maryland: American Physiological Society. pp. 467-520

Wiig H (1985) Comparison of methods for measurement of interstitial fluid pressure in cat skin/subcutis and muscle. Am J Physiol 249 (Heart Circ Physiol 18): H929-H944

Wiig H (1990) Evaluation of methodologies for measurement of interstitial fluid pressure (P_i): Physiological implications of recent P_i data. Crit Rev Biomed Eng 18: 27-54

Wiig H, Reed RK (1981) Compliance of the interstitial space in rats. II. Studies on skin. Acta Physiol Scand 113: 307-315

Wiig H, Reed RK (1985) Interstitial compliance and transcapillary Starling pressures in cat skin and skeletal muscle. Am J Physiol 248 (Heart Circ Physiol 17): H666-H673

Wiig H, Reed RK (1987) Volume-pressure relationship (compliance) of interstitium in dog skin and muscle. Am J Physiol 253 (Heart Circ Physiol 22): H291-H298

Wiig H, Reed RK, Aukland K (1987) Measurement of interstitial fluid pressure in dogs: evaluation of methods. Am J Physiol 253 (Heart Circ Physiol 22): H283-H290

SWELLING PRESSURE OF CARTILAGE: ROLES PLAYED BY PROTEOGLYCANS AND COLLAGEN

A. Maroudas, J. Mizrahi, E. Benaim, R. Schneiderman, G. Grushko
Julius Silver Institute, Department of Biomedical Engineering
Technion - Israel Institute of Technology
Haifa 32000
Israel

INTRODUCTION

Cartilaginous tissues such as articular cartilage and the intervertebral disc are called upon to function under very high pressures which they can do, thanks to the very special properties of their two major components, viz., the proteoglycans (PG) and collagen. The PG, a flexible polyelectrolyte of high fixed charge density has a high osmotic pressure and therefore a tendency to imbibe water and maintain tissue turgor while the collagen mesh, with its good tensile properties, resists undue swelling, thus enabling the proteoglycan-water mixture to exist as a concentrated solution (e.g. Fessler, 1960; Maroudas, 1973). The combination of the two components enables a cartilaginous tissue to exhibit flexibility and to withstand tensile stresses as well as high compressive loads (e.g. Maroudas, 1973; Maroudas, 1979; Weightman and Kempson, 1979). In Part I of the present chapter, we shall describe the organization of cartilaginous tissues and the factors which determine the levels of hydration under different conditions. In particular, we will consider in detail the role played by collagen and proteoglycans in determining the swelling pressure of cartilage. In Part II we shall describe how cartilage deforms when it is subjected to unconfined compression, with special reference to the factors resisting the change in shape and volume. We shall develop a mechanical model explicitly incorporating the osmotic pressure of the PG and the notion of two compartments. Finally, we shall show the relation based on experimental results between the applied pressure and the osmotic pressure in the compressed tissue.

NATO ASI Series, Vol. H 64
Mechanics of Swelling
Edited by T. K. Karalis

PART I: HYDRATION OF CARTILAGE: EFFECTS OF MATRIX ORGANIZATION, OSMOTIC PRESSURE OF THE PROTEOGLYCAN AND COLLAGEN TENSION

Two-compartment Model of Cartilage Matrix

When one considers the water content of a collagenous matrix one has to take into account the fact that the water is divided between at least two compartments: Figs. 1 and 2 show our two-compartment model of the cartilage matrix, based on the concepts of intra- and extrafibrillar spaces (Maroudas, 1990; Maroudas et al., 1991). Because of their size, the PG are confined to the extrafibrillar non-collagenous compartment. However, through their osmotic pressure acting across the surface of the collagen fibrils, they influence the amount of fluid retained within the fibrils.

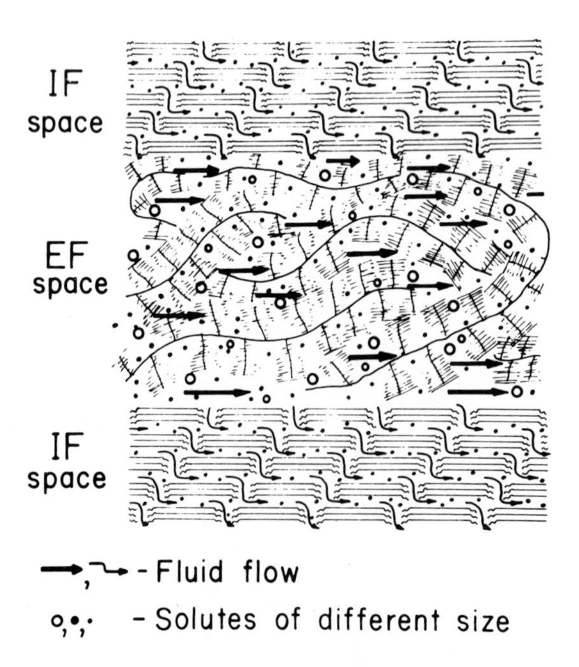

Figure 1: Two-compartment model of the cartilage matrix. (Reproduced from Maroudas et al., 1992)

The PG component is present in the matrix of cartilaginous tissues at very high concentrations, and is responsible for some of the most important physico-chemical and biomechanical properties of these tissues. However, it is not the overall concentration of the PG which is relevant, but rather their actual concentration in the extrafibrillar space. This effective concentration in turn depends on the proportions of water in the extra- and intrafibrillar compartments. The intrafibrillar compartment contains a small quantity of "bound" water. The rest consists of free water, the quantity of which is controlled in part by intermolecular repulsion forces, the precise nature of which is not clear at present, and in part by the osmotic pressure gradients between the outside and the inside of the fibrils. The state of tension of the fibril may also have an effect on the intrafibrillar hydration but this so far has not been investigated.

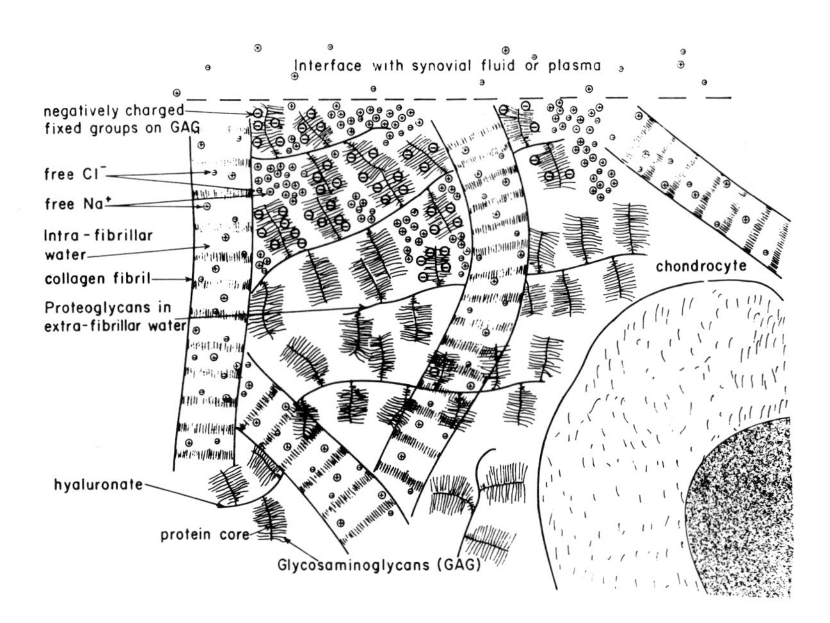

Figure 2: Schematic representation of the extra- and intrafibrillar compartments in the cartilage matrix.
(Reproduced from Maroudas and Urban, 1991)

It has been shown recently (Maroudas et al., 1991) that it is possible to quantify the relative proportions of water present in the two compartments under defined experimental conditions: these proportions vary and are themselves regulated by the osmotic pressure and hence the PG content in the extrafibrillar compartment. Fig. 3a shows the relationship between the intrafibrillar water, as calculated from X-ray diffraction data, and the osmotic pressure in the extrafibrillar space. Significant variations exist in the intrafibrillar water content over the range of osmotic pressures that correspond to the variations in PG content such as can occur in normal cartilage with distance from articular surface (e.g. Maroudas et al., 1969; Maroudas, 1979) and with topography (e.g. Grushko et al., 1989; Venn, 1979) or as a result of PG loss under pathological conditions. This range of values is further extended during weight-bearing when cartilage is subjected to pressures which lead to fluid loss from both compartments. Importantly, the extrafibrillar compartment constitutes the "outer phase" of the cartilage matrix. Thus, the interactions of the matrix with both the external environment (i.e. synovial fluid or blood) and with the chondrocytes are determined by the properties of this extrafibrillar compartment, in particular by the high osmotic pressure of the PG component. Provided the concentration of the PG is expressed on the basis of the extrafibrillar water, within which the PG are confined, the osmotic pressure exerted by the cartilage matrix can be deduced directly from the osmotic pressure of the proteoglycan solution at the same effective concentration.

PG Osmotic Pressure

For solutions of macromolecules, the relationship between the osmotic pressure of the solution and its concentration is in general highly non-ideal. The osmotic pressure can be expressed as a power series (Tombs and Peacock, 1974).

$$\frac{\pi_{osm}}{RT} = \frac{m}{M} + Bm^2 + Cm^3 + \ldots \quad (1)$$

where π_{osm} is the osmotic pressure,

R is the gas constant

T is absolute temperature,

M is the number average molecular weight of the macromolecule m is the concentration of macromolecules, and

B,C are the second and third virial coefficients of macromolecule.

In the case of polyelectrolyte molecules such as the proteoglycans, the second virial coefficient, B, includes terms depending on both the Donnan osmotic pressure and excluded volume contributions. These terms are additive (Tombs and Peacock, 1974). The third virial coefficient depends mainly on the Donnan contribution (Nichol et al., 1967).

In the case of PG solutions, it has been shown experimentally that the second term in eq. (1) is dominant and that the ionic contribution to the osmotic pressure accounts for some 80% of the total osmotic pressure π_{osm} of the PG (Urban et al., 1979).

Experimentally obtained, the relation between the osmotic pressure, π, and PG concentration, as expressed by fixed charge density (FCD) is given by Urban et al. (1979):

$$\pi_{osm} = B\ (FCD)^2 \tag{2}$$

According to recent data on the osmotic pressures of polyethylene glycol (PEG) solutions (see Fig. 4) (Maroudas and Grushko, 1990) and the data of Urban et al. (1979) for PG solutions, calibrated against the latter, the value of the coefficient B is 26.6 at 4°C and 29.0 at 37°C. These values of B correspond to π being expressed in atmospheres and FCD in mmol per cm^3 of solvent

It should be noted that the above equation was determined for solutions of extracted PG. Our assumption, however, is that PG within the tissue behaves in the same way as isolated PG in solution at the same concentration or FCD. Thus, we can make use of eq. (2) to calculate the osmotic pressure within cartilage provided FCD is expressed on the basis of extrafibrillar water. Fixed charge density in the extrafibrillar compartment can be calculated from the fixed charge density experimentally obtained for total tissue.

Effective Fixed Charge Density

The relationship between FCD obtained experimentally and calculated per total tissue water ($FCD_{overall}$) and the FCD in the extrafibrillar compartment, i.e. FCD_{EF} is given by the following expression:

$$FCD_{EF} = FCD_{overall} \left(\frac{W_{H_2O}}{W_{EF}} \right) = \frac{FCD_{overall} \times W_{H_2O}}{W_{H_2O} - W_{IF}} \qquad (3)$$

$$\text{where } W_{IF} = H_2O_{IF} \times C_{coll} \times W_0 \qquad (4)$$

In eqs. (3) and (4) the symbols have the following meanings:
W_0 is the total initial wet weight of the specimen; W_{H_2O} is the weight of total water in the specimen either initially or under pressure; W_{EF} is the weight of extrafibrillar water; C_{coll} is the weight of collagen per gram of initial specimen weight; $FCD_{overall}$ is the fixed charge density based on initial total weight of specimen water; and H_2O_{IF} is the weight of IF water per gram of dry collagen.

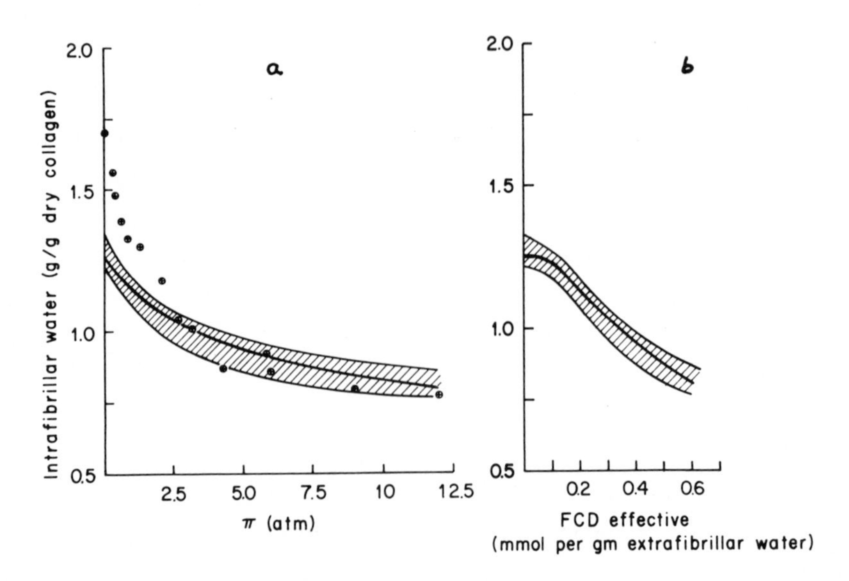

Figure 3: Variation of intrafibrillar water - (a) with osmotic pressure; (b) with effective FCD

The intrafibrillar water (W_{IF}) can be obtained from Fig. 3a for any given value of extrafibrillar osmotic pressure, π_{osm} (Maroudas et al., 1991).

Since π_{osm} is determined by FCD_{EF}, (see eq. 2), there is a unique relationship between the latter and W_{IF}, which is shown in Fig. 3b.

Factors Determining Cartilage Hydration

In the absence of an external load, matrix hydration is controlled by the balance between the osmotic pressure differential due to the proteoglycans, which leads to fluid imbibition, and the stress due to the tension in the collagen network, which tends to resist volume increases (e.g. Maroudas, 1976). If an external load is imposed upon the tissue, thus increasing the compressive stress, fluid will be lost from both the extra- and the intrafibrillar compartment until the concentration of the PG in the extrafibrillar space - and hence the osmotic pressure - has increased sufficiently to balance this new stress. When the external load is removed, the tissue will recover, i.e. it will "swell" until it returns to its initial hydration.

The above phenomena can be described in terms of the following relationships between the parameters involved: When cartilage is not under an external load, the swelling tendency of the PG is exactly balanced by the hydrostatic pressure created by the tensile stresses in the collagen network, P_c. This is expressed by the eq. (5)

$$\pi_{osm} = P_c, \tag{5}$$

When π_{osm} and P_c do not balance each other, the cartilage has a net swelling pressure (P_{SW}) which is defined as the applied external pressure (P_{app}) required to balance the difference between π_{osm} and P_c. Thus,

$$P_{app} = P_{SW} = \pi_{osm} - P_c, \tag{6}$$

where π_{osm} = osmotic pressure of the PG and P_c = hydrostatic pressure casued by the tensile stresses in the collagen network. P_{SW} as well as π_{osm} and P_c are functions of matrix hydration.

The parameter P_c decreases very rapidly as soon as the tissue begins to shrink in volume. When the decrease in volume is of the order of 5%, P_c becomes negligible in relation to π_{osm} (Maroudas et al., 1987). Under these conditions the tissue swelling pressure becomes equal to the osmotic pressure π_{osm} of the constituent PG (Maroudas & Bannon, 1981; Maroudas et al., 1987).

If $P_{app} > P_{SW}$, fluid is made to flow out of the tissue, the driving force for this process being the difference between P_{app} and P_{SW}. Fluid loss leads to a higher PG concentration per tissue volume, and hence an increased swelling pressure and lower permeability. Outward flow continues until the swelling pressure has increased enough to balance the applied load. Thus the equilibrium amount of fluid lost under a given load will depend on the composition of the matrix. When the external load is removed, the driving force for fluid reimbibition is the swelling pressure alone at the given hydration.

Since the only factors involved in controlling the overall level of hydration are the applied pressure, the collagen tension and the osmotic pressure differential due to the PG, any change in hydration, i.e., shrinking or swelling of cartilage, implies that one or more of the above three parameters have undergone a change. Thus, when isolated plugs of cartilage, rather than whole cartilage on the joint, are exposed to solution, the osmotic pressure differentials are increased (Maroudas, 1976; Maroudas and Venn, 1977). Therefore, upon excision cartilage will always have a tendency to swell. However, in normal adult human cartilage from large joints (e.g. femoral head and femoral condyle), the collagen network is inherently very stiff, and thus able to resist the higher osmotic pressure differentials created at the exposed interfaces. Hence, such cartilage swells by only a very limited amount upon excision. In contrast, fibrillated human femoral head or femoral condyle cartilage swells considerably upon excision from the joint and exposure to saline, due most probably to the loss of integrity of the collagen network, and hence decreased stiffness (e.g. Maroudas and Venn, 1977). Other cartilage, e.g., that from steer, swell upon excision, even when entirely intact. Here the reason is probably a differently organized collagen network, less stiff in itself and more dependent for its strength upon anchorage to bone.

Another reason for tissue swelling can be the removal of an external load. For instance, in the body the intervertebral disc is always under load. Even in the supine position, the pressures exerted on the lower lumbar discs by the tension in the muscles and ligaments is about 0.2 MN/m^2; in some positions the pressures can be as great as 1.2 MN/m^2. Thus, when the discs are removed from the spine and exposed to saline in the absence of load, they will swell considerably (Urban and Maroudas, 1981). The cartilage from facet joints, unlike human hip or knee cartilage, is under-hydrated when excised from the joint. The under-hydration is thought to reflect the permanent presence of stresses *in vivo* on some part of the facet joints, the position of the loaded site changing with time. Facet cartilage will thus swell when excised from the spine and exposed to physiological saline (Tobias et al., 1991).

The above examples illustrate how tissue swelling can arise under different circumstances, depending on which of the factors determining tissue hydration has undergone a change. In all cases, however, it is the osmotic pressure of the proteoglycans which endows cartilage with a tendency to swell; it should be noted that cartilage from which the PG have been completely removed does not re-swell after being compressed (Maroudas, unpublished data).

In the paragraphs which follow we shall describe experiments in which we compressed cartilage osmotically, using calibrated polyethylene glycol (PEG) solutins. As a result of these experiments we were able to compare the equilibrium swelling pressure of cartilage with the calculated osmotic pressure of the matrix PG.

Determination of Cartilage Swelling Pressure by Equlibrium Dialysis Against Polyethylene Glycol Solutions

The cartilage samples used originated from human femoral heads belonging to different age groups. Full-depth cylindrical plugs of cartilage (1-2 mm thick) were subjected to different osmotic pressures by immersion in calibrated solutions of PEG. The procedure is similar to that described by Maroudas and Urban (1983). In order to prevent penetration of PEG into the cartilage specimens, the samples after being initially weighed, were placed in fine-pore dialysis tubing (2000) and allowed to equilibrate in a PEG solution of desired concentration for a period of 24-48 h, depending on specimen thickness. The concentrations of PEG solutions used in this experiment were 25 and 30 g PEG per 100 g solvent, corresponding to applied pressures of 7 and 10 atm respectively. Once the equilibrium was reached, the sacs were lifted out of the PEG solutions, blotted well to remove any adherent moisture and the cartilage samples weighed. From the final and initial weight, it was possible to calculate for each slice the amount of water retained at equilibrium, corresponding to a given applied pressure. The dry weight was obtained by freeze drying the cartilage to constant weight. Knowing the initial fixed charge density, and the collagen content as well as the equilibrium water content, it was possible to calculate the effective FCD at the final equilibrium stage and hence the osmotic pressure.

Calibration of Polyethylene Glycol Solutions

Urban et al. (1979) calculated the osmotic pressures of PEG solutions using the virial coefficients reported by Edmond and Ogston (1968), at 25°C. These coefficients were used by Urban et al. (1979) for calculations at different temperatures by assuming that the virial coefficient remained constant over a range of 4-37°C. It was mentioned at the time that this assumption may not be exactly correct.

Empirical relations for 7°C and 30°C were recently published by Parsegian et al. (1986), who measured the osmotic pressure of PEG by direct membrane osmometry. Comparison of the two sets of data (Fig.4) shows considerable differences, especially at low temperatures, and leads to the conclusion that virial coefficients vary with temperature and extrapolation of results obtained at 25°C to other temperatures is not correct.

We ourselves repeated the calibration at 4°C, using a Diaflow cell and air pressure, and obtained the points also shown in Fig. 4 (unpublished data). It can be seen that our points fall very close to the curve obtained from Parsegian's data.

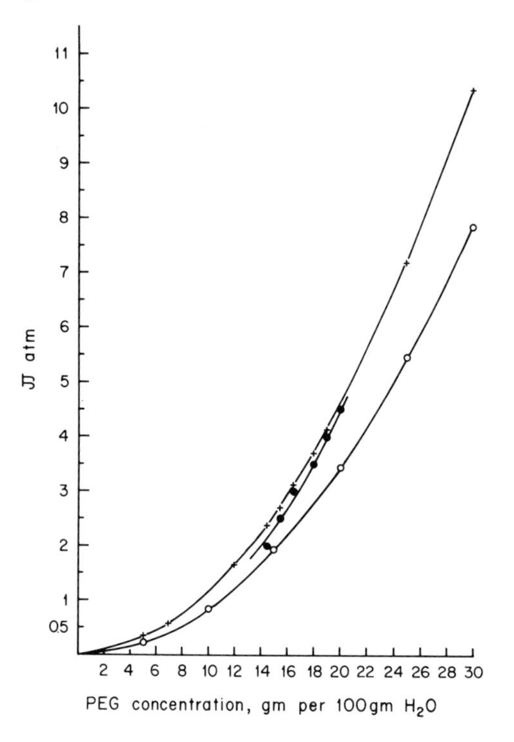

Figure 4: Calibration curve of osmotic pressure of polyethylene glycol solution in normal saline.
o - 7°C data of Urban based on Edmond & Ogston;
+ - 4°C data of Parsegian; o - 4°C, new calculation.

Comparison between Calculated Osmotic Pressure and Measured Swelling Pressure of Cartilage.

A comparison between measured swelling pressures and calculated values of osmotic pressure for cartilage of different age groups is given in Table 1. It should be noted that the original measurements of osmotic pressure of extracted PGs on which our calculations are based were obtained from a mean curve, not taking into account the systematic changes in the chondroitin sulfate: keratan sulfate: hyaluronan (CS:KS:HA) ratios which occur with age and which affect the mean interchange distances and hence π_{osm}. Clearly, it would be desirable to obtain separate curves of osmotic pressure for PG extracted from cartilage obtained from different age ranges. However, considering this shortcoming, the agreement between the two sets of values (within 10% except for the youngest age group) shows that once cartilage has been compressed and collagen tension released (P_c = 0), the swelling pressure in cartilage is indeed equal to the osmotic pressure of the constituent PG.

Table 1: Mean osmotic pressure of hip cartilage in different age groups: comparison between measured and calculated values.

Age (Years)	0.25 gm PEG per gm of H_2O 7 atm			0.30 gm PEG per gm of H_2O 10.1 atm		
	n	FCD_{eff}	π	n	FCD_{eff}	π
17 - 30	54	0.572 ± 0.022	8.73 ± 0.51	48	0.605 ± 0.018	9.76
30 - 45	30	0.538 ± 0.013	7.71 ± 0.35	48	0.589 ± 0.011	9.23 ± 0.34
50 - 70	28	0.509 ± 0.011	6.91 ± 0.29	33	0.590 ± 0.019	9.25 ± 0.59
70 - 81	36	0.507 ± 0.017	6.84 ± 0.04	33	0.605 ± 0.017	9.74 ± 0.23

Calculated Osmotic Pressure of Uncompressed Cartilage as a Function of Age

Table 2 shows the calculated values of osmotic pressure in uncompressed tissue, corresponding to the changes in the composition of

Table 2: Osmotic pressure for different age groups in uncompressed cartilage.

Age Range (years)	Water Content (per gm tissue)	Collagen Content (per gm tissue)	$FCD_{overall}$ (per gm tissue)	π (atm)	FCD_{Eff} (per gm extra-fibrillar water)
17-20	0.740	0.184	0.110	1.11	0.204
20-30	0.719	0.182	0.130	1.62	0.246
30-45	0.711	0.179	0.146	2.06	0.278
50-70	0.709	0.173	0.155	2.26	0.291
70-81	0.700	0.168	0.160	2.44	0.303

human articular cartilage with age. It can be seen that the osmotic pressure increases with age, thus endowing the tissue with a higher resistance to compression. Since, by virtue of eq. (2), the tensile strength of the collagen network, P_c, is equal to the osmotic pressure, we can conclude that P_c also increases with age, this of course only in the absence of degenerative changes.

PART II: MECHANICAL DEFORMATION IN RELATION TO COLLAGEN AND PG BEHAVIOUR

Based on preliminary results (Maroudas et al., 1987; Mizrahi et al., 1990), we have developed a model for cartilage deformation, in which one of our basic assumptions is that collagen fibrils cannot transmit compressive stresses. The meaning of this assumption is that in the loading (vertical) direction the contribution of the collagen network to load bearing decreases with increasing deformation: it becomes zero at a given deformation and remains zero at higher deformations. It should be noted that although at this stage there is no tension in the vertical fibrils, the horizontal fibrils are under tensile forces. As the specimen creeps, it decreases in volume and changes its shape. The latter change is accompanied by a change in orientation of the fibrils in the collagen network.

In order to test our model, we have developed a procedure to determine directly the changes in the lateral dimensions (Mizrahi et al., 1986), as well as in the thickness of the cartilage specimen during creep in unconfined compression. These two parameters can be combined to describe volume loss and change in shape. On the basis of such measurements, it is possible to analyze the deformation process and to attempt to clarify the respective roles of the collagen network and the proteoglycan-water solution in resisting the compressive stresses to which cartilage is subjected.

It should be noted, however, that we are not going to deal in this model with the forces responsible for the retention of fluid between the collagen molecules within the fibril although the intrafibrillar water content will be taken into account.

Experimental Apparatus

The experimental apparatus has already been described (Mizrahi et al., 1990). In summary, it consists of the following components. A loading apparatus is used for applying a step-loading compression to the specimen by means of a four-bar mechanism. During the experiment, the specimen is immersed in saline at 4°C in a transparent glass cell. The top surface of the specimen is compressed against a transparent rigid plunger, so that it can be observed throughout the creep phase. An optical system which consists of a microscope is used to view the deforming top surface to which a photo camera and a video camera are attached. A second video camera viewing the side of the specimen is used to monitor the changes in the shape and dimensions of the specimen's profile. A linear displacement transducer (LVDT) serves for measuring the thickness of the specimen. Finally, a computer is used to control the sequence of on-line data collection, to sample and to process the data obtained. A schematic description of the apparatus is given in Fig. 5.

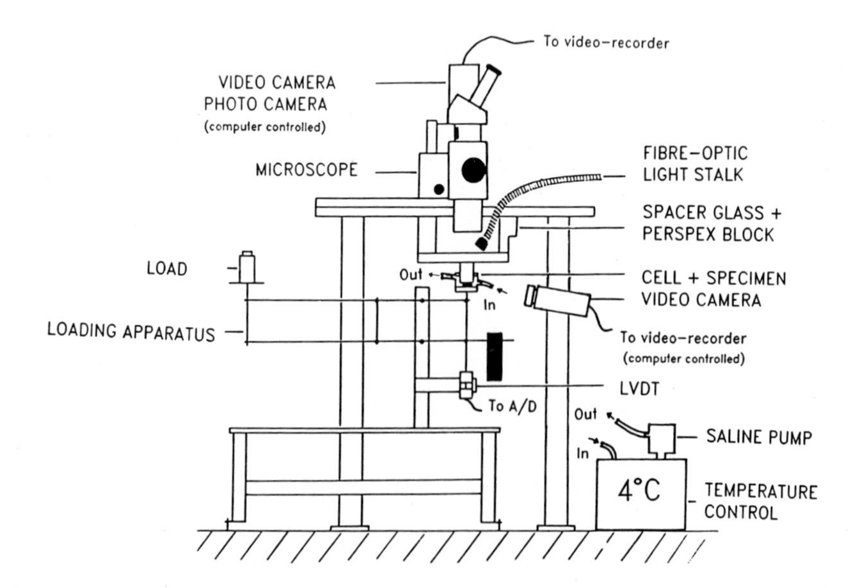

Figure 5: Diagram of the loading apparatus. (Reprinted from Mizrahi et al., 1990)

Procedure

Cylindrical cartilage specimens (c. 5 mm diameter) are cut, in our case from human hip and knee joints obtained at postmortem and from operations. Before testing, the initial dimensions are taken. Effective fixed charge density (FCD_{EF}) and hence osmotic pressure are calculated at all stages of creep as well as at the final equilibrium by methods described elsewhere (Maroudas and Grushko, 1990; Benaim et al., 1991). The above calculations are based on the volume changes that occur during creep, as well as on supplementary data determined from measurements made before and after the creep tests. These measurements include wet weight, dry weight, and overall FCD, as well as the collagen content.

Each specimen is loaded for 24 h, during which period its dimensions are measured. The dimensions measured include the diameters of both the top surface and the 'bulge' contour, the specimen thickness and the specimen profile.

The Model

The model to be presented in this paper refers to the "end states" of the creep process, i.e. the initial, instantaneous deformation and the final equilibrium state. The complete derivation of the model, including its dynamic part, i.e., the course of creep, is presented elsewhere (Benaim et al., 1991).

The deformation of cartilage in unconfined compression is non-homogeneous in the different layers and anisotropic (direction dependent). To simplify the formulation and solution of the equations of the model, the actual geometry of the deforming specimen was transformed into an exact cylindrical geometry as follows. Sections parallel to the top surface were first converted into equivalent circles, having the same area as those of the actual specimen, and the resulting solid of revolution was thereafter converted into a cylinder having the same volume and surface area as those of the actual specimen. In this way the average properties of the real sample and those of the equivalent specimen were made similar: hydration, osmotic pressure, permeability and flux. The equations to be presented will thus be written in cylindrical coordinates r, θ and z.

(a) Mass conservation

In the absence of fluid flow, the only non-trivial continuity equation reads:

$$\phi_{EF} = \frac{1}{J} (J - 1 + \phi_{EF_0} - (\omega_{IF} - \omega_{IF_0}) \phi(coll_0)) \tag{7}$$

where

$$\phi_i = \lim_{V \to 0} \frac{V_i}{V} \tag{7a}$$

ϕ_i is the volume ratio of component i to the total volume. Thus, ϕ_{EF} and ϕ_{coll} denote volume ratios for the extrafibrillar water and for the collagen, respectively. The subscript o stands for the initial state and ω_{IF} for intrafibrillar water, which is defined here as:

$$\omega_{IF} = \frac{\phi_{IF}}{\phi_{coll}} \tag{7b}$$

Note that both ϕ_{EF} and ω_{IF} can be derived from information described in Part I. J is defined as the ratio of the actual volume to the initial volume:

$$J = \lim_{Vo \to 0} \frac{V}{V_o} \tag{7c}$$

(b) Momentum conservation

Each of the components should be in separate equilibrium, yielding four equations, from which the following two can be obtained:

$$\underline{\nabla}^{\circ} \cdot (J\underline{F}^{-1} \cdot (\underline{\sigma} - P\underline{I}) = 0 \tag{8}$$

$$- \underline{K} \cdot \underline{\nabla}^{\circ} \cdot J\underline{F}^{-1} \cdot (P - \pi_{osm}) \, \underline{I} = J\underline{q} = 0 \tag{9}$$

where $\underline{\nabla}^{\circ}$ is the gradient operator, referring to the pre-deformation geometry; $\underline{\sigma}$ is the stress tensor; P and π_{osm} are the static and osmotic pressure, respectively; $\underline{I}$ is the unity tensor and $\underline{q}$ is the flux vector. $\underline{F}$ is the deformation gradient, given by:

$$\underline{F} = \underline{I} + \underline{U} \, \nabla^{\circ} \tag{9a}$$

where $\underline{U}$ is the deformation vector.

$\underline{K}$ is the overall permeability tensor given by the sum of the extra- and intrafibrillar permeability tensors:

$$\underline{K} = \underline{K}_{EF} + \underline{K}_{IF} \tag{9b}$$

The right hand side of eq. (9) is zero since in the stationary state the fluid flux $\underline{q}$ is zero.

(c) Constitutive equation for the collagen network

The three dimensional collagen network is assumed to consist of collagen fibers which run in a wound pattern, in a direction, generally normal to the articular surface (Broom, 1984; Maroudas et al. 1991). This direction is assumed to be maintained across most of the cartilage thickness, except for the superficial zone, where the majority of fibers tend to be tangential to the surface. The predominantly vertical fibers in the body of the tissue are connected together by lateral attachment nodes, forming structural units of a repetitive pattern. These units also include two families of fibers in planes parallel to the articular surface which are essential for the integrity of the network. This topic is discussed in detail elsewhere (Benaim et al., 1991). The analytical representation of the fiber trajectory is approximated as follows: in the vicinity of the attachment nodes the path is described by an elliptical arc and between the nodes the fibers are assumed to run in straight lines.

The constitutive model for the collagen network introduces two parameters as follows: the fibril stiffness C (in tension only) and a shape parameter α. This latter parameter describes the ability of the collagen network to change the angle between the families of the collagen fibrils whilst under load. Upon relating the deformation and stress in the fibrils to those in the tissue, we can derive a number of equations as shown below.

(d) Boundary conditions

For the top and bottom planes of the specimen

$$\int_{s^0} J\underline{e}_s^{\circ} \cdot \underline{F}^{-1} \cdot (\underline{\sigma} - P\underline{I})\, ds^{\circ} = \underline{F}_{appl} \qquad (10)$$

where $\underline{e}_s^{\circ}$ is the unit vector normal to the loaded surface s, expressed in the pre-deformation geometry, and $\underline{F}_{appl}$ is the applied loading force on that surface.

The circumferential wall of the specimen is stress-free and, at the same time it is in chemical equilibrium with the surrounding solution. Hence:

$$\underline{e}_a^{\circ} \cdot \underline{F}^{-1} \cdot (\underline{\sigma} - P\underline{I}) = 0 \qquad (11)$$

and

$$P - P_{ext} = P_{osm} - P_{osm\ ext} \qquad (12)$$

where the osmotic pressure of the surrounding solution is assumed constant.

Application of the model

We now refer to three different states of the specimen in the course of its deformation.

(1) The initial, unloaded state

The initial geometry of the specimen is assumed cylindrical.

If we denote in eq. (12) $P \equiv P - P_{ext}$ and $\pi_{osm} \equiv P_{osm} - P_{osm\ ext}$ we have, along the specimen's contour

$$P_o = \pi_{osm\ o} \tag{13}$$

and, from eq. (9), we also note that

$$\underline{\nabla}^{\circ} P_o = \underline{\nabla}^{\circ} \pi_{osm\ o} \tag{14a}$$

implying that eq. (13) holds for the points within the specimen.

From eq. (8) we conclude that

$$\underline{\nabla}^{\circ} \cdot \underline{\sigma} = 0 \tag{14b}$$

And since, in the absence of the external load, eq. (11) holds for all the boundaries of the specimen, it follows that the stress components, in polar coordinates, are

$$\sigma_{orr} = \sigma_{o\theta\theta} = \sigma_{ozz} = P_o = \pi_{osm\ o} \tag{14c}$$

(2) Instantaneous deformation following load application

The instantaneous deformation immediately follows load application, i.e. at $t = 0^+$, and is characterized by a very rapid rearrangement of the cartilage collagen network, with no net flow of the fluid relative to the matrix, i.e. $\underline{q} = 0$. Both the diameter and thickness deform instantaneously during this phase (Mizrahi et al., 1986). Thus, at constant volume:

$$J = (1 + e_{rr})^2 \ (1 + e_{zz}) = 1 \tag{15a}$$

This latter equation replaces the boundary conditions, expressed in eq. (12).

The system obtained includes eqs. (8), (9), (10), (11) and (15) and its solution should be of the form:

$$U_z = ze_{zz} \tag{16}$$

$$U_r = re_{rr} \tag{17}$$

where e_{zz} and e_{rr} are constants.

Since π_{osm} remains unchanged during the instantaneous deformation, i.e., $\pi_{osm} = \pi_{osm\ o}$, it follows that $\nabla \pi_{osm} = 0$, and also from eq. (9) that $°\underline{\nabla} \cdot JF^{-1} \cdot P\underline{I} = 0$. This implies that $°\underline{\nabla}P = 0$, since $J = 1$ and F^{-1} is constant.

The hydrostatic pressure P_i during the instantaneous deformation is assumed constant, hence $P = P_i$ gives from eq. (8).

$$°\underline{\nabla} \cdot JF^{-1} \cdot \underline{\sigma} = 0 \tag{18}$$

In components, eq. (18) reads:

$$\frac{\sigma_{rri}}{(1 + e_{rri})} = const_1 \tag{19}$$

$$\frac{\sigma_{zzi}^*}{(1 + e_{zzi})} = const_2 \tag{20}$$

where $\sigma_{zzi}^* \geq 0$, since the collagen fibrils cannot transmit compressive stresses. Note that σ_{zz} is related to P_c of eqs. (5) and (6).

The boundary condition expressed in eq. (10) yields

$$\int_0^{2\pi} \int_0^{Ro} \left[\frac{\sigma_{zzi}^*}{(1 + e_{zzi})} - \frac{P_i}{(1 + e_{zzi})} \right] rdr\, d\theta = F_{appl} \tag{21}$$

where F_{appl} is considered negative in compression.

We now define $(e_{zz})_{Ref}$ as the value of e_{zz} for which σ_{zz} becomes zero. Thus, for e_{zz} greater than $(e_{zz})_{Ref}$:

$$\frac{F_{appl}}{\pi R_o^{\,2}} = \frac{P_i}{(1 + e_{zz})} \tag{22a}$$

and for e_{zz} smaller than $(e_{zz})_{Ref}$, it can be shown that

$$\frac{F_{ext}}{\pi R_o^{\,2}} = Y_z\, e_{zz}\, (1 + e_{zz})\, (1 + 1/2\, e_{zz}) + P_o\, (1 + e_{zz}) - \frac{P_i}{(1 + e_{zz})} \tag{22b}$$

In most of the experiments performed, the applied load was such that the final deformation was large enough to satisfy eq. (22a).

From the boundary condition expressed in eq. (11)

$$\frac{\sigma_{rr}}{(1 + e_{rr})} - \frac{P_i}{(1 + e_{rr})} = 0 \tag{23}$$

by introducing the material property Y_r, we get

$$Y_r\, Eps_{rr}\, (1 + Eps_{rr})\, (1 + 1/2\, Eps_{rr}) + P_o\, (1 + Eps_{rr}) - \frac{P_i}{(1 + e_{rr})} = 0 \tag{24}$$

where $$Eps_{rr} = \Gamma\, (1 + e_{rr}) - 1 \tag{24a}$$

and $$\Gamma = \begin{cases} 1 & \sigma_{zz} > 0 \quad (e_{zz} < (e_{zz})_{Ref}) \\ [\,1 - \alpha\, (e_{zz}^{*})\, (2 + e_{zz}^{*})\,]^{-1/2} & ;\ \sigma_{zz} = 0 \end{cases} \tag{24b}$$

$$e_{zz}^{*} = \frac{1 + e_{zz}}{1 + (e_{zz})_{Ref}} - 1 \tag{25}$$

The material constant Y_r is given by

$$Y_r = \frac{\phi_{coll_o}}{\pi}\, C \tag{26}$$

(3) Deformation in the final equilibrium state

After equilibrium has been reached, the boundary condition in eq. (12) for the final deformation reads:

$$P_f = (\pi_{osm})_f \tag{27}$$

where f denotes the final state.

Assuming the deformation satisfies eq. (16) and (17), where e_{rr} and e_{zz} are constants, the osmotic pressure is uniform and thus $\pi_{osm} = (\pi_{osm})_f$ and also $\nabla\pi_{osm} = 0$. From eq. (9) it follows that P_f is therefore also uniform. Hence eq. (27) holds for all points of the specimen.

Equations (22) to (26) hold also for the final deformation, except that in this case P_i should be replaced by P_f.

Summary of the equations

Most of our measurements corresponded to a load causing σ_{zz} to become zero. Hence, our system of equations includes (22a) and (24), written for the instantaneous deformation, which together with two similar equations for the final deformation form a system of four algebraic equations for the four unknowns P_i, Y_r, α and P_f. In addition eq. (27) gives $(\pi osm)_f$ and eq. (26) enables us to determine the material constant C from Y_r.

Results

Diameter deformations

Two distinct phases characterize the deformation obtained: (i) the instantaneous phase, in which there is a marked change in shape, with no change in volume; and (ii) the creep phase, in which fluid is squeezed out of the specimen. In this phase, the volume decrease is characterized by a significant decrease in thickness, with a further increase in the surface area of the specimen. Average diameter deformations of the top surface in the instantaneous and final stages, are shown for typical specimens, in Table 3. The values presented are for the diameter of the 'equivalent' specimen. Two specimen configurations are compared: cartilage with its underlying bone and cartilage without bone and without the superficial layer. In this latter configuration, we note that the expansion of the specimen's surface is much higher, due to the absence of the constraining bone and the top layer of cartilage.

Table 3: Percentage instantaneous and final diameter deformations for the equivalent specimen. The results are of two specimen configurations: full depth cartilage speciments with the underlying bone and of specimens without bone and superficial layer. The applied pressure is 60 atm. Standard deviations are given in the parentheses.

Specimen Configuration	n =	Instantaneous Deformation (%)	Final Deformation (%)
With bone	11	1.69 (0.18)	4.99 (0.98)
Without bone and superficial layer	9	3.67 (0.83)	6.17 (3.1)

Average values of the elastic fibril parameter C and of the geometric parameter α are given in Table 4. These values were calculated from the instantaneous and final deformations, as explained earlier in the Model.

Table 4: Average values of C and α for the specimens tested. Standard deviations are given in aprentheses.

Specimen Configuration	n =	C (MN/m^2)	α
With bone	11	822 (161)	0.14 (0.03)
Without bone and superficial layer	9	252 (61)	0.17 (0.03)

The average value of the fibril stiffness parameter C was found lower in the specimens with bone than in those without bone and superficial layer. This difference was statistically significant ($p < 0.01$) and it suggests that the presence of these constraining surfaces has a stiffening effect on the collagen fibrils. The average value of network parameter α was higher in the unconstrained configuration of the specimen, as was to be expected. This difference, however, was not found significant.

Verification of the final osmotic pressure

Fig. 6 presents the comparison between applied pressure and PG osmotic pressure as calculated by methods described in Part I from the tissue composition and from the volume of the specimen at final equilibrium. In all experiments in which the mechanical loading pressure was higher than the initial osmotic pressure of the PG the final values of the osmotic pressure, π_{osm}, tended to coincide with the applied pressure as shown in Figure 6. This indicates that in the final equilibrium state the applied pressure is indeed resisted only by the osmotic pressure of the PG in the extrafibrillar space. As stated earlier, due to the retention of fluid between the collagen molecules, the fibrils themselves have a certain resistance to compressive loads. However, they do not contribute to the resistance of the extrafibrillar space to compression.

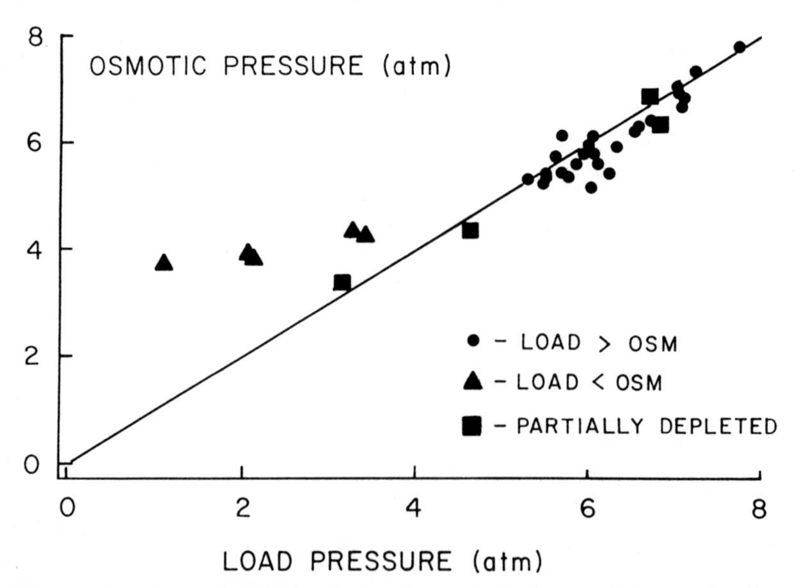

Figure 6: Calculated osmotic pressure vs. mechanically applied pressure at final equilibrium.

SUMMARY

The ability to quantify the distribution of water between the extra- and intrafibrillar spaces in cartilaginous tissues under different conditions renders it possible to relate the observed swelling equilibria of cartilage to the properties of the component proteoglycans and collagen. Knowing that the cartilage composition varies with age, we have been able to calculate the variation of osmotic pressure with age and to predict its effect on the resistance to compression. We have also described a model to analyse the behaviour of cartilage subjected to unconfined compression; the role played by the osmotic pressure is included in the model.

References

Benaim E, Mizrahi J, Maroudas A (1991) Shape and volume changes in cartilage during creep in unconfined compression. In preparation

Broom ND (1984) Further insights into the structural principles governing the function of articular cartilage. J Anat 139:275-294

Edmond E, Ogston AG (1968) Biochem J 1-9:569-576

Fessler JH (1960) A structural function of mucopolysaccharide in connective tissue. Biochem J 76:124

Grushko G, Schneiderman R, Maroudas A (1989) Some biochemical and biophysical parameters for the study of the pathogenesis of osteoarthritis: A comparison between the processes of ageing and degeneration in human hip cartilage. Conn Tiss Res 19:149-176

Maroudas A (1973) Physical-chemical properties of articular cartilage. In: Freeman MAR (ed) Adult articular cartilage. Pitman Medical, Tunbridge Wells, pp 131-170

Maroudas A (1976) Balance between swelling pressure and collagen tension in normal and degenerate cartilage. Nature 260:808-809

Maroudas A (1979) Physical-chemical properties of articular cartilage. In: Freeman MAR (ed) Adult articular cartilage, 2nd edn. Pitman Medical, Tunbridge Wells, pp 215-290

Maroudas A (1990) Tissue composition and organization. In: Maroudas E, Kuettner K (eds) Methods in cartilage research. Academic Press, London, pp 209-239

Maroudas A, Bannon C (1981) Measurement of swelling pressure in cartilage and comparison with the osmotic pressure of constituent proteoglycans. Biorheology 18:619-632

Maroudas A, Grushko G (1990) Measurement of swelling pressure of cartilage. In: Maroudas E, Kuettner K (eds) Methods in cartilage research. Academic Press, London, pp 298-301

Maroudas A, Venn M (1977) Swelling of normal and osteoarthritic femoral head cartilage. Ann Rheum Dis 36:399-406.

Maroudas A, Venn MF (1979) Biochemical and physico-chemical studies on osteoarthritic cartilage. In: Nuki G (ed) The aetiopathogenesis of osteoarthrosis. Pitman Medical, London

Maroudas A, Urban J (1983) In vitro and in vivo methods of studying articular cartilage and the intervertebral disc. In: Kunin, Simon (eds) Skeletal Research: An Experimental Approach, Vol. 2. Academic Press, New York, pp 135-182

Maroudas A, Urban J (1991) Articular cartilage and the intervertebral disc. In: Dulbecco R (ed) Encyclopedia of Human Biology, Vol. 1. Academic Press, New York, pp 365-370

Maroudas A, Mizrahi J, Ben Haim E, Ziv I (1987) Swelling pressure in cartilage. In: Staub NC, Hogg JC, Hargens AR (eds) Interstitial-lymphatic liquid and solute movement. Karger, Basel, pp 203-217

Maroudas A, Muir H, Wingham J (1969) The correlation of fixed negative charge with glycosaminoglycan content of human articular cartilage. Biochim Biophys Acta 177:492-500

Maroudas A, Schneiderman R, Popper O (1992) The role of water, proteoglycan and collagen in solute transport in cartilage. In: Kuettner KE, Schleyerbach R, Peyron JG, Hascall VC (eds) Articular cartilage ad osteoarthritis. Raven Press, New York (in press)

Maroudas A, Wachtel E, Grushko G, Katz EP, Weinberg P (1991) The effect of osmotic and mechanical pressures in water partitioning in articular cartilage. Biochim Biophys Acta 1073:285-294

Mizrahi J, Maroudas A, Benaim E (1990) Unconfined compression for studying cartilage creep. In: Maroudas E, Kuettner K (eds) Methods in cartilage research. Academic Press, London, pp 293-298

Mizrahi J, Maroudas A, Lanir Y, Ziv I, Webber TJ (1986) The instantaneous deformation of cartilage: effects of collagen fiber orientation and osmotic stress. J Biorheol 23:311-330

Nichol LW, Ogston AG, Preston BN (1967) The equilibrium sedimentation of hyaluronic acid and of two synthetic polymers. Biochem J 102:407

Parsegian VA, Rand RP, Ruller NL, Rau DC (eds) (1986) Methods in enzymology, Vol. 127. Academic Press, New York, 400-416

Tobias D, Ziv , Maroudas A (1991) Human Facet Cartilage: Swelling and some physico-chemical characteristics as a function of age: Part 1: Swelling of human facet joint cartilage. Spine, in press

Tombs MP, Peacock AR (1974) The osmotic pressure of biological macromolecules. Clarendon Press, Oxford

Urban J, Maroudas A (1981) Swelling of intervertebral disc in vitro. Conn Tiss Res 9:1-10

Urban J, Maroudas A, Bayliss MT, Dillon J (1979) Swelling pressures of proteoglycans at the concentrations found in cartilaginous tissues. Biorheology 16:447-464

Venn MF (1979) Chemical composition of human femoral head cartilage: Influence of topographical position and fibrillation. Annals of the Rheumatic Diseases 38:57

Weightman B, Kempson GE (1979) Load Carriage. In: Freeman MAR (ed) Adult articular cartilage. Pitman Medical, Tunbridge Wells, pp 291-332

CHANGES IN CARTILAGE OSMOTIC PRESSURE IN RESPONSE TO LOADS AND THEIR EFFECTS ON CHONDROCYTE METABOLISM

J.P.G. Urban and A. Hall

Physiology Laboratory
Oxford University, Parks Road
Oxford OX1 3PT, England

INTRODUCTION

Cartilage is a resilient, tough, connective tissue whose main functions are mechanical. It is rigid enough to provide structural support for tissues such as the ear, nose and larynx. Articular cartilage, found in skeletal joints is more deformable than bone, and is thus able to distribute the load and protect bone from mechanical forces and provide a low friction surface for articulating joints. In the spine, the cartilaginous intervertebral discs, as well as cushioning the vertebral bodies, impart flexibility to the spinal column, enabling it to bend and twist. The components of cartilage are made by the cartilage cells, the chondrocytes. In weight–bearing cartilages, these cells are subjected to an environment which can be dramatically altered by mechanical forces. This chapter describes the effects of mechanical stress on the osmotic environment of the chondrocyte, and shows how the metabolism of the cell is affected by these changes to the extracellular osmolarity.

A STRUCTURAL MODEL OF CARTILAGE

The mechanical properties of cartilage result from the properties and organisation of the macromolecules which make up its matrix, mainly collagen and proteoglycans. Collagen forms a fibrous network, which is inflated by the water imbibed by the polyanionic proteoglycans (Maroudas, 1979). How these components interact to form a structure which is able to carry high compressive loads, is demonstrated in a structural model (Broom and Marra, 1985).

This model consists of a string network inflated with balloons (Figure 1). String though strong in tension, is unable to form a rigid structure or carry compressive load and thus without the balloons the string network would collapse. On the other hand, without the string network, the balloons would fly apart. Together though, these form a structure which can be rigid and can also support and carry compressive loads. The rigidity of the structure and the manner in which it deforms under load will depend both on the nature of the network, and on the number of balloons and on their degree of inflation.

Figure 1. A structural model of cartilage (*from Broom and Marra, 1985, with permission*)

In cartilage, the collagen fibrillar network fulfils the function of the string network in the structural model. Like string, collagen is strong in tension but cannot support a compressive load. Proteoglycans and water together fulfil the function of the balloons. The osmotic properties of the proteoglycans which inflate cartilage and maintain turgor under high compressive loads (Maroudas, 1979), are thus vital to the functioning of cartilaginous tissues.

While this structural model is able to demonstrate how proteoglycans and collagen can form a load–bearing tissue, it is by no means an exact analogy for the structure of cartilaginous tissues. Water is not held in cartilage by an elastic membrane, but by osmotic pressure (Maroudas and Urban, 1981) and by the low hydraulic permeability of the tissue which arises because of the high concentration of proteoglycans (Maroudas, 1979; Mow *et al*, 1984; Eisenberg and Grodzinsky, 1987). Water is thus able to flow into and out of the tissue under pressure gradients. Proteoglycans are also not held in the tissue by a membrane but probably mostly through self-entanglement and entanglement with the collagen meshwork, rather through any specific interaction (Hascall and Hascall, 1981).

THE RELATIONSHIP BETWEEN FLUID CONTENT, OSMOTIC PRESSURE AND LOAD IN CARTILAGINOUS TISSUES

In unloaded cartilage, the proteoglycans, because of their high osmotic pressure, tend to imbibe fluid from a surrounding aqueous solution, setting the collagen network into tension. At equilibrium, when there is no net fluid loss or gain, the hydrostatic pressure which develops, balances the proteoglycan osmotic pressure in accordance with Starling's Law. The swelling pressure, defined as being equal in magnitude to the applied stress at fluid equilibrium is thus zero. Swelling pressure here results from the difference between proteoglycan osmotic pressure and hence hydrostatic pressure in the matrix, and the collagen network tension (Maroudas, 1979).

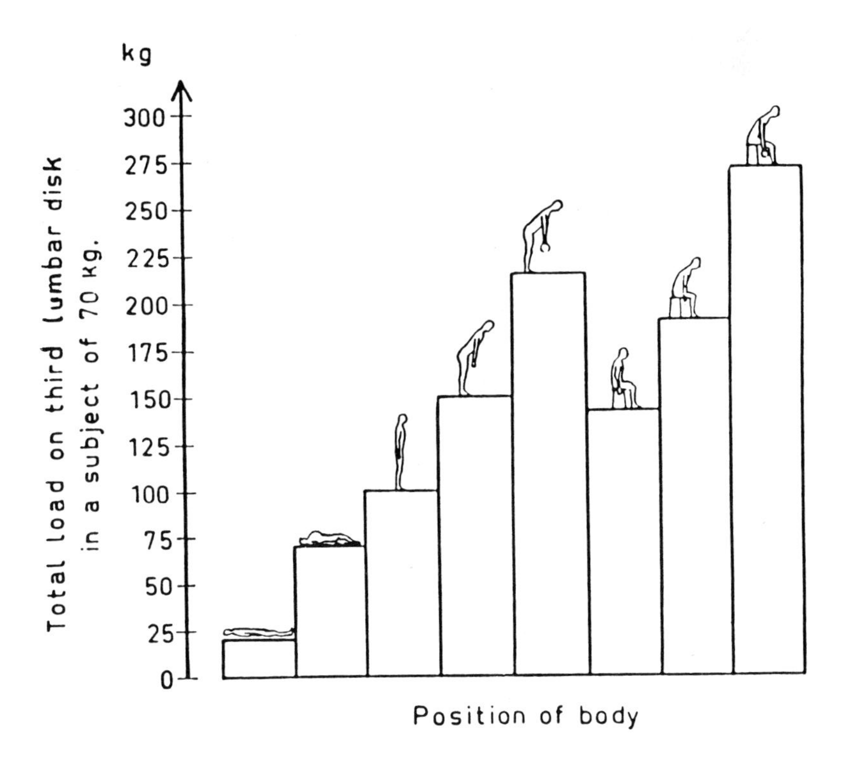

Figure 2. Variation of total load on the third lumbar disc with posture (Nachemson, 1975; with permission).

If a load is applied to such cartilage, the swelling pressure of the tissue is now less than the applied stress. Fluid is thus expressed from the tissue which increases proteoglycan concentration and hence osmotic pressure. Also as fluid is lost, the volume of the tissue decreases and the tensile stress in the collagen network falls. Both these processes lead to an increase in swelling pressure. Fluid expression continues until the swelling pressure has risen to balance the external stress.

Some cartilages, such as nasal cartilage, seldom have to carry a mechanical load. Articular cartilages, such as those in the hip and knee, are frequently loaded but when unloaded and at rest are in a pool of aqueous fluid. The collagen network of such cartilages is highly stressed, balancing a hydrostatic pressure of 0.1–0.2 MPa (Maroudas and Bannon, 1981). Because collagen is stiff, no significant swelling occurs in normal tissue (Maroudas, 1976) even if the osmotic pressure of proteoglycans is increased significantly.

Loads on the intervertebral disc

A different situation exists in the intervertebral disc. Intervertebral discs are always loaded *in vivo* both from body weight and from muscle and ligament forces. Even when lying relaxed the discs are under pressure of around 0.2 MPa (Nachemson, 1975). Thus the swelling pressure of the disc, unlike that of cartilages, is always greater than zero and the swelling pressure is similar in magnitude to the proteoglycan osmotic pressure (Urban and McMullin, 1987). The collagen network, especially in the nucleus, is thus not under tension, and not fully extended. The external load on the disc varies with posture (Figure 2).

The load increases three to four fold on standing upright and as much again if leaning forward or twisting. Throughout the day, as posture and activity alter, the discs are under varying mechanical loads which disturb the osmotic equilibrium in the tissue and hence result in fluid movement. However, in general, the load on the disc during the day's activity is much higher than that on the disc when lying down at night. Fluid thus tends to be expressed from the disc during the day to be reimbibed at night during rest. Because of the constantly changing loads on the disc with changes of posture and of muscle activity, complicated fluid flow patterns develop in the disc about which little is known (Simon *et al*, 1985).

The relationship between changes in load, disc swelling pressure and fluid movement is shown in Figure 3. The solid line represents the relationship between external stress and equilibrium fluid content; *i.e.*, it is the equivalent swelling pressure hydration curve of the tissue.

In this figure the disc is assumed to be at fluid equilibrium at point A, for example after a night's rest. The disc thus has a swelling pressure of P_A. On rising, the stress on the disc increases suddenly to point B; the disc is thus no longer at equilibrium since the applied stress is greater than the swelling pressure. Fluid flow in cartilage follows Darcy's law, and is proportional to the difference between the applied and equilibrium pressures, (P_B–P_A) at point B. It also depends on the hydraulic permeability coefficient of the tissue, which decreases steeply with fluid loss (Maroudas, 1979). Fluid loss is therefore fastest initially when both the pressure gradient and hydraulic permeability coefficient are at their maximum. As fluid is expressed, proteoglycan concentration rises, the swelling pressure of the tissue increases and the pressure gradient decreases as does the hydraulic permeability coefficient. Flow rate thus decreases rapidly as fluid is expressed limiting the amount of fluid lost from the tissue. Because of the size of the disc and the slow rate of fluid flow it is unlikely that osmotic equilibrium will be reached during a normal day's loading. If the load on the disc is removed at point C and the pressure returned to its original value of P_A, the tissue is now at D, but its swelling pressure is P_C. The disc, having lost fluid and increased its proteoglycan concentration, now lies below the equilibrium swelling pressure curve and fluid is imbibed to dilute the proteoglycans and reduce the swelling pressure. During swelling, the hydraulic permeability coefficient increases as the proteoglycan concentration falls; resistance to fluid flow thus decreases as equilibrium is approached. Swelling thus tends to be faster than fluid loss.

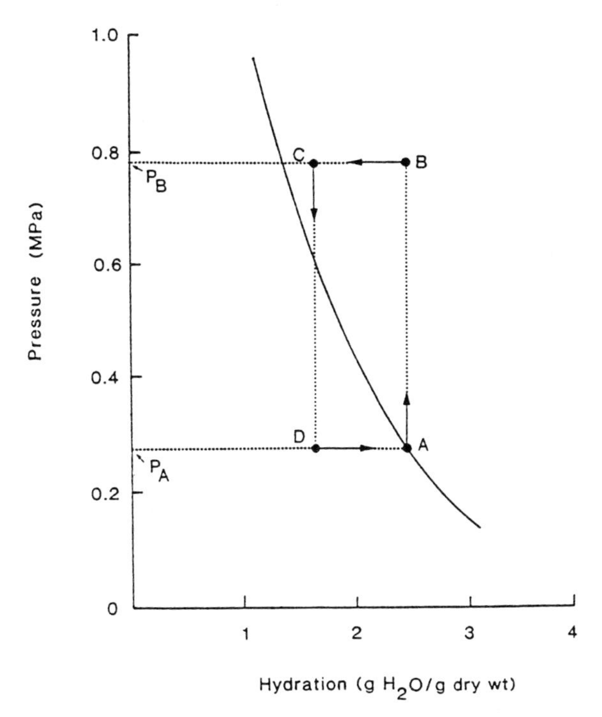

Figure 3. Schematic view of the changes in disc hydration with load history.

The amount of fluid exchanged between the disc and surrounding tissues during diurnal changes in load, is considerable. Adams and Hutton (1983), found that discs lost 11–25% of their fluid during a day. This fluid loss is partly responsible for the decrease in stature of about 17–20 millimetres which has been observed in the course of a day, and which varies with external loading (Eklund and Corlett, 1984). Alternatively, if load is removed from the disc *e.g.* during traction, or weightlessness in space, the discs imbibe fluid and swell; during a stay in Skylab an average height increase of 5 cm was recorded partly through this mechanism (Thornton *et al*, 1974). Thus the water content of any disc is not constant but varies with the applied load, and depends on loading history.

If the disc is excised from the body its swelling pressure is greater than the applied stress which is now zero, and if placed in aqueous solution, the tissue swells until further fluid imbibition is resisted by the collagen network. Since the collagen network of the disc is not under tension, the degree of swelling can be considerable and depends on the weave of the collagen network and on the osmotic pressure of the matrix. Such swelling causes many experimental artifacts (Urban and Maroudas, 1981). It can be prevented by incubating the disc in medium containing a high concentration of an osmotically active solute (usually polyethylene–glycol 20,000) which is excluded form the tissue, or alternatively by incubating the tissue under mechanical load. That the osmotic pressure arises mostly because of the Donnan effect on the charged proteoglycan can be seen by the effect of ionic strength on the extent and rate of swelling in similar disc slices. Slices swell most rapidly and to a greatest extent in low ionic strength aqueous solution and to a smaller extent in high salt; osmotically active uncharged solutes which enter the tissue freely such as sucrose, have little effect on swelling.

THE CELLS OF CARTILAGE

The functions of cartilage are primarily mechanical and depend on the properties of the major macromolecules, collagen and proteoglycan. These are turned over by the cells of cartilaginous tissues, the chondrocytes. Although the chondrocytes occupy only 1–5% of the tissue volume, and play no mechanical role their activity is essential for the health of the tissue (Stockwell, 1979). While collagen turnover is thought to be slow, proteoglycans are synthesised in cartilage throughout life. The rate at which proteoglycans are synthesised is governed to some extent by feedback from the surrounding extracellular matrix (Muir, 1981). It is not known how turnover is controlled by the cell. Specific cell–matrix interactions may play a role (Sommarin *et al*, 1989) and regulation may also be mediated by the osmotic environment of the chondrocyte (Bayliss *et al*, 1986; Schneiderman *et al*, 1986).

The extracellular environment of the chondrocyte

The chondrocytes *in vivo* are embedded in the proteoglycan–collagen matrix of cartilage. The proteoglycans are highly anionic macromolecules and their concentration around the cell determines the extracellular ionic environment of the chondrocyte; ion distributions in the matrix are governed by the Gibbs–Donnan equilibrium conditions (Maroudas, 1979; Grodzinsky, 1983). For cartilage in equilibrium with plasma or synovial fluid, the extracellular Na^+ is around 250–350 mM, free Ca^{++} may rise to 20 mM, and K^+ 10–15 mM. The anion concentration in contrast, is rather lower than in plasma with Cl^- being 60–90 mM, and the concentration of divalent anions such as SO_4^{--} being even lower. The chondrocyte thus lies in a rather unusual ionic environment surrounded by a high concentration of cations, a low concentration of anions and a total osmotic pressure arising mainly from these free ions of 350–450 mOsm (Urban and Hall, 1991)

The extracellular ionic and osmotic environment of the chondrocyte varies as fluid is expressed from the tissue under load. These changes are most dramatic in the intervertebral disc which can lose and regain up to 25% of its total fluid during each 24 hours. Since fluid is lost mainly from the extrafibrillar compartment in which the cells lie (Maroudas,1990), the effect on the cellular environment is even more extreme.

The changes in hydration which occur as the result of applying and external load to the joint or following swelling have a number of separate effects on the pericellular environment. Proteoglycan concentration itself increases if fluid is expressed from cartilage, and decreases if cartilage swells. Because of the polyelectrolyte nature of the proteoglycans an increase in proteoglycan concentration leads to an increase in the concentration of cations in the matrix and hence an increase in extracellular osmolarity. The reverse happens in swollen cartilage. Changes in the proteoglycan concentration may also affect the extracellular pH because of the effect on the distribution of hydrogen ions.

Influence of changes in hydration on cartilage metabolism

Changes in hydration of cartilage *in vitro* have been found to influence chondrocyte metabolism significantly. If cartilage is compressed either mechanically or osmotically and incubated in standard tissue culture medium, proteoglycan synthesis rate falls in direct proportion to the fall in hydration. Such results have been found in many different cartilages, both weight–bearing and non weight–bearing (Jones *et al*, 1982; Bayliss *et al*, 1986; Schneiderman *et al*, 1986; Gray *et al*, 1988; Urban and Bayliss, 1989; Sah *et al*, 1989).

This effect appears to arise, at least in part, from a change in extracellular osmolarity rather than from an increase in the concentration of macromolecules. In experiments where cartilage was incubated in medium of a low ionic strength the synthesis rate was much lower than in control cartilage incubated in standard medium even though the hydration and hence proteoglycan concentrations were equivalent (Urban and Bayliss, 1989). If the cartilage was then compressed the synthesis rate increased initially rather than decreased as the cartilage lost fluid (Figure 4); synthesis rates eventually decreased with further decrease in hydration but only after a significant fraction of the fluid had been lost.

The peak rate in this compressed cartilage incubated in low ionic strength, low osmolarity medium was similar to that of uncompressed cartilage incubated in a medium of standard ionic strength and osmolarity. This result suggests that the chondrocyte responds to extracellular ion concentrations or osmolarity rather than to the concentration of macromolecules as such.

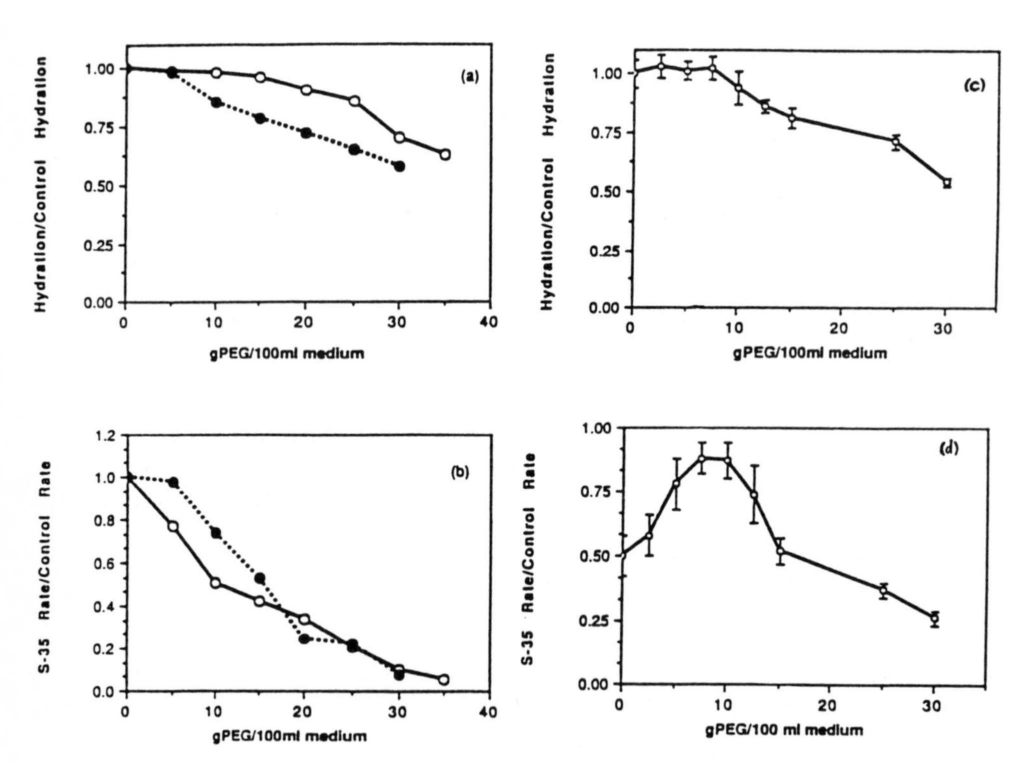

Figure 4. Effect of increasing polyethylene glycol (PEG) concentration on hydration, (a,c) and on synthesis rates relative to rates in standard medium with no PEG (b,d). In a,b the medium contained 140 mM NaCl, normal osmolarity; in c,d medium contained 70 mM NaCl, low osmolarity (from Urban and Bayliss, 1989)

Initially the extracellular osmolarity of the chondrocyte incubated in cartilage of low osmolarity containing only 70 mM NaCl was much lower than that of cartilage normal medium (140 mM NaCl) since by the ideal Donnan,

$$\Pi = (FCD^2+4.m^2)^{1/2} \qquad (1)$$

where II is the Donnan osmotic pressure, FCD is the fixed charge density and is directly proportional to the extrafibrillar proteoglycan concentration, and m is the concentration of solutes in the medium, mainly NaCl. As the tissue was compressed, proteoglycan concentration and hence FCD increased, cation concentrations around the cell increased through the Donnan effect, and the extracellular osmolarity rose (equation 1). Eventually the extracellular osmolarity increased to that found in control cartilage. Under these conditions though, $FCD_{compressed} > FCD_{control}$, while $m_{compressed} < m_{control}$. Synthesis rates fell as the tissue was further compressed and thus appear to correlate with changes in extracellular osmolarity rather than with changes in proteoglycan concentration.

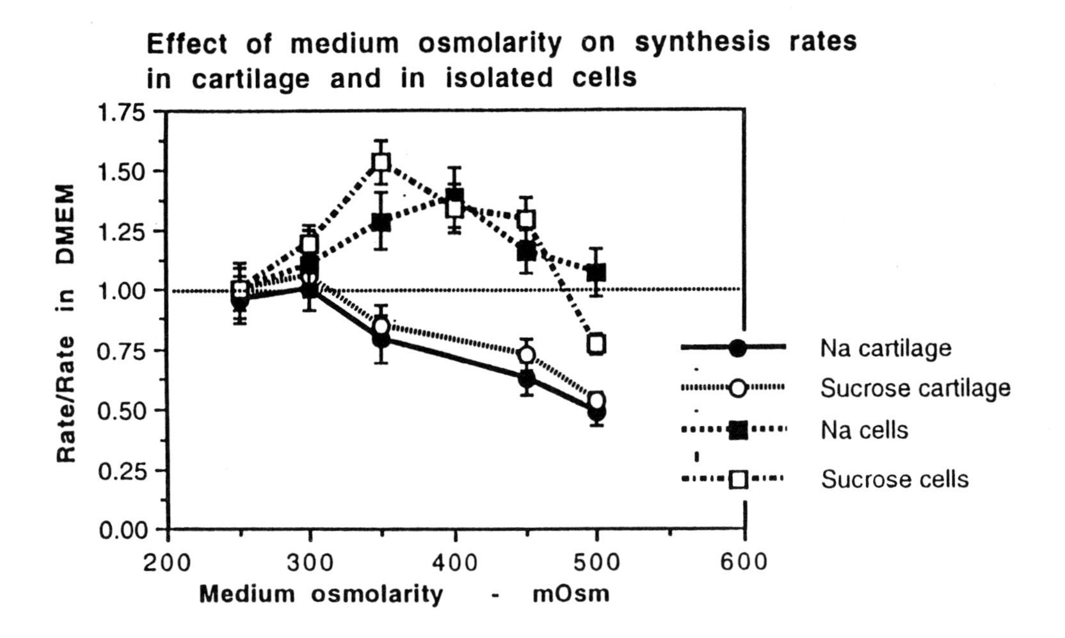

Figure 5. Effect of medium osmolarity on synthesis rates in cartilage slices and isolated chondrocytes.

This experiment has been performed in another way by adding extra ions or osmolites to tissue held at normal hydration (Figure 5). Under these conditions if the extracellular osmolarity is increased above control values or decreased below it, synthesis rates decrease even though hydration and hence extracellular proteoglycan concentrations remain constant (Schneiderman *et al*, 1986; Urban and Bayliss, 1989; Urban and Hall, 199). The effects were similar if either NaCl or sucrose were used to alter osmolarity, suggesting that it is extracellular osmolarity rather than ionic composition which has the major effects on synthesis rates.

Chondrocytes isolated from the matrix

If chondrocytes are isolated by digesting away the collagens and proteoglycans of the matrix using enzymes, and then suspended in standard culture medium, the chondrocytes are exposed to an environment which is very different to that they see in the tissue. The Donnan distribution no longer applies and both Na^+ and Cl^- have a similar concentration of around 140 mM, K^+ and Ca^{++} concentrations are considerably lower than in the tissue, and anion concentrations are higher. The extracellular osmolarity in culture medium is around 250–280 mOsm instead of 350–400 mOsm. In addition, chondrocytes are no longer attached to extracellular matrix components. It is thus not surprising that the rate of matrix synthesis in isolated chondrocytes was found to be considerably lower than that of chondrocytes in the tissue matrix (Urban and Hall, 1991). However if the extracellular osmolarity of the incubating medium was increased to 350–450 mOsm ie to the osmolarity the cell normally encounters in the tissue, synthesis rates increased 50–100% (Urban and Hall, 1991). Any further increase of osmolarity led to a fall in synthesis rates (Figure 5). It thus appears that chondrocytes isolated from cartilage respond to an extracellular osmolarity in a similar way to that of chondrocytes in the matrix.

A possible mechanism

If extracellular osmolarity around a chondrocyte is changed, as in all animal cells, cell volume is affected. The cell membrane is very permeable to water but essentially impermeable to most intracellular or extracellular osmolites except through specific transporters (Stein,1986). The cells thus shrink or swell in order to equilibrate osmotic pressure across the cell membrane by concentrating or diluting the intracellular osmotically active solutes (Hoffman and Simonsen, 1989). Such changes in volume are rapid. Some cells are able to lose or gain intracellular solutes and thus to regulate their volume to some extent. A change in external osmolarity will thus inevitably alter the concentration of intracellular solutes such as ions.

Freshly isolated bovine articular chondrocytes have a median diameter of 11.2 microns in standard tissue culture medium; this diameter appears to decrease by less than 2% over 2–4 hours. The median diameter falls to around 10.2–10.4 microns when the medium osmolarity is increased towards that present in the matrix (350–400 mOsm) and remains constant for at least 2–4 hrs. The diameter of chondrocytes *in situ* in unloaded bovine cartilage is around 10.1 microns (White, 1991), similar to their diameter in a medium of mOsm. The chondrocytes thus swell 30–40% when removed from the tissue matrix and placed in standard medium (Figure 6).

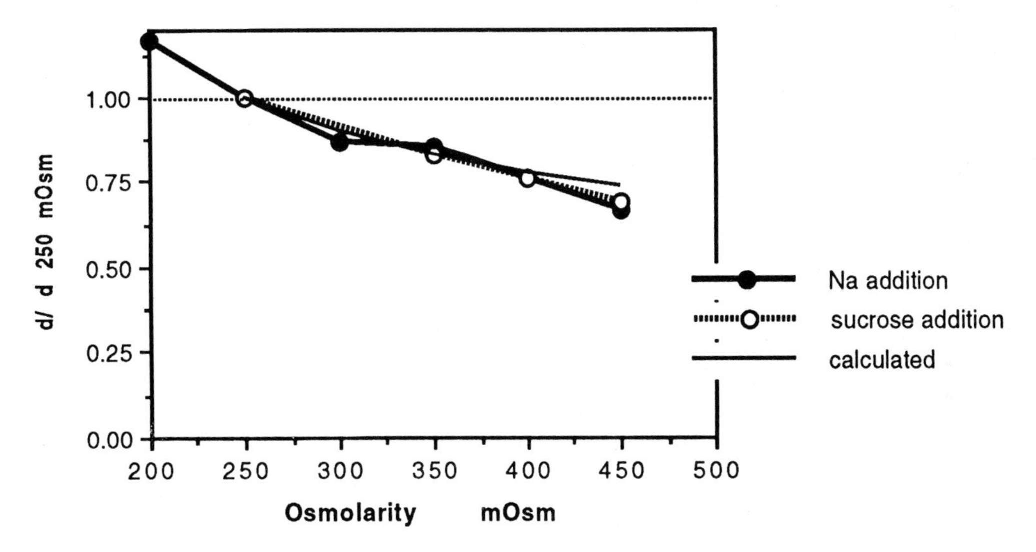

Figure 6. Effect of extracellular osmolarity on median volume of isolated chondrocytes measured using a coulter–counter. Measured values compared with those calculated from the Van't Hoff equation, using values for non–osmotic volume fraction in chondrocytes from data of McGann *et al*, (1988).

Such a volume change will intially lead to equivalent changes in the intracellular ionic composition of the chondrocyte. It is known that many enzymes involved in protein synthesis and energy metabolism are very sensitive to concentrations of intracellular ions particularly K^+, Na^+ and H^+ (Hoffman and Simonsen, 1989). Hence changes in their concentrations resulting from cell shrinkage or swelling would be expected to lead to changes in synthesis rates; if, in the tissue at rest, synthesis rates are operating at a optimum level, any changein intracellular ion concentration should lead to a decrease in rate. The cell may eventually be able to adapt to such changed conditions and indeed initial results suggest that chondrocyte metabolism can adapt to a wide range of extracellular osmolarities to some extent, given sufficient time (12–30 hours) (Urban and Hall, 1991).

CONCLUSIONS

The chondrocytes of cartilage are essential for maintaining and synthesising the matrix macromolecules which provide cartilage with its role of resisting mechanical stress and protecting tissues. Much of the behaviour of cartilage relies on the osmotic pressure developed by the charges on the polyanionic proteoglycans. In weight–bearing cartilages, fluid moving into and out of the tissue under load provides a changing osmotic pressure within the tissue. The chondrocytes of cartilage thus exist in an osmotic environment where the osmotic pressure may cycle through changes of up to 30% in each 24 hour cycle.

Such changes in the extracellular osmotic pressure of chondrocytes have an effect on chondrocyte volume whether it occurs in the tissue or in chondrocytes isolated from the matrix. Since the chondrocytes have no fast regulatory response such changes volume have a proportional change on the intracellular ionic composition of tissue and thus affect proteoglycan and other matrix synthesis rates. It is not known how these changes effect the overall regulation of matrix turnover in the tissue. However it is clear that the chondrocyte can tolerate such cyclical changes and can respond to them. If cartilage remains unloaded and these cyclical changes disappear, synthesis rates and matrix production are adversely affected in the long term.

Acknowledgements We thank the Arthritis and Rheumatism Council for supporting this work.

REFERENCES

Adams, M.A. and Hutton, W. (1983). The effect of posture on the fluid content of the lumbar intervertebral disc. Spine 8: 665–671.

Bayliss, M.T., Urban, J.P.G., Johnson, B. and Holms, S. (1986). In vitro method for measuring synthesis rates in the intervertebral disc. J. Othop. Res. 4: 10–17.

Broom, N.D. and Marra, D.L. (1985). New structural concepts of articular cartilage demonstrated with a physical model. Conn. Tiss. Res. 14: 1–8.

Eisenberg, S.R. and Grodzinsky, A.J. (1987). Kinetics of chemically induced non–equilibrium swelling of articular cartilage and corneal stroma. J. Biomech. Eng. 109: 79–89.

Eklund, J.A. and Corlett, E.N. (1984). Shrinkage as a measure of the effect of load on the spine. Spine 9: 189–194.

Gray M.L., Pizzanelli A.M., Gordzinsky AJ, Lee RC (1986).Mechanical and physicochemical determinants of chondrocyte biosynthetic response J Orthop Res 6, 777–792

Grodzinsky, A.J. (1983). Electromechanical and physicochemical properties of connective tissue. CRC Crit. Rev. Biomed. Eng. 9: 133–199.

Hascall V.C. and Hascall G.K. (1981) Proteoglycans In: Cell Biology of extracellular matrix Ed. ED Hay, Plenum Press, NY: 39–63

Hoffman, E.K. and Simonsen, L.O. (1989). Membrane mechanisms in volume and pH regulation in vertebrate cells. Physiol. Rev. 69: 315–382.

Jones, I.L., Klamfeldt, A. and Sandstrom, T. (1982). The effect of continuous mechanical pressure upon the turnover of articular cartilage proteoglycans in vitro. Clin. Orthop. 165: 283–289.

Maroudas, A. (1979). Physicochemical properties of articular cartilage In "Adult articular cartilage" Ed. M.A.R. Freeman, Pitman Medical, Tunbridge Wells. pp. 215–290.

Maroudas, A. (1976) Balance between swelling pressure and collagen tension in normal and degenerate cartilage. Nature 260: 808–809.

Maroudas, A. (1990). Different ways of expressing concentration of cartilage constituents with special reference to the tissue's organization and functional properties. In, "Methods in Cartilage Research." Eds. A. Maroudas and K.E. Kuettner. Academic Press, London.

Maroudas, A. and Bannon, C. (1981). Measurement of swelling pressure in cartilage and comparison with the osmotic pressure of constituent proteoglycans. Biorheology 18: 613–632.

Maroudas, A. and Urban, J.P.G. (1981). Swelling Pressure of Cartilaginous tissues In: Studies in Joint Disease I (eds) A Maroudas and E.J. Holborow, Pitman Medical, Tunbridge Wells

McGann, L.E., Stevenson, M., Muldrew, K. and Schachar, N. (1988). Kinetics of osmotic water movement in chondrocytes isolated from articular cartilage and applications to cryopreservation. J. Orthop. Res. 6: 109–115.

Mow, V.C., Holms, M.H. and Lai, W.H. (1984). Fluid transport and mechanical properties of articular cartilage: A review. J. Biomechanics 17: 377–394.

Muir, H. (1981). Chemistry of the ground substance of joint cartilage. In, "The Joints and Synovial Fluid, Vol. II. Ed. L. Sokoloff, Academic Press, New York.: 27–94.

Nachemson A. (1966). Load on the lumbar discs in different positions of the body Clin Orthop 45: 107–122

Sah, R.L., Kim, Y.L., Doong, J.Y.H., Grodzinsky, A.J., Plaas, A.H.K. and Sandy, J.D. (1989). Biosynthetic response of cartilage explants to dynamic compression. J. Othop. Res. 7: 619–639.

Schneiderman, R., Keret, D. and Maroudas, A. (1986). Effects of mechanical and osmotic pressure on the rate of glycosaminoglycan synthesis in the human adult femoral head cartilage. An in vitro study. J. Orthop. Res. 4: 393–408.

Simon B., Wu J.S.S., Carlton M.W. *et al.*, (1985). Poroelastaic dynamic structural models of rhesus spinal motion segments Spine 10: 494–505

Sommarin, Y., Larsson, T. and Heinegard, D. (1989). Chondrocyte matrix interactions. Exp.

Cell Res. 184: 181–192.

Stein, W.D. (1986) Transport and diffusion across cell membranes. Academic Press, London

Stockwell, R.A. (1979). Biology of cartilage cells. Cambridge University Press, Cambridge. pp. 148–163.

Thornton, W., Hoffler, W. and Rummel, J. (1974). Anthropometric changes and fluid shifts on Skylab. Presented Skylab Symposium, Aug. 28th.

Urban, J.P.G. and Bayliss, M.T. (1989). Regulation of proteoglycan synthesis rate in cartilage in vitro: influence of extracellular ionic composition. Biochim. Biophys. Acta 992: 59–65.

Urban, J.P.G. and Hall, A.C. (1991). Physical modifiers of cartilage metabolism. In "Articular Cartilage and Osteoarthritis," Ed. K.E. Kuettner. Raven Press, New York. In Press.

Urban, J.P.G. and Maroudas, A. (1981). Swelling of intervertebral disc in vitro. Conn. Tiss. Res. 9: 1–10.

Urban, J.P.G. and McMullin, (1988). Swelling pressure of the intervertebral disc Spine 13: 179–186.

White, N. (1991). Unpublished observations.

INTERSTITIAL MACROMOLECULES AND THE SWELLING PRESSURE OF LOOSE CONNECTIVE TISSUE

F.A. Meyer
Arthritis Research Unit
Ichilov Hospital
Tel Aviv 64239
Israel

INTRODUCTION

Interstitial (tissue) swelling pressure is an important parameter in the maintenance of homeostasis with respect to fluid balance and exchange between interstitium and blood and lymph. Pressure gradients between blood and interstitium and interstitium and lymph result in a steady state equilibrium which stabilizes tissue volume (largely determined by its water content) and the physicochemical environment in which cells can function normally (Zweifach and Silberberg, 1979; Bert and Pearce, 1984; Michel, 1984).

The swelling pressure of interstitium may arise from the osmotic pressure of the soluble macromolecules within the structured tissue space and/or hydrostatic pressure induced by stresses in the insoluble structural components. In order to account for the overall swelling pressure, therefore, it is necessary to consider not only the known osmotic pressures of the isolated macromolecules but also aspects of tissue composition and structure relating to the organisation and in situ effective concentrations of the tissue macromolecules and their intermolecular interactions.

The following is a review of work performed on Wharton's jelly of umbilical cord, a relatively simple loose connective tissue that contains an abundant interstitium. A study of its structure was made which was used as a basis for accounting for the swelling pressure of tissue.

TISSUE COMPOSITION

The interstitium of Wharton's jelly consists of an extracellular matrix (95% v/v) and cells (5% v/v). The extracellular matrix has a gel-like structure involving an insoluble three-dimensional fibrillar network within

NATO ASI Series, Vol. H 64
Mechanics of Swelling
Edited by T. K. Karalis

which soluble macromolecular components are distributed (Table 1). Water, including salts and low mol.wt. nutrients and metabolites ('water'), pervades the entire structure.

Table 1. Macromolecular Composition of Wharton's jelly

	% wet weight
Insoluble components	
Collagen	3.6
Glycoprotein microfibrils	0.3
Soluble components	
i) GAGs	
Hyaluronan	0.31
Chondroitin sulfate	0.14
ii) Plasma proteins	1.2

Ultrastructural, chemical and enzymatic studies (Meyer et al, 1983) indicate that the fibrillar network consists of two independent networks that interpenetrate one another (Fig. 1). The major network involves banded collagen fibrils (average diameter 39 ± 6 nm) and the secondary network involves glycoprotein microfibrils (diameter 13 nm) which have a beaded appearance. Elastin, often present in interstitial tissue, is absent from Wharton's jelly.

The soluble extracellular matrix macromolecules involve glycosaminoglycans (GAGs). Hyaluronan (HA), the major GAG present, is a linear chain macromolecule consisting of repeating disaccharide units which at neutral pH carry one charge/disaccharide. HA (Klein and Meyer, 1983) has a mol.wt. in the region of 10^7 D and in solution adopts a random coil configuration (diameter 600 nm). At tissue concentrations, HA molecules interpenetrate to form a fine pore meshwork (Comper and Laurent, 1978). In addition to HA, some chondroitin sulfate (CS) is also present. CS has a mol.wt. of $2x10^4$ and is a linear chain of repeating disaccharides which carry two charges/disaccharide. In its native form as proteoglycan the CS chains are linked to a common protein backbone in a 'bottle brush' structure (Mathews and Lozaitye, 1958). Proteoglycans have a mol.wt. of approx. 10^5-10^6 (Lowther et al, 1970; Heinegård et al, 1985) and being branched are more compact than the linear HA molecule. In addition to the extracellular matrix components, plasma

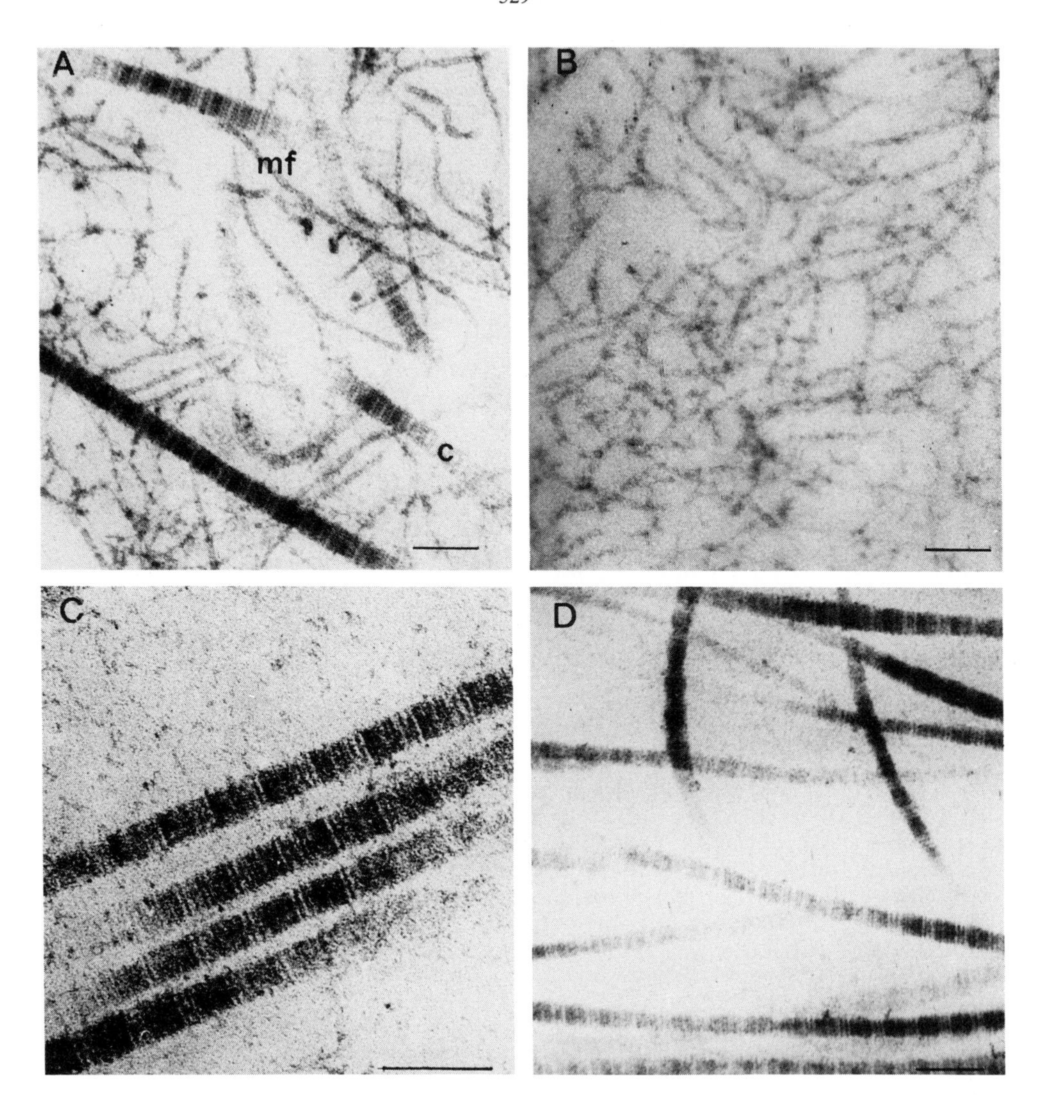

Fig. 1. Electron micrographs of fibrillar metworks in Wharton's jelly. In each case the starting tissue was the GAG-free tissue produced by digestion with testicular hyaluronidase. A: Intact networks of collagen (c) and glycoprotein microfibrils (mf). B: Glycoprotein microfibril network after removal of collagen by exhaustive digestion with bacterial collagenase. C: Break-up of the microfibrillar network on limited digestion with trypsin. D: Collagen network after removal of microfibril network by extensive digestion with Pronase. Bars indicate 100 nm. (From F.A. Meyer et al, 1983.)

protein macromolecules are present. A schematic diagram of tissue structure is shown in Fig. 2.

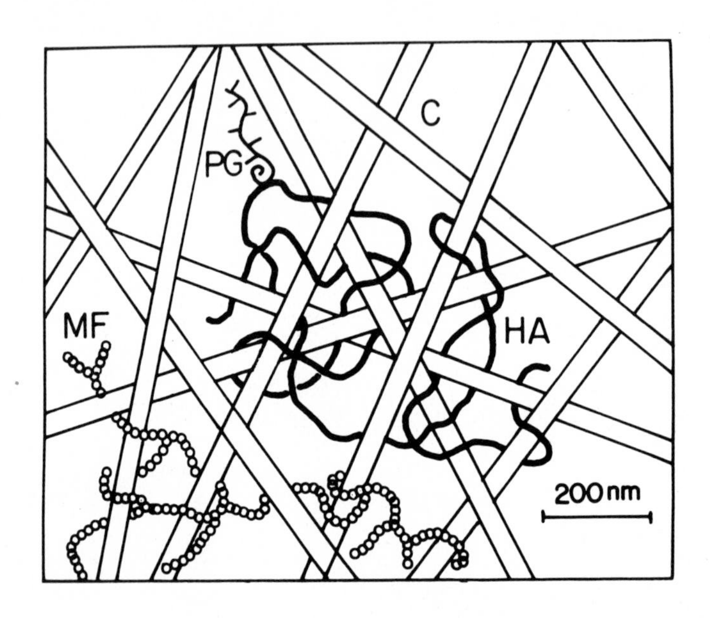

Fig. 2. Schematic illustration (approximately to scale) of the structural components within tissue. The components indicated are collagen (C), glycoprotein microfibrils (MF), hyaluronan (HA) and CS-proteoglycan (PG). For clarity, only part of the MF network and single molecules of HA and PG are shown.

Studies on tissue structure and the role of the individual extracellular matrix components have been facilitated by in situ enzymatic degradation of specific matrix components (Meyer and Silberberg, 1974; Meyer et al, 1977; Klein and Meyer, 1983; Meyer, 1983; Meyer et al, 1983). Chemical analyses and electron microscopy monitoring of enzyme action on umbilical cord slices demonstrated that testicular hyaluronidase causes the removal of GAGs while proteases cause the removal of CS and disruption of the glycoprotein microfibril network. Based on these findings, structural variants of tissue could be produced: i) intact fibrillar network minus GAGs (hyaluronidase treatment); ii) intact collagen fiber network (hyaluronidase followed by protease treatment); and iii) intact collagen network containing HA (protease treatment).

RETENTION OF SOLUBLE MACROMOLECULES IN TISSUE

The soluble macromolecules, GAGs and plasma proteins, may contribute osmotically to tissue swelling pressure by being selectively retained in the tissue space. The involvement of matrix structure in the retention of the soluble macromolecules was investigated by measuring their efflux rates from geometrically defined thin cylindrical slices confined at their in vivo volume in a porous cell (Klein and Meyer, 1983).

i) GAGs

Measurement of the rate of efflux of HA was found to be consistent with diffusional kinetics and a diffusion coefficient that is 1.1 ± 0.4% of the value in free solution. Preliminary experiments indicated that the efflux rate of CS-proteoglycan is similar, suggesting that the proteoglycan is held in tissue through association wtih HA. Although the formation of aggregates between HA and proteoglycan isolated from the present tissue have not been studied, such associations are well known in other connective tissues (Mörgelin et al, 1989). The significant reduction in diffusion coefficients of HA and proteoglycan indicates that these components are effectively immobilized and that they may be regarded together with the fibrillar networks as part of the fixed components of the tissue. By being immobilized, the tissue acts like an 'ideal' semi-permeable membrane for the GAGs which allows the generation of an osmotic contribution from these components.

Insights into the structural constraints on GAG mobility in tissue were revealed by using modified tissues. A study of HA diffusion through the collagen network alone indicated that its mobility increased somewhat to 10 ± 4% of that in free solution. Hence HA mobility in intact tissue is significantly, though not wholly, restricted by the collagen fiber network. The reduction in mobility through the collagen network was consistent with theoretical considerations whereby HA was assumed to be physically constrained within the network pores and that movement was restricted to reptational translation (De Gennes, 1971), whereby the flexible HA molecular chain threads its way through the pores of the collagen fiber network. For intact tissue where the pore size is further reduced by the additional presence of the glycoprotein microfibril network, the predicted diffusion agreed within a factor of four with the experimental result. An even closer correspondence between theory and experiment for the intact tissue might be expected, however, by the attachment of proteoglycan appendages to the HA chain which would further impede reptational movement. It is also possible that the immobilization of HA in intact tissue may be influenced in part by a HA/proteoglycan and proteoglycan/collagen (Brown and Vogel, 1989) interaction. The present results indicate that the GAGs are effectively held in tissue primarily by physical constrainment within the pores of the insoluble fiber network. This is consistent with previous studies (Fessler, 1960) showing that the GAGs of Wharton's jelly can be readily extracted on physical disruption of the fibrillar network.

ii) Plasma Proteins

In contrast to the immobilization of GAGs within tissue, plasma proteins were not retained by tissue. Efflux data indicated that these diffused out of tissue with a diffusion coefficient that was 55 ± 10% of that for free diffusion. This reduction could be accounted for by theoretical predictions based on diffusion through the fibrillar network and through GAG solutions. Although plasma proteins cannot be regarded as part of the fixed tissue structure, they nevertheless may make an osmotic contribution in vivo via the endothelium lining which acts as a selective membrane that hampers plasma protein but not 'water' movement (Taylor and Granger, 1984).

COMPARTMENTALIZATION AND THE EFFECTIVE CONCENTRATION OF SOLUBLE MACROMOLECULES IN TISSUE

The space available in tissue for the molecularly disperse soluble components, plasma protein and GAGs, will determine their effective concentration in tissue and hence their osmotic pressure (Meyer, 1983). For plasma proteins the space available is dictated by the exclusion properties of the GAGs, fibrils and cells and hence is given by the exclusion behavior of the intact tissue. On the other hand, the space available for the GAGs is determined by exclusion due to the fibrils and cells which can be determined on GAG-free tissues. The available space for a molecule is dependent both on the self-volume of the excluding components and on spaces created by the spatial organization of the network components, whose dimensions are smaller than those of the molecule (Ogston, 1958). Exclusion, therefore, is dependent on molecular size and will be relatively large for compact globular proteins such as albumin (radius 3.55 nm) as compared with open-chain macromolecules such as the GAGs (chain radius 0.35 nm) whose chains can penetrate into smaller spaces.

The excluded volume fraction (= 1 - available volume fraction) of the tissues held at the in vivo volume by confinement in porous cells was determined for a series of spherical non-interacting probe molecules of differing radii. After equilibration between probe solution and tissue, the concentration of the probe in the total tissue volume was compared with that in the solution and the ratio (partition coefficient) used to calculate the excluded volume fraction (Fig. 3).

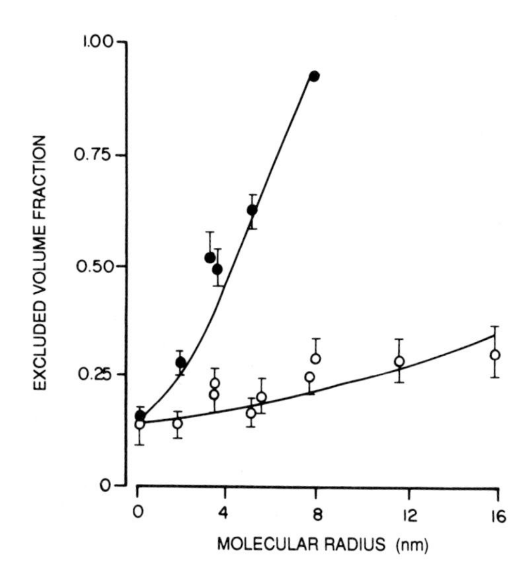

Fig. 3. Excluded volume fraction of intact tissue (●) and GAG-free tissue (o) for spherical molecules of varying radii. Equilibration was with sorbitol, myoglobin, albumin, transferrin, catalase and thyroglobulin of hydrodynamic radii (in nm) of 0.36, 1.98, 3.55, 3.82, 5.2 and 8.17, respectively. For GAG-free tissue, equilibration was also performed with dextran fractions of radii 5.5, 8.0, 11.7 and 16.0 nm. The solid lines represent the exclusion predicted using a cylinder/sphere geometric model. (Redrawn from F.A. Meyer, 1983.)

In the intact tissue the excluded volume fraction is markedly dependent on molecular size. For albumin, which represents the bulk of the diffusible plasma proteins in tissue, the excluded volume fraction is 0.55. This corresponds to an available space which is 45% of the tissue volume and hence to an effective concentration in tissue which is 2.22 times higher than that based on total tissue volume. The difference between the data for intact and

GAG-free tissues gives the exclusion due to the GAGs. It is evident that exclusion of albumin is due both to the GAGs and to the insoluble elements in the tissue.

The excluded volume fraction for the GAGs can be obtained from data for the GAG-free hyaluronidase-treated tissue. The data show only a slight dependence on molecular size and that the excluded volume fraction of the GAGs (chain radius 0.35 nm) is essentially given by the self-volume fraction (zero radius) of the insoluble tissue components. An estimate of 0.14 is indicated which corresponds to an available space for GAGs in 86% of the tissue volume and hence an effective concentration which is 1.16 time higher than that estimated from the total tissue volume. The coincidence of the curves for the treated and untreated tissues at zero probe size is consistent with the self-volume fraction of the GAGs in tissue being negligible (0.003) in relation to total tissue volume.

The solid curves in Fig. 3 represent the exclusion predicted by theory based on the composition of the tissues, the molecular characteristics of the matrix macromolecules and the self-volume fraction of the cells in the tissue. The calculations are based on a cylinder/sphere steric exclusion model (Ogston, 1958), where the GAG chains, collagen and microfibrils are regarded as cylinders and the spherical probes as spheres. Good agreement is seen between the predicted curves and the experimental data. This suggests, inter-alia, that the GAGs in tissue behave in a manner comparable to their state in free solution and hence are present in tissue in a similar molecularly dispersed state, and that the probes used interact physically but not chemically with the extracellular matrix. Extrapolation of the predicted curves indicate a pore radius of 18 nm for intact tissue and of 110 nm for the hyaluronidase-treated tissue (not shown). In the present treatment fibrillar volume was assumed to be unaffected by removal of the GAGs. In cartilage, however, osmotic effects of GAGs on intrafibrillar water have been shown to cause changes in fibrillar volume and hence to influence available tissue volume (Katz et al, 1986). These effects are unlikely to be of significance in the present case since the concentration of GAGs in Wharton's jelly is too low to exert such effects.

TISSUE SWELLING PRESSURE

Tissue swelling pressure may be expected to involve osmotic contributions from the soluble extracellular matrix components (GAGs) and plasma proteins and a hydrostatic contribution from the insoluble fibrillar network. An osmotic contribution from the latter is unlikely in view of the small surface/bulk ratio of the fibrils. The contribution of the extracellular matrix macromolecules was obtained using tissue from which plasma proteins had first been allowed to diffuse out.

i) Extracellular Matrix Components

Tissue slices placed in open contact with a physiological buffer solution will swell to twice their in vivo volume (Fig. 4). The osmotic pressure of

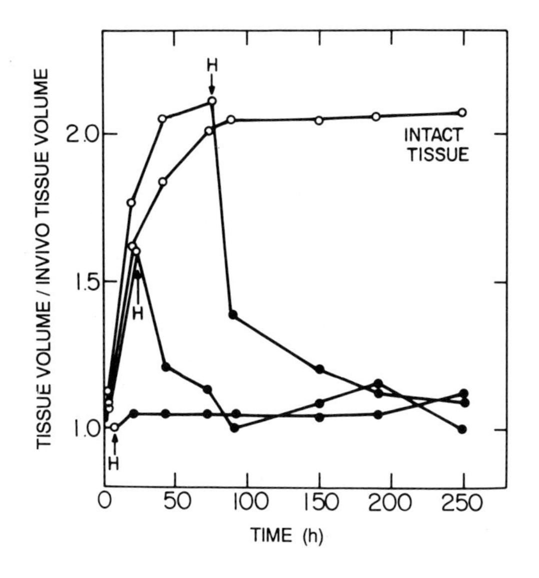

Fig. 4. Time dependence of swelling to equilibrium in physiological buffer of intact tissue (o), and of tissue treated with hyaluronidase (H) at time zero (●) and at two subsequent times (●). (Redrawn from F.A. Meyer and A. Silberberg, 1974.)

the GAGs is responsible for this swelling since tissue treated to remove the GAGs no longer swells. Tissue volume, then, remains close to its in vivo volume (Fig. 4), indicating that the fibrillar network when at the in vivo volume of tissue does not exert a hydrostatic pressure. When intact tissue swells, however, elastic stresses are built up in the fibrillar network, and upon removal of the GAGs from swollen tissue, the fibrillar network returns to its unstressed state close to the in vivo volume of tissue (Fig. 4). The elastic stresses that develop upon swelling limit the GAG-induced swelling and equilibrium is reached when the osmotic pressure of the GAGs is balanced by the elastic stresses in the network.

These findings suggest that at close to the in vivo volume of tissue, tissue swelling pressure is given only by the osmotic pressure of the GAGs and that the fibrils are in a zero-stress state at this volume. Upon swelling or deswelling, however, the fibrils may make a net hydrostatic contribution due to the elastic stresses that develop on volume change.

The zero-stress state of the fibrillar network close to the in vivo volume of tissue may be attributed to an interaction between the collagen and glycoprotein microfibril networks. Experiments with GAG-free tissues (Fig. 5) indicate that treatment with a protease to break up the glycoprotein microfibril network causes the collagen network to expand, suggesting that the latter is contrained by the former. On the other hand, if the collagen network is subjected to collagenase attack, expansion does not occur.

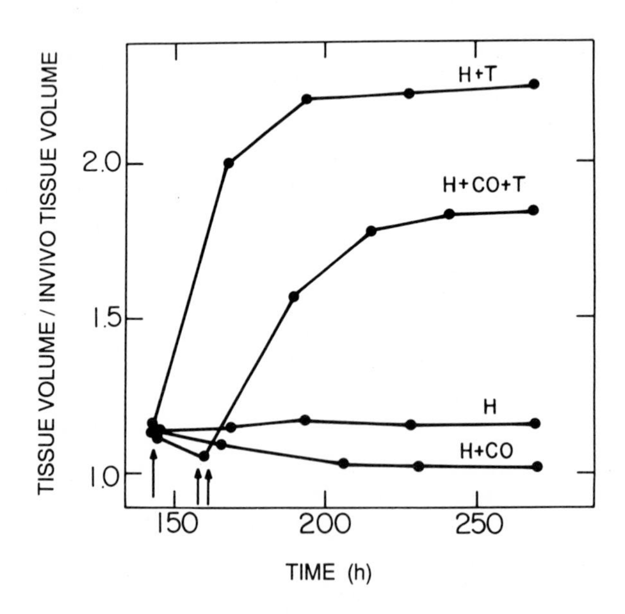

Fig. 5. Effect of treatment of GAG-free (hyaluronidase-treated) tissue (H) with trypsin (H + T), bacterial collagenase (H+CO) and with both enzymes (H + CO + T) on equilibrium swelling in physiological buffer solution. Treatment with H was at time zero and with second (CO or T) and third (T) enzymes at the times indicated by the single and double arrows, respectively. Similar results were obtained when trypsin was replaced by chymotrypsin, elastase or Pronase. (From F.A. Meyer et al., 1983).

Collagenase in this study solubilizes 46% of the collagen which corresponds to the loss of the outer three molecular layers from the collagen fibrils, since digestion is uniform and was performed after diffusion of the enzyme into the tissue. Upon subsequent breakup of the microfibril network, the weakened collagen network is still capable of expansion, though to a lesser extent (Fig. 5). Interaction between the networks is likely to be mechanical rather than chemical since collagenase attack at the collagen fibril surface might be expected to disrupt chemical links between interacting fibrillar surfaces. Furthermore, some mechanical coupling might be expected by the mutual interpenetration of independent networks of relatively bulky fibrils. Modulation of collagen fiber contortions, therefore, appears to be a role of the glycoprotein microfibrils.

To determine the contribution of the various extracellular matrix components to tissue swelling pressure, measurements were performed on intact and modified tissues as a function of tissue hydration (Meyer, 1983). Tissue slices were allowed to swell or deswell to equilibrium against an external osmotically active solution. Since swelling pressures are relatively low compared, e.g., with cartilage (Maroudas and Grushko, 1990), it was possible to use dilute solutions of HA of high molecular weight as calibrants. Such solutions are convenient to use and their osmotic pressure is known (Silpananta et al, 1968). Although tissue was in direct contact with HA solutions, HA entry from the external solution into intact tissue did not occur over the time scale for solvent equilibration. For enzyme- modified tissues from which GAGs had been removed, some entry of HA was indicated by a superimposed swelling phase. To correct for this, the swelling phase was subtracted from the data.

The swelling pressure of intact tissue and the individual contributions of the extracellular macromolecules as a function of tissue volume is shown in Fig. 6. The results show that the swelling pressure of tissue over much of the hydration range is given mainly by the osmotic pressure of the GAGs. At the point where the curves for intact tissue and GAGs cross, which is close to the in vivo volume of tissue, the swelling pressure is entirely due to the GAGs. At this volume, as discussed above, the combined fibrillar network of collagen and glycoprotein microfibrils is in an unstressed configuration (zero net pressure) with no tendency to expand or contract. This is consistent with the swelling pressure curve of the fibrillar system which shows that at close to the in vivo volume of tissue, the fibrillar network exerts a net zero pressure. When the fibrillar system is expanded or contracted, elastic stresses are generated which respectively produce a positive hydrostatic pressure that tends to expel solvent and a negative hydrostatic pressure that tends to imbibe solvent.

The pressure contribution of the fibrillar network consists of the individual contributions of the collagen and glycoprotein microfibril networks.

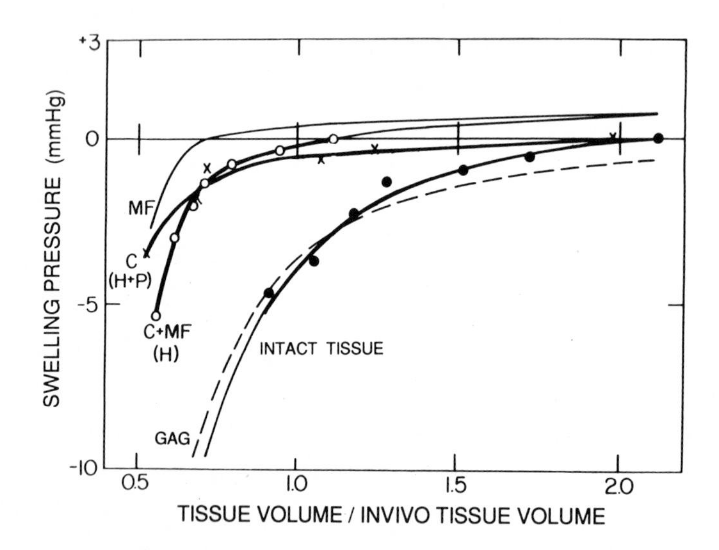

Fig. 6. Swelling pressure of intact tissue and the individual contributions of glycosaminoglycans (GAG), collagen (C) and microfibrils (MF) as a function of tissue volume. Experiments were performed using tissues from which diffusible proteins were removed. The pressure is given in mmHg, above or below atmospheric pressure (regarded as zero pressure) and corresponds respectively to a tendency to express or to imbibe solvent. The curve (C + MF) represents hyaluronidase-digested tissue (H) and the curve (C) represents tissue after hyaluronidase and Pronase digestion (H + P). The curve for GAG (dashed line) was derived using data from Fig. 7, as described in the text. The solid curves were constructed from experimental data (●, o or x) and/or by difference using the experimental and GAG curves. (From F.A. Meyer, 1983.)

The collagen curve shows that the pressure exerted by the collagen network is negative over the entire range of tissue hydration studied and hence is in a state of compressive stress. The microfibril curve, on the other hand, crosses over from a negative to a positive pressure indicating that the stresses developed in this network change direction in going from compression to expansion. This transition point is reflected in the curve for the combined fibrillar network which undergoes a transition at close to the in vivo volume of the tissue and represents the configuration where the net pressure of the combined network is zero.

The curve for the osmotic pressure of the tissue GAGs in Fig. 6 was estimated from the effective concentration of the GAGs in the extrafibrillar, extracellular space of tissue as a function of hydration. Using the data for swelling equilibrium of intact tissue against external HA solution, a plot of the effective GAG concentration in tissue against the external HA concentration at swelling equilibrium was made (Fig. 7). At the tissue volume (arrow) where the fibrillar system is at zero pressure (essentially the in vivo volume of tissue) the swelling pressure is due to the GAGs alone. At

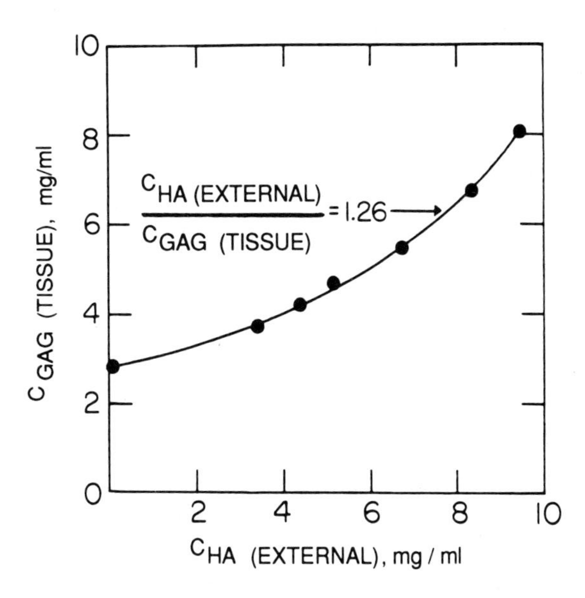

Fig. 7. Plot of GAG concentration, C_{GAG}, in the extrafibrillar, extracellular space of intact tissue against external HA concentration, C_{HA}, at swelling equilibrium. The arrow corresponds to the volume of tissue at which the overall fibrillar system is in its zero-pressure configuration. (From F.A. Meyer, 1983).

this point their pressure is balanced externally by a concentration of HA that is 1.26 times higher than that of the GAGs. This value is consistent with that expected for the osmotic activity of a 70% HA/30% CS mixture that constitutes the tissue GAGs. The osmotic pressure of HA and proteoglycan solutions are predominantly due to Donnan counter-ion effects (Brace and Guyton, 1979; Maroudas and Urban, 1980, Maroudas et al, 1988). For GAG concentrations involved in tissue, ideal Donnan behavior applies, as the fixed charge density of the GAGs (approximately 0.019 M) is very much smaller than the physiological microion concentration (approximately 0.15 M) of the buffer solution. Hence, CS, which carries two charges/disaccharide, has an osmotic pressure approximately twice that of HA which carries only one charge/disaccharide. Hence the osmotic pressure for the 70% HA/30% CS mix of tissue GAGs is equivalent to that of an HA solution at a 1.30 fold concentration, consistent with experimental findings. Use of this equivalence was made in deriving the osmotic pressure curve for the tissue GAGs in Fig. 6. It may be seen (Fig. 6) that this curve, derived from the effective concentration of the GAGs in the extrafibrillar, extracellular space of tissue, is in accord with the GAG contribution indicated by swelling pressure measurements for tissue with and without GAGs.

ii) Plasma Proteins

The osmotic pressure of the plasma proteins in intact tissue is given by their effective concentration in tissue. Their concentration based on total tissue volume is 12 mg/ml (Klein and Meyer, 1983). Assuming these proteins to be albumin (the major diffusible plasma protein in tissue) then localization of the protein to the available albumin space in tissue (Fig. 3) would give an effective protein concentration of 26.6 mg/ml which corresponds to an osmotic pressure of -8.9 mmHg (Landis and Pappenheimer, 1963) at the in vivo volume of tissue.

CONCLUSIONS

The behavior of tissue on swelling and deswelling is similar to that of a gel in that there is an interplay between the osmotic and elastic forces. In conventional gels, however, the stress-bearing network system is also the osmotically active component, whereas in tissue, because of the insolubility of the fibrillar network, these roles are separated such that the latter produce the elastic stresses and the molecularly disperse components the osmotic effects. Our studies indicate that at volumes close to the in vivo volume of tissue, the elastic forces become zero since the fibrillar system is in its net zero-stress configuration. At this volume, therefore, only the molecularly disperse components make a contribution to tissue swelling pressure. A similar situation may apply to other tissues as well, e.g., skin (Wiederhielm, 1981), heart valves (Meyer, 1971), articular cartilage (Maroudas and Banon, 1981) and intervertebral disc (Urban and McMullin, 1985). Although the hydrostatic pressure of the insoluble fibrillar system was not measured in these tissues, the measured swelling pressure of tissue was comparable with estimates of the osmotic pressure contribution expected from the molecularly disperse components.

At the in vivo volume of tissue, osmotic pressures of -4 mmHg for the GAGs and -8.9 mmHg for plasma proteins are indicated which are consistent with those expected from their effective concentrations in tissue. These estimates are comparable to values for the swelling pressure measured by others on umbilical cord of -12 mmHg (Granger et al, 1975) and -5.6 ± 3.7 mmHg (Snashall, 1977) in the presence and absence, respectively, of diffusible proteins. The osmotic pressures for the plasma proteins given above are those expected across an 'ideal' semipermeable membrane which is selective for protein but not for 'water'. In vivo, however, their osmotic pressure may be considerably lower since the membrane involved (the endothelium) is in general only partially selective for proteins (Taylor and Granger, 1984). On the other hand, an 'ideal' semipermeable membrane may be assumed for the GAGs since, as was shown, they are effectively trapped within the fibrillar network which therefore provides an 'ideal' semipermeable membrane for these components.

The finding that the unstressed configuration of the fibrillar network system is close to the in vivo volume of tissue suggests that the fibrils provide the basic skeletal shape and volume of tissue in vivo while the molecularly disperse components pressurize the system by adjusting 'water' chemical potential so that a steady state balance with blood and lymph is achieved. How this is done is not known; however it is possible that tissue cells, by sensing stress in the fibrillar network, can respond by modulating GAG synthesis to relieve the stress. In this regard, it is noteworthy that mechanical stress can induce effects on the rate of GAG synthesis by chondrocytes (De Witt et al, 1983; Schneiderman et al 1986).

REFERENCES

Bert JL, Pearce RH (1984) The interstitium and microvascular exchange. In: Renkin EM, Michel CC, Geiger SR (eds) The cardiovascular system. Microcirculation. American Physiological Society, Bethesda (Handbook of Physiology, sect. 2, vol. IV, chap 12, p 521)

Brace RA, Guyton AC (1979) Interstitial fluid pressure: capsule, free fluid, gel fluid, and gel absorption pressure in subcutaneous tissue. Microvasc Res 18:217-228

Brown DC, Vogel KG (1989) Characteristics of the in vitro interaction of a small proteoglycan (PG II) of bovine tendon with Type I collagen. Matrix 9:468-478

Comper WD, Laurent TC (1978) Physiological function of connective tissue polysaccharides. Physiol Rev 58:255-315

De Gennes PG (1971) Reptation of a polymer chain in the presence of fixed obstacles. J Chem Phys 55:572-579

De Witt MT, Handley CJ, Oakes BW, Lowther DA (1983) In vitro response of chondrocytes to mechanical loading. The effect of short term mechanical tension. Connect Tissue Res 12:97-109

Fessler JH (1960) A structural function of mucopolysaccharides in connective tissue. Biochem J 76:124-132

Granger HJ (1981) Physicochemical properties of the extracellular matrix. In: Hargens AR (ed) Tissue fluid pressure and composition, Williams & Wilkins, Baltimore, p 43

Heinegård D, Bjorne-Persson A, Coster L, Franzen A, Gardell S, Malmstrom A, Paulsson M, Sandfalk R, Vogel K (1985) The core proteins of large and small interstitial proteoglycans from various connective tissues form distinct subgroups. Biochem J 230:181-194

Katz EP, Wachtel EJ, Maroudas A (1986) Extrafibrillar proteoglycans osmotically regulate the molecular packing of cartilage collagen . Biochim Biophys Acta 882:136-139

Klein J, Meyer FA (1983) Tissue structure and macromolecular diffusion in umbilical cord. Immobilization of endogenous hyaluronic acid. Biochim Biophys Acta 755:400-411

Landis EM, Pappenheimer JR (1963) Exchange of substances through the capillary walls. In: Hamilton WF (ed) Circulation. American Physiological Society, Washington DC (Handbook of Physiology, sect. 2, vol. II, chap. 29, p 961)

Lowther DA, Preston BN, Meyer FA (1970) Isolation and properties of chondroitin sulphates from bovine heart valves. Biochem J 118:595-601

Maroudas A, Bannon C (1981) Measurement of swelling pressure in cartilage and comparison with the osmotic pressure of constituent proteoglycans. Biorheology 18:619-632

Maroudas A, Grushko G (1990) Measurement of swelling pressure of cartilage. In: Maroudas A, Kuettner K (eds) Methods in cartilage research, Academic Press, London, p 298

Maroudas A, Urban JPG (1980) Swelling pressures of cartilaginous tissues. In: Maroudas A, Holborrow EJ (eds) Studies in joint disease. Pitman Medical, Tunbridge Wells, Kent, p 87

Maroudas A, Weinberg PD, Parker KH, Winlove CP (1988) The distribution and diffusivities of small ions in chondroitin sulphate, hyaluronate and some proteoglycan solutions. Biophys Chem 32:257-270

Mathews MB, Lozaitye I (1958) Sodium chondroitin sulfate-protein complexes of cartilage. 1. Molecular weight and shape. Arch Biochem Biophys 74: 158-174

Meyer FA (1971) A biochemical and biophysical approach to the structure of heart valves. PhD Thesis, Monash University, Clayton, Australia
Meyer FA (1983) Macromolecular basis of globular protein exclusion and of swelling pressure in loose connective tissue (umbilical cord). Biochim Biophys Acta 755:388-399
Meyer FA, Koblentz M, Silberberg A (1977) Structural investigation of loose connective tissue by using a series of dextran fractions as non-interacting macromolecular probes. Biochem J 161:285-291
Meyer FA, Laver-Rudich Z, Tanenbaum R (1983) Evidence for a mechanical coupling of glycoprotein microfibrils with collagen fibrils in Wharton's jelly. Biochim Biophys Acta 75:376-387
Meyer FA, Silberberg A (1974) In vitro study of the influence of some factors important for any physicochemical characterization of loose connective tissue in the microcirculation. Microvasc Res 8:263-273
Michel CC (1984) Fluid movements through capillary walls. In: Renkin EM, Michel CC, Geiger SR (eds) The cardiovascular system. Microcirculation. American Physiological Society, Bethesda (Handbook of Physiology, sect. 2, vol. IV, chap 9, p 375)
Mörgelin M, Paulsson M, Malström A, Heinegård D (1989) Shared and distinct structural features of interstitial proteoglycans from different bovine tissues revealed by electron microscopy. J Biol Chem 264:12080-12090
Ogston AG (1958) The spaces in a uniform random suspension of fibers. Trans Faraday Soc 54:1754-1757
Schneiderman R, Keret D, Maroudas A (1986) Effect of mechanical and osmotic pressure on the rate of glycosaminoglycan synthesis in the human adult femoral head cartilage. An in vitro study. J Orthop Res 4:393-408
Silpananta P, Dunstone JR, Ogston AG (1968) Fractionation of a hyaluronic acid preparation in a density gradient. Biochem J 109:43-50
Snashall PD (1977) Mucopolysaccharide osmotic pressure in the measurement of interstitial pressure. Am J Physiol 232:608-616
Taylor AE, Granger DN (1984) Exchange of macromolecules across the microcirculation. In: Renkin EM, Michel CC, Geiger SR (eds) The cardiovascular system. Microcirculation. American Physiological Society, Bethesda (Handbook of Physiology, sect. 2, vol. IV, chap 11, p 467)
Urban JPG, McMullin JF (1985) Swelling pressure of the intervertebral disc. Influence of proteoglycan and collagen contents. Biorheology 22:145-157
Wiederhielm CA (1981) The tissue pressure controversy, a semantic dilemma. In: Hargens AR (ed) Tissue fluid pressure and composition, Williams & Wilkins, Baltimore, p 21
Zweifach BW, Silberberg A (1979) The interstitial-lymphatic flow system. In: Guyton AC, Young DB (eds) International review of physiology. Cardiovascular physiology III, Vol. 18, University Park Press, Baltimore, p 215

A MIXTURE APPROACH TO THE MECHANICS OF THE HUMAN INTERVERTEBRAL DISC

H. Snijders[1], J. Huyghe[1,2], P. Willems[1], M. Drost[1], J. Janssen[1,2], A. Huson[1,2]

[1]Dept. of Movement Sciences
University of Limburg
P.O. Box 616
Maastricht
The Netherlands

[2]Dept. of Mechanical Engineering
Eindhoven University of Technology
P.O. Box 513
Eindhoven
The Netherlands

Summary

A finite deformation model for charged hydrated tissues, such as the intervertebral tissue, is developed. The model predicts not only the response to a mechanical load but also to a chemical load. This response is characterised by ion diffusion, fluid flow relative to the solid matrix and osmotic pressure. The reasonable agreement between experiment and prediction of the model supports its physical basis.

Notation

a	scalar	$\underline{I}$	unit second order tensor
$\vec{a}$	vector	$\vec{a} \circ \vec{b}$	dot vector product
$\underline{a}$	second order tensor	$\vec{a}\vec{b}$	dyadic vector product
$\underline{a}^{-1}$	inverse of $\underline{a}$	$\underline{a} \circ \underline{b}$	dot matrix product
$\underline{a}^{c}$	conjugate of $\underline{a}$	$\underline{a} : \underline{b}$	double dot matrix product

Introduction

Intervertebral disc tissue consists of a collagen and elastine fiber network inbedded in a hydrated proteoglycan (PG) matrix. Small nutrients and ions are dissolved within the tissue. Because of the entanglement of the PG and the fiber network only the interstitial fluid may flow. The PG are ionized and because they are relatively stagnant, "osmotic" effects are important. Deformation of the tissue can be achieved either by mechanical or chemical loading. The overall respons which occurs in cartilaginous tissues, is the result of four different, simultaneously occurring physical processes, each with a different time constant. (1) The diffusion of mobile ions. A typical value for the time constant of diffusion for intervertebral disc tissue is 1500 s (specimen thickness 1 mm) (Maroudas, 1975). (2) Specific binding of ions to the PG-chain, resulting in limiting the diffusion will be neglected. This is valid for the diffusion of NaCl (Maroudas, 1975). (3) The large deformation of the fiber network and the ground substance with relative fluid flow into or out of the tissue. In the limiting case of small deformation a typical value of the time constant is 12000 s (specimen thickness 1 mm) (Best et al., 1989). (4) Readjustment of the local electrical fields within the matrix, due to the volume change of the PG-matrix or the change of ion content. Typical values are 1 ns (Grodzinsky et al., 1981). This charge relaxation is never rate-limiting. On the basis of the estimation of the time constants we

NATO ASI Series, Vol. H 64
Mechanics of Swelling
Edited by T. K. Karalis

developed a mechanical model for intervertebral disc tissue including "osmotic" effects, ion diffusion and large matrix deformation combined with relative fluid flow.

In order to describe this complex behaviour we make use of the theory of mixtures. Starting with the kinematic relationships and the general balance laws, specific assumptions for intervertebral disc tissue will be incorporated and constitutive restrictions based on the entropy principle are derived. The theory is compared with experimental results.

Suffix α stands for an arbitrary constituent, while the suffix f, s and i stands for respectively fluid, solid and ions.

Kinematics

The theory presented is based on the theory of mixtures by Bowen (1980) and Müller (1985). The basic assumption is that the mixture may be viewed upon as a superposition of single continua, each following its own motion. This implies an averaging procedure within an elementary volume. During this averaging procedure both mathematical and physical requirements should be adopted to define averaged macroscopic quantities in terms of the microscopic quantities. Details of such an averaging procedure can be found in the literature (Whitaker, 1969; Slattery, 1972; Hassanizadeh and Gray, 1979). Every property γ has a 'true' and an 'apparent' value. The true property of the αth constituent $\mathring{\gamma}^\alpha$ is defined as the volume integral of γ^α over the averaging volume devided by the true volume V^α. The apparent property or bulk volume averaged value γ^α is integrated in the same way but is devided by the mixture volume V. By these definitions it is clear that $\gamma^\alpha = n^\alpha \mathring{\gamma}^\alpha$, with $n^\alpha = V^\alpha/V$ being the volume fraction of constituent α. So by means of the averaging procedure each constituent is spread over the mixture volume. Unless otherwise stated all properties are apparent. It is assumed that at any time t there exists an one-to-one continuous and differentiable relationship between the current position vector $\vec{x}^\alpha$ and the initial position vector $\vec{x}_0^\alpha$ of a constituent:

$$\vec{x} = \vec{x}^\alpha = \vec{\psi}^\alpha (\vec{x}_0^\alpha, t) \tag{1}$$

We assume that within each averaging volume the different constituents are homogeneously distributed. Therefore all current position vectors $\vec{x}^\alpha$ coincide for all different phases; which is generally not true for the corresponding reference vectors $\vec{x}_0^\alpha$.

For the definition of the material time derivative it is important with which constituent the observer is moving:

$$\frac{D^\alpha}{Dt}\gamma = \frac{\partial\gamma}{\partial t} + \vec{\nabla}\gamma \circ \vec{v}^\alpha \tag{2}$$

$\vec{\nabla}$ represents the gradient operator with respect to the current configuration and is constituent independent; $\vec{v}^\alpha$ the velocity of the αth constituent. The αth material time derivative may be transformed to the material derivative following another arbitrary constituent β by:

$$\frac{D^{\beta}}{Dt}\gamma - \frac{D^{\alpha}}{Dt}\gamma = (\vec{v}^{\beta} - \vec{v}^{\alpha}) \circ \vec{\nabla}\gamma \qquad (3)$$

Because the constituents are regarded as continua all other kinematic quantities, such as the velocity or the deformation gradient are formulated in the same way as in a single continuum theory. Therefore their definition is omitted here.

Balance laws

Mass balance. The local mass balance for the αth constituent is given by:

$$\frac{D^{\alpha}}{Dt}\varrho^{\alpha} + \varrho^{\alpha}\vec{\nabla}\circ\vec{v}^{\alpha} = c^{\alpha} \quad or \quad \dot{\varrho}^{\alpha}+\varrho^{\alpha}\vec{\nabla}\circ\vec{v}^{\alpha}=c^{\alpha} \qquad (4)$$

The mass supply from the other constituents is accounted for by the interaction term c^{α}. If chemical reactions or phase transitions occur, this term is nonzero. However the mass balance of the mixture leads to:

$$\sum_{\alpha} c^{\alpha}=0 \qquad (5)$$

Within intervertebral disc tissue three different constituents may be distinguished. The elastine and collagen fibers and the proteoglycan ground matrix are considered to be an intrinsically incompressible solid. This means that the true density Γ^s is constant. The second constituent is the fluid, which is also intrinsically incompressible. The fluid and solid constituent are immiscible. Volume change of the mixture is achieved only by squeezing fluid out of the tissue. The third constituent consists of the small nutrients and ions. Although many different ions and nutrients are involved, they are treated as one constituent: the ionphase. The volume of the ions is neglected compared to the volume occupied by the solid and the fluid. The density of the fluid and solid constituent (ρ^f and ρ^s) is expressed as a function of the volume fractions (n^f and n^s) and the intrinsic density (Γ^f and Γ^s):

$$\varrho^s = n^s\Gamma^s \quad and \quad \varrho^f = n^f\Gamma^f \qquad (6)$$

As the tissue remains hydrated the sum of the volume fractions equals unity: $n^s + n^f = 1$. No chemical reaction takes place ($c^{\alpha}=0$), so the mass balances for the three constituents are:

$$\begin{aligned} &solid\text{:}\ \dot{n}^s+n^s\vec{\nabla}\circ\vec{v}^s=0 \\ &fluid\text{:}\ \dot{n}^f+n^f\vec{\nabla}\circ\vec{v}^f=0 \\ &ions\text{:}\ \dot{\varrho}^i+\varrho^i\vec{\nabla}\circ\vec{v}^i=0 \end{aligned} \qquad (7)$$

The ionized proteoglycans affect the equations in two ways: their volume contribution is incorporated in the mass balance of the solid while their electrical charge influences the ion flux. Summation of the mass balance of the solid and fluid (7.1 and 7.2) and eliminating the local time derivates yields the mass balance of the mixture:

$$\vec{\nabla}\circ\vec{v}^s+\vec{\nabla}\circ(n^f(\vec{v}^f-\vec{v}^s))=0 \tag{8}$$

Balance of momentum. The balance of momentum for the αth constituent is given by:

$$\varrho^\alpha \frac{D^\alpha}{Dt}\vec{v}^\alpha = \vec{\nabla}\circ\underline{\sigma}^\alpha + \varrho^\alpha\vec{q}^\alpha + \vec{\pi}^\alpha \tag{9}$$

with:

$\underline{\sigma}^\alpha$ partial Cauchy stress tensor

$\vec{q}^\alpha$ body force

$\vec{\pi}^\alpha$ momentum supplied by the other constituents

The restriction on the interaction terms, found by making use of the momentum balance for the mixture, is given by:

$$\sum_\alpha \vec{\pi}^\alpha + c^\alpha\vec{v}^\alpha = \vec{0} \tag{10}$$

Because we consider only slow deformation rates or transient swelling behaviour, the inertial effects are neglected. The body force is also neglected. This results in:

$$\begin{aligned} &\textit{solid:} && \vec{\nabla}\circ\underline{\sigma}^s + \vec{\pi}^s = \vec{0} \\ &\textit{fluid:} && \vec{\nabla}\circ\underline{\sigma}^f + \vec{\pi}^f = \vec{0} \\ &\textit{ions:} && \vec{\nabla}\circ\underline{\sigma}^i + \vec{\pi}^i = \vec{0} \\ &\textit{interaction:} && \vec{\pi}^s + \vec{\pi}^f + \vec{\pi}^i = \vec{0} \end{aligned} \tag{11}$$

Balance of moment of momentum. The balance of moment of momentum for the αth constituent is given by:

$$\vec{m}^\alpha - \vec{s}^\alpha = \vec{0} \tag{12}$$

with

$\vec{m}^\alpha$ the moment of momentum supplied by the other constituents

$\vec{s}^\alpha$ axial vector of the skew-symmetric tensor $\underline{\sigma}^\alpha$ - $(\underline{\sigma}^\alpha)^c$.

A non-zero $\vec{m}^\alpha$ implies a non-symmetric partial Cauchy stress tensor. Until proof of the contrary, we assume $\vec{m}^\alpha$ equal to zero, and hence the partial Cauchy stress to be symmetric.

Balance of energy or first axiom of thermodynamics. The energy balance for the αth constituent is given by:

$$\varrho^{\alpha}\dot{\varepsilon}^{\alpha} = \underline{\sigma}^{\alpha}:\vec{\nabla}\vec{v}^{\alpha} - \vec{\nabla}\circ\vec{J}^{\alpha} + \varrho^{\alpha}r^{\alpha} + g^{\alpha} \tag{13}$$

with

ε^{α}	specific internal energy
$\vec{J}^{\alpha}$	heat flux
r^{α}	external specific heat supply
g^{α}	heat supply by the other constituents

Again the balance of the mixture is used to find the restriction on the interaction terms:

$$\sum_{\alpha} g^{\alpha} + \vec{\pi}^{\alpha}\circ\vec{v}^{\alpha} + c^{\alpha}(\varepsilon^{\alpha} + \frac{1}{2}\vec{v}^{\alpha}\circ\vec{v}^{\alpha}) = 0 \tag{14}$$

Constitutive relations

Entropy inequality or second axiom of thermodynamics. The mass, momentum and energy balance form a set of coupled differential equations. Even with the right initial and boundary conditions this set of equations cannot be solved because the constitutive behaviour is unknown. The constitutive theory for mixtures based on the second axiom of thermodynamics (entropy inequality) has been a main point of discussion in the literature: the main dispute concerns the relationship between the mixture properties and the constituents properties. It can be shown that for 'simple' mixtures no essential differences result from application of either theory (Atkin and Craine, 1976). In this context the formulation for the entropy inequality according to Müller (1985) will be used. For each constituent it is hypothesized that:

$$\varrho^{\alpha}\dot{\eta}^{\alpha} + \vec{\nabla}\circ\vec{\theta}^{\alpha} = \xi^{\alpha} \tag{15}$$

with

η^{α}	specific entropy
$\vec{\theta}^{\alpha}$	total entropy flux
ξ^{α}	entropy production

For the purpose of deriving the restrictions on the constitutive relations, it is unneccessary to include the external heat supply, because these restrictions are unaffected by the presence of this supply. Because the entropy production may be caused by entropy exchange with other constituents, the classical demand of a positive entropy production is not claimed for each individual constituent. Their sum, which is the total entropy production on the mixture scale, must be positive or zero:

$$\sum_{\alpha} \varrho^{\alpha}\,\dot{\eta}^{\alpha} + \vec{\nabla}\circ\vec{\theta}^{\alpha} = \sum_{\alpha}\xi^{\alpha} \geq 0 \tag{16}$$

By means of the introduction of the Helmholtz free energy A ($A=\varepsilon-T.\eta$, T the absolute temperature) it is possible to combine the first and second axiom of thermodynamics. All experimental conditions which are considered in a following paragraph and the in vivo situation are assumed isothermal: each constituent has the same constant temperature. With no chemical reactions (c^α=0) the resulting entropy inequality for the mixture is given by:

$$\sum_\alpha -\varrho^\alpha \dot{A}^\alpha + \underline{\sigma}^\alpha : \vec{\nabla}\vec{v}^\alpha + \vec{\pi}^\alpha \circ \vec{v}^\alpha \geq 0 \tag{17}$$

The constitutive relationships should be such that for any state of the mixture, complying with the kinematic restrictions (7.3) and (8) the inequality (17) holds. When incorporating these restrictions into the inequality we find:

$$\begin{aligned} &-\varrho^i \dot{A}^i - \varrho^s \dot{A}^s - \varrho^f \dot{A}^f + \underline{\sigma}^i : \vec{\nabla}\vec{v}^i + \underline{\sigma}^f : \vec{\nabla}\vec{v}^f + \underline{\sigma}^s : \vec{\nabla}\vec{v}^s \\ &+ \vec{\pi}^i \circ \vec{v}^i + \vec{\pi}^f \circ \vec{v}^f + \vec{\pi}^s \circ \vec{v}^s + \lambda\{\vec{\nabla} \circ \vec{v}^s + \vec{\nabla} \circ (n^f(\vec{v}^f - \vec{v}^s))\} \\ &+ \kappa(\dot{\varrho}^i + \varrho^i \vec{\nabla} \circ \vec{v}^i) \geq 0 \end{aligned} \tag{18}$$

The Lagrange multipliers κ and λ are unknown and have to be identified by the elaboration of the extended entropy inequality (18).

Constitutive restrictions. By means of the extended entropy inequality (18) it is possible to find restrictions on the constitutive relations for the dependent variables:

$$A^s,\ A^f,\ A^i,\ \underline{\sigma}^s,\ \underline{\sigma}^f,\ \underline{\sigma}^i,\ \vec{\pi}^s,\ \vec{\pi}^f,\ \vec{\pi}^i,\ \lambda,\ \kappa \tag{19}$$

This list can be found by just counting the number of unknowns in the balance laws. The dependent variables may be functions of the primary unknowns: mass densities, displacements or their derivatives. Based on experimental results or even physical intuition a set of independent variables has to be chosen. The constitutive restrictions found, are only valid and consistent within this framework. By choosing another set different results may be obtained. For the set of independent variables we choose:

$$\underline{E}^s,\ \varrho^f,\ \vec{\nabla}\varrho^f,\ \varrho^i,\ \vec{\nabla}\varrho^i,\ \vec{\omega},\ \vec{v} \tag{20}$$

with

$$\begin{aligned} \vec{\omega} &= \vec{v}^f - \vec{v}^i \\ \vec{v} &= \vec{v}^s - \vec{v}^i \\ \vec{u} &= \vec{v}^f - \vec{v}^s = \vec{\omega} - \vec{v} \end{aligned} \tag{21}$$

The Green-Lagrange strain tensor $\underline{E}^s$ accounts for the deformation of the solid constituent. The mass density ρ^s is solely determined by the reference value ρ_0^s and the volume change of the porous solid. This volume change can be expressed in terms of the invariants of $\underline{E}^s$. Therefore ρ^s is not included in the list of independent variables. The fluid constituent is assumed to be macroscopically nonviscous; hence the deformation rate tensor has not been chosen as an independent variable. The microscopic viscous effect is incorporated by the relative velocity $\vec{u}$. Diffusion effects are included by means of the relative velocities $\vec{\omega}$ and $\vec{v}$. The buoyancy effect as defined by Bowen (1980) is included by $\vec{\nabla}\varrho^f$; ϱ^i and $\vec{\nabla}\varrho^i$ allow for modelling ionic effects.

The principle of equipresence requires that all dependent variables depend on all independent variables. By postulating:

$$
\begin{aligned}
A^{i} &= A^{i}(\varrho^{f}, \varrho^{i}) \\
A^{s} &= A^{s}(\underline{E}^{s}, \varrho^{f}, \varrho^{i}) \\
A^{f} &= A^{f}(\varrho^{f}, \varrho^{i}) \\
\vec{\pi}^{f} &= \vec{\pi}^{f}(\varrho^{f}, \varrho^{i}, \vec{\nabla}\varrho^{f}, \vec{\nabla}\varrho^{i}, \vec{v}, \vec{\omega}) \\
\vec{\pi}^{s} &= \vec{\pi}^{s}(\varrho^{f}, \varrho^{i}, \vec{\nabla}\varrho^{f}, \vec{\nabla}\varrho^{i}, \vec{v}, \vec{\omega}) \\
\vec{\pi}^{i} &= \vec{\pi}^{i}(\varrho^{f}, \varrho^{i}, \vec{\nabla}\varrho^{f}, \vec{\nabla}\varrho^{i}, \vec{v}, \vec{\omega}) \\
\kappa &= \kappa(\varrho^{f}, \varrho^{i}, \vec{\nabla}\varrho^{f}, \vec{\nabla}\varrho^{i}, \vec{v}, \vec{\omega}) \\
\lambda &= \lambda(\varrho^{f}, \varrho^{i}, \vec{\nabla}\varrho^{f}, \vec{\nabla}\varrho^{i}, \vec{v}, \vec{\omega})
\end{aligned} \tag{22}
$$

this principle is approximated. However it can be shown that this choice is an early simplification, which may be obtained from the full set of constitutive relations (Oomens et al., 1987). The constitutive relations (22) should not violate the extended entropy inquality (18). The chain rule of differentiation is used to evaluate the time derivatives:

$$
\begin{aligned}
\dot{A}^{i} &= \frac{\partial A^{i}}{\partial \varrho^{i}}\dot{\varrho}^{i} + \frac{\partial A^{i}}{\partial \varrho^{f}}\frac{D^{i}}{Dt}(\varrho^{f}) \\
\dot{A}^{f} &= \frac{\partial A^{f}}{\partial \varrho^{i}}\frac{D^{f}}{Dt}(\varrho^{i}) + \frac{\partial A^{f}}{\partial \varrho^{f}}\dot{\varrho}^{f} \\
\dot{A}^{s} &= \frac{\partial A^{s}}{\partial \underline{E}^{s}} : \underline{\dot{E}}^{s} + \frac{\partial A^{s}}{\partial \varrho^{f}}\frac{D^{s}}{Dt}(\varrho^{f}) + \frac{\partial A^{s}}{\partial \varrho^{i}}\frac{D^{s}}{Dt}(\varrho^{i})
\end{aligned} \tag{23}
$$

The material time derivative of a property of a constituent different from the phase associated with the derivative is transformed by eq. (3). Making use of the symmetry of $\partial A^{s}/\partial \underline{E}^{s}$ it can be shown that:

$$
\frac{\partial A^{s}}{\partial \underline{E}^{s}} : \underline{\dot{E}}^{s} = \underline{F} \circ \frac{\partial A^{s}}{\partial \underline{E}^{s}} \circ \underline{F}^{c} : \vec{\nabla}\vec{v}^{s} \tag{24}
$$

Substituting these results and eq. (11.4) in the extended entropy inequality (18) and after rearranging, results into:

$$
\begin{aligned}
&[-\varrho^{s}\frac{\partial A^{s}}{\partial \varrho^{i}} - \varrho^{i}\frac{\partial A^{i}}{\partial \varrho^{i}} - \varrho^{f}\frac{\partial A^{f}}{\partial \varrho^{i}} + \kappa]\,\dot{\varrho}^{i} + \\
&[-\varrho^{s}\frac{\partial A^{s}}{\partial \varrho^{f}} - \varrho^{i}\frac{\partial A^{i}}{\partial \varrho^{f}} - \varrho^{f}\frac{\partial A^{f}}{\partial \varrho^{f}}]\,\dot{\varrho}^{f} + \\
&[-\varrho^{s}(\underline{F} \circ \frac{\partial A^{s}}{\partial \underline{E}^{s}} \circ \underline{F}^{c}) + n^{s}\lambda \underline{I} + \underline{\sigma}^{s}] : \vec{\nabla}\vec{v}^{s} + \\
&[\kappa\varrho^{i}\underline{I} + \underline{\sigma}^{i}] : \vec{\nabla}\vec{v}^{i} + [n^{f}\lambda\underline{I} + \underline{\sigma}^{f}] : \vec{\nabla}\vec{v}^{f} + \\
&(\varrho^{s}\frac{\partial A^{s}}{\partial \varrho^{f}}\vec{\nabla}\varrho^{f} + \lambda\vec{\nabla}n^{f}) \circ \vec{u} - \\
&(\varrho^{s}\frac{\partial A^{s}}{\partial \varrho^{i}}\vec{\nabla}\varrho^{i} + \vec{\pi}^{s}) \circ \vec{v} - \\
&(\vec{\pi}^{f} + \varrho^{f}\frac{\partial A^{f}}{\partial \varrho^{i}}\vec{\nabla}\varrho^{i} - \varrho^{i}\frac{\partial A^{i}}{\partial \varrho^{f}}\vec{\nabla}\varrho^{f}) \circ \vec{\omega} \geq 0
\end{aligned} \tag{25}
$$

Insertion of the constitutive relations (22) into the extended entropy inequality (18) leads to an inequality that is partly linear in the derivatives:

$$\dot{\varrho}^i,\ \dot{\varrho}^f,\ \vec{\nabla}\vec{v}^s,\ \vec{\nabla}\vec{v}^f,\ \vec{\nabla}\vec{v}^i \tag{26}$$

The inequality must hold for all primary fields: $\varrho^s, \varrho^f, \varrho^i, \vec{v}^f, \vec{v}^s, \vec{v}^i$. It must also hold for arbitrary values of the above mentioned derivatives (26). By choosing arbitrary values for the derivatives one defines a path, either in space or time for the primary fields. To ensure that the inequality holds, the primary fields should therefore comply with the condition that terms inside the square brackets vanish independently:

$$\begin{aligned}
&\kappa = \varrho^s \frac{\partial A^s}{\partial \varrho^i} + \varrho^f \frac{\partial A^f}{\partial \varrho^i} + \varrho^i \frac{\partial A^i}{\partial \varrho^i}\\
&\varrho^i \frac{\partial A^i}{\partial \varrho^f} + \varrho^f \frac{\partial A^f}{\partial \varrho^f} + \varrho^s \frac{\partial A^s}{\partial \varrho^f} = 0\\
&\underline{\sigma}^i = -\varrho^i \kappa \underline{I}\\
&\underline{\sigma}^s = -\lambda n^s \underline{I} + \varrho^s \underline{F} \circ \frac{\partial A^s}{\partial \underline{E}^s} \circ \underline{F}^c\\
&\underline{\sigma}^f = -\lambda n^f \underline{I}
\end{aligned} \tag{27}$$

leaving as inequality:

$$\begin{aligned}
&(\varrho^s \frac{\partial A^s}{\partial \varrho^f} \vec{\nabla}\varrho^f + \lambda \vec{\nabla} n^f) \circ \vec{u}\\
&-(\varrho^s \frac{\partial A^s}{\partial \varrho^i} \vec{\nabla}\varrho^i + \vec{\pi}^s) \circ \vec{v}\\
&-(\vec{\pi}^f + \varrho^f \frac{\partial A^F}{\partial \varrho^i} \vec{\nabla}\varrho^i - \varrho^i \frac{\partial A^i}{\partial \varrho^f} \vec{\nabla}\varrho^f) \circ \vec{\omega} \geq 0
\end{aligned} \tag{28}$$

By choosing:

$$\begin{aligned}
&\varrho^s \frac{\partial A^s}{\partial \varrho^f} \vec{\nabla}\varrho^f + \lambda \vec{\nabla} n^f = \underline{B}_u \circ \vec{u}\\
&\varrho^s \frac{\partial A^s}{\partial \varrho^i} \vec{\nabla}\varrho^i + \vec{\pi}^s = -\underline{B}_v \circ \vec{v}\\
&\vec{\pi}^f + \varrho^f \frac{\partial A^f}{\partial \varrho^i} \vec{\nabla}\varrho^i - \varrho^i \frac{\partial A^i}{\partial \varrho^f} \vec{\nabla}\varrho^f = -\underline{B}_\omega \circ \vec{\omega}
\end{aligned} \tag{29}$$

and $\underline{B}_u$, $\underline{B}_v$ and $\underline{B}_\omega$ as semi-positive definite tensors inequality (28) is always fulfilled. These tensors are not independent because of the relation between the three relative velocities (21). However we keep them separated because of the ease of the derivations in the subsequent paragraphs.

Identification of Lagrange multipliers. As stated before all physical properties are averaged values. The fluid is modelled as macroscopically non-viscous. The constitutive relation for such a barotropic fluid is given by:

$$\underline{\Sigma}^f = -p\underline{I} \tag{30}$$

with p the hydrodynamic pressure.

The Cauchy stress tensor of the fluid $\underline{\sigma}^f$, used in our constitutive theory is the bulk volume average of $\underline{\Sigma}^f$:

$$\underline{\sigma}^f = \tilde{\underline{\Sigma}}^f = \frac{1}{V_e}\int_{V_e^f}\underline{\Sigma}^f dV = \frac{1}{V_e}\int_{V_e^f} -p\underline{II}dV \tag{31}$$

The size of the elementary averaging volume V_e is determined by the distribution of the constituents present within the volume. Intervertebral disc tissue is mainly a mixture on a molecular scale. So the averaging volume is very small compared to a tissue sample. Within this volume it is permissible to assume that the hydrodynamic pressure is constant:

$$\underline{\sigma}^f = -\frac{V_e^f}{V_e}p\underline{II} = -n^f p\underline{II}$$

Comparing this result with the constitutive relationship found by making use of the entropy inequality (27.5) shows that the Lagrange multiplier λ is the hydrodynamic pressure p. The constitutive relationship for the Cauchy stress of the solid is also determined by λ (27.4). Deformation of the solid matrix is achieved either by the hydrodynamic pressure or the contact between the solid particles.

$$\underline{\sigma}^s = \tilde{\underline{\Sigma}}^s = \frac{1}{V_e}\int_{V_e^s}\underline{\Sigma}^s dV = \frac{1}{V_e}\int_{V_e^s}(\underline{\Sigma}_c^s - p\underline{II})dV = \underline{\sigma}_{eff} - n^s p\underline{II} \tag{33}$$

with

$\underline{\Sigma}_c^s$ contact stress

$\underline{\sigma}_{eff}$ effective stress

Comparing this result with the outcome of the entropy inequality (27.4) yields:

$$\begin{aligned}\underline{\sigma}_{eff} &= \varrho^s\underline{F}\circ\frac{\partial A^s}{\partial \underline{E}^s}\circ\underline{F}^c \\ \lambda &= p\end{aligned} \tag{34}$$

Again λ equals the hydrodynamic pressure. Within physico-chemical theories (Bowen, 1980) it is common to define a chemical potential tensor by:

$$\underline{K}^\alpha = A^\alpha\underline{II} - \frac{\underline{\sigma}^\alpha}{\varrho^\alpha} \tag{35}$$

Substituting the constitutive relationship for $\underline{\sigma}^i$ (26.3) and (26.5) gives:

$$\begin{aligned}\underline{K}^i &= A^i\underline{II} + \kappa\underline{II} \\ \underline{K}^f &= A^f\underline{II} + p/\Gamma^f\underline{II}\end{aligned} \tag{36}$$

Two conclusions may be drawn. It is obvious that the chemical potential tensors $\underline{K}^i$ and $\underline{K}^f$ are scalar functions: μ^i and μ^f. The Lagrange multiplier κ is the difference between the chemical potential and the Helmholtz free energy of the salt. By neglecting the volume of the ions, expressions for the Helmholtz free energy as a function of the composition may be found in the physico-chemical literature (Richards, 1980):

$$\begin{aligned} A^i &= A_0^i + \frac{RT}{M^+}\ln(f^+x^+) + \frac{RT}{M^-}\ln(f^-x^-) \\ A^f &= A_0^f + \frac{RT}{M^f}\ln(f^f x^f) \end{aligned} \qquad \textbf{(37)}$$

with

A_0^α Helmholtz free energy in the reference state

R universal gas constant

T absolute temerpature

M^α mol mass

f^α activity coefficient

x^α mol fraction

The activity coefficient is a function of the composition of the tissue. The mole fraction x^α is defined as the number of moles of constituent α (m^α), devided by the total number of moles in the tissue:

$$x^\alpha = \frac{m^\alpha}{\sum_\alpha m^\alpha} \qquad \sum_\alpha x^\alpha = 1 \qquad \textbf{(38)}$$

The Helmholtz free energy of the ions is split into those of the anions (positive charge) and cations (negative charge). The mole fraction of the anions does not equal the mole fraction of the cations because of the presence of ionized proteoglycans.

The momentum interactions. To complete the constitutive theory closed expressions for the momentum interaction terms $\vec{\pi}^i$, $\vec{\pi}^s$ *and* $\vec{\pi}^i$ should be obtained. Substituting eq. (27.2) in eq. (29.3) and within this result substituting (29.1) and by making use of eq. (21.3) gives and expression for $\vec{\pi}^f$:

$$\vec{\pi}^f = -\underline{B}_\omega \circ \vec{\omega} - \underline{B}_u \circ \vec{u} + p\vec{\nabla} n^f - \varrho^f \vec{\nabla} A^f \qquad \textbf{(39)}$$

The momentum exchange of the fluid is respectively caused by the relative flow of the ions and the solid matrix (often called the Stokes drag), the buoyancy force and the gradient of the Helmholtz free energy. Starting with eq. (11.4) and substituting the sum of (29.2) and (29.3) and by making use of (21.1), gives an expression for $\vec{\pi}^i$:

$$\vec{\pi}^i = -(\vec{\pi}^f + \vec{\pi}^s) = \underline{B}_v \circ \vec{v} + \underline{B}_\omega \circ \vec{\omega} + \eta \vec{\nabla} \varrho^i - \varrho^i \vec{\nabla} A^i \qquad \textbf{(40)}$$

The momentum exchange of the ions can be described in the same mathematical way as the momentum exchange of the fluid. An expression for $\vec{\pi}^s$ can be obtained by combining eq. (39) and eq. (40).
Substituting the relationships found for the momentum exchange of the ions and fluid (38 and 39) in the respective balances of momentum (11.2 and 11.3) and making use of the constitutive relationships for $\underline{\sigma}^f$ and

$\underline{\sigma}^i$ (27.3 and 27.5) and the definition of the chemical potential (36), yields:

$$-\varrho^i\vec{\nabla}\mu^i + \underline{B}_\omega \circ \vec{\omega} + \underline{B}_v \circ \vec{v} = \vec{0}$$
$$-\varrho^f\vec{\nabla}\mu^f - \underline{B}_\omega \circ \vec{\omega} - \underline{B}_u \circ \vec{u} = \vec{0} \tag{41}$$

Adding these two equations (41) and by assuming the diffusion within the solid matrix to be negligible compared to the diffusion within the fluid, yields:

$$\underline{B}_v \approx \mathbf{0}$$
$$-\underline{B}_u \circ \vec{u} = \varrho^i\vec{\nabla}\mu^i + \varrho^f\vec{\nabla}\mu^f \tag{42}$$

This relation may be viewed upon as an extended form of Darcy's law: relative fluid flow is caused by the gradient of the chemical potential of the ions and fluid. Substituting the definition of the chemical potential (36) and the expression for the Helmholtz free energy (37) results into:

$$-\underline{B}_u \circ \vec{u} = n^f\{\vec{\nabla}p + \frac{\varrho^i}{\varrho^f}\Gamma^f\vec{\nabla}\eta - RT[(1+H) - \frac{\varrho^i}{\varrho^f}\frac{M^f}{M^+}(1+F)]\vec{\nabla}C^+$$
$$-RT[(1+H) - \frac{\varrho^i}{\varrho^f}\frac{M^f}{M^-}(1+G)]\vec{\nabla}C^+\} \tag{43}$$

where C^+ and C^- are moles per unit fluid volume. The coefficients F, G and H are functions of the ion concentrations. For dilute solutions they tend to zero, which is also true for the quotient of ϱ^i/ϱ^f. Both in the in vivo situation and in all experimental conditions which are considered in a following paragraph, the ion concentrations are less than 1 mol/l, and no external electrical field is applied. So in a first approximation intervertebral disc tissue is a dilute solution in which charge neutrality is maintained:

$$C^- + C^{PG} = C^+ \tag{44}$$

with C^{PG} proteoglycan concentration.

Due to these assumptions (42) Darcy's law is expressed as:

$$n^f(\vec{v}^f - \vec{v}^s) = -\underline{K} \circ \{\vec{\nabla}p - RT(2\vec{\nabla}C^- + \vec{\nabla}C^{PG})\} \tag{45}$$

with $\underline{K} = (n^f)^2\underline{B}_u^{-1}$ the permeability tensor.

Relative fluid flow is governed by a gradient of both a hydrodynamic pressure and an ion concentration. On the other hand no fluid flow may occur even with a pressure gradient if this is compensated by the ion-concentration gradient. In the absence of the ions the classical form of Darcy's law is obtained.

Mass balance of the ions. Finally we will examine the mass balance of the ion constituent again. The ion velocity $\vec{v}^i$ in the mass balance can be eliminated by using the modified balance of momentum (41.1), resulting in the diffusion equation:

$$\frac{\partial \varrho^i}{\partial t} = \vec{\nabla} \circ \{\varrho^{i^2}\underline{B}_\omega^{-1} \circ \vec{\nabla}\mu^i - \varrho^i\vec{v}^f\} \tag{46}$$

The total ion flux includes diffusion caused by the gradient of the chemical potential and convection by the macroscopic fluid flow. In a similar way as used by the derivation of Darcy's law the gradient of the chemical potential can be expressed in terms of the ion concentrations. Substituting this result in eq. (46) and by using the charge neutrality (44) gives one diffusion equation governing both the diffusion of the mobile C^--ions and the stagnant C^{PG}-ions. The diffusion of the stagnant proteoglycan chains is neglected compared to the diffusion of the anions, resulting into:

$$\frac{\partial C^-}{\partial t} = \frac{1}{n^f}\vec{\nabla}\circ(\underline{D}\circ\vec{\nabla}C^-) - \vec{v}^f\circ\vec{\nabla}C^- \tag{47}$$

with $\underline{D} = RT\varrho^{i^2}\frac{M^f}{M^+M^-}\frac{\underline{B}_{\omega}^{-1}}{\Gamma^f}$ the diffusion tensor.

This relationship is used as the mass balance for the mobile ions.

Experimental methods

Two year old Bouvier des Flandres dogs are sacrified. The lumbar spine with the two adjacent thoracal vertebrae and the sacrum are dissected and within 1 to 2 hours after death frozen at -80°C. Individual motion segments (½ vertebra - disc - ½ vertebra) are sawed by a bandsaw and stored at -80°C. Samples are prepared batchwise as needed for testing while cooling with liquid nitrogen (-196°C). The segments are sawed in square pieces of 6 mm. Each motion segment renders one sample of nucleus and 3 or 4 samples of annulus. Characterization of the tissue is done visually. Mixed specimens are discarded. The location of each specimen is recorded. The frozen square specimen is turned to a diameter of 4 mm. The testing specimen is cut from the midzone of the cylinder. The two remaining parts are stored for histological and biochemical analysis. The sample thickness is approximately 0.9 mm.

Uniaxial confined compression experiments are performed in a special testing chamber, diameter 4 mm, height 4 mm. The bottom of the chamber is a ceramic porous filter (pore size between 16 and 40 μm), while the wall is an impervous metal ring. Mechanical loading is achieved by an impervous metal shaft. The shaft is connected to Schaevitz DC operated LVDT interfaced by a Labmaster AD converter to an IBM-AT. Through the filter the sample is in contact with a bathing solution. By altering the ionic-strength of the solution it is possible to load the specimen in a chemical way. The bathing solution is recirculated through the filter, thus minimizing the influence of stagnant films on the transport of ionic solutes into the tissue. The temperature of the solution is controlled to 27±1°C.

The frozen specimen is placed in the testing chamber and the shaft is lowered. While the specimen thaws it is loaded with a preload of 0.5 N, so a good fit in the chamber is achieved. A vacuum pump (underpressure 85.0 kPa) frees the filter from air, thus optimizing the contact between the bathing solution and the sample. After 10 seconds the bathing solution, deionized water flows into the filter. The specimen is allowed to swell until equilibrium is reached. After equilibrium the load is raised to 2.0 N and the specimen is compressed. After the new equilibrium is reached the load is altered to the original preload and the specimen swells again. It reaches almost the same swelling displacement of the first equilibrium, indicating that the loss of PG-subunits and degeneration of the tissue is minimal. During the experiment the displacement of the shaft is automatically recorded. The sample frequency is 0.1 Hz. In order to avoid stick phenomena the testing apparatus is shaken during 1.0 second, 0.5 second after each sample is taken.

Based on the experimental results from the swelling phase the material parameters are calculated by means of an unbaised minium-variance parameter estimation (Hendriks, 1991). Fig. 1 shows the experimental swelling displacement and the theoretical curve using the converged material parameters. These values are within the physiological range. The values found for both the compressive stiffness (0.206 MPa) and the permeability (0.665 10^{-16} m^4/Ns) are lower then those reported in the literature (Best et al., 1989)

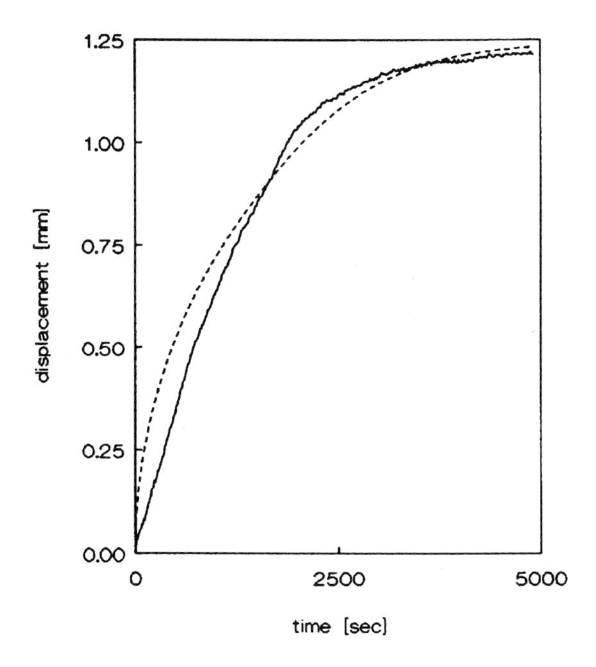

Fig. 1 Swelling displacement versus time both experimental (——) and theoretical (- - -). Bathing solution deionized water, initial thickness 0.889 mm.

Conclusions

A theory has been developed to describe the swelling behaviour of intervertebral disc tissue. Swelling of the tissue is initiated by a change of the chemical potential of both the solvent and the solutes. The resulting "osmotic" pressure for an externally applied load accounts for the fluid flow and the tissue deformation. Based on experimental data a preliminary validation of the model has been done. Model and experiment show a reasonable similarity.

References

Atkin RJ, Crain RE (1976) Continuum theories of mixtures: Basic theory and historical development. Quart J Mech Appl Math 29:209-243

Best BA, Guilak F, Weidenbaum M, Mow VC (1989) Compressive stiffness and permeability of intervertebral disc tissues: variation with radial position, region and level. Proc Winter Annual Meeting

ASME, San Francisco

Bowen RM (1980) Incompressible porous media models by use of the theory of mixtures. Int J Eng Sci 18:1129-1148

Grodzinsky AJ, Roth V, Meyers E, Grossman WD, Mow VC (1981) The significance of electromechanical and osmotic forces in the nonequilibrium swelling behaviour of articular cartilage in tension. J Biomech Eng 103:221-231

Hassanizadeh M, Gray WG (1979) General conservation equotations for multi phase systems: 1 Average procedure. Adv Water Resources 2:131

Hendriks MAN (1991) Identification of the mechanical behaviour of solid materials. PhD thesis Eindhoven University of Technology, The Netherlands

Maroudas A (1975) Biophysical chemistry of cartilaginous tissues with special reference to solute and fluid transport. Biorheology 12:233-248

Muller I (1985) Thermodynamics. Pitman Advanced Publishing Program, Boston 1-202

Oomens CWJ, Campen DH van, Grootenboer HJ (1987) A mixture approach to the mechanics of skin. J Biomech 20:877-885

Richards EG (1980) An introduction to physical properties of large molecules in solution. Cambridge University Press, Cambridge

Slattery JC (1972) Momentum, energy and mass transfer in continua. McGraw-Hill, New York

Whitaker S (1969) Advances in the theory of fluid motion in porous media. Ind Engng Chem 61:14

BLOOD - TISSUE FLUID EXCHANGE - TRANSPORT THROUGH DEFORMABLE, SWELLABLE POROUS SYSTEMS.

A. Silberberg
Weizmann Institute of Science, Rehovot 76100, Israel.

INTRODUCTION

A living system in thermodynamic equilibrium is a contradiction in terms. A system in equilibrium would be dead. On the other hand, if, after some embryonic period, a living system reaches its prime development then, as often celebrated in song and dance, it becomes our foremost objective to maintain this state, as a steady state, for as long as possible.

In a steady state, perfect or otherwise, there has to be an input of products to replace materials that have deteriorated and gone to waste. In a steady state there has also to be an input of energy to maintain the potential gradients upon which it depends. Anything eliminated is either freshly supplied from the outside, or newly made. Any energy put in is converted into work, chemical products or heat.

For this to function well a highly effective transport system is required. In human beings the major supply line is through the blood stream. This links all the organs of the system. No living cell of the body can afford to be too strongly distanced. The metabolically required materials are extracted from the contents of the gastrointestinal tract, enter the blood, and are then delivered and selectively absorbed by the target cells. The organs and all the cells of the system eliminate their waste by discharging it into the blood. Blood is purified and regenerated in the kidney and freshly stocked with oxygen and delivered of its carbon dioxide in the lungs.

Blood communicates chemically with its surroundings. Water and practically all the components of blood plasma pass through the blood vessel wall and enter the extravascular space. (Guyton 1966, Michel 1985, Michel 1988). With this two purposes are achieved. The blood stream supplies and collects materials from the cells distributed throughout the system and maintains, as much as possible, the environment surrounding the cells in a well established homeostatic state characterized by a given water content, a given pH, a given salt concentration and a given temperature (Zweifach and Silberberg 1979). It is the blood/tissue exchange process which establishes the steady state conditions required by the cells to function normally and which controls the way in which the various components distribute themselves in flow between blood and its surroundings.

Among many other things that can go awry, failure, sometimes unavoidable failure, to restrict the entry of water causes edema, i.e. causes the tissue, through which the extravascular flow is directed, to swell up. As much as possible the system tries to avoid such departures from steady state, but we are never far from losing control.

THE ESSENTIAL CHARACTERISTICS OF BLOOD, THE VESSEL WALL, AND TISSUE.

Blood is a suspension of cells. Roughly half the volume of blood is taken up by cells. The major cell type is the red cell. Red cells store and release oxygen. The cells are suspended in a medium which contains salt and a variety of sugars, urea and other low molecular weight solutes. It also contains many proteins but the most abundant of these by far is serum albumin (Guyton 1966).

The vessel wall is composed of a layer, a pavement of contiguous endothelial cells which are encapsulated in a stress-bearing membrane the so called basement membrane. Depending upon the diameter and pressure in the vessel this inner structure may be surrounded by further layers which give strength and support to the vessel. The finest, the narrowest and the most prevalent of blood vessels, the capillaries, essentially involve an endothelial cell layer and a basement membrane only.

The swellable, deformable tissue into which the organs and cells of the system are built and which "connects" between them is based on a network of load bearing collagen fibers, a system of highly extensible elastin fibers and some space filling highly branched supermolecular structures of enormous size generally going under the name of proteoglycans. The backbone is hyaluronic acid, a huge charged polysaccharide chain to which a system of glycoprotein molecules is attached each in itself an extremely high molecular weight structure involving a protein backbone and another system of polysaccharide side chains of varying length, fairly tightly spaced along the protein cores. These oversized molecules are stable only because they are supported and mechanically protected by the collagen network (Meyer 1983).

While proteoglycans are very soluble in water, collagen and elastin do not dissolve at all. The proteoglycans are thus the swelling agents and the collagen and elastin networks are the mechanically resisting elements which limit swelling. There do not appear to be any chemical links between the proteoglycan molecules and the collagen or elastin. They are just very heavily entangled with each other. A solution containing water, albumin and salt, but indeed also just about all the components of plasma, constitutes the medium with which the network is filled.

The passage from blood to connective tissue is across the vessel wall (Figure 1). For water the important path is between the endothelial cells through clefts held open by what appears to be a system of posts (Silberberg 1988 a and b). All low molecular weight material and much of the albumin also crosses this way, but specific receptor mediated pathways through the endothelial cells exist as well.

The protein which leaves the blood via the vessel wall cannot be reabsorbed directly into the blood. For thermodynamic reasons, as we shall show, this is impossible. Protein transferred from blood to tissue is returned to blood in the form of lymph, a protein solution which is formed spontaneously by oozing out of the tissue into so-called terminal lymph vessels from which the lymph is then pumped and actively returned to the blood stream (Figure 1).

Some of the water, however, is returned to the blood from tissue directly. While on the arterial side the high blood pressure in the capillary forces the water out, on the venous side the driving force has reversed itself (Zweifach and Silberberg 1979).

To sum up and suppressing many important details we have water and protein transfer to tissue occurring mainly in the capillaries. These have the largest combined surface area and being also the narrowest

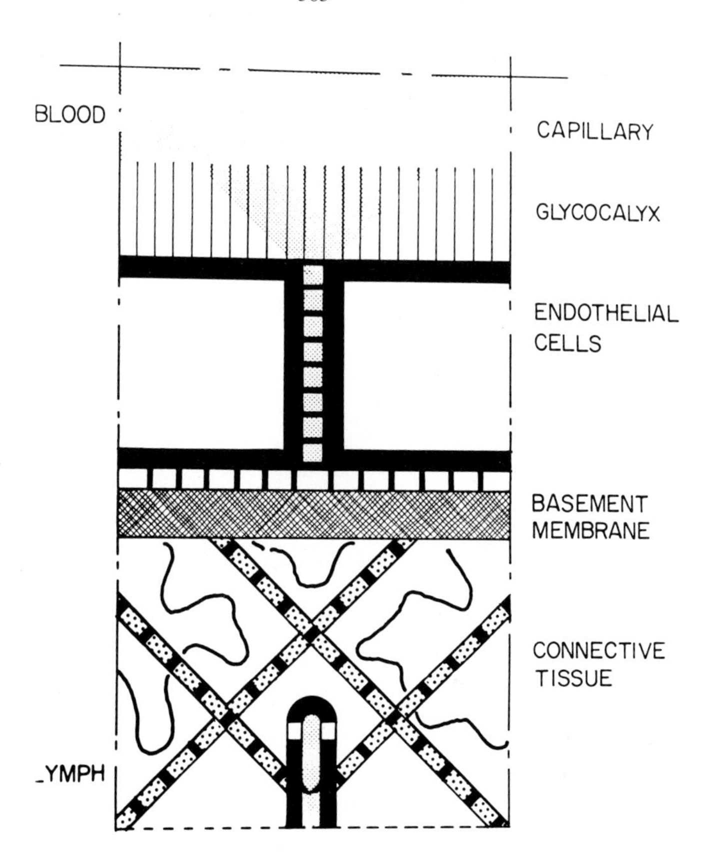

Figure 1: Blood to Lymph Transport.

(Highly schematized drawing with length scales grossly distorted). Capillaries are some 6μ in diameter and pressure along them varies from 35-55 mm Hg to 5-10 mm Hg. The glycocalyx is now believed to be some 1μ thick. It is probably composed of extended strands some 40Å in diameter and some 160Å apart. The clefts, due to tortuosity, probably present paths some 50 to 100μ in length. They are about 180Å wide. The posts are perhaps 30Å in diameter and are some 300Å apart. The basement membrane is a loosely woven, but very strong, web some 500Å in thickness. Collagen fibrils are about 600Å in diameter. The terminal lymphatic is a vessel (with large gaps between endothelial cells) generally larger in diameter than the blood capillary. A system of one way valves forces the lymph upwards as the vessels are squeezed and released by normal muscular activity.

element in the blood cycle have the highest rate of pressure drop along them. The capillary wall is a reflective membrane for protein but one which albumin can cross. Flow out is mainly on the hydrostatic pressure difference, but this is modified by the high protein content of plasma (~ 7% albumin) as compared with the relatively low albumin content of tissue (~ 2-3% albumin). Plasma has an osmotic pressure, due to its high molecular weight solutes, of about 25 mm Hg, connective tissue only 12 mm Hg of which quite a lot is contributed by the proteoglycan part of the network. Between tissue and blood there are thus 13 mm Hg (actually somewhat less) in osmotic pressure difference which tend to favor returning water to the blood stream. In regions where the pressure in the capillary is 35-55 mm Hg these 13 mm Hg are easily overwhelmed, but on the venous side, where blood pressure is very low, water is reabsorbed into the blood.

The filtration flow, the reabsorption flow and the lymph flow combine to maintain the tissue in steady state.

This said let us stress that we have only barely sketched out the complexity of the system. There are many more low molecular weight components than ions, urea and sugars in solution and hundereds of proteins and protein assemblies. The vessel wall is far more complex, the interaction with blood cells is more complex, the organization of tissue varies and involves a series of components that have not been mentioned. We have ignored metabolism and material turnover, cell division and cell death. Under these circumstances then, will it be even remotely possible to bring models to bear upon the case, models that would not only be adequately accurate in portraying this vast array of biologically active species, but also adequately simple to allow one to obtain a reasonable insight and a semi-quantitative description at least .

TRANSPORT ACROSS POROUS DEFORMABLE MEDIA

The case of transport from blood to lymph in steady state can be treated within the framework of irreversible thermodynamics if the following basic description can be considered to apply (Silberberg 1989):

There are (m-1) mobile species and a matrix component m. The matrix component is linked to some reference frame and does not move in steady state. The other components i are distributed according to a volume fraction field $\phi_i(\mathbf{x})$. In addition at each locality $\mathbf{x}$ component i possesses a mean velocity $\mathbf{v}_i(\mathbf{x})$. It is assumed that all velocities are directed into the x_1-direction and that the other components of $\mathbf{v}_i$ in the 2 and 3 directions are zero. There is thus a flux $\mathbf{J}_i$ of i across the system into the x_1-direction given by

$$\mathbf{J}_i = \phi_i(\mathbf{x})\mathbf{v}_i(\mathbf{x}) \qquad ; \qquad i = 1, ..., (m-1) \qquad (1)$$

which in steady state must be the same at all $\mathbf{x}$. As said

$$\mathbf{J}_m = 0 \qquad (2)$$

because $\mathbf{v}_m$ is taken to be zero.

The forces which drive these fluxes are the gradients of the chemical potentials of the m components. Chemical potentials are usually defined per molecule (or per mole) of i, but here it is more convenient to express them in terms of unit volume of i. The system and its components are assumed to be incompressible so that unit volume of i refers to the same amount of i everywhere. Let us call this gradient of chemical potential per unit volume, in the case of component i, $\mathbf{X}_i(\mathbf{x})$.

Flow is taken to be slow and hence $\mathbf{X}_i$ is opposed only by the drag on the molecules of i (moving at their average velocity $\mathbf{v}_i$) which derives from the (m-1) types of molecules j ($j \neq i$) each moving with their velocity $\mathbf{v}_j$. These drag, or friction, forces $\mathbf{F}_{ij}$, in first order approximation, are proportional to $(\mathbf{v}_i - \mathbf{v}_j)$:

$$\mathbf{F}_{ij}(\mathbf{x}) = -[r_{ij}(\mathbf{x})/\phi_i(\mathbf{x})][\mathbf{v}_i(\mathbf{x}) - \mathbf{v}_j(\mathbf{x})] \quad : \tag{3}$$

The parameters r_{ij}, defined by equation (3), characterize the level of the local mechanical interaction of species i in its movement relative to j. We thus have the following m equations of motion:

$$\mathbf{X}_i(\mathbf{x}) + \sum_j \mathbf{F}_{ij}(\mathbf{x}) = 0, \qquad i = 1,\ldots,m \, . \tag{4}$$

Under isothermal conditions we can express $\mathbf{X}_i$ by

$$\mathbf{X}_i(\mathbf{x}) = -(-\nabla)\,\mathbf{Z}_i + (\phi_1/\phi_i)(-\nabla)\pi_i \, . \tag{5}$$

In equation (5), $\mathbf{Z}_i$ is the stress tensor, per unit volume of i, at the position $\mathbf{x}$ and

$$(-\nabla)\pi_i = (\phi_i/\phi_1 V_i) k_B T (-\nabla) \ln a_i \quad , \tag{6}$$

where V_i is the molecular volume of i, k_B is the Boltzmann constant, T is absolute temperature and a_i is the activity, i.e. the effective concentration, of i at $\mathbf{x}$.

We now assume that the overall stress tensor $\boldsymbol{Z}$ at $\mathbf{x}$ is given by the mixing rule

$$\boldsymbol{Z} = \sum_i \phi_i \boldsymbol{Z}_i \,, \tag{7}$$

where local mechanical equilibrium requires that

$$\sum_i \phi_i(-\nabla)\boldsymbol{Z}_i = 0. \tag{8}$$

The stress tensor $\boldsymbol{Z}_i$ for unit volume of component i can also be written

$$\boldsymbol{Z}_i = -p\boldsymbol{I} + \boldsymbol{Z}_i' \quad , \tag{9}$$

where p is the hydrostatic pressure and $\boldsymbol{Z}_i'$ is the local extra stress tensor per unit volume of i. The extra stress tensor $\boldsymbol{Z}_i'$ is non-zero only when i = m, i.e. only in the case of the solid-like matrix component m. The hydrostatic pressure p, in an incompressible medium, is defined only up to some constant.

We now replace the system by an infinite array of lamellar phases of infinitesimal thickness, dx_1, along the x_1-axis, but of infinite dimensions in the x_2- and x_3-directions. Each such lamellar phase is considered to be in internal thermodynamic equilibrium. The gradients of chemical potential which exist in reality in the x_1-direction are ignored because dx_1 is taken to be small enough.

Hence each composition along the x_1 axis corresponds to an equilibrium state and the Gibbs-Duhem relation for internal equilibrium will apply. This, in terms of the variables π_i introduced above, is, at constant temperature, expressed by

$$\sum_i (-\nabla)\pi_i = 0 \qquad (10)$$

Furthermore, since the law of detailed balance must apply to the drag between molecules i and j, we have

$$r_{ij} = r_{ji} \quad . \qquad (11)$$

An overall balance equation thus follows when equations (8), (10) and (11) are substituted into (4) and the result is summed. We find that the equation

$$\sum_i \phi_i(\mathbf{x})\mathbf{X}_i(\mathbf{x}) = 0 \qquad (12)$$

is obeyed by the system in steady state.

Equation (4) is a system of m linear equations for the forces $\mathbf{X}_i$ in terms of the m fluxes $\mathbf{J}_i$. Only m-1 of these equations, only m-1 forces $\mathbf{X}_i$ and only m-1 fluxes $\mathbf{J}_i$, are independent. We can easily invert this system of m-1 equations and derive expressions for $\mathbf{J}_i$. The evaluation of the fluxes depends upon our knowing the functional dependence of the friction parameters r_{ij} on the local concentrations, the local structure and on certain rheological parameters such as for example the viscosity of the solvent. So far we have only the most rudimentary models for r_{ij}, theoretically, and experimentally the r_{ij}, are far from easy to measure.

If m = 2 and there is only a matrix and a solvent there is only one friction coefficient in the problem. This is the coefficient r_{1m} which can be related to the Darcy coefficient, or the sedimentation constant, or the

$$\eta = \phi_1 r_{1m} / [\phi_2 r_{1m} + (\phi_1+\phi_2) r_{12}] \tag{18}$$

$$\sigma = [r_{2m} - (\phi_2/\phi_1)\, r_{1m}]/[r_{2m} + [\phi_2/\phi_1]^2\, r_{1m} + [(\phi_1+\phi_2)/\phi_1]^2 r_{12}] \tag{19}$$

It turns out to be more informative to write the fluxes $\mathbf{J}_1$ and $\mathbf{J}_2$ in terms of the three transport parameters L_p, σ and η, instead of the three friction coefficients r_{1m}, r_{2m} and r_{12}. L_p is essentially the Darcy coefficient and σ is the reflection coefficient of the matrix for the solute familiar from discussions of membrane function.

It is most important to remember that these parameters are functions of concentration and that they vary from place to place. Equations (13) to (19) are local equations and L_p, σ and η are not constants, but have values which change from point to point. Any individual local value is, however, not easily determined. In general only overall values are measured, i.e. values which are averages for a certain path along x_1.

It should also be stressed that pressure and the extra stresses in the membrane component are linked by equation (8). Substituting (9) we thus have

$$(-\nabla)p = \phi_m(-\nabla)\, \mathbf{Z}_m' \tag{20}$$

where the extra stress tensor $\mathbf{Z}_m'$ in turn depends upon the equation of state which links the stress to the strain, i.e. the deformation of the matrix m, specifically the gradient of ϕ_m, to the stresses composing $(-\nabla)\mathbf{Z}_m'$. The gradients in ϕ_2 and ϕ_m are also the functions which determine $(-\nabla)\pi_2$ and $(-\nabla)\pi_m$. It is thus the two concentration distributions $\phi_2(\mathbf{x})$ and $\phi_m(\mathbf{x})$ which solve the problem. The functions of

ϕ_2 and ϕ_m which express the other parameters introduced must of course also be known. We generally have some good idea about the forms which $\phi_2(\mathbf{x})$ and $\phi_m(\mathbf{x})$ will take and can thus guess at the answer. Even if a fully quantitative answer cannot be given, a qualitative result can generally be reached and is often adequately enlighting since we know that the coefficients r_{ij} must be positive.

Examining equation (13) for $\mathbf{J}_2$, for example, it can be shown that since $(1-\sigma)$, L_p, $\mathbf{X}_V$, η and $\mathbf{X}_D$ are all positive in the direction where flow of solute is from blood to lymph, $\mathbf{J}_2$ cannot change sign and that solute flow from lymph (tissue) to blood is impossible.

BLOOD TO LYMPH TRANSPORT AS A THREE COMPONENT SYSTEM

In applying the foregoing formalism to blood to lymph transport it is first of all necessary to assume that there are no active elements in the pathway. In other words, we must assume that the only sources of potential are the hydrostatic pressure of blood, as put in by the heart, and the protein concentration of the plasma, as adjusted by the kidney.

For certain proteins this will not be true. For these there is a receptor based pathway taking them across the endothelial cells. These molecules are thus specifically "pumped" across making use of chemical energy. We shall ignore these effects and hope that the amounts so transferred can be neglected in comparison with the main fluxes of solvent and solute going through the clefts.

We must now build up (divide the system into) three groups of species, a matrix, a solvent and a solute.

There is really no problem building up the matrix component, since all the materials constituting the network, either by virtue of chemical bonding, or by entanglement, share the same zero velocity. It does not matter if the composition of the matrix is extremely complex and that at some localities one or other of the matrix composing substances may have zero concentration. What is important is that nowhere along the flow path should any of the matrix components detach themselves from the matrix and begin to have a non-zero velocity. Let us remember that these assumptions have to apply from the inside of the blood vessel along the entire path until the interior of the terminal lymphatic is reached (Figure 1).

For the solvent, obviously based on water, we can assume, although with lesser confidence, that many of the components dissolved in plasma will move with the water as though chemically linked to it, i.e., here now, at the same concentration and at the same velocity. We shall assume indeed that all plasma components of low molecular weight are of this kind, are strongly linked to water and have the same median velocity as water. We know that this is not entirely true, but only in this way can we create a "solvent" component 1 of fixed composition. For the matrix component we had $\mathbf{v}_m = 0$ and thus $\mathbf{J}_m = 0$ irrespective of how the contributions to $\mathbf{J}_m$ were distributed over $\mathbf{x}$. For $\mathbf{J}_1$ the relative concentrations have to be the same for all the species composing 1.

For the "solute" we are left with all the high molecular weight diffusible solutes in blood. Only a handwaving argument can help us here and we are fortunate, as already said, that serum albumin is so much more abundant than the polysaccharides and the proteins, other than albumin, present in blood. One can look upon these other proteins as "hangers on"! Whatever happens to serum albumin will happen to them, or more precisely formulated: "the serum albumin distribution fixes the gradients and the parameter values", what happens to the other proteins (hopefully) makes no difference to the state of the system. This is set by serum albumin.

The cells suspended in blood we ignore entirely since for our purposes these components do not partake in the lymph creating flow.

We have thus defined three components and in running through the structural elements between blood and lymph in Figure 1 we see that everywhere this subdivision applies and is respected. It turns out, however, that the matrix, through which the serum albumin solution winds, is somewhat more complicated than we have so far indicated. The luminal surface of the endothelium is covered by a grafted glycoprotein layer which we now believe to be as much as much as 1μ thick (Silberberg 1990). It is most likely composed of glycoproteins (long protein chains with frequent short sugar side chains) which are embedded in the endothelial cell surface and constitute what is called the glycocalyx (Born and Palinski 1985). In our view the glycocalyx is the real selective barrier to the serum albumin component (Silberberg 1990). Its local reflection coefficient σ is very high. The flow through the clefts between the cells and the subendothelial layers is made possible because these spaces are crossed by frequent very stiff cross bridges, or posts, which prevent their collapse (Silberberg 1988 a and b). The posts rigidify the whole channel system and the blood pressure acting over the endothelial surface is communicated to the cell's interior and from there, via the posts, to the basement membrane. The outward force is compensated by a hoop stress in the basement membrane to which the posts in the subendothelial layer are attached. The posts are some 300Å apart in the 180Å wide clefts and offer very little hindrance to serum albumin (a serum albumin molecule is about 40Å in diameter). The reflection coefficient in the cleft, in the sub-endothelium and the basement membrane is taken to be very small.

There are large concentration and pressure gradients in the glycocalyx, particularly in the region where it sits over a cleft, and there is a large pressure gradient down the cleft. In the sub-endothelium, in the basement membrane and across into the tissue space, on the other hand, these gradients essentially disappear. The reason is the enormous increase in aspect area for the flow. This goes up by a factor of 300, or perhaps even more, and what were finite gradients $(-\nabla)p$, $(-\nabla)\pi_2$ and $(-\nabla)\pi$ in the cleft and in the glycocalyx are now all essentially zero. We can look upon connective tissue, in zeroth approximation, as in equilibrium with the fluid in the subendothelial space and with the fluid in the terminal lymphatic (a tube made up of an endothelial cell layer held open by tethers between this lymphatic endothelium and some structural elements in the system frame).

For equilibrium between sub-endothelium and connective tissue we put $\mathbf{X}_V$ and $\mathbf{X}_D$ (equations (15) and (16)) equal to zero and interpret $(-\nabla)p$, $(-\nabla)\pi_2$ and $(-\nabla)\pi_m$ as the differences Δp, $\Delta\pi_2$ and $\Delta\pi_m$ across the interface between connective tissue and sub-endothelium. We thus obtain the relationships

$$\Delta p = \phi_1/[\phi_1+\phi_2]\,\Delta\pi_m \approx \Delta\pi_m \quad ; \qquad \phi_2 << \phi_1 ,$$

and

$$\Delta p = -\,[\phi_1/\phi_2]\,\Delta\pi_2 \,.$$

where $\Delta\pi_m$ refers to the difference in the "matrix" contribution to permeability. If m = 3 and there is in addition to the matrix a "solute" in solution in the solvent, there are three independent parameters r_{ij} to consider. For m = 4 there are six and this is already too many. For practical purposes m = 3 is the point where we must stop.

How then is it possible, in the case of a very complex system involving perhaps hundreds of different chemical species, to get any insight from the approach outlined. The answer is that the analysis only partially depends upon the full chemical complexity. The only components we have to distinguish from each other are the components which move with a significantly different velocity. All species which move with the same velocity can be grouped into one component from the point of view of the transport analysis. We only have to make sure that zero velocity difference in one set of localities does not become a finite difference at some other location along the path.

We shall show below with what justification we can form three groups of components in the blood to lymph case. Here, first of all, we will write down some important equations for the three component case:

$$\mathbf{J}_2 = [\phi_2/(\phi_1+\phi_2)](1-\sigma)L_p[\mathbf{X}_V + \eta\mathbf{X}_D] \tag{13}$$

$$\mathbf{J}_1 + \mathbf{J}_2 = L_p[\mathbf{X}_V - \sigma\mathbf{X}_D] \tag{14}$$

$$\mathbf{X}_V = (-\nabla)p - [\phi_1/(\phi_1+\phi_2)](-\nabla)\pi_m \tag{15}$$

$$\mathbf{X}_D = (-\nabla)\pi_2 + [\phi_2/(\phi_1+\phi_2)](-\nabla)\pi_m \tag{16}$$

$$L_p = [\phi_1(\phi_1+\phi_2)/r_{1m}]\ [\eta/(\sigma+\eta)] \tag{17}$$

osmotic pressure and $\Delta\pi_2$ to the corresponding difference in albumin contribution. We know, from one well investigated case at least, that $\Delta\pi_m$ is about 6 mm Hg (Meyer 1983). Hence Δp will be some 6 mm Hg and since ϕ_1/ϕ_2 is very large an expected small, negative $\Delta\pi_2$ is consistent with a finite positive Δp. In earlier work we could show that p equals atmospheric pressure inside unswollen connective tissue (Meyer and Silberberg 1974). Hence the pressure in the sub-endothelial space, or in any other open cavity equilibrated with connective tissue should be about -6 mm Hg. This is the value found by Guyton in his specially implanted capsules (Guyton 1966). To the extent that higher pressures are measured the "open" space into which the pressure measuring device is inserted probably contain some unsuspected matrix component, or some large macromolecular species which are excluded from the connective tissue space. Such may be the case in terminal lymphatics where pressures of the order zero are said to exist (Zweifach and Silberberg 1979).

Of course it is only an approximation to view the connective tissue space as in equilibrium with sub-endothelium. There really are gradients but these are kept low. Any disturbance is, therefore, slow to act and changes in water content, the build-up and disappearance of edema, the swelling and deswelling of connective tissue, are very slow processes.

REFERENCES

Born GVR, Palinski W (1985) Unusually high concentration of sialic acids on the surface of vascular endothelia. Br J exp Path 66:543-549.

Guyton AC (1966) Textbook of medical physiology, 3rd edn. Saunders, Philadelphia.

Meyer FA (1983) Macromolecular basis of globular protein exclusion and of swelling pressure in loose connective tissue (umbilical cord). Biochim Biophys Acta 755:388-399.

Meyer FA, Silberberg A (1974) In vitro study of the influence of some factors important for any physicochemical characterization of loose connective tissue in the microcirculation. Microvascular Res 8:263-273.

Michel CC (1985) Vascular permeability - the consequence of Malpighi's hypothesis. Int J Microcirc Clin & Exp 4:265-284.

Michel CC (1988) Capillary permeability and how it may change. J Physiol 404:1-29.

Silberberg A (1988a) Structure of the interendothelial cell cleft. Biorheology 25:303-318.

Silberberg A (1988b) Vessel wall permeability: ultrastructure of the pathway, site of selectivity, pressure and concentration distribution in filtration and reabsorption flow through the interendothelial cleft. Int J Microcir Clin & Exp, Special Issue:S76.

Silberberg A (1989) Transport through deformable matrices. Biorheology 26:291-313

Silberberg A (1990) The physiological role of a thick gel-like layer over the endothelial surface. Int J Microcirc Clin & Exp, Special Issue:S204.

Zweifach BW, Silberberg A (1979) The interstitial-lymphatic flow system. In: Guyton AC, Young DB (eds) International review of physiology, cardiovascular physiology III. University Park Press, Baltimore, 215-260.

RHEOLOGY OF CONNECTIVE TISSUES AS SWELLING FIBROUS STRUCTURES

Y. Lanir
Department of Biomedical Engineering and
The Julius Silver Institute of Biomedical Engineering Sciences
Technion–Israel Institute of Technology
Haifa, 32000, Israel

1. INTRODUCTION

The most important functions of connective tissues is to maintain structural integrity and allow transport of chemicals and water. This is achieved through their multiphase structure which consists primarily of solid components and fluid matrix.

The solid components include (in various proportions) networks of collagen, elastin and structural glycoproteins. The liquid phase contains (apart from water), freely diffusing solutes of low molecular weight, slowly diffusing solutes (predominantly serum albumin), and very large molecules (with molecular weight which may exceed 10^8 daltons) of soluble proteoglycans (PG) aggregates. They in turn consist of hyaluronic acid backbone attached to proteoglycan subunits. These subunits contain a large number of sulfated polysaccharides – the glycosaminoglycans (GAG). The GAG molecules have a high density of negatively charged fixed groups which give rise to imbalance in the concentration of freely diffusible electrolytes. Electro–neutrality requires that positive cations be in approximation to the fixed negative groups. Hence the tissue space has a higher concentration of electrolytes than do the isotonic body fluids. This imbalance in concentration gives rise to the swelling potential and maintains the tissues in a hydrated state.

The rheology of tissues has important bearing on their mechanical functioning. Practical analysis of the tissue's response to external loading is possible if the various phases can be grouped into very few "components". For this approach to be valid, constituents of a single component must all have the same velocity [Silberberg, 1982; Silberberg, 1989].

NATO ASI Series, Vol. H 64
Mechanics of Swelling
Edited by T. K. Karalis

Swelling tissues have been considered as tricomponent systems [Mow et al., 1990; Silberberg, 1982; Silberberg, 1989] consisting of solid, liquid solvent (fluid) and solutes. In tissues, like cartilage, the PG aggregates are effectively immobilized within the fibrous network due to both their very large volumes and the dense structure of the fibrils. Under *in vivo* conditions a tissue is in equilibrium with its local ionic environment so that the rapidly moving small solutes by themselves do not give rise to concentration gradients. Hence, *in vivo* the tissue can be considered as a bicomponents system in which the fibrous networks and the attached GAG with their accompanying small solutes are the "solid" while the interstitial water with the rest of the small solutes are the "fluid".

An essential characteristic of connective tissues is their fibrous structure. This has important bearing on their mechanical response which has for the most part hitherto not been considered in modeling work.

In the present study features of the tissue's fibrous structure will be incorporated into a biocomponent *in vivo* theory for connective tissues. The goal is to establish a theoretical framework for the analysis and prediction of the tissue's rheological behavior, including the effects of swelling. The present formulation extends an earlier theory for small deformations [Lanir, 1987] into the domain of finite deformations.

2. NOTATION

The matrix and solvent will be referred to as solid and fluid and be designated the subscripts s and f respectively. The subscript i stands for both s and f (i = s,f).

d_i True density of component i (mass per unit volume of the pure component).
C_i Concentration expressed as moles per unit tissue volume.
ρ_i Concentration in the tissue (mass per unit tissue volume).
ρ_i^o Concentration in the reference configuration.
ϕ_i Volume fraction of component i in the tissue.
ϕ_i^o Volume fraction in reference configuration.
$\mathbf{V}_i$ Velocity of component i.
$\mathbf{q}$ Fluid filtration flux.
σ_i The solid Cauchy stress.
σ_i^* The solid extra Cauchy stress.

p The pore fluid pressure.
π_i Concentration pressure.
S The Second Piolla–Kirchoff stress in the solid.
e The solid Green–Lagrange strain.
s Uniaxial Second–Piolla Kirchoff stress in a tissue fiber ensemble.
ε Axial strain in a fiber.
$R(\phi,\theta)$ Orientation density distribution for the tissue fibers in the reference configuration
$\mathbf{\Phi}_i$ Viscous resistance to flow of component i.
$\underline{V}_i$ Partial molar volume of component i.

3. MASS BALANCE

The volume fraction ϕ_i and densities ρ_i and d_i are interrelated by:

$$\phi_i = \rho_i / d_i \qquad i = s,f \tag{3.1}$$

The two components occupy the whole volume. Hence

$$\phi_s + \phi_f = 1.0 \tag{3.2}$$

Connective tissues are incompressible, i.e. both solid and fluid are incompressible. Hence:

$$d_s = \text{const.} \qquad d_f = \text{const.} \tag{3.3}$$

In the absence of chemical reactions, the mass of each component is conserved, so that

$$\partial\rho_i/\partial t + \nabla\cdot(\rho_i \mathbf{V}_i) = 0 \qquad i = s,f \tag{3.4}$$

which in view of Eqs. (3.1) and (3.3) can be written as:

$$\partial\phi_i/\partial t + \nabla\cdot(\phi_i \mathbf{V}_i) = 0 \qquad i = s,f \tag{3.5}$$

where ∇ is the del operator referred to the deformed state, and $\mathbf{V}_i$ is the component velocity.

By combining Eqs. (3.5) for the solid and fluid, it is readily obtained that

$$\nabla \cdot (\phi_s \mathbf{V}_s + \phi_f \mathbf{V}_f) = 0 \tag{3.6}$$

Eq. (3.6) was derived earlier [e.g. Kenyon, 1976]. It can be written in an alternative form as:

$$\nabla \cdot (\mathbf{V}_s + \mathbf{q}) = 0 \tag{3.7}$$

where the filtration flux **q** is defined by:

$$\mathbf{q} = \phi_f(\mathbf{V}_f - \mathbf{V}_s) \tag{3.8}$$

4. EQUATIONS OF MOTION

In most problems related to connective tissues, the inertia and acceleration forces are negligible, and there are no body forces. Under these conditions the equations of motion for the solid and fluid are [Bowen, 1980]:

$$\nabla \cdot (\sigma_i) + \hat{\mathbf{p}}_i = 0 \tag{4.1}$$

where σ_i is the component's bulk averaged (partial) Cauchy's stress and $\hat{\mathbf{p}}_i$ is the local momentum supply which represents local interactions between components. Hence:

$$\hat{\mathbf{p}}_s + \hat{\mathbf{p}}_f = 0 \tag{4.2}$$

The momentum supply vectors are postulated to consist of three parts as follows:

$$\hat{\mathbf{p}}_i = P\, \nabla(\phi_i) + \phi_i\, \mathbf{\Phi}_i + \phi_i\, \nabla(\pi_i) \tag{4.3}$$

where P is the fluid pore pressure, $\mathbf{\Phi}_i$ is the component's viscous filtration resistance and π_i is the strain dependent concentration pressure which is an expression of the concentration dependent portion of the component's chemical potential. Its concentration dependence is a constitutive law to be established experimentally (e.g. [Maroudas and Bannon, 1981]). The term $P\nabla(\phi_i)$ arises from the incompressibility constraint [Bowen, 1980].

Under *in vivo* conditions, the connective tissue space is in balance with its local ionic environment. The Gibbs–Duhem equation states that under these conditions

$$C_s\,\nabla(\mu_s^c) + C_f\,\nabla(\mu_f^c) = 0 \qquad (4.4)$$

where C_i are the concentrations and μ_i^c are the concentration dependent portions of the chemical potentials. The concentration pressures π_i are defined by

$$\nabla(\pi_i) = \nabla(\mu_i^c)\,/\,\underline{V}_i \qquad (4.5)$$

where $\underline{V}_i$ are the partial molar volumes. Since $C_i\,\underline{V}_i = \phi_i$, Eq. (4.3) can be introduced into Eq. (4.2) resulting in:

$$\phi_s\,\nabla(\pi_s) + \phi_f\,\nabla(\pi_f) = 0 \qquad (4.6)$$

thus establishing the interactive nature of the concentration pressures. In the following π_f will be designated as π.

The viscous resistance forces are interactive as well, i.e., the viscous forces which the solid and fluid exert on each other are equal in magnitude and opposite in directions. Hence

$$\phi_s\,\mathbf{\Phi}_s + \phi_f\,\mathbf{\Phi}_f = 0 \qquad (4.7)$$

In connective tissues the fluid filtration flux relative to the solid matrix is very slow (due to the very low permeability). Hence it is assumed that the filtration is governed by Darcy's Law in which the permeability is strain dependent. Darcy's law states that

$$\mathbf{\Phi}_f = -\boldsymbol{\alpha}\cdot\mathbf{q} \qquad (4.8)$$

where $\boldsymbol{\alpha}$ is the strain dependent resistivity matrix.

Under the quasi–static conditions which prevail in connective tissues, the solid and fluid partial Cauchy stresses are assumed to be [e.g. Holmes, 1986]

$$\boldsymbol{\sigma}_s = \boldsymbol{\sigma}_s^* - \phi_s\,\mathrm{P}\,\mathbf{I} \qquad (4.9a)$$

$$\sigma_f = -\phi_f \, P \, \mathbf{I} \tag{4.9b}$$

where $\boldsymbol{\sigma}_s^*$ is the extra Cauchy stress in the solid, which is that portion of the stress which depends on the solid deformation. It is related to in more detail in the following section.

By introducing Eqs. (4.3), (4.8) and (4.9) into Eq. (4.1), the equations of motion for the fluid becomes as follows:

$$\mathbf{q} = -\mathbf{K}\cdot\nabla(P - \pi) \quad , \qquad \mathbf{K} = \alpha^{-1} \tag{4.10}$$

where **K** is the permeability matrix. Its strain dependence is a constitutive law which must be established experimentally (e.g. [Lai et al., 1981]).

Summing up the two Eq. (4.1) while using Eq. (4.2) and (4.9) results in the equation of motion for the whole tissue as follows:

$$\nabla\cdot\boldsymbol{\sigma}_s^* - \nabla\,(P) = 0 \tag{4.11}$$

By using Eq. (4.11), Eq. (4.10) becomes:

$$\mathbf{q} = \mathbf{K}\cdot(\nabla\cdot\boldsymbol{\sigma}_{sw}) \tag{4.12}$$

where $\boldsymbol{\sigma}_{sw}$ is the tissue swelling stress [Lanir, 1983],

$$\boldsymbol{\sigma}_{sw} = \pi\,\mathbf{I} - \boldsymbol{\sigma}_s^* \tag{4.13}$$

The swelling stress represents the fluid (and solid) total chemical potential since by Eq. (4.12) it is the driving force for fluid–solid relative motion. The name swelling reflects the fact that a tissue will swell or deswell if $\boldsymbol{\sigma}_{sw} \neq 0$.

5. CAUCHY STRESS IN THE TISSUE FIBROUS NETWORK

The extra Cauchy's stress $\boldsymbol{\sigma}_s^*$ is related to the response of the fibrous network to the overall tissue deformation. In the following, structural features of the tissue fibers will be utilized to develop a three–dimensional finite deformation constitutive law for the solid extra Cauchy stress. The formulation follows the general theory for fibrous tissues [Lanir, 1983]. For simplicity of presentation, the tissue is assumed to have just one type of elastic fibers (e.g. collagen in cartilage). The extension to more than one type of fibers and to the viscoelastic case is straightforward.

Since the wetted fibers are considered elastic, then the whole fibrous skeleton is elastic as well. The Kirchoff's (second Piolla–Kirchoff) stress in an elastic body is derived from a strain energy function W as:

$$\mathbf{S} = dW/d\mathbf{e} \qquad (5.1)$$

where **e** is the Green strain. Under isothermal conditions W is a function of the strain alone. It is equal to the elastic energy stored in the tissue per unit undeformed volume. In fibrous structures of tissues, this energy is equal to the sum of the strain energies of all the fibers. The strain energy of an elastic fiber $w(\varepsilon)$ is a function of the fiber's axial strain ε and characterizes the fiber's uniaxial response. The uniaxial stress–strain relationship for the fiber is

$$s(\varepsilon) = dw/d\varepsilon \qquad (5.2)$$

The three–dimensional fibrous network of a tissue can be characterized by three parameters. All relate to the fibers contained in a representative elementary volume. They are: The volume density of the fibers in the tissue space in the reference state – ϕ_S^o ; the uniaxial material law of the fiber – $w(\varepsilon)$; and the fibers' orientation density distribution in the reference state. The latter is designated by $R(\phi,\theta)$ where ϕ and θ are the two polar angles. Hence within a representative elementary volume of the tissue, for each direction in space (characterized by the dual value ϕ,θ) the proportion of all fibers oriented at angles between ϕ,θ and $\phi + d\theta$, $\theta + d\theta$ is $R(\phi,\theta)d\phi d\theta$.

The function $R(\phi,\theta)$ satisfies the normalizing condition:

$$\int_0^{\pi/2} \int_0^{\pi} R(\phi,\theta) \cdot d\phi \cdot d\theta = 1.0 \qquad (5.3)$$

In a nonhomogeneous tissue, the orientation distribution of the fibers is nonconstant. In the cartilage, for example, the collagen fibers are parallel to surface at the outer layer but are inclined nearly normally to the surface in the deeper layers. As a consequence, the material law for the cartilage is expected to be nonhomogeneous, reflecting the variable structure. The orientation distribution of fibers has been measured in the cartilage (Hukins and Aspden, 1989) and in the skin (Sachs, 1988).

The total strain energy of the tissue's fibers (W) in a representative elementary volume is thus expressed by

$$W = \int_{\phi} \int_{\theta} \phi_s^o \, R(\phi,\theta) \, w(\varepsilon) \, d\phi d\theta \qquad (5.4)$$

where integration is performed over all possible orientations (values of ϕ and θ).

The axial strain ε of each fiber depends on its orientation. Under affine deformation it can be obtained by tensorial transformation of the overall tissue strain **e**. Hence $\varepsilon = \varepsilon\,(\mathbf{e}, \phi,\theta)$ and the derivative $\partial/\partial\mathbf{e}$ in Eq. (5.1) can be expressed by the chain rule as:

$$\partial/\partial\mathbf{e} = (\partial/\partial\varepsilon)\,(\partial\varepsilon/\partial\mathbf{e}) \qquad (5.5)$$

so that the stress in Eq. (5.1) is given by:

$$\mathbf{S} = \phi_s^o \int_{\phi} \int_{\theta} R(\phi,\theta)\,(\partial w/\partial\varepsilon)\,(\partial\varepsilon/\partial\mathbf{e})\, d\phi d\theta \qquad (5.6)$$

where the derivative $\partial\varepsilon/\partial\mathbf{e}$ arises from the tensorial transformation $\varepsilon = \varepsilon(\mathbf{e},\phi,\theta)$.

In view of Eq. (5.2) the three–dimensional constitutive law (5.6) can be expressed as:

$$\mathbf{S} = \int_{\phi} \int_{\theta} R(\phi,\theta)\, s(\varepsilon)\, \partial\varepsilon/\partial\mathbf{e}\, d\phi d\theta \qquad (5.7)$$

The extra Cauchy stress for the tissue's solid σ_s^* is related to the Kirchoff stress **S** by:

$$\sigma_s^* = J^{-1} \cdot \mathbf{F} \cdot \mathbf{S} \cdot \mathbf{F}^T \qquad (5.8)$$

where **F** is the deformation gradient and $J = \det(\mathbf{F})$.

6. THE BOUNDARY CONDITIONS

The boundary conditions for the driving forces in an incompressible solid–fluid mixture at a singular surface relate to the jump in total stress and total chemical potential [Liu, 1980]. In the present case these jump conditions can be expressed as follows [Lanir, 1983]:

$$[\,(\sigma_s^* - P\,\mathbf{I})\,]\cdot\mathbf{n} = 0 \tag{6.1}$$

where **n** is a normal unit vector and a square bracket indicates the change in the argument across the surface.

Under equilibrium the jump condition for the fluid is:

$$[\,(P - \pi)\,] = 0 \tag{6.2}$$

Direct substitution of Eq. (6.2) in Eq. (6.1) yields:

$$[\,\sigma_{sw}\,]\cdot\mathbf{n} = 0 \tag{6.3}$$

Eq. (6.3), together with Eq. (4.12), imply that a swelling tissue will be in equilibrium if and only if the swelling stress at each point in the tissue is in balance with the external loading applied on the boundary. Otherwise, the tissue will swell or deswell until equilibrium is obtained.

The kinematical jump condition can be expressed via the diffusive velocities $\mathbf{U}_i = \mathbf{V}_i - \mathbf{V}_i^*$ (i=s,f) where $\mathbf{V}_i^*$ is the velocity of the singular surface. The jump condition is [Liu, 1980]:

$$[\,\rho_s\,\mathbf{U}_s + \rho_f\,\mathbf{U}_f\,]\cdot\mathbf{n} = 0 \tag{6.4}$$

If the boundary surface is permeable only to the fluid, then $\mathbf{U}_s = 0$ at the surface and the condition (6.4) reduces to

$$[\,\rho_f\,\mathbf{U}_f\,]\cdot\mathbf{n} = 0 \tag{6.5}$$

REFERENCES

Bowen EM (1980) Incompressible porous media models by use of the theory of mixtures. Int J Engng Sci 18:1129–1148

Holmes MH (1986) Finite deformation of soft tissue: Analysis of a mixture model in uni–axial compression ASME Trans J Biomechanical Eng 108:372–389

Hukins DWL and Aspden RM (1989) Fibre reinforcing in articular cartilage. In: Yettram AL (ed) Material Properties and Stress Analysis, Manchester Univ. Press, Manchester, pp 44–69

Kenyon DE (1976) The theory of an incompressible solid–fluid mixture. Arch Ration Mech Analysis 62:131–147

Lai WM, Roth V, Mow VC (1981) Effect of nonlinear permeability and rate of compression on the stress behavior of articular cartilage. ASME Trans J Biomech Eng 103:61–66

Lanir Y (1983) Constitutive equation for fibrous connective tissues. J Biomechanics 16: 1–12

Lanir Y (1987) Biorheology and fluid flux in swelling tissues. I. Biocomponent theory for small deformations, including concentration effects. Biorheology 24:173–187

Liu IS (1980) On chemical potential and compressible porous media. J de Mecanique 19:327–342

Maroudas A and Bannon C (1981) Measurement of swelling pressure in cartilage and comparison with the osmotic pressure of constituent proteoglycans. Biorheology 18:619–632

Mow VC, Lai WM and Hou JS (1990) Triphasic theory for swelling properties of hydrated charged soft biological tissues. Appl Mech Rev 43:5134–5141

Sachs M (1988) Focus on materials with scattered light. Research and Development, September, pp 74–78

Silberberg A (1982) The mechanics and thermodynamics of separation flow through porous molecularly disperse, solid media – The Poiseulle Lecture 1981. Biorheology 111–127

Silberberg A (1989) Transport through deformable matrices. Biorheology 26:291–313

THE INFLUENCE OF BOUNDARY LAYER EFFECTS ON ATHEROGENESIS IN DIALYSIS TREATMENT PATIENTS

Pallotti Giovani, Pettazzoni P.

Department of Physics, Faculty of Medicine
Bologna University, Via Ranzani 5, 40127, Italy

Cianciolo G., Coli' L. Campieri C., Stefoni S.

Department of Physics, Faculty of Medicine,
Bologna University, Italy

Slawomirski M. R.

Institute Gornictwa Naftowego
Krakow, Poland

INTRODUCTION

Cardiovascular diseases represent at present the most frequent cause of death in patients subjected to regular dialysis treatment sessions (*DTS*). Statistically, the risk of death from myocardial infraction in a haemodialysis population aged 35 to 54 is 20 times greater than among the normal population. These complications are strictly related to accelerated atherosclerosis, one of the main complications in patients subjected to *DTS*.

Pathology in patients subjected to *DTS* is still by no means clear. Several factors seem to be implicated in contributing to accelerated atherosclerosis. Many lipidic abnormalities such as hypertrigliceridemia, low *HDL* cholesterol and others are generally associated with uraemia. Hypertrigliceridemia observed in patients subjected to *DTS* is essentially due to an increase in triglyceride *TG* contained in very low density in lipoproteins *VLDL*, intermediate density lipoproteins *IDL* and low density lipoproteins *LDL*. It has been shown to be the main or autonomous risk factor in the development of atherosclerosis; *HDL* cholesterol being a reliable prognostic index of coronary artery disease.

The low blood level of *HDL* and the lecithin–cholesterol–acyltransferase *LCAT*

NATO ASI Series, Vol. H 64
Mechanics of Swelling
Edited by T. K. Karalis

deficiency may impair the process of reverse cholesterol transport from lipid stores to *HDL*, contributing to its accumulation in the arterial wall. However, although the central role that the lipoprotein plays in metabolism it is not known, one can say that certain functional and quantitative changes in apoproteins could increase the atherogenicity of lipoproteins themselves.

Several other atherogenic factors, pathologically retained or produced during an extracorporeal session, may determine endothelial injuries or damages in the underlying intima; the latter once exposed. The serotonin released by the platelet stimulates directly platelet aggregation and may aggravate the immune mediated endothelial cell injury; increasing intimal alterations.

Platelet Derived Growth Factor *PDGF1*, a specific protein released by platelet α–granules as a result of blood/artificial materials during extracorporeal circulation, stimulates intima proliferation. Other possible factors that may play a role in endothelial or intima injury include the parathyroid hormone, the heparin employed for anticoagulation of the extracorporeal circuit, hepatic lipase deficiency, the dialysate acetate buffer, free oxygen radicle and others.

It is highly probable that the increase in concentration of the blood *LDL* particles during *DTS*, is directly related to the intradialytic haemoconcentration, increasing the risk of particle sedimentation on the arterial wall leading to the development of atheroma.

Since knowledge on Haemorheology inclines us to suspect the importance of the change in the blood flow properties related to haemoconcentration as the mechanism behind this vascular complication, certain aspects of this problem are explained from the behaviour of the boundary layer during dialysis sessions. In the present study a discussion of the particles' sedimentation theory as a function of the intermittent haemoconcentration, in the search for a possible new approach to the pathogenesis in patients subjected to dialysis treatment sessions accelerated atherosclerosis.

MATHEMATICAL CLARIFICATION

The boundary layer is a subdomain of the flow located on the artery wall. The flow in the boundary is dominated by viscous effects whereas outside the boundary layer it behaves as quasi fluid. Figure 1 presents the distribution of velocity near the beginning of the artery. It is quite different from the well known Poiseuille distribution which develops over a long distance from the artery inlet (several dozen times the artery diameter). At the artery wall the blood velocity u is equal to zero, increasing progressively with distance from the wall up to velocity U which is typical of the flow outside the boundary layer.

Starting from the beginning of the artery the thickness of the boundary layer δ increases with distance according to the following formula

$$\delta = C\sqrt{\frac{\eta x}{\rho U}} \tag{1}$$

where η is the blood viscosity, ρ is the blood density, x is the distance from the beginning

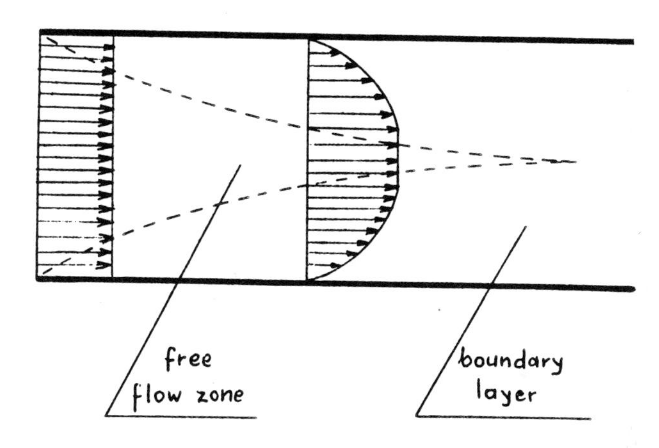

Figure 1. Velocity distribution near the beginning of the artery

of the artery and C is a constant value. Consequently, the boundary layer develops with distance x; starting from a certain point it occupies the entire cross–section of the artery (Fig.2). Note that the interface between boundary layer and remaining flow zone is arbitrary, since the velocity of flow u asymptotically tends to U, and during the flow there exists a permanent mass transfer between the boundary layer and the remaining flow zone.

At the beginning of an artery the flow in the boundary layer is laminar. Since the thickness of the boundary layer increases with distance, in certain cases the Reynolds number for the boundary layer

$$R = \frac{\rho \upsilon \delta}{\eta} = C\upsilon\sqrt{\frac{\rho x}{\eta U}} \tag{2}$$

exceeds the critical value at the defined point, staring from the point where turbulence in the boundary layer may be observed (Fig.2). The blood speed outside the boundary layer is high, and therefore the probability that *LDL* particles transported in this flow zone may be "trapped" by the artery wall is low. Thus, *LDL* particles which constitute the atheroma are previously intercepted by the boundary layer. The flow in the boundary layer appropriate for *LDL* sedimentation should be laminar. In turbulent motion the oscillatory and random fluctuations in speed lead to exceptionally irregular *LDL* particle paths and to an intense mixing of the blood which makes sedimentation almost impossible.

The haemoconcentration accompanying the dialysis process modifies the behaviour of the boundary layer in the artery. The main reason for this is the increase in blood viscosity

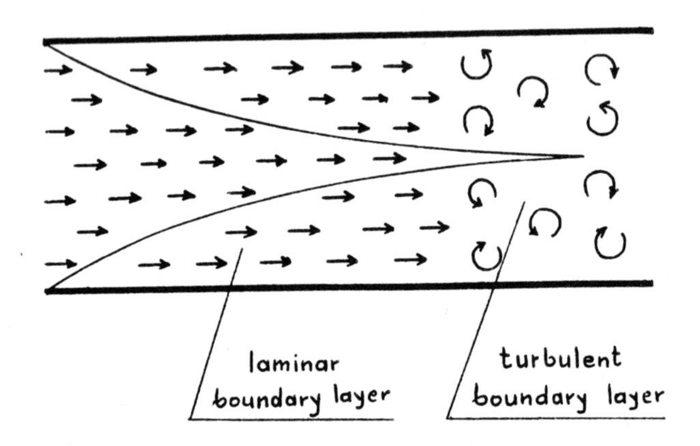

Figure 2. Boundary layer developed in the distance x; starting from a certain point until the point where the turbulence occupies the entire cross–section of the artery

η. According to equation (1), the thickness of the boundary layer increases and consequently the zone from which *LDL* particles may be intercepted extends. Moreover, the length of the flow zone outside the boundary layer is reduced. The increase in viscosity also reduces the magnitude of the Reynolds number for boundary layer R (Fig. 3). As a result, turbulent oscillation is damped and the turbulization of the boundary layer is displaced to a long way from the beginning of the artery.

All the factors mentioned above increase the volume of the laminar boundary layer. The increase in *LDL* concentration in the blood, associated with the simultaneous extension of the laminar boundary layer, increases the possibility of sedimentation of lipoproteins on the wall of an artery leading to the development of atherosclerosis.

CONCLUSIONS

The periodic application of the artificial kidney in uraemic patients and the consequent haemoconcentration represent an acknowledged atherogenic factor above all in dialysis patients with a high interdialytic body weight increase and consequent high ultrafiltration rate

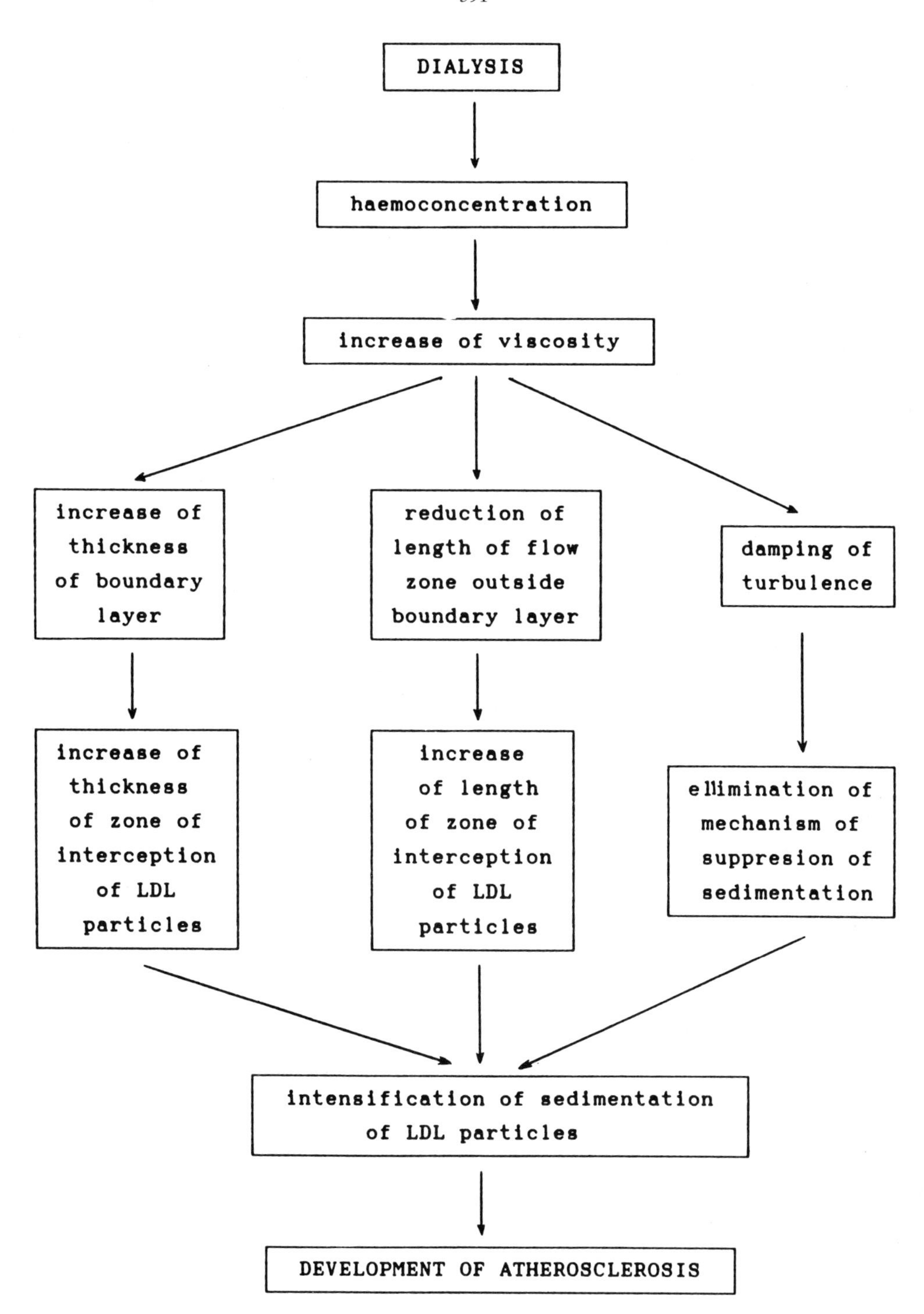

Figure 3. Diagram showing the factors which contribute to atherogenesis

or with high values of plasmatic viscosity (↑↑Hmt, ↑↑plasmatic protein, ↑↑plasmatic lipoproteins). The pathogenesis of accelerated atherosclerosis in dialysis patients is multifactorial. The sedimentation theory represents a possible new approach in analyzing the problem.

REFERENCES

Bagdate J.D., Porte D., Bierman E.L., (1968) Hypertriglyceridemia a metabolic consequence of chronic renal failure, N. Engl. J. Med. 279: 181.

Brunzell J.D., Albers J.J., Hass L.B., (1977) Prevalence of serum lipid abnormalities in chronic haemodialysis, Metabolism 26: 903.

Burke J.F., Francos G.G., Moore L.L., (1978) Accelerated atherosclerosis in chronic dialysis patients: Another look, Nephron 21: 181.

Lindner A, Charra B, Sherrard D.J., Scribner B.H., (1974) accelerated atherosclerosis in prolonged maintenance haemodialysis. N.Engl.J.Med. 290: 697.

Howarth L., (1959) Laminar Boundary Layers, Handbuch der Physik, herausgegeben von S. Flugge and C. Truesdell, Bd. VIII/1, Springer, Berling, Gottingen, Heidelberg.

Landau L.D., Lifshitz E.M., (1959) Fluid Mechanics, Pergamon, New York.

Pallotti G., Pettazzoni P., Stefoni S., Ciancolo G., Bonomini V., (1989) Renal Arterial Stenosis in Renal Transplantation: A Mathematical Pathogenetic Hypothesis and its possible Clinical Implications. In: Proceedings 2nd International Symposium on Biofluid Mechanics and Biorheology.

Schlichting H., (1958) Grenzschicht–Theorie, Braun, Karlsruhe.

Slawomirski M.R., Coli' L., Bonomini V., Pallotti G., (1987) Accelerated Arteriosclerosis in Dialysis Patients. A Possible Role of the Sedimented Particles. In: Proceedings 8th International Conference of the Cardiovascular System Dynamics Society, 4–7 October 1987, Osaka, Japan.

Slawomirski M.R., Pallotti G., Stefoni S., Coli' L., Bonomini V. (1989) A Mathematical approach to the development of Arteriosclerosis in Dialysis Patients. In: Proceedings 2nd International Symposium on Biofluid Mechanics and Biorheology.

Part 5

Blisters, Forces Between Particles, Phase Transitions of Gels and Flow in Deformable Media

BLISTERS

P.G. de Gennes

Physique de la Matière Condensée
Collège de France
11 place Marcelin–Berthelot
75231 Paris Cedex 05
France

Abstract. *When a liquid sheet (monolayer or bilayer) is deposited on a solid surface and covered by a liquid, many types of defects can occur: among these are the blisters, which in typical cases are filled with water. Blisters may be generated by: (a) thermal agitation of a flexible sheet, (b) excess area frozen in by the deposition process, (c) mismatch conditions between the solid and a crystallized form of the lipid. The thermal blisters are expected to be small, mobile, and difficult to see by force microscopy (because they can be moved out of the field). They can give rise to an exponential force between two adjacent plates (this however, is not an explanation for the exponential forces observed in liquid/water lamellae, without any solid support). The blisters may transport water parallel to the solid surface, with a well defined diffusion coefficient.*

I. Blister types

We often wish to transform a solid surface by surfactant deposition. This may not always generate an ideal "carpet" (fig.1). (a) If we hydrophobize a polar solid with a <u>monolayer</u> and cover it by an oil, it will often happen that ambient water will diffuse and swell the interface between the solid and the polar heads.... We call this a "water in oil blister" (W/O B). (b) If

NATO ASI Series, Vol. H 64
Mechanics of Swelling
Edited by T. K. Karalis

we cover a polar solid by a single bilayer, we can think of a "water in water" blister (W/W B). Also, apart from this classification, we should distinguish between flexible lipid layers and ordered lipid layers: in the latter case, we can have a certain mismatch between the area per polar head layers: in the latter case, we can have a certain mismatch between the area per polar head at the moment of deposition, and the optimal area in the ordered phase: this could generate fold defects as shown on fig.1c.

The case of W/W B with flexible sheets, generated by thermal agitation is discussed in ref[1] and this is summarized in section II. In section III the dynamics of these blisters is analyzed qualitatively. In section IV we think more generally of the practical effects of any type of blisters.

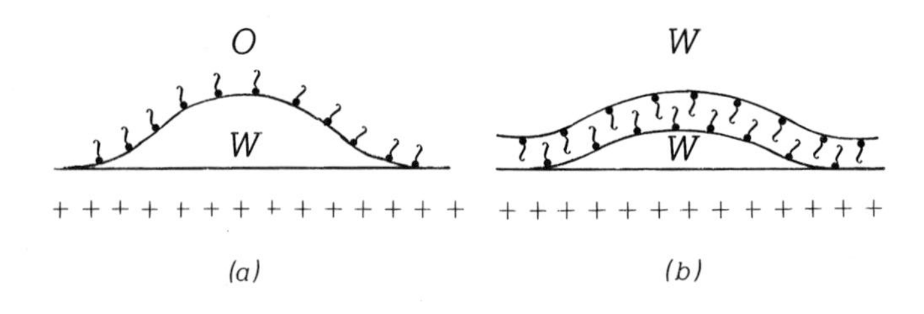

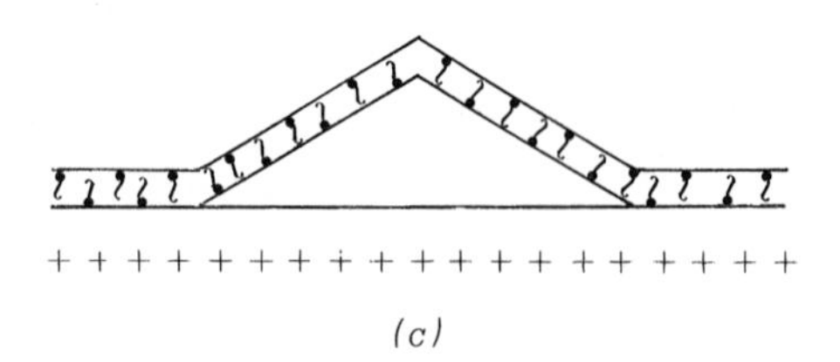

Figure 1. (a) "Water on oil blister". (b) "Water in water blister". (c) Blister in a crystalline bilayer with mismatch.

II. Thermal blisters: equilibrium properties

The geometry is shown in fig.(2). Physically the major ingredients are: (1) an unbinding energy Δ (required to separate one cm^2 of sheet from the solid), and (2) a rigidity constant K for the sheet. Notice that the sheet is assumed to have optimized its area per polar head without ant surface tension. In the presence of tension, blisters become much less probable. Then the free energy f of a blister of height Z and radius X is of the order

$$f = X^2 \left[K \frac{Z^2}{X^4} + \Delta \right] \tag{1}$$

and the optimum corresponds to

$$X^2 = l Z \tag{2}$$

where

$$l = \left(\frac{K}{\Delta} \right)^{1/2} \tag{3}$$

represents a typical curvature radius near the apex (fig.2). The whole continuum description makes sense only if Δ is small (say of order 1 erg./cm^2) so that l becomes relatively large. Inserting (2) into (1) we find that the blister free energy

$$f \cong \Delta l Z \tag{4}$$

is linear in height! This remarkable property implies that the probability of finding a blister of height Z is exponential

$$p(Z) = \text{const.} \exp\left(-\frac{Z}{\lambda}\right) \tag{5}$$

where

$$\lambda = \frac{kT}{\Delta l} = \frac{kT}{(K\Delta)^{1/2}} \quad (\sim 5 \text{ Å}) \tag{6}$$

and we emphasize again that this remarkable exponential law is a specific feature of sheets with zero tension.

III. Blister dynamics

Consider a thermal blister (of height $Z = X^2/l$) moving at constant velocity V and assume that the dominant losses are due to the internal fluid with viscosity η. The velocity gradients are of the order V/Z (since the velocity on the solid vanishes) and the dissipation reads

$$T\dot{S} \equiv FV = X^2 Z\eta\left(\frac{V}{Z}\right)^2 = \eta l V^2. \tag{7}$$

Thus the friction force is

$$F = \eta l V \tag{8}$$

and, by Einstein's relation, the diffusion coefficient scales according to

$$D = \frac{kT}{\eta l} \tag{9}$$

which is independent of the thickness Z, and essentially equal to the Stokes Einstein diffusion of a sphere of radius l in a fluid of viscosity η.

The estimate (9) is correct if there are no important losses associated with the detachment or attachment of the lipid to the surface. In the opposite limit of strong adhesion, very different laws would hold.

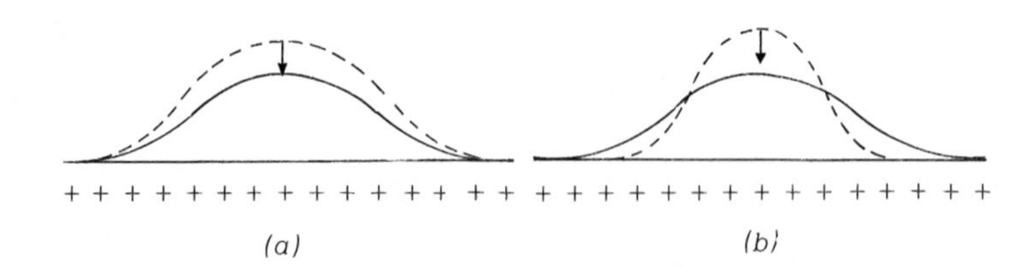

Figure 2. (a) The deflation process (at constant area), (b) first fluctuation mode (at constant volume).

Deflection time of a W/W blister (fig.2a)

Let us now take the opposite viewpoint and consider the effects of water permeability on a W/W blister. We call this permeability Λ defining it *via*

$$V_n = \Lambda p_{12} \tag{10}$$

where V_n is the water flow normal to the bilayer and p_{12} is the pressure drop across the bilayer.

Here the process corresponds to X_o = constant (no change in adhesion) and Z decreases with time. The free energy results from eq.(1)

$$f \cong \Delta l Z \tag{11}$$

and the dissipation is

$$-\Delta l V_n = T\dot{S} = \frac{V_n^2}{\lambda} X_o^2 \tag{12}$$

Thus we arrive at

$$Z\frac{dZ}{dt} = -D \cong -\Lambda\Delta \tag{13}$$

where D has the dimensions of a diffusion coefficient. This gives a deflation law of the form

$$Z^2(o) - Z^2(t) = 2Dt \tag{14}$$

or a deflation time

$$\tau_o \cong \frac{Z_o^2}{\Lambda\Delta}. \tag{15}$$

Of course, apart from this deflation trend, a Brownian motion will tend to restore a larger water content inside the blister; leading ultimately to the equilibrium Boltzmann distribution of Eq.(5). However, blisters of thickness $Z_o >> \lambda$ will follow eq.(14).

Deformation modes of a sealed blister (Fig.2a)

A W/O B essentially fluctuates at fixed water content. This also holds for a W/W B, if the permeability of water through the bilayer is weak. Let us consider such a "sealed" blister of original size X_o and $Z_o = X_o^2/l$, and assume that it is deformed: $X_o \rightarrow X_o + \delta X$ and

$Z_o \rightarrow Z_o + \delta Z$, with

$$\delta f \sim \frac{K}{X_o^2} \delta Z^2 \tag{16}$$

The related restoring pressures (which are negative near X = 0 and positive near $X = X_o$, for $\delta Z>0$) are thus given by

$$\delta p X_o^2 \sim \frac{\partial}{\partial(\delta Z)} \delta f = \frac{K}{X_o^2} \delta Z \tag{17}$$

The pressure gradients $\partial p/\partial x$ are then of order $\delta p/X_o$, and they drive a viscous flow V_x

$$\frac{\delta p}{X_o} \sim \frac{\eta V_x}{Z_o^2} \tag{18}$$

The volume conservation gives

$$\frac{V_x}{X_o} \sim \frac{\delta Z_o}{Z_o} = -\frac{1}{Z_o \tau_v} \delta Z \tag{19}$$

where τ_v is the relaxation time of our process.

Inserting (19) into (17) and considering (18) plus the optimum condition, we arrive at a relaxation rate independent of Z_o

$$\frac{1}{\tau_v} \simeq \frac{K}{\eta l^3} \tag{20}$$

IV. Concluding remarks

1) It is remarkable to observe that thermal blisters have an exponential distribution of thickness (eq.5), if the sheet has zero tension. It is then possible to conceive that thermal W/W blisters contribute to the hydrophillic repulsions observed between two mica plates, both covered by a bilayer, which are known to be exponential[(2)]. This possibility was discussed in ref.(1). However, this feature may not be very important for various reasons: (a) Many of the data on hydrophillic repulsions come from osmotic deswelling of lamellar lipids[(3)], where there is no solid substrate and no blisters that we can think of. (b) Many other, independent sources of exponential repulsions can, and have been invoked[(3)(4)].

1) On the other hand W/O blisters (of non–thermal origin) may be an important

feature of hydrophobized layers exposed to humidity. Very anomalous, long range, attractions have been found in systems of this type. It is true that blisters have not been observed in a recent study using the force microscope [5]: but they may well be "ironed out" by the probe.

3) Many other features may become important, if blisters (thermal or non-thermal) do exist in practice: among those are the (opposed) processes of coalescence and fission. Note, however, that for the thermal blisters of eq (9), the diffusion coefficient is independent of these processes.

Acknowledgements

These ideas have greatly benefitted from written or oral discussions with E. Evans, J. Israelashvili, A. Parsegian, P. Pincus - but none of them should be held responsible for the - very tentative - conclusions.

V. REFERENCES

[1] De Gennes P.-G., Pincus P. (1990) Comptes Rend. Acad. Sci. (Paris) 310: 697-700

[2] Clunie J.S., Goodman J.F., Symons P.C. (1967) Nature 216: 1203
Le Neveu D.M., Rand R.P., Parsegian V.A. (1976) Nature 259: 600

[3] Rand R.P., Parsegian V.A. (1989) Biochimica Biophysica Acta 988: 351-376

[4] Israelashvili J., Wennerstrom H. (1990) Langmuir 6: 873-876

[5] Evans E. (1991) Private Communication

ORIGIN OF SHORT-RANGE FORCES IN WATER BETWEEN CLAY SURFACES AND LIPID BILAYERS

Jacob Israelachvili

Department of Chemical & Nuclear Engineering, and Materials Department, University of California, Santa Barbara, California 93106, USA.

The forces between clays surfaces, surfactant micelles, lipid bilayers and biomembranes in water include the expected attractive van der Waals and repulsive electric double-layer forces (DLVO forces). In the case of the rigid crystalline surfaces of clays, there are additional short-range forces due to water structure and hydrogen-bond correlations. The force associated with water structure is generally oscillatory, while that associated with the H-bond network is monotonically repulsive (between hydrophilic surfaces) or monotonically attractive (between hydrophobic surfaces). These two forces are still not understood theoretically.

Amphiphilic structures such as micelles and bilayers also have an additional non-DLVO short-range repulsive force in water which was also originally thought to be purely due to hydration, that is, water or H-bond structuring. However, it now appears that their origin may be completely different. Because of the dynamic (fluid-like) nature of these interfaces additional entropic "thermal fluctuation" forces arise from the overlap of thermally excited surface modes. Four such repulsive forces have been identified: *undulation, peristaltic, protrusion* and *headgroup overlap* forces. The first two arise from collective motions of bilayers or membranes and can be described in terms of their continuum elastic bending and compressibility moduli, respectively. The last two arise from molecular-scale fluctuations of molecules bobbing in and out of the surfaces (protrusion force), and a similar osmotic repulsion between overlapping mobile headgroups (headgroup overlap force). It is shown that both these two forces are expected to decay roughly exponentially with distance with characteristic decay lengths in water in the range 0.1 to above 0.3 nm, which is exactly within the measured range.

An assessment is made of the relative contributions of DLVO forces, entropic forces and genuine hydration forces between clay and amphiphilic surfaces. It is concluded that the short-range repulsive forces between amphiphilic surfaces are not primarily due to water structure (they occur in other liquids than water) but rather to entropic or osmotic effects, and that they are more akin to the "steric" repulsion between polymer-covered surfaces. With this new interpretation of these forces many aspects of interbilayer interactions may now be understood.

Introduction: historical origin of "hydration" forces

The two classical forces of colloid science are the attractive van der Waals and repulsive electric "double-layer" forces. These two forces form the basis if the the so-called DLVO theory [Derjaguin & Landau, 1941; Verwey & Overbeek, 1948] which has been experimentally verified in direct force measurements between surfaces across a variety of liquids, both aqueous and non-aqueous [Israelachvili, 1985a].

At small distances, below about 5 to 10 solvent diameters the continuum treatment breaks down, and both theory and experiment indicate a transition to an oscillatory type of force

NATO ASI Series, Vol. H 64
Mechanics of Swelling
Edited by T. K. Karalis

[Henderson & Lozada-Cassou, 1986; Israelachvili & McGuiggan, 1988]. Figures 1 and 2 show the forces measured between two mica surfaces in various aqueous electrolyte solutions. At low concentrations the forces are well described by the DLVO theory at all separations down to molecular contact (in the "primary minimum"). At high concentrations, as more hydrated cations bind to the surfaces, an additional non-DLVO force comes in which has both a monotonic and an oscillatory component (Fig. 1, inset). The oscillations have a periodicity of about 0.25 nm — the diameter of a water molecule. These oscillatory solvation or hydration forces are intimately connected to the geometry (size and shape) of the solvent molecules, and they are now reasonably well understood, at least for the case of two *smooth, rigid* walls interacting across simple liquids including water (see Fig. 2, inset).

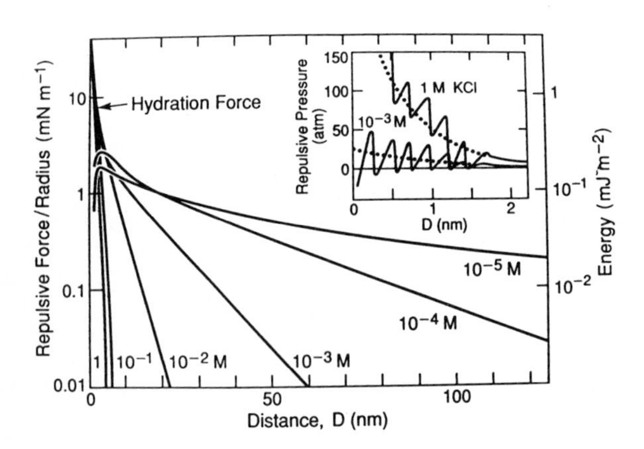

Fig. 1 *Measured forces and pressures between mica surfaces in aqueous KCl solutions. In 10^{-5} M and 10^{-4} M solutions the forces obey DLVO theory, but at higher concentrations an additional "hydration" force comes in that has both an oscillatory and a monotonically repulsive component (inset). [Modified from Israelachvili & Pashley, 1983.]*

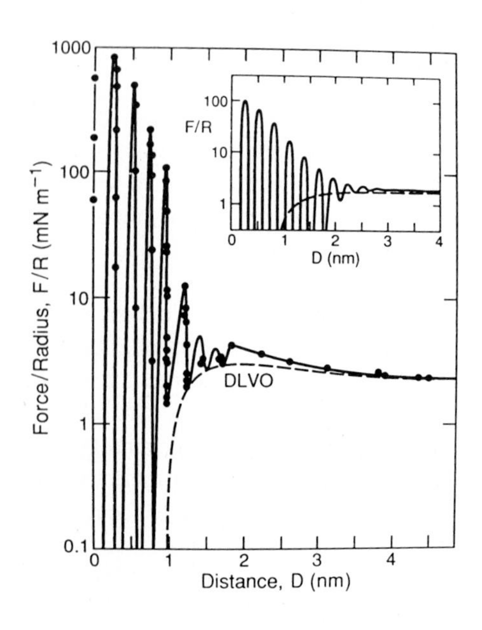

Fig. 2 *Detailed force-profile in 10^{-3} M KCl (Fig. 1). Inset: theoretical force computed by Henderson and Lozada-Cassou (1986).*

The origin of the additional monotonic component is, however, not understood. It appears to arise only in aqueous solutions. Experimental measurements indicate that the monotonic repulsion decays roughly exponentially with distance, D, and so may be expresssed as

$$F = F_0 e^{-D/\lambda_0} \qquad (1)$$

where the sign and magnitude of F_0 depends on the "hydrophilicity" or "hydrophobicity" of the surfaces. Between hydrophilic surfaces the force is repulsive, F_0 is positive, and the decay length λ_0 varies between 0.6 and 1.2 nm (Pashley, 1982). Between hydrophobic surfaces the force is attractive, F_0 is negative, and $\lambda_0 \approx 1.0$ nm [Israelachvili & McGuiggan, 1988].

Additional short-range repulsive forces also arise between amphiphilic, i.e., surfactant and lipid, structures such as micelles, microemulsions, bilayers and biological membranes. Langmuir (1933, 1938) was one of the first to consider the forces between surfactant molecules and identified three types of repulsive forces between them: First, there is a purely entropic (osmotic) repulsion $P=\Sigma nkT$ that, by analogy to Boyle's Law $PV=RT$ for a gas, takes the following two-dimensional (2-D) form for a monolayer: $\Pi A=RT$. When excluded volume and headgroup area effects are taken into account, the above two equations become the well-known equations of state in 3-D and 2-D:

$$P = \frac{RT}{V-b} \quad \textit{for a gas, and} \quad \Pi = \frac{RT}{A-A_o} \quad \textit{for a monolayer.} \qquad (2)$$

Equation (2) was proposed by Langmuir (1933) as a suitable Equation of State for monolayers and, like the van der Waals Equation of State, is still used today.

Second, there is the well-known repulsive electrostatic "double-layer" force whenever the headgroups are charged, though as was first pointed out by Langmuir (1938) this interaction, too, must be considered as essentially entropic since it is the osmotic repulsion between the ions that is responsible for the net repulsion — the purely electrostatic contribution to the interaction being *attractive*.

Third, there are repulsive "hydration" forces, which Langmuir (1938) considered to arise from " ... the intense electric field [which] draws the dipole water molecules into the spaces between the particles and so holds them apart." These forces were invoked to explain the observed short-range repulsion between oppositely charged "coacervates" (micelles, giant vesicles, microemulsion droplets, etc.) in water. It is not clear why Langmuir did not consider the possibility that here, too, the net electrostatic interaction involving charges and dipoles is also bound to be attractive (so long as the system is overall electroneutral). Nor is it clear why Langmuir only considered the osmotic repulsion between the quasi-free ions at water-surfactant interfaces and not that between the mobile molecules themselves. Given that most of the diffuse double-layer counterions are located within 2-5 Å of a charged surface [Israelachvili, 1985a] and still give rise to a double-layer force that can extend hundreds of ångstroms, a similar force arising from molecular density fluctuations near surfaces might have been anticipated.

Later, when non-DLVO repulsive forces were measured between other hydrophilic surfaces (e.g. of solid colloidal particles, clays, silica) it was assumed that these too are due to water structure. These forces were called "structural" or "hydration" forces [Jordine, 1973; van Olphen, 1977] and were believed to be the same as the repulsive forces between amphiphilic surfaces. Hydration forces are believed to arise whenever surfaces have a strong affinity for water (i.e., are hydrophilic) which is in turn related to the way the surfaces induce order in, or alter the structure of, the adjacent water molecules. These forces are now routinely invoked to explain any unexpected repulsion and/or swelling of colloids, clays, soap films and amphiphilic structures in water.

However, apart from their approximately exponential distance-dependence, the "hydration" forces between the rigid, crystalline surfaces of clays such as mica are actually very different from those between the soft, amorphous or fluidlike surfaces of micelles and bilayers. Three qualitative differences may be noted: (i) the non-DLVO repulsive forces between clay surfaces have an oscillatory component while those between bilayers do not; (ii) the forces between clays and mica appear unique to aqueous solvents, while those between bilayers are not (and are usually even stronger in other liquids — see Fig. 5 below); and (iii) the forces between mica and silica have typical decay lengths of about 1.0 nm, while those between lipid bilayers are closer to 0.2 nm (see below). As will be argued in this paper, these two interactions have different origins, with the force between amphiphilic surfaces not directly related to water structure at all.

Early measurements of the forces between lecithin and other uncharged bilayers in aqueous solutions [LeNeveu *et al.* 1976; Parsegian *et al.* 1979; Lis *et al.* 1982] showed that at distances D below about 1-3 nm they decay approximately exponentially with distance, as given by Eq. (1), with a "characteristic" decay length, λ_o, roughly equal to the size of the water molecule. This gave further support to the notion that this force is due to water structure, and theoretical work on the subject [Marcelja & Radic, 1976; Gruen & Marcelja, 1983; Schiby & Ruckenstein, 1983; Attard & Batchelor, 1988] tended to confirm the possible existence of an exponentially repulsive force arising from the decaying "polarization" of water molecules by surfaces. However, the decay length was not easy to derive theoretically and had to be assumed or fitted, though it seemed conceivable that it could be close to the size of a water molecule.

Subsequent Molecular Dynamics simulations [Kjellander & Marcelja, 1985a, 1985b] failed to predict the expected monotonically decaying force. Instead, with surfaces modelled on lecithin and mica, only decaying oscillatory profiles were obtained. Could the monotonically repulsive forces between fluid-like amphiphilic surfaces be a smeared-out oscillatory force, due to the thermal motion of the mobile head-groups? After all, in the computer simulations the lecithin headgroups were kept *immobilized*, whereas those of real surfaces are highly mobile. There was no definitive answer to this question until very recently when Granfeldt and Miklavic (1991) carried out a computer simulation of two *thermally mobile* lecithin bilayers interacting with each other. They found that the entropic interactions between the thermally fluctuating surface groups leads to an exponentially repulsive force with a decay length of about 0.2 nm — very close to the measured value. We return to consider these theoretical matters again later.

Recent measurements of repulsive forces between bilayers

Experimentally, too, the data looks increasingly less consistent with a hydration origin for these forces, but more consistent with an entropic one. Five key examples will now be described; these also provide a convenient background for the ensuing theoretical analysis of thermal fluctuation forces.

Range of repulsive forces

In the earlier force measurements by LeNeveu *et al.* (1976) and Lis *et al.* (1982) the distances between bilayer surfaces were inferred from the repeat spacings of the multibilayers as measured using x-rays. More recently, McIntosh and Simon (1986) showed how the aqueous gap separation could be determined more directly from the measured electron density profiles across the multibilayers. They re-evaluated the older data and also measured the force-distance profiles for a variety of surfactant and lipid bilayers in both aqueous and nonaqueous solvents. One of their conclusions is that the earlier experiments had significantly overestimated both the range and decay lengths of the respulsive forces. This is illustrated in Fig. 3 which shows the force between bilayers of the pure uncharged zwitterionic lipid dipalmitoyl-phosphatidylcholine (DPPC) in the fluid state in water. The range of the repulsion is about 1.17 nm below which it is roughly exponential with a decay length of $\lambda_o \approx 0.14$ nm (this may be compared with the earlier data of Lis *et al.* (1982) which gave a range of 3.3 nm and a decay length of 0.22 nm).

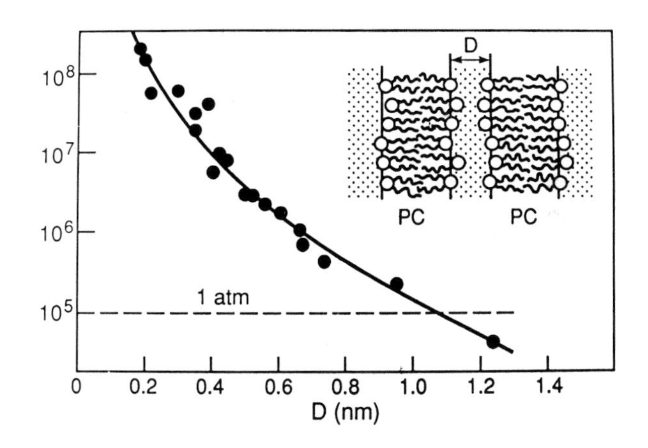

Fig. 3 *Repulsive force profile between fluid-state DPPC bilayers in water [from McIntosh & Simon, 1986].*

Likewise, for egg-PC, McIntosh *et al.* (1989) obtained a range of 1.54 nm and $\lambda_o \approx 0.17$ nm, compared to 2.75 nm and $\lambda_o \approx 0.26$ nm obtained by Lis *et al.* (1982). Thus, the range of these forces is less than 2 nm, rather than 2.0 - 3.5 nm as previously supposed. With the repulsive forces now considered to extend no more than 2 nm (less than 1 nm per surface), they are now well within the range of steric forces expected to arise from the overlap of thermally excited surface groups, discussed in the next theoretical section.

A purely hydration origin for these forces also looks increasingly unlikely when we consider that measured values for λ_o currently range from 0.08 nm to 0.64 nm depending on the lipid system [Lis *et al.* 1982; Lyle & Tiddy, 1986; Marsh, 1989]. Even allowing for the different ways that the experimental data are processed to obtain these decay lengths, such a large variation no longer appears to correlate λ_o with the size of the water molecule, or even with some characteristic property of water — something that has always been tacitly assumed for hydration forces.

Shape of repulsive potentials

Most experimentally measured force-laws are automatically fitted to an exponential function. However, on close examination of the data it is clear that the force often becomes more repulsive at smaller separations (See Fig. 3 and McIntosh *et al.* 1988). As shown in the next section, this is exactly as expected from theories based on entropic repulsion where as $D \rightarrow 0$ the force must always tend to infinity according to $F \propto 1/D$.

Another interesting aspect concerning the measured forces between bilayers is that these are always monotonic, whereas both theory and experiments on the hydration forces between smooth surfaces are always oscillatory. The fact that oscillatory forces have so far never been seen in the interactions of bilayers brings into question the idea that the monotonically repulsive forces that *have* been measured are due to solvent structuring effects.

Effect of solvent

Recent experiments with egg-PC bilayers in different solvents [Persson & Bergenståhl, 1985; McIntosh *et al.* 1989] show that exponentially repulsive forces having similar magnitudes and decay lengths also exist in liquids other than water. This is illustrated in Fig. 4 for egg-PC in three different solvents: water, formamide and 1:3-propanediol.

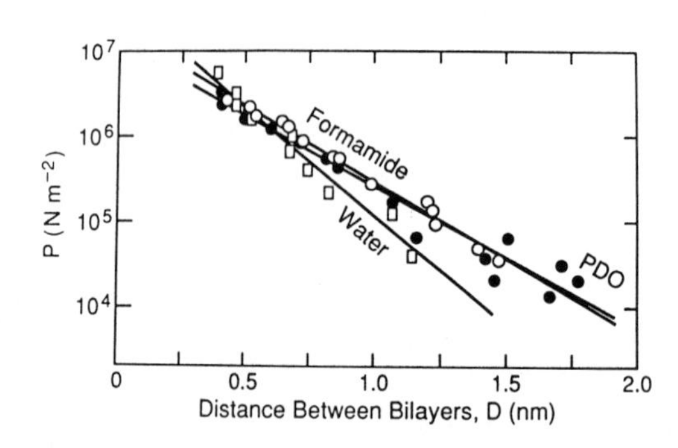

Fig. 4 *Repulsive force profiles between egg-lecithin (egg-PC) bilayers in water, formamide and 1:3-propanediol or PDO [from McIntosh et al. 1989].*

These forces are therefore not unique to water. Since entropic or osmotic do not depend on some unique property of the solvent, the measured forces could well have a "nonspecific" entropic origin. Indeed, the range of the forces between egg-PC increases as we go from water to formamide to propanediol, exactly the opposite from what would be expected if solvent structuring were responsible. On the other hand, the range increases with decreasing interfacial energy of the hydrocarbon-solvent interfaces, which is precisely what would be expected for entropic forces (see below).

Effect of temperature

As shown in Fig. 5, the repulsion between bilayers in water generally increases with temperature [Marra & Israelachvili, 1985; Israelachvili, 1985b]. In addition, many surfactant bilayers do not swell at all when in the solid-crystalline state but do swell as soon as the temperature is raised to above the liquid-crystalline state [Tiddy, 1980]. Within the hydration model this trend would suggest that the water structure increases with increasing temperature. This is very unlikely: with increasing temperature, as the amphiphilic molecules become less ordered, one would expect the same to occur for the water molecules in the gap between them. This latter observation again suggests that these forces are of thermal origin.

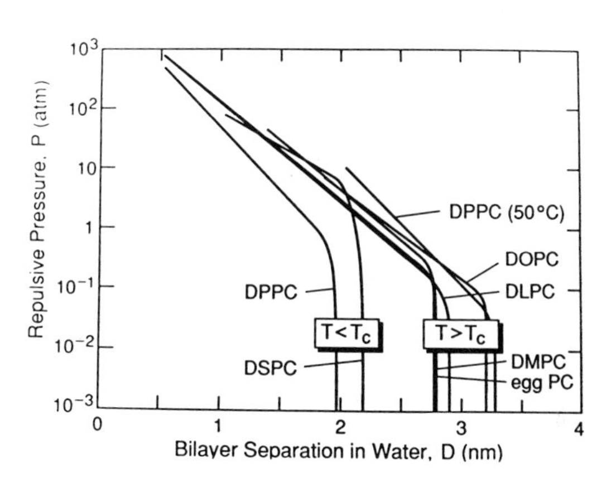

Fig. 5
Repulsive force profiles between various lipid bilayers in water [from Israelachvili, 1985b].

Effect of acyl chains

The decay lengths for single-chained lysolipids are about twice the values for double-chained lipids. For example, $\lambda_0 \approx 0.45$ nm for lyso-PS and 0.64 nm for lyso-PC, compared to less than half these values for the double-chained homologues [Marsh, 1989]. This is entirely consistent with an entropic interaction, but not with a hydration force which would suggest similarity of forces so long as the headgroups and headgroup areas remain unchanged.

Effect of headgroup

Steric forces arising from headgroup overlap are expected to increase with the length and hydrophilicity of the headgroups. This is precisely what is observed experimentally [Lyle & Tiddy, 1986; Homola & Robertson, 1976]. Figure 6 shows the measured forces between monolayers of the surfactant polyethylene-oxide ($C_{12}EO_4$) in water, together with theoretical fits based on the Alexander-de Gennes theory for two interacting "polymer brush" layers (see later). It is clear that in this case the forces can be well accounted for in terms of a simple entropic repulsion of overlapping headgroups without invoking water structuring effects.

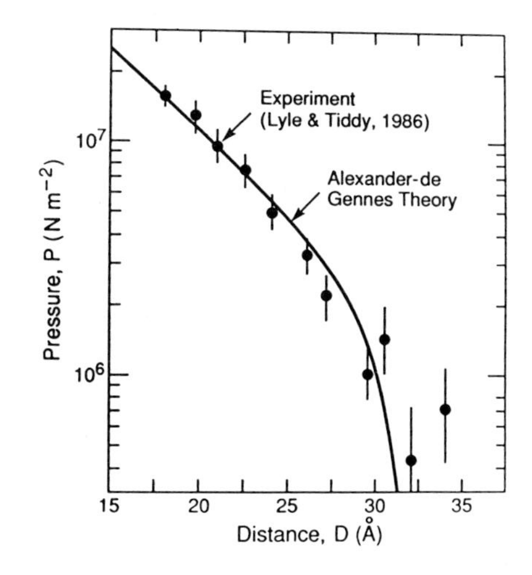

Fig. 6
Measured forces between bilayers of the poly(ethylene-oxide) surfactant $C_{12}EO_4$ in water [from Lyle & Tiddy, 1986]. The solid line is the fitted Alexander-de Gennes Equation, Eq. (13), for overlapping headgroups modelled as polymer "brush" layers, using L=16Å (4Å per EO group), s=9.3Å, and T=25°C.

As will be argued below, headgroup overlap repulsion together with protrusion repulsion probably dominate the short-range repulsion between the surfaces of micelles, bilayers and other self-assembled amphiphilic structures in different solvents.

Theory of Thermal fluctuation forces

In one of his classic papers on attractive and repulsive forces, Langmuir (1938) criticised "the past use of [potential] energy diagrams" because they ignored "the effect of the thermal agitation" of ions in the solution. He proposed that "instead of potential energy ... the osmotic pressure be used, which includes these previously neglected factors". Langmuir also demonstarted that it is the osmotic pressure between ions that was responsible for the "double-layer" repulsion between charged colloids and "coacervates" in solution. Here it is argued that the osmotic pressure between thermally mobile molecules and molecular groups [Pfeiffer *et al.* 1989] has likewise been ignored, and that this contributes an additional and often dominant term to the short-range force between amphiphilic surfaces in solution.

There are at least four different types of repulsive "thermal fluctuation" forces between amphiphilic aggregates that are entropic in origin, i.e., having the same origin as osmotic pressure or the pressure of a perfect gas. The molecular origin of these forces is illustrated in Fig. 7. None of these forces should exist between hard surfaces (e.g., solid colloidal particles), but between amphiphilic bilayers we shall see that both their magnitude and distance dependence quantitatively account for the measured repulsions.

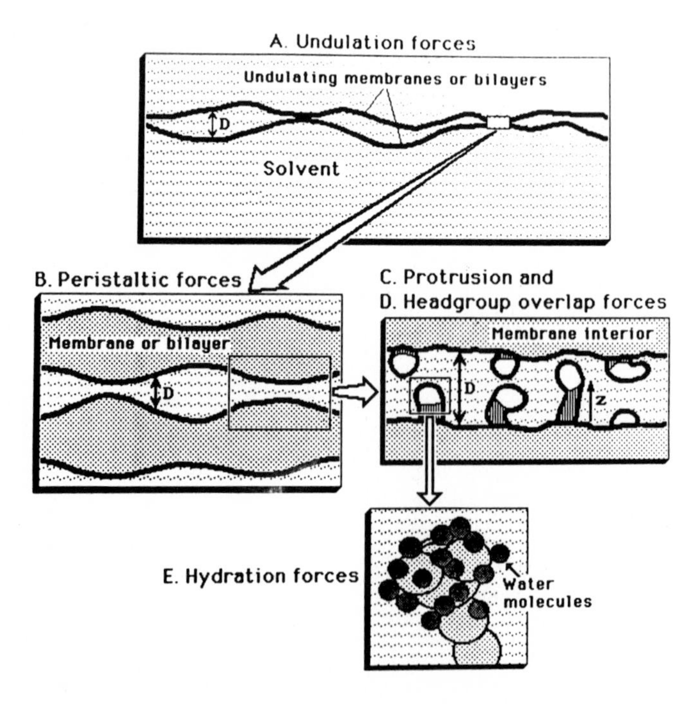

Fig. 7 *Four types of thermal fluctuation forces between amphiphilic surfaces such as surfactant and lipid bilayers.*

Pure osmotic repulsion

Before we consider the various repulsive forces in complex fluid systems it is well to recall the very large osmotic pressure that a collection of non-interacting particles will have in solution. This is simply given by Eq. (2). If we define the mean interparticle or intersolute separation by D, Eq. (2) for a gas may be expressed as

$$P = \frac{RT}{(V-b)} \approx \frac{kT}{D^3}. \tag{3}$$

Figure 8 shows a plot of $P=kT/D^3$ over the range from D=0.3 nm to D=1.6 nm. If a straight line is drawn through the points one obtains the quasi-exponential force-law $F=F_o e^{-D/\lambda}$ of decay length $\lambda_o \approx 2.7$Å and magnitude $F_o \approx 2\times10^9$ dyne/cm² — both of which are typical of fitted values for a whole range of different amphiphilic systems.

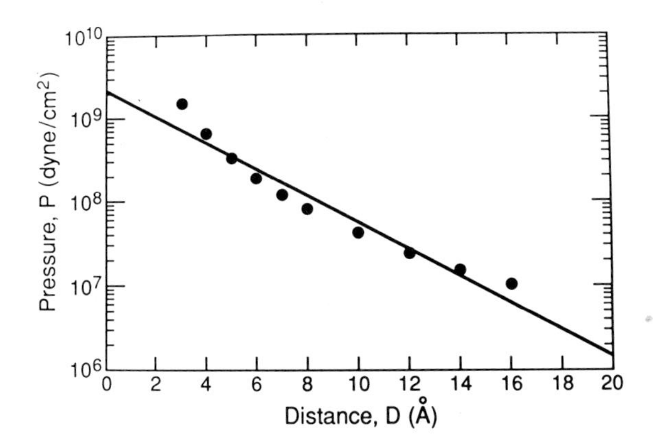

Fig. 8
Ideal osmotic pressure (or Boyle's Law): P=kT/V plotted as P against mean interparticle separation distance $D=V^{1/3}$. The fitted straight line is $P=2\times10^9\, e^{-D/2.7Å}$ dyne/cm², and is independent of the solvent.

The curve in Fig. 8 shows the maximum repulsive force (osmotic limit) that may be expected for any osmotic pressure measurement where the dissolved solute molecules or particles *do not interact with each other*, i.e., where there is no real force between the particles. The mere measurement of such a repulsion is therefore no indication of the existence of an interparticle force, whether attractive or repulsive. In complex fluid systems there will always be an osmotic pressure between the fluidlike structures or aggregates, but this will modified from that given by Eq. (3). The question is: what is this modified force?

First, in the limit of small separations the force should always tend to the osmotic limit given by $P=\Sigma nkT$, where Σn is the number density of counterions or mobile molecular species. For example, it is well known that in the limit of small separations the double-layer repulsion between two surfaces of constant charge density Γ (electronic charges per unit area) as given by the Poisson-Boltzmann equation is independent of salt concentration and given by [Israelachvili, 1985a]

$$P = 2\Gamma kT / D \tag{4}$$

Note that this is the expected osmotic limit for an ideal gas of ions (or any noninteracting particles) of volume density $\Gamma/\frac{1}{2}D$ confined uniformly a gap of thickness D. For small D, Eq. (4) predicts forces very similar to those predicted by Eq. (3). Figure 9 shows an example of this effect. Here the measured repulsion between two bilayers of the charged lipid PG is shown, together with the theoretically expected repulsion based on the full Poisson-Boltmann equation. The agreement between the two is remarkable, and yet these short-range forces, as well as those between the $C_{12}EO_4$ bilayers shown in Fig. 6, were considered to be due entirely to "hydration" effects.

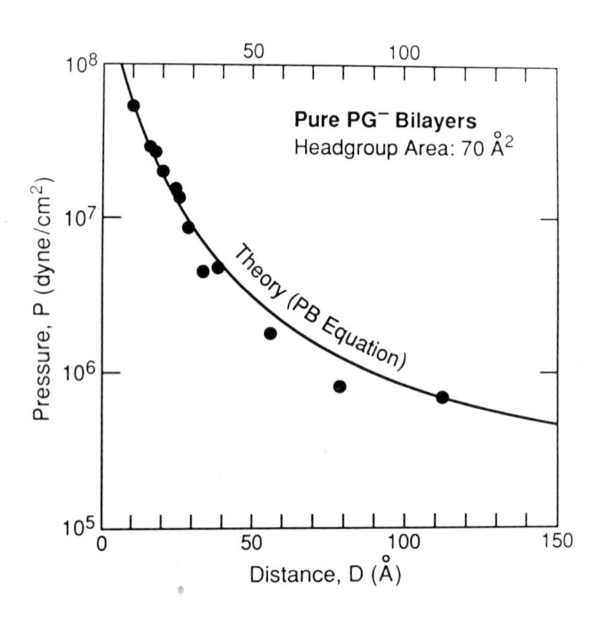

Fig. 9 *Theoretical force-profile arising from the electrostatic repulsion between two charged surfaces in water (charge density: 1e per 70Å^2) and the measured forces between two charged phosphatidylglycerol (PG) bilayers with the same headgroup area in water. [Theoretical calculations based on Poisson-Boltzmann (PB) equation, as described by Israelachvili (1985a, Ch. 12). Experimental points: Cowley et al. (1978), Fig. 4] Note the similarity of the forces to the pure osmotic pressure (Fig. 8) for D<20Å.*

Clearly, regardless of the system, so long as there are mobile molecular or ionic species near the surfaces, there will always be a steep short-range repulsion of similar decay and magnitude (at least on a log scale).†

---

† The repulsive forces between bilayers have been linked to apparently similar short-range forces measured between oriented DNA molecules [Rau *et al.* 1984; Podgornik *et al.* 1989]. Since DNA molecules are charged in water the net repulsive force between them is a combination of at least three separate components: a repulsive electrostatic "double-layer" force, a thermal fluctuation force due to molecular motions, and a genuine hydration force. At small separations the magnitude of the double-layer force between any two surfaces of constant charge is expected to be independent of salt and determined by the osmotic pressure between the trapped counterions which, as Figs 8-10 show, is very similar to the thermal fluctuation forces arising from molecular motions in bilayers. The presence of at least three repulsive forces, and the difficulty of defining D=0 for the DNA double-helix, makes it difficult to say anything definitive about the separate contributions of these three forces in DNA systems.

The undulation force

Historically, the first thermal fluctuation force to be considered was the "undulation" force, first proposed by Helfrich (Helfrich, 1978; Helfrich & Servuss, 1984; Servuss & Helfrich, 1989). All elastic sheets, including fluid bilayers, have thermal undulations whose amplitude and density increase with increasing temperature, T, and decreasing bilayer bending modulus, K_b. Helfrich showed that for two bilayers under no external tension at a mean distance D apart (Fig. 7A) the force per unit area between them is repulsive and given by

$$F \approx \frac{3\pi^2 (kT)^2}{64 K_b D^3}. \tag{5}$$

The undulation force is essentially an entropic force arising from the confinement of thermally excited waves (undulation modes) into a smaller region of space as two membranes approach each other. The undulation force has been measured and the inverse third distance dependence verified experimentally by Safinya *et al.* (1986), McIntosh *et al.* (1989), and Abillon and Perez (1990).

The Peristaltic force

In addition to bending fluctuations, bilayers or membranes also undergo peristaltic (or squeezing) fluctuations wherein the thickness of the membrane fluctuates about the mean thickness (Fig. 7B). It can be shown that for two membranes whose area expansion or compressibility modulus is K_a, the peristaltic force between them is

$$F \approx \frac{2(kT)^2}{\pi^2 K_a D^5}. \tag{6}$$

Note that in contrast to the undulation force, which is in terms of the membranes' bending modulus, K_b, the peristaltic force depends on the expansion modulus, K_a. These two elastic properties are quite different and have different dimensions, though they are not necessarily totally independent of each other.

The Protrusion force

The undulation and peristaltic forces, given by Eqs (5) and (6), are valid for large wavelength fluctuations where bilayers can be treated in terms of their continuum elastic properties. These equations should not apply to thermal excitations occuring at the molecular level, i.e., when the lateral dimensions of the excited modes approach molecular dimension, σ, as shown in Fig. 7C. These molecular-scale protrusions and the additional thermal fluctuation force they give rise to must be treated in terms of a molecular rather than a continuum framework. In this subsection we consider the protrusions of whole molecules, while in the next subsection we consider the protrusions headgroups.

The protrusion of a surfactant or lipid molecule out of a micelle or bilayer (Fig. 7C) is generally unfavourable because of the positive hydrophobic energy associated with the increased hydrocarbon-water contact area. In a first approximation, this energy increases linearly with the distance, z, the molecule protrudes into the water. This gives rise to a molecular "protrusion" potential given by

$$\Phi(z) = \alpha z \tag{7}$$

where the interaction parameter α is in units of J/m. The density profile of molecules protruding into the aqueous phase is therefore expected to decay exponentially according to

$$\rho_z = \rho_o e^{-\alpha z_i / kT} = \rho_o e^{-z_i / \lambda} \qquad \textit{where} \qquad \lambda = kT / \alpha \tag{8}$$

is the characteristic "protrusion decay length". Equation (8) was first used by Aniansson and others [Aniansson, 1978; Aniansson *et al.* 1976] to analyse the protrusion dynamics of surfactant molecules in micelles and their exchange rates with the monomers in the bulk. They arrived at a value for the interaction parameter of $\alpha = 3 \times 10^{-11}$ J/m, which corresponds to a decay length of $\lambda = kT/\alpha = 0.14$ nm at 25°C. We return to consider other data relevant to both single-chained and double-chained amphiphiles in the following section.

When two such surfaces approach each other, they experience a repulsive force as their protruding groups are forced back. This "protrusion" overlap force is entirely entropic or osmotic in origin and is given by [Israelachvili & Wennerström, 1990]

$$F = \frac{\Gamma \alpha (D/\lambda) e^{-D/\lambda}}{\left[1 - (1 + D/\lambda) e^{-D/\lambda}\right]}. \tag{9}$$

where Γ is the surface number density of protrusions, in units of m^{-2}. In the distance regime between 1 and 10 decay lengths, the protrusion force as given by Eq. (9) varies roughly exponentially and is adequately given by

$$F \approx 2.7 \Gamma \alpha e^{-D/\lambda_o} \qquad \textit{where} \quad \lambda_o = 1.15\lambda = 1.15kT/\alpha \tag{10}$$

while at smaller separations there is a steep upturn in the force which diverges according to

$$F(D \to 0) = 2\Gamma kT / D \qquad \textit{for } D < \lambda_o. \tag{11}$$

Note that this is the same as Eq. (4) and is expected osmotic limit for an ideal gas of protrusions of density $2\Gamma/D$ confined within a gap of thickness D.

The unfavourable energy of a finger-like protrusion (Fig. 7C) is related to the *excess* area it exposes, $\pi\sigma z$, multiplied by the interfacial energy, γ_i. Accordingly, α and λ_o may also be expressed as

$$\alpha = \pi \sigma \gamma_i$$

$$\textit{and} \qquad \lambda_o = 1.15kT/\alpha = 1.15kT/\pi\sigma\gamma_i. \tag{12}$$

Headgroup overlap forces

The headgroups of many surfactants and lipid molecules are longer than the chains. They also protrude into the aqueous phase, but because they are hydrophilic and miscible with the solvent phase the energetics and density profiles associated with these protrusions are different from those of protruding hydrophobic chains, discussed above. There are no simple general expressions to describe the dynamics and density profiles of headgroups, and so far only computer simulations have been able to theoretically probe this highly complex region of amphiphilic surfaces [Leermakers & Scheutjens, 1988; Granfeldt & Miklavic, 1991]. However, in a first approximation we may consider a flexible headgroup as behaving like an end-grafted polymer exposed to a theta-solvent. Given that the mean separation between headgroups is typically 0.6-0.9 nm and that this is also about how much they extend into the solution, we may expect to use the theory of polymer brushes to decribe headgroup-headgroup interactions.

The Alexander-de Gennes theory gives for the repulsion between two end-grafted polmer layers (de Gennes, 1987)

$$F = \frac{kT}{s^3}\left[\left(\frac{2L}{D}\right)^{9/4} - \left(\frac{D}{2L}\right)^{3/4}\right] \qquad \textit{for } D < 2L \tag{13}$$

where L is the thickness of each polymer layer or "brush" and where s is the mean distance between grafting sites (corresponding to the distance between headgroups, so that $\Gamma = 1/s^2$).

Equation (13) assumes that L>s and applies in the distance regime D>2L. Also, there is an unknown numerical factor at the front so that the exact magnitude of F is unknown even though the shape of the force curve is. In Fig.6 the Alexander-de Gennes equation nicely fits the experimental points for $C_{12}EO_4$ bilayers using reasonable values for L and s. Since L>s in this case it appears that the headgroups are long enough to be considered as quasi-brush layers.

Quantitative estimates of thermal fluctuation forces

Figure 10 shows theoretical plots of the undulation, peristaltic, protrusion and headgroup overlap forces between two amphiphilic surfaces such as two lecithin bilayers in water at 25°C. The attractive van der Waals force has also been included for comparison. Electrostatic, hydrophobic, and hydration forces have not been included; these only arise under certain conditions whereas the former five forces should be quite general. After describing how each curve was computed the results will be discussed in the light of current experimental data.

Van der Waals forces

The attractive van der Waals force or pressure between two bilayers is given by

$$F = -A / 6\pi D^3 \tag{14}$$

where A is the Hamaker constant, which from the Lifshitz theory is expected to be 4-5 $\times$ 10^{-21} J for two interacting bilayers, though a value of 1.3 $\times$ 10^{-21} J was measured between two lecithin bilayers [Marra & Israelachvili, 1985]. In Fig. 10 the van der Waals pressure is plotted in the range 0.2 to 2.8 nm using Eq. (14) with A = 5 $\times$ 10^{-21} J, which should be an upper limit for the strength of the van der Waals attraction. Equation (14) is expected to remain valid out to bilayer separations of about 2.5-3.0 nm, above which the force decays faster than $1/D^3$ due to the finite thickness of bilayers and retardation effects. Thus at the larger separations the plotted van der Waals force is an upper bound.

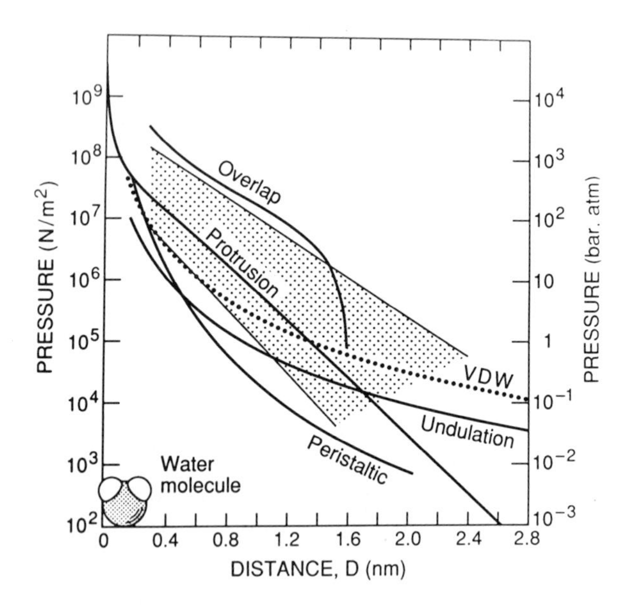

Fig. 10 *Expected repulsive steric forces and attractive van der Waals force profiles for two lipid bilayers in water in the fluid state.*

Undulation forces

The repulsive undulation force of Eq. (5) has the same distance dependence ($P \propto 1/D^3$) but is of opposite sign to the attractive van der Waals force. The only parameter needed to calculate the undulation force is the bending modulus, K_b. Experiments show that typical values for lecithin bilayers are in the range 9×10^{-20} to 2×10^{-19} J [Cevc & Marsh, 1987; Marsh, 1990]. Accordingly, a value of 10^{-19} J (10^{-12} erg) was used in the calculation. The range of the undulation force is expected to exceed that of all other forces and to become the dominant interaction at large distances (in the absence of other long-range forces such as electrostatic and hydrophobic forces). At smaller separations, however, this continuum, mean-field interaction is expected to become replaced by the protrusion and headgroup overlap forces.

Peristaltic forces

The peristaltic force depends on the area expansion modulus of bilayers, K_a, which has been found to lie in the range 100-200 mJ/m² for a variety of bilayers [Cevc & Marsh, 1987; Marsh, 1990]. Accordingly, a value of $K_a = 150$ mJ/m² (150 erg/cm²) has been used in Eq. (6) to obtain the peristaltic curve of Fig. 10. Equation (6) is not expected to apply at very small separations, but neither should it apply at separations greater than some fraction of the bilayer thickness (see Fig. 7B). The maximum range over which the peristaltic force, as given by Eq. (6), is estimated to apply is between 0.2 and 2.0 nm, as plotted in Fig. 10.

Protrusion forces

Two parameters define the protrusion force of Eqs (9) and (10): the interaction parameter, α (in J m^{-1}), and the density of protrusions, Γ (in m^{-2}). For amphiphilic molecules in a bilayer in water, α would normally be the "hydrophobic" energy per unit length needed to extend the headgroup or a hydrocarbon chain into the aqueous phase. The hydrophobic energy of a pure hydrocarbon chain in water (with no headgroup) is obtained from the solubility of alkanes in water [Tanford, 1980] and is about 6.3×10^{-21} J/mole per CH_2 group (900 cal/mole/CH_2) at 25°C. Since the length of a CH_2 group along a chain is 0.1265 nm, this value corresponds to $\alpha = 5.0\times10^{-11}$ J/m, and thus to a protrusion decay length of $\lambda = kT/\alpha = 0.08$ nm.

In the case of surfactant molecules in micelles, the hydrophobic energy per CH_2 group or per unit length can be determined from the CMC. The hydrophobicity of an amphiphilic chain is less than that of a pure alkane chain due to the proximity of the hydrophilic headgroup and the higher chain ordering within micelles. Typical values for α are $\sim3\times10^{-11}$ J/m ($\lambda \approx 1.4$ nm) for single-chained surfactants and $\sim5\times10^{-11}$ J/m ($\lambda \approx 1$ nm) for double-chained surfactants (see Table 1). The reason why α for double-chained amphiphiles is less than twice that for single-chains is because the two monomer chains associate in water which reduces their contact area with water [Tanford, 1980; Cevc & Marsh, 1987].

Table 1. Protrusion energy per unit length for single- and double-chained surfactants in micelles as determined from CMCs.†

Surfactant (lipid)	Protrusion energy, α (10^{-11} J m^{-1})	
	Single-chained	Double-chained
alkyl sulphates	2.3	3.7
alkyl dimethyl ammonium chlorides	2.1	4.1
phosphatidylcholines (lecithins)	3.5	5.4

† CMC values from Tanford (1980), Shinoda *et al.* (1963), Mukerjee & Mysels (1970), and Cevc & Marsh (1987).

Note that α may also be estimated from the diameter σ and interfacial energy γ_i of the protruding groups via Eq. (12). For a surfactant chain of diameter σ=0.4 nm, using $\gamma_i \approx 20$ mJ/m^2 as determined for bilayer lamellar phases in water [Jönsson & Wennerstrom, 1981; Parsegian, 1966], we obtain $\alpha = \pi\sigma\gamma_i = 2.5\times10^{-11}$ J/m.

Turning now to *bilayers,* Aniansson (1978) concluded that the protrusion energy of a surfactant in a bilayer would be less, and λ more, than in a micelle. Note that in bilayers composed of double-chained surfactants the protrusion of a headgroup now involves pulling out two chains for which the value of α is higher than for a single chain (Table 1). Thus we should expect shorter decay lengths for aggregates composed of double-chained lipids. However, both chains do not always have to protrude together. Isolated chains can still protrude on their own. Indeed, since the protrusion energy of isolated, single chains remains unchanged these are now energetically more favourable and therefore more probable than pair-protrusions. This suggests that in bilayers composed of double-chained lipids, individual chains are likely to protrude farther than the headgroups (the second chain remaining in the bilayer). This is consistent with the recent theoretical analysis of lecithin bilayers by Leermakers & Scheutjens (1988, Fig. 5), who also concluded that the density profiles of the headgroups and individual chains, or "tails", fall off more or less exponentially over a distance comparable with the length of the molecules. The computed decay lengths were $\lambda \approx 0.2$-0.3 nm for the headgroups and $\lambda \approx 0.3$-0.4 nm for the tails, which corresponds to $\alpha \approx 1.0$-2.0×10^{-11} J/m. Further theoretical work by Leermakers (private communication) based on the same theoretical model indicates that the repulsive force between two bilayers should be roughly exponential with a decay length of about 0.3 nm, and that it arises from *both* entropic headgroup-headgroup and headgroup-tail protrusion repulsions. These findings are consistent with the recent Monte Carlo simulations of Granfeldt and Miklavic (1991) who computed a decay length for lecithin of about 0.2 nm.

From the above analysis of the current literature on protrusion it appears that realistic α values for bilayers in water should range from above 5×10^{-11} J/m to below 1.5×10^{-11} J/m. This corresponds to λ_o values in the range 0.08 to 0.3 nm — a range that exactly matches that obtained in force measurements (a value of $\alpha \approx 2.5 \times 10^{-11}$ J/m, corresponding to $\lambda_o = 0.18$ nm, was used in the calculation of the protrusion force shown in Fig. 10).

Concerning the values for the density of protrusion sites, Γ, we may infer these from the known headgroup areas of amphiphilic structures. These range from 0.4 to 0.7 nm^2 for single and double chained surfactants in micelles and bilayers, corresponding to n between 2.5 and 1.4×10^{18} m^{-2}. A value of $\Gamma = 2 \times 10^{18}$ m^{-2} (corresponding to one protrusion per 50$Å^2$) was used in Fig. 10. Note that the protrusion force given by Eq. (9) is not expected to extend beyond about 3-4 nm, corresponding to two fully extended hydrocarbon chain lengths.

Headgroup overlap forces

Fig. 10 also shows the overalap force as determined from Eq. (13) using $L = s = 0.8$ nm. These values correspond to a headgroup area of 0.64 nm^2 (64$Å^2$) and to a headgroup layer thickness of 0.8 nm — both being typical values expected for lecithins and other surfactant/lipid molecules in bilayers. Longer headgroups should have longer-ranged forces, as illustrated in Fig. 6 where $L = 1.6$ nm. The combined effects of protrusion and overlap forces are discussed below.

Comparison of theoretical and experimental forces

From Fig. 10, it appears that protrusion and overlap forces are expected to dominate the repulsive interaction between bilayers at small separations (less than 1.5 nm), beyond which the undulation repulsion and van der Waals attraction take over. However, for very long headgroups the overlap force may dominate the interaction at all separations. The peristaltic force is always small compared to these forces.

Since the protrusion and overlap forces appear to dominate at small separations where most interbilayer force measurements have been made, it is worth characterizing these forces further. Figure 10 shows that both these forces have a near exponential distance dependence with magnitudes and decay lengths close to those measured for a wide variety of lipid bilayers [Lis *et al.* 1982; McIntosh & simon, 1986; Lyle & Tiddy, 1986; Rand & Parsegian, 1989; McIntosh *et al.* 1989; Marsh, 1989]. For example, for the much studied egg-PC, the measured values for λ_o range from 0.17 to 0.26 nm, while the values for F_o range over two orders of magnitude, from 4×10^7 to 4×10^9 N/m^2, depending on how the experimental data are interpreted [McIntosh & simon, 1986; Rand & Parsegian, 1989; McIntosh *et al.* 1989]. These two limits are shown by the upper and lower shaded boundaries in Fig. 10.

As will now be shown, protrusion and overlap forces can account for many hitherto unexplained phenomena, either quantitatively or qualitatively. They also rationalize a number of trends that were always difficult to reconcile with the hydration model, particularly the effects of non-aqueous solvents, chain number, and temperature on these forces:

(1) The theory predicts that these forces should occur in nonaqueous solvents. In addition, Eq. (12) predicts that λ_o should increase as the interfacial energy (γ_i) of the hydrocarbon-solvent decreases. This trend has been found in the forces measured between lecithin bilayers in water, formamide and 1,3-propanediol [McIntosh *et al.* 1989], where the measured decay lengths are: $\lambda_o = 0.17$ nm ($\gamma_i = 51$ mJ/m^2), 0.24 nm ($\gamma_i = 32$ mJ/m^2), and 0.26 nm ($\gamma_i = 21$ mJ/m^2), respectively, paralleling the decreasing interfacial energies of these liquids.

(2) Equation (12) suggests that longer decay lengths should arise between single-chained bilayers than double-chained bilayers. This is precisely what is observed: the three highest decay lengths so far measured are all for lyso-lipids, viz. >3.0Å for $C_{12}E_3$ [Lyle & Tiddy, 1986], 4.5Å for lyso-PC, and 6.4Å for lyso-PS [Marsh, 1989], while double-chained lipids such as PE, DLPC, DMPC, DPPC and egg-PC (all in the fluid state) generally have less than half these decay lengths.

(3) Equations (9) and (10) predict an increased repulsion with increasing temperature, as is generally observed [Fig 5; Israelachvili, 1985b; Marra & Israelachvili, 1985]. Such a trend would also be expected for other entropic forces but not for forces arising from solvent structuring effects (unless one accepts the possibility of increased structure/ordering at higher temperatures).

(4) For most of the measured force profiles low decay lengths are generally associated with high pre-exponential factors [McIntosh *et al.* 1989]. Equation (10) readily accounts for this trend following the reciprocal relation between α and λ_o.

(5) Equations (9) and (13) predict that the repulsive forces are not simple exponential functions of distance. In particular, they predict a steep upturn in the repulsion at very small D (see Fig. 10) which has been observed in many systems where forces have been measured down to separations below 3-5Å [McIntosh *et al.* 1988]. These same two equations also predict that the forces decay more rapidly to zero beyond some finite distance corresponding to the fully extended chain and headgroup lengths. This may explain the abrupt drop in the forces that is often observed just prior to the potential well [Lis *et al.* 1982].

(6) Since the interfacial energy of an *unsaturated* hydrocarbon-water interface is less than that of a saturated hc-water interface [Landolt-Börnstein, Vol. II, No. 3, page 462], Eq. (12) predicts larger decay lengths with unsaturated and biological lipids, as has been observed. For example, Marsh (1989) measured λ_o=2.2-2.3Å for di-C(18:0)PC, di-C(16:0)PC and di-C(14:0)PC, but obtained λ_o=3.1Å for di-C(18:1)PC and λ_o=2.8Å for egg-PC, all at 22°C.

(7) For bilayers in the fluid state having the same type of chains but different headgroups, we expect the main effects to come from changes in the headgroup overlap forces. For longer headgroups (e.g., C_mEO_n with n>3) we should expect longer-ranged forces, as observed [Homola & Robertson, 1976], while with shorter headgroups (e.g., PE) we should expect shorter-ranged forces. We may also expect headgroup overlap forces to be suppressed by altering the solution conditions for the headgroup segments, for example, on going from "good" to "poor" solvent conditions. This could be achieved by going through the theta-temperature for nonionic surfactants, or by changing the ionic strength and pH for ionic and zwitterionic surfactants.

Interplay of Thermal Fluctuation, Hydration, DLVO and Hydrophobic Forces

At small separations, below λ_o, both the protrusion and headgroup overlap forces tend to the osmotic limit given by Eq. (11). This result also suggests a way of treating genuine solvation/hydration effects. In a first approximation one may simply consider the effect of nonelectrostatically "bound" water molecules (Fig. 7E) as increasing the effective size of the headgroups. This is the conventional approach used in electrolyte solution theory. Accordingly, by analogy with the van der Waals equation of state, Eq. (2) would become

$$F = \frac{kT}{(D/2\Gamma - b)} = \frac{2\Gamma kT}{(D - D_o)}, \quad where \quad D_o = 2\Gamma b = 8\Gamma v, \tag{15}$$

where v is the excluded "hydrated" volume per protruding group. Note that the above result suggests that, in a first approximation, the effect of finite headgroup size and hydration effects on the forces could simply be treated by shifting the origin from $D = 0$ to $D = D_o$. The same conclusion was arrived at by Aniansson (1978) regarding the effects of protrusions on shifting out the plane of origin of repulsive electrostatic double-layer forces. This procedure would therefore be equivalent to adding an effective "hydrated" layer of thickness $\frac{1}{2}D_o = 4\Gamma v$ per surface. For example, assuming 10 "bound" water molecules per headgroup each of volume 30 $\times 10^{-30}$ m^3, and using $\Gamma = 2 \times 10^{18}$ m^{-2}, we obtain a hydrated layer thickness of 0.24 nm per surface, i.e. the effective plane of origin of the forces or interaction potential is shifted out to $D_o = 0.5$ nm.

It is not obvious whether the two DLVO forces, i.e., the van der Waals and electric double-layer forces, can be considered as being simply additive with the thermal fluctuation forces. The fact that both the amphiphilic molecules and their (charged) headgroups are mobile and sample quite a large volume of space suggests that the plane of origin of both these interactions is not well defined. It is also not obvious whether the different fluctuation forces are additive with each other and/or how much correlation exists between all these forces.

These matters can probably only be resolved by complex theoretical modelling and computer simulations. Initial results by Leermakers & Scheutjens (1988) and Granfeldt & Miklavic (1991) suggest that the picture presented here is essentially correct and that protrusion and overlap forces, acting together, result in an overall exponentially repulsive force having a range of 1-2 nm and a decay length of 0.1-0.4 nm.

Finally, concerning the role of attractive long-range hydrophobic forces, these have been shown to act more or less independently of the repulsive steric forces (Helm *et al.* 1989).

When bilayers are stressed so that they expose an increased hydrophobic area to the aqueous phase, the emerging hydrophobic attraction leads to a dramatic increase in their adhesion and, beyond a certain strain, to bilayer fusion [Helm *et al.* 1989; Bailey *et al.* 1990]. This effect manifestes itself in a spontaneous "jump" of the two surfaces from a finite separation into contact which is immediately followed by fusion. The sudden break in the repulsive curve is reminiscent of the onset of a two-phase region in the Van der Waals equation of State where at a certain point the volume "collapses" at fixed pressure. The sudden collapse of two bilayers into each other may also be viewed in this way, and probably describable by simply adding an attractive distance-dependent term to Eq. (15) which will lead to such an instability. If this approach is correct, the fusion of bilayers and membranes may be viewed as a first order phase transition.

Conclusions

From the good agreement obtained between experiment and theory without any need to invoke any additional structural-hydration interactions, it is concluded that the short-range repulsion between amphiphilic surfaces is mainly of entropic origin. Whether water structure provides an additional long-range "hydration" force between *mobile* amphiphilic interfaces as it does between *rigid, crystalline* surfaces such as silicate surfaces [Israelachvili, 1985a, 1985b; van Olphen, 1977] remains an open question.

It is clear that more carefull thought must be given before using terms such as "structured water", "water of hydration", "hydration forces", etc. For example, it is no longer obvious that the amount of water uptake during the swelling of a lyotropic lamellar phase can be unambiguously associated with the "structured water" or "hydration" of the headgroups or surfaces.

Acknowledgements

I thank Frans Leermakers, Christiane Helm, Wolfgang Knoll, Joe Zasadzinski and Håkan Wennerström for their helpful comments, and the National Science Foundation for supporting this work under grant number CTS-9015537.

References

Abillon, O., and Perez, E. (1990). *J. Phys. (France)* **51**, 2543-2556.
Aniansson, E. A. G. (1978). *J. Phys. Chem.*, **82**, 2805-2808.
Aniansson, E. A. G., Wall, S. N., Almgren, M., Hoffmann, H., Kielmann, I., Ulbricht, W., Zana, R., Lang, J., and Tondre, C. (1976). *J. Phys. Chem.* **80**, 905-922.
Attard, P., and Batchelor, M. T. (1988). *Chem. Phys. Lett.* **149**, 206.
Bailey, S. M., Chiruvolu, S., Israelachvili, J. N., and Zasadzinski, J. A. N. (1990). *Langmuir* **6**, 1326-1329.
Cevc, G., and Marsh, D. (1987). "Phospholipid Bilayers," Wiley, New York.
Cowley, A. C., Fuller, N. L., Rand, R. P. and Parsegian, V. A. (1978). *Biochemistry* **17**, 3163.
de Gennes, P. G. (1987). *Adv. Colloid Interface Sci.* **27**, 189-209.
Derjaguin, B. V., and Landau, L. (1941). *Acta Physicochim.* URSS **14**, 633-662.
Granfeldt, and Miklavik, (1991). *J. Phys. Chem.* (in press)
Gruen, D. W. R., and Marcelja, S. (1983). *J. Chem. Soc. Faraday Trans. II* **79**, 225.
Helfrich, W. (1978). *Z. Naturforsch.* **33a**, 305-315.
Helfrich, W., and Servuss, R.M. (1984). *Il Nuovo Cimento* **3D**, 137.

Helm, C. A., Israelachvili, J. N., and McGuiggan, P. M. (1989). *Science* **246**, 919-922.
Henderson, D., and Lozada-Cassou, M. (1986). *J. Colloid Interface Sci.* **114**, 180-183.
Homola, A., and Robertson, A. A. (1976). *J. Colloid Interface Sci.* **54**, 286-498.
Israelachvili, J.N. (1985a). "Intermolecular and Surface Forces," Academic Press, London & New York.
Israelachvili, J. N. (1985b). *Chemica Scripta* **25**, 7-14.
Israelachvili, J. N., and Pashley, R. M. (1983). *Nature* **306**, 249-250.
Israelachvili, J.N., and McGuiggan, P.M. (1988). *Science* **241**, 795.
Israelachvili, J. N., and Wennerström, H. (1990). *Langmuir* **6**, No. 4, 873-876.
Jordine, E. St. A. (1973). *J. Coll. Interface Sci.* **45**, 435.
Jönsson, B., and Wennerström, H. (1981). *J. Colloid Interface Sci.* **80**, 482-496.
Kjellander, R., and Marcelja, S. (1985a). *Chem. Phys. Letters* **120**, 393-396.
Kjellander, R., and Marcelja, S. (1985b). *Chemica Scripta* **25**, 73-80.
Kwok, R. and Evans, E. (1981). *Biophys. J.* **35**, 637-652.
Langmuir, I. (1933). *J. Chem. Phys.* **1**, 756-776.
Langmuir, I. (1938). *J. Chem. Phys.* **6**, 873.
Leermakers, F. A. M., and Scheutjens, J. M. H. M. (1988). *J. Chem. Phys.* **89**, 3264.
LeNeveu, D.M., Rand, R.P. and Parsegian, V.A. (1976). *Nature* **259**, 601.
Lis, L.J., McAlister, M., Fuller, N.L., Rand, R.P., and Parsegian, V.A. (1982). *Biophys. J.* **37**, 657.
Lyle, I. G., and Tiddy, G. J. T. (1986). *Chem. Phys. Letts* **124**, 432-436.
Marcelja, S., and Radic, N. (1976). *Chem. Phys. Lett.* **42**, 129-130.
Marra, J., and Israelachvili, J. N. (1985). *Biochemistry* **24**, 4608-4618.
Marsh, D. (1989). *Biophys. J.* **55**, 1093.
Marsh, D. (1990). CRC "Handbook of Lipid Bilayers", CRC Press, Boca Raton, Florida.
McIntosh, T.J, Magid, A.D., and Simon, S.A. (1989). *Biochemistry* **28**, 7904.
McIntosh, T.J., and Simon, S.A. (1986). *Biochemistry* **25**, 4058.
McIntosh, T.J., Magid, A.D., and Simon, S.A. (1988). *Biochemistry* **26**, 7325.
Mukerjee, P., and Mysels, K.J. (1970). "Critical Micelle Concentrations of Aqueous Systems," Nat. Bur. Stand (U.S.), *Nat. Stand. Ref. Data. Ser., No.* **36**.
Parsegian, V. A. (1966). *Trans. Faraday Soc.* **62**, 848-860.
Parsegian, V.A., Fuller, N.L., and Rand, R.P. (1979). *Proc. Natl. Acad. Sci. USA* **76**, 2750.
Pashley, R. M. (1982). *Adv. Colloid Interface Sci.* **16**, 57-63.
Persson, P. K. T., and Bergenståhl, B. A. (1985). *Biophys. J.* **47**, 743-746.
Pfeiffer, W., Henkel, Th., Sackman, E., Knoll, W., and Richter, D. (1989). *Europhys. Lett.* **8**, 201-206.
Podgornik, R., Rau, D. C., and Parsegian, V. A. (1989). *Macromolecules* **22**, 1780.
Rand, R.P., and Parsegian, V.A. (1989). *Biochim. Biophys. Acta.* **988**, 351-376.
Rau, D. C., Lee B., and Parsegian, V. A. (1984). *Proc. Natl. Acad. Sci. USA*, **81**, 2621.
Safinya, C.R., Roux, D., Smith, G.S., Sinha, S.K., Dimon, P., Clark, N.A., and Bellocq, A.M. (1986). *Phys. Rev. Lett.* **57**, 2718-2721.
Schiby, D., and Ruckenstein, E. (1983). *Chem. Phys. Lett.* **95**, 435-438.
Servuss, R.M., and Helfrich, W. (1989). *J. Phys. France* **50**, 809-827.
Shinoda, K., Nakagawa, T., Tamamushi B., and Isemura, T. (1963). "Colloidal Surfactants," Academic Press, New York & London.
Tanford, C. (1980). "The Hydrophobic Effect", 2nd Edition, Wiley, New York.
Tiddy, G. J. T. (1980). *Phys. Rep.* **57**, 1-46.
van Olphen, H. (1977). "An Introduction to Clay Colloid Chemistry," 2nd ed., ch. 10, Wiley, New York.
Verwey, E. J. W., and Overbeek, J. Th. G. (1948). "Theory of Stability of Lyophobic Colloids," Elsevier, Amsterdam.

SWELLING FROM THE PERSPECTIVE OF MOLECULAR ASSEMBLIES AND SINGLE FUNCTIONING BIOMOLECULES

V. Adrian Parsegian
National Institutes of Health
Bethesda, MD 20892 USA

R. Peter Rand
Biological Sciences
Brock University
St. Catharines, Ont. L2S 3A1 Canada

Donald C. Rau
National Institutes of Health
Bethesda, MD 20892 USA

Given the origins of life in an aqueous milieu and the importance of the internal aqueous environment, it is no surprise that the hydration and swelling of biomolecules is a central feature of their organization and function. Through direct measurement of intermolecular forces in aqueous solutions, we have come to recognize new ways in which nature achieves the controlled swelling essential to living matter. In this paper we will outline how we see that nature may make use of similar hydration/dehydration energetics in controlling single-molecular kinetics as she does in controlling molecular assembly. We summarize what seem to be universal principal features of hydration. In many different kinds of systems these features allow us to see how the swelling energetics of living matter not only compares with that in inanimate systems but also can be quantitatively probed with great sensitivity by using osmotic stress.

Because molecular hydration and consequent intermolecular hydration forces are of paramount importance in the immediate vicinity

NATO ASI Series, Vol. H 64
Mechanics of Swelling
Edited by T. K. Karalis

of individual molecules and in the crucial last Angstroms of their separation when approaching contact, these two expressions of swelling will be the main theme of this report. To emphasize generality, we will begin with the equations for the free energies and derivative forces. We will then describe the methods of force and energy measurement together with a review of the variety of membranes and molecules that have been studied so far. From this perspective, then, it is possible to see new directions in thinking that emerge from direct force and energy measurement.

THE COMMON LANGUAGE OF HYDRATION SWELLING

Intermembrane forces

One can speak of the work of transfer of water from any structural array in terms of a chemical potential of water or equivalently of an osmotic stress Π on the array as a function of a structural parameter. Between parallel membranes, (LeNeveu et al. 1976, 1977, Parsegian et al 1979, Lis et al 1982, Rand 1981, Rand and Parsegian 1989, Rand et al 1990, McIntosh and Simon 1986a,b, McIntosh et al. 1988) the hydration pressure or force per unit area varies essentially exponentially with a decay distance lambda, λ, such that one may write

$$\Pi = P = P_o \exp(-d_w/\lambda)$$

where $\log_{10}(P_o)$ can be from 9 to 12 with P_o in dyne/cm^2 and λ can be from 1 to 2.5 Angstroms; d_w is membrane separation. Integration of this pressure produces an energy per unit area

$$E = \lambda \cdot P$$

with respect to infinite separation, an energy that can be on the order of 100 ergs/cm^2 for polar surfaces brought to anhydrous contact.

These hydration energies of contact can be instructively converted into chemical units. Consider, for example, the energy of a 1 nm by 1 nm patch of membrane. The energy of 100 ergs/cm^2 becomes

$$100 \text{ ergs}/10^{14} \text{ nm}^2 = 10^{-12} \text{ ergs}/\text{ nm}^2$$

In units of kT (1 kT at room T = 4 x 10^{-14} ergs) the energy becomes

$$25 \text{ kT}/\text{nm}^2$$

and recalling that kT per particle turns into RT = 0.6 kcal per mole of particles, we have

15 kcal/mole per nm^2 patch of membrane.

This area, 1 nm^2, is on the order of the area occupied by one or two phospholipid molecules in membranes. This energy, 15 kcal/mole, is already on the order of the high-energy bonds involved in the utilization or production of ATP.

Intermolecular Forces

Measurements of forces between parallel linear molecules in well- defined arrays also show exponential variation (Rau et al 1984, Rau and Parsegian 1987,1990). One can again speak of the work of transfer of water from such an array in terms of a chemical potential of water or equivalently of an osmotic stress Π on the lattice as a function of a lattice parameter such as their interaxial distance d_i.

$$\Pi = P = P_o \exp(-d_i/\lambda)$$

In cases of purely repulsive interaction between parallel linear molecules, the decay distance λ can range from 2.5 to 3.5 Angstroms. In those (usual) cases where the molecules pack into a hexagonal array the osmotic stress of the lattice can be converted to a force per unit length f between parallel molecules by the relation

$$f(d_i) = P(d_i)\ d_i / \sqrt{3}$$

(Rau et al. 1984). In either case, since $\lambda << d_i$, both force per unit length and osmotic stress vary essentially exponentially.

Free Energy of Assembly

In both the two-dimensional and the three-dimensional geometry of molecular arrays, the free energy of assembly is measured as the integral of osmotic stress vs. solvent volume V.

$$\Delta G = \quad \Pi\, d V$$

It is worth noting in this connection that the temperature derivative of Π gives the temperature derivative of free energy and is thus a measure of the entropy changes with molecular separation (Prouty et al. 1985).

Chemical Free Energies.

We want to draw the direct connection between the hydration energies of these structural arrays and the hydration energy of a single molecule undergoing a transition reaction between states. To do this, it is particularly instructive to translate the free energy changes from force measurements as described above into the terms of free energies that are inferred from simple reactions. For the simplest possible case,

$$A <==> B$$

we speak of an equilibrium constant

$$K = [B]/[A]$$

and a "standard free energy of reaction"

$$\Delta G = -RT \ln K$$

"Reaction" in this sense might be the "allosteric" switching of a protein between two states often written

$$T <= => R$$

for the Tense and Relaxed states,

or the opening and closing of a trans-membrane ionic channel

$$C <= => O.$$

Now if the transition between states incurs association with small molecules, the free energy of transition is changed by an amount equal to the number of small molecules added times the chemical potential of the small species

$$A + nW <= => B$$

$$K = [B]/([A]\ [W]^n)$$

$$\Delta G = -RT \ln K = -RT \ln ([B]/[A]) - nRT \ln [W]$$

More accurately speaking, the function RT ln [W] should really be the variable part of the chemical potential of species W, and the concentration [W] the activity of substrate W, a_W. The relation is then

$$\Delta G = -RT\ln ([B]/[A]) - n\ RT \ln a_W$$

We now combine this consideration of single molecules and their change in state in the A/B transformation with the idea of pressure volume work. This is done by considering W in the above reaction to be water, and recognizing that there can be an identifiable hydration part to this molecular transition since the numbers of waters bound to A and B are different by n. That difference will be seen as an osmotic sensitivity of the equilibrium balance of two states A and B, since a change in water activity, a_W, made through a change in osmotic pressure of the medium, will shift the balance between A and B. Specifically, an osmotic part of ΔG will go as $\Pi\Delta V$, where ΔV is the volume of water bound or unbound in the reaction against osmotic pressure Π. Specifically,

$$\Pi\Delta V = - n\ RT \ln a_w$$

and

$$\Delta G = \text{constant} = -RT\ln([B]/[A]) + \Pi\Delta V.$$

Therefore

$$\Delta \ln([B]/[A]) / \Delta \Pi \propto \Delta V / RT.$$

And so a measurement of how ln([B]/[A]) changes with the osmotic pressure of the system, Π, provides a direct measure of the change in associated water volume, ΔV, in the molecular transition.

By so considering both individual molecules and molecular arrays in parallel language, it becomes clear that the physics of swelling is in principle the same at both levels. We turn now to the actual measurement of swelling pressures and free energies of several biological systems that have now been characterized in this way.

HOW TO PROBE MOLECULAR SWELLING

We recognize that in the area of measurement of forces between surfaces, alternative methods, such as the surface force apparatus (Israelachvili and Adams 1978, Pashley et al 1986), sometimes coincide with the original measurements made with osmotic stress (Horn et al 1988). But measurements of forces between materials and molecules that are bound to rigid surfaces are not equivalent to measurements of forces between free surfaces, nor of forces between free molecules, nor of interactions within molecules. In comparison, the strategy of using osmotic stress allows the system and molecules to be measured in their free form. Since the swelling of any material in water is an osmotic process, i.e. a transfer of water between the system of interest and a reference phase of known water activity, the system should be free to swell simply by controlling water activity in its vicinity. The osmotic stress (OS) method, amply described elsewhere (Parsegian et al 1986), was first used for direct force measurement between assembled molecules (LeNeveu et al 1977). It has since been used to observe the change in configurational motion of interacting macromolecules as a function of molecular separation wherein entropic forces of configuration couple with direct interactions to enhance the action of each (Podgornik et al 1989). But more recently osmotic stress has been used to change the state of individually functioning proteins. The method, in brief, is to bring the system of interest into thermodynamic equilibrium with a separate 'reservoir' phase whose volume is in vast excess to that of the system under observation and whose water activity, pH, temperature, and small solute activities are all strictly controlled. The system so thermodynamically equilibrated can then be observed further with every available experimental approach. For example, using X-ray diffraction by ordered assemblies, a marriage of thermodynamics and structure, one can obtain molecular dimensions and intermolecular spacings with accuracies often better than Angstroms. With probes of ionic channel conductance or of protein activity, one extracts the osmotically sensitive part of the underlying structural transformation and consequently the change in the volume of associated water as the system goes between active and inactive forms.

We will now demonstrate the common features of swelling at these different levels in osmotic stress measurements on several kinds of experimental systems.

Ion Channels in Membranes

Perhaps the molecular swelling most easily conceived is that of a ionic channel that spans a low dielectric lipid bilayer or cell membrane. The ability to observe one molecular channel or conducting unit at a time, shows the protein switching between open and closed states corresponding to changes in ability to conduct ions.

In the OS strategy, the opening of the channel is made to occur in the presence of neutral or charged molecules so large that they are excluded from channel entry and transport. Effectively, there is a semi-permeable membrane across the mouth of the channel. Excluded species then act osmotically against the aqueous space that is created when the channel is opened. N.B. the solute- excluding space so created need not be identical with the space through which ions move during transport.

In the language of reaction-equilibrium as described above, there is an osmotic work

$$\Pi \, \Delta V$$

required for the act of opening the channel. A measure of the ratio of times that the channel is open and closed is thereby a measure of the osmotic work. Specifically

$$\ln \left([open]/[closed]\right) \infty \; -\Pi \, \Delta \, V/kT).$$

How this ratio changes with the osmotic pressure of excluded solutes gives Δ V.

$$\Delta \, (\ln K)/ \, \Delta \, (\Pi) \;=\; - \, \Delta V \, / \, kT.$$

(where we are using molecular unit kT rather than molar RT = N_0kT because we want to compute a volume change per single molecule).

Δ V yields the actual volume of water associated with the opening and closing of the channel!

A schematic of the experiment is shown in Figure 1. The osmotic stress method was first applied to the large Voltage Dependent Anion Channel (VDAC), also known as "mitochondrial porin", from the outer membrane of mitochondria (Zimmerberg and Parsegian 1986). The protein was placed in an artificial bilayer membrane (BLM) and the conductance of a single molecule observed between electrodes placed in adjoining chambers.

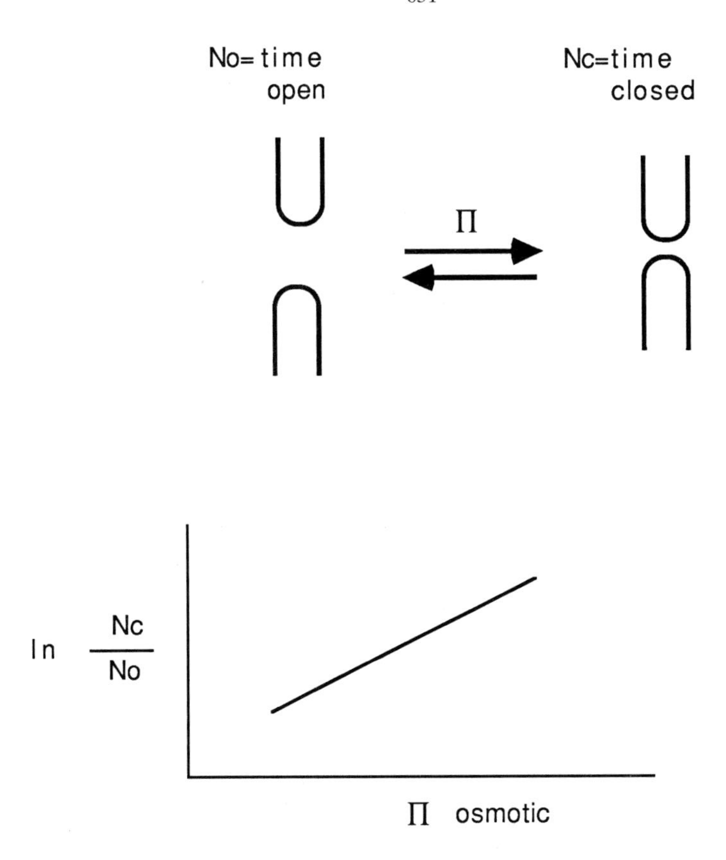

Figure 1. Schematic illustration showing the osmotic stress strategy used to measure the volume of water associated with the gating of the membrane channel.

The osmotically measured volume of water to open VDAC was 2 to 4 x 10^4 cubic Angstroms, a significant fraction of the total channel volume of 2 to 8 x 10^4 cubic Angstroms inferred from the channel length and diameter. (Channel diameter is estimated from the single channel conductivity (Benz et al 1985), or from the size of excluded solutes or from image reconstruction (Manella et al 1984).) The qualitatively important result is that this channel "gating" seems to involve significant changes in channel structure. The act of opening and closing is not the result of inserting a small plug or closing a turnstile that would show minimal osmotic sensitivity. The individual channel seems to swell and shrink by about 700 water molecules! (Zimmerberg and Parsegian 1986)

Perhaps of greater significance, even though the channel might stay in the open or closed state for minutes, the statistics gathered over very long observation times show that the occupation of these states follows the Boltzmann statistics implicit in the equations stated above. Thermodynamics holds not only between but also within states whose lifetimes are much longer than all characteristic molecular times of vibration, rotation and even helix coil transition.

Qualitatively similar results have been seen in the four other channels that have now been osmotically stressed: the potassium channel from the squid giant axon (Zimmerberg et al 1990)), the sodium channel from the crayfish giant axon (Starkus et al 1991)), alpha-toxin from bacteria (Kasianowicz et al 1991), and the peptide channel alamethicin (Bezrukov et al 1991)). The volume of water in each case seems to be a major part of the water volume inferred from open channel size.

This view of channel gating suggests an unusual interplay between specificity of channel ion permeability and control of channel activity. When one realizes that most channels are specific to the species that can pass through them, one is forced immediately to recognize that most channels are under osmotic stress from the excluded species. This natural osmotic stress could be a significant factor in the control of those channels with large internal aqueous volume.

Isolated Molecules in Solution

From single channels in membranes it is conceptually a small step to individual molecules in solution. There again, any change in molecular shape can incur both changes in exposure of protein area to the bathing solution and, often, changes in the volume of aqueous cavities between components of tetrameric or oligomeric molecules. In principle, any of the seemingly ubiquitous "allosteric" transitions of functioning proteins are susceptible to osmotic stress, in the test tube or in the native cell itself.

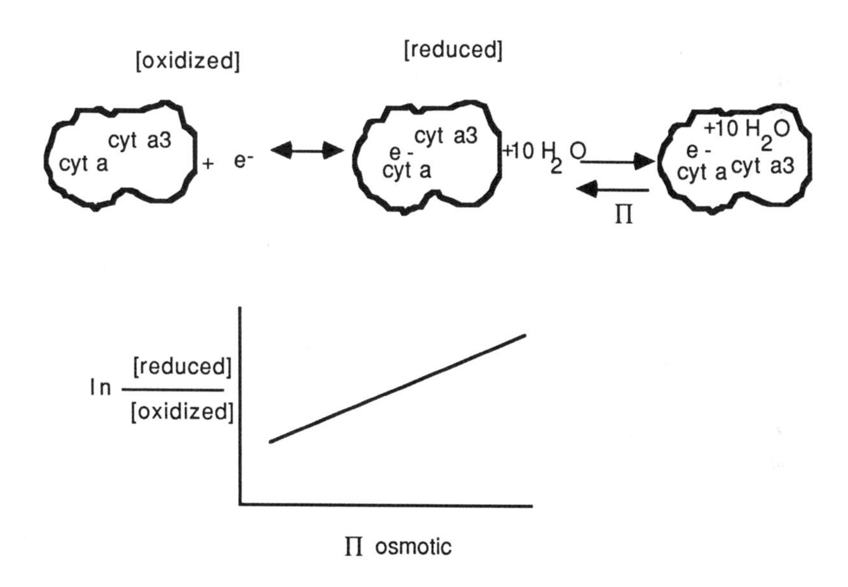

Figure 2. Schematic illustration showing the osmotic stress strategy used to measure the volume of water associated with the process of reduction of cytochrome a in cytochrome oxidase. The reaction kinetics are actually steady state.

The first successful application of osmotic stress to a functioning protein was that of Kornblatt and Hui Bon Hoa (1990). They reduced cytochrome c oxidase in the presence of a variety of different solutes used to adjust Π to find that a balance between the fully oxidized state and the first reduced state was osmotically sensitive. As described above, the quantitative relation between osmotic pressure and the reduced/oxidized ratio showed that a volume equal to some 10 water molecules must become associated with the protein along with the entry of an electron. Furthermore it appears that subsequent intramolecular electron transport requires these ten waters to dissociate from the protein. From the energetic perspective of a charge going from a highly polarizable water to low polarizable protein, it is not surprising that the electron needs water on its inbound journey. But from the perspective of protein mechanism it is newsworthy that the 'apparently futile' water cycle must reflect a swelling and shrinking of the protein as it conducts charges through itself.

Hemoglobin seems to act like a molecular osmometer (Colombo et al 1991). A schematic of the OS method applied to its oxygen binding activity is shown in Figure 3.

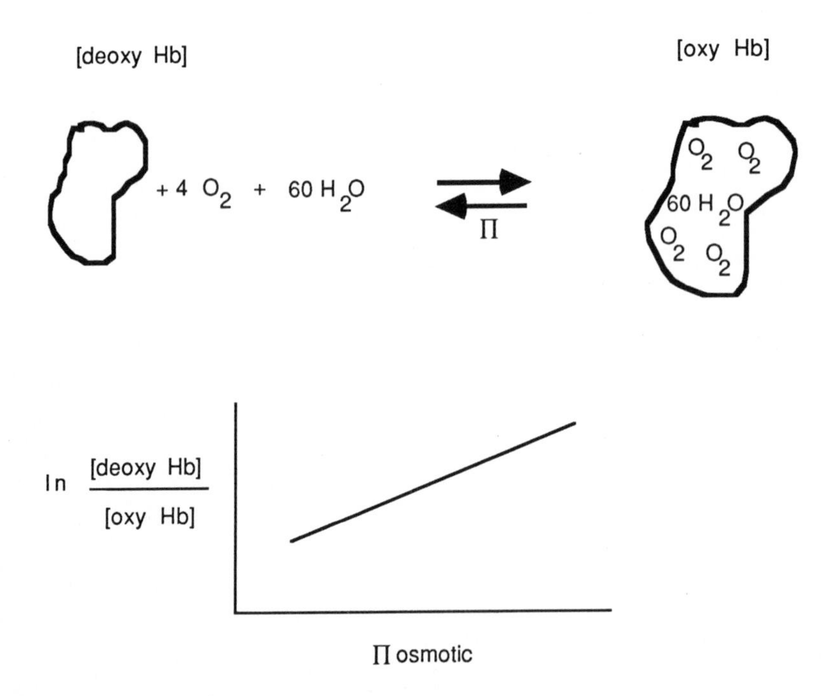

Figure 3. Schematic illustration showing the osmotic stress strategy used to measure the volume of water associated with the process of oxygenation of hemoglobin.

It has long been known from protein structure that there is a 500 to 700 square Angstrom increase in surface exposed to solution when the tetramer switches from the "tense" deoxygenated form, with its tight bridges between polar groups on facing surfaces, to the "relaxed" oxygenated form. What one learns from the application of osmotic stress to hemoglobin is that some 60 solute-excluding water molecules associate with the protein during the switching event.

In the language of equilibrium reactions, we might write

$$T + 60H_2O \Leftarrow\Rightarrow R{\cdot}60H_2O$$

to emphasize the addition of water. Or including the four oxygens even write

$$T + 4\,O_2 + 60H_2O \Leftarrow\Rightarrow R{\cdot}4O_2{\cdot}60H_2O$$

One could think of these waters as ligands whose action effecting the transition is formally (and maybe physically) analogous to ligands such as oxygen, protons, various anions and controller molecules such as diphosphoglycerate (DPG). Since the sensitivity of a macromolecule to the activity of an associating ligand is directly proportional to their required number, the fact of 60 waters compared to four oxygens or one DPG should cause one to re-think the importance of solvation in controlling oxygen uptake. This re-consideration takes us on the one hand to the purely phenomenological, to the fact that lowering the activity of water shifts the oxygen dissociation curve down; fewer oxygens bind at a given oxygen pressure. Hydration can be a major source of energy that changes the oxygen binding affinity of heme groups by 100 or 1000 times when the molecule makes its deoxy <= => oxy switch. Ever since x-ray diffraction showed little difference in the two forms for the structure around the heme binding site, it has been a puzzle where the energy came in to change oxygen binding affinity. The best thinking on this seems to say that the energy is delocalized over a large volume of protein (Hopfield 1973). The large energy of hydrating the protein surface, implicit in the fact of 60 newly bound water molecules, may be the missing ingredient. The attractive pulling forces of intramolecular bridging polar groups in the deoxy form might become repulsive spreading forces of hydration when those same groups are exposed to water. Just how large the energies are might be gauged from the measured work of surface dehydration to which we turn our attention now.

Direct Measurement of Forces between Molecules

Polysaccharides

As has been repeatedly noted, the most striking feature of forces between membranes or molecules approaching contact is their exponential variation with distance, essentially independent of ionic strength.

A recent study of forces between stiff Xanthan polysaccharides (Rau et al 1990) aligned into well-defined arrays showed a 3.3 Angstrom decay distance independent of whether the bathing medium contained 0.1 to 1. 0 Molar NaCl, or 0.05 to 0.1 Molar $MgCl_2$ or Ca rather than Mg ions. Independence of salt concentration and type is all the more striking when one recalls that Xanthan bears charged pyruvate groups. Removal of pyruvate merely weakens the force between helices but has no effect on the spatial decay constant. Another polysaccharide, schizophyllan, showed similar spatial variation but even weaker forces.

For their ability to absorb water, to control viscosity, to maintain the integrity of neighboring cells, polysaccharides are found throughout the plant and animal kingdoms. In most cases these molecules are flexible, and their swelling a combination of polymer entropic, electrostatic, and hydration forces. The difficulty of accounting for so many factors simultaneously has made difficult the identification of each. Now, with the results of forces measured between stiff molecules in various ionic solutions, it is possible to separate the contribution of direct solvation from the entropic and electrostatic.

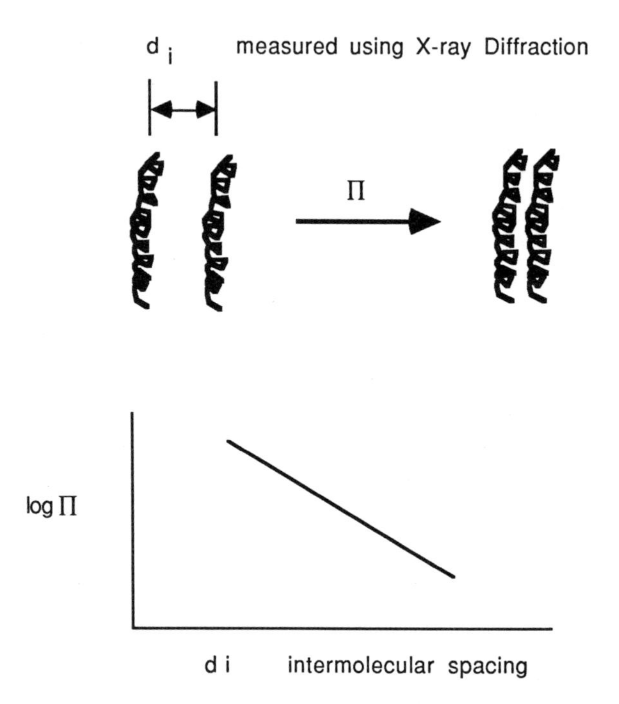

Figure 4. Schematic illustration showing the osmotic stress strategy used to measure the force between organized polymers in solution as it varies with polymer distance.

Double Helical DNA

The molecular interaction most studied to date is the repulsion and attraction of DNA. In simple salts, repulsion is again exponential with 2.8 to 3.3 Angstrom decay rates (Rau et al. 1984). The coefficient of the force seems to vary with the type of cation while electrostatic double layer repulsion is low enough to suggest that the helix is largely neutralized by ion association.

The tight packaging of DNA in cells seems to require the action of condensing species that seem to hold the double helices together. In vitro experiments with several different kinds of condensing agents show the DNA precipitating into well-defined arrays but with equilibrium separations of 5 to 15 Angstroms between molecular surfaces. We have measured the forces encountered in bringing the molecules closer together than these 5 to 15 Angstrom spontaneous separations and have realized that the operative attractive force is also a force of hydration or, rather, also a structural force with its origin in the perturbation of the arrangement of water around the DNA helix.

The measured strength of this attractive hydration force, and its sensitivity to the identity and activity of condensing agent make it a likely candidate for explaining the strong fit of complementary surfaces. This old idea, is brought to new life with a quantitative idea of the actual nature of the attractive forces directly measured (Rau and Parsegian submitted).

Since the force between two molecules is a derivative of the free energy of bringing them together, and since the temperature derivative of a free energy is an entropy, by measuring the temperature dependence of the inter-helical force, it has been possible to extract the Entropic and the Enthalpic components of the force. Remarkably, both of these components too show exponential variation with separation. For the best-studied case, Mn-DNA, the entropy of water release from between two helices is POSITIVE (Leikin et al submitted). That is evidence for the idea that water is to some extent immobilized in the vicinity of the molecular surface relative to its condition in water solution.

Entropically inspired forces have long been recognized as a key factor in macromolecular organization (Kauzmann 1959), usually in terms of "hydrophobic" forces around non-polar surfaces. The DNA results suggest that water entropy is also important in considering the interaction of polar molecules. In fact, the possibilities of specifying and modifying polar surfaces are so great that we believe now that entropic forces involving hydration are a key factor in molecular assembly.

HYDRATION OF FREE SURFACES

From the beginning of their systematic measurement some 15 years ago, no body of force data has been better elaborated than that of interactions between amphiphilic assemblies, particularly between phospholipid bilayer membranes (for a recent review see Rand and Parsegian 1989). Amphiphilic molecules by their nature self-assemble in water into structures that separate polar group/water compartments from hydrophobic compartments. The aqueous spaces swell and shrink under the combined influence of all the forces that stabilize the molecular assemblies as well as those that determine their association. Osmotic stress of those spaces combined with a measure of the structural consequences tells of those forces. Since the subject has been so thoroughly reviewed, we need mention only the general points here.

Direct Measurement of Interbilayer Forces

For every single one of the several dozen systems looked at, bilayer interactions are dominated at close range by exponential hydration repulsion which can be described by the relation given at the beginning of this exposition and shown in Figure 5. Within this universal behavior, a number of specific differences can be measured, reflected in the coefficients and decay lengths, and that have been documented in detail (Rand and Parsegian 1989).

In general, electrostatic repulsion and van der Waals attraction contribute little to the interaction energy when the surfaces are closer than about 20-30 Å. Much of the design and concern of these studies has been to understand the cause of hydration repulsion. Polar group identity, polar group methylation, the physical state of the hydrocarbon chain, chain heterogeneity, mixing of lipid species, all appear to affect hydration repulsion. One general and large influence on the magnitude of hydration repulsion is methylation of the surface. The addition of a single methyl group or of a small fraction of methylated species to the surface molecules causes a disproportionate increase in hydration repulsion and bilayer swelling or separation. The ability of surface molecules to form intermolecular hydrogen bonds appears to diminish considerably their ability to perturb vicinal water and propagate that perturbance to neighbouring bilayers.

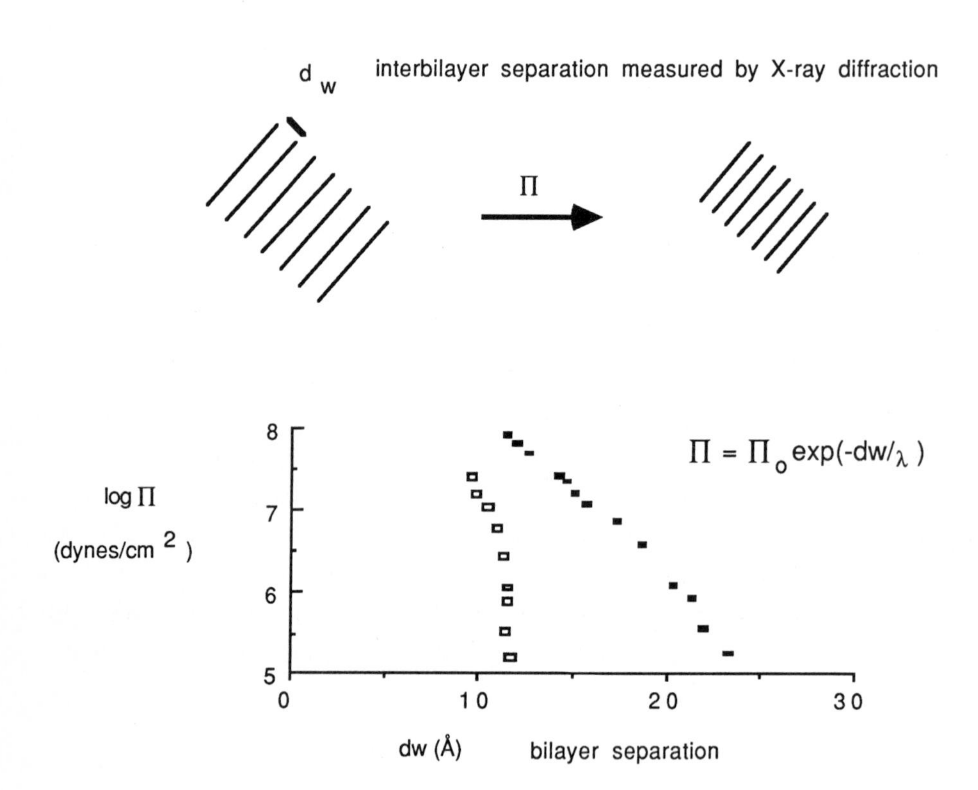

Figure 5. Schematic illustration showing the osmotic stress strategy used to measure the net repulsive force between membranes as it varies with membrane separation.
Closed symbols are stearoyloleoylphosphatidylcholine, open symbols are palmitoyloleoylphosphatidylethanolamine.

Correlation of the strength of hydration forces with surface dipole potentials (Simon and McIntosh 1989) has raised the difficulty of separating the cause and effect of these two, since perturbed water would be expected to contribute to both. As in the studies with DNA one is led to an overall picture in which specific polar group/water interactions determine the coefficient of the hydration forces, attractive and repulsive, which then dominate all interactions within 10-20Å separation.

While hydration forces dominate at short range, it is the equilibrium separations in excess water that really define the 'degree of swelling' in these systems. And here it is the other forces against which the shorter range hydration repulsion is balanced that determines the final state.

Van der Waals attraction or hydration attractive forces are normally balanced by hydration repulsion to determine the equilibrium swelling of electrostatically neutral bilayers. But observed differences among these systems are not easily attributed to one or other of these forces (Rand et al 1988). But van der Waals attraction or hydration attraction is capable of producing collapse to the near dehydrated state only in those instances where hydration repulsion is nearly eliminated, for example as seen in the cases of divalent-cation-induced aggregation of acidic phospholipids (Kachar et al 1986, Coorssen and Rand 1988), reminiscent of the action of condensing agents on DNA.

One force that has been developed to a fuller appreciation in the past few years is that caused by thermal fluctuations of bilayers (Helfrich 1978). When the net forces between bilayers are weak enough to allow bilayer thermal motion the consequent steric forces of mutual collision enhance the separation of neutral bilayers (Safinya et al 1989). And through the interplay between these configurational and electrostatic forces, decay rates of membrane repulsion are substantially modified (Evans and Parsegian (1986). Otherwise low levels of electrostatic repulsion alone result in indefinitely large swelling or separation of charged lipid bilayers and to the formation of closed vesicles.

Hydration of non-bilayer assemblies

The OS method has also been used to probe the energetics of non-bilayer assemblies of phospholipids, particularly the inverted H_{II} phase. In this structure parallel tubes of water, whose axes are arranged on a hexagonal net, are bounded by the polar groups of the phospholipids whose tails fill the intervening space. Such an experiment is shown in Figure 6.

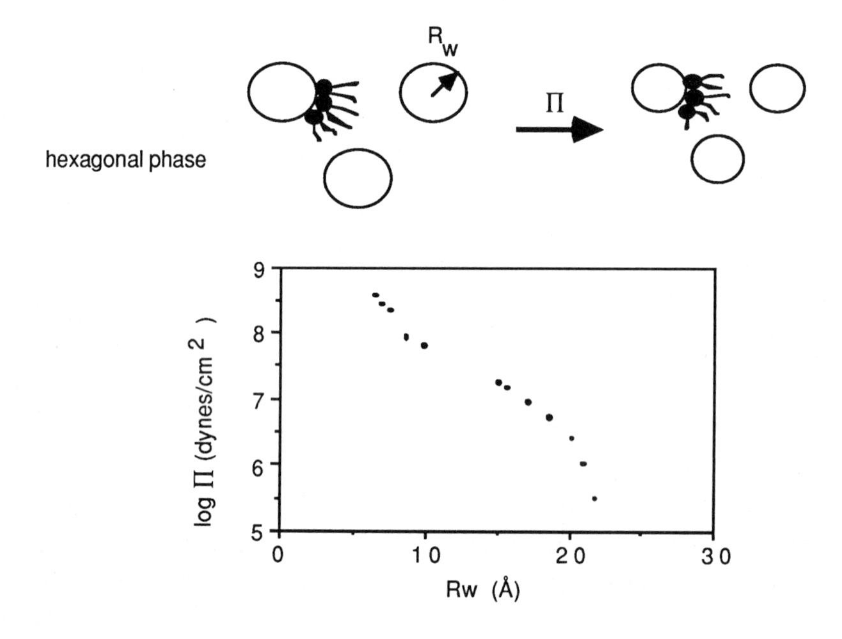

Figure 6. Schematic illustration showing the osmotic stress strategy used determine the relation between structural dimensions of the aqueous spaces and the applied stress.

The monolayer of lipid has been seen as having a spontaneous radius of curvature which it will take on in excess water and alkane which relax hydration and hydrocarbon chain stresses (Kirk and Gruner 1985). This radius then defines the degree of swelling, or water/lipid ratio, and it can be greatly increased by the addition of those lipids with larger or more hydrated polar groups. What is remarkable is that the hexagonal phases of larger spontaneous radii of curvature take up much more water than does the equivalent lamellar phase (Rand et al 1990). This extra drive for hydration, above that required to directly hydrate the polar groups, is understood in terms of the monolayer curvature energy that creates large water cavities required by the large spontaneous radius. Extra hydration in this sense is driven not by the polar groups alone but by the energetics of the whole molecular ensemble. As we point out below, contributions to those energetics are very poorly understood.

Osmotic stress of the H_{II} phase shrinks the aqueous compartment necessarily bending the lipid monolayers encircling the aqueous channels. The osmotic work can be translated into monolayer and bilayer bending moduli (Rand et al 1990) and yields a bilayer value of about 1.5 X 10^{-12} ergs, in good agreement with measurements from bilayer mechanics (Bo and Waugh 1989). One instructive consequence of dehydration of the H_{ii} phase is the structural observation that as the monolayer curls the molecular area at the polar end of the molecule decreases as the hydrocarbon end increases. An intermediate pivotal position, very near the polar group/hydrocarbon interface, undergoes little change in area and serves as a reference position for measuring monolayer curvature. It is seen as a position in the lipid monolayer where lateral interactions of the constituent molecules balance.

Hydration Driven Phase transitions

The seminal work of Luzzati (1968) demonstrated the rich polymorphism, of which the multilamellar and H_{II} phases are only two structures, of amphiphilic systems based on water content. Ever since, water-induced phase transitions in an immensely varied field have been successfully understood and rationalized in terms of water increasing polar group molecular area. So it was a surprise recently to observe an H_{II} - lamellar - H_{II} transition sequence resulting from successive hydration of dioleoylphosphatidylethanolamine (Gawrisch et al 1991). Furthermore, an estimate of the cost of uncurling the monolayer in the formation of bilayers based on the bending modulus described above, far exceeds the osmotic work that actually produced the transition. Ηυμβλινγ λεσσονσ χομε φρομ τηε υνεξπεχτεδ. This transition sequence can successfully be accounted for on simple thermodynamical principles (Gawrisch et al 1991). But what is made clear is that within these structures there are many geometry-dependent free energy contributions which we simply do not yet understand. Again, just as direct hydration is only one contribution to that free energy, the final degree of swelling is dependent on all those other contributions that determine the final conformation of the aqueous space.

SUMMARY

From single molecules going between well-defined states to amorphous polymers filling and lubricating space, the swelling of biological materials is intimately connected with the actual hydration of the macromolecular surface. This theme of hydration as an organizing principle gives us a new logic for thinking about biological processes. It also gives us a way to relate the exquisitely controlled physical and chemical properties of bio-materials with the rougher behavior of inanimate systems.

The fact that biological material appear to control hydration through biochemical modification of interacting surfaces or by the adsorption of specifically designed molecular ligands leads us to recognize that hydration is a dynamic event reflecting the simultaneous exercise of physical, chemical and cell biological processes. Hydration energies are large; the energy differences that occur in controlled transitions can be very small.

The free energetic consequence of water removal on the free energy of transition need not be that great. Recall that the 0.15 ionic Molarity of mammalian blood translates into an osmotic pressure of the order of 5 to 10 atmospheres or 5 to 10 x 10^6 erg/cm^3. One way to comprehend the meaning of this is to say that removal of a 5000 cubic Angstrom volume of solute-inaccessible water will "cost" kT per molecule or RT per mole of protein.

Or consider even the situation where there is a near-zero difference in the energy in the free energy of transition ΔG between states. Say, for example, that without an osmotic term the equilibrium constant between forms A and B equals 1, $\Delta G = 0$. How big a change does a $\Pi\,\Delta V$ term need to be to change the 50-50 distribution between A and B to 60-40?

$$\Pi\,\Delta V = RT\ \ln\,(60/40) = 0.4\ RT$$

or

$$\Delta\ V = 2000 \text{ cubic Angstroms,}$$

still a rather large volume.

The part of the energy of hydration that counts is what occurs in exposing new surface to the solution. For example, take a conservative estimate of 10 erg/cm^2, an order of magnitude smaller than what is typically measured with lipid bilayers. Consider a 10 square Angstrom area occupied by one water of hydration, 100 square Angstroms for 10 water molecules. A hydration energy of 10 $erg/cm^2 \times 10^{-14}\ cm^2 = 10^{-13}$ erg or 4 kT, just for this additional small area.

These energies, then, can translate into large stresses on solvent-exposed surfaces. It has been known for a very long time that hydration forces between surfaces couple with forces of lateral deformation within each surface (Parsegian et al 1979).

For a functioning protein, exposure of new surface to water does not only mean a large energy of solvation but also a significant energy for stressing the protein surface. This change in the balance of forces is likely to provide a key to seeing how "allosterism", hydration, and changes in protein function can be connected through the fact of molecular swelling.

Acknowledgements: RPR acknowledges the financial support of the Natural Sciences and Engineering Research Council of Canada.

REFERENCES

Benz, R., Ludwig, O., de Pinto, V., Palmieri, F. (1985) in Achievements and Perspectives of Mitochondria Research. eds. Quagliariello, E., Slater, E. C., Saccone, C., Palmieri, F., Kroon, A. M. 1:317-327.

Bezrukov, S. M., Vodyanoy, I. (1991) Biophys. J. 59:457a.

Bo, L. and R. E. Waugh (1989) Biophys. J. 55:509-517

Coorssen, J. and R. P. Rand (1988) Studia Biophysica 127:53-60

Cowley, A. C., Fuller, N. L. Rand, R. P., Parsegian, V. A. (1978) Biochemistry 17:3163-3168.

Colombo, M. F., Rau, D. C., Parsegian, V. A. (1991) Biophys. J. 59: 611a.

Gawrisch, K., Parsegian, V. A., Hajduk, D. A., Tate, M. W., Gruner, S. M., Fuller, N. L. and R. P. Rand (1991) Biophys, J. 59:548a.

Gawrisch, K., Zimmerberg, J., Rushton, D., Rand, R. P. and N. L. Fuller (1990) Biophys. J. 57:35a.

Evans E. and Parsegian,V. A. (1986) Proc. Nat. Acad. Sci. 83:7132-7136.

Gruner, S.M. Proc. Natl. Acad. Sci. USA 82,(1985),3665-3669.

Gruner, S. M., Parsegian, V. A. and Rand. R. P. (1986) Discussions Chem. Soc. 81:29-37.

Helfrich,W. (1978) Z.Naturforsch.33a:305-315.

Hopfield, J. J. (1973) J. Mol. Biol. 77:207-222.

Horn,R.G., Israelachvili,J.N., Marra,J., Parsegian,V.A., Rand,R.P. (I988). Biophys. J.

Israelachvili,J.N., Adams,G.E. (1978) J. Chem. Society Faraday Trans 1, 74:975-1001.

Kachar, B., Fuller, N. L. and R. P. Rand (1986) Biophys. J. 50:779-788.

Kasianowicz, J. J., Moore, C. R., Zimmerberg, J., Pasternak, C., Bashford, C. L., Parsegian, V. A. (1991) Biophys. J. 59:458a.

Kauzmann,W. (1959) Adv. Protein Chemistry 14:1-63.

Kirk, G. L. and S. M. Gruner (1985) J. Phys. (Les Ulis, Fr.) 46:761-769.

Kornblatt, J. A., Hoa, G. H. B. (1990) Biochemistry 29:9370-9376.

Leikin, S., Rau, D. C., Parsegian, V. A. (submitted Phys. rev A)

LeNeveu, D. M, Rand, R. P, Parsegian, V. A, Gingell, D. (1977) Biophys. J. 18:209-230.

LeNeveu, D. M, Rand, R. P, Parsegian, V. A. (1976) Nature 259:601-603.

Lis, L. J., McAlister, M., Fuller, N. L., Rand, R. P., Parsegian, V. A., (1982). Biophys.J. 37:657-666.

Loosley-Millman, M. E., Rand, R. P., Parsegian, V. A. (1982). Biophys. J. 40:221-232.

Luzzati, V. (1968) in Biological Membranes (Chapman, D. Ed), pp.71-123, Academic Press, New York.

Manella, C. A., Radermacher, M., Frank, F. (1984) Proc. 42nd Electron Micros. Soc. Am. p 664.
McIntosh, T. J. and Simon, S. A. (1986) Biochemistry 25:4058-4066.
McIntosh, T. J. and Simon, S. A. (1986) Biochemistry 25:4948-4952.
McIntosh,T.J., Magid, A.D., Simon,S.A. (1988) . Biochemistry 26: 7325-7332.
Parsegian, V. A., Fuller, N. L., Rand, R. P. (1979). Proc. Nat. Acad. Sci. USA 76:2750-2754.
Parsegian, V.A., Rand, R.P., Fuller, N.L. and Rau, D.C. (1986) in Methods in Enzymology. Vol. 127. Biomembranes; Protons and Water: Structure and Translocation (Packer, L., ed.), pp. 400-416, Academic Press, New York
Pashley, R.M.,McGuiggan, P.M.,Ninham, B.W., Brady, J. and Evans, D.F. (1986) J. Phys. Chem. 90:1637-1642.
Podgornik,R., Rau,D.C., Parsegian,V.A. (1989) Macromolecules 22:1780-1786.
Prouty, M. S., Schecter, A. N., Parsegian, V. A. (1985) J. Mol. Biol. 184:517.
Rand, R. P., Fuller, N. L., Lis, L. J. (1979) Nature 279:258-260.
Rand, R. P. N.L. Fuller, S.M. Gruner, V.A. Parsegian (1990) Biochemistry 29:76-87
Rand, R. P. and V. A. Parsegian (1989) Biochim. Biophys. Acta 988:351-376.
Rand, R. P. and V. A. Parsegian (1986) Ann. Rev. Physiology 48:201-212.
Rand, R.P. (1981). Ann. Rev. Biophys. Bioeng. 10:277-314.
Rand, R.P., Fuller, N.L., Parsegian, V.A., Rau, D.C. (1988) Biochemistry 27:7711-7722.
Rau, D. C., Lee, B. K., Parsegian, V. A. (1984) Proc. Nat Acad. Sci. USA. 81:2621-2625.
Rau,D.C. and Parsegian,V.A. (l987) Biophys. J. 51:503
Rau, D. C., Parsegian, V. A. (1990) Science 249:1278-1281.
Safinya, C. R., Smith, E. B., Roux, D., Smith, G. S. (1989) Phys. Rev. Letts. 62:1134-1137.
Simon, S.A. and T. J. McIntosh (1989) Proc. Natl. Acad. Sci. USA 86:9263-9267.
Simon, S.A., McIntosh, J., Magid, A.D. (in press) J. Colloid & Interface Science.
Starkus, J. G., Rayner, M. D., Ruben, P. C., Alicata, D. A. (1991) Biophys. J. 59:71a.
Zimmerberg, J., Benzanilla F., Parsegian, V. A. (1990) Biophys. J. 57:1049-1064.
Zimmerberg, J., Parsegian, V. A. (1986) Nature 323:36-39.

ATTRACTIVE DOUBLE LAYER INTERACTION: FROM CALCIUM CLAYS TO BIOLOGICAL MEMBRANES

S. Marčelja
Department of Applied Mathematics
Research School of Physical Sciences
The Australian National University
GPO Box 4, Canberra ACT 2601, Australia

Introduction

In most practical cases, the electrical double layer interaction is repulsive. In the popular theory of the electrical double layer, based on the Poisson-Boltzmann (PB) equation, the calculated interaction is always repulsive. Yet in several very important naturally occurring systems, the double layer contribution to the interaction between the surfaces or interaction between particles is attractive.

The attractive interaction originates from the correlation in the positions of ions between the charged surfaces. It can be thought of as the slow (zero frequency) component of the Van der Waals interaction. Although the concept is not new (Mahanty and Ninham 1976), it was never thoroughly investigated because the magnitude of the interaction is difficult to calculate.

Within the past five years, accurate results were obtained by Monte Carlo simulation (Guldbrand *et al.* 1984, Bratko *et al.* 1986) and by hypernetted chain (HNC) integral equations of the liquid state theory applied to the non-uniform ionic fluid (Kjellander and Marčelja 1984-1988, Kjellander *et al.* 1988, 1990). The calculations used the so called primitive model of the electrolytes, where ions are approximated as charged hard spheres and the solvent as a dielectric continuum. The resulting interaction contains two contributions neglected in the PB theory: The ion correlation attraction and the hard core repulsion. The overall interaction is either attractive or repulsive, with the attractive regime found at small surface separations when counterions have a valency of 2 or higher.

The HNC theoretical scheme and different accuracy checks have been described in detail in our earlier publications. In brief, several detailed comparisons with Monte Carlo simulations (Bratko *et al.* 1986, Kjellander and Marčelja 1984,1986, Valleau *et al.* 1991) of the same systems show that the HNC approximation is remarkably accurate in describing the behaviour of primitive model electrolytes, even in the vicinity of interfaces.

Unfortunately, the calculation of the interaction between charged surfaces in primitive model electrolytes only partially solves the problem of

NATO ASI Series, Vol. H 64
Mechanics of Swelling
Edited by T. K. Karalis

the interaction of surfaces in real aqueous electrolytes. There are other non-electrostatic contributions to the interaction, notably the oscillatory contribution originating from the size of the solvent molecules and the repulsive contribution originating from the hydration of ions at the surfaces. From the experimental experience with the surface force apparatus, those contributions seem to be approximately additive to the double layer force.

In this brief report we shall consider the significance of the attractive double layer interaction in two important examples: the swelling of clays and the fusion of biological membranes. Some of the results for clays have been previously published, while those for membranes are new or even preliminary (Figs. 1 and 8). A review of most results until 1988 is available in Marčelja (1990).

Double layer attraction and the swelling of clays

Consider the double layer interaction of two highly charged surfaces in various mixed electrolytes (Fig. 1). At 60Å^2/unit charge this example corresponds to natural vermiculite clays. In 0.1M 1:1 electrolytes the HNC and the PB theory give very similar results. This would also be true for more dilute 1:1 electrolytes. The small differences between the HNC and PB pressures follow from the attractive electrostatic correlation term in the former (hence the repulsion is weaker than in the PB case) and from the repulsive core interaction which is effective only at smaller separations.

As 1mM of 2:1 electrolyte is added, divalent counterions preferentially enter the double layer. The ion correlation interaction rapidly increases, and in this example at 90Å^2/unit charge balances the repulsion at the separation of 4 Å.

With 10mM of 2:1 electrolyte, the ion correlation attraction dominates at small separations, where the double layer interaction has turned attractive. By this stage, most of the ions in the double layer are divalent counterions and in the separation range shown in the figure the interaction is very similar to that obtained in pure 2:1 electrolytes or even the system which contains only divalent counterions and no added electrolyte.

Like all theoretical results presented in this report, the data were obtained by applying the HNC approximation to the nonuniform electrolyte between the surfaces.

The effect of varying electrolyte concentration on the double layer attraction is illustrated in Fig. 2. With the increasing concentration the attractive region becomes deeper and extends to closer separations, but there are no major qualitative changes.

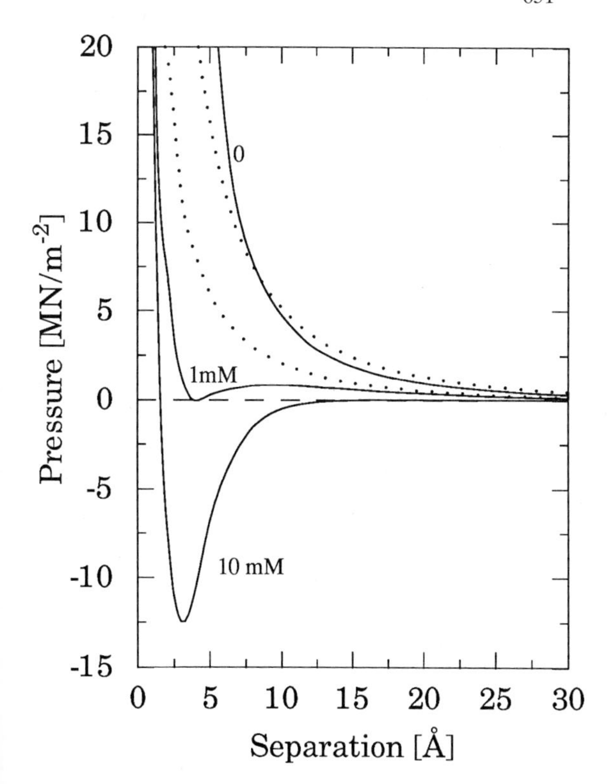

Fig.1. Calculated double layer interaction between two charged planar surfaces in mixed 1:1 and 2:1 electrolytes. The full lines give HNC results for a 0.1M 1:1 electrolyte with 2:1 electrolyte added as marked in the figure. The ion radius is 2.125 Å and the surface charge density is one elementary charge per 60 $Å^2$ for the cases of 0 and 10 mM and one elementary charge per 90 $Å^2$ for 1 mM. The dotted lines are the PB results corresponding to the cases of 0 and 10mM.
In all figures with the exception of Fig. 5 a convention is adopted where separation is measured between the centers of the ions touching the surface. The actual surfaces are one ion diameter further apart. An example of a feasible system in our HNC studies is shown in Fig. 6.
In Figs. 1-5 and 7 the surface charge is assumed to be continuous and there are no electrostatic images (i.e. the dielectric constant is taken as 78.36 for both the solution and the surfaces).

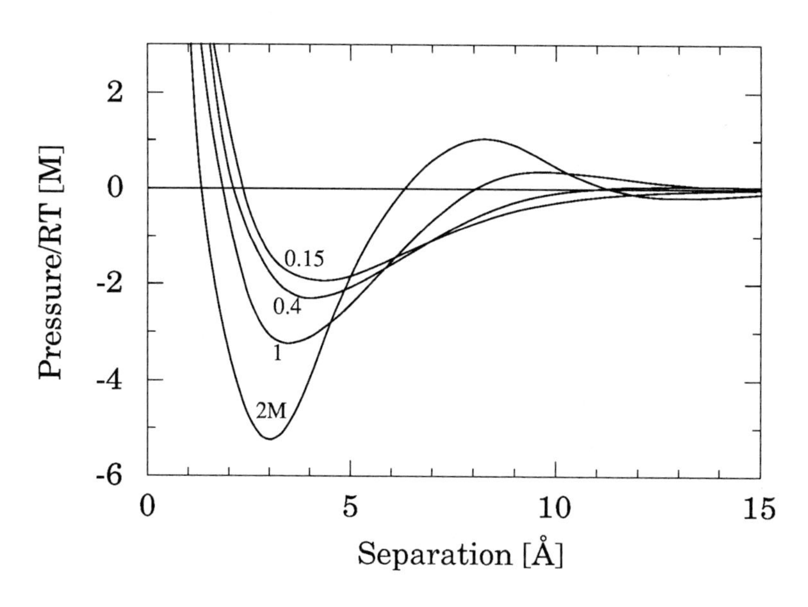

Fig. 2. Calculated double layer interaction between two charged planar surfaces with a surface charge density of one elementary charge per 60 $Å^2$ across a 2:1 electrolyte solution. The bulk concentrations of the electrolyte are marked in the figure. Data from Kjellander *et al.* (1990).

The calculations like those shown in Fig. 2 have been performed for many different values of surface charge and separation. The system studied in most detail consists of charged planar surfaces with only counterions present in the solution. The results can be summarized in a contour plot of Fig. 3. It is seen that the region of attractive double layer interaction is obtained for large surface charge and small separations. Corresponding plots for 2:1 or 2:2 electrolyte solutions are qualitatively very similar.

The comparison of the above calculations with the experiment is encouraging but far from decisive. The attractive part of the double layer interaction cannot be easily studied in the surface force apparatus, because the surfaces jump over the attractive region until they encounter repulsion or come into contact. Just in the last month, modifications of the apparatus, due to John Parker, have overcome this limitations with a feedback control of the surface movement. Until new results become available later in the year, our only indication of the attractive region is a jump of the surfaces in the apparatus.

As an example, Fig. 4 shows a measurement (Kjellander *et al.* 1990) of the force between two molecularly smooth mica surfaces in 0.15M $CaCl_2$. A jump between about 1.8 and 0.6 nm indicates attractive region. A comparison with the calculated electrostatic interaction shows that this range approximately corresponds to the region of attraction predicted from the double layer theory (Fig. 3). It also shows that the comparison between the theory and the experiment is still rather poor, with the hydration force and the oscillations due to water structure outside the reach of the primitive model calculations.

Over the past few years we have studied the implications of the attractive double layer interaction for the swelling behaviour of clays (Kjellander *et al.* 1988, 1990) . The obvious expectation is that Ca clays should not swell because the double layer interaction is attractive, while Na clays should swell. It is indeed well known that calcium clays do not swell, a fact which could not be explained on the basis of the older theories.

The actual situation with Na (or other monovalent counterion) clays is however more subtle. The double layer repulsion (which in a more accurate HNC calculation is weaker than predicted by the PB theory) is similar in magnitude to the Van der Waals attraction. The theory therefore predicts a very small interaction and can no longer be considered reliable. For example, Na montmorillonite swells for electrolyte concentrations lower than 0.3M.

For calcium montmorillonites and vermiculites, we find that the attractive minimum obtained by the double layer calculations with the addition of conventional Van der Waals force is very close to the observed surface separations (Fig.5). The agreement is not perfect - the measured separations must coincide with the minima of the oscillatory structure imposed by the molecular size of the solvent.

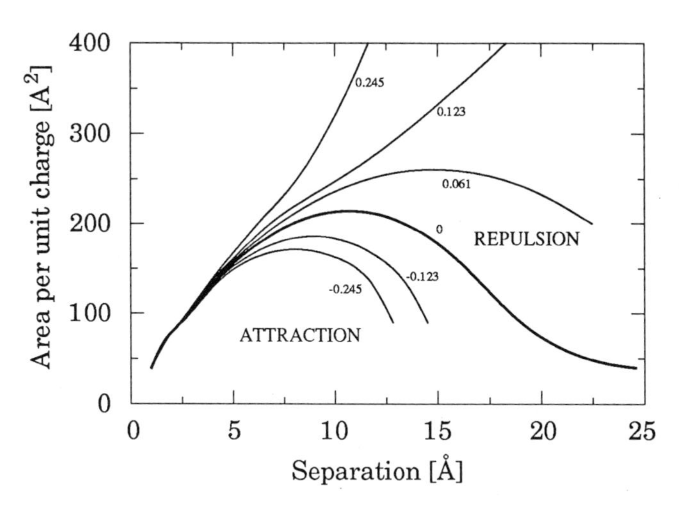

Fig. 3. Boundary between the attraction and repulsion in electrical double layers containing only divalent counterions. The pressure is equal to zero on the heavy line; other contours are marked with the corresponding value of the pressure in MNm^{-2}. The counterion radius was chosen as 2.125 Å. For very high values of surface charge, counterion radius affects the results and below some 60 Å^2 per unit charge the results are no longer universal.

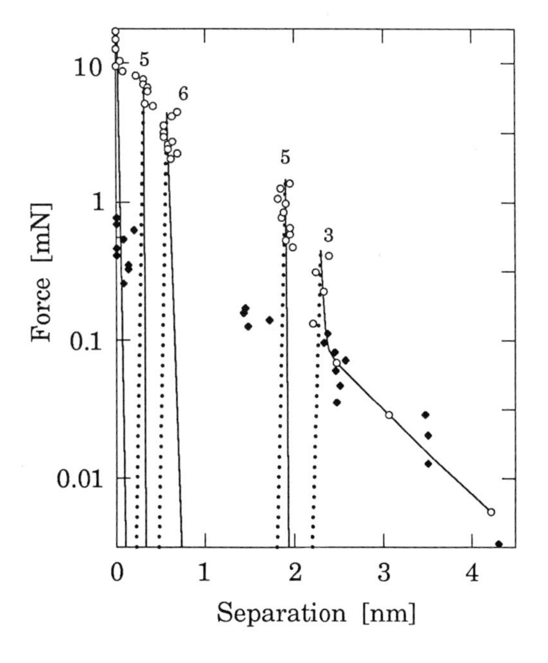

Fig. 4. Forces measured between mica surfaces in 0.15M $CaCl_2$ at *p*H values of 5.7 (O) and 4.9 (♦). At *p*H 5.7, the oscillatory force was first seen at about 2.2 nm but a gap in the structure was observed between about 1.8 nm and 0.7 nm. The gap corresponds to the expected attractive regime of the interaction, which cannot be measured in the surface force apparatus. The numbers shown at the maxima are estimated pressures between the surfaces in units of MNm^{-2}. From Kjellander *et al.* (1990).

A great deal of work must be done before we fully understand the swelling of clays. The practically most important case of swelling in mixed electrolytes (Quirk 1986) is only beginning to be studied. We also find that the size of the hydrated ion is more important than it was originally assumed, and it may control the entry of electrolyte into the narrow slit between the clay platelets. Finally, the calculations including the full solvent structure are an order of magnitude more complex, but some encouraging initial steps have already been taken by several groups.

Adhesion and fusion of biological membranes

Another situation where the double layer force may be very important is the interaction of biological membranes. For example, in neuromuscular junctions the fusion of vesicles containing neurotransmiter with the cell membrane is controlled by the release of small amounts of calcium. In model systems, it is well known that bilayer vesicles containing around 10% of negatively charged lipids (similar to the fraction of charged lipids in biological membranes) readily fuse in the presence of small (for example 1mM) amounts of calcium.

Membrane fusion is a two-step process, first involving membrane adhesion and then fusion. Mg is just effective as Ca in inducing membrane adhesion, but it does not promote membrane fusion.

Could the electrostatic interaction be a controlling factor in membrane adhesion? The surface charge of biological membranes is much smaller than was the case with clays and the double layer interaction is repulsive even in the presence of divalent counterions (Fig. 5). But it is experimentally very well established that Ca or Mg ions adsorb to model bilayer membranes. In specialised biological membranes, Ca receptors further increase the binding affinity by several orders of magnitude.

The adhesion is most easily studied in model membranes, where the strength of the adsorption has been measured by fluorescence, electrophoresis, and other techniques. Let us consider the interaction of such model membranes in the presence of divalent counterions. In the first example we show the interaction of weakly charged surfaces without the added electrolyte, where the counterions lower their energy by 4 kT when adsorbed to the surface. The geometry of the model system is shown in Fig. 6, and the resulting pressure in Fig. 7. The example also illustrates the effect of larger ion radius (in this case 3Å), discrete surface charges and the electrostatic images.

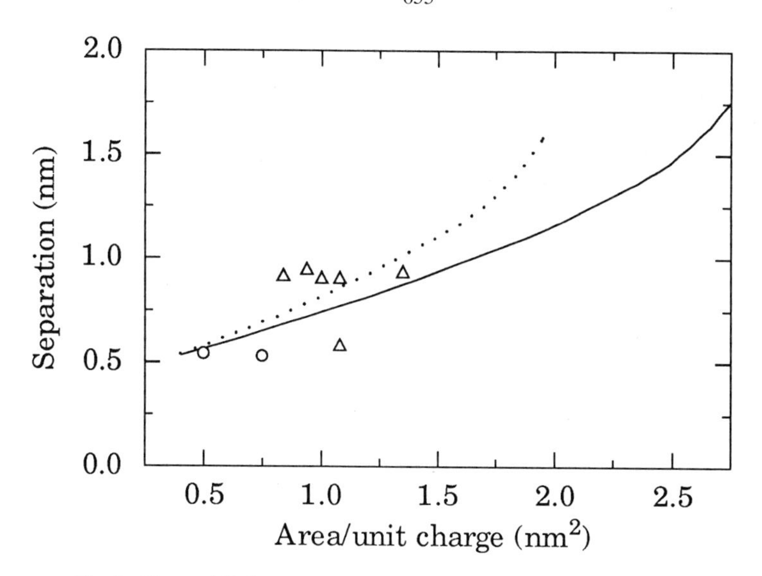

Fig. 5. The equilibrium plate separation as a function of the area per unit surface charge for calcium clays. Full line: the separation calculated for the primitive model electrolyte with the addition of the conventional Van der Waals interaction; Dotted line: same, but without the Van der Waals interaction. The experimental separations for various Ca-vermiculites (O) and Ca-smectites (Δ) were obtained from the X-ray diffraction data assuming the plate thickness of 0.96 nm. The separation in this figure is measured between the actual clay surfaces.
The experimental separations take discrete values due to solvent size effects. The solvent size and the hydration forces are not included in the theoretical model. After Kjellander *et al.* (1988).

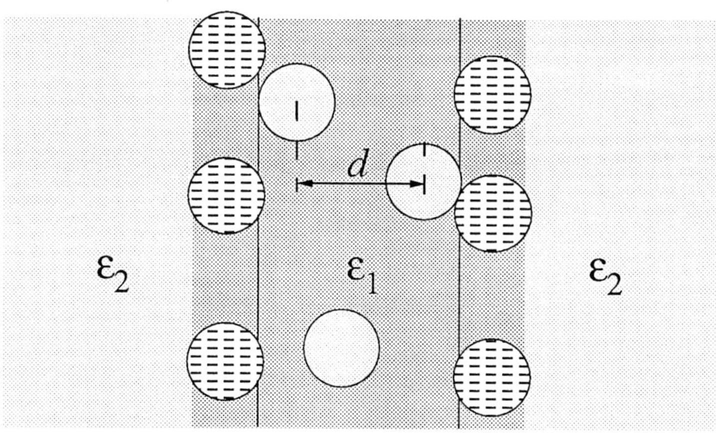

Fig. 6. Geometry of the model double layer system whose interaction is shown in Fig. 7. The ion radius for both discrete monovalent surface charges and divalent counterions is 3Å. Surface charge groups can move in the plane of the surface and their centers are 2.5Å inside the surface. The dielectric constants are ε_1 = 78.358 and ε_2 = 3. Higher dielectric constant region extends 5Å beyond the surfaces. The convention for measuring the separation d is shown in the figure. Ions can move within the distance d between the surfaces; when they adsorb to the surface their energy is lowered by 4 kT.

In the absence of adsorption and at larger surface separations the interaction is much weaker than the corresponding PB result. More results on this weaker repulsion were presented by Kjellander *et al.* (1990).

With typical rather moderate adsorption, the double layer repulsion is turned into an electrostatic attraction. The attractive region would not be expected in the older theories or without counterion adsorption. The attraction is not very strong, its order of magnitude being comparable to the Van der Waals force. Of course, in a broader sense the correlation attraction is a part of the Van der Waals force (cf. the Introduction), and hence this result is not surprising.

The effect of ion size is seen as the very steep rise in the calculated pressure at the separation of 3.5Å. At that separation, the counterions adsorbed on one of the surfaces come into contact with discrete surface charge groups on the other surface and the core repulsion consequently rapidly increases. This equilibrium position for the electrostatic interaction is thus reached at a surface separation where the adsorbed layers are already in contact. Other than core size effects, discreteness of surface charge has very little effect on the overall interaction (Langner *et al.* 1990). The effect of electrostatic images is not very strong. At small surface separations they add a weak repulsive term to the overall pressure.

In the last figure we keep a typical physiological concentration of 0.1M 1:1 electrolyte in the background and examine the effect of adding small amounts of divalent counterions. The surface charge density is taken as uniform, with the value of one charge per 2.5 nm^2, and the electrostatic image potential is not included. The adsorption of Ca to model membranes formed from phosphatidylcholine (PC) and phosphatidylserine (PS) was studied by McLaughlin *et al.* (1981). We find that we can reproduce the surface charge neutralisation measured in their electrophoresis experiments on vesicles formed from PC and PS in a 5:1 ratio by assuming that the adsorption energy for the divalent counterion is 6 kT.

It can be seen from Fig. 8. that at 250$Å^2$/unit charge the addition of 1mM of a divalent ion reverses the electrostatic interaction between surfaces from repulsion to attraction. The electrostatic attraction is similar in magnitude to the Van der Waals force, and the total attractive force can certainly lead to membrane adhesion. With specialised biological membranes concentrations of Ca as small as several μM will induce fusion, because even at such low concentrations Ca will rapidly bind to its receptors in the membrane.

As we have not considered the discrete solvent or the hydration force the problem of membrane adhesion is certainly more complicated than our model can admit. Nevertheless, as a result of this investigation we do know that in the presence of divalent counterions the attractive double layer interaction favours contact between weakly charged adsorbing surfaces.

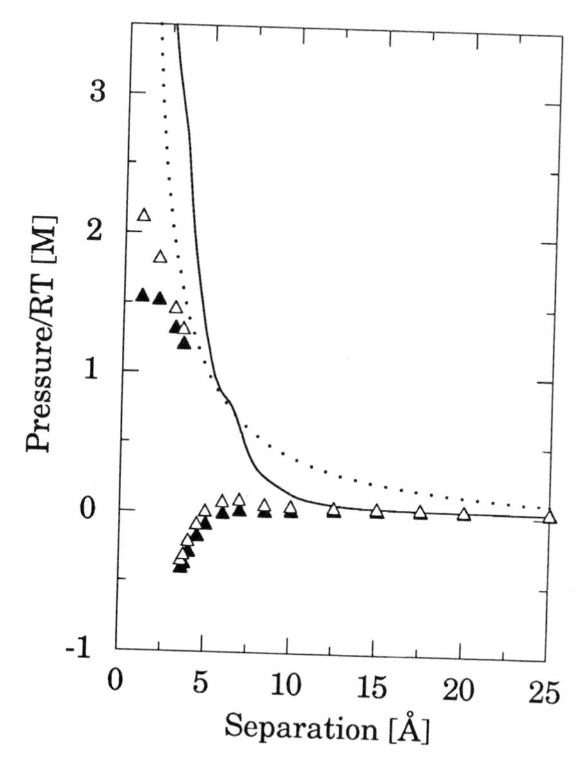

Fig. 7. Double layer interaction between two weakly charged planar surfaces in the presence of counterion adsorption. The area per unit surface charge is 2.5 nm^2, and the divalent counterion can adsorb to the surface with the adsorption energy of 4 kT. In this example the surface charges are discrete with a radius of 3Å and placed as shown in the previous figure. There is no electrolyte and the counterion radius is 3Å. (Similar results are obtained with dilute 2:1 electrolytes).

(–) HNC pressure without adsorption;
(···) PB pressure without adsorption;
(Δ) HNC pressure in the presence of adsorption;
(Δ) same, but with electrostatic images.

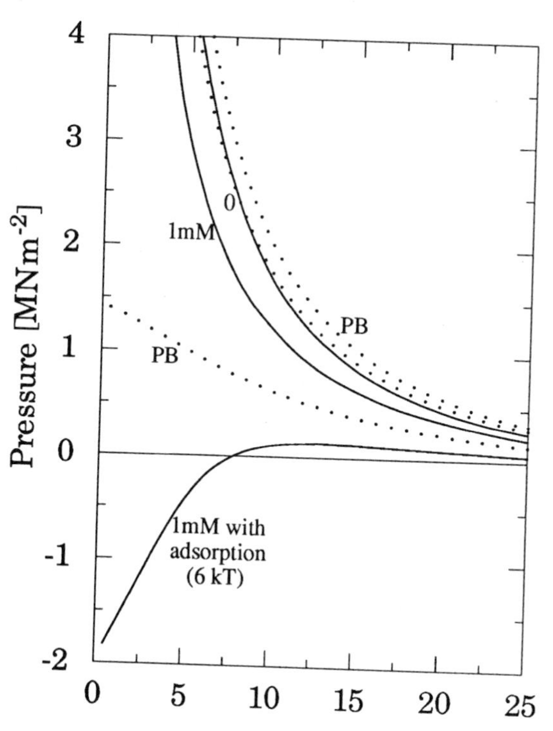

Fig. 8. Interaction of weakly charged adsorbing surfaces in mixed electrolytes as a model for bilayer adhesion. In the background we have 0.1M 1:1 electrolyte which does not adsorb to the surfaces (top line). The surfaces carry one charge per 2.5 nm^2. Upon of addition of 1mM 2:1 electrolyte, the repulsion is decreased. However, if divalent counterions adsorb to the surfaces the interaction turns attractive. The value of 6 kT for the adsorption energy was obtained from a comparison of the adsorbed charge density with electrophoresis data for PC/PG membranes (McLaughlin *et al.* 1981). The three dotted curves are the respective PB results.

Acknowledgment: I am indebted to Roland Kjellander, Richard Pashley and Jim Quirk for many developments and discussions during our studies of attractive double layer interaction.

References

Bratko D, Jönsson B, Wennerström H (1986) Electrical double layer interactions with image charges. Chem. Phys. Lett. 128: 449-454

Guldbrand L, Jönsson B, Wennerström H, Linse P (1984) Electrical double layer forces. A Monte Carlo study. J. Chem. Phys. 80: 2221-2228

Kjellander R, Marčelja S (1984) Correlation and image charge effects in electric double layers. Chem. Phys. Letters 112:49-53 ; 114:124(E)(1985)

Kjellander R, Marčelja S (1985) Inhomogenous Coulomb fluids with image interactions between planar surfaces. J. Chem. Phys. 82: 2122-2135

Kjellander R, Marčelja S (1986) Interaction of charged surfaces in electrolyte solutions. Chem. Phys. Letters 127:402-407

Kjellander R, Marčelja S (1988) Surface interactions in simple electrolytes. J. Physique 49:1009-1015

Kjellander R, Marčelja S, Quirk JP (1988) Attractive double-layer interaction between calcium clay particles. J. Colloid Interface Sci. 126: 194-211

Kjellander R, Marčelja S, Pashley RM, Quirk JP (1990) A theoretical and experimental study of forces between charged mica surfaces in aqueous CaCl2 solutions. J. Chem. Phys., 92: 4399-4407

Langner M, Cafiso D, Marčelja S, McLaughlin S (1990) Electrostatics of phosphoinositide bilayer membranes. Biophys. J. 57: 335-349

Mahanty J, Ninham BW (1976) Dispersion Forces. Academic, New York

Marčelja S (1990) Interactions between Interfaces in Liquids. In: Charvolin J, Joanny JF, Zinn-Justin J (eds) Liquids at Interfaces (Les Houches, Session XLVIII, 1988), North-Holland, Amsterdam, p 100

McLaughlin S, Mulrine N, Gresalfi T, Vaio G, McLaughlin A (1981) Adsorption of divalent cations to bilayer membranes containing phosphatidylserine. J. Gen. Physiol. 77: 445-473

Quirk JP (1986) Soil permeability in relation to sodicity and salinity. Phil. Trans. R. Soc. Lond. A316: 297-317

Valleau JP, Ivkov R, Torrie GM (1991) Colloid Stability: The forces between charges surfaces in an electrolyte. J. Phys. Chem. (in press)

HYDRATION AND STERIC INTERACTIONS BETWEEN LIPID MEMBRANES

Sidney A. Simon[1], Thomas J. McIntosh[2], and Alan D. Magid[2]
Departments of Neurobiology[1] and Cell Biology[2]
Duke University Medical Center
Durham, North Carolina 27710

INTRODUCTION

The interactions between molecules in solution are of fundamental importance in surface chemistry, physics of the liquid state, and biology. Knowledge of the range and magnitude of these interactions is important in such diverse fields as the stabilization of colloids, enzyme kinetics, lubrication, and the interaction between macromolecules such as polymers and carbohydrates (Israelachvili, 1985). Life processes depend on the ability of biological molecules to specifically associate with the same or different compounds so that cellular processes can be carefully regulated. To exhibit the specificity or selectivity that occurs in chemical and biological interactions, molecules must approach to within a few Ångstroms of each other. At these distances, there are several attractive and repulsive interactions present. The two principle repulsive short-range pressures between polar surfaces in solution are the *hydration* and *steric* pressures. Here we present our work on the these two short-range repulsive pressures acting between lipid membrane surfaces. The hydration and steric pressures are critical to many properties of both cell and model membranes. For example, these pressures are important in the numerous biological processes where membranes associate, such as cell-cell recognition, synaptic transmission, and protein secretion, and they also determine to a large extent the hydration, aggregation, and fusogenic properties of lipid bilayers.

The hydration pressure (or more generally, the solvation pressure) forms a major barrier to the close approach of two hydrophilic surfaces. For a variety of lipid bilayer membranes it has been found that the hydration pressure, P_h, decays exponentially with increasing fluid separation between bilayers, d_f, such that $P_h = P_o \cdot \exp(-d_f/\lambda)$ (Parsegian et al., 1979; McIntosh & Simon, 1986; Rand & Parsegian, 1989). A goal of our work is to determine those properties of the solvent and the membrane surface that are responsible for the

NATO ASI Series, Vol. H 64
Mechanics of Swelling
Edited by T. K. Karalis

magnitude (P_o) and decay length (λ) of the hydration pressure.

Considerably less experimental information is available concerning steric pressures between membrane surfaces, even though steric repulsion is important in many cell-cell and receptor-ligand interactions. Here we present measurements of the steric pressure between phosphatidylcholine and phosphatidylcholine:cholesterol bilayers.

METHODS

Pressure versus distance relations were obtained by applying known osmotic pressures to lipid multilayers and measuring the distance between adjacent bilayers at each applied pressure by analysis of X-ray diffraction data, as described by McIntosh and Simon (1986). Osmotic pressures in the range of 1.1×10^5 to 3.2×10^7 dyn/cm^2 were applied to unoriented multiwalled liposomes by incubation with large neutral polymers such as polyvinylpyrrolidone (PVP) or dextran (Parsegian et al., 1979). Pressures in the range of 2.8×10^7 to 2.3×10^9 dyn/cm^2 were applied to oriented lipid multilayers through the vapor phase by established procedures (Parsegian et al., 1979; McIntosh et al., 1987). Measurements of the dipole potential (V) were made for monolayers with 0.01 M KCl as a subphase as described in detail in Simon et al., (1988). The values of dipole potential represent the differences in the potential of the subphase surface in the presence and absence of the monolayer.

RESULTS AND DISCUSSION

The applied osmotic pressure, P, versus lamellar repeat period relationship for egg phosphatidylcholine (EPC) multilayers in the liquid-crystalline phase is shown in Figure 1. For repeat periods from 60 Å to 52 Å the data points can be fit quite closely with a straight line. However, there is a sharp upward break in the plot at repeat period values of about 51 Å and at an applied pressure of $\ln P \approx 17$.

The lamellar repeat period represents the total width of each unit cell, and includes the sum of bilayer thickness and the fluid space between adjacent bilayers. To determine the width of the bilayer and the fluid space, we calculated electron density profiles from the x-ray diffraction patterns obtained at each value of applied pressure.

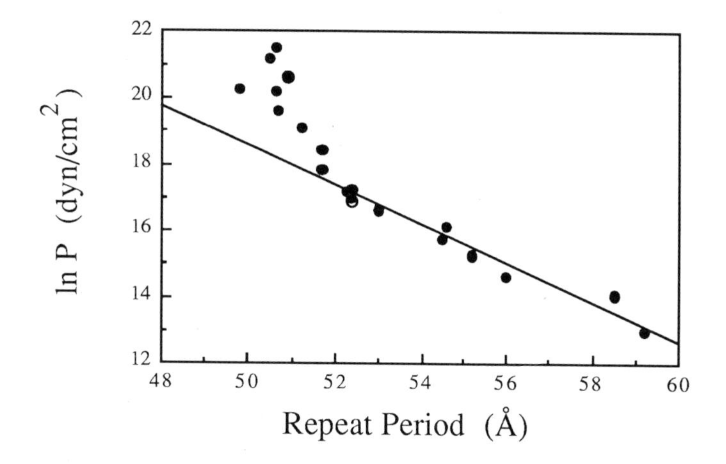

Figure 1. Natural logarithm of applied pressure plotted versus lamellar repeat period for EPC bilayers. The straight line is a least squares fit to the data for $12.5 < \ln P < 17.5$.

Figure 2 shows typical electron density profiles for EPC multilayers which were subjected to osmotic pressures by incubation in three concentrations of the large neutral polymer polyvinylpyrrolidone (PVP). For each profile, the geometric center of one bilayer is at the origin and the high density peak at 18 Å corresponds to the lipid head group of that bilayer. The high density peaks positioned at about 32 Å for 60% PVP, at about 38 Å for 20% PVP, and at about 44 Å for 0% PVP correspond to the head group peaks from the apposing bilayer. The distance between head group peaks from adjacent bilayers decreases with increasing PVP concentration. Thus water is removed from between adjacent bilayers with increasing osmotic pressure.

We operationally define the bilayer width as the total thickness of the bilayer assuming that the head group conformation is the same as it is in single crystals of dimyristoylphosphatidylcholine, that is parallel to the bilayer surface (Pearson and Pascher, 1979). Therefore, the edge of the bilayer lies about 5 Å outward from the center of the high density peaks in the electron density profiles (McIntosh & Simon, 1986; McIntosh et al., 1987; McIntosh et al., 1989a&b).

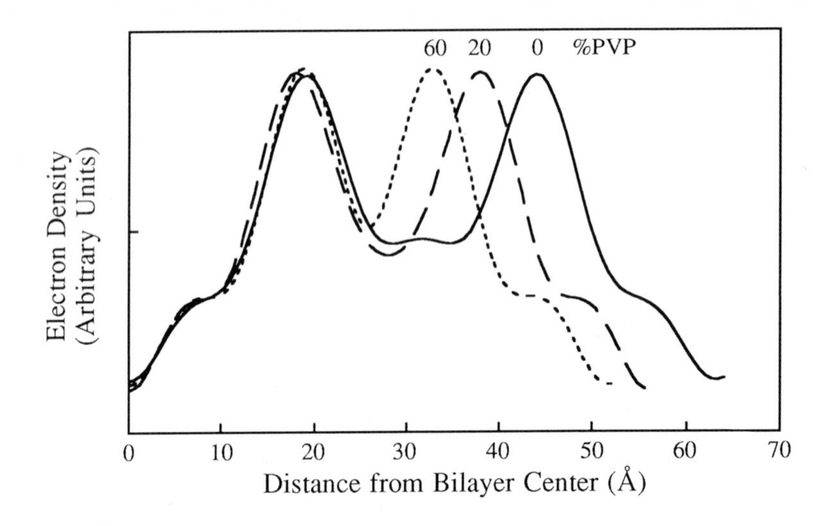

Figure 2. Electron density profiles of EPC bilayers in excess water (solid line), 20% PVP (dashed line), and 60% PVP (dotted line).

Using electron density profiles and the above definition for the water/bilayer interface, we obtained the applied pressure-bilayer separation relationship for EPC bilayers shown in Figure 3. Again, there are two distinct regions in the pressure-bilayer separation relationship. That is, for pressures such that $\ln P < 17.5$, or bilayer separations > 4 Å, the data points can be fit quite closely with a single straight line, such that $P = P_o \cdot \exp(-d_f/\lambda)$, where $P_o = 4.0 \times 10^8$ dyn/cm^2 and $\lambda = 1.7$ Å. This exponential decrease in pressure with increasing fluid spacing has been attributed to the hydration pressure. For $\ln P >$ 17.5 and bilayer separation < 4 Å, the pressure decays much more rapidly with increasing bilayer separation, so that the pressure decays exponentially with increasing d_f with a decay length of about 0.6 Å. As discussed below, we attribute this part of the pressure-separation relationship to steric pressure between head groups from apposing bilayers.

First, let us consider the low pressure region of Figure 3, where the hydration pressure is the dominant interaction. For EPC the hydration pressure decays exponentially with increasing bilayer separation such that $P_h = P_o \cdot \exp(-d_f/\lambda)$, where P_o is the magnitude and λ is the decay length. This relationship holds for a number for a number of lipid bilayers including: phosphatidylcholines in their gel, liquid-crystalline,

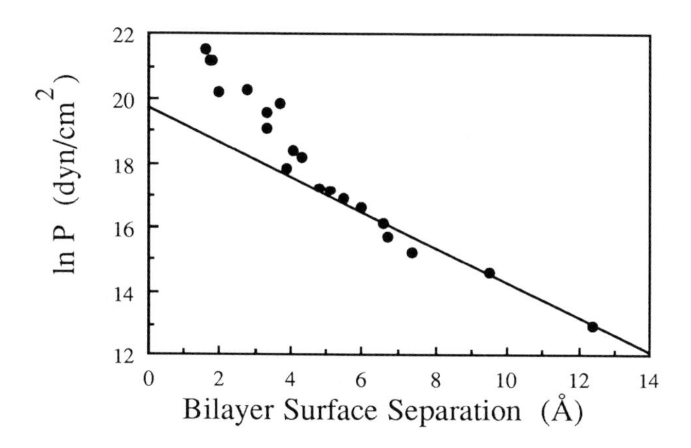

Figure 3. Natural logarithm of applied pressure plotted versus the separation between adjacent EPC bilayers. The solid line is a least-squares fit (r^2 = 0.97) to the data or bilayer separations greater than 4 Å.

and interdigitated phases, phosphatidylcholine:cholesterol bilayers, monoglycerides, and phosphatidylglycerol (McIntosh & Simon, 1986; McIntosh et al., 1987; Simon et al., 1988; Simon & McIntosh, 1989; McIntosh et al., 1989a,b,&c; Simon et al., 1991). We are interested in determining the factors which govern P_o and λ for these amphiphiles. For these lipids with water as the solvent, λ ranges from 1.2 - 2.1 Å (Simon & McIntosh, 1989). For phosphatidylcholines, we found that the decay length does not depend on the lipid's acyl chain length (Simon et al., 1988). We also investigated how the decay length varies with other solvents, namely formamide and 1,3-propanediol (1,3-PDO), whose dimensions, as well dipole moments, dielectric constants and surface tensions are quite different than water (McIntosh et al., 1989b). Plots of ln P versus d_f provide values for λ and P_o for each solvent. For both EPC and equimolar EPC:cholesterol bilayers, the decay length was found to increase with increasing size of the solvent molecule. For equimolar EPC:cholesterol bilayers, λ was 2.1 Å, 2.9 Å, and 3.1 Å in water, formamide, and 1,3-PDO, respectively (McIntosh et al., 1989b).

Since λ does not vary monotonically with either the static dielectric constant or bulk dipole moment of the solvent, the importance of these two parameters in determining the range of the solvation pressure must be second order. However, the decay length does vary monotonically with the molecular weight of the solvent. As a first approximation of the packing of the solvent molecules in the direction perpendicular to the bilayer surface, we use the cube root of the number of solvent molecules per volume ($n^{1/3}$) in bulk solution, where $n = N\rho/M_w$, where N is Avogadro's number, ρ is the solvent density, and M_w is the solvent molecular weight. This calculation makes no assumptions about the shapes of the solvent molecules. For EPC:cholesterol multilayers in water, formamide, and 1,3-PDO, a plot of the decay length versus $1/n^{1/3}$ yields a straight line which has a slope of 0.66 and passes through the origin (Figure 4). The factor of 0.66 might represent a measure of the packing of the solvent molecules in the interbilayer space. For example, if the solvent molecules were spherical and hexagonally close packed then $r = 0.56/n^{1/3}$, where r is the "radius" of the solvent molecule. This implies that the decay of the solvation pressure, at least for the three solvents studied here, is approximately equal to the "radius" of the solvent molecule calculated assuming spherical symmetry and hexagonal close packing of the solvent molecules. Theoretical models for the solvation pressure that have λ proportional to the size of the solvent molecules (Schiby & Ruckenstein, 1983; Attard & Batchelor, 1988) are consistent with these data.

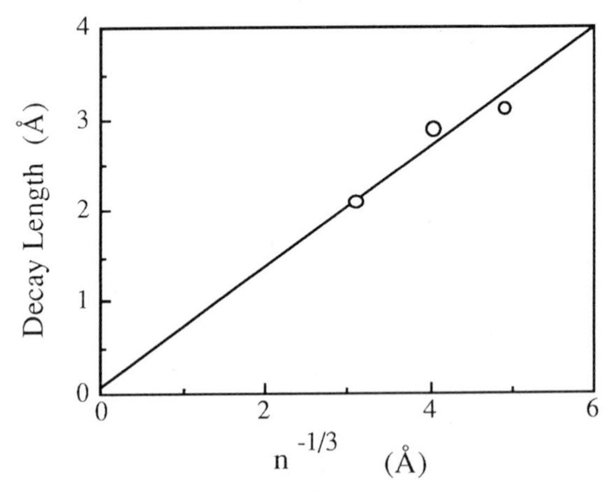

Figure 4. Decay length of solvation pressure plotted versus the inverse cube root of the number of solvent molecules per unit volume. This figure is taken from McIntosh et al. (1989b) and reprinted with permission of the American Chemical Society.

Next, we consider the factors that determine the magnitude of the hydration pressure, P_o. Cevc and Marsh (1985) derived the following expression for the magnitude of the hydration pressure:

$$P_o = 2\chi\ (\psi(0)/\lambda)^2 \qquad \text{Eqn. (1)}$$

where $\psi(0)$ is the "hydration potential" at $d_f = 0$ and χ is orientational susceptibility of the solvent and is equal to $\varepsilon_o(\varepsilon - 1)/\varepsilon$ where ε is the solvent's bulk dielectric constant and ε_o is the permittivity of free space. That is, according to Cevc and Marsh the magnitude of the hydration pressure is proportional to the square of the potential that arises from the perpendicular components of the "multipole surface charge densities" the lipid polar head groups.

We have performed experiments to test the validity of equation (1) by equating the hydration potential, $\psi(0)$, to the dipole potential, V, measured for monolayers in equilibrium with liposomes (Simon & McIntosh, 1989). The measured dipole potential includes contributions from all fixed charges and all oriented dipoles and multipoles, of both the lipid *and* solvent molecules. This identification predicts that

$$P_o = 2\chi\,(V/\lambda)^2 \qquad \text{Eqn. (2)}$$

For zwitterionic and uncharged lipids V can arise only from the vector sum of perpendicular components of the dipole (and multipole) moments of the lipid *and* solvent molecules in the bilayer. To vary V, we used lipids with different polar head groups (phospholipids, monoglycerides), gel and liquid crystalline phase lipids, lecithins with different amounts of cholesterol, and lipids with different solvents (Simon & McIntosh, 1989). Figure 5 shows P_o as obtained by x-ray diffraction measurements plotted versus $2\chi\,(V/\lambda)^2$.

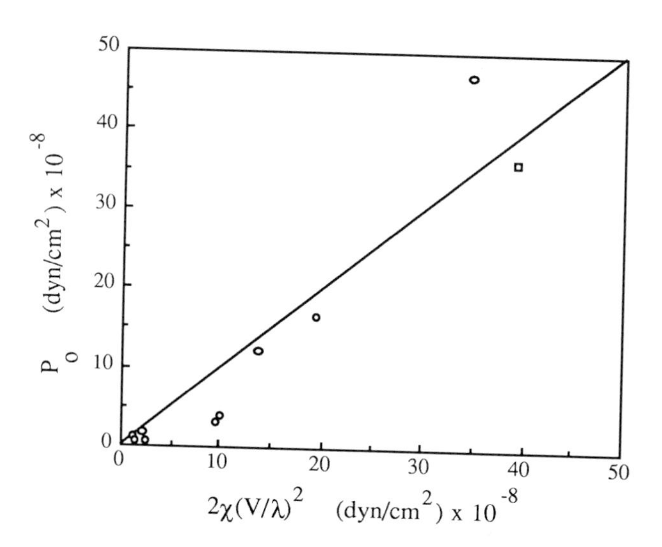

Figure 5. P_o versus the quantity $2\chi\,(V/\lambda)^2$. The circles represent data from phosphatidylcholine and phosphatidylcholine:cholesterol bilayers and the square represents results from phosphatidyl-dimethylethanolamine bilayers. The straight line is a plot of the theoretical prediction $P_o = 2\chi\,(V/\lambda)^2$. This figure is reproduced from the *Biophysical Journal,* Simon et al., (1991) by copyright permission of the Biophysical Society.

As can be seen from the data in Figure 5, there is a good correlation between the magnitude of the hydration pressure and the quantity $2\chi(V/\lambda)^2$ obtained from measurements of the Volta potential. The close agreement between the measured and predicted values of P_o supports our choice of the edge of the bilayer as the plane of origin of the hydration pressure.

The upward break in the pressure-separation curve at ln P ≈ 17.5, d_f ~ 5 Å seen in Figure 3 arises from the onset of steric repulsion between apposing lipid head groups. The onset of steric repulsion at this fluid separation is consistent with NMR data (Hauser, 1981) which indicate that the phosphocholine group rotates from its preferred position, approximately parallel to the bilayer, to another position where it extends 2 to 3 Å farther into the fluid space. Thus, the trimethylammonium groups from apposing bilayers would be expected to come into steric hindrance at d_f ≈ 4 to 6 Å. Local variations in bilayer thickness, or "breathing" modes, would tend to "soften" the edge of the steric barrier barrier and could extend the range of steric interactions. The work to overcome this steric pressure is approximately the work to omit one degree of freedom of the configurational entropy of the polar head group (McIntosh et al., 1987).

These data also suggest that the extent of this upward break should depend on the volume fraction of PC head groups at the hydrocarbon/water interface. To quantitatively test this hypothesis, we systematically varied the volume fraction of interfacial PC head groups by incorporating various concentrations of cholesterol into EPC bilayers (McIntosh et al., 1989a). Cholesterol is an amphipathic molecule with a small polar head group and a molecular area of about 38 Å^2. Therefore incorporation of cholesterol into EPC bilayers separates adjacent PC head groups and decreases their volume fraction at the bilayer/water interface.

Figure 6 shows a plot of pressure versus bilayer separation for EPC and equimolar EPC:cholesterol bilayers. The solid line corresponds to a least-squares fit to the the equimolar EPC:cholesterol data. For ln P < 17.5, the data points for both EPC and equimolar EPC:cholesterol fall quite closely to this line, indicating that the hydration pressure is similar for both systems. However, the pressure-separation data are quite different for ln P > 18. As noted above, for EPC bilayers, the data points deviate upward from the straight line due to steric hindrance between the bulky PC head groups from apposing bilayers. For equimolar EPC:cholesterol bilayers the data points fall closely to the same straight line up to the highest applied pressures (ln P = 21.5) so that there is a larger decrease in d_f at high applied pressure for EPC:cholesterol bilayers than for EPC bilayers. For example, at the highest applied pressure (ln P = 21.5) the value of d_f is about 6 Å smaller for equimolar EPC:cholesterol bilayer than for EPC bilayers. In fact, the values of d_f are negative for equimolar EPC:cholesterol bilayers at the highest applied pressures, meaning that the PC head groups from apposing bilayers have interpenetrated.

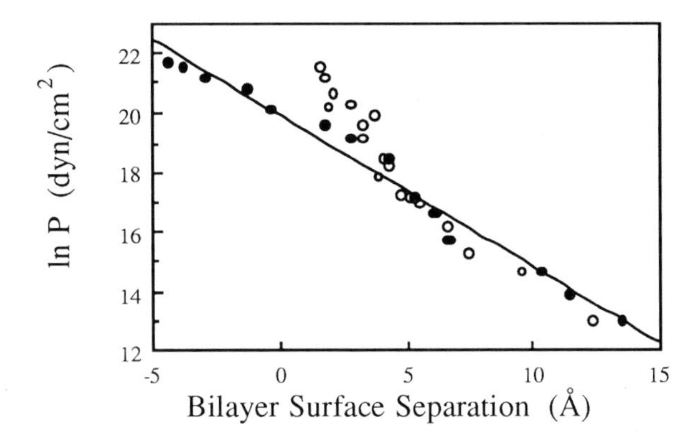

Figure 6. Natural logarithm of applied pressure plotted versus the separation between adjacent bilayer surface for EPC (open circles) and equimolar EPC:cholesterol bilayers (solid circles). The solid line is a linear least-squares fit to the equimolar EPC:cholesterol data.

These observations can be explained in terms of a simple model in which cholesterol spreads the EPC molecules apart in the plane of each bilayer, reducing the volume fraction of PC head groups at the interface, thereby decreasing steric repulsion between adjacent bilayers. For equimolar EPC:cholesterol bilayers the EPC molecules are spread far enough apart that head groups from apposing bilayers can interpenetrate (McIntosh et al., 1989a). In plots of ln P versus bilayer separation, data points for EPC bilayers containing 0.2 and 0.33 mole fraction cholesterol fall in between the extreme cases of 0 and 0.5 mole fraction cholesterol (McIntosh et al., 1989a. This indicates that the magnitude of steric repulsion depends on the amount of cholesterol in the bilayer, or on the volume fraction of PC head groups at the interface (McIntosh et al., 1989a).

SUMMARY

Two short-range pressures primarily govern the fluid spaces between electrically neutral lipid bilayers--steric pressure and hydration pressure. Although the origin of these pressures is not well understood for rough, fluid surfaces, we have found parameters that influence the range and magnitude of these pressures. The magnitude of the hydration pressure depends on the square of the Volta potential and its range depends on the packing density of the solvent molecules, whereas the magnitude of the steric pressure depends on the volume fraction of polar head groups at the membrane/water interface and its range is a function of the extension of the lipid head groups into the fluid space.

REFERENCES:

Attard P, Batchelor MT (1988) A mechanism for the hydration force demonstrated in a model system. *Chem. Phys. Lett.* **149**, 206-211.

Cevc G, Marsh D (1985) Hydration of noncharged lipid bilayer membranes. Theory and experiments with phosphatidylethanolamine. *Biophys. J* **47**: 21-32.

Hauser, H. (1981) The polar group conformation of 1,2-dialkyl phosphatidylcholines. An NMR study. *Biochim. Biophys. Acta* **646**: 203-210.

Israelachvili, JN (1985) *Intermolecular and Surface Forces with Applications to Colloidal and Biological Systems,* pages 1-43 Academic Press, London.

McIntosh TJ, Magid AD, Simon SA (1987) Steric repulsion between phosphatidylcholine bilayers. *Biochemistry* **26**: 7325-7332

McIntosh TJ, Magid AD, Simon SA. (1989a) Cholesterol modifies the short-range repulsive interactions between phosphatidylcholine membranes. *Biochemistry* **28**: 17-25.

McIntosh TJ, Magid AD, Simon SA (1989b) Range of the solvation pressure between lipid membranes: dependence on the packing density of solvent molecules. *Biochemistry* **28**: 7904-7912.

McIntosh TJ, Magid AD, Simon SA (1989c) Repulsive interactions between uncharged bilayers. Hydration and fluctuation pressures for monoglycerides. *Biophys. J.* **55**: 897-904.

McIntosh TJ, Simon SA (1986) The hydration force and bilayer deformation: a reevaluation. *Biochemistry* **25**: 4058-4066

McIntosh TJ, Simon SA, Dilger JP (1989) in *Water Transport in Biological Membranes* (ed. Benga, G.), CRC Press, Inc., Boca Raton, FL.

Parsegian VA, Fuller N, Rand RP (1979) Measured work of deformation and repulsion of lecithin bilayers. *Proc. Nat. Acad. Sci. USA* **76**: 2750-2754.

Pearson SH, Pascher I (1979) The molecular structure of lecithin dihydrate. *Nature (London)* **281**: 499-501.

Rand RP, Parsegian VA (1989) Hydration forces between phospholipid bilayers. *Biochim. Biophys. Acta* **988**: 351-376.

Schiby D, Ruckenstein E (1983) The role of the polarization layers in hydration forces. *Chem. Phys. Lett.* **95**: 435-438.

Simon SA, Fink CA, Kenworthy AK, McIntosh TJ (1991) The hydration pressure between lipid bilayers: a comparison of measurements using x-ray diffraction and calorimetry. *Biophys. J.* **59**: 538-546.

Simon SA,. McIntosh TJ (1989) Magnitude of the solvation pressure depends on dipole potential. *Proc. Natl. Acad. Sci. USA* **86**: 9263-9267.

Simon SA, McIntosh TJ, Magid AD (1988) Magnitude and range of the hydration pressure between lecithin bilayers as a function of head group density. *J. Colloid Interface Sci.* **126**: 74-83.

POLYMER GEL PHASE TRANSITION: THE MOLECULAR MECHANISM OF PRODUCT RELEASE IN MUCIN SECRETION?

Pedro Verdugo, Ingrith Deyrup-Olsen[**], Arthur W. Martin[**], and Daniel L. Luchtel[*]
Center for Bioengineering, Depts of Environmental Health[*] and Zoology[**]
University of Washington WD-12
Seattle, Washington 98195
USA

ABSTRACT

Mucus is a polymer gel resulting from the annealing and hydration of a variety of secretory products. Mucins which are the principal polymeric species in the mucus matrix, are condensed inside secretory granules and undergo massive swelling upon release. Upon exocytosis from goblet cells mucins swell following a typical first order kinetics that is characteristic of polymer gels (Verdugo, 1984; Tanaka and Fillmore, 1979). The swelling of mucus, which is the leading parameter for the control of mucus rheology, is driven by a Donnan potential (Tam and Verdugo, 1981). However the condensation of mucins inside the secretory granule is still not well understood. An important condition for mucin condensation is the shielding of their polyionic charges which depends upon the presence of large amounts of Ca^{2+} inside the secretory granule (Verdugo et. al., 1987). Although the cationic shield provided by calcium is expected, it is not sufficient to explain a specific molecular mechanism of mucins condensation. We have proposed that the condensation and decondensation of secretory mucins could be explained by a polymer gel phase transition mechanism. The experiments reported here were designed to test if the native polymer network contained in the giant mucin granule of the slug *Ariolimax columbianus* can undergo polymer gel phase transition. Isolated granules were demembranated and suspended in solutions containing different water/glycerol ratios. The volume of the mucin polymer network was monitored by video-microscopy. Video recordings were stored in magnetic tape. Results indicate that the mucin network can expand as much as 600 fold. This volume change is reversible and characteristically discontinuous demonstrating the typical features of a polymer gel phase transition: The critical solvent/non-solvent ratio is 74/26% water/glycerol (pH 7, 20°C), and varies with pH, temperature and Ca^{2+} concentration.

The present results are consistent with the idea that mucins must be in condensed phase in the secretory granule. Their charges being shielded by Ca^{2+}. Product release upon exocytosis could result from a polymer gel phase transition whereby the mucin polymer network goes from a condensed phase to an expanded hydrated phase.

NATO ASI Series, Vol. H 64
Mechanics of Swelling
Edited by T. K. Karalis

INTRODUCTION

Depending upon the temperature and solvent composition polymer gels can undergo discrete and reversible bistable volume changes that have the characteristic features of a critical phenomenon. Changes in the volume of the gel reflect a transition between two states or phases distinguished by two correspondingly different molecular configurations of the gel's polymer network, which can be either condensed, in low affinity solvents, or expanded, in high affinity solvents. The volume change can amount to several hundred-fold. The transition between these two phases is typically discontinuous and take place at a critical solvent/non-solvent ratio. In artificial polyionic hydrogels the critical solvent/non-solvent ratio depends upon pH, the concentration of electrolytes, and the temperature of the solvent/non-solvent mix (Tanaka, 1981).

Although phase transition has been well characterized in artificial polymer gels, the application of this principle in natural gels, and in particular to explain the mechanism of condensation and decondensation in secretion has been only recently investigated (Verdugo, 1986; Fernandez et al., 1991).

Mucins, the principal constituent of the mucus gel polymer network, are condensed in secretory granules and expand by hydration upon release from secretory cells (Verdugo 1984). Phase transition may provide a novel conceptual framework to understand the molecular mechanism of mucin condensation during storage in secretory granules, and its subsequent decondensation upon exocytosis. The experiments described here were aimed at establishing whether the native condensed mucins polymer network stored in secretory granules can undergo polymer gel phase transition.

METHODS

The giant secretory granule of the slug *Ariolimax columbianus* was used as experimental model for this investigation. The condensed polymer network of slug mucus can be extracted intact from demembranated giant slug granules and represent a unique model to study the physical properties of a native polymer gel network. Giant mucus granules are about the size of an erithrocyte (see Fig 1). Their mechanism of release from secretory cells of the slug skin is not clearly understood. However, intact isolated giant mucus granules can be normally found in the fresh mucus trails that the slug leaves behind upon locomotion.

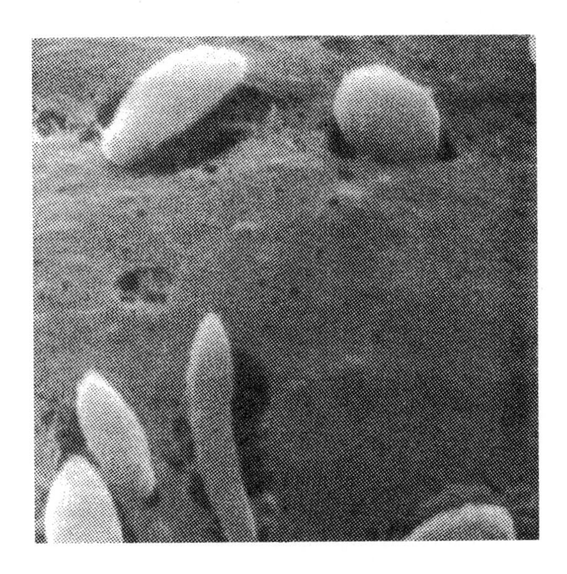

Figure.1. Scanning electron micrographs of quick-frozen freeze-dried mucin secretory granules of the slug *Ariolimax columbianus*. These giant mucus granules have the shape of a rugby ball measuring about 10 to 15 μm in their longer axis, and about 5 to 7 μm in the minor axis.

Slight mechanical shearing can induce these granules to release their secretory contents in an explosive fashion smearing in less than 40 msec. a plug of mucus gel several hundred times their original size.

Granules were collected from slugs following the procedure published elsewhere (Deyrup-Olsen et al, 1983). Briefly, the posterior body of the slug foot is dissected, forming a sac that is then attached to a glass tube, mounted vertically, and filled with Ringer solution. Gentle mechanical or electrical stimulation of the surface of the foot yields a suspension of intact secretory granules as those illustrated in Fig 1.

In the present studies a drop of the granule-containing suspension secreted by the slug foot was deposited between two cover slips separated by 300 μm spacers, forming a tunnel chamber with a total volume of 30 mm^3. After 5 minutes the chamber was perfused with 1 ml of a solution containing 50 mM $CaCl_2$ buffered at pH 7 with HEPES/MOPS buffer. The temperature of the chamber was set at 20°C using a closed loop proportional temperature servo controller. Following perfusion, the granules that became spontaneously bound to the glass surface remained in the chamber while those in suspension were washed out. A field containing 6 to 7 granules was imaged in a video camera using phase contrast microscopy at a total magnification of 300x. Video images were recorded in magnetic tape and further digitized frame by frame in a computer.

Experimental protocols were designed to verify whether the mucin polymer network of the mucus gel can undergo phase transition, and to investigate the effect of temperature, pH and Ca^{2+} concentration on the critical solvent/non-solvent ratio.

The experiments consisted in equilibrating the condensed demembranated granules in solutions containing different water/glycerol ratios. The chamber was first perfused with 100% glycerol, while the granules were continuously recorded on video tape at a sample rate of 60 frames/second. Glycerol in this case served both to keep the polymer matrix condensed, and also to solubilize the membrane of the secretory granules, leaving the polymer network of mucins condensed and devoid of the granular membrane.

The condensed demembranated mucin polymer networks remained bound to the cover slip of the chamber, and were then equilibrated in solutions containing increasing buffer/glycerol ratios. The effect of Ca^{2+}, pH, and temperature on the critical solvent/non solvent ratio was studied by changing the concentration of $CaCl_2$, pH, and temperature of the buffer/glycerol mixture. The effect of Ca^{2+} was studied at 1 μM, 20 mM and 50 mM $CaCl_2$, while pH was varied from 6 to 8 in solutions containing 20 mM $CaCl_2$. The effect of temperature was investigated at 20° and 30°C in mucin networks equilibrated in buffer/glycerol mixtures containing 20 mM $CaCl_2$.

RESULTS

The experimental outcome of the present studies indicate that the native mucin polymer network of giant mucin granules can indeed undergo a typical polymer gel phase transition. The expansion/condensation of the mucin network observed as a result of variations in the solvent/non-solvent ratio was characteristically discontinuous.

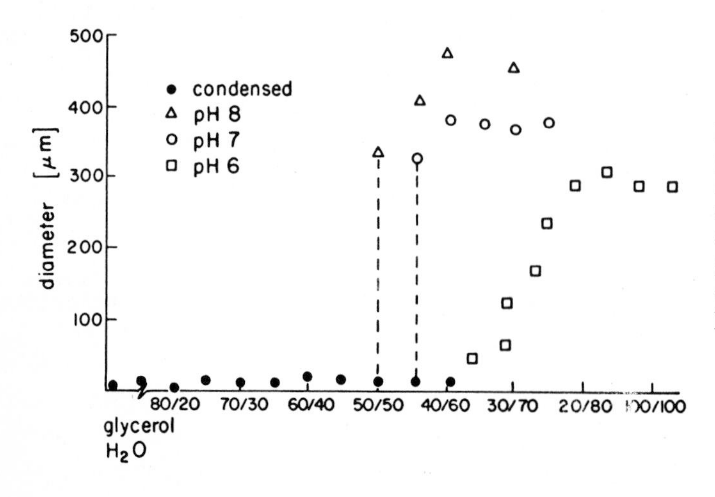

Figure. 2. Polymer gel phase transition of the native demembranated mucin polymer matrix of a giant secretory granules of *Ariolimax columbianus*. Effect of pH.

The corresponding volume expansion amounted to several hundred-fold. As predicted by the theory of phase transition of polymer gels , small (<1%) changes in buffer/glycerol ratio around the critical ratio, or slight changes in temperature (<1°C) resulted the recondensation or re-expansion of the mucus gel.

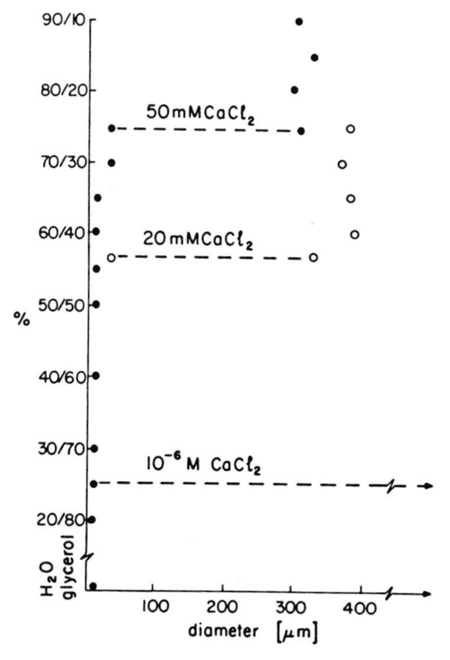

Figure. 3. Polymer gel phase transition of the native demembranated mucin polymer matrix of a giant secretory granules of the slug *Ariolimax columbianus*. Effect of Ca^{2+} concentration.

As in other polyionic polymer gels, variations of pH and Ca^{2+} concentration in the the buffer resulted in a typical change of both the critical water/glycerol ratio, and the magnitude of the expanded volume of the polymer network (Figs. 2 and 3).

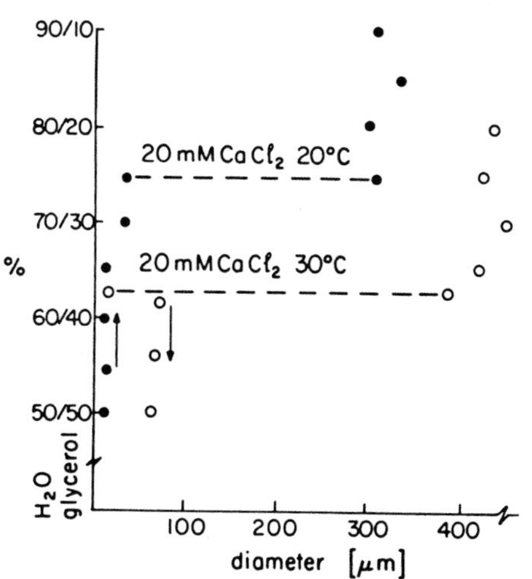

Figure. 4. Polymer gel phase transition in a native demembranated mucin polymer matrix of giant secretory granules of the terrestrial slug *Ariolimax columbianus*. Effect of temperature. The arrows indicate decondensation-recondensation cycle.

Notice also that phase transition of the the mucin network is reversible, and show a characteristic temperature dependance (Fig. 4). However, due to the tangled, rather than crosslinked nature of the polymer network of mucus (Lee et al., 1977; Verdugo et al., 1983), the recondensation does not return to the original size.

DISCUSSION

Mucus is a ubiquitous polymer hydrogel that functions as a protective coat on the skin and mucosa of a broad range of species. Polymer gels are made out of a polymer network and a solvent. In mucus the solvent is water while its polymer matrix is composed of long chain glycoproteins called mucins which are tangled together forming a randomly woven, highly polyionic network (Verdugo et al., 1983). Mucins produced by mammalian goblet cells *in vitro* undergo massive swelling upon release from secretory granules. Thus, mucins are condensed in the granule and expand by hydration during and after exocytosis to form the mucus gel (Verdugo, 1984). However, the molecular mechanism of condensation of mucin, during storage in secretory granules, and its expansion, upon release from the cell, had only recently begun to be examined. The hydration of mucins during release from goblet cells follows a typical first order kinetic (Verdugo, 1984; Aitken et al., 1989). The application of Tanaka's linear theory of swelling of polymer gels (Tanaka and Fillmore 1979) revealed that during release, the mucin polymer network exhibits diffusivities about one order of magnitude higher than those measured by dynamic laser scattering in swollen mucus (Verdugo et al., 1987). These, as well as previous observations on the Donnan properties of mucus (Tam and Verdugo, 1981), prompted the idea that the exocytotic swelling of secretory granules might be powered by a Donnan potential, rather than by simple concentration-driven osmotic gradients as it has been thought before (Pollard et al. 1984; Finkelstein et al., 1986; Holz, 1986; Whitaker and Zimmerberg, 1987). We further proposed that the condensed pre-secretory and the hydrated post-secretory forms of the mucus gel might correspond to the two characteristic phases of a polymer gel that undergoes a typical phase transition upon release from the cell.

The experiments reported here were designed to test the ability of mucin gels to undergo polymer gel phase transition. Results show that the mucin polymer network isolated from giant secretory granules of the slug *Ariolimax columbianus* can indeed undergo a typical, reversible polymer gel phase transition.

A "jack-in-the-box" model for exocytosis.

The present observations, together with the findings of high concentrations of Ca^{2+} inside secretory granules, and that Ca^{2+} release precedes the bursting of giant granules (Verdugo et al., 1987), suggest a novel molecular mechanism to explain mucin release, and perhaps the mechanism of exocytosis in other secretory cells. The basic feature of this idea (see Fig. 5) is that inside the secretory granule there must be a cation-shielded polyion in condensed phase. Phase transition of the polyion would result in the decondensation, driving the exocytosis of the granular content.

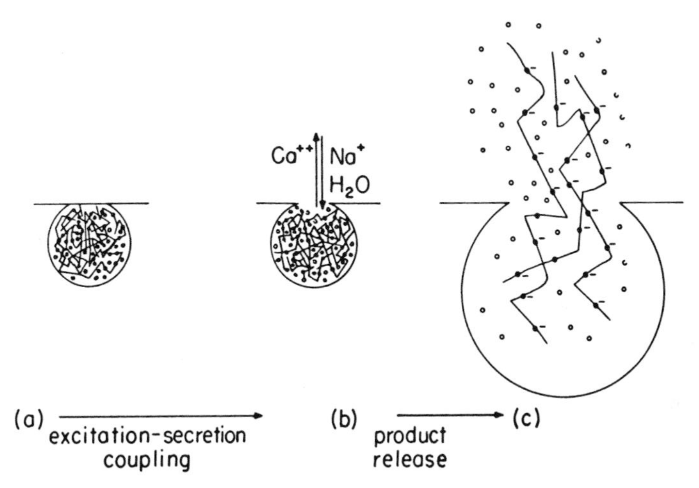

Figure 5.- Mechanisms of mucin exocytosis. By analogy to the sequence of activation in muscle contraction, in exocytosis it is convenient to separate the stimulus coupling steps from the actual mechanical phenomena that drive product release. The stimulus-coupling steps encompass the events that start with receptor activation that lead to the docking of the secretory granule to the plasma membrane, the formation of a secretory pore, and the switching of the pore to high ionic conductance. Once a high conductance channel is established between intragranular and extracellular compartments, n^{+}/Ca^{+2} ion exchange follows, triggering a phase transition which turns the mucin polymer from a condensed to a hydrated phase. At this stage electrochemical energy stored in the mucin polymer network is transformed into mechanical energy, via Donnan swelling, driving the final release of secretory products.

When intragranular divalent cations are exchanged by extracellular monovalent, a "jack-in-the-box" mechanism is turned on whereby the condensed gel undergo phase transition, and the polymer chains quickly unfold, expanding the polymer network and its final release from the cell. Although osmotic swelling has been proposed in the past as the driving force for exocytosis in other secretory cells (Pollard et al. 1984; Finkelstein et al., 1986; Holz, 1986; Whitaker and Zimmerberg, 1987), the kinetics of expansion in mucin granules is much too fast to be explained by a simple concentration-driven osmotic drive. This is particularly striking in the giant secretory granule of the terrestrial slug where the volume of the mucus network can change about 600 fold in about 40 msec (Verdugo et al., 1987). High rates of exocytotic discharge have also been observed in the statocyst in *Paramecium caudatum* (Garofalo and Satir,1984), and particularly in the Nematocyst of *Hydra attenuata* where the linear expansion can reach velocities of about 2 m/sec (Holstein and Tardent, 1984). In these particular secretory systems, however, the expansion of the granular content is highly anisotropic (vectored in a linear direction) as compared to the characteristic isotropic expansion observed in other secretory systems like mucin release or in mast cell exocytosis.

The only other secretory cell where a polymer gel phase transition mechanism has been recently investigated is in mast cell granules. In that system the exocytosed heparin network can be readily recondensed by 50 mM divalent histamine at pH 3. These conditions mimic those normally found inside the granule. Also histamine-condensed heparin networks remain condensed even when they are equilibrated in distilled water, virtually ruling out the idea of a simple osmotic mechanism to explain the exocytotic swelling in mast cells. However, small concentrations of Na^{+} (<10 mM) in the medium can readily trigger swelling of histame-recondensed heparin networks (Fernandez et al., 1991). As shown by Ohmine and Tanaka (1982), divalent cations are much more effective than monovalent cations in inducing polymer gel condensation. For instance, the critical concentration of non-solvent to sustain condensation of polyionic polyacrylamide is about 4000 fold smaller for equimolar concentrations of divalent as compared to monovalent cations. Thus, the ion exchange properties of mucins and heparine which have been well documented (Crowther and Marriot, 1983; Forstner and Forstner, 1975; Uvnas and Aborg, 1977), can provide a convenient mechanism to explain the initiation of phase transition in mucin secreting cells, and in mast cell.

The proposed "jack-in-the-box" paradigm whereby a polymer gel phase transition mechanism is triggered by ion exchange between the intragranular polymer network and the extracellular space could as well be operating in exocytosis in other secretory cells. Indeed, the two fundamental components for such a system to function , namely, a condensed polyanion and a complementary shielding cation, seem to be present in a variety of secretory granules. While in mucin granules the polyion is mucin and the shielding cations are Ca^{2+} and/or Mg^{2+} (Verdugo et al., 1977; Sasaki et al., 1983; Roomans et al., 1986; Isutzu et al., 1986), the corresponding pair in mast cells is the polyionic heparin and the cation is histamine. In chromaffin cells the condensed polyion is the highly acidic protein chromogranin A, and the shielding cations is probably H^+, and/or catecholamines (Winkler and Westhead, 1980). The parathyroid granules also contains a negatively charged polyion called protein I (SP-I), and the cation is Ca^{2+} (Cohen et al, 1981). The presence of cationic moieties in secretory granules has been investigated only in a few cells. However, highly polyanionic proteins of similar molecular weight and showing immunoreactivity with chromogranin A and SP-I have been found in secretory granules of a broad variety of cells, including the anterior pituitary, the thyroid, pancreatic islets, gastric antrum, and celiac, and mesenteric sympathetic ganglia (Cohen et al, 1981). Although the function of this ubiquitous group of proteins has not been established, the idea of a polymer gel phase transition to explain condensation-decondensation of secretory products identifies specific roles for these polyions and the corresponding cations found in secretory granules, indicating testable predictions to study the molecular mechanisms that drive exocytosis.

The detailed physiologic mechanisms that control phase transition in secretory mucins remain to be explored. The findings reported here open a novel alternative of high predictive power to explain the mechanisms of storage and release in secretion.

ACKNOWLWDGEMENTS

This work was supported by grants HL 38494 and RR-07096 from the National Institute of Healt., and grant R 010 7 01 from the Cystic Fibrosis Foundation.

REFERENCES

Aitken ML, Verdugo P (1989) Donnan mechanism of mucin release and conditioning in goblet cells: the role of polyions. In: Chantler EN (ed). Mucus and Related Topics. Plenum Press, New York, PP. 73-80.

Cohen DV, Morrissey JJ, Hamilton JW, Shofstall RE, *et.al* (1981) Isolation and partial characterization of secretory protein I from bovine parathyroid glands. Biochemistry 20:4135-4140

Crowther RS, Marriot C (1983) Counter-ion binding to mucus glycoproteins. J Pharm Phamacol 36:21-26.

Deyrup-Olson I, Luchtel DL, Martin AW (1983) Components of mucus of terrestrial slugs (Gastropoda) Am J Physiol 245:R448-R452.

Fernandez JM, Villalon M, Verdugo P (1991) Reversible condensation of mast cell secretory products in vitro. Biophys J 59:1022-1027

Finkelstein A, Zimmerberg J, Cohen FS (1986). Osmotic swelling of vesicles: its role in the fusion of vesicles with planar phospholipid bilayers membranes and its possible role in exocytosis. Annu Rev Physiol 48:163-174.

Forstner JF, Forstner GG (1975) Calcium binding to intestinal goblet cell mucin. Biochim Biophys Acta 386:283-292.

Garofalo R. and Stir B (1984) Paramecium secretory granule content: quantitative study on in vitro expanssion and its regulation by Ca and pH J. Cell Biol. 99: 2193-2199.

Holstein T., Tardent P. (1984) An ultrahigh-speed analysis of exocytosis: nematocyst discharge. Science 223:830-833

Holz RW (1986) The role of osmotic forces in exocytosis from adrenal chromaffin cells. Ann Rev Physiol 48:175-189

Izutsu K, Johnson D, Schubert M, Wang E, Ramsey B. *et al* (1985) Electron microprobe analysis of human labial gland secretory granules in cystic fibrosis. J Clin Invest 75:1951-1956

Lee W I, Verdugo P, Blandau R J (1977) Molecular arrangement of cervical mucus: a re-evaluation based on laser scattering spectroscopy. Gynec Invest 8:254-266

Ohmine I, Tanaka T (1982) Salt effects on the phase transition of ionic gels. J Chem Phys 77:5725-5729.

Pollard HB, Pazoles CJ, Creutz CE, Scott HJ, Zinder O, Hotchkiss A (1984). An osmotic mechanism for exocytosis from dissociated chromaffin cells. J Biol Chem 259:1114-1121.

Roomans GM, von Euler AM, Muller RM, and Gilljman H (1986) X-ray microanalysis of goblet cells in bronquial epithelium of patients with cystic fibrosis. J Submicroc Cytol 18:613-615

Sasaki S, Nakagaki I, Mori H, Imai Y (1983) Intracellular calcium store and transport of elements in acinar cells of the salivary gland determined by electron probe x-ray microanalysis. Jap J Physiol 33:69-83

Tam PY, Verdugo P (1981) Control of mucus hydration as a Donnan equilibrium process. Nature 292:340-342

Tanaka T, Fillmore DJ (1979) Kinetic of swelling of gel. J Chem Phys 70:1214-1218

Tanaka T. (1981) Gels. Sci Am 244:124-138

Uvnas B, Aborg CH (1977) On the cation exchanger properties of rat mast cell granule and their storage of histamine. Acta Physiol Scand 100:309-314.

Verdugo P, Tam PY, Butler J (1983) Conformational structure of respiratory mucus studied by laser correlation spectroscopy. Biorheology 20:223-230

Verdugo P (1984) Hydration kinetics of exocytosed mucins in cultured secretory cells of the rabbit trachea: a new model. In: Nugent J, O'Connor (eds), Mucus and Mucosa (Ciba Foundation Symposium Vol. 109). Pitman, London, pp. 212-234.

Verdugo P (1986) Polymer gel phase transition: a novel mechanism of product storage and release in mucin secretion. Biophys J 49:231a

Verdugo P, Deyrup-Olsen L, Aitken M. Villalón M, Johnson D. (1987) Molecular mechanism of mucin secretion: the role of intragranular charge shielding. J Dent Res 66:506-508

Whitaker M, Zimmerberg J (1987) Inhibition of secretory granule discharge during exocytosis in sea urchin eggs by polymer solutions. J Physiol 389:527-539.

Winkler H, Westhead E (1980) The molecular organization of adrenal chromaffin granules. Neuroscience 5:1803-1924

Phase Transitions of Gels

Toyoichi Tanaka, Masahiko Annaka, Franck Ilmain, Kazuhiro Ishii
Etsuo Kokufuta, Atsushi Suzuki, and Masayuki Tokita

Department of Physics and Center for Materials Science and Engineering
Massachusetts Institute of Technology
Cambridge, Massachusetts 02139

Polymer gels are known to exist in two distinct phases, swollen and collapsed. Volume transition occurs between the phases either continuously or discontinuously in response to chemical and physical stimuli such as temperature, solvent composition, pH, ionic composition, electric field, light, and particular molecules. For a gel to undergo the phase transition, it is necessary that polymers interact with each other through both repulsive and attractive interactions and the balance of competing interactions has to be modified by various stimuli. The phase behavior of a gel, therefore, crucially depends on the nature of interactions between polymers. Recently new phases and volume transitions between them have been discovered in some gels. Detailed examination of the gel phase behavior provides a deep insight into the polymer-polymer interactions and configurations of polymers. The knowledge on physical and chemical fundamentals of gel phase transition will play a role as guiding principles for a wide variety of technological applications of gels as functional elements.

Introduction

During the last decade gel research has made remarkable progress owing to the strenuous efforts by researchers from many different fields. Since ancient times gels have been closely affiliated with our daily life. From the early 1940's numerous synthetic gels have been designed and developed for various usages *(1,2)*. What, then, has made the gel research very popular again? The purpose of this article is to provide an answer to this question *(3)*.

NATO ASI Series, Vol. H 64
Mechanics of Swelling
Edited by T. K. Karalis

Gels are found everywhere. In our bodies, the cornea, the vitreous, and the connective tissues are gels. The surfaces of the internal tracts such as the stomach and lung are covered with gels. The basement membranes for the kidney and blood vessels are also gels. These membranes are believed to play a fundamental role in the transport of water and solute molecules.

Disposable diapers and sanitary napkins use gels as super water absorbents. For these industries it is crucial to understand the physical and chemical principles that govern the degree and speed of gel swelling in the physiological conditions.

Sheets of gels are developed that tightly wrap fresh fishes and meats. The gels keep them moist but absorb unnecessary excess water, and are useful for efficient transportations and storages.

Gels are used for agricultural purposes as retainers of water and solutes.

Gels are indispensable materials as molecular sieves to separate molecules according to size in gel permeation chromatography and electrophoresis.

Soft contact lenses, artificial lenses, artificial vitreous, and materials used in plastic surgeries are made of gels. Gels are also used as control delivery systems for drugs and perfumes.

Builders mix a concrete powder with water. It is difficult to mix a powder and a liquid homogeneously, and it is necessary to put an excess amount of water for a thorough homogeneous mixing. Recently a method has been developed where gels rather than water are used. Mixing of two powders is much easier than mixing powder and water and requires much less of water. After thorough mixing the gel gradually shrinks due to the pH change squeezing out water into the concrete powder uniformly, thus a concrete with an extremely high density can be obtained. Its hardness and strength are similar to those of rock.

The gels play a vital role in the fields of medicine, foods, chemical, agricultural, and other industries. The list of gels and their usages would be too long to be fully described in this paper. Readers may refer to some recently published books on applications *(1-2,8-9)*.

Structure of gels

The unique structure allows gels to be useful in various applications *(10,11)*. Gels are cross-linked network of polymers swollen with a liquid. For example, two thirds of the human body consists of water, and almost all the biological reactions and other activities occur in water. Water needs a container to retain its shape as an entity; nature invented two kinds of containers, cells and gels. A cell membrane provides a boundary to water, whereas a polymer network incorporates water in its intersticious space with its affinity due to interaction energy and polymer entropy. These containers allow water to retain its shape.

The history of the science and technology of membranes is long. Recent revival of the technology of the Langmuir-Blogget films allowed remarkable progress in the field. Fruitful outcome in the membrane technology is expected to flourish during the next decade. Incidentally gel research is also expected to grow rapidly during the next decade.

As our society becomes richer and more sophisticated, and as we increasingly recognize that the natural resources are not unlimited, materials with better quality and higher functional performance become more wanted and needed. Soft and gentle materials are beginning to replace some of the hard mechanical materials in various industries. Recent progress in biology and polymer sciences that is unveiling the mystery of marvelous functions of biological molecules and organelles promises new development in gel technologies. All these factors bring us to realize the importance and urgent needs of establishing gel sciences and technologies *(1)*.

The principles of phase transitions and critical phenomena in polymer gels are essential for the understanding of gels and of polymer systems in general *(12)*. These phenomena provide an experimental tool with which to explore the fundamentals underlying the molecular interactions and recognitions in natural and synthetic polymers.

Historical Background

Theoretical works

Study of the phase transition of gels can be traced back to the theoretical work by Dusek and Patterson in 1968, who suggested the possibility of a discontinuous volume change of a gel when an external stress is imposed upon it *(13)*. Using the Flory-Huggins equation of state, these researchers realized that with the presence of an external force, Maxwell's loop appears in the isobar of the gel. A similar phenomenon in a single polymer chain, known as the coil-globule transition, has been theoretically studied by Ptitsen and Eizner *(14)*, and Lifshitz, Grosberg, and Khokhlov *(15)*, and deGennes *(16)*. In a sense a gel phase transition may be considered as a macroscopic manifestation of a coil-globule transition.

Collective diffusion of gels

On the experimental side the first relevant event may be the observation of the so-called collective diffusion, which was found in 1973 using quasielastic light scattering on acrylamide gels *(17)*. Light is scattered by the density fluctuations of polymer network or the phonons in the polymer network. The phonon would propagate without water, but in water the phonon is completely overdamped and the motion becomes a diffusion process. The equation for the displacement **u** of a point in the gel whose average location was x is given by:

$$\rho \frac{\partial^2 \vec{u}(\vec{x},t)}{\partial t^2} = K \frac{\partial^2 \vec{u}}{\partial \vec{x}^2} - f \frac{\partial \vec{u}}{\partial t} \tag{1}$$

The left hand side represents the density ρ times acceleration, the first in the rhs is the elastic restoring force. K denotes the longitudinal elastic modulus. These two terms constitute the sound wave equation within an elastic medium, which, in the case of a gel, is the polymer network . The last term is the viscous friction between the network and water. In most gels the inertia term is much smaller than the friction and elastic terms and can be neglected, and the equation becomes that of a diffusion:

$$\rho \frac{\partial^2 \vec{u}}{\partial t^2} = K \frac{\partial^2 \vec{u}}{\partial \vec{x}^2} - f \frac{\partial \vec{u}}{\partial t} \cong 0 \qquad \Rightarrow \qquad \frac{\partial \vec{u}}{\partial t} = \frac{K}{f} \frac{\partial^2 \vec{u}}{\partial \vec{x}^2} \tag{2}$$

It is interesting that polymers undergo a diffusion process even though they are all cross-linked into a network. This process is, therefore, called the collective diffusion of a gel.

However, the notion that an elastic medium obeys the diffusion equation when it moves in a viscous fluid is more general and universal than the diffusion of isolated molecules in water. For example, the diffusion of ink in water obeys the diffusion equation. Without water the ink solute can be considered as a gas whose elastic constant and friction is given by using Stokes formula:

$$K = nkT \qquad \text{and} \qquad f = n 6\pi\eta r \tag{3}$$

where n is the number density of the ink molecules, η is the water viscosity, and r is the radius of the ink molecule. From these two equations the Stokes-Einstein formula is derived for the diffusion coefficient of particle

$$D = \frac{K}{f} = \frac{nkT}{n6\pi\eta r} = \frac{kT}{6\pi\eta r} \qquad (4)$$

Thus the diffusion of ink molecules can be derived from the equation (2).

The light scattering study confirmed the diffusion process in gels. For example, the mode decayed in proportion to the square of the wave-vector ($\sim q^2$) as theoretically predicted from the equation (2). The collective diffusion mode has been confirmed in various gel systems. Munch and colleagues showed an interesting demonstration on the collective nature of the diffusion of a polymer network. These researchers measured a solution containing polymers of a same chemical structure but with two different molecular weights *(18)*. The two diffusion coefficients were observed, each corresponding to each molecular weight. When the polymers were cross-linked the correlation function of scattered light became a single exponential and the diffusion coefficient became a single value. This observation revealed the essence of nature of collectivity of the diffusion process of all the connected polymers.

Viscoelastic parameters were determined using light scattering technique on various gels *(19)*. The collective diffusion theory has been improved by taking into account of the counter flow of water, the effect of shear modulus, and shape dependence *(19-22)*.

Critical behavior of gels

In 1977 the critical phenomena were discovered while the light intensity scattered from a gel was being measured as a function of temperature *(23)*. As the temperature was lowered, both the intensity and the fluctuation time of scattered light increased and appeared to diverge at a special temperature. The phenomenon was explained as the critical density fluctuations of polymer network, although the polymers were cross-linked *(24-25)*. The critical behavior was not along the critical isobar, but the spinodal line was reached under super cooled conditions.

Phase transition of gels

The light scattering study of the acrylamide gels in water showed that the fluctuations of the polymer network diverged at minus 17°C. This finding raised a question of ice formation although such possibility was carefully checked and eliminated by the measurement of refractive index of the gel. Such a question could be answered if the spinodal temperature were raised to much above the freezing temperature. The gels were placed in acetone-water mixtures with different concentrations to find a proper solvent in which the opacification of the gel occurred at room temperature. The next day half of the gels were swollen, and the remaining gels were collapsed, meaning the gel volume changed discontinuously as a function of acetone concentration.

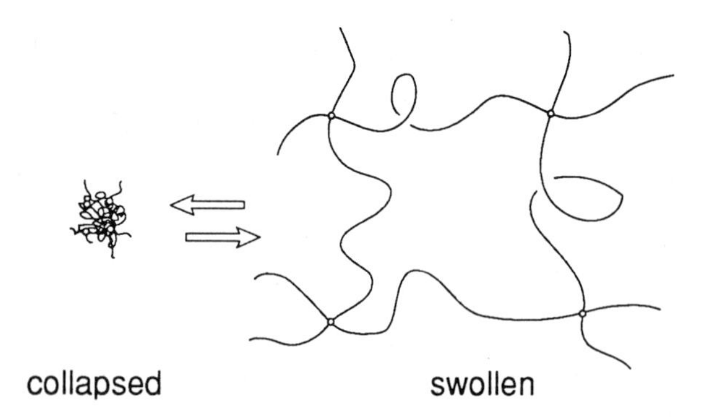

The experiments were repeated but were not reproducible: new acrylamide gels were made with various recipes and their swelling curves were determined as a function of acetone concentration, but they were all continuous. It was later recognized that the gels that showed the discontinuous transition were old ones, that is, gels made a month before and left within tubes in which they were prepared. Repeated experiments were all carried out on "new" gels, and thus they underwent a continuous transition.

Later the difference of new and old gels was identified as ionization which induced an excess osmotic pressure within the gels leading to the discontinuous transition. Hydrolysis was gradually taking place in the gel in a mildly high pH solution. This explanation was experimentally proven by artificially hydrolyzing the gel and observing the increase in the discontinuity of the volume transition *(26)*. It was also shown theoretically that when the osmotic pressure due to counter ions was added to the equation of state of a gel, Maxwell's loop appeared in the swelling curves (temperature vs volume) and a discontinuous transition emerged. As the ionization increased the transition temperature was predicted to increase and the volume change at the transition was found to be larger.

Phase Transitions of Gel

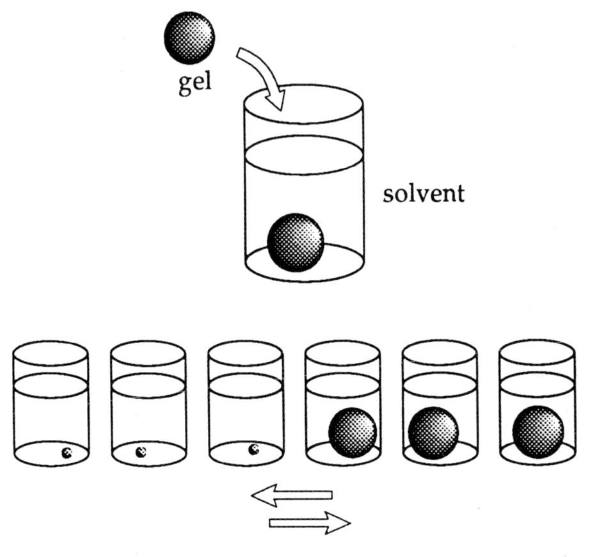

Temperature, Solvent composition, pH, Ions
Electric field, UV, Light
Specific molecules, or chemicals

The figure below shows the swelling curves of acrylamide gels immersed in acetone water mixtures with different volume compositions *(26)*. The gel is swollen in water and gradually shrinks as acetone is added to the solution. The days shown in the graphs indicate the length of hydrolysis of the acrylamide gels in the pH 10 solution of tetramethylethylenediamine, the accelerator of the gelation of acrylamide.

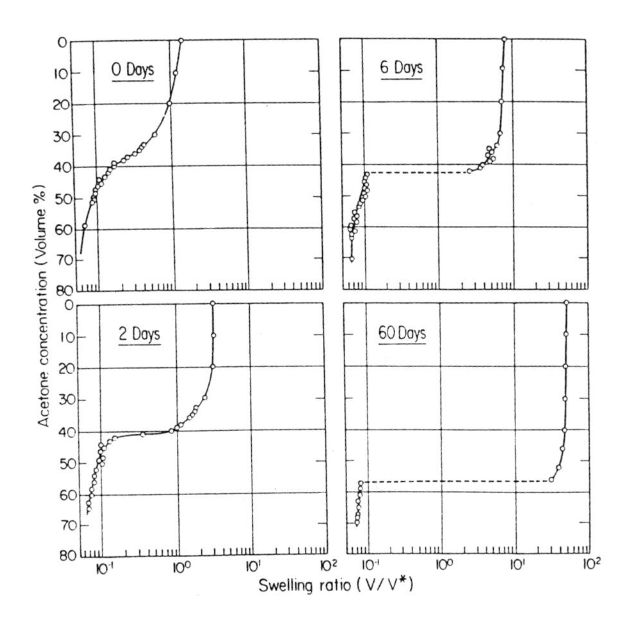

As the hydrolysis proceeds the transition acetone concentration becomes higher and the volume change at the transition increases. The volume change at the transition reaches 500 times at 60 days hydrolysis. A special hydrolysis duration exists at which the transition becomes discontinuous for the first time. The special condition of temperature, solvent composition and gel composition, at which the swollen phase and collapsed phase merge corresponds to the critical point.

Phase transition is universally observed in gels

The set of the swelling curves shown in the figure has a strong resemblance to the set of volume-temperature curves of water. Water becomes vapor at 100°C and changes its volume seventeen hundred times at one atmosphere. As the pressure decreases the vaporization temperature is lowered and the volume change at the transition becomes large. (This phenomenon is why a mountain climber needs a pressurized cooker to boil rice at high altitudes.) The ionization causes a similar effect on the gel phase transition. In gels the ionization adds an internal pressure, which is opposite to an external pressure. Ionization of a gel has a similar effect as decreasing the external pressure in water-vapor transition. Theoretically the swelling curve is given by minimizing the free energy per polymer segment with respect to polymer density or equivalently by the zero osmotic pressure on the gel. The Flory-Huggins formula gives the free energy in the form *(10,11,23)*:

$$F = F_{\text{rubber}} + F_{\text{counter-ion}} + F_{\text{virial}}$$

$$\frac{F_{\text{rubber}}}{VkT} \sim \nu_e \left[\left(\frac{\rho}{\rho_o} \right)^{1/3} + \left(\frac{\rho}{\rho_o} \right) \ln \left(\frac{\rho}{\rho_o} \right) \right]$$

$$\frac{F_{\text{counter-ion}}}{VkT} \sim f \nu_e \left(\frac{\rho}{\rho_o} \right) \ln \left(\frac{\rho}{\rho_o} \right)$$

$$\frac{F_{\text{virial}}}{VkT} \sim \left[v + \left(\frac{f}{\rho_o} \right)^2 \right] \rho^2 + w\rho^3 + x\rho^4 + \cdots \qquad (5)$$

where F_{rubber} represents the free energy of rubber elasticity, $F_{\text{counter-ion}}$ represents the counter ion osmotic pressure, and F_{virial} denotes the set of virial terms including the charge-charge repulsion. The curves calculated using this formula satisfactorily describes the phenomena qualitatively. Thus, the phase transition of a gel may be considered a phenomenon similar to the gas-liquid phase transition. The latter occurs in the group of separated molecules, whereas in the former all the molecules are connected in the form of a network.

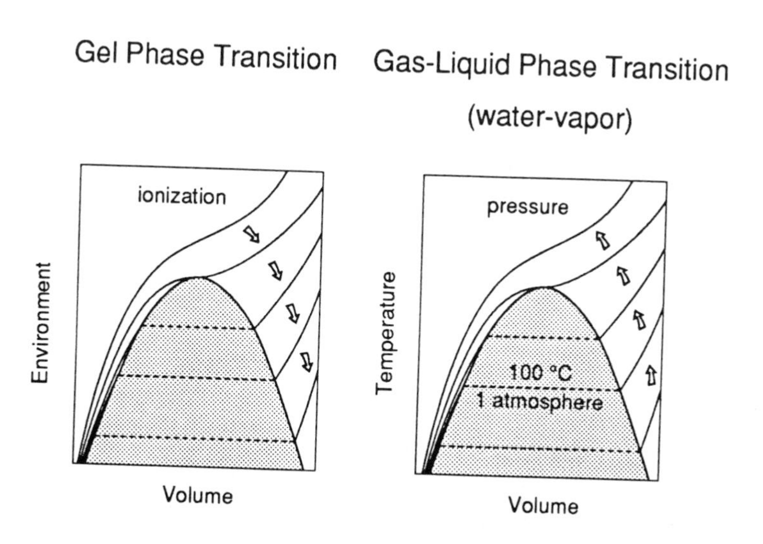

The analogy between the water-vapor transition and gel transition indicates that the gel transition should be observed in any gel, since the gas-liquid transition is also universal to any materials. The universality of the gel phase transition was confirmed by the observations of phase transition in gels of various synthetic polymers. Most of the synthetic polymers in which the transition was found had the vinyl chain backbone. As is known only three types of polymers exist in nature: polypeptides, polysaccharides, and polynucleotides. These polymers have backbone structures totally different from each other and from vinyl polymers. Amiya chose a representative from each one of these biopolymer groups and cross-linked them by covalent bonds. Gels made of gelatin, agarose, and DNA exhibited a phase transition *(27)*. Many synthetic gels have also been studied and the universality of the phase transition in gels seems to have been well established *(28-32)*. It may now be concluded that the phase transition of gels should be universal to any gel.

Theories of phase transition of gels have been extensively improved. Readers should refer to references such as *(33-36)*.

Examples of gel phase transition in the biological world

Now that the phase transition is universally observed in gels, it is natural to ask whether nature makes use of the phenomenon in biological activities. Verdugo found a fascinating example where the gel phase transition is indeed utilized in biological activity *(37)*. Slug mucin, which is stored in the body in an extremely compact form, swells more than 1000 times when secreted out of the body and absorbs water in its neighborhood. By doing so the slugs are able to keep water and maintain the moist environment necessary for survival. The idea is also used for recently available disposable diapers. The question has been raised by some biologists concerning how slugs can create such highly dense mucin for storage in the body full of water. A slug might need a significant osmotic pressure, thus a large energy to do so.

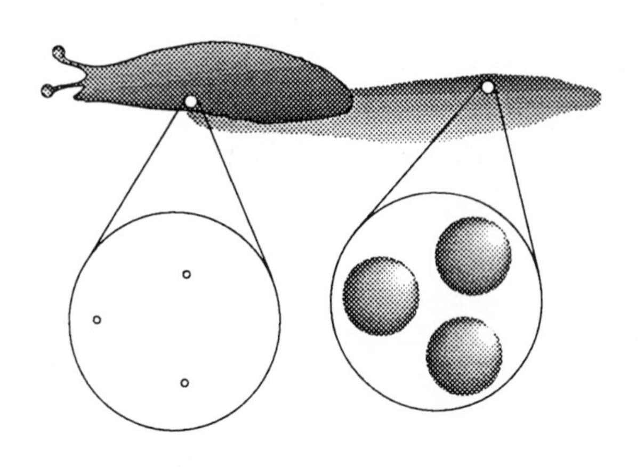

Verdugo and his colleagues found that the mucin undergoes phase transition in acetone-water mixtures, and by adding calcium ions. Namely slug does not need much energy to squeeze the mucins, but only have to increase calcium concentration around mucins. Such mechanisms seem to be used for other animals such as the eight-eyed eels in Japan, and probably for the mucins that cover our internal ducts. This finding is not only amusing but seems to have a profound implication for biological activities. It is interesting to note how sophisticated are the slug's method that utilize gels near the phase transition threshold and thus in the reversible condition, compared to the super water absorbents in diapers which are used only in the irreversible processes.

Verdugo further found that the delivery of histamines is carried out by phase transition of granular mucus gels *(38)*. The granules contain highly concentrated histamine molecules, which are released upon swelling transition after the granules are secreted out of a mast cell. The gel granules are collapsed with the presence of high concentration of histamine for recycling.

Some researchers consider that the phase transition of gels plays a vital role in muscular contraction and relaxation *(39)*, and also in neural excitation and signal transmission *(40)*. It is known that a cytoskeletal gel is needed for membrane excitation, and the gel changes its volume upon polarization and repolarization *(41)*.

Diminishing of gel-solvent friction

According to the study of dynamic light scattering of acrylamide gel, the friction between the water and the polymer network decreases reversibly and diminishes as the gel approaches to a critical point or spinodal line *(23)*. The physical picture is that the density fluctuations become large as the critical point is approached, which create effectively large pores as determined by the correlation length. The correlation length diverges and the friction diminishes at the critical point. Tokita found the decrease of gel-solvent friction by three orders of magnitude in poly(N-isopropylacrylamide) gel in water *(47-48)*.

Fundamental biological interactions and gel phase transitions

Four fundamental interactions play an essential role in biological nature in determining the structures and specific functions of macromolecules and their assemblies: hydrogen bonding; hydrophobic interactions; van der Waals interactions; and ionic interactions *(49)*. Molecular recognition that occurs between DNA strands, antigen-antibody, receptor-hormone, enzyme-substrate, and actin-myosin filaments are all achieved by exquisitely coordinating these forces in space. The magnitude, temperature dependence, and behavior in aqueous environment are totally different among the fundamental forces, and it is those differences that allow the variety in the biological functions created by these forces. Understanding how these forces determine the phase behaviors of polymers is of fundamental importance.

Gel phase transition is a result of a competitive balance between a repulsive force that acts to expand the polymer network and an attractive force that acts to shrink the network. The most effective repulsive force is the electrostatic interaction between the polymer charges of the same kind, which can be imposed upon a gel by introducing ionization into the network. The osmotic pressure by counter ions adds to the expanding pressure. The attractive interactions can be van der Waals, hydrophobic interaction, ion-ion with opposite kinds, and hydrogen bonding. The phase transition was discovered in gels induced by all the fundamental forces.

van der Waals interaction

The partially hydrolyzed acrylamide gel undergoes phase transition in acetone water mixtures *(26)*. The main polymer-polymer affinity is due to the van der Waals interaction. It was necessary to add acetone, a nonpolar poor solvent, to water in order to increase the attraction to a sufficiently large value to induce the transition. The

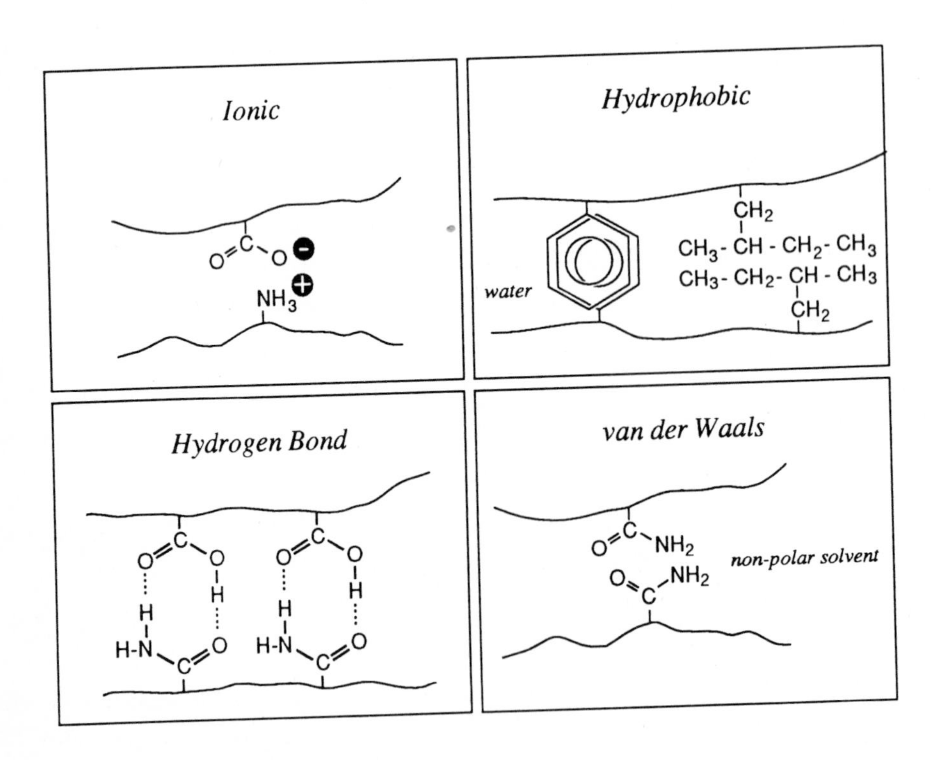

transition is also observed when the temperature is varied while the solvent composition is fixed near the transition threshold at room temperature. The gel swells at higher temperatures and shrinks at lower temperatures.

Hydrophobic interaction

In an attempt to find a gel that undergoes a volume phase transition in pure water, rather than acetone-water mixture, Hirokawa studied gels which had side groups with more hydrophobicity than that of acrylamide *(50)*. In the figure below the volumes of copolymer gels of N-isopropylacrylamide and sodium acrylate in water are plotted as a function of temperature as determined by Hirotsu, Hirokawa, and Tanaka *(51)*. This gel is unique in that the solvent is a simple liquid in contrast to the previous cases where mixtures of solvents were necessary to observe the phase transition. Without ionizable sodium acrylate, the gel undergoes a discontinuous volume change by eight times at T=33.0°C. As the sodium acrylate concentration increases, the transition temperature rises and the volume change becomes larger.

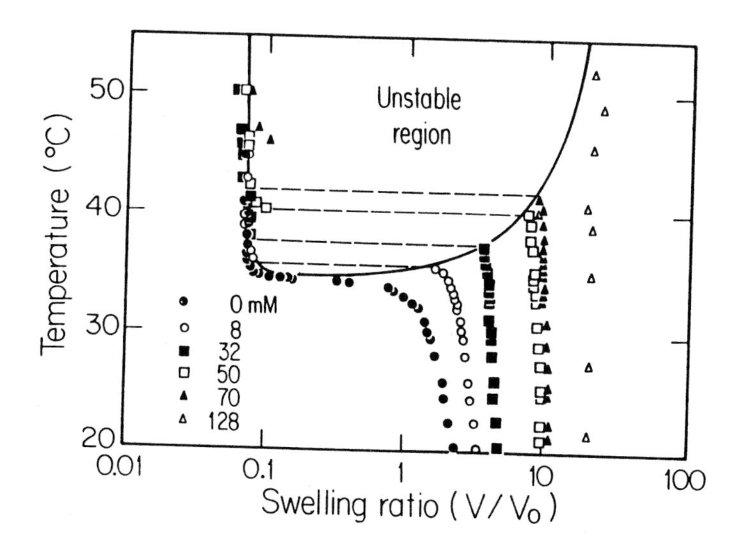

It is interesting to observe that the gel swells at lower temperatures and collapses at higher temperatures. This temperature dependence, which is opposite to the transition induced by van der Waals interaction, is due to the hydrophobic interaction of the gel and water. The water molecules in the vicinity of hydrophobic polymer chains have more ordered structures, and thus a lower entropy, than those away from the polymers. At higher temperatures the polymer network shrinks and becomes more ordered, but the water molecules excluded from the polymer network become less ordered. As a whole the gel collapses amount to a higher entropy of the entire gel system as should be. More detailed theory and experiments have been carried out *(52-54)*.

Hydrogen boding

We now look for a gel that undergoes phase transition in pure water, but in the temperature dependence opposite to that of hydrophobic interaction. Such a phase transition has been found in an interpenetrating polymer network (IPN) consisting of two independent networks intermingled each other *(55)*. One network is poly(acrylic acid) and the other poly(acrylamide). The gel was originally designed and developed by Okano and his colleagues *(56-59)*. They found that the gel is shrunken at low temperatures in water, and the volume increases as temperature rose. There was a sharp but continuous volume change at about 30°C. They identified the main interaction to be hydrogen bonding and also pointed out the importance of the so-called "zipper" effect, which described the cooperative nature of the interaction between two polymers *(56)*. Such polycomplexation phenomena were studied extensively in solutions of various polymer pairs *(60-64)*.

By slightly ionizing the gel, Ilmain, Tanaka, and Kokufuta succeeded in inducing the discontinuous volume transition of the IPN in pure water *(55)*. The transition temperature was approximately 20°C and a large hysteresis was observed.

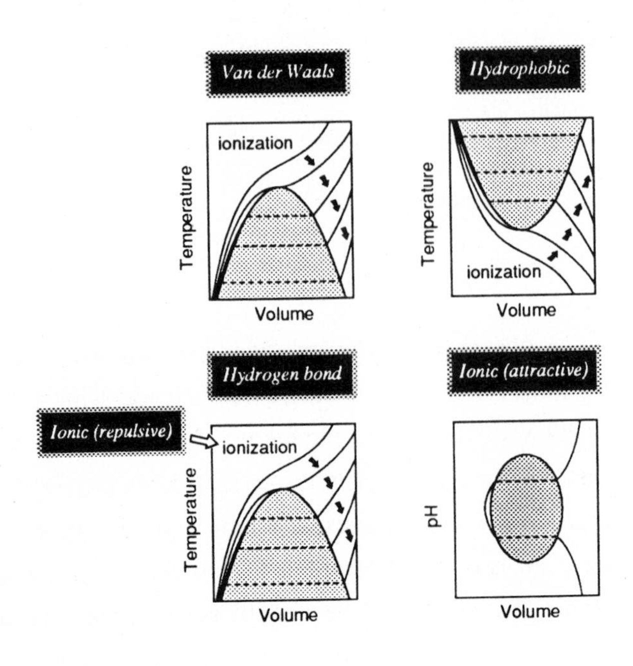

<u>Electrostatic attraction</u>

Polymers having both cationic and anionic groups are called polyampholytes. The polymers can be either positively or negatively charged, and repel each other over short ranges, but attract over long ranges *(65-69)*. Myoga and Katayama studied such a gel and observed that at neutral pH's the gel is indeed shrunken and at both higher and lower pH's *(70)* it is swollen. In the neutral pH's both cations and anions are ionized, and they attract each other, thus the gel shrinks. Otherwise the one of the ionizable groups is neutralized, while the other remains ionized, and the gel swells. The volume change was gradual and continuous.

Recently Annaka found a phase transition in a polyampholyte gel of copolymers of acrylic acid and methacrylamidopropylammonium chloride *(71)*. Phase transitions at higher pH and at lower pH are both accompanied by large pH hystereses.

These observations of phase transitions driven mainly by each one of the fundamental biological interactions allow us to draw a general picture of how polymers interact with each other through these interactions. It is important to note that each interaction is strong enough to induce a collapse but weak enough to allow swelling transition near the body temperature. The differences in the nature of each interaction are clearly revealed in the phase transition, particularly in the temperature dependence and in the size of hysteresis. Nature uses those differences to create extremely effective and specific molecular recognition mechanisms. It will be interesting to study the phase behavior of gels where combinations of these fundamental interactions are introduced and their balance is varied by changing variables such as temperature, solvent composition, and pH. We shall show in the next section that such a study is now revealing a fascinating new aspect in the phase behavior of gels.

New phases

We have seen that each of the four fundamental interactions induces a transition between two different phases in water near body temperature. What will happen if two or more interactions are combined within a gel? Would a gel, and in general a polymer system, have new phases other than just swollen or collapsed phases?

Such phases are known to exist in the biological world. For example, every protein has a unique and stable structure. Configuration of a natural protein must be at its free energy minimum separated by free energy barriers from other possible configurations. By thermodynamic definition the protein configuration should be a phase. Similarly an antibody forms a pair only with its antigen. Any other combination does not lead to a pair. Configuration of such a pair should thus be separated by the free energy barrier from other possible combinations. Again by definition the antibody-abtigen should be a phase.

However, no such phases were observed in synthetic polymer systems. It was thought that only biopolymers, not synthetic polymers, may have such phases. It took billions of years for nature to select some 10^6 proteins for biological activities, whereas theoretically 20^{400} variety is possible for polypeptides assuming each consists of 400 amino acids of 20 kinds. The number of actual proteins is negligible compared to the theoretically possible number of polypeptides. Proteins could be an exception. However, new phases have recently been found by Annaka in copolymer gels consisting of cationic and anionic groups that form interpolymer hydrogen bondings *(71)*. As temperature or pH is varied, the gel changes its volume discontinuously in water among many phases distinguished by different volumes. The number of phases and the transition thresholds depend on the the ratio of the cationic and anionic monomers in the polymer network. Each phase can be reached by following a different way of changing pH. New phases have also been discovered by Ishii in a gel made of homogeneous polymers where polymers are interacting through hydrogen bonding and hydrophobic interactions *(72)*.

The internal structures of these new phases are not yet understood. The finding demonstrated the possibility that the marvelous functions and structures, that were considered to be available only to biopolymers, may be achieved by ordinary synthetic polymers. It is a democratization of polymers.

On the other side, it will introduce into polymer science the concept and principles of specificity, individuality, and diversity: Macroscopic behaviors of polymers are far richer than one used to think and crucially depend on how the fundamental interactions are coordinated within the polymers.

Some future directions

Gel research is wide and deep. We have learned a great deal from the analogy of gels and other systems such as magnetic systems and gas-liquid systems. During the earlier stage of research, the detailed characteristics of gels were purposefully ignored to advance gel research into a rigorous physics. Such efforts indeed allowed a better and more general understanding of various aspects of gels.

However, gels have their own unique features. For example, gels have shear elasticity that polymer solutions do not have. Because of the shear elasticity, for example, the kinetics of gel shrinking and diffusion of ink have a fundamental difference. It was mentioned earlier in this paper that a gel swelling and shrinking obeys the diffusion equation. But it is actually necessary to add one more equation to fully describe the kinetic process, namely, the overall energy due to shear deformation should always be minimized. This is because the relaxation of shear deformation does not require relative motion between the network and water. This additional condition makes the diffusion process dependent not only on the size of a gel but on its shape. For example, Li predicted *(73)* that the relative values of effective diffusion coefficient of sphere : cylinder : slab = 1 : 2/3 : 1/3. This prediction was confirmed experimentally and interpreted as follows. A spherical gel diffuses in all the three dimensions. A cylindrical gel can diffuse only in the radial direction, since there should be no relative motion between water and network along the axis. The diffusion along the radius is shared by the axial shear relaxation and thus becomes two-thirds that of a spherical gel. A slab of gel can diffuse only along the thickness, not within the plain. The diffusion should be shared by two other dimensions, and the overall diffusion becomes one third of the sphere.

Because of the shear elasticity, a gel can have a wide variety of patterns during an extensive swelling or shrinking processes *(74)*. Permanent structural inhomogeneities are also specific to gels, and to identify their origin is an important task.

Universality classes of the phase transition of gels may well be dependent on the attractive interaction that drives the transition. For example, the hydrogen bond requires a specific orientation for the two molecules, and thus new order parameters such as hydrogen bond density and orientation will be needed to describe the equation of state. Then the universality class of gels having hydrogen bonding as a key attractive interaction may be different from that of the hydrophobic gel, which is the only gel system whose universality class has been identified.

Finding principles associated with the uniqueness of gels is thus one of the key directions.

A second important direction for future research is to establish the principles underlying the individuality or specificity of gels. Namely, the principles should be established that predict the phase behavior of gels from the knowledge of their chemical structures. As shown in the recent studies of new phases of gels, it is clear that small changes in chemical composition induces a drastic change in the macroscopic phase properties of gels. This area has to be explored through combined efforts of experiments and theory *(75-80)*.

Finally, a rapid progress has been underway in technological applications of polymer gels, including chemical design, synthesis, fabrication in various length scales from μm to cm, and development of devices.

References

This review overlaps with contents given in the ACS Book on *Polyelecrtolyte Gels* edited by Ronald Harland and Robert Prud'homme.

(1) DeRossi, D., Kajiwara, K., Osada, Y., and Yamauchi A., *Polymer Gels* Plenum Press, New York, 1991

(2) Yamauchi, A., (Editor), *Organic Polymer Gels (Japanese)*, Gakkai Publisher, Tokyo, 1990

(3) In this article only gels cross-linked by permanent chemical bondings will be considered. There are numerous reviews for physically cross-linked gels. Refer, for example *(4-7)*.

(4) Kramer, O. (Editor), *Biological and Synthetic Polymer Networks*, Elsevier Appl. Sci., London, 1988

(5) Clark, A. H. and Ross-Murphy, S. B., Adv. Polym. Sci., **1987**, *83*, 60

(6) Stepto, R.F.T. and Haward, R. N., *Development in Polymerization - 3. Network Formation and Cyclization*, Appl. Sci., London and New York, 1982

(7) Stauffer, D., *Introduction to Percolation Theory*, Taylor and Francis, London, 1985

(8) Masuda, F., *Superabsorbent Polymers (in Japanese)*, Kyoritsu Press, Tokyo, 1987

(9) Fushimi, T., *Ideas of Applications of Superabsorbent Polymers (in Japanese)*, Kogyo Chosakai, Tokyo, 1990

(10) Flory, P. J., *Principles of Polymer Chemistry* Cornell University Press, Ithaca, NY, 1957

(11) DeGennes, P. G., *Scaling Concepts in Polymer Physics*, Cornell University, Ithaca, New York, 1979

(12) Tanaka, T., *Sci. Am.* , **1981**, *244*, 110.

(13) Dusek, K. and Patterson, D., J. Polym. Sci., **1968**, *Part A-2, 6*, 1209.

(14) Ptitsen, O. B.. and Eizner, Yu. E., Biofizika, **1965**, *10*, 3

(15) Lifshitz, I. M., Grosberg, A. Yu., and Khokhlov, A., R., Rev. Modern Phys., **1978**, *50*, 683-713

(16) DeGennes, P. G., Phys. Lett., **1972**, *A38*, 339
(17) Tanaka, T., Hocker, L. O., Benedek, G. B., J. Chem. Phys., **1973**, *59*, 5151
(18) Munch, J. P., Candau, S. J., Duplessix, R., Picot, C., Herz, J., and Benoit, R., J. Polym. Sci. **1976**, *14*, 1097
(19) Geissler, E. and Hecht, A. M., Macromolecules, **1980**, *13*, 1276, , **1981**, *14*, 185
(20) Doi, M. and Edwards, S., *The Theory of Polymer Dynamics*, Oxford Univ. Press, Clarendon, 1980
(21) Onuki, A., Phys. Rev. A, **1989**, *39*, 2308
(22) Peters, A. and Candau, S. J., Macromolecules, **1986**, *19*, 1952, **1988**, *21*, 2278
(23) Tanaka, T., Ishiwata, S., and Ishimoto, C., Phys. Rev. Lett., **1977**, *39*, 474
(24) Tanaka, T.; Phys. Rev. Lett., **1978**, *40*, 820
(25) Hochberg, A., Tanaka, T., and Nicoli, D., Phys. Rev. Lett., **1979**, *43*, 217
(26) Tanaka, T., Fillmore, D.J., Sun, S.-T., Nishio, I., Swislow, G., and Shah, A., *Phys. Rev. Lett.*, **1980**, *45*, 1636
(27) Amiya, T. and Tanaka, T., Macromolecules, **1987**, *20* 1162
(28) Katayama, T. and Ohta, A., Macromolecules, **1985**, *18*, 2781
(29) Siegel, R. A. and Firestone, B. A., Macromolecules, **1988**, *21*, 3254
(30) Ilavsky, M., Macromolecules, **1978**, *59*, 5151
(31) Hrouz, J., Ilavsky, M., Ulbrich, K., and Kopecek, J., Europ. Polym. J., **1981**, *17* 361
(32) Ilavsky, M., Hrouz, J., and Ulbrich, K., Polym. Bull., **1982**, *7* 107
(33) Khokhlov, A. R., Polymer **1980**, *21*, 376
(34) Grosberg, A. Yu. and Khokhlov, A. R., *Statistical Physics of Macromolecules*, Nauka, Moscow, 1989
(35) Grosberg, A. Yu., Nechaev, S. K., Shakhnovich, E. I., J. de Phys. (Paris), **1988**, *49*, 2095
(36) Khokhlov, A. R. and Nechaev, S. K., Phys. Lett., **1985**, *112-A*, 156
(37) Verdugo, P., Biophys. J., **1986**, *49*, 231
(38) Verdugo, P., Annu. Rev. Physiol., **1990**, *52*, 157
(39) Shay, J. W., *Cell and Muscle Mobility*, Plenum Press, New York, 1984
(40) Tasaki, I., *Physiology and Electrochemistry of Nerve Fibers*, Academic Press, New York, 1982
(41) Arai, T. and Matsumoto, G., J. Neurochem., **1988**, *51*, 1825, **1989**, *52*, 93
(42) Wilson, K. G. and Kogut, J. B., Physics Reports, **1974**, *12c*, 77
(43) Stanley, H. E., *Introduction to Phase Transition and Critical Phenomena*, Oxford Univ. Press, Oxford, 1971

(44) Li, Y. and Tanaka, T., J. Chem. Phys., **1989**, *90*, 5161

(45) Onuki, A., Phys, Rev. A., **1988**, *38*, 2192

(46) Onuki, A., J. Phys. Soc. Japan, **1988**, *56*, 699 and 1868

(47) Tokita, M., Tanaka, T., J. Chem. Phys., **1991**, *in press*

(48) Tokita, M., Tanaka, T., *in press*

(49) Lehninger, A. L., I., *Principles of Biochemistry,* Worth Publishers, New York, 1982

(50) Hirokawa, Y. and Tanaka, T., J. Chem. Phys., **1984**, *81*, 6379

(51) Hirotsu, S., Hirokawa, Y., and Tanaka, T. J. Chem. Phys., **1987**, *87*, 1392

(52) Otake, K., Tsuji, T., Konno, M., and Saito, S., J. Chem. Eng. Japan, **1988**, *21*, 443

(53) Inomata, H., Yagi, Y., and Saito, S., Macromolecules, **1990**, *29*, 4887

(54) Ichita, H., Miyano, Y., Kiyota, y., and Nakano, Y., *Proc. Gel Conference in Tokyo*, Polymer Soc. Japan, pp.92-93, 1991

(55) Ilmain, F., Tanaka, T., and Kokufuta, E., Nature, **1991**, *349*, 400

(56) Okano, T., Bae, Y. H., and Kim, S.W., in *Modulated Control Release System, (Ed. by Kost, J.)*, CRC Press, *(in press)*

(57) Okano, T., Bae, Y. H., Jacobs, H., and Kim, S.W., J. Controlled Release, **1990**, *11*, 255

(58) Bae, Y. H., Okano, T., and Kim, S.W., Makromol. Chem. Rapid Commun., **1987**, *8*, 481, **1988**, *9*, 185

(59) Bae, Y. H., Okano, T., and Kim, S.W., J. Controlled Release, **1989**, *9*, 271

(60) Baranovsky, Yu. V., Litmanovich, A. A., Papisov, I. M., Kabanov, V.A., Europ. Polym. J., **1981**, *17*, 969

(61) Eustace, D. J., Siano, D. B., and Drake, E. N., J. Appl. Polym. Sci., **1988**, *35*, 707

(62) Osada, Y., J. Polym. Sci. Polym. Chem. Ed. **1979**, *17*, 3485

(63) Tsuchida, E. and Abe, K., *Interaction between Macromolecules in Solution,* Springer-Verlag, Berlin (1982)

(64) Abe, K. and Koide, M., Macromolecules, **1977**, *10*, 1259

(65) Edwards, S., King, P. R., and Pincus, P., Ferroelectrics, **1980**, *30*, 3

(66) Qian, C. and Kholodenko, A. L., J. Chem. Phys., **1988**, *89*, 5273

(67) Khokhlov. A. R., and Kachaturian, K. A., Polymer, **1982**, *23*, 1742

(68) Higgs, P., and Joanny, J-F., J. Chem. Phys., **1991**, *94*, 1543

(69) Kantor, Y. and Kardar, M., (to be published) show that a polyampholyte has a

repulsive interaction in the short range.

(70) Myoga, A., and Katayama, S., Polym. Prep. Japan, **1987**, *36*, 2852

(71) Annaka, M. and Tanaka, T., (to be published)

(72) Ishii, K. and Tanaka, T. (to be published)

(73) Li, Y. and Tanaka, T., J. Chem. Phys., **1990**, *92,* 1365

(74) Onuki, A., *Formation, Dynamics, and Statistics of Patterns, (Edited by Kawasaki, K.)*, World Sci. Singapore, 1989
Sekimoto, K. and Kawasaki, K., J. Phys. Soc. Japan, **1987**, *56*, 2997

(75) Kantor, Y. and Kardar, M., (to be published) show that a polyampholyte has a repulsive interaction in the short range.

(76) Golubovic, L., and Lubensky, T., Phys. Rev. Lett., **1989**, *63*, 1082

(77) Shakhnovich, E. I., and Gutin, A. M., Europhys. Lett., **1989**, *8,* 327

(78) Goldbart, P., and Goldfeld, N., Phys. Rev. Lett., **1987**, *58*, 2676

(79) Umemoto, S., Itoh, Y., Okui, N., and Sakai, T.,
Reports on Progress in Polym. Phys. in Japan, **1988**, *31*, 295

(80) Umemoto, S., Okui, N., and Sakai, T., *Polymer Gels,* pp257-270,
Plenum Press, New York (1991)

POWER GENERATION BY MACROMOLECULAR POROUS GELS

M. Suzuki, M. Matsuzawa*, M. Saito* and T. Tateishi
Biomechanics Division
Department of Applied Physics
and Information Science
Mechanical Engineering Laboratory
1-2 Namiki, Tsukuba 305 Japan

Introduction

From the view point of mechanical engineering, polymer gels are being expected as a muscle-like soft linear actuator which is awaited in the fields of robotics and medical welfare instruments. The basic idea originates from Kuhn and Katchalsky's work on polyelectrolyte gels in 1950. Up to this point various kinds of polymer gels have been studied as chemomechanical energy conversion materials. Some reviews are now available, such as by Osada[1987]. However gels have never been applied to any actuator uses. Some of the main reasons are as follows. First the response is slow, second the contraction force is still not enough, thirdly energy source is not familiar compared to usual actuator's and fourthly the control method has not been established. Animal muscle generates contraction tension about 0.5MPa within 30ms by neural control. Many of recent studies are relating to volume phase transition phenomena of poly(acryl amide) gel[Tanaka,1978], poly(N-isopropyl acrylamide) gel [Hirokawa,1984] and poly(vinyl methyl ether) gels [Hirasa, 1989]. The volume phase transition in gel was first pointed out and studied by Tanaka and his coworkers[1978,1979,1984]. As an actuator material poly(acrylonitril) gel fibers were treated with a heat process and hydrolysis to get amphoteric hydrogel fibers by Okui[1987], which showed high tensile strength and large shrinkage. On the other hand in 1982 Nanbu found that poly(vinyl alcohol) (PVA) aqueous solution more than 5w% becomes a strong hydrogel by repetitive freezing and thawing. Then one of the authors have tried to enhance chemomechanical properties of PVA hydrogel by intermixing polyelectrolytes before repetitive freezings. It was so successful that one could expect amphoteric PVA hydrogel as an artificial muscle.[Suzuki, 1986, 1987,1989,1991] The purpose of this study is to clarify how this material can become closer to muscle in the mechanical active properties.

1.Preparation of Amphoteric PVA Hydrogel

---

*University of Tsukuba, 2-1 Tennodai, Tsukuba 305 Japan

NATO ASI Series, Vol. H 64
Mechanics of Swelling
Edited by T. K. Karalis

A uniform aqueous solution was obtained by mixing poly(vinyl alcohol) (Kurare 117H; Mw:74,800; saponification>99.6%), poly(acrylic acid) (-COOH, SP2; Mw: 170,000), and poly(allyl amine) hydrochloride ($-NH_3Cl$, Nittobo; Mw: 60,000) at monomeric concentrations of 1.74 M, 0.245 M and 0.26 M, respectively. With some spacers the solution was sandwiched between two plates of optically ground heat resistant glass of 10 mm thick, to form a liquid layer of constant thickness. Amphoteric PVA hydrogel films were then obtained by subjecting the solution to repetitive freezing and thawing. The temperature range was between-50 degC and room temperature around 23 degC, with temperature

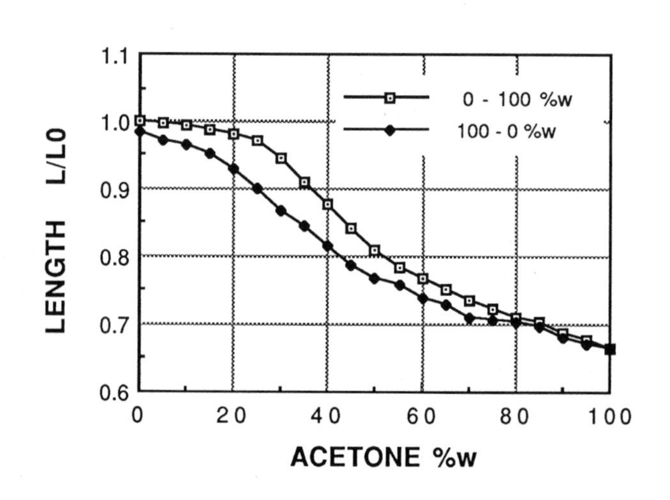

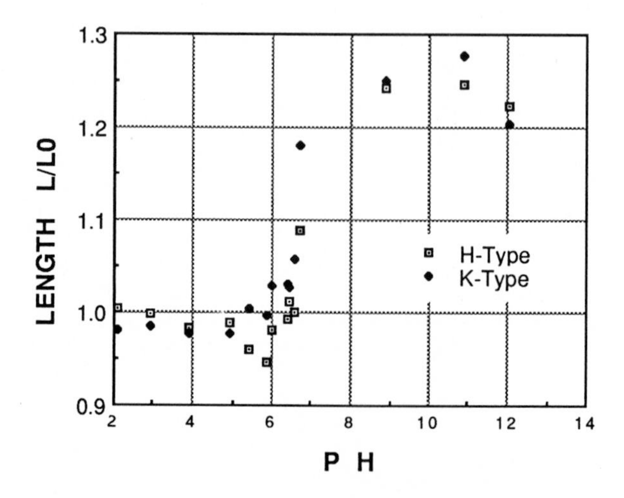

Fig.1 Equilibrium Swelling of Amphoteric PVA Hydrogel in a:water/acetone mixture, b:HCl/KOH aqueous solution. H-Type was brought from HCl 1N and K-Type from KOH 1N.

falling and rising at a rate of 3.0 deg/min just before the freezing and thawing points. Some gel films were rendered anisotropic by stretching uniaxially. The properties of this anisotropic materials were already reported in the previous paper. [Suzuki,1989,1991] To get finished samples, the gel films were washed with 1N KOH, 1N HCl, and water three times.

The equilibrium swelling characteristics of these gels in a mixed solvent of acetone and water are shown in Fig.1a and in HCl/KOH aqueous solution without extra salts in Fig.1b. Inside the amphoteric PVA gel ionization of polyelectrolytes occurs, for example poly(acrylic acid) does not dissociate in acidic solution and poly(allylamine) does not dissociate in alkali solution. Therefore the gel swells both in acidic and basic solutions and shrinks in neutral pH. In acetone those polyelectrolytes cannnot dissociate so that they associate each other because the affinity between solvent and polyelectrolytes are low. It causes gel contraction in acetone. In the following experiments were made with acetone/water solvent.

To examine the effect of freezing rate on the internal structure and the mechanical properties of hydrogel we have made three types of gel film A,B and C by changing the freezing rate 3.78, 0.0766 and 0.0329 deg/s, respectively at the freezing point.

2. Mechanical Properties of PVA hydrogel

2.1 Measurement of modulus and poisson ratio

To characterize the mechanical properties of gel at least we have to measure modulus and poisson ratio first. After a gel film was dipped in a solvent for a sufficient time to reach its equilibrium state the film was pulled in the longitudinal direction for the modulus measurement by using the apparatus shown in Fig.2. At one end of the bar a weight was hung as well as the gel film at the other end. The solvent was a mixture of water and acetone. The Young's modulus of gel film was measured for each acetone concentration. The modulus changed from 0.2MPa in water to 10MPa in acetone as shown in Fig.3. The Young's moduli for gels A, B and C were 0.049,0.101 and,0.245MPa, respectively.

The poisson ratio was measured both from the width change by taking photographs and from the thickness with Fig.5 system. The measurement of poisson ratio from the width change was difficult over the whole range of acetone concentration. At the high acetone concentration region the film became hard and wavy, on the other hand at the low acetone concentration the film was so soft to fold itself by small perturbations. Although there are big error bars one can see two maxima around 25w% and 65w% of acetone content as shown in Fig.4. The values of poisson ratio were much smaller than 0.5. Fig.5 shows the apparatus to detect the thickness change of hydrogel film. The gel film is set between two pins, one of which is fixed at the bottom end of balance bar(TB) and the other at the end of fixed bar(FB). Simultaneously the gel film was subjected to a longitudinal tention with LB bar. The thickness change of the gel film causes the angle change of the balance bar(TB) which can be

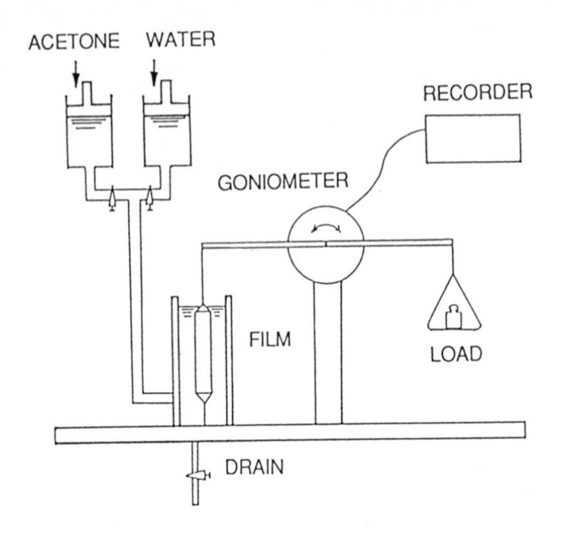

Fig.2 Apparatus to measure static and dynamic change of film length. Acetone or Water was injected to the cell with one of the pressurized syringes. When the film contracts the angle change of the bar was electrially detected.

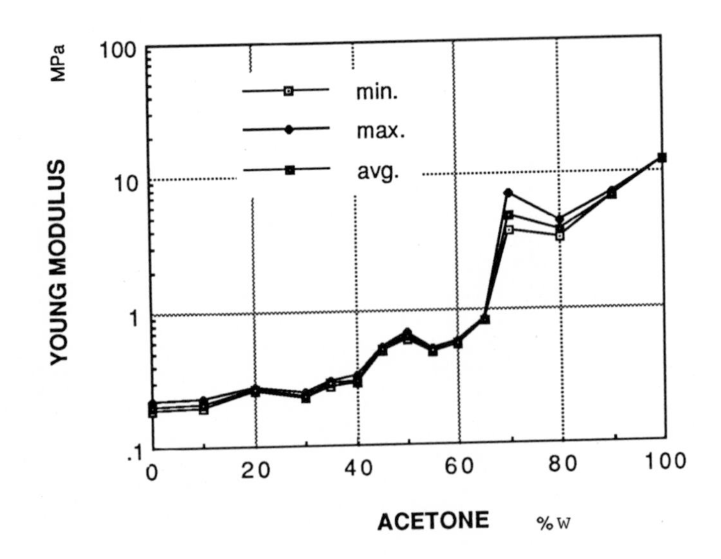

Fig.3 Young's modulus vs. acetone concentration in acetone/water mixed solvent.

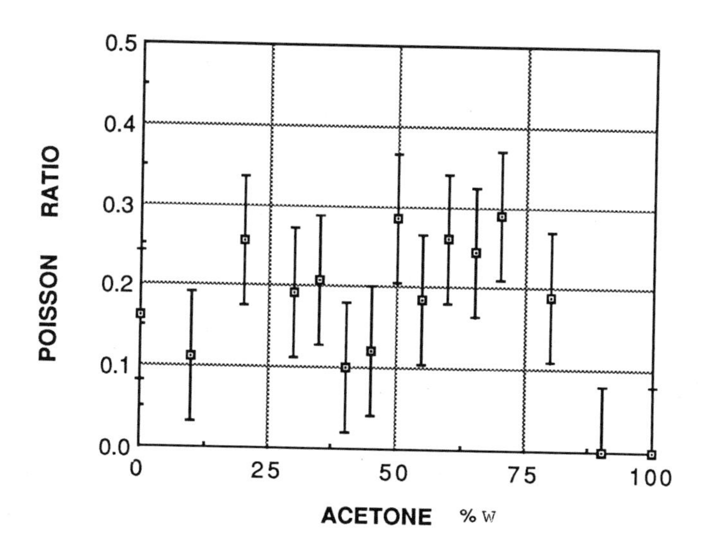

Fig.4 Poisson ratio vs. acetone concentration. This relation was obtained from width change of gel film by photographs.

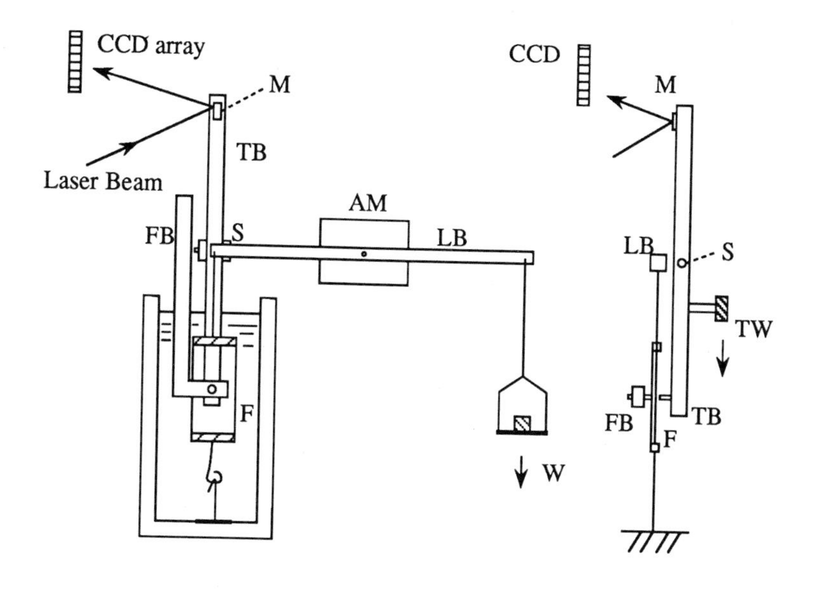

Fig.5 Apparatus to measure thickness change of gel film. AM:anglemeter, LB:bar to apply a longitudinal tension,W:weight, TB:bar to detect thickness change, M:mirror, S:fulcrum, FB:fixed bar,F:gel film, TW: weight to apply preload on gel film through pins.

detected with a CCD array and a laser beam reflection technique. This system could easily resolve 100nm in thickness change. Fig.6 shows the dynamic change of film thickness when the solvent was exchanged from water to acetone or vice versa. The preload applied to the pin on TB of 1mm diameter can be varied from 0.08g to 1.0g with a weight(TW). From this, zero-preload deformation was obtained after extraporation. Fig. 7 shows the poisson ratio obtained with this method in water. The value was about 0.08 below the longitudinal load of 0.12MPa, which was close to or lower than that obtained from width change.

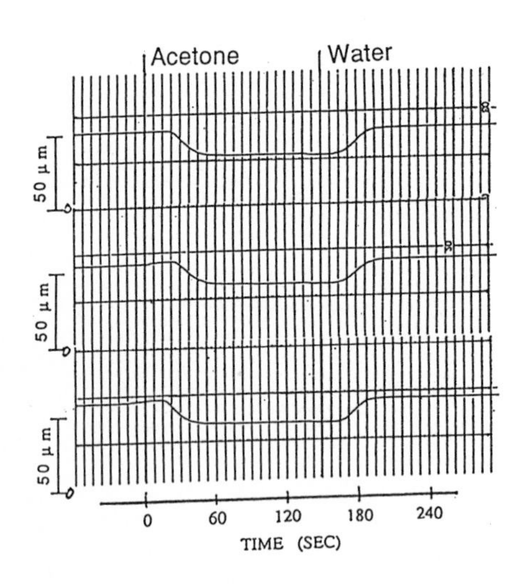

Fig.6 Thickness change due to solvent exchange. Applied preload was ,from the top, 0.52, 0.11,0.08g, respectively.

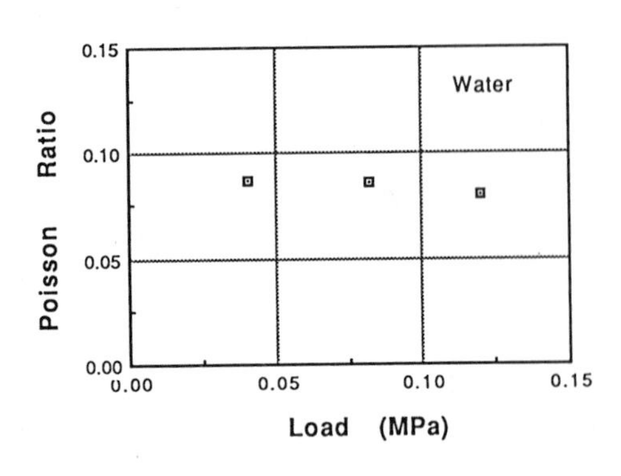

Fig.7 Poisson ratio obtained from thickness change with the apparatus in Fig.5 in water.

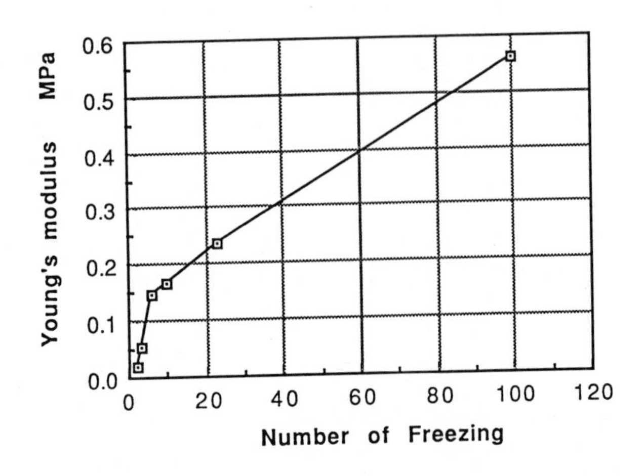

Fig.8 Relation between elasticity and number of freezing times.

2.2 Elasticity increase due to repetitive freezing

As increasing the number of freezing times the elasticity (Young's modulus) increased from 0.02MPa up to 0.56MPa as shown in Fig.8. It is attributed to increased physical crosslinking due to cooperative hydrogen bonds between polyvinylalcohol chains. There is a distinction in the slope around six times freezing above which the increasing slope of elesticity decreased.

3. Structural Observation by SEM

The hydrogel films were first quickly dipped in liquid nitrogen precooled below -208 degC by evacuation, then cut in it. The samples frozen were removed on a copper block precooled at -15 degC, then evacuated for a day below 1Torr in order to evaporate water directly from ice to vapor. By this treatment the structural deformation of hydrogel network due to surface tention was avoided. The samples were then coated with gold and observed with a usual scanning electron microscopy technique. Fig.9 shows the cross sectional views of hydrogel films A,B and C. As increasing the freezing rate the pore size became smaller and smaller. For the slowly frozen gels there were porous region and dense region which should lead to slow response.

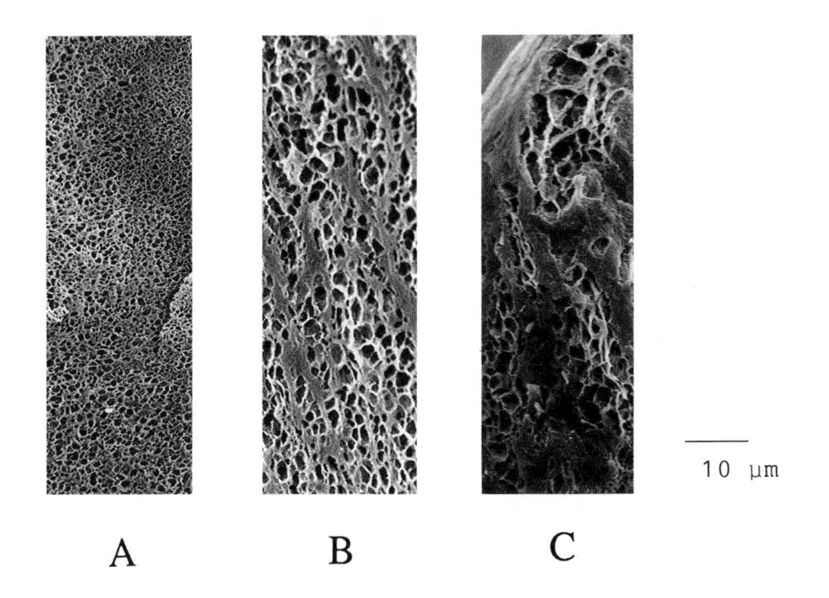

Fig.9 Cross sectional views of gel A,B and C obtained with scanning electron microscopy. Freezing rates were, A:3.78, B:0.0766, C:0.0329 in deg/s.

4. Mechanical Power Generation by PVA Hydrogels
4.1 Dynamic response of chemomechanical reaction

The response time of the chemomechanical volume change of hydrogel was thought to be governed by the motion of equation of network segment if the chemical reaction inside the network were very fast. Since the pore size of gel network is so small the inertia term in the equation of motion can be neglected as suggested by Tanaka[1973,1979]. It gives the collective diffusion equation for the segment displacement. From this relation the characteristic time constant is proportional to the square of the typical size and to the inverse of collective diffusion coefficient D which is given by the modulus divided by the friction. In order to improve the chemomechanical response one may take several ways realizing 1) thinner film or fiber, 2) high elasticity material, 3) low friction material, 4) porous structure material and so on.

The effect of the thickness of gel film on the response is shown in Fig.10. When the thickness was 10 micrometers the contraction response time was about 0.2s which is almost same speed as animal skeletal muscle contraction.

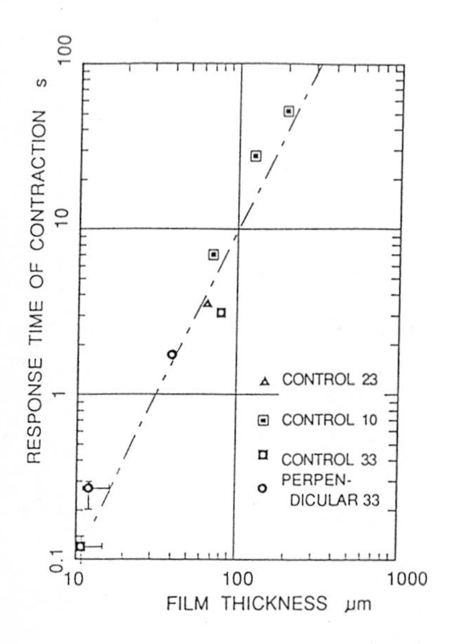

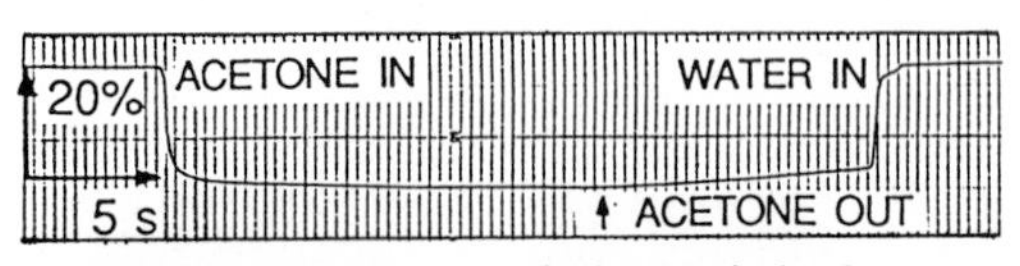

Fig.10 Relation between contraction response time vs. film thickness.

To obtain a high elasticity material one may utilize the method as mentioned in section 2.2. Fig.11 shows the result of contraction ratio vs. load as a function of freezing times. The higher the elasticity the higher is the contraction ratio under loading. However the response was not clearly improved.

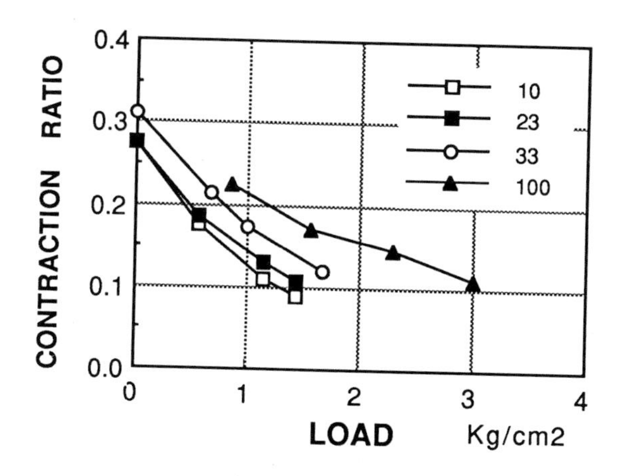

Fig.11 Contraction ratio vs.load as a function of number of freezing times.

Porous structure is one substantial answer to realize a low friction and high elasticity material. As mentioned in section 1 and section 3 by changing the freezing rate the porous structure of PVA hydrogel could be controlled. Fig.12 shows the typical contraction curves for A,B and C. Under loading condition the largest initial contraction speed was found in gel C but the longest settling time was also from gel C. This result tells us that the larger the pore size and the thinner the dense region the faster is the response.

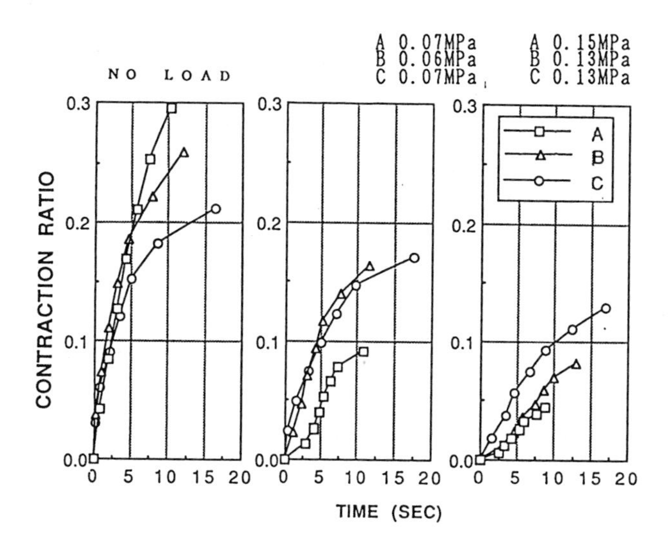

Fig.12 Contraction curves for A,B and C under several loading conditions. The curves are indicated with time and were terminated at 7/8 of final contraction ratio.

4.2 Power generation by amphoteric PVA hydrogels

The mechanical power generated by hydrogel films was evaluated from contraction force multiplied by contraction speed. The speed v was obtained from the time t at 50% of total length change(l'-l) as v = (l'-l)/2t. Fig.13 shows the generated power density for various amphoteric PVA gel films. In the case of the film of 10 micrometers thick could generate as high as 0.1 watt/g which is more than frog muscle's. The freezing rate of this gel was 0.05 deg/s. In Fig.12 the group of numbers means thickness, number of freezing times, anisotropy(-/no,+/yes) and the water content of hydrogel at the initial solution state before gelation, respectively.

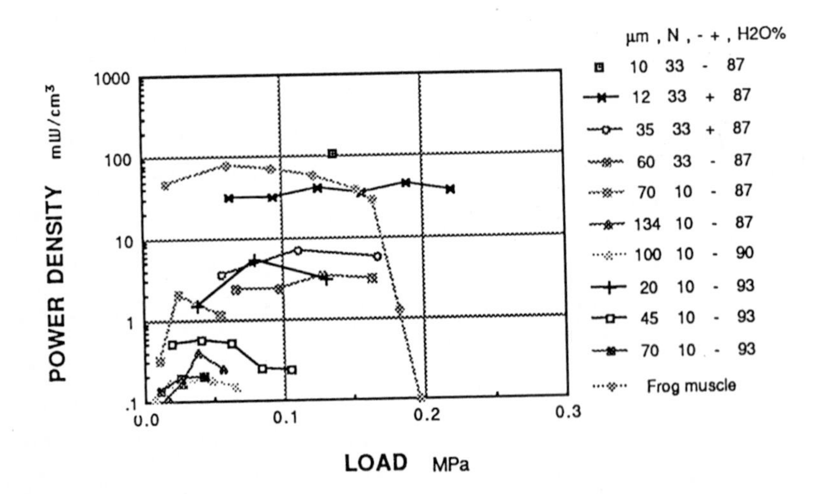

Fig.13 Generated power density vs. load for various gel films made in diferent conditions. Power density of frog muscle is indicated for comparison's sake.

Conclusion

In this study we have examined several ways to improve the chemomechanical actuator material. It was ascertained that a high elasticity and low friction material can generate large power density, especially when the gel had a suitable pore structure. These conditions were well satisfied with amphoteric PVA hydrogel made by means of repetitive freezing process. Since the response time was strongly affected by the typical size the directions of the material developement should be to combine micronization of material and micropore structure control technique. To realize an actuator device methods of energy input and integration of material unit should be developed in the near future.

Acknowledgement

The authors would like to express their sincere thanks to Dr. S. Fujishige for his important suggestion, to Mrs.Y. Satoh in carrying out the experiment, to the members of Biomechanics division, and also to Dr. Y. Osada, Dr. A. Yamauchi, Dr.O.Hirasa and Dr. T.Tanaka for fruitful discussions. The authors also thanks to Ms.K.Saitoh of MEL who kindly cooperated to take cryo-SEM photographs. This study was supported by AIST and STA.

References

Hirasa O,Morishita Y,Onomura R,Ichijo H,Yamauchi A(1989) Preparation and mechanical properties of thermo-responsive fibrous hydrogels made from poly(vinyl methyl ether)s(in Japanese).Kobunshi Ronbunshu 46:11-661

Hirokawa H,Tanaka T(1984)Volume phase transition in a nonionic gel.J Chem Phys 81:12-6379

Kuhn W,Hargitay B,Katchalsky A,Eisenberg H(1950)Reversible Dilatation and Contraction by Changing the State of Ionization of High Polymer Acid Networks.Nature 165:4196-514

Nanbu M(1983)High water content rubber(in Japanese).Polymer Appl 32:11-523

Okui N,Umemoto S,Matumura T,itoh Y,Sakai T(1987)High speed reversible contraction gel fiber.In:Chem Soc Japan(ed) IUPAC CHEMRAWN VI,Tokyo,pIIK05

Osada Y(1987) Conversion of Chemical into Mechanical Energy by Synthetic Polymers.In:Olive S,Henrici G(ed)Advances in Polymer Science82,Springer,Berlin Heidelberg,p1

Suzuki M,Tateishi T,Ushida T,Fujishige S(1986) An Artificial Muscle of Polyvinyl Alcohol Hydrogel Composites, Biorheology 23:3-274

Suzuki M(1987)Responsive Mechanochemical Actuator Materials by PVA Hydrogel.Chem Soc Japan(ed) IUPAC CHEMRAWN VI,Tokyo,pIB11

Suzuki M(1989)Amphoteric Poly(vinyl alcohol) Hydrogel as a Material of Artificial Muscle(in Japanese).Kobunshi Ronbunshu 46:10-603

Suzuki M(1991)Amphoteric Polyvinyl Alcohol Hydrogel and Electro hydrodynamic Control Method for Artificial Muscles.In:DeRossi D,Osada Y,Yamauchi A,Kajiwara K(ed)Polymer Gels,Plenum, New York,p221

Tanaka T,Hocker LO,Benedek GB(1973)Spectrum of light scattered from a viscoelastic gel.Phys Rev Lett 59:9-5151

Tanaka T,Fillmore DJ(1979)Kinetics of swelling of gels.J Chem Phys 70:3-1214

Tanaka T(1978)Collapse of Gels and the Critical Endpoint.J Chem Phys 40:12-820

SWELLING AND CONSOLIDATION OF A MUSCLE-LIKE HYDROGEL

P. J. Basser
Biomedical Engineering and Instrumentation Program
National Institutes of Health
Bethesda, MD 20892
USA

Introduction:

Polyelectrolyte gels and soft tissue have many common characteristics. Because of their muscle-like compliance and the ease with which they convert electrical or chemical energy into mechanical work (Kuhn, et al., 1950), gel actuators for robotics and gelatinous artificial muscles for prosthetic devices are under development. However, long mechanical response times and ease of fracture of these hydrogels have impeded their production.

Recently, a synthetic hydrogel was developed that contracts as rapidly and forcefully as skeletal muscle (Suzuki, 1989a, b). It is an interconnected network of dense hydrogen-bonded polymer fibers consisting of polyvinyl alcohol (PVA) and polyallylamine (PAllyA). Its open-cell structure gives the gel rigidity and strength. Scanning electron micrographs reveal that this amphoteric gel's network architecture is both homogeneous and isotropic (Suzuki, 1989a, b). The fiber network defines macroscopic pores through which interstitial solvent flows when the gel is either compressed or stretched. As such, this gel is poroelastic; it consolidates as soils do when it is loaded.

Still, this hydrogel exhibits chemical activity; it swells or contracts in response to changes in pH or solvent composition. Figure 1 illustrates a chemically induced shape change caused when acetone is exchanged for water as the gel's interstitial solvent. The equilibrium fractional change in gel network volume, $(V-V_0)/V_0$, varies markedly with concentration of acetone in the interstitial solvent.

To exploit these phenomena (consolidation and chemical activity) in devices such as artificial muscles, more must be known about their origin. The artificial muscle designer would like to predict the dynamic and steady-state response of the gel--in particular, the time course and distribution of mechanical stress during stimulation--and the velocity and displacement distributions when the material is loaded or unloaded.

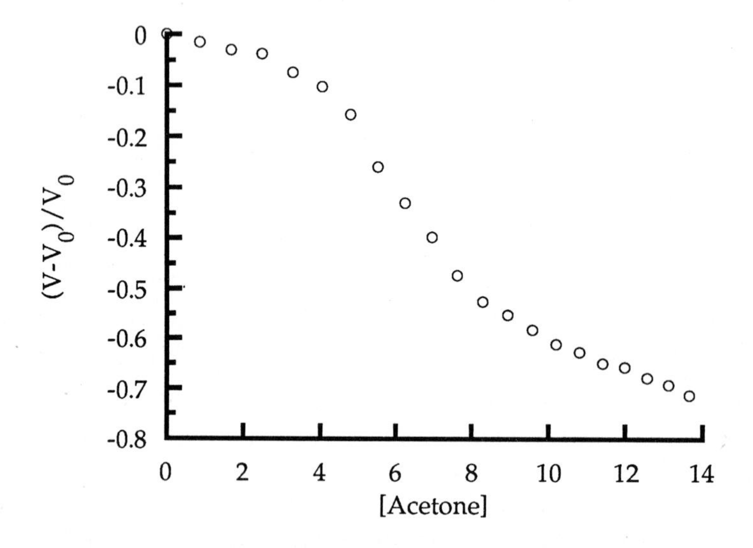

Figure 1. Network dilatation vs. molar concentration of acetone. Replotted from (Suzuki, 1989b).

A model describing the dynamics of shape change of the gel must not only explain its "passive" poroelastic behavior (consolidation) and its "active" chemical response (swelling and contraction); it should also identify the processes that limit the rate of swelling and contraction. In Figure 2, it is apparent that the time of contraction scales with the square of the thickness of the sample, suggesting that the rate-limiting process has a diffusive character.

Still, we must discover which of the following diffusion processes actually limits the rate. Network consolidation, solvent transport into the pores and into the fiber strands, and chemical reaction within the strands all have a diffusive character. It is well known that consolidation of a linear poroelastic network is described by a diffusion equation. Solvent infiltrates the voids in the polymer network by diffusion and advection. Once within a pore, solvent infiltrates the

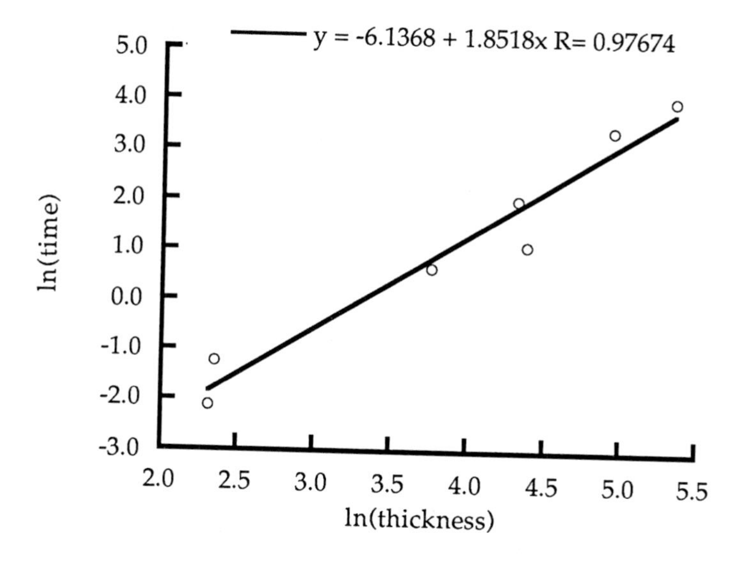

Figure 2. ln(contraction time) vs. ln(gel thickness). Replotted from (Suzuki, 1989b).

polymer fibers by diffusion. Even chemical reactions that cause shape changes in the polymer may appear to be diffusive. Identifying the rate-limiting step in this sequence is one goal of this modelling effort.

Model of solvent mediated shape change of the poroelastic gel network:

It is prudent to construct a continuum model of the dynamics of the poroelastic medium. At this scale of description we can incorporate microstructural features of the gel, yet predict its macroscopic behavior in different experimental protocols. The model presented here has a linear constitutive law; Darcy's law of fluid flow through porous media; conservation of mass of fluid and solid that make up the gel; conservation of linear momentum (F = ma); a solvent transport equation (in which the total flux is sum of diffusive and advective fluxes); and finally, a reaction/diffusion equation describing solvent transport within the fibers and fiber rearrangement. Each of these equations is described below:

The constitutive law of network:

The polymer network obeys a linear constitutive law (i.e., the relationship between stress and strain):

$$\tau_{ij} = 2\,G\,\varepsilon_{ij} + \lambda\left(\varepsilon_{kk} - \varepsilon_{kko}\right)\delta_{ij} - \sigma\,\delta_{ij} \tag{1}$$

Above, τ_{ij} is the total stress tensor of the poroelastic material and ε_{ij} is the gel network strain tensor:

$$\varepsilon_{ij} = \frac{1}{2}\left(\frac{\partial u_i}{\partial x_j} + \frac{\partial u_j}{\partial x_i}\right) \quad , \tag{2}$$

ε_{kk} is the network dilatation, u_i is the network displacement in the x_i direction, σ is the hydrostatic pressure in the pores, β is the molar concentration of acetone, $G(\beta_{eq})$ is the shear modulus, $\lambda(\beta_{eq})$ is the Lamé constant of the consolidated network, and $\varepsilon_{kko}(\beta_{eq})$ is the reference network dilatation.

These three material "constants" are actually functions of the equilibrium concentration of acetone; they can be estimated from free-swelling or other experiments. Eisenberg and Grodzinsky (Eisenberg and Grodzinsky, 1985) present a similar constitutive law for cartilage, but they use an isotropic chemical stress (analogous to a thermal stress) to describe the effect of changing the local chemical environment on the total stress.

Mechanical equilibrium of the poroelastic medium:

The three equations of mechanical equilibrium of the poroelastic medium are restatements of Newton's second law of motion, $\mathbf{F} = m\,\mathbf{a}$, written in each coordinate direction. For an isotropic, poroelastic medium, they are:

$$\rho\,\frac{\partial^2 u_i}{\partial t^2} = \frac{\partial \tau_{ij}}{\partial x_j} = 0 \tag{3}$$

where ρ is the mean density of the medium. Substituting the constitutive law and the definition of infinitesimal strain into the equation of conservation of momentum, these three equilibrium equations become:

$$0 = G \frac{\partial^2 u_i}{\partial x_j \partial x_j} + (G + \lambda) \frac{\partial \varepsilon_{kk}}{\partial x_i} - \frac{\partial \sigma}{\partial x_i}$$

$$+ \ 2 \frac{\partial G}{\partial x_j} \varepsilon_{ij} + \frac{\partial \lambda}{\partial x_i} \varepsilon_{kk} - \frac{\partial (\lambda \varepsilon_{kk0})}{\partial x_i} \quad (4)$$

shear modulus gradient — Lamé constant gradient — "chemical stress" gradient

The top line of Eq. (4) is simply Navier's equation as derived by Biot (Biot, 1941; Biot, 1955). The remaining three terms are new; they arise because the material "constants" are now variables that depend on chemical concentration and position. Additional forces may develop in the poroelastic medium when there are gradients in material properties, just as fluid stresses develop when there are gradients in surface tension. Gradients in shear modulus permit both shear and normal strains to produce a stress, but gradients in Lamé constant only permit isotropic network dilatation to produce a stress. Finally, gradients in the chemical stress $\lambda \varepsilon_{kk0}$ (that Eisenberg and Grodzinsky lump into a single term (Eisenberg and Grodzinsky, 1985)), also produce network stress.

Conservation of mass of the poroelastic medium:

The equation of conservation of mass, when applied to an incremental volume element of the poroelastic medium is (Biot, 1941; Biot, 1955):

$$- \frac{\partial V_i}{\partial x_i} = \frac{\partial \varepsilon_{kk}}{\partial t} \quad (5)$$

where V_i is the flux of the interstitial solvent relative to the solid network in the x_i direction. It has been assumed above that each constituent is incompressible.

Solvent flow within the network:

Darcy's law is used to describe the drag on the elastic network caused by solvent flow (Biot, 1941; Biot, 1955):

$$V_i = -\kappa \frac{\partial \sigma}{\partial x_i} \tag{6}$$

Above, κ is the hydraulic permeability of the network, which also may depend upon position and concentration.

Consolidation equation of the medium:

The equations of conservation of mass and momentum and Darcy's law, when solved together, yield a non-linear "consolidation" equation. To see this, we consider an experiment in which the gel sample swells only in the z direction:

$$u_x = \frac{\partial u_x}{\partial z} = \varepsilon_{xz} = u_y = \frac{\partial u_y}{\partial z} = \varepsilon_{yz} = 0 \quad , \tag{7}$$

so that the deformation and chemical transport depend only on z. The new consolidation equation becomes:

$$\frac{\partial \varepsilon_{zz}}{\partial t} = (2G+\lambda)\,\kappa\,\frac{\partial^2 \varepsilon_{zz}}{\partial^2 z} + g\left(2G+\lambda;\, \varepsilon_{zz0};\, \lambda;\, \kappa;\, \varepsilon_{zz}\right)\frac{\partial \beta}{\partial z} + h\left(2G+\lambda;\, \varepsilon_{zz0};\, \lambda;\, \kappa;\, \varepsilon_{zz}\right)\frac{\partial^2 \beta}{\partial z^2} \quad . \tag{8}$$

Biot named the coefficient of the first term on the right-hand side of Eq. (8) the consolidation constant, c (Biot, 1941):

$$c = (2G+\lambda)\,\kappa \quad . \tag{9}$$

He recognized its importance as the effective diffusivity of the strain field (Biot, 1941). There are new terms in the mechanical swelling equation that depend on the gradients and the Laplacian of the chemical concentration; however, when the material constants are independent of chemical concentration, Biot's original consolidation equation is recovered.

Transport of solvent within voids:

The flux of the acetone/water mixture within the gel, J_i, is the result of diffusive and advective motion (Crank, 1975):

$$J_i = -D \frac{\partial \beta}{\partial x_i} + \beta V_i \quad , \tag{10}$$

where D is the diffusivity of the acetone in water.

The equation of conservation of mass of the solvent is:

$$\frac{\partial \beta}{\partial t} = -\frac{\partial J_i}{\partial x_i} \quad . \tag{11}$$

For constrained swelling, these two equations yield:

$$\frac{\partial \beta}{\partial t} = D \frac{\partial^2 \beta}{\partial^2 z} + f\left(2G+\lambda; \varepsilon_{zz_0}; \lambda; \kappa; \varepsilon_{zz}\right) \frac{\partial \beta}{\partial z}$$

$$+ w\left(2G+\lambda; \varepsilon_{zz_0}; \lambda; \kappa; \varepsilon_{zz}\right) \frac{\partial^2 \beta}{\partial z^2} + q\left(2G+\lambda; \varepsilon_{zz_0}; \lambda; \kappa; \varepsilon_{zz}\right) \beta \,. \tag{12}$$

This advection/diffusion equation contains quantities that depend on the swelling of the material.

Transport of solvent within fibers:

The most "phenomenological" aspect of this model is the way reaction and diffusion within the fiber bundles are treated. Since little is known about the diffusivity of the individual fibers and the reaction mechanisms that ultimately lead to the mechanical readjustment of individual fibers, a set of simple, first-order rate equations are proposed that allow the material constants to relax to their steady-state values in a prescribed time. The equations are given by:

$$\frac{\partial \varepsilon_{kk_0}}{\partial t} = -k\left(\varepsilon_{kk_0} - \varepsilon_{kk_{0eq}}\right) \tag{13a}$$

$$\frac{\partial \lambda}{\partial t} = -k\left(\lambda - \lambda_{eq}\right) \tag{13b}$$

$$\frac{\partial G}{\partial t} = -k\left(G - G_{eq}\right) \tag{13c}$$

$$\frac{\partial \kappa}{\partial t} = -k\left(\kappa - \kappa_{eq}\right) \tag{13d}$$

In the absence of additional information, the rate constants, k, are all assumed to be the same. If transport in the fibers is diffusion-limited, then k can be expressed as a function of the polymer fiber radius, a (which can be estimated from SEMs (Suzuki, 1989)) and the diffusivity of the solvent within a fiber, D_f. By comparing the analytical solution to diffusion of a solute into an infinite cylinder (Crank, 1975) with the analytical solution to a first-order rate equation, like the ones shown above, an expression for k can be found:

$$k = D_f \alpha_1^2 \qquad (1/\text{sec}) \tag{14}$$

where α_1 is the smallest positive root of $J_0(\alpha_1 a) = 0$. For a fiber with radius a = 10^{-4} cm, $\alpha_1 = 24048.3$ cm^{-1}. If $D_f \sim 10^{-7}$ cm^2/sec in the condensed polymer fiber, then k ~ 58 sec^{-1}. This estimate for the reaction rate is close to the observed times of mechanical readjustment in thin polymer sheets (Suzuki, 1989).

The nature of mechanochemical coupling:

The diagram below illustrates the intimate coupling that exists between mechanical and chemical processes in the gel. We glean this understanding from the equations of network consolidation/swelling and solvent transport. Any change in the chemical composition of the external bath has immediate mechanical consequences. The superficial layers of the gel begin to contract or swell, resulting in advection of the solvent. Meanwhile, the solvent diffuses into the interstitial spaces. New regions deeper in the interior now begin to change shape. The cycle continues until a new steady-state configuration is achieved throughout the gel.

Initial and boundary conditions:

To solve the system of partial differential equations, both initial and boundary conditions must be prescribed. In this problem the hydrostatic pressure is assumed to be continuous at the gel/bath interface, so no external traction is applied at the free surfaces of the gel. The acetone concentration is also prescribed in the external bath.

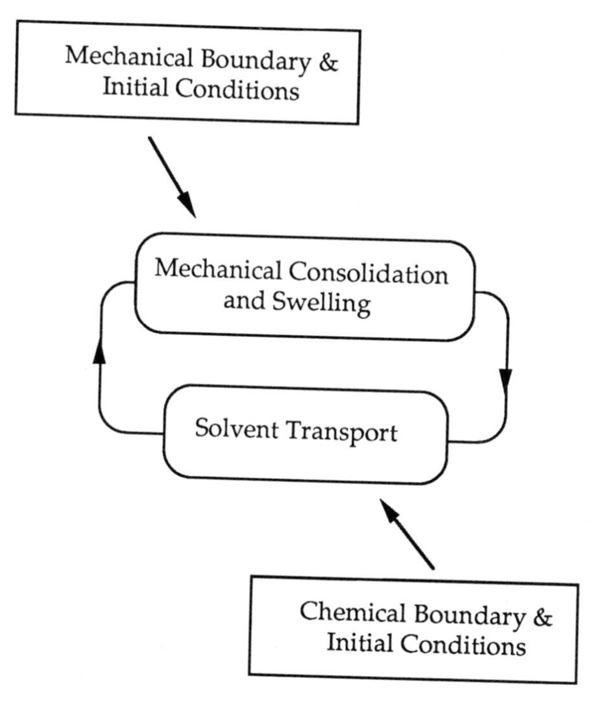

Figure 3. Proposed coupling between mechanical and chemical processes

Concentration and strain profiles within gel strip:

Using the model we can calculate the response of the thin sheet in a solvent-exchange experiment. In Figure 4, the concentration in moles/liter is given as a function of time and width coordinate, z. The concentration is uniform at t=0 before the sample (that is assumed to have equilibrated in a bath of acetone) is suddenly put in contact with water. The initial decrease in concentration is most pronounced at the periphery, where the solvent is most readily exchanged with water. After about 0.1 seconds, the acetone has been flushed from the gel. The gel now has equilibrated with the surrounding water bath.

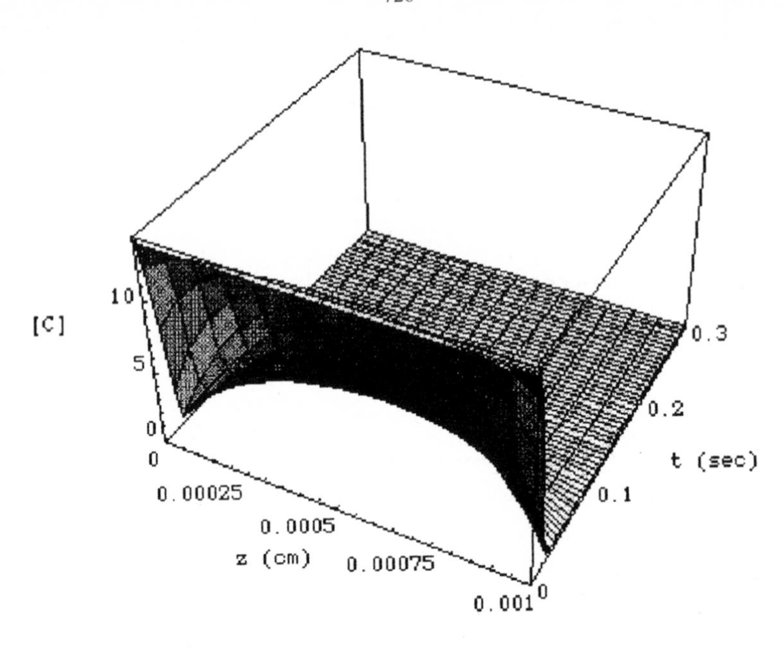

Figure 4. Concentration profile within gel strip during swelling

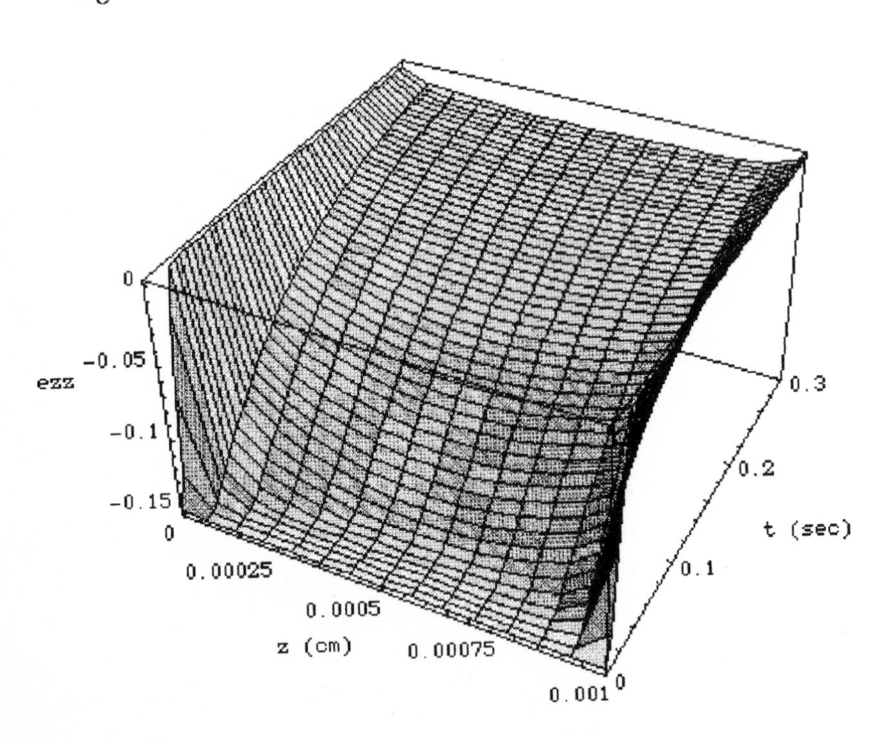

Figure 5. Strain profile within gel strip during swelling

In the same experiment the initial strain is everywhere negative at t=0, indicating that the sample is compressed with respect to its free state in water. Strain goes to zero almost instantly at the periphery, but only slowly in the center of the sample. The mechanical time constant for swelling is significantly longer than the chemical time constant.

Dimensional Analysis:

Once the model is prescribed and the range of parameter values has been determined from experiments, we can normalize the equations and determine dimensionless groups that help to identify rate-limiting processes and to interpret the numerical results of the model. One dimensionless group, π_1, is the ratio of the chemical and mechanical diffusivities. This quantity also can be seen as the ratio of the consolidation and diffusive time constants, i.e.,

$$\pi_1 = \frac{D}{c} = \frac{D}{(2G+\lambda)\kappa} = \frac{\tau_c}{\tau_d} \sim 10^{-3} \quad . \tag{15}$$

This equation shows that the time of consolidation for the network is substantially shorter than the diffusion time of the solvent; therefore, with respect to chemical diffusion, consolidation can be treated in the steady state.

The relative importance of diffusion and advection can be estimated from the Peclet number (Pe), which is defined as the ratio of the these two fluxes per unit volume. It also can be interpreted as the ratio of the characteristic times for diffusion and advection, i.e.,

$$\pi_2 = \mathrm{Pe} = \frac{G\kappa\varepsilon_0}{D} = \frac{\tau_d}{\tau_a} \sim 50 \quad . \tag{16}$$

This ratio shows that the time of advection is also shorter than the diffusion of the solvent. Even so, the time of diffusion, which is the longest of the three processes, can be estimated directly for a gel strip of thickness δ:

$$\tau_d \sim \frac{\delta^2}{D} \quad . \tag{17}$$

For $\delta \sim 10^{-3}$ cm, $\tau_d \sim 10^{-5}$ seconds. This estimate is three orders of magnitude too small to explain the observed time of mechanical readjustment for samples of this thickness, which are about 10^{-2} seconds. This simple analysis suggests either that the rate-limiting process must take place within the fibers themselves (such as by diffusion or reaction) or that it is caused by some mechanism not considered in this analysis.

Utility of a mathematical model:

A model derived from basic conservation laws provides a reasonable framework for examination of the dynamics of swelling and shrinking of a gel. The model helps identify the important physical variables and material constants that need to be measured. Numerical solutions to the underlying equations and dimensional analysis can then be used to identify the rate-limiting process. In addition, this model yields an unexpected prediction: The rate of swelling is expected to be higher than the rate of contraction. The model also predicts that the rate of swelling or contraction should scale with the square of sample thickness, and that it can be used with different boundary and initial conditions to represent or mimic different experimental conditions.

Still, this model has deficiencies. First, it contains several parameters that are difficult to measure. It cannot be used to predict accurately the gel dynamics when network deformations are large. It does not treat the "moving boundary" condition at the gel/solvent interface that would accompany a large deformation.

Such a model could be used to predict the response of gels to various chemical stimuli just as the response of skeletal muscle to excitation is described. Experimental protocols that can be described include washout experiments, in which stress is measured under isometric conditions or displacement is measured under isotonic conditions. The length following quick-release experiments could also be predicted.

References:

Biot MA (1941) General theory of three-dimensional consolidation. J Appl Phys 12(2):155-164

Biot MA (1955) Theory of elasticity and consolidation for a porous anisotropic solid. J Appl Phys 26(2):182-185

Crank J, (1975) The mathematics of diffusion, 2nd. Ed., Oxford University Press, Oxford

Eisenberg SR, Grodzinsky AJ, (1985) Electromechanical micromodel of articular cartilage. In: Butler D, Hung TK, Mates RE (eds), 1985 Biomechanics Symposium. A.S.M.E., New York, 85-89

Kuhn W, Hargitay B, Katchalsky A, Eisenberg H (1950) Reversible dilation and contraction by changing the state of ionization of high-polymer acid networks. Nature 165(514-516

Suzuki M, (1989a) Amphoteric polyvinyl alcohol hydrogel and electro-hydrodynamic control method for artificial muscles. In: DeRossi D, Kajiwara K, Osada Y,Yamauchi A (eds), Polymer gels: Fundamental and biomedical applications. Plenum Press, New York, 221-246

Suzuki M (1989b) Amphoteric polyvinyl alcohol hydrogel as a material of artificial muscle. Kobunshi Ronbunshu 46(10):603-611

Dynamics of osmotic fluid flow

George Oster
Departments of Molecular & Cellular Biology, and Entomology, University of California, Berkeley, CA 94720

Charles S. Peskin
Courant Institute of Mathematical Sciences, 251 Mercer Street, New York, NY 10012

Introduction

The classical thermodynamic treatment of osmotic pressure is quite sufficient to compute most quantities of interest without reference to any molecular model (see, for example, Finkelstein, 1987). However, one's intuition is always aided by molecular models, and a microscopic picture can be quite useful when thinking about osmotic flow in unfamiliar situations. Therefore, we offer here a surprisingly simple and intuitive molecular model for describing solvent flow driven by osmotically generated pressure gradients. We will show that osmotic pressure arises from the external force on solute particles which is transmitted to the solute by viscous drag. This simple picture enables us to propose a simulation method that can be used to compute the osmotic flows associated with quite complex situations arising in polymer and gel swelling.

A microscopic model for osmotic pressure

We begin with the simplest situation of an ideal solute in water. The molecular picture we shall adopt is the following. A population of N noninteracting particles of mass m immersed in water is driven by Brownian motion. The motions of the particles and the fluid are coupled by simple hydrodynamic drag. Let $\mathbf{X}_k$ be the location of the kth particle. The equation of motion for a particle in a fluid flowing with velocity $\mathbf{u}$ is:

$$m\frac{d^2\mathbf{X}_k}{dt^2} = -\underbrace{\nabla\phi(\mathbf{X}_k)}_{\text{External force on particle k}} + \underbrace{\beta\left[\bar{\mathbf{u}}(\mathbf{X}_k(t),t) - \frac{d\mathbf{X}_k}{dt}\right]}_{\text{Frictional force between particle k and fluid with velocity } \bar{\mathbf{u}}} + \underbrace{\mathbf{R}_k(t)}_{\text{Random force exerted on particle k by the fluid}} \qquad [1]$$

Here ϕ is a potential function that defines all of the forces acting on the particle; in the simplest case, $-\nabla\phi$ is the external force acting on the particle. β is the coefficient of the frictional drag between the fluid and

NATO ASI Series, Vol. H 64
Mechanics of Swelling
Edited by T. K. Karalis

the particle, which depends on their relative velocity and the shape of the particle (e.g. for a sphere, β is given by Stokes law, $\beta = 4\pi\mu a$, where a is the particle radius and μ the fluid viscosity). $\overline{\mathbf{u}}$ is a weighted average fluid velocity in the neighborhood of the particle:

$$\overline{u}(\mathbf{X}_k,t) = \int u(\mathbf{x},t)\delta_d(\mathbf{x}-\mathbf{X}_k)d\mathbf{x} \qquad [2]$$

where $\mathbf{x}$ locates a point in the fluid, and the function δ_d defines the average fluid velocity in the neighborhood of the particle. The reason for performing such an average instead of just evaluating the fluid velocity at the particle will appear below. The random force, $\mathbf{R}(t)$ has the following statistics:

$$\langle \mathbf{R}_k(t)\rangle = 0, \quad \langle R_{jp}(t)R_{kq}(t')\rangle = 2k_B T\beta\delta_{jk}\delta_{pq}(t-t') \qquad [3]$$

where k_B is Boltzmann's constant, T the absolute temperature, and δ_{jk} and $\delta_{pq}(t-t')$ are the Kroneker and Dirac delta functions, respectively. Equation [3] is the usual definition of "white" noise employed in Langevin dynamical equations such as [1] (Chandrasekhar, 1943).

The particle motions are coupled to the local fluid flow via the incompressible Navier-Stokes equations:

$$\rho\left(\frac{\partial \mathbf{u}}{\partial t} + \mathbf{u}\cdot\nabla\mathbf{u}\right) + \nabla p = \mu\nabla^2\mathbf{u} - \underbrace{\sum_{k=1}^{N}\delta_d(\mathbf{x}-\mathbf{X}_k(t))\left[\beta\left(\overline{u}(\mathbf{X}_k(t),t) - \frac{d\mathbf{X}_k}{dt}\right) + \mathbf{R}_k(t)\right]}_{\text{Frictional and random forces exerted on the fluid by the particles}} \qquad [4]$$

$$\nabla\cdot\mathbf{u} = 0 \qquad \text{incompressibility} \qquad [5]$$

where the last term in [4] is the equal and opposite force applied to the fluid by the particles in the neighborhood defined by the weighting function δ_d . That is, the viscous drag force exerted by the fluid on each particle is felt equally, but in the opposite direction, by the fluid. Also, the random force acting on the particles arises from the statistical fluctuations in the fluid near the particle, and so the same force must act reciprocally on the fluid volume surrounding the particle. If this force were applied at a point, it would produce an infinite velocity there. Instead, we spread the force out over a fluid volume of diameter d by making use of the weighting function $\delta_d(\mathbf{x} - \mathbf{X}_k)$. Once we have decided to spread the force in this way, conservation of energy demands that we also average the fluid velocity using the same weighting function. This is the reason for the use of the spatially averaged velocity given by Equation [2] as the convection velocity in Equation [1].

Equations [1] - [5] constitute our "microscopic" model. It is not really microscopic, since we describe the solvent flow by continuum equations; however, molecular dynamics simulations have shown that the Navier- Stokes equations are approximately valid even for flow of 1Å particles flowing through a 3.2Å channel (Levitt, 1973).

For the situations of biological interest we can neglect the particles' inertia, so that the equation for the particles is:

$$\nabla\phi(\mathbf{X}_k) = \beta\left(\bar{\mathbf{u}}(\mathbf{X}_k(t),t) - \frac{d\mathbf{X}_k}{dt}\right) + \mathbf{R}_k(t) \qquad [6a]$$

or

$$\boxed{\beta\frac{d\mathbf{X}_k}{dt} = -\nabla\phi(\mathbf{X}_k) + \beta\bar{\mathbf{u}}(\mathbf{X}_k(t),t) + \mathbf{R}_k(t)} \qquad [6b]$$

Now [6a] can be used to eliminate $\mathbf{R}_k(t)$ from the fluid equations:

$$\rho\left(\frac{\partial\mathbf{u}}{\partial t} + \mathbf{u}\cdot\nabla\mathbf{u}\right) + \nabla p = \mu\nabla^2\mathbf{u} - \sum_{k=1}^{N}\delta_d(\mathbf{x} - \mathbf{X}_k(t))[\nabla\phi(\mathbf{X}_k)] \qquad [7]$$

Notice that the only randomness in the fluid equations is the location of the particles, and that the forces, $-\nabla\phi$, felt by the particles are now transferred to the fluid as an additional force term.

To see the consequences this, consider the case where the fluid inertia is also neglected, so that the fluid equations become simply

$$\boxed{\nabla^2\mathbf{u} = \frac{1}{\mu}\left[\nabla p + \sum_{k=1}^{N}\delta_d(\mathbf{x} - \mathbf{X}_k(t))[\nabla\phi(\mathbf{X}_k)]\right]} \qquad [8]$$

It may appear strange that the frictional coupling coefficient, β, doesn't appear in Equation [8]. Note, however, that Equation [8] must be solved simultaneously with Equation [6] for the particle locations, $\mathbf{X}_k(t)$, and the fluid velocity field, $\mathbf{u}(\mathbf{x},t)$. The limiting situation where $\beta \rightarrow 0$ is singular: with no drag, the particles move at infinite velocity, and so neglect of the particle mass is not possible. For any finite particle drag, all of the force, $-\nabla\phi$, on the particle is transmitted to the fluid, and so the fluid equations do not contain β.

Imagine, for example, a collection of particles released in the center of a vesicle (e.g. a liposome) bounded by a semipermeable membrane. The only force acting on the particles is the repulsion of the membrane, which is not felt until the particles are very close to the vesicle walls. Thus nothing interesting happens until the particles diffuse to the membrane. Once the particles feel the repulsive force of the membrane, they transfer this force to the fluid by viscous drag. This drag force is in the same direction as the membrane repulsion, and it raises the pressure inside the vesicle with respect to the outside. To analyze this situation further, we adopt a macroscopic view of the membrane and consider it as uniformly porous, i.e., we ignore the discreteness of the pores. Then the whole situation is spherically symmetric, and it is easy to see that the (incompressible) fluid cannot move. Nevertheless, the semipermeable membrane will be driven outward through the static fluid by the high internal pressure until it stretches sufficiently to balance the internal pressure. (Of course, from the standpoint of the membrane, this looks exactly like fluid flowing through it.) Thus the vesicle will swell until it reaches an equilibrium state, or bursts.

Equilibrium: van't Hoff's law

At equilibrium, the fluid velocity vanishes, $\mathbf{u} = 0$, and Equation [8] reduces to

$$\boxed{\nabla p = -\sum_{k=1}^{N} \delta_d(\mathbf{x} - \mathbf{X}_k)\nabla\phi(\mathbf{X}_k)} \qquad [9]$$

The right hand side of [9] is a force field which is generated by applying the external force acting on each particle to the fluid which surrounds that particle. The pressure, p, generated by this force field is just the osmotic pressure. This can be seen as follows.

Consider the situation shown in Figure 1: two fluid compartments separated by a semipermeable membrane. We introduce into the compartment on the left N solute molecules, and wait until equilibrium is reached. To analyze the equilibrium situation, we take the ensemble average of both sides of Equation [9]. Recall that the ensemble average of any function,f(X), is defined by

$$\langle f(X) \rangle \equiv \int f(X)P(X)dX,$$

where P(X) is the probability density function of the random variable X. In our case, the random variables are the locations of the particles, $\mathbf{X}_k(t)$. At equilibrium these all have the same probability density function $P(\mathbf{X}) \equiv c(\mathbf{X}) / N$, where N is the total number of particles and $c(\mathbf{X})$ is their equilibrium concentration (number per unit volume). At equilbrium, $c(\mathbf{x})$ obeys the Boltzmann distribution

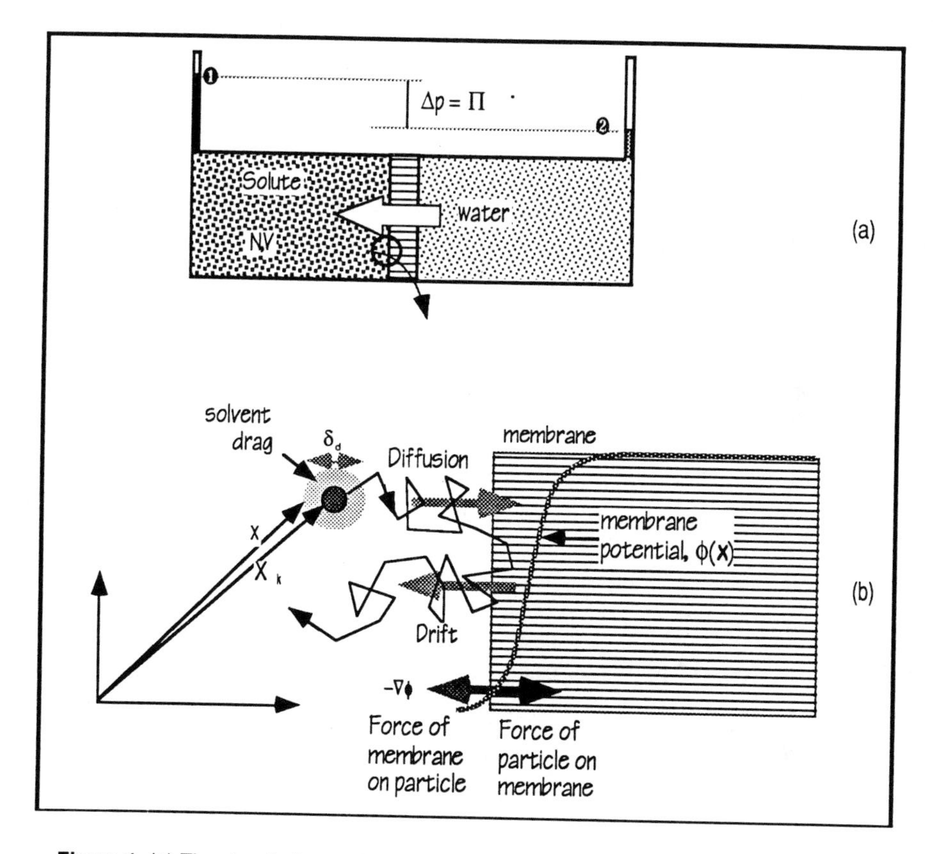

Figure 1. (a) The classical osmotic flow experiment. A semipermeable membrane separates two compartments of volume V; N particles of solute are introduced into the left compartment. The resulting osmotic flow induces a hydrostatic pressure head, Δp, just equal to the osmotic pressure, Π.

(b) A microscopic view of the trajectory of particles near the membrane. The random diffusion of particles is biased by the membrane repulsion which induces a drift component directed away from the membrane. Therefore, the particles drag more fluid to the left than to the right until mechanical equilibrium is achieved.

$$c(\mathbf{x}) = c_0 e^{-\frac{\phi(\mathbf{x})}{k_B T}} \qquad [10]$$

where c_0 is the concentration at a location where $\phi = 0$. Taking the gradient of both sides of this equation and rearranging the result, we get

$$k_B T \nabla c = -c \nabla \phi \qquad [11]$$

Taking the ensemble average of Equation [9] in the manner described above results in

$$\langle \nabla p(\mathbf{x}) \rangle = \sum_{k=1}^{N} \int \delta_d(\mathbf{x} - \mathbf{X})(-\nabla \phi(\mathbf{X})) \frac{c(\mathbf{x})}{N} d\mathbf{X} \quad [12]$$

Since the quantity being summed is independent of N, the sum of N terms just cancels the factor 1/N, and we are left with

$$\langle \nabla p(x) \rangle = \int \delta_d(\mathbf{x} - \mathbf{X})(-\nabla \phi(\mathbf{X})) c(\mathbf{x}) d\mathbf{X} \quad [13]$$

At this point, we make two approximations. First, since $p(\mathbf{x})$ is a macroscopic variable, we assume it is close to its ensemble average. Thus, we replace $\langle p(\mathbf{x}) \rangle$ by $p(\mathbf{x})$. Second, since ϕ and c are also macroscopic, and hence slowly varying in comparison to d, we may replace δ_d by the Dirac δ-function itself. This makes it possible to evaluate the integral in Equation [13]. The result is

$$\nabla p(x) = -\nabla \phi(\mathbf{x}) c(\mathbf{x}) \quad [14]$$

Finally Equations [11] and [14] may be combined to yield

$$\nabla p(x) = k_B T \nabla c(\mathbf{x}) \quad [15]$$

If the membrane is completely impermeable to solute and if there is no solute to the right of the membrane in Fig. 1, we integrate Equation [15] across the membrane to obtain

$$\Delta p = k_B T c = k_B T \frac{N}{V} \quad [16a]$$

where V is the volume in which the solute is distributed. Multiplying numerator and denominator on the right-hand side of Equation [16a] by Avogadro's number, $N_A = R/k_B$, we obtain van't Hoff's law for the osmotic pressure of a dilute solution (Finkelstein, 1987):

$$\boxed{\Pi = RTC} \quad [16b]$$

where Π is the osmotic pressure and $C = N/N_A V$ is the molar concentration of solute. The reasons why we get an osmotic pressure law appropriate to a dilute solution are: (1) our model particles do not interact, and

(2) the particles are treated as points, so that there is no excluded volume effect (despite the presence of the parameter d) in the simple form of the model that we have presented so far.

The picture of osmotic pressure that emerges from this model is shown in Figure 1b. According to Equation [9] particles diffusing close to the membrane have their random motions biased away from the membrane by the force $-\nabla\phi$, and this force exerted on the particles by the membrane is transferred to the fluid. The membrane pushes on the particles—or, equivalently, the particles push on the membrane—and the particles drag fluid away from the membrane. Until equilibrium is reached the particles drag more fluid away from the membrane than towards it. At equilibrium this drag force on the fluid near the membrane is still present, but it is balanced by a pressure gradient that develops near the membrane. Seen from this viewpoint, osmotic pressure is generated by biasing the random motions of the solute particles, which induces a viscous drag of the fluid by the particles.

Nonequilibrium and nonideal osmotic flows

The analysis of the microscopic model that we have just done for the equilibrium case generalizes readily to the nonequilibrium situation. The key result of the foregoing analysis was that the external forces exerted on the solute particles (e.g., by a membrane) are transferred to the fluid (see Equations [8] and [9]). This same result holds in non-equilibrium situations. Thus, to obtain macroscopic equations corresponding to our microscopic model, we need only add an external force density to the fluid equations of the form $c(\mathbf{x},t)\,(-\nabla\phi(\mathbf{x},t)$, since $c(\mathbf{x},t)$ is the local density of solute particles and $(-\nabla\,\phi(\mathbf{x},t)$ is the external force per particle. In the case where fluid inertia may be neglected, the macroscopic equations are:

$$\nabla p = \mu\nabla^2\mathbf{u} - c\nabla\phi \quad \text{(Stokes equations)} \qquad [17]$$

$$\nabla\cdot\mathbf{u} = 0 \qquad \text{(Incompressibility)} \qquad [18]$$

Here the local body force in the fluid equations is just the force per particle times the local particle concentration. These equations must be supplemented by the solute particle conservation equation, which takes the form:

$$\frac{\partial c}{\partial t} = -\nabla \cdot \left[\underbrace{-\frac{kT}{\beta}\nabla c}_{\text{Diffusion}} + \underbrace{uc}_{\text{Convection}} + \underbrace{\frac{-\nabla\phi}{\beta}c}_{\text{Drift}} \right] \quad \text{Conservation of particles} \tag{19}$$

Note that the velocity multiplying c in Equation [19] is a sum of two terms: the macroscopic convection velocity, **u**, and the drift velocity relative to the fluid generated by the external force per particle: $\mathbf{v}_{drift} = -\nabla\phi/\beta$. At equilibrium, both the fluid velocity and the solute flux (the quantity in square brackets in Equation [19]) must be zero. Thus,

$$c\nabla\phi = -kT\nabla c \tag{20}$$

which, when substituted into the fluid equations yields:

$$\nabla p = kT\nabla c \tag{21}$$

as before.

It is not easy to generalize the macroscopic equations derived above to cover the case of interacting particles (although it is easy to include chemical reactions by simply adding a reaction term to Equation [19] and a separate force term to Equation [17] for each species). The advantage of the microscopic equations, however, is that they can describe not only ideal solutions, where the particles do not interact, but also polymer solutions where the particles are coupled. For example, instead of placing the particles inside a vesicle, let them be polymerized into a chain. The equations for polymer osmotic pressure are obtained by adding the inter-particle forces to the potential in Equation [1]. Neglecting inertial forces, the equations for a flexible polymer inside a vesicle take the following form (Figure 2a).

$$\underbrace{\beta_k \frac{d\mathbf{X}_k}{dt}}_{\text{Viscous drag}} = \underbrace{-\frac{\partial \Phi}{\partial \mathbf{X}_k}(\mathbf{X}_1,\ldots,\mathbf{X}_N)}_{\text{Forces on kth monomer}} + \underbrace{\mathbf{R}_k(t)}_{\text{Random force}} \tag{22}$$

Where the potential, Φ, is given , for example, by:

$$\Phi(\mathbf{X}_1,\cdots,\mathbf{X}_N) = \underbrace{\sum_{i=1}^{n}\phi(\mathbf{X}_k)}_{\text{Membrane potential}} + \underbrace{\sum_{k=1}^{N-1}\frac{k_0}{2}\left(|\mathbf{X}_{k+1}-\mathbf{X}_k|-D_0\right)^2}_{\text{Extensional rigidity of the polymer}} + \underbrace{\sum_{k=2}^{N-1} S_k \frac{k_1}{2}|\mathbf{X}_{k+1}+\mathbf{X}_{k-1}-2\mathbf{X}_k|^2}_{\text{Bending rigidity of the polymer}} \tag{23}$$

The forces acting on the particles now include those holding the chain together. These forces are directed (on the average) towards the center of mass of the polymer so as to oppose the diffusive dispersal of the monomers that would result from the random force if the monomers were not linked together into a polymer chain. Just as in the simpler case of independent solute particles (see Equation [8]), these forces are transferred to the fluid in the same direction. Since the fluid is incompressible, there can be no net flow into the center of the polymer coil, but there will be an increase in hydrostatic pressure at the center as the polymer hydrates. If the particles are contained inside a vesicle, then some of the force they formerly imparted to the membrane (and thus to the fluid in the opposite direction) is now transmitted through the interparticle bonds. Then the net force on the fluid at the membrane is less than when the monomer particles were uncoupled, and so the osmotic pressure of the polymer is less than that of its independent monomer particles. How much does the length and flexibility of the polymer chain affect the osmotic effect (i.e. the solvent drag) of each monomer? This is an interesting question that we hope to answer through computer simulation.

If several charged polymers are crosslinked into a gel, the contribution of the counterions to the swelling pressure can be computed as a Donnan equilibrium (Figure 2b). This result can be obtained from Equation [9] as well by letting the particles be the mobile counterions and setting the force acting on them, $-\nabla\phi$, equal to the Donnan potential. That is, the Donnan potential acts as an electrostatic membrane around the gel which pushes on the particles, and so imparts forces to the solvent. There are many interesting dynamic situations that one can address with the osmotic model. For example, it will be intriguing to see if our osmotic equations produce the same swelling dyanamics as predicted by Tanaka's well accepted diffusion equation (Tanaka and Fillmore, 1979). The Brownian drag model predicts that the counterions would generate an increased hydrostatic pressure inside the gel as they attempt to pump incompressible fluid into the gel interior. The Tanaka model, however, does not predict this effect, although it fits the swelling kinetics quite well. In principle, this increase in internal hydrostatic pressure could be measured as a test of the model's validity.

Discussion

The physical origins of osmotic flow have been the subject of considerable controversy (e.g. Hammel, 1979; Kiil, 1989). It is not our purpose to enter into that debate here. Rather we offer a simple dynamical model for computing the osmotically driven fluid flow in a variety of complex, nonequilibrium situations. Intuitively, the model is appealing, for one can interpret osmotic flows and pressures entirely in terms of the viscous drag of fluid by Brownian particles. The key idea is that the solute particles are excited by Brownian forces generated by the fluid in their immediate neighborhood and that they are simultaneously retarded by viscous drag of the same fluid. It follows that the fluid sees these same forces equally and in the opposite direction. Consequently, since we neglect the mass of the solute particles, any external force applied to the particles—for example, repulsion from a semi-permeable membrane—is transferred to the fluid by viscous drag. That is, the membrane acts on the fluid using the particles as its agent.

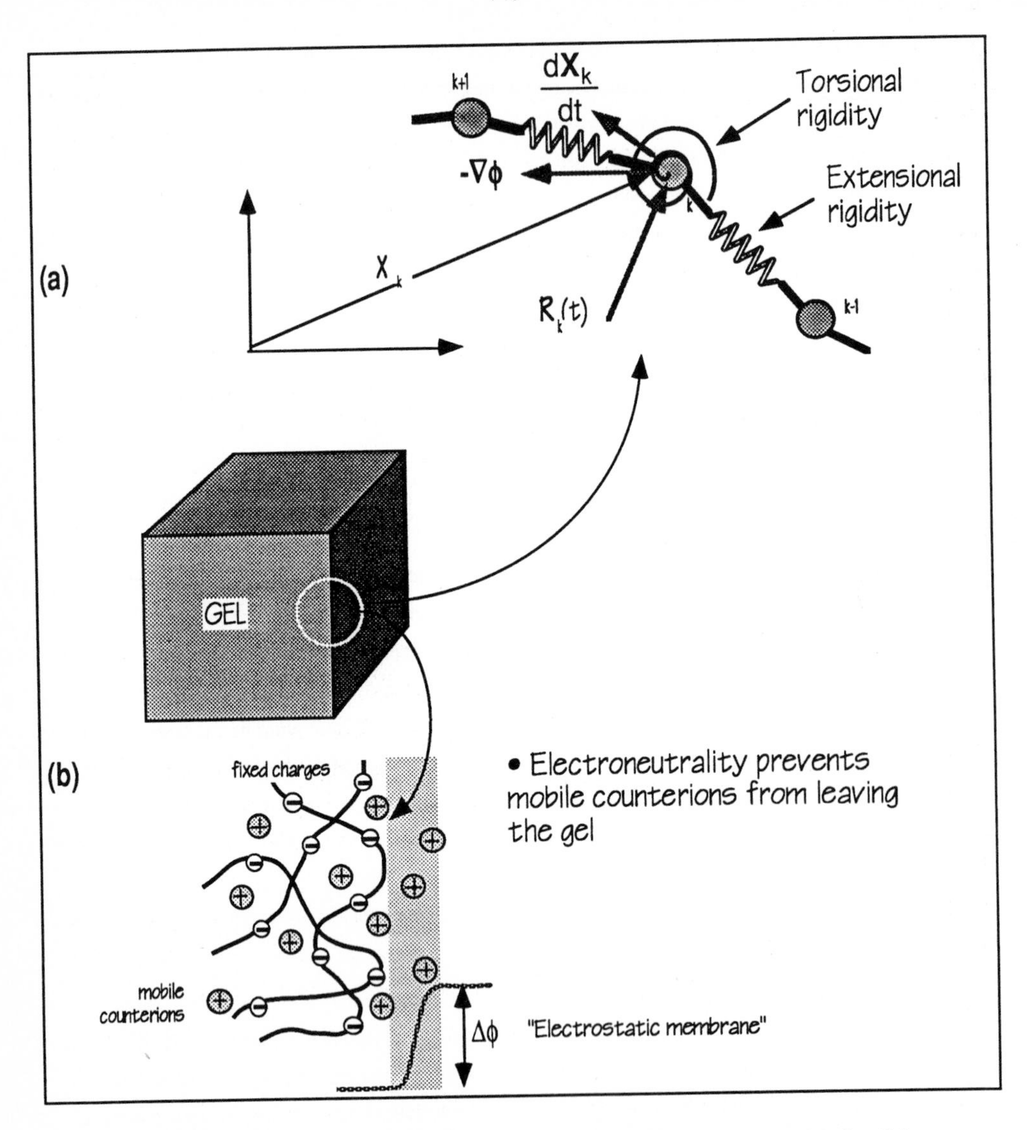

Figure 2. (a) Solute molecules joined into an elastic polymer. Since we neglect inertial forces, the velocity vector of the kth monomer, dX/dt, is proportional to the resultant of the random force, R(t), the external force, $-\nabla\phi$, and the internal torsional and extensional elastic forces. This will not be the same as the velocity of the monomer in isolation, and so the solvent dragged by the kth monomer will be smaller, on the average, when it is a member of a polymer than when it acts alone.

(b) Because of the constraint of electroneutrality, a polyelectrolyte gel generates a Donnan potential at its surface which acts as an "electrostatic membrane" confining the mobile counterions. This increases the hydrostatic pressure inside the gel, contributing to its swelling pressure.

In its simplest form, this microscopic model of osmotic pressure involves four parameters: (i) the fluid viscosity, μ; (ii) the drag coefficient, β, between particles and fluid, which depends on the size and shape of the particles; (iii) the absolute temperature, T, which influences the intensity of the random force (see Equation [3]), and (iv) the diameter, d, of the local region of fluid that that interacts with each solute particle (one might also vary the shape of the function δ_d, but the diameter of its support is presumably its most important characteristic). Of course β,μ, and T are physical parameters which can be measured or calculated for each situation. The meaning of the parameter d is less clear, however. It is tempting to equate it with the diameter of the solute particle, but it remains to be seen whether the physical consequences of changes in the effective particle diameter can be simulated by varying d. For polymer molecules there are additional parameters such as bending rigidity that must be given to completly specify the model.

The model equations reduce to the usual equations at equilibrium. They generalize easily to quite complex situations where particles interact with one another, for example if they are polymerized and/or crosslinked into a gel. Therefore, one can compute the effect of polymer mechanical properties such as bending or extensional stiffness on osmotic pressure and flow. In principle, this permits one to compute the effects of coupling mechanical and osmotic stresses, as measured by Ito and his colleagues (Ito et al., 1989; Ito et al., 1987; Suzuki et al., 1989). In order to carry out such caclulations, however, one must solve a rather large set of molecular dynamical equations[1]. Fortunately, computers of sufficient speed are now available for this task. Moreover, recent progress in molecular dynamics algorithms has made it possible to take large enough time steps to make this task a reasonable undertaking (Peskin & Schlick, 1989; Schlick & Peskin, 1989; Peskin, 1990), and the computational apparatus for coupling immersed elastic particles to a viscous incompressible fluid has already been developed in a different context (Fogelson & Peskin, 1988). We shall report on such calculations in subsequent publications. For now, however, we present this picture of osmotic flow in the hopes that it will provide both a new intuitive view of the phenomena as well as a practical method for computing osmotic flows and pressures in complex situations.

---

[1] Note, however, that molecular dynamics is applied only to the solute particles; the solvent is modeled macroscopically.

Acknowledgements

GO was supported by NSF Grant MCS-8110557. CSP was supported by NSF Grant CHE-900-2416. Both GO and CSP acknowledge the support provided by MacArthur Foundation Fellowships. GO acknowledges the hospitality of the Neurosciences Institute during which part of this work was performed. Finally, we would like to thank our colleagues for their thoughtful comments and discussions, especially Nicolas Cordova, Alan Finkelstein, Robert Macey, Garrett Odell, and Ken Spring. These colleagues, should not, however, be held responsible for the (possibly heretical) views expressed herein.

References

Chandrasekhar, S. (1943). Stochastic problems in physics and astronomy. Rev. Mod. Phys. **15**:1-89.

Doi, M. S. Edwards. (1986). *The Theory of Polymer Dynamics.* New York: Oxford University Press.

Finkelstein, A. (1987). *Water Movement Through Lipid Bilayers, Pores, and Plasma Membranes. Theory and Reality.* 4. New York: John Wiley & Sons.

Fogelson, A.L., C.S. Peskin (1988). A fast numerical method for solving the three-dimensional Stokes' equations in the presence of suspended particles. J. Comput. Phys. **79**:50-69.

Hammel, H. (1979). Forum on osmosis. I. Osmosis: diminished solvent activity or enhanced solvent tension? Amer. J. Physiol. **237**:R95-R107.

Ito, T., M. Yamazaki S. Ohnishi (1989). Poly(ethylene glycol)-induced shrinkage of Sephadex gel. A model system for quantitative analysis of osmoelastic coupling. Biophys J. **56**:707-11.

Ito, T., K.S. Zaner T.P. Stossel (1987). Nonideality of volume flows and phase transitions of F-actin solutions in response to osmotic stress. Biophys J. **51**:745-53.

Kiil, F. (1989). Molecular mechanisms of osmosis. Amer. J. Physiol. **256**:R801-R808.

Levitt, D. (1973). Kinetics of diffusion and convection in 3.2Å pores. Exact solution by computer simulation. Biophys. J. **13**:

Peskin, C. (1990). Analysis of the backward-Euler/Langevin method for molecular dynamics. Comm. Pure Appl. Math. **XLIII**:599-645.

Peskin, C.S., T. Schlick (1989). Molecular dynamics by the backward-Euler method. Comm. Pure Appl. Math. **XLII**:1001-1031.

Schlick T, C.S. Peskin (1989): Can classical equations simulate quantum-mechanical behavior: a molecular dynamics investigation of a diatomic molecule with a Morse potential. Commun. Pure \& Appl. Math. **42**:1141-1163.

Suzuki, A., M. Yamazaki T. Ito (1989). Osmoelastic coupling in biological structures: formation of parallel bundles of actin filaments in a crystalline-like structure caused by osmotic stress. Biochemistry. **28**:6513-8.

Tanaka, T. D. Fillmore (1979). Kinetics of swelling of gels. J. Chem. Phys. **70**:1214-18.

MECHANISM OF OSMOTIC FLOW

Wayne D. Comper and R.P.W. Williams
Biochemistry Department
Monash University
Clayton
Victoria, 3168
Australia

Introduction

Osmotic flow of water is an important biological transport process that occurs across cell membranes, across basement membranes and in extracellular matrices of connective tissues. The osmotic flow process is the basis, too, for swelling in biological systems. In many cases in biology the osmotically active solute is often nonideal macromolecular one. The nonideality (which yields parameters which vary nonlinearly with concentration) influences osmotic pressure through excluded volume polymer interactions and, for charged polymers through the influence of active counterions on the ambient simple electrolyte concentration. Nonideal effects are also manifested in dynamic parameters, particularly hydrodynamic frictional terms that describe the viscous dissipation of water over the surface of the polymer chain.

Osmotic flow is often considered in a system where a semipermeable membrane (permeable to solvent but not to solvent) separates a compartment containing a binary solution of solvent and solute and a compartment containing solvent. Water flow proceeds in the direction against the osmotic activity gradient or down the chemical potential gradient of water. The osmotic water flow is regarded as being directly proportional to the difference in osmotic pressure across the membrane. The proportionality constant has formerly been regarded in the literature as a solution-independent quantity which may be related to the geometry and structure of the membrane.

Prior to our initial studies in 1987 (Williams and Comper, 1987) there had

NATO ASI Series, Vol. H 64
Mechanics of Swelling
Edited by T. K. Karalis

hitherto been no account made of the influence of the dynamic properties of the osmotically active solute in the osmotic flow-membrane system. This was inspite of the compelling reasons to do so especially when considering nonmembranous systems. A solute concentration gradient across a free liquid boundary will give rise to solute diffusion which could equally well be described by water diffusion (an osmotic volume flow) in the opposite direction. In fact, the diffusion coefficient of water and that of the solute in a volume-fixed frame of reference are identical (Bearman, 1961). A more familiar system associated with the osmotic flow process is the swelling of gels. The gels imbibe water through the difference in osmotic pressure between the gel phase and the external solvent phase. It has been demonstrated that the rate of osmotic water flow into the gel is governed by the diffusion of the gel-polymer network and that this is identical with polymer diffusion in semi dilute polymer solutions (Buckley and Berger, 1962; Tanaka and Filmore, 1979; Candau et al., 1982). (Semi dilute polymer solutions are characterised by concentrations where there are continuous, entropic, molecular interactions between molecules to form a transient molecular network that manifests molecular weight independent properties.) The osmotic flow studies described below are with semi dilute solutions of polymers which vary in molecular weight, chemical composition, charge, hydrodynamic frictional coefficient and osmotic activity together with well characterised membranes with either tortuous pores or straight-through cylindrical pores of known radius.

Osmotic Flow in Solution (Binary Systems)

The diffusional relaxation of a solute is accompanied by a corresponding relaxation in the chemical activity of water. This volume flow of water is essentially an osmotic flow - arising from the gradient in chemical potential of water or osmotic pressure of solute. The translational diffusional process in the isothermal system is related to the concentration gradient of solute through Fick's law such that

$$J_1 = -D_1(\text{grad } c_1) \tag{1}$$

where J_1 is the flux of component 1 and c_1 the concentration of component 1 (moles per unit volume). The relationship between the molecular flux of component 1 and its concentration gradient is given by the translational (or mutual) diffusion coefficient

D_1. The diffusion coefficient in a binary system for a volume-fixed frame (designated by the subscript v) has been derived by Bearman (1961) for component 1 and component 2 (solvent) as

$$(D_1)_v = (D_2)_v = (1-\phi_1)(c_1/f_{12})(\partial\mu_1/\partial c_1)_{T,p} \tag{2}$$

where ϕ_1 is the volume fraction of 1, T is the temperature, p the pressure, μ_1 the chemical potential of 1 and f_{12} is the frictional factor between 1 mole of solute and water defined as (Spiegler, 1958)

$$-(\partial\mu_1/\partial x) = f_{12}\,(u_1 - u_2) \tag{3}$$

where u_i is the velocity of i.

(a)

Solute ⇌ Solvent volume exchange

free boundary

(b)

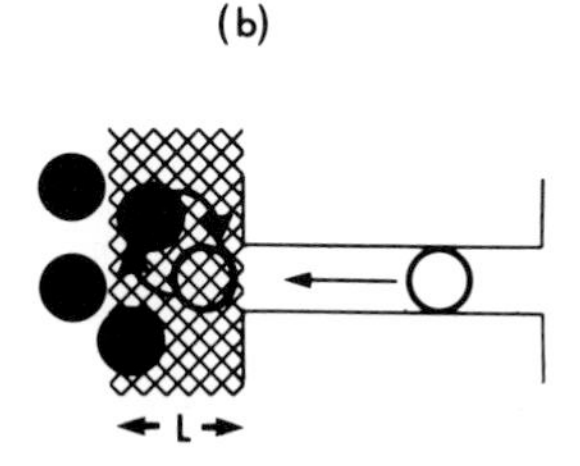

Figure 1. Schematic diagram of long time averaged solute (filled circles) - solvent (empty circles) volume exchange a) across a free boundary (dashed line) and b) in a thin layer on the solution side of the semipermeable membrane.

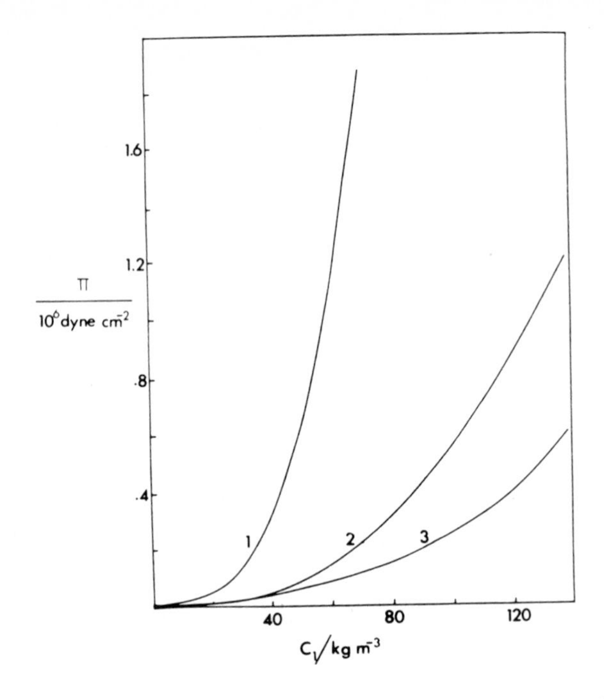

Figure 2. Osmotic pressures of polymers in thermodynamic equilibrium with PBS (phosphate buffered saline which consists of 0.14 mol dm^{-3} NaCl, 2.68 x 10^{-3} mol dm^{-3} KCl, 1.5 x 10^{-3} KH_2PO_4 and 8.1 x 10^{-3} Na_2HPO_4, pH 7.5) as determined from sedimentation-diffusion data for PGS (curve 1), dextran sulfate T500 ($M_w \approx 500000$) (curve 2), and dextran T500 ($M_w \approx 500000$) (curve 3) (from Williams and Comper, 1990 with permission of the publishers).

Derivation of the translational diffusion coefficient corresponding to common experimental standard conditions of constant temperature and chemical potential of solvent μ_2 (Kurata, 1982; Comper et al., 1986) is

$$(D_1)_v = RT(1-\phi_1)^2(M_1/f_{12})(\partial\Pi^*/\partial C_1)_{T,\mu_2} \qquad (4)$$

$$(\partial\Pi^*/\partial C_1)_{T,\mu_2} = (1/M_1) + 2A_2C_1 + 3A_3C_1^2 + \ldots \qquad (5)$$

where R is the universal gas constant, M_1 is the molecular weight of component 1, Π^* the osmotic pressure at constant T and μ_2, A_2 and A_3 are the standard osmotic second and third virial coefficients respectively and C_1 is the concentration of component 1 in mass units. The diffusion coefficient in equation 4 is then a composite function of an hydrodynamic parameter associated with the frictional coefficient and a thermodynamic equilibrium term associated with osmotic pressure. Therefore, the diffusional flux of solute generates a reverse flow of solvent which is driven by the osmotic pressure gradient ie., the volume flux associated with this diffusional process will be exactly balanced by an opposite volume flux of solvent (Figure 1) so that

$$J_1V_1 = - J_2V_2 \quad (6)$$

$$=RT(1-\phi_1)^2 v_1(M_1/f_{12})(\partial\Pi^*/\partial x) \quad (7)$$

where V_i is the partial molar volume and v_i the partial specific volume if i ($V_i=v_iM_i$). The right hand term of equation 7 could be thought of as an osmotically driven flow in free solution or the swelling dilution of component 1; the rate of which will vary as $\sqrt{(\text{time})}$.

Many studies have been performed in our laboratory on the mutual diffusion coefficients (defined by equation 4) on various polymers. This includes the polymers that we consider in this study namely dextran (a linear polymer of glucose-sugar residues with a low degree of branching) (Comper et al., 1986), dextran sulfate (Comper and Zamparo, 1989) and Swarm rat chondrosarcoma proteoglycan subunit (PGS) (Comper and Williams, 1987; Comper and Zamparo, 1990) (isolated from Swarm rat chondrosarcoma and is a molecule of approximately 2.6×10^6 in molecular weight consisting of pendant chondroitin sulfate chains (almost 90% by mass of the whole molecule) ($20\text{-}30 \times 10^4$) covalently bound to a protein core - a similar polysaccharide exists in cartilagenous tissue and acts to control tissue hydration and tissue biomechanics). In summary, the results of these studies have demonstrated that the terms D_1, Π^* and M_1/f_{12} generally exhibit a marked concentration dependence in the semi dilute regime. While quantitative differences exist for the various polymers these terms do approach molecular weight independence demonstrating that critical segments of the polymer chain are responsible for determining translational mobility (for the proteoglycan subunit its behaviour in the semi dilute regime is determined primarily by segments of its constituent chondroitin sulfate chains). The exact size of these segments is unknown at present, although they will be less than 10000 in molecular weight for the polymers studied here. Generally, too, there is a marked increase in the mutual diffusion coefficient with concentration - this is due to the predominance of the osmotic term (Figure 2) over the hydrodynamic frictional term (Figure 3).

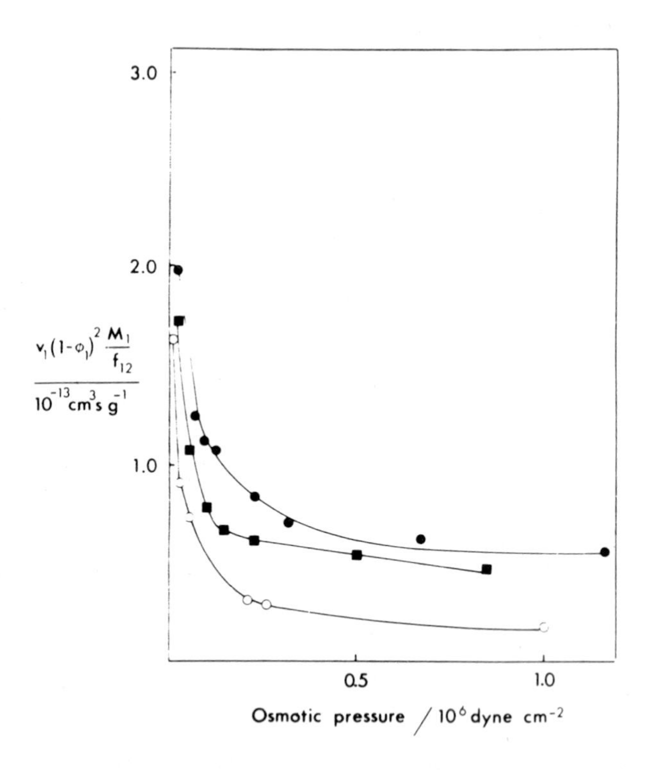

Figure 3. Variation of $v_1(1-\phi_1)^2M_1/f_{12}$ from equation 7 as determined from sedimentation studies as a function of osmotic pressure for dextran T500 (■), dextran sulfate T500 (•) and (○) PGS in PBS (from Williams and Comper, 1990 with permission of publishers).

Membrane-Related Phenomena (Binary and Pseudo Binary Systems)

1) Membranes impermeable to solute

We regard that in the solution in contact with the membrane there will be a very thin layer, extending from the membrane pore to a distance L into the solution where the solute concentration (or chemical potential) drops from its solution value to zero at the membrane pore/solution interface (Figure 1). It is the relative movement of water in relation to solute within this layer, at the mouth of the pore, that will govern osmotic flow and ultrafiltration. We had previously demonstrated that polymer transport across a free boundary where high polymer concentration gradients exist proceeds by a diffusion process (Van Damme et al., 1982) - such gradients would operate at the membrane pore. This model forms the basis for a new interpretation of membrane hydrodynamics (Williams and Comper 1987, 1990; Zamparo and Comper, 1989; Comper and Williams, 1990).

In a membrane system, osmotic flow occurs when a solute dissolved in water (a solution) is separated from solvent by a membrane permeable to solvent but not solute. Flow of solvent will spontaneously occur from the solvent compartment into the solution compartment. The driving force for solvent flow will be the tendency of the solute concentration gradient to relax in the thin layer adjacent to the membrane. It is exactly the same force that governs mutual diffusion in free solution as discussed *vide supra*. In this case it was pointed out that a solute concentration gradient across a free liquid boundary will give rise to a solute diffusion which could equally well be described by water diffusion (volume flow) in the opposite direction. This water flow is driven by identical osmotic pressure gradients that drive osmotic flow across membranes. Yet we know that in the free boundary system the diffusion coefficient of water and that of the solute in a volume fixed frame of reference are identical (Bearman, 1961) ie., the rate of osmotic flow across a free boundary is governed by solute diffusion.

Osmotic flow across a membrane has been modelled in our studies on a series of flows consisting of 1) water flow across the membrane, and 2) the diffusional exchange of osmotically active solute with solvent in the thin layer on the solution side of the membrane - it is the exchange process that draws water through the membrane. This can be described then as two resistances namely R_s associated with the solution layer and R_m associated with the membrane acting in series to control flow. For a given osmotic pressure $\Delta\Pi$ then the volume flow is given by

$$J_v=(1/(R_s + R_m))\,\Delta\Pi \tag{8}$$

In the experimental systems studied here with porous membranes and polymer solutions it is evident that the polymers have such high concentration dependent resistance that the solution layer adjacent to the membrane becomes rate limiting.

Considerable controversy has previously existed as to the mechanism of osmosis with most interpretations recognising flow through the membrane as the effective kinetic process (Berkely and Hartley, 1910; Kedem and Katchalsky, 1961; Heyer et al., 1969). The kinetic behaviour of the solute has not generally been recognised at all. However, some investigators have previously acknowledged the importance of the solution-exchange layer adjacent to the membrane (Ray 1960; Dainty 1965; Soodak and Iberall, 1978).

The usual relationship between osmotic volume flow, J_v, of water and the osmotic pressure difference ($\Delta\Pi$) across the membrane is

$$J_v = - L_p^{\circ}\Delta\Pi \quad (9)$$

where L_p° is the osmotic permeability coefficient which has formerly been regarded as a constant describing the interaction of water with the membrane. Prior to our studies, there had been no account made of the influence of the strong concentration dependent dynamic properties of the osmotically active solute on the magnitude of L_p° Rather, any deviation from ideal behavior of water - membrane interaction was commonly ascribed to an 'unstirred boundary layer effects' (Heyer et al., 1969). We have derived (Williams and Comper 1990) the following expression for L_p° which takes into account the series interaction model of water with the membrane and solute such that

$$L_p^{\circ} = 1/[(1/L_p) + (L/v_1(1-\phi_1)^2 A_{eff} M_1/f_{12})] \quad (10)$$

where L_p is the hydraulic permeability coefficient in the absence of solute and embodies the water-membrane interaction term. [1]

---

[1]Osmotic flows were measured in a well stirred (>200rpm) bivalve cell through either directly monitoring volume flow or solution density as a function of time. The flows were linear in the first 30 min.

The water-solute interaction will be included in the M_1/f_{12} term. The membrane area A_{eff} is the effective area of the pores in the membrane and L represents the distance over which the osmotic solute-solvent exchange process takes place in front of the

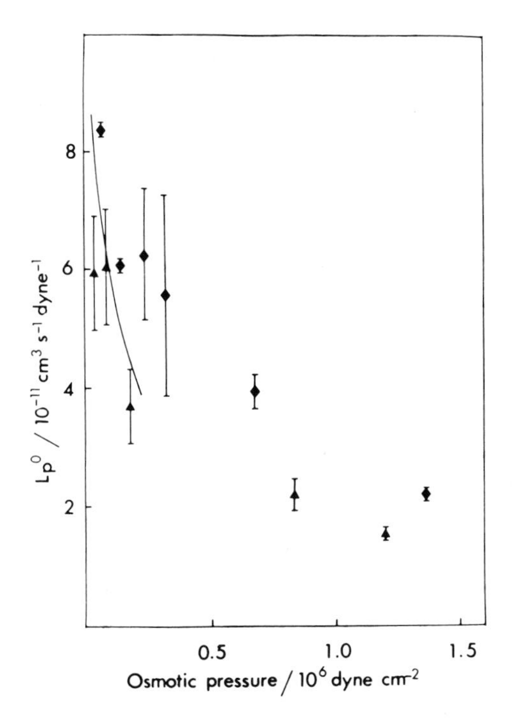

Figure 4a. The variation of the osmotic permeability coefficient L_p° measured for osmotic flows across semipermeable tortuous polyester membrane as a function of osmotic pressure for dextran ($M_w \sim 70000$) for stirred (♦) and unstirred (▲) as compared to previous data from an earlier model osmotic flow cell (Williams and Comper (1987). L_p for the membrane in solvent only was 10.97×10^{-11} cm³ dyne⁻¹s⁻¹ (from Williams and Comper, 1990 with permission of publishers).

pore (Figure 1). Any involvement of solute-membrane interaction in modifying the exchange process will be included in the A_{eff}/L term.

For osmotically active polymer solutions at concentrations greater that 10 mg ml⁻¹ the second term on the rhs of equation 10 far outweighs the first term governing water-membrane interactions for all types of commercially available membranes studied (Williams and Comper, 1987; 1990) (Figure 4a & b). This demonstrates that the limiting factor governing the rate of osmotic flow is the frictional term M_1/f_{12}. The data in Figure 4a & b also demonstrates that L_p° decreases rapidly with osmotic pressure and that this is mirrored in a similar fashion in the variation of the frictional factor (as determined from sedimentation studies) with osmotic pressure (Figure 3).

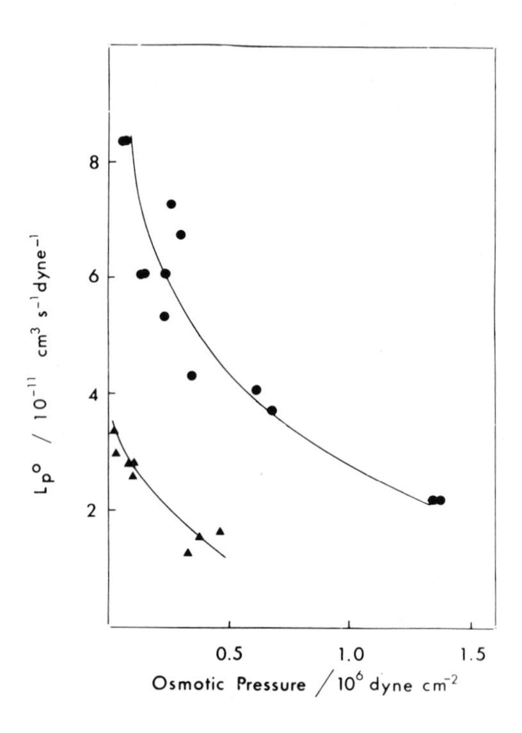

Figure 4b. The variation of the osmotic permeability coefficient $L_p°$ measured for osmotic flows across semipermeable tortuous polyester membrane as a function of osmotic pressure for PGS (▲) and dextran sulfate (•) in thermodynamic equilibrium with PBS. L_p for the membrane in solvent only was 10.97×10^{-11} cm^3 $dyne^{-1}s^{-1}$ (from Williams and Comper, 1990 with permission of publishers).

The relative differences between dextran, dextran sulfate and PGS in terms of $L_p°$ are also exemplified in the M_1/f_{12} function in Figure 3. The possible exception is the relatively low values of $L_p°$ for dextran sulfate which could be rationalised on the basis that the membrane may not be totally impermeable to this solute as the preparation is extremely heterogeneous with M_n/M_w ratios less than 0.5. (Our initial studies (Williams and Comper,1987) also demonstrated the molecular weight independence of $L_p°$ for a series of dextran fractions of different molecular weight in semi dilute solution which is in accord with the molecular weight dependence of the

M_1/f_{12} function.) The results of evaluation of the A_{eff}/L term in equation 10 from data in Figures 3 and 4 is shown in Figure 5 (represented as $K(=A_{eff}/L)$). In general we find a marked similarity in the magnitude of K for all three polymers studied with values in the range of 0.4-0.6. Similar values of K have been obtained from ultrafiltration studies of dextran and proteoglycan (Zamparo and Comper, 1989).

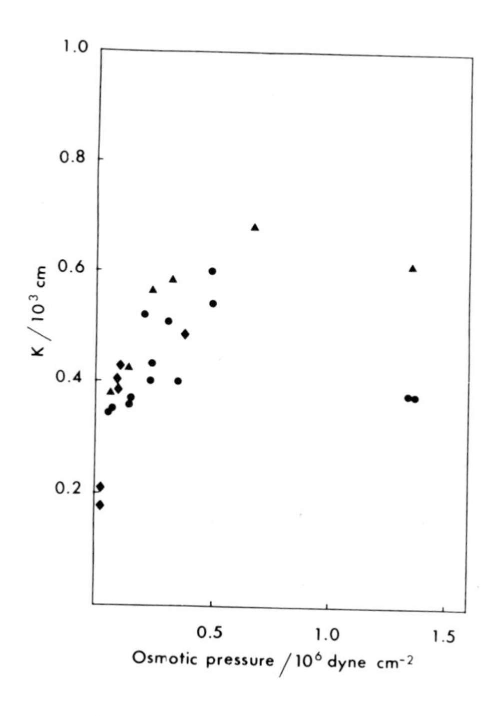

Figure 5. Variation of the parameter $K(=A_{eff}/L)$ from equation 10 as a function of solute osmotic pressure for dextran sulfate (•), PGS (♦) and dextran (▴) (from Willia ms and Comper, 1990 with permission of publishers).

While it is commonly recognised that the colligative properties of the polyelectrolytes, such as dextran sulfate and PGS, are governed mainly by their constituent micro-counterions, the kinetics of osmotic flow is determined by the frictional factor of the polymers which, as shown above, resides essentially with the dynamic interaction of water with segments of the polymer chain. These segments, characterised by their glycosidic linkage, and chain conformation (Comper and Zamparo), therefore are critical in the manifestation of osmotic pressure; reducing the mobility of these units may have an influence on the osmotic properties of the polymer.

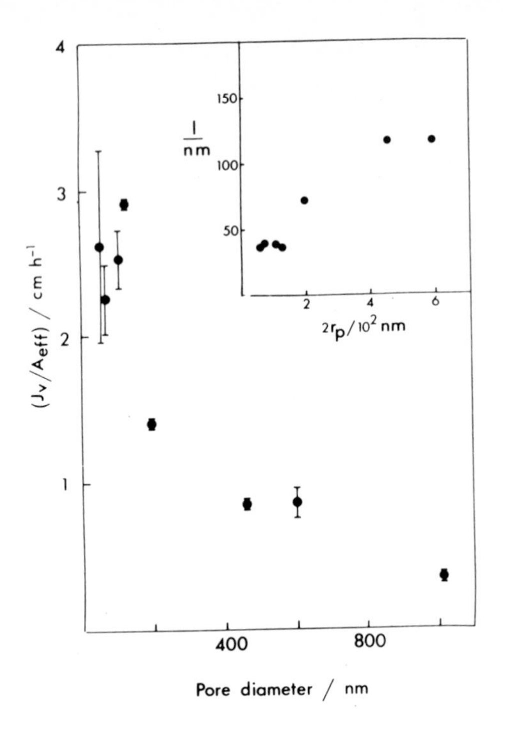

Figure 6. The osmotic flow J_v/A_{eff} obtained over a 1h period generated by 19.8 mg ml $^{-1}$ PGS across well characterised Nuclepore straight capillary pore membranes of known pore diameter ($2r_p$) and effective pore area. The corresponding distance L in front of the pore ,calculated from equation 10, is shown in the inset.From Comper and Williams (1990).

In studies of osmotic flow across membranes with straight cylindrical pores of known effective area, the thin layer distance L for a 20 mg ml $^{-1}$ PGS solution was estimated to be the order of a molecular dimension namely 40nm (Comper and Williams 1990). This value was consistently obtained for membranes of different pore size that were impermeable to the proteoglycan (Figure 6). This would also suggest that in spite of varying membrane surface exposed at the pore, and therefore area available for adsorbtion of proteoglycan, this did not affect the consistency of the results. For the membranes impermeable to the proteoglycan the variation of the flux is to r_p^2 which indicates that the process is diffusion-controlled. If we take the common assumption that the coefficient in equation 9 is directly related to Poiseuille's law then the flows would be related to r_p^4. This is clearly not the case as shown in Figure 7. This data also reinforces the fact that the limiting region for osmotic flow is the thin exchange solution layer in front of the pore and not the membrane. The latter would have to be an order of magnitude thicker to have a significant effect on flow.

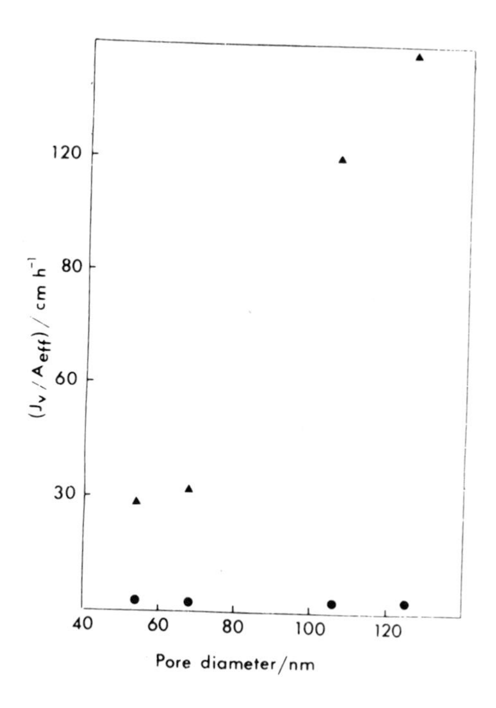

Figure 7. Osmotic flow represented as J_v/A_{eff} (●) and its predicted value (▲) calculated from Poiseuille's law as function of membrane pore diameter. Membrane thicknesses were in the range of 6μ to 10μ (from Comper and Williams, 1990 with permission of publishers).

2) Membranes with partial permeability to solute.

For permeable membranes the solute will partition itself at the membrane pore. The simplest approach is to regard the solute as an effective sphere of radius r_1 that is sterically excluded from the pore of radius r_p. This gives a partition coefficient (λ)

$$\lambda=(1-\alpha)^2=C_1'/C_1 \tag{11}$$

where $\alpha=r_1/r_p$ and $C_1'$and C_1 are the concentrations of component 1 just inside the membrane pore and in the bulk solution respectively. This approach regards processes at the membrane-solution interface as critical in determining the degree of osmosis. For osmotically active solute initially on one side of the membrane only, it will then be the corresponding difference in osmotic pressure across the mouth of the pore $\Pi(C_1)$-$\Pi(C_1')$ that will drive osmotic flow. This will be generated by molecules excluded by the pore and undergo diffusional exchange with solvent, with a frictional coefficient corresponding to the concentration C_1 over the characteristic distance L in front of the pore. It is assumed that all other molecules will enter the pore and do not become osmotically active in generating flows across the membrane. The volume flow generated by diffusion of solute in the pore is also assumed to be negligible. With these considerations, we can write the volume flow as

$$J_v = [\Pi^*(C_1) - \Pi^*(C_1')]/[(1/L_p(C_1') + (f_{12}(C_1)L/v_1(1-\phi_1)^2M_1A_{eff})] \qquad (12)$$

where $L_p(C_1')$ is the hydraulic permeability of the solution assumed to be at concentration C_1' through the membrane. Efforts aimed at predicting proteoglycan-generated osmotic flow through permeable membranes using equation 12 are shown in Figure 8.

It is of interest to examine whether the data give meaningful values of proteoglycan size and how flow is predicted to vary with r_p. It is evident that reasonable agreement is apparent for molecular radui in the range of 37nm up to 60nm although the analysis is not accurate enough to nominate a particular effective size of the material. It is apparent, however, that the viscosity component embodied in the L_p term is insignificant in relation to the frictional coefficient term in equation 12. Further, there is a strong dependence predicted of the J_v/A_{eff} term in relation to the variation of r_p especially in the range of r_p from 125 nm to 194 nm. This is essentially derived from the marked nonideality of the osmotic pressure term.

Significance for cartilage swelling

The finite dimension of the thin layer governing solute-solvent exchange that drives osmosis would also indicate that in collagen fiber networks in tissues that retain proteoglycan, this range of dimensions be available for the proteoglycan to move about in order to generate osmotic flow.

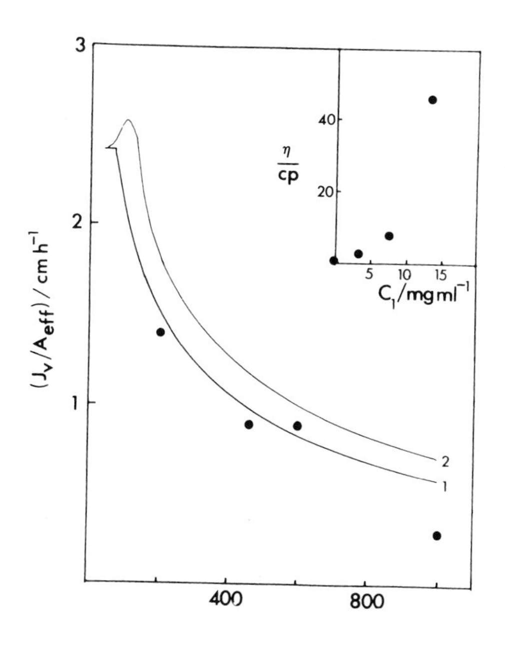

Figure 8. Predicted osmotic flow for PGS at 19.8 mg ml^{-1} from equation 12 with L=37 nm and $L_p(C_1')$from Poiseuille's law and the PGS viscosity data in the inset. The values have been calculated for r_1=37 nm (one half the average molecular distance between PGS molecules at 19.8 mg ml^{-1}) (curve 1) and r_1=60 nm (curve 2) (from Comper and Williams, 1990 with permission of publishers).

Therefore, in understanding the mechanism of osmotic flow recognition must be made of, extracellular matrix microorganisation of proteoglycans that enable segmental mobility, and the finite distances over which solute-solvent exchange takes place. Clearly a recognition of the role of collagen and its coexistance with proteoglycans must be made although not much is known yet of the dynamic properties of collagen network. This model also realises the important function of distance over which the internal osmotic pressure can operate in relation to the swelling pressure of the whole system. It was shown *vide supra* that in a membrane-constrained system a distance of 40nm was required for the proteoglycan chondroitin sulfate segments to move. Any immobilisation of the chondroitin sulfate chain segments themselves would have a drastic influence on the osmotic activity of the proteoglycan.

Osmotic Equilibration in Ternary Systems

We have established that for a given osmotic pressure the lower the frictional coefficient (or faster the effective diffusive mobility) then the greater the rate of osmotic flow. It was pointed out by Comper et al (1990) that this provides an interesting situation as to the mechanism of osmotic equilibration between two solutions of identical osmotic pressure but different binary frictional coefficients. For the ternary case we redefine component nomenclature as 1=solute 1, 2=solute 2, and 3=solvent. The osmotic equilibration will the involve a membrane impermeable to solutes 1 and 2 which are placed on either side of the membrane. The equations for ternary diffusion are

$$-J_1=D_{11}(\partial c_1/\partial x) + D_{12}(\partial c_2/\partial x) \qquad (13)$$

$$-J_2=D_{21}(\partial c_1/\partial x) + D_{22}(\partial c_2/\partial x) \qquad (14)$$

where D_{ij} are the ternary diffusion coefficients. These equations can be reexpressed in terms of binary frictional coefficients and osmotic pressures of the two solutions on either side of the membrane Π_1 and Π_2 such that for dilute solutions (Miller, 1959; Comper et al,. 1984)

$$J_1=(RT/\xi)[(f_{23}+f_{21})(\partial\Pi_1/\partial x_1) + f_{21}(\partial\Pi_2/\partial x_2)] \qquad (15)$$

$$J_2=(RT/\xi)[f_{12})(\partial\Pi_1/\partial x_1) + (f_{13}+f_{12})(\partial\Pi_2/\partial x_2)] \qquad (16)$$

where

$$\xi=-(f_{13}f_{23} + f_{13}f_{21} + f_{12}f_{23}) \qquad (17)$$

At equilibrium we have zero volume flow and equal osmotic pressures so that

$$J_1V_1=-J_2V_2 \text{ and } \partial\Pi_1/\partial x_1=\partial\Pi_2/\partial x_2$$

When these conditions are used in combination with equations 15 and 16 we get a finite value of the cross frictional term existing across the membrane such that

$$f_{12}=-(V_1 f_{23} + V_2 f_{13})/2(V_1(c_1/c_2)+V_2) \qquad (18)$$

This result suggests that there are thin layers on either side of the membrane, where solute-solvent exchange occurs, where molecules will be distinct from their counterparts in the bulk solutions due to the influence of the cross membrane frictional coefficient.

Objections to the Series Resistance Osmotic Flow Model

It had been previously stated in the literature that a major factor that will influence the magnitude of osmotic flow is the presence of an apparent unstirred layer (or osmotic boundary layer) on the solution side of the membrane. This layer represents a thin film of fluid on the solution side of the membrane which is thought to have a lower solute concentration than the bulk solution due to the fact that the osmotic flow of water would advect solute away from the membrane. The layer would give rise to a lower effective osmotic pressure and flow on the basis of solute concentration in the bulk solution. Further, the magnitude of the flow may depend on the efficiency of the stirring motion in the solution to prevent the unstirred layer from forming. Under these circumstances, the ideal flow would yield a concentration - independent L_p° which has commonly been described in terms of parameters associated with the membrane structure. At present, there is no experimental evidence or established theoretical models which demonstrates the nature of the concentration gradients of solute adjacent to the membrane.

It is clear from our studies that even in efficiently stirred systems with no boundary layers the concept of the series resistance would still apply. In fact an underlying feature of the interpretation adopted is that stirring effects are small in our experimental system. These conclusions have come from a number of experiments which have been designed to test the effect of stirring including modifying the stirring design (to have stirrers situated 0.5nm from membrane surface), varying stirring speed (up to 200 rpm) , and linearity of flows from 30 s to 60 min. The vertical position of the membrane in the osmotic flow experiment would also make boundary layers gravitationally unstable with the tendency to convect away any concentration gradients. In contrast to the advecting (of solute) osmotic boundary layer, the diffusional solute-solvent exchange layer driving flow as proposed in our model will tend to promote solute towards the membrane. This represents a clear distinction from the unstirred layer model. It is clear that while we cannot absolutely disprove that stirring or lack of it is not affecting our results we can rationalise that it is not a major factor governing them. The marked variation in L_p° is in accord with the variation in the M_1/f_{12} function. Notably too, all experiments to test the model with materials with different osmotic pressures and frictional coefficients consistently behaved with the framework of the model although there are some reservations about this. Firstly, data from the tortuous polyester membranes have been interpreted in terms of a K function (= A_{eff}/L). While there seems to be reasonable agreement of the K values between the different polymer preparations used here there is no *a priori* reason for this particularly in terms of factors that may govern the magnitude of L which we presume is related to the characteristic jump length of the solute. However, we do find it reassuring that for dextran fractions of different molecular weights the same K value is obtained which seems to be in accord with the molecular weight dependence of the hydrodynamic frictional coefficient (in mass units) and that similar values of K are obtained for proteoglycan and dextran from ultrafiltration (Zamparo and Comper,1989). THe latter finding would confirm Darcy's law for the system. Secondly, evaluation of L from well characterised straight through cylindrical pore membranes gave consistent results with the model but there is still a need to establish from independent measurements the dynamics of the polymer near a membrane pore. Studies with the cylindrical pore membranes also demonstrated the possibility of transient networks in semi dilute solution having an osmotic effect in relation to relatively large pore membranes (not discussed in this article but see Comper and Williams, 1990).

Acknowledgements

This work was supported by grants from the Australian Research Council and Monash University Special Research Section.

References

Bearman RJ (1961) On the molecular basis of some theories of diffusion. J Phys

Chem 73:1961-1968

Berkeley Earl of, Hartley EG (1910) Dynamic osmotic pressures. Proc Roy Soc (Lond) Series A 82:271-275

Buckley DJ, Berger M (1962) The swelling of polymer systems in solvents II Mathematics of diffusion. J Polym Sci 56:175-188

Candau S, Bastide J, Delsanti M (1982) Structural, elastic, and dynamic properties of swollen polymer networks. Adv Polym Sci 44:27-71

Comper WD, Checkley, GJ, Preston BN (1984) Kinetics of multicomponent transport by structured flow in polymer solutions 5. Ternary diffusion in the system dextran-poly (vinyl pyrrolidone) - water. J Phys Chem 88:1068-1076

Comper WD, Preston BN, Davies P (1986) The approach of dextran mutual diffusion coefficients to molecular weight independence in semi dilute solutions of polydisperse dextran fractions. J Phys Chem 90:128-132

Comper WD, Williams RPW (1987) Hydrodynamics of concentrated proteoglycan solutions. J Biol Chem 262:13464-13471

Comper WD, Williams RPW (1990) Osmotic flow caused by chondroitin sulfate proteoglycan across well defined Nuclepore membranes. Biophys Chem 36:215-222

Comper WD, Williams RPW, Zamparo O (1990) Water transport in extracellular matrices. Connective Tissue Res 25:3-16

Comper WD, Zamparo O (1990) The hydrodynamic properties of connective tissue polysaccharides. Biochem J 269: 561-564

Dainty J (1965) Osmotic flow. In: The state and movement of water in living organisms. Symp Soc Exp Biol XIX (Federation Proc) University Press, Cambridge, p75

Heyer E, Cass A, Mauro A (1969-70) A demonstration of the effect of permeant and impermeant solutes, and unstirred boundary layers on osmotic flow. Yale J Biol and Med 42:139-153

Kedem O, Katchalsky A (1961) Physical interpretation of the phenomenological coefficients of membrane permeability. J Gen Physiol 45:143-179

Kurata M (1982) Thermodynamics of polymer solutions, Hardwood Academic, Chur London New York

Miller DG (1959) Ternary isothermal diffusion and the validity of the Onsager reciprocity relations. J Phys Chem 63:570-578

Ray PM (1960) On the theory of osmotic water movement. Plant Physiol 35:783-795

Spiegler KS (1958) Transport processes in ionic membranes. Trans Faraday Soc 54:1408-1428

Soodak H, Iberall A (1978) Osmosis, diffusion, convection. Am J Physiol 235:R3-R17

Tanaka T, Fillmore D (1970) Kinetics of swelling of gels. J Chem Phys 70:1214-1218

Urban JPG, Maroudas A, Bayliss MT, Dillon J (1979) Swelling pressures of proteoglycans at concentrations found in cartilagenous tissues. Biorheology 16:447-464

Van Damme M-P, Comper WD, Preston BN (1982) Experimental measurements of polymer unidirectional fluxes in polymer and solvent systems with non-zero chemical-potential gradients. J Chem Soc Faraday Trans I 78:3357-3367

Williams RPW, Comper WD (1987) Osmotic flow caused by non-ideal macromolecular solutes. J Phys Chem 91:3443-3448

Williams RPW, Comper WD (1990) Osmotic flow caused by polyelectrolytes. Biophys Chem 36:223-231

Zamparo O, Comper WD (1989) Hydraulic conductivity of chondroitin sulfate proteoglycan solutions. Arch Biochem Biophys 274:259-269

Part 6

Developments in Various Techniques

CELL PROTRUSION FORMATION BY EXTERNAL FORCE

Sergey V. Popov

Department of Biological Sciences
Columbia University
New York 10027
USA

Leonid B. Margolis

Belozersky Laboratory of Physico–Chemical Biology
Moscow 119899
USSR

SUMMARY

A method of applying a mechanical force to the plasma membrane of cells is described. The force is generated by an alternating current (*AC*) electrical field (*EF*) of high frequency. It is applied to the plasma membrane and is directed outwards. This force was sufficient to generate morphologically normal cell protrusions in mouse embryo and *3T3* fibroblasts. Specific inhibitors of actin polymerization were not able to prevent generation of protrusions in an electrical field. Organization of the cytoskeleton inside the processes was examined using a platinum replica method. Bundles of microfilaments morphologically similar to those observed during normal physiological spreading were found in electrical field–generated protrusions. Local application of electrical field to polarized cells demonstrated that different areas of the cell surface differ in their ability to form processes under the action of membrane–applied force.

1. INTRODUCTION

Most of the cells of an organism are attached to each other or to non–cellular substrate with the help of the processes (or cell protrusions). The nature of the force pushing the membrane outward, as well as the factors determining the morphology of the protrusions are unclear (Trinkaus, 1985).

NATO ASI Series, Vol. H 64
Mechanics of Swelling
Edited by T. K. Karalis

In normal physiological conditions bundles of actin microfilaments are found in protrusions (Small, 1981) and it has been proposed that the force pushing the membrane outward is generated by actin polymerization into microfilaments (*viz.* Vasiliev, 1985). An alternative hypothesis considers the protrusions to be induced by the forces of other origin *e.g.*, osmotic (Oster and Pelerson, 1987).

To investigate the mechanism of the formation of cells protrusions a model system is described in the present study, based on the phenomenon of dielectrophoresis. We demonstrate that the membrane–applied force generated by the application of an electrical field is sufficient to produce native–like protrusions with normal morphology and normal organization of the cytoskeleton–even in conditions where the cytoskeletal activity is inhibited.

2. MATERIAL AND METHODS

Cells. Primary mouse embryo fibroblasts were prepared according to the method of Domnima *et al* (1982). *NIH 3T3* fibroblasts were obtained from the American Tissue Culture Collection. Cells were maintained in Dulbecco's modified Eagle's medium containing *10%* fetal calf serum. Before electrical field treatment cells were transferred to buffered sucrose (*290 mM sucrose, 1 mM Hepes–NaOH, pH 7.4*).

Cell treatment by electrical field. Cells were treated by an electrical field as previously described (Popov and Margolis, 1988). *50–100 μl* of a cell suspension in buffered sucrose were placed on the surface of the coverslip in the gap between electrodes (Fig. 1). The cells were treated by an electrical field (frequency *1 MHz*, voltage *6–8 V* with a gap between electrodes of *60–80 μm*) at room temperature for *20 s* to *5 min*.

We also treated with an electrical field spread on the surface of the coverslip fibroblasts (Fig.2). An electrical field was applied *via* tungsten microelectrodes with *2–5 μm* diameter tips. Electrodes were placed *10–20 μm* from the cell surface and the electrical field was applied for *10–60 s* (voltage *7–8 V*, frequency *1 MHz*).

A force is generated during cell treatment with an *AC* electrical field in low conducting medium. The force is applied to the plasma membrane and to a thin (*1–2 nm*) cytoplasmic layer and is directed outwards (Pohl, 1978; Pastushenko *et al*, 1985). For the typical intensities of electrical field used in the present study (*1–10 KV/cm*), the magnitude of the membrane–applied force is 10^1–10^2 N/m^2 (Margolis and Popov, 1988).

Electron microscopy. Cells were fixed for scanning electron microscopy by *2.5 %*

glutaraldehyde in phosphate–buffered saline or in buffered sucrose. The cells were dehydrated in acetone, critical point dried and examined in a Hitachi *405 A* scanning electron microscope.

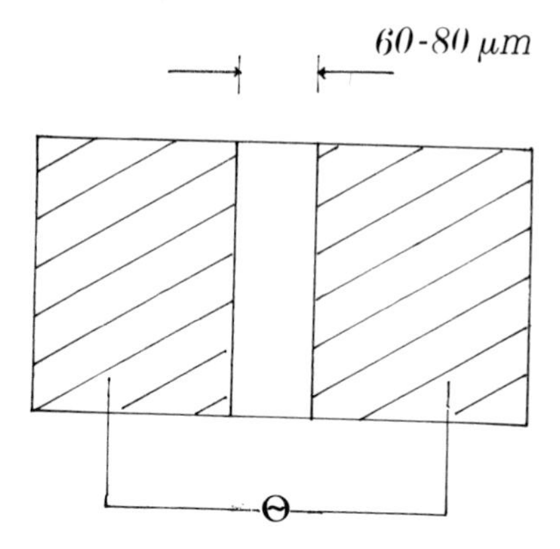

Figure 1. An experimental chamber. A coverslip was covered with aluminum films of 0.1 μm thick. The metal films were used as a pair of electrodes. The gap between two electrodes was 60–80 μm. Cells suspended in buffered sucrose were treated with alternative current electrical field for 20 s to 5 min.

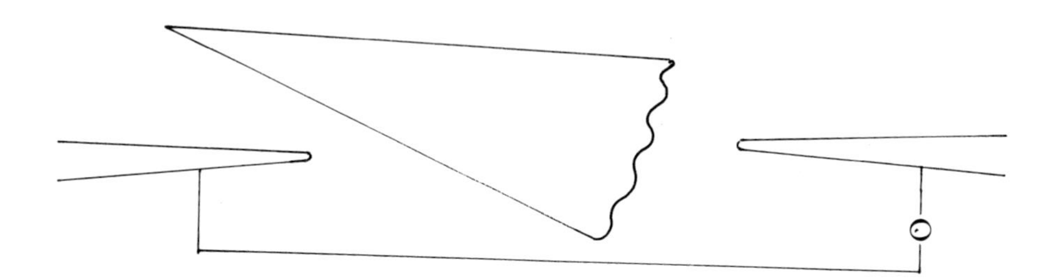

Figure 2. An alternative current field was locally applied to the fibroblasts placed on the surface of a coverslip by 2 tungsten microelectodes (tip diameter 2–5 μm).

Platinum replicas of *EF*–treated fibroblasts were prepared according to Svitkina *et al* (1984). Briefly, Triton *X–100*–extracted cells were rotary shadowed with platinum and strengthened with carbon. The replicas were examined at *80 KV* in a Jeol–*100C* electron microscope.

Immunoelectron microscopy of platinum replicas was performed according to Popov *et al* (1991). Triton *X–100*–extracted cells were incubated with rabbit anti–actin antibodies followed by incubation with secondary antibodies conjugated to *10 nm* colloidal gold particles.

3. RESULTS

Cell process formation. After the electrical field was switched on, a drop of the cell suspension in buffered sucrose solution was placed in the middle of the chamber (Fig.1). The suspended cells were concentrated on the coverslip along the edges of the metal electrodes, in the area of maximal field strength. The observed behaviour of the cells is in full agreement both with the theory of the dielectrophoresis and previously reported experimental data (Pastushenko *et al*, 1985; Popov and Margolis, 1988).

After electrical field application for *20 s* to *5 min* cells were fixed and examined in scanning electron microscope. A system of cell protrusions, connecting the cell body with the electrode was found (Fig.3). The protrusions were formed only on the parts of the membrane facing the electrode. Most of them were cylindrical in shape, *0.2–0.5 μm* in diameter and up to *15 μm* long. These cylindrical processes are morphologically similar to the filopodia processes that form during the initial stage of cell spreading under normal physiological conditions.

No processes at all were formed by fibroblasts during a *5 min* incubation in sucrose solution without electrical field application, indicating that the electrical field is the immediate cause of cell process formation.

Distribution of the process over the cell body *i.e.,* their formation at sites of the membrane experiencing the largest force, suggest that the observed are possibly generated directly under the action of a membrane–applied force. For the typical intensities of electrical field used in these experiments, the membrane–applied force is estimated to be in the range of 10^1–10^2 N/m^2 (Margolis and Popov, 1988). This force is enough to induce bending of the lipid bilayer and to form cell protrusion (Bo and Waugh, 1989).

Previously we have demonstrated that other possible effects of electrical field– electrical breakdown of the plasma membrane, electrophoresis of receptors on the cell surface and heating do not contribute to the phenomenon of cell protrusion formation (Popov and Margolis, 1988).

It could be suggested that the role of an electrical field in our experiments is to bring the cells in close contact with the substrata and to accelerate normal (cytoskeleton dependent) generation of cell protrusions. To understand the role of cell cytoskeleton in the protrusion formation in an electrical field we pretreated fibroblasts with cytochalsin B to inhibit actin polymerization, with colcemide to destroy microtubule organization of the cells, and with

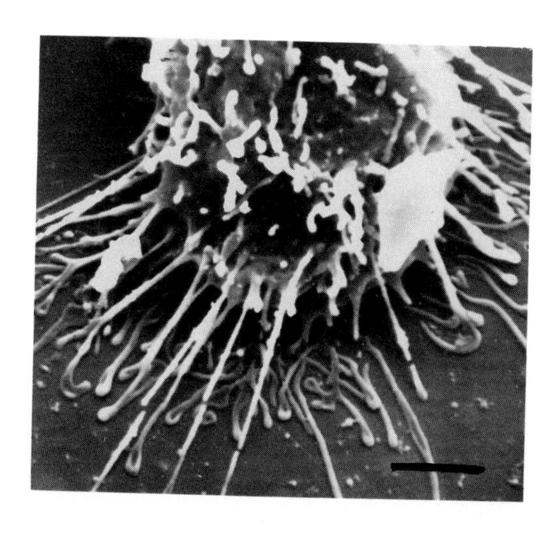

Figure 3. Scanning electron microphotograph of mouse embryo fibroblast fixed after 5 min application of the electrical field. The cylindrical protrusions are directed from the cell body to the electrode, which occupies the lower part of the photograph. Bar, 3 μm.

inhibitors of *ATP* synthesis (sodium azide and carbonyl–m–chlorphenylhydrazon (to block *ATP*–dependent reorganizations of the cytoskeleton. In normal physiological conditions, each of these treatments significantly inhibited formation of cell processes. Scanning electron microscopy demonstrated that contrary to normal physiological conditions the mechanical force in the dielectrophoretic chamber generates the cell surface protrusions independently of these treatments. Neither the morphology of the processes nor their distribution over the cell body was affected by the drug treatments.

Thus, it can be concluded that in our experiments morphologically normal cell protrusions are formed directly under the action of membrane–applied force without active (*ATP*–dependent) reorganization of the cytoskeleton.

Organization of the cytoskeleton in the processes. To understand how the deformation of the cell membrane induced by electrical field–generated force affects cytoskeleton organization, the internal structure of the protrusions was investigated using a platinum replica method. A system of cytoskeletal elements connecting the cell body with the electrode was revealed inside the protrusions. Bundles of filaments, corresponding to the cylindrical processes were found inside the protrusions (Fig.4). The system of filaments observed after a *5 min* application of electrical field was morphologically similar to the one formed by control cells in the early stages of spreading under normal physiological conditions. In the latter case the filaments are composed of *F*–actin. To find out the nature of filaments inside electrical field–generated protrusions, we stained the cells with *FITC*–phalloidin and used the method of immunoelectron microscopy of platinum replicas. Both of these approaches demonstrated that the main component of filaments inside protrusions is *F*–actin.

Two mechanisms of formation of actin bundles inside electrical field–generated protrusions can be proposed. (1) Actin filaments, linked to the plasma membrane, passively follow it and organize in the bundles directly under the action of membrane–applied force. (2) The processes are free from F–actin bundles immediately after their formation. Eventually G–actin *ATP*–dependently polymerize into bundles of microfilaments.

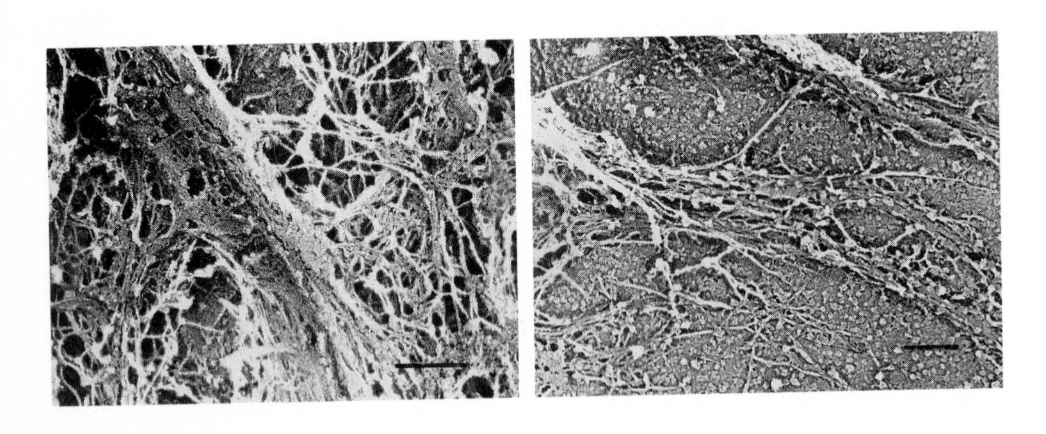

Figure 4. Platinum replicas of cytoskeleton of mouse embryo fibroblasts. Cells were fixed after a 5 min. application of electrical field. (a) A bundle of filaments at the distal part of the lamella. (b) Bundles of filaments inside the distal processes. The filaments are oriented parallel to the axis of cylindrical protrusion. Bar, 0.2 μm.

To distinguish between these mechanisms we applied electrical field to the cells for *20 sec* and immediately fixed them for phalloidin staining or electron microscopy. Both fluorescent microscopy and platinum replica method demonstrated the presence of F–actin bundles inside filopodia protrusion, implicating in favour of the first mechanism.

In another set of experiments we pretreated the cells with sodium azide or carbonyl–m–chlorphenylhydrazone to inhibit *ATP*–dependent polymerization of actin into microfilaments. As in the case of cells not treated with these drugs, actin filaments oriented along the axis of cylindrical protrusion were revealed by the platinum replica method.

Taken together, results of these experiments demonstrate that actin filaments inside filopodia can be organized into bundles directly under the action of membrane–applied force, without *ATP*–dependent actin polymerization.

Formation of cell processes at different parts of the cell membrane. To investigate the potential for different regions of fibroblasts to form processes under the action of force, we applied an electrical field locally to different sites of the membrane and examined morphological changes in cells by light microscopy.

A few seconds after electrical field application protrusions were generated on that part of the membrane closest to the electrode. The processes were cylindrical; they were directed towards the tip of the electrode and reached it in approximately *30 sec* (Fig.5).

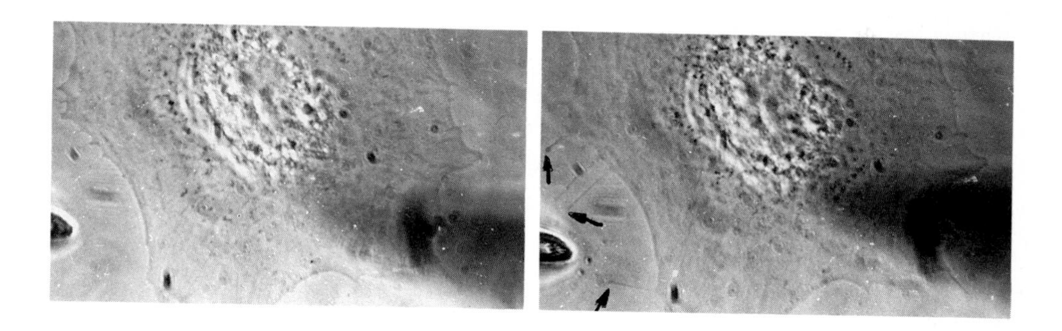

Figure 5. A cell after 3 h incubation on the surface of a coverslip (a) 30 sec after electrical field application (b) Cylindrical processes are formed at the electrode–facing part of the cell (arrows). They are directed from the cell to the tip of the electrode. Bar, 20 μm.

After a few hours of incubation on the surface of the coverslip the cells polarized with the formation of active and stable edges. Protrusive activity of the cell at the stable edge stopped then cell protrusions were formed only at the lamella region of the cell. By placing the electrodes close to certain parts of the membrane, we were able to apply the force locally to the lamella region of the cell and to stable edge. Under electrical field treatment the processes were formed only at the active edge (where they form normally) but not at the stable edge.

4. DISCUSSION

The mechanism of cell protrusion formation remains unclear. Even the most basic question about the role of actin inside the protrusions is unresolved. The main experimental difficulty is that up to now it was impossible to "separate" formation of protrusions from polymerization of actin inside them. Bundles of filaments were always found in protrusions, suggesting that the membrane is pushed outward by actin polymerization. A model system described in the present study provides a tool to separate the contributions of actin polymerization and membrane–applied force to cell protrusion formation.

The experimental phenomenon that we observe (e.g., the electrical field–generated force produce cell protrusions in animal cells) is surprising in itself. Indeed, it was demonstrated that the force P applied to the unit area of the cell membrane in the electrical field can be estimated as

$$P=\epsilon_o\epsilon_w\frac{E^2}{2}$$

where ε_o is the permittivity of vacuum, ε_w is the dielectric constant of water and E is the intensity of electrical field near the cell surface (Margolis and Popov, 1988). The voltage at the plasma membrane can be found by

$$U=\frac{\epsilon_w}{\epsilon_m}Eh$$

where ε_m is the dielectric constant of the plasma membrane and h is its thickness. Since the breakdown voltage U_{max} of the plasma membrane is of the order of *1 V* (Zimmermann, 1982), the force P_{max} applied to the cell surface in high frequency electrical field cannot exceed

$$P_{max}=\frac{\epsilon_m^2\epsilon_o}{2\epsilon_w}\left(\frac{U_{max}}{h}\right)^2\sim10^3N/m^2.$$

This force is enough to deform lipid bilayer. But since fibroblasts, as well as other animal cells have a rigid cortical layer of cytoskeletal elements connected to the plasma membrane, it is generally accepted that a typical animal cell is a few orders of magnitude more rigid than a liposome (Bray *et al*, 1986; Pasternak and Elson, 1985). For this reason it seems *a priori* unlikely that electrical field–generated force can induce significant deformation of the surface of a fibroblast.

We did not observe *global* deformation of fibroblasts in an electrical field: cells remained spherical during application of an electrical field. But as it is clearly demonstrated by electron microscopy, cell surface deformed *locally*–forming cylindrical cell processes up to *15 μm* long. Morphologically these processes were strikingly similar to that formed during spreading of the cells under normal physiological conditions. As demonstrated by experiments with inhibitors of cytoskeleton activity, the protrusions were generated directly by membrane–applied force without active participation of the cytoskeleton.

In both electrical field–generated protrusions and in the outgrowths formed during spreading in normal physiological conditions, bundles of actin filaments oriented along the axis of the filopodia were observed.

Moreover, the processes can be generated by the membrane–applied force only at those sites of the cell, where they form under normal physiological conditions (leading lamella) but not at the stable edge of the cell.

The results of our experiments demonstrate that a surprisingly small (of the order of 10^1–10^2 N/m^2) membrane–applied force is sufficient to reproduce the whole repertoire of cell surface protrusive activity without *ATP*–dependent reorganizations of the cytoskeleton:

- Protrusions of normal morphology are generated;
- Cell cytoskeleton in the protrusions is organized in the native way:
- The processes are induced at those sites of the polarized cells, where they form in normal physiological conditions.

Acknowledgement. We thank Drs. Y. A. Chizmadzev and L. V. Chernomordik for useful discussions.

5. REFERENCES

Bo L., Waugh R.E. (1989) Determination of bilayer membrane stiffness by tether formation from giant, thin–walled vesicles. Biophys. J., 55: 509–517.

Bray D., Heath J., Moss D. (1986). The membrane–associated "cortex" of animal cells: its structure and mechanical properties. J. Cell Sci. Suppl., 4: 71–88.

Domnina L.V., Gelfand V.I., Ivanova O.Y., Leonova E.V., Vasiliev J.M., Gelfand I.M. (1982) Effect of small doses of cytochalasins on fibroblasts: preferential changes of active edges and focal contacts. Proc. Natl. Acad. Sci. USA, 79: 7754–7757

Margolis L.B., Popov S.V. (1988) Generation of cell processes in high frequency electrical field. Biolectrochem Bioenerget., 20: 143–154.

Oster G.F., Perelson A.S. (1987) The physics of cell motility. J. Cell Sci. Suppl., 8:35–54

Pasternak C., Elson E.L. (1985) Lymphocyte mechanical response triggered by cross–linking surface receptors. J. Cell Biol., 100: 860–872

Pastushenko V.Ph., Kuzmin P.I., Chizmadzhev Yu.A. (1985) Dielectrophoresis and electrorotation–a unified theory of spherically symmetrical cells. Stud. Biophys., 110: 51–57

Pohl H.A. (1978) Dielectrophoresis. Cambridge University Press, Cambridge.

Popov S.V., Margolis L.B. (1988) Formation of cell outgrowths by external force:a model study. J. Cell Sci., 90: 379–389

Popov S.V., Svitkina T.M., Margolis L.B., Tsong T.Y. (to be published) Mechanism of cell protrusion formation in electrical field: the role of actin. Biochim. Biophys. Acta.

Small J.V. (1981) Organization of actin in the leading edge of cultured cells: influence of osmium tetroxide and dehydration on the ultrastructure of actin meshworks. J. Cell. Biol., 91: 695–705

Svitkina T.M., Shevelev A.A., Bershadsky A.A., Gelfand V.I. (1984) Cytoskeleton of mouse embryo fibroblasts. Electron microscopy of platinum replicas. Eur. J. Cell Biol., 34: 64–74

Trinkaus J.B. (1985) Protrusive activity of the cell surface and the initiation of cell movement during morphogenesis. Exp. Biol. Med., 10: 130–173

Vasiliev J.M. (1985) Spreading of non–transformed and transformed cells. Biochim. Biophys. Acta, 780: 21–65

Zimmermann U. (1982) Electric field–mediated fusion and related electrical phenomena. Biochim. Biophys. Acta, 694: 227–277

AUTHOR AND SUBJECT INDEX

N

O

P

Q

R

S

T

U

V

W

X

Y

Z

Printing: Druckerei Zechner, Speyer
Binding: Buchbinderei Schäffer, Grünstadt

NATO ASI Series H

NATO ASI Series H

Vol. 22: Neural Development and Regeneration. Cellular and Molecular Aspects.
Edited by A. Gorio, J. R. Perez-Polo, J. de Vellis, and B. Haber. 711 pages. 1988.

Vol. 23: The Semiotics of Cellular Communication in the Immune System.
Edited by E.E. Sercarz, F. Celada, N.A. Mitchison, and T. Tada. 326 pages. 1988.

Vol. 24: Bacteria, Complement and the Phagocytic Cell.
Edited by F. C. Cabello und C. Pruzzo. 372 pages. 1988.

Vol. 25: Nicotinic Acetylcholine Receptors in the Nervous System.
Edited by F. Clementi, C. Gotti, and E. Sher. 424 pages. 1988.

Vol. 26: Cell to Cell Signals in Mammalian Development.
Edited by S.W de Laat, J.G. Bluemink, and C.L. Mummery. 322 pages. 1989.

Vol. 27: Phytotoxins and Plant Pathogenesis.
Edited by A. Graniti, R. D. Durbin, and A. Ballio. 508 pages. 1989.

Vol. 28: Vascular Wilt Diseases of Plants. Basic Studies and Control.
Edited by E. C. Tjamos and C. H. Beckman. 590 pages. 1989.

Vol. 29: Receptors, Membrane Transport and Signal Transduction.
Edited by A. E. Evangelopoulos, J. P. Changeux, L. Packer, T. G. Sotiroudis, and K.W.A. Wirtz. 387 pages. 1989.

Vol. 30: Effects of Mineral Dusts on Cells.
Edited by B.T. Mossman and R.O. Begin. 470 pages. 1989.

Vol. 31: Neurobiology of the Inner Retina.
Edited by R. Weiler and N.N. Osborne. 529 pages. 1989.

Vol. 32: Molecular Biology of Neuroreceptors and Ion Channels.
Edited by A. Maelicke. 675 pages. 1989.

Vol. 33: Regulatory Mechanisms of Neuron to Vessel Communication in Brain.
Edited by F. Battaini, S. Govoni, M.S. Magnoni, and M. Trabucchi. 416 pages. 1989.

Vol. 34: VectorsasTools for the Study of Normal and Abnormal Growth and Differentiation.
Edited by H. Lother, R. Dernick, and W. Ostertag. 477 pages. 1989.

Vol. 35: Cell Separation in Plants: Physiology, Biochemistry and Molecular Biology.
Edited by D. J. Osborne and M. B. Jackson. 449 pages. 1989.

Vol. 36: Signal Molecules in Plants and Plant-Microbe Interactions.
Edited by B.J.J. Lugtenberg. 425 pages. 1989.

Vol. 37: Tin-Based Antitumour Drugs. Edited by M. Gielen. 226 pages. 1990.

Vol. 38: The Molecular Biology of Autoimmune Disease.
Edited by A.G. Demaine, J-P. Banga, and A.M. McGregor. 404 pages. 1990.

Vol. 39: Chemosensory Information Processing.
Edited by D. Schild. 403 pages. 1990.

Vol. 40: Dynamics and Biogenesis of Membranes.
Edited by J. A. F. Op den Kamp. 367 pages. 1990.

Vol. 41: Recognition and Response in Plant-Virus Interactions.
Edited by R. S. S. Fraser. 467 pages. 1990.

NATO ASI Series H

NATO ASI Series H

Vol. 62: Endocytosis. From Cell Biology to Health, Disease and Therapie.
Edited by P. J. Courtoy. 547 pages. 1992.

Vol. 63: Dynamics of Membrane Assembly.
Edited by J. A. F. Op den Kamp. 402 pages. 1992.

Vol. 64: Mechanics of Swelling. From Clays to Living Cells and Tissues.
Edited by T. K. Karalis. 802 pages. 1992.